Transmission and Population Genetics

A Conceptual Approach

FOURTH EDITION

Benjamin A. Pierce

Southwestern University

W. H. Freeman and Company
New York

Publisher: Kate Ahr Parker
Executive Editor: Susan Winslow
Development Editor: Lisa Samols
Senior Project Editor: Georgia Lee Hadler
Manuscript Editor: Patricia Zimmerman
Art Director: Diana Blume
Illustrations: Dragonfly Media Group
Illustration Coordinator: Janice Donnola
Photo Editor: Ted Szczepanski
Photo Researcher: Elyse Rieder
Production Coordinator: Paul Rohloff
Media Editor: Aaron Gass
Supplements Editor: Anna Bristow
Associate Director of Marketing: Debbie Clare
Composition: Preparé
Printing and Binding: RR Donnelley

Library of Congress Control Number: 2010943169

© 2012, 2009 by W. H. Freeman and Company. All rights reserved.

ISBN-13: 978-1-4292-5494-6
ISBN-10: 1-4292-5494-7

Printed in the United States of America

First printing

W. H. Freeman and Company
41 Madison Avenue
New York, NY 10010
Houndsmills, Basingstoke RG21 6XS. England

www.whfreeman.com

To my parents, Rush and Amanda Pierce;

my children, Sarah and Michael Pierce;

*and my genetic partner, friend, and soul mate
for 30 years, Marlene Tyrrell*

Contents in Brief

1	Introduction to Genetics	1
2	Chromosomes and Cellular Reproduction	15
3	Basic Principles of Heredity	43
4	Sex Determination and Sex-Linked Characteristics	73
5	Extensions and Modifications of Basic Principles	99
6	Pedigree Analysis, Applications, and Genetic Testing	135
7	Linkage, Recombination, and Eukaryotic Gene Mapping	161
8	Bacterial and Viral Genetic Systems	203
9	Chromosome Variation	239
10	Gene Mutations and DNA Repair	271
11	Quantitative Genetics	303
12	Population Genetics	337
13	Evolutionary Genetics	365
	Reference Guide to Model Genetic Organisms	A1

Contents

Letter from the Author xv

Preface xvi

Chapter 1 Introduction to Genetics 1

ALBINISM IN THE HOPIS 1

1.1 Genetics Is Important to Us Individually, to Society, and to the Study of Biology 2
- The Role of Genetics in Biology 4
- Genetic Diversity and Evolution 4
- Divisions of Genetics 5
- Model Genetic Organisms 5

1.2 Humans Have Been Using Genetics for Thousands of Years 7
- The Early Use and Understanding of Heredity 7
- The Rise of the Science of Genetics 9
- The Future of Genetics 10

1.3 A Few Fundamental Concepts Are Important for the Start of Our Journey into Genetics 11

Chapter 2 Chromosomes and Cellular Reproduction 15

THE BLIND MEN'S RIDDLE 15

2.1 Prokaryotic and Eukaryotic Cells Differ in a Number of Genetic Characteristics 17

2.2 Cell Reproduction Requires the Copying of the Genetic Material, Separation of the Copies, and Cell Division 18
- Prokaryotic Cell Reproduction 18
- Eukaryotic Cell Reproduction 18
- The Cell Cycle and Mitosis 21
- Genetic Consequences of the Cell Cycle 24

Connecting Concepts: Counting Chromosomes and DNA Molecules 25

2.3 Sexual Reproduction Produces Genetic Variation Through the Process of Meiosis 25
- Meiosis 26
- Sources of Genetic Variation in Meiosis 29

Connecting Concepts: Mitosis and Meiosis Compared 31

The Separation of Sister Chromatids and Homologous Chromosomes 31
Meiosis in the Life Cycles of Animals and Plants 33

Chapter 3 Basic Principles of Heredity 43

THE GENETICS OF RED HAIR 43

3.1 Gregor Mendel Discovered the Basic Principles of Heredity 44
- Mendel's Success 45
- Genetic Terminology 46

3.2 Monohybrid Crosses Reveal the Principle of Segregation and the Concept of Dominance 47
- What Monohybrid Crosses Reveal 48

Connecting Concepts: Relating Genetic Crosses to Meiosis 49

- Predicting the Outcomes of Genetic Crosses 51
- The Testcross 55
- Genetic Symbols 55

Connecting Concepts: Ratios in Simple Crosses 55

3.3 Dihybrid Crosses Reveal the Principle of Independent Assortment 56
- Dihybrid Crosses 56
- The Principle of Independent Assortment 56
- Relating the Principle of Independent Assortment to Meiosis 57
- Applying Probability and the Branch Diagram to Dihybrid Crosses 57
- The Dihybrid Testcross 59

3.4 Observed Ratios of Progeny May Deviate from Expected Ratios by Chance 61
- The Goodness-of-Fit Chi-Square Test 61

Chapter 4 Sex Determination and Sex-Linked Characteristics 73

THE STRANGE CASE OF PLATYPUS SEX 73

4.1 Sex Is Determined by a Number of Different Mechanisms 74
- Chromosomal Sex-Determining Systems 75
- Genic Sex Determination 77

Environmental Sex Determination 77
Sex Determination in *Drosophila melanogaster* 78
Sex Determination in Humans 79

4.2 Sex-Linked Characteristics Are Determined by Genes on the Sex Chromosomes 81
X-Linked White Eyes in *Drosophila* 81
Nondisjunction and the Chromosome Theory of Inheritance 82
X-Linked Color Blindness in Humans 84
Symbols for X-Linked Genes 85
Z-Linked Characteristics 85
Y-Linked Characteristics 86

Connecting Concepts: Recognizing Sex-Linked Inheritance 88

4.3 Dosage Compensation Equalizes the Amount of Protein Produced by X-Linked Genes in Males and Females 88
Lyon Hypothesis 89
Mechanism of Random X Inactivation 90
Dosage Imbalance Between X-Linked Genes and Autosomal Genes 90

Chapter 5 Extensions and Modifications of Basic Principles 99

CUÉNOT'S ODD YELLOW MICE 99

5.1 Additional Factors at a Single Locus Can Affect the Results of Genetic Crosses 100
Types of Dominance 100
Penetrance and Expressivity 103
Lethal Alleles 103
Multiple Alleles 104

5.2 Gene Interaction Takes Place When Genes at Multiple Loci Determine a Single Phenotype 106
Gene Interaction That Produces Novel Phenotypes 106
Gene Interaction with Epistasis 107

Connecting Concepts: Interpreting Ratios Produced by Gene Interaction 111

Complementation: Determining Whether Mutations Are at the Same Locus or at Different Loci 113
The Complex Genetics of Coat Color in Dogs 113

5.3 Sex Influences the Inheritance and Expression of Genes in a Variety of Ways 115
Sex-Influenced and Sex-Limited Characteristics 115
Cytoplasmic Inheritance 117

Genetic Maternal Effect 119
Genomic Imprinting 120

5.4 Anticipation Is the Stronger or Earlier Expression of Traits in Succeeding Generations 122

5.5 The Expression of a Genotype May Be Affected by Environmental Effects 123
Environmental Effects on the Phenotype 123
The Inheritance of Continuous Characteristics 124

Chapter 6 Pedigree Analysis, Applications, and Genetic Testing 135

HUTCHINSON–GILFORD SYNDROME AND THE SECRET OF AGING 135

6.1 The Study of Genetics in Humans Is Constrained by Special Features of Human Biology and Culture 136

6.2 Geneticists Often Use Pedigrees to Study the Inheritance of Characteristics in Humans 137
Symbols Used in Pedigrees 137
Analysis of Pedigrees 137
Autosomal Recessive Traits 138
Autosomal Dominant Traits 139
X-Linked Recessive Traits 139
X-Linked Dominant Traits 141
Y-Linked Traits 142

6.3 Studying Twins and Adoptions Can Help Assess the Importance of Genes and Environment 143
Types of Twins 143
Concordance in Twins 144
A Twin Study of Asthma 145
Adoption Studies 146

6.4 Genetic Counseling and Genetic Testing Provide Information to Those Concerned about Genetic Diseases and Traits 146
Genetic Counseling 146
Genetic Testing 148
Interpreting Genetic Tests 152
Direct-to-Consumer Genetic Testing 153
Genetic Discrimination and Privacy 153

6.5 Comparison of Human and Chimpanzee Genomes Is Helping to Reveal Genes That Make Humans Unique 153

Chapter 7 Linkage, Recombination, and Eukaryotic Gene Mapping 161

LINKED GENES AND BALD HEADS 161

7.1 Linked Genes Do Not Assort Independently 162

7.2 Linked Genes Segregate Together and Crossing Over Produces Recombination Between Them 163
- Notation for Crosses with Linkage 164
- Complete Linkage Compared with Independent Assortment 164
- Crossing Over with Linked Genes 166
- Calculating Recombination Frequency 167
- Coupling and Repulsion 168

Connecting Concepts: Relating Independent Assortment, Linkage, and Crossing Over 169

- Evidence for the Physical Basis of Recombination 170
- Predicting the Outcomes of Crosses with Linked Genes 171
- Testing for Independent Assortment 172
- Gene Mapping with Recombination Frequencies 174
- Constructing a Genetic Map with the Use of Two-Point Testcrosses 175

7.3 A Three-Point Testcross Can Be Used to Map Three Linked Genes 176
- Constructing a Genetic Map with the Three-Point Testcross 177

Connecting Concepts: Stepping Through the Three-Point Cross 182

- Effect of Multiple Crossovers 184
- Mapping Human Genes 185
- Mapping with Molecular Markers 186
- Locating Genes with Genomewide Association Studies 186

7.4 Physical-Mapping Methods Are Used to Determine the Physical Positions of Genes on Particular Chromosomes 187
- Deletion Mapping 188
- Somatic-Cell Hybridization 188
- Physical Chromosome Mapping Through Molecular Analysis 190

7.5 Recombination Rates Exhibit Extensive Variation 191

Chapter 8 Bacterial and Viral Genetic Systems 203

LIFE IN A BACTERIAL WORLD 203

8.1 Genetic Analysis of Bacteria Requires Special Methods 204
- Bacterial Diversity 204
- Techniques for the Study of Bacteria 205
- The Bacterial Genome 206
- Plasmids 206

8.2 Bacteria Exchange Genes Through Conjugation, Transformation, and Transduction 208
- Conjugation 208
- Natural Gene Transfer and Antibiotic Resistance 215
- Transformation in Bacteria 216
- Bacterial Genome Sequences 218
- Horizontal Gene Transfer 218

8.3 Viruses Are Simple Replicating Systems Amenable to Genetic Analysis 219
- Techniques for the Study of Bacteriophages 219
- Transduction: Using Phages to Map Bacterial Genes 220

Connecting Concepts: Three Methods for Mapping Bacterial Genes 223

- Gene Mapping in Phages 223
- Fine-Structure Analysis of Bacteriophage Genes 224
- RNA Viruses 227
- Human Immunodeficiency Virus and AIDS 227
- Influenza Virus 229

Chapter 9 Chromosome Variation 239

TRISOMY 21 AND THE DOWN-SYNDROME CRITICAL REGION 239

9.1 Chromosome Mutations Include Rearrangements, Aneuploids, and Polyploids 240
- Chromosome Morphology 240
- Types of Chromosome Mutations 241

9.2 Chromosome Rearrangements Alter Chromosome Structure 242
- Duplications 242
- Deletions 244
- Inversions 246
- Translocations 248
- Fragile Sites 251
- Copy-Number Variations 252

9.3 Aneuploidy Is an Increase or Decrease in the Number of Individual Chromosomes 252
- Types of Aneuploidy 252
- Effects of Aneuploidy 252

Aneuploidy in Humans 254
Uniparental Disomy 257
Mosaicism 257

9.4 Polyploidy Is the Presence of More than Two Sets of Chromosomes 258
Autopolyploidy 258
Allopolyploidy 260
The Significance of Polyploidy 261

9.5 Chromosome Variation Plays an Important Role in Evolution 262

Chapter 10 Gene Mutations and DNA Repair 271

A FLY WITHOUT A HEART 271

10.1 Mutations Are Inherited Alterations in the DNA Sequence 272
The Importance of Mutations 272
Categories of Mutations 272
Types of Gene Mutations 273
Phenotypic Effects of Mutations 275
Suppressor Mutations 276
Mutation Rates 280

10.2 Mutations Are Potentially Caused by a Number of Different Natural and Unnatural Factors 281
Spontaneous Replication Errors 281
Spontaneous Chemical Changes 283
Chemically Induced Mutations 284
Radiation 287

10.3 Mutations Are the Focus of Intense Study by Geneticists 288
Detecting Mutations with the Ames Test 288
Radiation Exposure in Humans 288

10.4 A Number of Pathways Repair Changes in DNA 290
Mismatch Repair 291
Direct Repair 292
Base-Excision Repair 292
Nucleotide-Excision Repair 293

Connecting Concepts: The Basic Pathway of DNA Repair 294

Repair of Double-Strand Breaks 294
Translesion DNA Polymerases 294
Genetic Diseases and Faulty DNA Repair 295

Chapter 11 Quantitative Genetics 303

CORN OIL AND QUANTITATIVE GENETICS 303

11.1 Quantitative Characteristics Vary Continuously and Many Are Influenced by Alleles at Multiple Loci 304
The Relation Between Genotype and Phenotype 305
Types of Quantitative Characteristics 306
Polygenic Inheritance 306
Kernel Color in Wheat 307
Determining Gene Number for a Polygenic Characteristic 308

11.2 Statistical Methods Are Required for Analyzing Quantitative Characteristics 309
Distributions 309
Samples and Populations 310
The Mean 310
The Variance and Standard Deviation 311
Correlation 312
Regression 313
Applying Statistics to the Study of a Polygenic Characteristic 315

11.3 Heritability Is Used to Estimate the Proportion of Variation in a Trait That Is Genetic 316
Phenotypic Variance 316
Types of Heritability 318
Calculating Heritability 318
The Limitations of Heritability 320
Locating Genes That Affect Quantitative Characteristics 322

11.4 Genetically Variable Traits Change in Response to Selection 324
Predicting the Response to Selection 324
Limits to Selection Response 326
Correlated Responses 327

Chapter 12 Population Genetics 337

GENETIC RESCUE OF BIGHORN SHEEP 337

12.1 Genotypic and Allelic Frequencies Are Used to Describe the Gene Pool of a Population 338
Calculating Genotypic Frequencies 339
Calculating Allelic Frequencies 339

12.2 The Hardy–Weinberg Law Describes the Effect of Reproduction on Genotypic and Allelic Frequencies 341

Genotypic Frequencies at Hardy–Weinberg Equilibrium 341
Closer Examination of the Assumptions of the Hardy–Weinberg Law 342
Implications of the Hardy–Weinberg Law 342
Extensions of the Hardy–Weinberg Law 343
Testing for Hardy–Weinberg Proportions 343
Estimating Allelic Frequencies with the Hardy–Weinberg Law 344

12.3 Nonrandom Mating Affects the Genotypic Frequencies of a Population 345

12.4 Several Evolutionary Forces Potentially Cause Changes in Allelic Frequencies 348
Mutation 348
Migration 349
Genetic Drift 350
Natural Selection 353

Connecting Concepts: The General Effects of Forces That Change Allelic Frequencies 358

Chapter 13 Evolutionary Genetics 365

TASTER GENES IN SPITTING APES 365

13.1 Organisms Evolve Through Genetic Change Taking Place Within Populations 366

13.2 Many Natural Populations Contain High Levels of Genetic Variation 367
Molecular Variation 368
Protein Variation 368
DNA Sequence Variation 370

13.3 New Species Arise Through the Evolution of Reproductive Isolation 373
The Biological Species Concept 373
Reproductive Isolating Mechanisms 373
Modes of Speciation 375
Genetic Differentiation Associated with Speciation 379

13.4 The Evolutionary History of a Group of Organisms Can Be Reconstructed by Studying Changes in Homologous Characteristics 380
The Alignment of Homologous Sequences 381
The Construction of Phylogenetic Trees 381

13.5 Patterns of Evolution Are Revealed by Changes at the Molecular Level 382
Rates of Molecular Evolution 382

The Molecular Clock 384
Genome Evolution 384

Reference Guide to Model Genetic Organisms A1
The Fruit Fly *Drosophilia melanogaster* A2
The Bacterium *Escherichia coli* A4
The Nematode Worm *Caenorhabditis elegans* A6
The Plant *Arabidopsis thaliana* A8
The Mouse *Mus musculus* A10
The Yeast *Saccharomyces cerevisiae* A12

Glossary B1

Answers to Selected Questions and Problems C1

Index D1

Letter from the Author

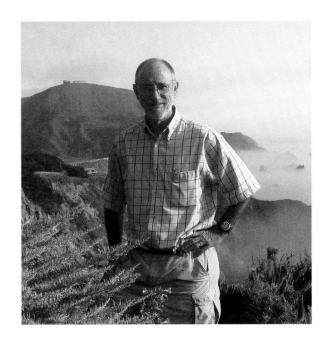

One of my passions in life is teaching. Probably like many of you, I've been inspired by past teachers. Miss Amos, my high-school English teacher, demonstrated the importance of having excitement for the subject that you teach. My organic chemistry professor, Harold Jesky, taught me that students learn best when they are motivated. And my first genetics professor, Ray Canham, taught me that the key to learning genetics is to focus on concepts and problem solving. All are important lessons that I learned from my teachers—lessons that I've tried to incorporate into my own teaching and into my textbook.

Seventeen years ago, I set out to write a new genetics textbook. My vision was to create a book that conveys the excitement of genetics, that motivates students, and that focuses on concepts and problem solving. Those were the original goals of the first edition of *Genetics: A Conceptual Approach* and they remain the core features of this fourth edition of the book.

In this book, I've tried to share some of what I've learned in my 30 years of teaching genetics. I provide advice and encouragement at places where students often have difficulty, and I tell stories of the people, places, and experiments of genetics—past and present—to keep the subject relevant, interesting, and alive. My goal is to help you learn the necessary details, concepts, and problem-solving skills while encouraging you to see the elegance and beauty of the discipline.

At Southwestern University, my office door is always open, and my students often drop by to share their own approaches to learning, things that they have read about genetics, and their experiences, concerns, and triumphs. I learn as much from my students as they learn from me, and I would love to learn from you—by email (pierceb@southwestern.edu), by telephone (512-863-1974), or in person (Southwestern University, Georgetown, Texas).

Ben Pierce
PROFESSOR OF BIOLOGY AND
HOLDER OF THE LILLIAN NELSON PRATT CHAIR
SOUTHWESTERN UNIVERSITY

Preface

Transmission and Population Genetics was developed to address the needs of courses that focus solely on those topics. Like its parent text, *Genetics* 4e, this edition of *Transmission and Population Genetics* strives to help students uncover major concepts of genetics and make connections among those concepts so as to have a fuller understanding of genetics. This conceptual and holistic approach to genetics has proved to be effective in the three preceding editions of this book. After taking part in class testing of those editions and of a sample chapter of the current edition, students say that they come away with a deeper and more complete understanding of genetics, thanks to the accessible writing style, simple and instructive illustrations, and useful pedagogical features throughout the book.

Hallmark Features

- **Key Concepts and Connections** Throughout the book, I've included features to help students focus on the major concepts of each topic.
  - *Concept boxes* throughout each chapter summarize the key points of preceding sections. *Concept Checks* ask students to pause for a moment and make sure that they understand the take-home message. Concept Checks are in multiple-choice and short-answer format, and answers are listed at the end of each chapter.
  - *Connecting Concepts* sections draw on concepts presented in several sections or several chapters to help students see how different topics of genetics relate to one another. These sections compare and contrast processes or integrate ideas across chapters to create an overarching, big picture of genetics. All major concepts are listed in the *Concepts Summary* at the end of each chapter.

> **CONCEPTS**
> The process of transformation indicates that some substance—the transforming principle—is capable of genetically altering bacteria. Avery, MacLeod, and McCarty demonstrated that the transforming principle is DNA, providing the first evidence that DNA is the genetic material.
>
> ✔ **CONCEPT CHECK 3**
> If Avery, MacLeod, and McCarty had found that samples of heat-killed bacteria treated with RNase and DNase transformed bacteria, but samples treated with protease did not, what conclusion would they have made?
> a. Protease carries out transformation.
> b. RNA and DNA are the genetic materials.
> c. Protein is the genetic material.
> d. RNase and DNase are necessary for transformation.

- **Accessibility** The welcoming and conversational writing style of this book has long been one of its most successful features for both students and instructors. In addition to carefully walking students through each major concept of genetics, I invite them into the topic with an **introductory story**. These stories include relevant examples of disease or other biological phenomena to give students a sample of what they'll be learning in a chapter. More than a third of the introductory stories in this edition are new.
- **Clear, Simple Illustration Program** I have worked closely with illustrators to create attractive and instructive illustrations, which have proved to be an effective learning tool for students. Each illlustration was carefully rendered to highlight main points and to step the reader through experiments and processes. Most illustrations include textual content that walks students through the graphical presentation. Illustrations of experiments reinforce the scientific method by first proposing a hypothesis, then pointing out the methods and results, and ending with a conclusion that reinforces concepts explained in the text.

I liked the amount of concept reinforcement that is in the chapter. The concept check questions seem like they would be very useful in helping students understand the material by getting them to stop and think about what they just read. —William Seemer, Student, University of North Florida

The style of writing is easy to follow and is directed at college-level students. I appreciate the use of real-world language, and examples that can be found in daily life. —Shannon Forshee, Student, Eckerd College

The figures and balloon dialogue really make the concepts easier to understand. Visualization is really important to me when I am studying. —Jamie Adams, Student, Arkansas State University

- **Emphasis on Problem Solving** One of the things that I've learned in my 30 years of teaching is that students learn genetics best through problem solving. Working through an example, equation, or experiment helps students see concepts in action and reinforces the ideas explained in the text. In the book, I help students develop problem-solving skills in a number of ways. **Worked Problems** follow the presentation of difficult quantitative concepts. Walking through a problem and solution within the text reinforces what the student has just read. New **Problem Links** spread throughout each chapter point to end-of-chapter problems that students can work to test their understanding of the material that they have just read. I provide a wide range of end-of-chapter problems, organized by chapter section and split into Comprehension, Application Questions and Problems, and Challenge Questions. Some of these questions draw on examples from published papers and are marked by a data analysis icon.

> *I liked the addition of the yellow tags telling you what problems to do that go along with the section. I also really liked that a worked problem was included in the text and not just at the end. I felt like that helped my problem-solving skills while I was reading.* —Alexandra Reynolds, Student, California Polytechnic State University, San Luis Obispo

NEW Problem Link

is the probability of Y^+y ($1/2$) multiplied by the probability of C^+c ($1/2$), or $1/4$. The probability of each progeny genotype resulting from the testcross is:

Progeny genotype	Probability at each locus		Overall probability	Phenotype
$Y^+y\ C^+c$	$1/2 \times 1/2$	=	$1/4$	red peppers
$Y^+y\ cc$	$1/2 \times 1/2$	=	$1/4$	peach peppers
$yy\ C^+c$	$1/2 \times 1/2$	=	$1/4$	orange peppers
$yy\ cc$	$1/2 \times 1/2$	=	$1/4$	cream peppers

When you work problems with gene interaction, it is especially important to determine the probabilities of single-locus genotypes and to multiply the probabilities of *genotypes*, not phenotypes, because the phenotypes cannot be determined without considering the effects of the genotypes at all the contributing loci. **TRY PROBLEM 25**

Gene Interaction with Epistasis

Sometimes the effect of gene interaction is that one gene masks (hides) the effect of another gene at a different locus, a phenomenon known as **epistasis**. In the examples of genic

New to the Fourth Edition

The fourth edition builds on successful features of the preceding editions of *Genetics: A Conceptual Approach* while providing an up-to-date look at the field.

- **More Help with Problem Solving** I encourage students to practice applying the concepts that they've just learned with new **Problem Links**, spread throughout each chapter. Each link points to an end-of-chapter problem that addresses the concept just discussed, so that students can immediately test their understanding of the concept. These problem links, in addition to the Concept Check questions, provide valuable just-in-time practice, enabling students to monitor their own progress.

- **Updated Coverage** The fourth edition addresses recent discoveries in the field of genetics, corresponding to our ever-changing understanding of inheritance, the molecular nature of genetic information, and genetic evolution. **Epigenetics**, an exciting new area of genetics, has been given extended and updated coverage in this edition. Additional updates include:

- genetic-conflict hypothesis of genomic imprinting (Chapter 5)
- epigenetics and the development of queen bees (Chapter 5)
- interpreting genetic tests (Chapter 6)
- direct-to-consumer genetic tests (Chapter 6)
- genomewide association studies (Chapter 7)
- horizontal gene transfer (Chapter 8)
- rapid evolution of influenza viruses (Chapter 8)
- segmental duplications (Chapter 9)
- copy-number variations (Chapter 9)
- repair of double-strand breaks (Chapter 11)

8.33 New strains of influenza virus are created by reassortment of genetic material from different strains.

NEW Introductory Stories

- **More Flexibility** Model genetic organisms are now presented at the end of the book in a **Reference Guide to Model Genetic Organisms**. Each model organism is presented in a two-page spread that summarizes major points about the organism's life cycle, genome, and uses as a model organism. This new organization makes it easier to teach model genetic organisms as a single unit at any time in the course or to cover individual organisms in the context of particular studies throughout the course.

- **More Connections** New **summary tables** throughout the book help students make connections between concepts. I use tables to make side-by-side comparisons of similar processes, to summarize the steps and players in complex biological pathways, and to help visually organize important genes, molecules, or ideas.

- **More Relevance** Each chapter begins with a brief **introductory story** that illustrates the relevance of a genetic concept that students will learn in the chapter. These stories—a favorite feature of past editions—give students a glimpse of what's going on in the field of genetics today and help to draw the reader into the chapter. New introductory topics are "The Strange Case of Platypus Sex," "Linked Genes and Bald Heads," "Life in a Bacterial World," and "Corn Oil and Quantitative Genetics." New end-of-chapter problems specifically address concepts discussed in most introductory stories, both old and new.

Media and Supplements

The complete package of media resources and supplements is designed to provide instructors and students with the most innovative tools to aid in a broad variety of teaching and learning approaches—including e-learning. All the available resources are fully integrated with the textbook's style and goals, enabling students to connect concepts in genetics and think like geneticists as well as develop their problem-solving skills. Students and instructors using *Transmission and Population Genetics* will have access to all of the supplements that accompany the parent text *Genetics* 4e.

GENETICS PORTAL
http://courses.bfwpub.com/pierce4e

GeneticsPortal is a dynamic, fully integrated learning environment that brings together all of our teaching and learning resources in one place. It features problem-solving videos, animations of difficult-to-visualize concepts, and our new **problem-solving engine**—an engaging tool that contains all of the end-of-chapter questions and problems in the textbook, converted into multiple-choice problems, including illustrations from the textbook and drop-down menus. Easy-to-use assessment tracking and grading tools enable instructors to assign problems for practice, as homework, quizzes, or tests.

NEW Genetics Portal

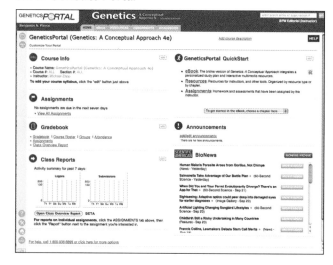

Some of the GeneticsPortal features are as follows:

- Hundreds of self-graded end-of-chapter practice problems allow students to fill in Punnett Squares, construct genetic maps, and calculate probabilities in multiple-choice versions.

- Step-by-step problem-solving videos walk through the specific process to solve select problems.

- Animations and activities to help students visualize genetics.

- A personalized calendar, an announcement center, and communication tools all in one place to help instructors manage the course.

GeneticsPortal also includes the fully interactive **eBook** and all of the Student and Instructor Resources that are on the **Book Companion Website**. The GeneticsPortal is included with all new copies of the fourth edition of *Genetics: A Conceptual Approach*.

eBook
http://ebooks.bfwpub.com/pierce4e

The eBook is a completely integrated electronic version of the textbook—the ultimate hybrid of textbook and media. Problems and resources from the printed textbook are incorporated throughout the eBook along with Check Your Understanding questions that are linked to specific sections of the textbook, to ensure that students can easily review specific concepts. The eBook is available as a stand-alone textbook (sold for approximately 60% of the retail price of the printed textbook) or packaged with the printed textbook at a discounted rate. The eBook version that accompanies *Transmission and Population Genetics* includes all 26 chapters of the parent text, *Genetics* 4e. The eBook enables students to:

eBook

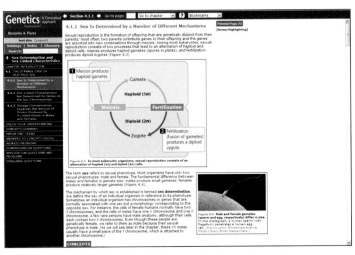

- Access the complete book and its electronic study tools from any internet-connected computer by using a standard Web browser;
- Navigate quickly to any section or subsection of the book or any page number of the printed book;
- Add their own bookmarks, notes, and highlighting;
- Access all the fully integrated media resources associated with the book, including the Interactive Animations and the Problem-Solving Videos (described on p. xx);
- Review quizzes and personal notes to help prepare for exams; and
- Search the entire eBook instantly, including the index and spoken glossary.

Instructors teaching from the eBook can assign either the entire textbook or a **custom version** that includes only the chapters that correspond to their syllabi. They can choose to add notes to any page of the eBook and share these notes with their students. These notes may include text, Web links, animations, or photographs. Also available is a **CourseSmart eBook**.

Instructor Resources

Instructors are provided with a comprehensive set of teaching tools, carefully developed to support lecture and individual teaching styles.

On the Book Companion Site
www.whfreeman.com/pierce4e

- All **Textbook Images and Tables** are offered as high-resolution JPEG files in PowerPoint. Each image has been fully optimized to increase type sizes and adjust color saturation. These images have been tested in a large lecture hall to ensure maximum clarity and visibility.
- **Layered or Active PowerPoints** deconstruct key concepts, sequences, and processes, step-by-step in a visual format, allowing instructors to present complex ideas in clear, manageable chunks.
- **Clicker Questions** allow instructors to integrate active learning in the classroom and to assess students' understanding of key concepts during lectures. Available in Microsoft Word and PowerPoint, numerous questions are based on the Concepts Check questions featured in the textbook.
- The **Test Bank** has been prepared by Brian W. Schwartz, Columbus State University; Bradley Hersh, Allegheny College; Paul K. Small, Eureka College; Gregory Copenhaver, University of North Carolina at Chapel Hill; Rodney Mauricio, University of Georgia; and Ravinshankar Palanivelu, University of Arizona. It contains 50 questions per chapter, including multiple-choice, true-or-false, and short-answer questions, and has been updated for the fourth edition with new problems. The Test Bank is available on the Instructor's Resource DVD as well as on

the Book Companion Site as chapter-by-chapter Microsoft Word files that are easy to download, edit, and print. A computerized test bank also is available.

- **Lecture Connection PowerPoint Presentations** for each chapter have been developed to minimize preparation time for new users of the book. These files offer suggested lectures including key illustrations and summaries that instructors can adapt to their teaching styles.
- The **Solution and Problem-Solving Manual** (described below) is available to download as pdf files.
- The **Instructor's Resource DVD** contains all of the resources on the Book Companion Site, including text images in PowerPoint slides and as high-resolution JPEG files, all animations, the Solutions and Problem-Solving Manual, and the Test Bank in Microsoft Word format.
- **Blackboard** and **WebCT** cartridges are available and include the Test Bank and end-of-chapter questions in multiple-choice and fill-in-the-blank format.

The most popular images in the textbook are also available as **Overhead Transparencies**, which have been optimized for maximum visibility in large lecture halls.

Student Resources

Students are provided with media designed to help them fully understand genetic concepts and improve their problem-solving ability.

- **Solutions and Problem-Solving Manual** by Jung Choi, Georgia Institute of Technology, and Mark McCallum, Pfeiffer University, contains complete answers and worked-out solutions to all questions and problems in the textbook. The manual has been found to be an indispensable tool for success by students and has been reviewed extensively by instructors for the fourth edition. (ISBN: 1-4292-3254-4)

On the Book Companion Site
www.whfreeman.com/pierce4e

Interactive Animations

- **Problem-Solving Videos** developed by Susan Elrod, California Polytechnic State University, San Luis Obispo, offer students valuable help by reviewing basic problem-solving strategies. Students learn first-hand how to deconstruct difficult problems in genetics by using a set of questioning techniques. The problem-solving videos demonstrate in a step-by-step manner how these strategies can be applied to the in-text worked problems and selected end-of-chapter problems in many of the chapters.
- **Interactive Animations/Podcasts** illuminate important concepts in genetics. These tutorials help students understand key processes in genetics by outlining these processes in a step-by-step manner. The tutorials are also available on the eBook. Podcasts adapted from the tutorial presentations in the following list are available for download from the Book Companion Site and from the eBook. Students can review important genetics processes and concepts at their convenience by downloading the animations to their MP3 players. The major animated concepts are:

  2.1 Cell Cycle and Mitosis
  2.2 Meiosis
  2.3 Genetic Variation in Meiosis
  3.1 Genetic Crosses Including Multiple Loci
  4.1 X-Linked Inheritance
  7.1 Determining Gene Order by Three-Point Cross
  8.1 Bacterial Conjugation
  18.1 DNA Mutations
  25.1 The Hardy–Weinberg Law and the Effects of Inbreeding and Natural Selection

Also available for students is *Genetics: A Conceptual Approach*, Fourth Edition (ISBN: 1-4292-3250-1). *Genetics*, Fourth Edition, is also available in looseleaf (ISBN: 1-4292-3251-X) at a reduced cost.

Acknowledgments

I am indebted to the thousands of genetics students who have filled my classes in the past 30 years, first at Connecticut College, then at Baylor University, and now at Southwestern University. The intelligence, enthusiasm, curiosity, and humor of these students have been a source of motivation and pleasure throughout my professional life. From them I have learned much about the art of teaching and the subject of genetics. I am also indebted to my genetics teachers and mentors, Dr. Raymond Canham and Dr. Jeffry Mitton, for introducing me to genetics and encouraging me to be a lifelong learner and scholar.

Five years ago, Southwestern University provided me with the opportunity to return to teaching and research at a small undergraduate university, and I have greatly enjoyed the experience. The small classes, close interaction of students and faculty, and integration of teaching and research have made working at Southwestern fun and rewarding. My colleagues in the Biology Department continually sustain me with friendship and advice. I thank James Hunt, Provost of Southwestern University and Dean of the Brown College, for valued friendship, collegiality, and support.

I have been blessed with an outstanding team of colleagues at W. H. Freeman and Company. Publisher Kate Ahr Parker provided support for this new edition. Executive Editor Susan Winslow expertly shepherded the project, providing coordination, creative ideas, encouragement, and support throughout the project. Developmental Editor Lisa Samols was my daily partner in crafting this edition. Her organizational skills, creative insight, and superior editing were—as always—outstanding. Importantly, Lisa also kept me motivated and on schedule with a positive attitude and good humor.

Patty Zimmerman, an outstanding manuscript editor, kept close watch on details and contributed valuable editorial suggestions. Patty has worked with me on all four editions of *Genetics: A Conceptual Approach*, as well as *Genetics Essentials*; her editorial touch resonates throughout the book. Senior Project Editor Georgia Lee Hadler at W. H. Freeman expertly managed the production of this fourth edition, as well as all preceding editions. I thank Craig Durant at Dragonfly Media Group for creating and revising the book's illustrations and Bill Page and Janice Donnola for coordinating the illustration program. Thanks to Paul Rohloff at W. H. Freeman and Pietro Paolo Adinolfi at Preparé for coordinating the composition and manufacturing phases of production. Diana Blume developed the book's design and the cover for this edition. I thank Ted Szczepanski and Elyse Rieder for photo research. Anna Bristow did an outstanding job of managing the supplements and assisting with the editorial development of the book. Aaron Gass and Ashley Joseph coordinated the excellent multimedia package that accompanies the book. I am grateful to Jung Choi and Mark McCallum for writing solutions to new end-of-chapter problems. Brian W. Schwartz, Bradley Hersh, Paul K. Small, Gregory Copenhaver, Rodney Mauricio, and Ravinshankar Palanivelu developed the Test Bank. Debbie Clare brought energy, creative ideas, and much fun to the marketing of the book.

I am grateful to the W. H. Freeman sales representatives, regional managers, and regional sales specialists, who introduce my book to genetic instructors throughout world. I have greatly enjoyed working with this sales staff, whose expertise, hard work, and good service are responsible for the success of Freeman books.

A number of colleagues served as reviewers of this book, kindly lending me their technical expertise and teaching experience. Their assistance is gratefully acknowledged. Any remaining errors are entirely my own.

Marlene Tyrrell—my spouse and best friend for 30 years—and our children Michael and Sarah provide love, support, and inspiration for everything that I do.

My gratitude goes to the reviewers of this new edition of *Genetics: A Conceptual Approach*.

Ann Aguanno
Marymount Manhattan College

Andrea Bailey
Brookhaven College

Paul W. Bates
University of Minnesota

Bethany V. Bowling
Northern Kentucky University

Natalie Bronstein
Mercy College

Gerald L. Buldak
Loyola University Chicago

J. Aaron Cassill
University of Texas at San Antonio

Maria V. Cattell
University of Colorado

Henry C. Chang
Purdue University

Mary C. Colavito
Santa Monica College

John F. Davis

Kimberley Dej
McMaster University

Steve H. Denison
Eckerd College

Laura E. DeWald
Western Carolina University

Moon Draper
University of Texas at Austin

Richard Duhrkopf
Baylor University

Cheryld L. Emmons
Alfred University

Michele Engel
University of Colorado Denver

Victor Fet
Marshall University

David W. Foltz
Louisiana State University

Wayne Forrester
Indiana University

Robert G. Fowler
San Jose State University

Laura Frost
Point Park University

Maria Gallo
University of Florida

Alex Georgakilas
East Carolina University

William D. Gilliland
DePaul University

Jack R. Girton
Iowa State University

Doreen R. Glodowski
Rutgers University

Elliott S. Goldstein
Arizona State University

Jessica L. Goldstein
Barnard College

Steven W. Gorsich
Central Michigan University

Liu He
University of Denver

Stephen C. Hedman
University of Minnesota Duluth

Bradley Hersh
Allegheny College

Carina Endres Howell
Lock Haven University of Pennsylvania

Colin Hughes
Florida Atlantic University

Dena Johnson
Tarrant County College NW

David H. Kass
Eastern Michigan University

Margaret Kovach
The University of Tennessee at Chattanooga

Brian Kreiser
University of Southern Mississippi

Catherine Kunst
University of Colorado at Denver

Mary Rose Lamb
University of Puget Sound

Haiying Liang
Clemson University

William J. Mackay
Edinboro University of Pennsylvania

Cindy S. Malone
California State University at Northridge

Jessica L. Moore
University of South Florida

Daniel W. Moser
Kansas State University

Mary Rengo Murnik
Ferris State University

Sang-Chul Nam
Baylor University

Ann V. Paterson
Williams Baptist College

Helen Piontkivska
Kent State University

Michael Lee Robinson
Miami University

Katherine T. Schmeidler
Irvine Valley College

Rodney J. Scott
Wheaton College

Barbara B. Sears
Michigan State University

Barkur S. Shastry
Oakland University

Mark Shotwell
Slippery Rock University

Anoop S. Sindhu
University of Saskatchewan

Paul K. Small
Eureka College

Thomas Smith
Ave Maria University

Lara Soowal
University of California at San Diego

Ruth Sporer
Rutgers University

Nanette van Loon
Borough of Manhattan Community College

Sarah Ward
Colorado State University

Lise D. Wilson
Siena College

Kathleen Wood
University of Mary Hardin-Baylor

Malcolm Zellars
Georgia State University

Zhongyuan Zuo
University of Denver

1 Introduction to Genetics

Hopi bowl, early twentieth century. Albinism, a genetic condition, arises with high frequency among the Hopi people and occupies a special place in the Hopi culture. [The Newark Museum/Art Resource, NY.]

ALBINISM IN THE HOPIS

Rising a thousand feet above the desert floor, Black Mesa dominates the horizon of the Enchanted Desert and provides a familiar landmark for travelers passing through northeastern Arizona. Not only is Black Mesa a prominent geological feature, but, more significantly, it is the ancestral home of the Hopi Native Americans. Fingers of the mesa reach out into the desert, and alongside or on top of each finger is a Hopi village. Most of the villages are quite small, having only a few dozen inhabitants, but they are incredibly old. One village, Oraibi, has existed on Black Mesa since 1150 A.D. and is the oldest continuously occupied settlement in North America.

In 1900, Aleš Hrdlička, an anthropologist and physician working for the American Museum of Natural History, visited the Hopi villages of Black Mesa and reported a startling discovery. Among the Hopis were 11 white persons—not Caucasians, but actually white Hopi Native Americans. These persons had a genetic condition known as albinism (**Figure 1.1**).

Albinism is caused by a defect in one of the enzymes required to produce melanin, the pigment that darkens our skin, hair, and eyes. People with albinism don't produce melanin or they produce only small amounts of it and, consequently, have white hair, light skin, and no pigment in the irises of their eyes. Melanin normally protects the DNA of skin cells from the damaging effects of ultraviolet radiation in sunlight, and melanin's presence in the developing eye is essential for proper eyesight.

The genetic basis of albinism was first described by the English physician Archibald Garrod, who recognized in 1908 that the condition was inherited as an autosomal recessive trait, meaning that a person must receive two copies of an albino mutation—one from each parent—to have albinism. In recent years, the molecular natures of the mutations that lead to albinism have been elucidated. Albinism in humans is caused by defects in any one of several different genes that control the synthesis and storage of melanin; many different types of mutations can occur at each gene, any one of which may lead to albinism. The form of albinism found in the Hopis is most likely oculocutaneous albinism type 2, due to a defect in the *OCA2* gene on chromosome 15.

The Hopis are not unique in having albinos among the members of their tribe. Albinism is found in almost all human ethnic groups and is described in ancient writings; it has probably been present since humankind's beginnings. What is unique about the

1.1 Albinism among the Hopi Native Americans. In this photograph, taken about 1900, the Hopi girl in the center has albinism. [The Field Museum/Charles Carpenter.]

Hopis is the high frequency of albinism. In most human groups, albinism is rare, present in only about 1 in 20,000 persons. In the villages on Black Mesa, it reaches a frequency of 1 in 200, a hundred times as frequent as in most other populations.

Why is albinism so frequent among the Hopi Native Americans? The answer to this question is not completely known, but geneticists who have studied albinism in the Hopis speculate that the high frequency of the albino gene is related to the special place that albinism occupied in the Hopi culture. For much of their history, the Hopis considered members of their tribe with albinism to be important and special. People with albinism were considered pretty, clean, and intelligent. Having a number of people with albinism in one's village was considered a good sign, a symbol that the people of the village contained particularly pure Hopi blood. Albinos performed in Hopi ceremonies and assumed positions of leadership within the tribe, often becoming chiefs, healers, and religious leaders.

Hopi albinos were also given special treatment in everyday activities. The Hopis farmed small garden plots at the foot of Black Mesa for centuries. Every day throughout the growing season, the men of the tribe trekked to the base of Black Mesa and spent much of the day in the bright southwestern sunlight tending their corn and vegetables. With little or no melanin pigment in their skin, people with albinism are extremely susceptible to sunburn and have increased incidences of skin cancer when exposed to the sun. Furthermore, many don't see well in bright sunlight. But the male Hopis with albinism were excused from this normal male labor and allowed to remain behind in the village with the women of the tribe, performing other duties.

Geneticists have suggested that these special considerations given to albino members of the tribe are partly responsible for the high frequency of albinism among the Hopis. Throughout the growing season, the albino men were the only male members of the tribe in the village during the day with all the women and, thus, they enjoyed a mating advantage, which helped to spread their albino genes. In addition, the special considerations given to albino Hopis allowed them to avoid the detrimental effects of albinism—increased skin cancer and poor eyesight. The small size of the Hopi tribe probably also played a role by allowing chance to increase the frequency of the albino gene. Regardless of the factors that led to the high frequency of albinism, the Hopis clearly respected and valued the members of their tribe who possessed this particular trait. Unfortunately, people with genetic conditions in many societies are often subject to discrimination and prejudice.

TRY PROBLEMS 1 AND 25 →

Genetics is one of the most rapidly advancing fields of science, with important new discoveries reported every month. Pick up almost any major newspaper or news magazine and chances are that you will see articles related to genetics: the completion of another genome, such as that of the platypus; the discovery of genes that affect major diseases, including multiple sclerosis, depression, and cancer; a report of DNA analyzed from long-extinct animals such as the woolly mammoth; and the identification of genes that affect skin pigmentation, height, and learning ability in humans. Even among the advertisements, one is likely to see genetics: ads for genetic testing to determine paternity, one's ancestry, and susceptibility to diseases and disorders. These new findings and applications of genetics often have significant economic and ethical implications, making the study of genetics relevant, timely, and interesting.

This chapter introduces you to genetics and reviews some concepts that you may have encountered briefly in a biology course. We begin by considering the importance of genetics to each of us, to society at large, and to students of biology. We then turn to the history of genetics, how the field as a whole developed. The final part of the chapter presents some fundamental terms and principles of genetics that are used throughout the book.

1.1 Genetics Is Important to Us Individually, to Society, and to the Study of Biology

Albinism among the Hopis illustrates the important role that genes play in our lives. This one genetic defect, among the 20,000 genes that humans possess, completely changes the life of a Hopi who possesses it. It alters his or her occupation, role in Hopi society, and relations with other members of the tribe. We all possess genes that influence our lives in significant ways. Genes affect our height, weight, hair color, and skin pigmentation. They affect our suscepti-

Introduction to Genetics 3

1.2 Genes influence susceptibility to many diseases and disorders. (a) An X-ray of the hand of a person suffering from diastrophic dysplasia (bottom), a hereditary growth disorder that results in curved bones, short limbs, and hand deformities, compared with an X-ray of a normal hand (top). (b) This disorder is due to a defect in a gene on chromosome 5. Braces indicate regions on chromosome 5 where genes giving rise to other disorders are located. [Part a: (top) Biophoto Associates/Science Source/Photo Researchers; (bottom) courtesy of Eric Lander, Whitehead Institute, MIT.]

1.3 In the Green Revolution, genetic techniques were used to develop new high-yielding strains of crops. (Left) Norman Borlaug, a leader in the development of new strains of wheat that led to the Green Revolution. Borlaug was awarded the Nobel Peace Prize in 1970. (Right) Modern, high-yielding rice plant (left) and traditional rice plant (right). [Left: UPI/Corbis-Bettman. Right: IRRI.]

bility to many diseases and disorders (**Figure 1.2**) and even contribute to our intelligence and personality. Genes are fundamental to who and what we are.

Although the science of genetics is relatively new compared with sciences such as astronomy and chemistry, people have understood the hereditary nature of traits and have practiced genetics for thousands of years. The rise of agriculture began when people started to apply genetic principles to the domestication of plants and animals. Today, the major crops and animals used in agriculture are quite different from their wild progenitors, having undergone extensive genetic alterations that increase their yields and provide many desirable traits, such as disease and pest resistance, special nutritional qualities, and characteristics that facilitate harvest. The Green Revolution, which expanded food production throughout the world in the 1950s and 1960s, relied heavily on the application of genetics (**Figure 1.3**). Today, genetically engineered corn, soybeans, and other crops constitute a significant proportion of all the food produced worldwide.

The pharmaceutical industry is another area in which genetics plays an important role. Numerous drugs and food additives are synthesized by fungi and bacteria that have been genetically manipulated to make them efficient producers of these substances. The biotechnology industry employs molecular genetic techniques to develop and mass-produce substances of commercial value. Growth hormone, insulin, and clotting factor are now produced commercially by genetically engineered bacteria (**Figure 1.4**). Genetics has also been used to produce bacteria that remove minerals from ore, break down toxic chemicals, and inhibit damaging frost formation on crop plants.

Genetics plays a critical role in medicine. Physicians recognize that many diseases and disorders have a hereditary component, including rare genetic disorders such as sickle-cell anemia and Huntington disease as well as many common diseases such as asthma, diabetes, and hypertension. Advances in genetics have resulted in important insights into the nature of diseases such as cancer and in the development

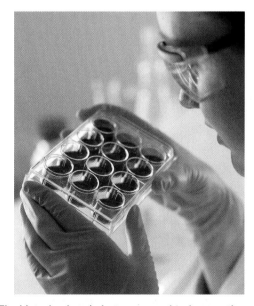

1.4 The biotechnology industry uses molecular genetic methods to produce substances of economic value.
[Andrew Brooks/Corbis.]

of diagnostic tests such as those that identify pathogens and defective genes. Gene therapy—the direct alteration of genes to treat human diseases—has now been administered to thousands of patients.

The Role of Genetics in Biology

Although an understanding of genetics is important to all people, it is critical to the student of biology. Genetics provides one of biology's unifying principles: all organisms use genetic systems that have a number of features in common. Genetics also undergirds the study of many other biological disciplines. Evolution, for example, is genetic change taking place through time; so the study of evolution requires an understanding of genetics. Developmental biology relies heavily on genetics: tissues and organs develop through the regulated expression of genes (**Figure 1.5**). Even such fields as taxonomy, ecology, and animal behavior are making increasing use of genetic methods. The study of almost any field of biology or medicine is incomplete without a thorough understanding of genes and genetic methods.

Genetic Diversity and Evolution

Life on Earth exists in a tremendous array of forms and features in almost every conceivable environment. Life is also characterized by adaptation: many organisms are exquisitely suited to the environment in which they are found. The history of life is a chronicle of new forms of life emerging, old forms disappearing, and existing forms changing.

Despite their tremendous diversity, living organisms have an important feature in common: all use similar genetic systems. A complete set of genetic instructions for any organism is its **genome**, and all genomes are encoded in nucleic acids—either DNA or RNA. The coding system for genomic information also is common to all life: genetic instructions are in the same format and, with rare exceptions, the code words are identical. Likewise, the process by which genetic information is copied and decoded is remarkably similar for all forms of life. These common features of heredity suggest that all life on Earth evolved from the same primordial ancestor that arose between 3.5 billion and 4 billion years ago. Biologist Richard Dawkins describes life as a river of DNA that runs through time, connecting all organisms past and present.

That all organisms have similar genetic systems means that the study of one organism's genes reveals principles that apply to other organisms. Investigations of how bacterial DNA is copied (replicated), for example, provide information that applies to the replication of human DNA. It also means that genes will function in foreign cells, which makes genetic engineering possible. Unfortunately, these similar genetic systems are also the basis for diseases such as AIDS (acquired immune deficiency syndrome), in which viral genes are able to function—sometimes with alarming efficiency—in human cells.

Life's diversity and adaptation are products of evolution, which is simply genetic change through time. Evolution is a two-step process: first, inherited differences arise randomly and, then, the proportion of individuals with particular differences increases or decreases. Genetic variation is therefore the foundation of all evolutionary change and is ultimately the basis of all life as we know it. Furthermore, techniques of molecular genetics are now routinely used to decipher evolutionary relationships among organisms; for example, recent analysis of DNA isolated from Neanderthal fossils has yielded new information concerning the relationship between Neanderthals and modern humans. Genetics, the study of genetic variation, is critical to understanding the past, present, and future of life. **TRY PROBLEM 17**

1.5 The key to development lies in the regulation of gene expression. This early fruit-fly embryo illustrates the localized expression of the *engrailed* gene, which helps determine the development of body segments in the adult fly. [Stephen Paddock Digital Image Gallery.]

> **CONCEPTS**
>
> Heredity affects many of our physical features as well as our susceptibility to many diseases and disorders. Genetics contributes to advances in agriculture, pharmaceuticals, and medicine and is fundamental to modern biology. All organisms use similar genetic systems, and genetic variation is the foundation of the diversity of all life.
>
> ✔ **CONCEPT CHECK 1**
>
> What are some of the implications of all organisms having similar genetic systems?
>
> a. That all life forms are genetically related
> b. That research findings on one organism's gene function can often be applied to other organisms
> c. That genes from one organism can often exist and thrive in another organism
> d. All of the above

Divisions of Genetics

The study of genetics consists of three major subdisciplines: transmission genetics, molecular genetics, and population genetics (**Figure 1.6**). Also known as classical genetics, **transmission genetics** encompasses the basic principles of heredity and how traits are passed from one generation to the next. This area addresses the relation between chromosomes and heredity, the arrangement of genes on chromosomes, and gene mapping. Here, the focus is on the individual organism—how an individual organism inherits its genetic makeup and how it passes its genes to the next generation.

Molecular genetics concerns the chemical nature of the gene itself: how genetic information is encoded, replicated, and expressed. It includes the cellular processes of replication, transcription, and translation (by which genetic information is transferred from one molecule to another) and gene regulation (the processes that control the expression of genetic information). The focus in molecular genetics is the gene, its structure, organization, and function.

Population genetics explores the genetic composition of groups of individual members of the same species (populations) and how that composition changes geographically and with the passage of time. Because evolution is genetic change, population genetics is fundamentally the study of evolution. The focus of population genetics is the group of genes found in a population.

Division of the study of genetics into these three groups is convenient and traditional, but we should recognize that the fields overlap and that each major subdivision can be further divided into a number of more-specialized fields, such as chromosomal genetics, biochemical genetics, quantitative genetics, and so forth. Alternatively, genetics can be subdivided by organism (fruit fly, corn, or bacterial genetics), and each of these organisms can be studied at the level of transmission, molecular, and population genetics. Modern genetics is an extremely broad field, encompassing many interrelated subdisciplines and specializations.

TRY PROBLEM 18 →

Model Genetic Organisms

Through the years, genetic studies have been conducted on thousands of different species, including almost all major groups of bacteria, fungi, protists, plants, and animals. Nevertheless, a few species have emerged as **model genetic organisms**—organisms having characteristics that make them particularly useful for genetic analysis and about which a tremendous amount of genetic information has accumulated. Six model organisms that have been the subject of intensive genetic study are: *Drosophila melanogaster*, the fruit fly; *Escherichia coli*, a bacterium present in the gut of humans and other mammals; *Caenorhabditis elegans*, a nematode worm; *Arabidopsis thaliana*, the thale-cress plant; *Mus musculus*, the house mouse; and *Saccharomyces cerevisiae*, baker's yeast (**Figure 1.7**). These species are the organisms of choice for many genetic researchers, and their genomes were sequenced as a part of the Human Genome Project. The life cyles and genetic characteristics of these model genetic organisms are described in more detail in the Reference Guide to Model Genetic Organisms located at the end of the book.

At first glance, this group of lowly and sometimes despised creatures might seem unlikely candidates for model organisms. However, all possess life cycles and traits that make them particularly suitable for genetic study, including a short generation time, large but manageable numbers of progeny, adaptability to a laboratory environment, and the ability to be housed and propagated inexpensively. Other species that are frequently the subjects of genetic research and considered genetic models include *Neurospora crassa* (bread mold), *Zea mays* (corn), *Danio rerio* (zebrafish), and *Xenopus laevis* (clawed frog). Although not generally considered a genetic model, humans also have been subjected to intensive genetic scrutiny; special techniques for the genetic analysis of humans are discussed in Chapter 6.

1.6 Genetics can be subdivided into three interrelated fields.
[Top left: Junior's Bildarchiv/Alamy. Top right: Mona file M0214602tif. Bottom: J. Alcock/Visuals Unlimited.]

Drosophila melanogaster
Fruit fly

Escherichia coli
Bacterium

Caenorhabditis elegans
Nematode

1.7 Model genetic organisms are species having features that make them useful for genetic analysis. [Part a: SPL/Photo Researchers. Part b: Gary Gaugler/Visuals Unlimited. Part c: Natalie Pujol/Visuals Unlimited. Part d: Peggy Greb/ARS. Part e: Joel Page/AP. Part f: T. E. Adams/Visuals Unlimited.]

The value of model genetic organisms is illustrated by the use of zebrafish to identify genes that affect skin pigmentation in humans. For many years, geneticists have recognized that differences in pigmentation among human ethnic groups are genetic (**Figure 1.8a**), but the genes causing these differences were largely unknown. The zebrafish has recently become an important model in genetic studies because it is a small vertebrate that produces many offspring and is easy to rear in the laboratory. The mutant zebrafish called *golden* has light pigmentation due to the presence of fewer, smaller, and less-dense pigment-containing structures called melanosomes in its cells (**Figure 1.8b**). Light skin in humans is similarly due to fewer and less-dense melanosomes in pigment-containing cells.

Keith Cheng and his colleagues at Pennsylvania State University College of Medicine hypothesized that light skin in humans might result from a mutation that is similar to the *golden* mutation in zebrafish. Taking advantage of the ease with which zebrafish can be manipulated in the laboratory, they isolated and sequenced the gene responsible for the *golden* mutation and found that it encodes a protein that takes part in calcium uptake by melanosomes. They then searched a database of all known human genes and found a similar gene called *SLC24A5*, which encodes the same function in human cells. When they examined human populations, they found that light-skinned Europeans typically possessed one form of this gene, whereas darker-skinned Africans, Eastern Asians, and Native Americans usually possessed a different form of the gene. Many other genes also affect pigmentation in humans, as illustrated by mutations in the *OCA2* gene that produce albinism among the Hopi Native Americans (discussed in the introduction to this chapter). Nevertheless, *SLC24A5* appears to be responsible for 24% to 38% of the differences in pigmentation between Africans and Europeans. This example illustrates the power of model organisms in genetic research.

(a)

(b)

Normal zebrafish

Golden mutant

1.8 The zebrafish, a genetic model organism, has been instrumental in helping to identify genes encoding pigmentation differences among humans. (a) Human ethnic groups differ in degree of skin pigmentation. (b) The zebrafish *golden* mutation is caused by a gene that controls the amount of melanin pigment in melanosomes. [Part a: PhotoDisc. Part b: K. Cheng/J. Gittlen, Cancer Research Foundation, Pennsylvania State College of Medicine.]

(d)
Arabidopsis thaliana
Thale-cress plant

(e)
Mus musculus
House mouse

(f)
Saccharomyces cerevisiae
Baker's yeast

CONCEPTS

The three major divisions of genetics are transmission genetics, molecular genetics, and population genetics. Transmission genetics examines the principles of heredity; molecular genetics deals with the gene and the cellular processes by which genetic information is transferred and expressed; population genetics concerns the genetic composition of groups of organisms and how that composition changes geographically and with the passage of time. Model genetic organisms are species that have received special emphasis in genetic research; they have characteristics that make them useful for genetic analysis.

✔ CONCEPT CHECK 2

Would the horse make a good model genetic organism? Why or why not?

1.2 Humans Have Been Using Genetics for Thousands of Years

Although the science of genetics is young—almost entirely a product of the past 100 years or so—people have been using genetic principles for thousands of years.

The Early Use and Understanding of Heredity

The first evidence that people understood and applied the principles of heredity in earlier times is found in the domestication of plants and animals, which began between approximately 10,000 and 12,000 years ago in the Middle East. The first domesticated organisms included wheat, peas, lentils, barley, dogs, goats, and sheep (**Figure 1.9a**). By 4000 years ago, sophisticated genetic techniques were already in

1.9 Ancient peoples practiced genetic techniques in agriculture. (Left) Modern wheat, with larger and more numerous seeds that do not scatter before harvest, was produced by interbreeding at least three different wild species. (Right) Assyrian bas-relief sculpture showing artificial pollination of date palms at the time of King Assurnasirpalli II, who reigned from 883 to 859 B.C. [Left: Scott Bauer/ARS/USDA. Right: The Metropolitan Museum of Art, gift of John D. Rockefeller, Jr., 1932. (32.143.3) Photograph © 1996 Metropolitan Museum of Art.]

use in the Middle East. Assyrians and Babylonians developed several hundred varieties of date palms that differed in fruit size, color, taste, and time of ripening (**Figure 1.9b**). Other crops and domesticated animals were developed by cultures in Asia, Africa, and the Americas in the same period.

Ancient writings demonstrate that early humans were also aware of their own heredity. Hindu sacred writings dating to 2000 years ago attribute many traits to the father and suggest that differences between siblings are produced by the mother. The Talmud, the Jewish book of religious laws based on oral traditions dating back thousands of years, presents an uncannily accurate understanding of the inheritance of hemophilia. It directs that, if a woman bears two sons who die of bleeding after circumcision, any additional sons that she bears should not be circumcised; nor should the sons of her sisters be circumcised. This advice accurately corresponds to the X-linked pattern of inheritance of hemophilia (discussed further in Chapter 6).

The ancient Greeks gave careful consideration to human reproduction and heredity. Greek philosophers developed the concept of **pangenesis**, in which specific particles, later called gemmules, carry information from various parts of the body to the reproductive organs, from which they are passed to the embryo at the moment of conception (**Figure 1.10**). Although incorrect, the concept of pangenesis was highly influential and persisted until the late 1800s.

Pangenesis led the ancient Greeks to propose the notion of the **inheritance of acquired characteristics**, in which traits acquired in a person's lifetime become incorporated into that person's hereditary information and are passed on to offspring; for example, people who developed musical ability through diligent study would produce children who are innately endowed with musical ability. The notion of the inheritance of acquired characteristics also is no longer accepted, but it remained popular through the twentieth century.

Although the ancient Romans contributed little to an understanding of human heredity, they successfully developed a number of techniques for animal and plant breeding; the techniques were based on trial and error rather than any general concept of heredity. Little new information was added to the understanding of genetics in the next 1000 years.

Dutch eyeglass makers began to put together simple microscopes in the late 1500s, enabling Robert Hooke (1635–1703) to discover cells in 1665. Microscopes provided naturalists with new and exciting vistas on life, and perhaps excessive enthusiasm for this new world of the very small gave rise to the idea of **preformationism**. According to preformationism, inside the egg or sperm there exists a fully formed miniature adult, a *homunculus*, which simply enlarges in the course of development (**Figure 1.11**). Preformationism meant that all traits were inherited from

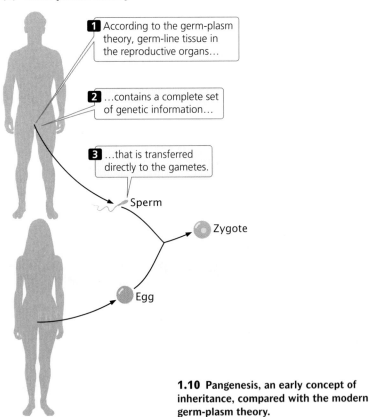

1.10 Pangenesis, an early concept of inheritance, compared with the modern germ-plasm theory.

1.11 Preformationists in the seventeenth and eighteenth centuries believed that sperm or eggs contained fully formed humans (the homunculus). Shown here is a drawing of a homunculus inside a sperm. [Science VU/Visuals Unlimited.]

only one parent—from the father if the homunculus was in the sperm or from the mother if it was in the egg. Although many observations suggested that offspring possess a mixture of traits from both parents, preformationism remained a popular concept throughout much of the seventeenth and eighteenth centuries.

Another early notion of heredity was **blending inheritance**, which proposed that offspring are a blend, or mixture, of parental traits. This idea suggested that the genetic material itself blends, much as blue and yellow pigments blend to make green paint. Once blended, genetic differences could not be separated out in future generations, just as green paint cannot be separated out into blue and yellow pigments. Some traits do *appear* to exhibit blending inheritance; however, we realize today that individual genes do not blend.

The Rise of the Science of Genetics

In 1676, Nehemiah Grew (1641–1712) reported that plants reproduce sexually by using pollen from the male sex cells. With this information, a number of botanists began to experiment with crossing plants and creating hybrids, including Gregor Mendel (1822–1884; **Figure 1.12**), who went on to discover the basic principles of heredity. Mendel's conclusions, which were not widely known in the scientific community for 35 years, laid the foundation for our modern understanding of heredity, and he is generally recognized today as the father of genetics.

Developments in cytology (the study of cells) in the 1800s had a strong influence on genetics. Robert Brown (1773–1858) described the cell nucleus in 1833. Building on the work of others, Matthias Jacob Schleiden (1804–1881) and Theodor Schwann (1810–1882) proposed the concept of the **cell theory** in 1839. According to this theory, all life is composed of cells, cells arise only from preexisting cells, and the cell is the fundamental unit of structure and function in living organisms. Biologists interested in heredity began to examine cells to see what took place in the course of cell reproduction. Walther Flemming (1843–1905) observed the division of chromosomes in 1879 and published a superb description of mitosis. By 1885, it was generally recognized that the nucleus contained the hereditary information.

Charles Darwin (1809–1882), one of the most influential biologists of the nineteenth century, put forth the theory of evolution through natural selection and published his ideas in *On the Origin of Species* in 1859. Darwin recognized that heredity was fundamental to evolution, and he conducted extensive genetic crosses with pigeons and other organisms. However, he never understood the nature of inheritance, and this lack of understanding was a major omission in his theory of evolution.

In the last half of the nineteenth century, cytologists demonstrated that the nucleus had a role in fertilization. Near the close of the nineteenth century, August Weismann (1834–1914) finally laid to rest the notion of the inheritance of acquired characteristics. He cut off the tails of mice for 22 consecutive generations and showed that the tail length in descendants remained stubbornly long. Weismann proposed the **germ-plasm theory**, which holds that the cells in the reproductive organs carry a complete set of genetic information that is passed to the egg and sperm (see Figure 1.10b).

1.12 Gregor Mendel was the founder of modern genetics. Mendel first discovered the principles of heredity by crossing different varieties of pea plants and analyzing the transmission of traits in subsequent generations. [Hulton Archive/Getty Images.]

The year 1900 was a watershed in the history of genetics. Gregor Mendel's pivotal 1866 publication on experiments with pea plants, which revealed the principles of heredity, was rediscovered, as considered in more detail in Chapter 3. The significance of his conclusions was recognized, and other biologists immediately began to conduct similar genetic studies on mice, chickens, and other organisms. The results of these investigations showed that many traits indeed follow Mendel's rules. Some of the early concepts of heredity are summarized in **Table 1.1**.

After the acceptance of Mendel's theory of heredity, Walter Sutton (1877–1916) proposed in 1902 that genes, the units of inheritance, are located on chromosomes. Thomas Hunt Morgan (1866–1945) discovered the first genetic mutant of fruit flies in 1910 and used fruit flies to unravel many details of transmission genetics. Ronald A. Fisher (1890–1962), John B. S. Haldane (1892–1964), and Sewall Wright (1889–1988) laid the foundation for population genetics in the 1930s by synthesizing Mendelian genetics and evolutionary theory.

Geneticists began to use bacteria and viruses in the 1940s; the rapid reproduction and simple genetic systems of these organisms allowed detailed study of the organization and structure of genes. At about this same time, evidence accumulated that DNA was the repository of genetic information.

1.13 The human genome was completely sequenced in 2003. A chromatograph of a small portion of the human genome. [Science Museum/SSPL.]

James Watson (b. 1928) and Francis Crick (1916–2004), along with Maurice Wilkins (1916–2004) and Rosalind Franklin (1920–1958), described the three-dimensional structure of DNA in 1953, ushering in the era of molecular genetics.

By 1966, the chemical structure of DNA and the system by which it determines the amino acid sequence of proteins had been worked out. Advances in molecular genetics led to the first recombinant DNA experiments in 1973, which touched off another revolution in genetic research. Walter Gilbert (b. 1932) and Frederick Sanger (b. 1918) developed methods for sequencing DNA in 1977. The polymerase chain reaction, a technique for quickly amplifying tiny amounts of DNA, was developed by Kary Mullis (b. 1944) and others in 1983. In 1990, gene therapy was used for the first time to treat human genetic disease in the United States, and the Human Genome Project was launched. By 1995, the first complete DNA sequence of a free-living organism—the bacterium *Haemophilus influenzae*—was determined, and the first complete sequence of a eukaryotic organism (yeast) was reported a year later. A rough draft of the human genome sequence was reported in 2000, with the sequence essentially completed in 2003, ushering in a new era in genetics (**Figure 1.13**). Today, the genomes of numerous organisms are being sequenced, analyzed, and compared. TRY PROBLEMS 22 AND 23

The Future of Genetics

Numerous advances in genetics are being made today, and genetics remains at the forefront of biological research. New, rapid methods for sequencing DNA are being used to sequence the genomes of numerous species, from bacteria to elephants, and the information content of genetics is increasing at a rapid pace. New details about gene structure and function are continually expanding our knowledge of how genetic information is encoded and how it specifies traits. These findings are redefining what a gene is.

The power of new methods to identify and analyze genes is illustrated by recent genetic studies of heart attacks in

Table 1.1 Early concepts of heredity

Concept	Proposed	Correct or Incorrect
Pangenesis	Genetic information travels from different parts of the body to reproductive organs.	Incorrect
Inheritance of acquired characteristics	Acquired traits become incorporated into hereditary information.	Incorrect
Preformationism	Miniature organism resides in sex cells, and all traits are inherited from one parent.	Incorrect
Blending inheritance	Genes blend and mix.	Incorrect
Germ-plasm theory	All cells contain a complete set of genetic information.	Correct
Cell theory	All life is composed of cells, and cells arise only from cells.	Correct
Mendelian inheritance	Traits are inherited in accord with defined principles.	Correct

humans. Physicians have long recognized that heart attacks run in families, but finding specific genes that contribute to an increased risk of a heart attack has, until recently, been difficult. In 2009, an international team of geneticists examined the DNA of 26,000 people in 10 countries for single differences in the DNA (called single-nucleotide polymorphisms, or SNPS) that might be associated with an increased risk of myocardial infarction. This study and other similar studies identified several new genes that affect the risk of coronary artery disease and early heart attacks. These findings may make it possible to identify persons who are predisposed to heart attack, allowing early intervention that might prevent the attacks from occurring. Analyses of SNPS are helping to locate genes that affect all types of traits, from eye color and height to glaucoma and heart attacks.

Information about sequence differences among organisms is also a source of new insights about evolution. For example, recent analysis of DNA sequences at 30 genes has revealed that all living cats can trace their ancestry to a pantherlike cat living in Southeast Asia about 11 million years ago and that all living cats can be divided into eight groups or lineages.

In recent years, our understanding of the role of RNA in many cellular processes has expanded greatly; RNA has a role in many aspects of gene function. The discovery in the late 1990s of tiny RNA molecules called small interfering RNAs and micro RNAs led to the recognition that these molecules play central roles in gene expression and development. Today, recognition of the importance of alterations of DNA and chromosome structure that do not include the base sequence of the DNA is increasing. Many such alterations, called epigenetic changes, are stable and affect the expression of traits. New genetic microchips that simultaneously analyze thousands of RNA molecules are providing information about the activities of thousands of genes in a given cell, allowing a detailed picture of how cells respond to external signals, environmental stresses, and disease states such as cancer. In the emerging field of proteomics, powerful computer programs are being used to model the structure and function of proteins from DNA sequence information. All of this information provides us with a better understanding of numerous biological processes and evolutionary relationships. The flood of new genetic information requires the continuous development of sophisticated computer programs to store, retrieve, compare, and analyze genetic data and has given rise to the field of bioinformatics, a merging of molecular biology and computer science.

A number of companies and researchers are racing to develop the technology for sequencing the entire genome of a single person for less than $1000. As the cost of sequencing decreases, the focus of DNA-sequencing efforts will shift from the genomes of different species to individual differences within species. In the not-too-distant future, each person will likely possess a copy of his or her entire genome sequence, which can be used to assess the risk of acquiring various diseases and to tailor their treatment should they arise. The use of genetics in agriculture will contine to improve the productivity of domestic crops and animals, helping to feed the future world population. This ever-widening scope of genetics will raise significant ethical, social, and economic issues.

This brief overview of the history of genetics is not intended to be comprehensive; rather it is designed to provide a sense of the accelerating pace of advances in genetics. In the chapters to come, we will learn more about the experiments and the scientists who helped shape the discipline of genetics.

CONCEPTS

Humans first applied genetics to the domestication of plants and animals between 10,000 and 12,000 years ago. Developments in plant hybridization and cytology in the eighteenth and nineteenth centuries laid the foundation for the field of genetics today. After Mendel's work was rediscovered in 1900, the science of genetics developed rapidly and today is one of the most active areas of science.

✔ CONCEPT CHECK 3

How did developments in cytology in the nineteenth century contribute to our modern understanding of genetics?

1.3 A Few Fundamental Concepts Are Important for the Start of Our Journey into Genetics

Undoubtedly, you learned some genetic principles in other biology classes. Let's take a few moments to review some fundamental genetic concepts.

Cells are of two basic types: eukaryotic and prokaryotic. Structurally, cells consist of two basic types, although, evolutionarily, the story is more complex (see Chapter 2). Prokaryotic cells lack a nuclear membrane and possess no membrane-bounded cell organelles, whereas eukaryotic cells are more complex, possessing a nucleus and membrane-bounded organelles such as chloroplasts and mitochondria.

The gene is the fundamental unit of heredity. The precise way in which a gene is defined often varies, depending on the biological context. At the simplest level, we can think of a gene as a unit of information that encodes a genetic characteristic. We will enlarge this definition as we learn more about what genes are and how they function.

Genes come in multiple forms called alleles. A gene that specifies a characteristic may exist in several forms, called alleles. For example, a gene for coat color in cats may exist as an allele that encodes black fur or as an allele that encodes orange fur.

Genes confer phenotypes. One of the most important concepts in genetics is the distinction between traits and genes. Traits are not inherited directly. Rather, genes are inherited and, along with environmental factors, determine the expression of traits. The genetic information that an individual organism possesses is its genotype; the trait is its phenotype. For example, albinism seen in some Hopis is a phenotype, the information in *OCA2* genes that causes albinism is the genotype.

Genetic information is carried in DNA and RNA. Genetic information is encoded in the molecular structure of nucleic acids, which come in two types: deoxyribonucleic acid (DNA) and ribonucleic acid (RNA). Nucleic acids are polymers consisting of repeating units called nucleotides; each nucleotide consists of a sugar, a phosphate, and a nitrogenous base. The nitrogenous bases in DNA are of four types: adenine (A), cytosine (C), guanine (G), and thymine (T). The sequence of these bases encodes genetic information. DNA consists of two complementary nucleotide strands. Most organisms carry their genetic information in DNA, but a few viruses carry it in RNA. The four nitrogenous bases of RNA are adenine, cytosine, guanine, and uracil (U).

Genes are located on chromosomes. The vehicles of genetic information within a cell are chromosomes (**Figure 1.14**), which consist of DNA and associated proteins. The cells of each species have a characteristic number of chromosomes; for example, bacterial cells normally possess a single chromosome; human cells possess 46; pigeon cells possess 80. Each chromosome carries a large number of genes.

Chromosomes separate through the processes of mitosis and meiosis. The processes of mitosis and meiosis ensure that a complete set of an organism's chromosomes exists in each cell resulting from cell division. Mitosis is the separation of chromosomes in the division of somatic (nonsex) cells. Meiosis is the pairing and separation of chromosomes in the division of sex cells to produce gametes (reproductive cells).

Genetic information is transferred from DNA to RNA to protein. Many genes encode traits by specifying the structure of proteins. Genetic information is first transcribed from DNA into RNA, and then RNA is translated into the amino acid sequence of a protein.

1.14 Genes are carried on chromosomes.

Mutations are permanent changes in genetic information that can be passed from cell to cell or from parent to offspring. Gene mutations affect the genetic information of only a single gene; chromosome mutations alter the number or the structure of chromosomes and therefore usually affect many genes.

Some traits are affected by multiple factors. Some traits are affected by multiple genes that interact in complex ways with environmental factors. Human height, for example, is affected by hundreds of genes as well as environmental factors such as nutrition.

Evolution is genetic change. Evolution can be viewed as a two-step process: first, genetic variation arises and, second, some genetic variants increase in frequency, whereas other variants decrease in frequency. **TRY PROBLEM 24 →**

CONCEPTS SUMMARY

- Genetics is central to the life of every person: it influences a person's physical features, susceptibility to numerous diseases, personality, and intelligence.
- Genetics plays important roles in agriculture, the pharmaceutical industry, and medicine. It is central to the study of biology.
- All organisms use similar genetic systems. Genetic variation is the foundation of evolution and is critical to understanding all life.
- The study of genetics can be broadly divided into transmission genetics, molecular genetics, and population genetics.
- Model genetic organisms are species about which much genetic information exists because they have characteristics that make them particularly amenable to genetic analysis.

- The use of genetics by humans began with the domestication of plants and animals.
- Ancient Greeks developed the concepts of pangenesis and the inheritance of acquired characteristics. Ancient Romans developed practical measures for the breeding of plants and animals.
- Preformationism suggested that a person inherits all of his or her traits from one parent. Blending inheritance proposed that offspring possess a mixture of the parental traits.
- By studying the offspring of crosses between varieties of peas, Gregor Mendel discovered the principles of heredity. Developments in cytology in the nineteenth century led to the understanding that the cell nucleus is the site of heredity.
- In 1900, Mendel's principles of heredity were rediscovered. Population genetics was established in the early 1930s,

followed closely by biochemical genetics and bacterial and viral genetics. The structure of DNA was discovered in 1953, stimulating the rise of molecular genetics.
- Cells are of two basic types: prokaryotic and eukaryotic.
- The genes that determine a trait are termed the genotype; the trait that they produce is the phenotype.
- Genes are located on chromosomes, which are made up of nucleic acids and proteins and are partitioned into daughter cells through the process of mitosis or meiosis.
- Genetic information is expressed through the transfer of information from DNA to RNA to proteins.
- Evolution requires genetic change in populations.

IMPORTANT TERMS

genome (p. 4)
transmission genetics (p. 5)
molecular genetics (p. 5)
population genetics (p. 5)
model genetic organism (p. 5)
pangenesis (p. 8)
inheritance of acquired characteristics (p. 8)
preformationism (p. 8)
blending inheritance (p. 9)
cell theory (p. 9)
germ-plasm theory (p. 9)

ANSWERS TO CONCEPT CHECKS

1. d
2. No, because horses are expensive to house, feed, and propagate, they have too few progeny, and their generation time is too long.
3. Developments in cytology in the 1800s led to the identification of parts of the cell, including the cell nucleus and chromosomes. The cell theory focused the attention of biologists on the cell, eventually leading to the conclusion that the nucleus contains the hereditary information.

COMPREHENSION QUESTIONS

Answers to questions and problems preceded by an asterisk can be found at the end of the book.

Section 1.1

*1. How does the Hopi culture contribute to the high incidence of albinism among members of the Hopi tribe?

2. Outline some of the ways in which genetics is important to each of us.

*3. Give at least three examples of the role of genetics in society today.

4. Briefly explain why genetics is crucial to modern biology.

*5. List the three traditional subdisciplines of genetics and summarize what each covers.

6. What are some characteristics of model genetic organisms that make them useful for genetic studies?

Section 1.2

7. When and where did agriculture first arise? What role did genetics play in the development of the first domesticated plants and animals?

*8. Outline the notion of pangenesis and explain how it differs from the germ-plasm theory.

9. What does the concept of the inheritance of acquired characteristics propose and how is it related to the notion of pangenesis?

*10. What is preformationism? What did it have to say about how traits are inherited?

11. Define blending inheritance and contrast it with preformationism.

12. How did developments in botany in the seventeenth and eighteenth centuries contribute to the rise of modern genetics?

*13. Who first discovered the basic principles that laid the foundation for our modern understanding of heredity?

14. List some advances in genetics made in the twentieth century.

Section 1.3

15. What are the two basic cell types (from a structural perspective) and how do they differ?

*16. Outline the relations between genes, DNA, and chromosomes.

APPLICATION QUESTIONS AND PROBLEMS

Section 1.1

17. What is the relation between genetics and evolution?

*18. For each of the following genetic topics, indicate whether it focuses on transmission genetics, molecular genetics, or population genetics.

a. Analysis of pedigrees to determine the probability of someone inheriting a trait

b. Study of people on a small island to determine why a genetic form of asthma is prevalent on the island

c. Effect of nonrandom mating on the distribution of genotypes among a group of animals

d. Examination of the nucleotide sequences found at the ends of chromosomes
e. Mechanisms that ensure a high degree of accuracy in DNA replication
f. Study of how the inheritance of traits encoded by genes on sex chromosomes (sex-linked traits) differs from the inheritance of traits encoded by genes on nonsex chromosomes (autosomal traits)

19. Describe some of the ways in which your own genetic makeup affects you as a person. Be as specific as you can.

20. Describe at least one trait that appears to run in your family (appears in multiple members of the family). Does this trait run in your family because it is an inherited trait or because it is caused by environmental factors that are common to family members? How might you distinguish between these possibilities?

Section 1.2

*21. Genetics is said to be both a very old science and a very young science. Explain what is meant by this statement.

22. Match the description (*a* through *d*) with the correct theory or concept listed below.

    Preformationism
    Pangenesis
    Germ-plasm theory
    Inheritance of acquired characteristics

    a. Each reproductive cell contains a complete set of genetic information.
    b. All traits are inherited from one parent.
    c. Genetic information may be altered by the use of a characteristic.
    d. Cells of different tissues contain different genetic information.

*23. Compare and contrast the following ideas about inheritance.
    a. Pangenesis and germ-plasm theory
    b. Preformationism and blending inheritance
    c. The inheritance of acquired characteristics and our modern theory of heredity

Section 1.3

*24. Compare and contrast the following terms:
    a. Eukaryotic and prokaryotic cells
    b. Gene and allele
    c. Genotype and phenotype
    d. DNA and RNA
    e. DNA and chromosome

CHALLENGE QUESTIONS

Introduction

25. The type of albinism that arises with high frequency among Hopi Native Americans (discussed in the introduction to this chapter) is most likely oculocutaneous albinism type 2, due to a defect in the *OCA2* gene on chromosome 15. Do some research on the Internet to determine how the phenotype of this type of albinism differs from phenotypes of other forms of albinism in humans and which genes take part. Hint: Visit the Online Mendelian Inheritance in Man Web site (http://www.ncbi.nlm.nih.gov/omim/) and search the database for albinism.

Section 1.1

26. We now know as much or more about the genetics of humans as we know about that of any other organism, and humans are the focus of many genetic studies. Should humans be considered a model genetic organism? Why or why not?

Section 1.3

*27. Suppose that life exists elsewhere in the universe. All life must contain some type of genetic information, but alien genomes might not consist of nucleic acids and have the same features as those found in the genomes of life on Earth. What might be the common features of all genomes, no matter where they exist?

28. Choose one of the ethical or social issues in parts *a* through *e* and give your opinion on the issue. For background information, you might read one of the articles on ethics marked with an asterisk in the Suggested Readings section for Chapter 1 at www.whfreeman.com/pierce4e.

    a. Should a person's genetic makeup be used in determining his or her eligibility for life insurance?
    b. Should biotechnology companies be able to patent newly sequenced genes?
    c. Should gene therapy be used on people?
    d. Should genetic testing be made available for inherited conditions for which there is no treatment or cure?
    e. Should governments outlaw the cloning of people?

*29. A 45-year old women undergoes genetic testing and discovers that she is at high risk for developing colon cancer and Alzheimer disease. Because her children have 50% of her genes, they also may have increased risk of these diseases. Does she have a moral or legal obligation to tell her children and other close relatives about the results of her genetic testing?

*30. Suppose that you could undergo genetic testing at age 18 for susceptibility to a genetic disease that would not appear until middle age and has no available treatment.

    a. What would be some of the possible reasons for having such a genetic test and some of the possible reasons for not having the test?
    b. Would you personally want to be tested? Explain your reasoning.

2 Chromosomes and Cellular Reproduction

Chromosomes in mitosis, the process through which each new cell receives a complete copy of the genetic material. [Photograph by Conly L. Reider/Biological Photo Service.]

THE BLIND MEN'S RIDDLE

In a well-known riddle, two blind men by chance enter a department store at the same time, go to the same counter, and both order five pairs of socks, each pair a different color. The sales clerk is so befuddled by this strange coincidence that he places all ten pairs (two black pairs, two blue pairs, two gray pairs, two brown pairs, and two green pairs) into a single shopping bag and gives the bag with all ten pairs to one blind man and an empty bag to the other. The two blind men happen to meet on the street outside, where they discover that one of their bags contains all ten pairs of socks. How do the blind men, without seeing and without any outside help, sort out the socks so that each man goes home with exactly five pairs of different colored socks? Can you come up with a solution to the riddle?

By an interesting coincidence, cells have the same dilemma as that of the blind men in the riddle. Most organisms possess two sets of genetic information, one set inherited from each parent. Before cell division, the DNA in each chromosome replicates; after replication, there are two copies—called sister chromatids—of each chromosome. At the end of cell division, it is critical that each new cell receives a complete copy of the genetic material, just as each blind man needed to go home with a complete set of socks.

The solution to the riddle is simple. Socks are sold as pairs; the two socks of a pair are typically connected by a thread. As a pair is removed from the bag, the men each grasp a different sock of the pair and pull in opposite directions. When the socks are pulled tight, it is easy for one of the men to take a pocket knife and cut the thread connecting the pair. Each man then deposits his single sock in his own bag. At the end of the process, each man's bag will contain exactly two black socks, two blue socks, two gray socks, two brown socks, and two green socks.*

Remarkably, cells employ a similar solution for separating their chromosomes into new daughter cells. As we will learn in this chapter, the replicated chromosomes line up at the center of a cell undergoing division and, like the socks in the riddle, the sister chromatids of each chromosome are pulled in opposite directions. Like the thread connecting two socks of a pair, a molecule called cohesin holds the sister chromatids together until severed

---

*This analogy is adapted from K. Nasmyth. Disseminating the genome: joining, resolving, and separating sister chromatids, during mitosis and meiosis. *Annual Review of Genetics* 35:673–745, 2001.

by a molecular knife called separase. The two resulting chromosomes separate and the cell divides, ensuring that a complete set of chromosomes is deposited in each cell.

In this analogy, the blind men and cells differ in one critical regard: if the blind men make a mistake, one man ends up with an extra sock and the other is a sock short, but no great harm results. The same cannot be said for human cells. Errors in chromosome separation, producing cells with too many or too few chromosomes, are frequently catastrophic, leading to cancer, miscarriage, or—in some cases—a child with severe handicaps.

This chapter explores the process of cell reproduction and how a complete set of genetic information is transmitted to new cells. In prokaryotic cells, reproduction is relatively simple, because prokaryotic cells possess a single chromosome. In eukaryotic cells, multiple chromosomes must be copied and distributed to each of the new cells, and so cell reproduction is more complex. Cell division in eukaryotes takes place through mitosis or meiosis, processes that serve as the foundation for much of genetics.

Grasping mitosis and meiosis requires more than simply memorizing the sequences of events that take place in each stage, although these events are important. The key is to understand how genetic information is apportioned in the course of cell reproduction through a dynamic interplay of DNA synthesis, chromosome movement, and cell division. These processes bring about the transmission of genetic information and are the basis of similarities and differences between parents and progeny.

	Prokaryotic cells	Eukaryotic cells
Nucleus	Absent	Present
Cell diameter	Relatively small, from 1 to 10 μm	Relatively large, from 10 to 100 μm
Genome	Usually one circular DNA molecule	Multiple linear DNA molecules
DNA	Not complexed with histones in eubacteria; some histones in archaea	Complexed with histones
Amount of DNA	Relatively small	Relatively large
Membrane-bounded organelles	Absent	Present
Cytoskeleton	Absent	Present

2.1 Prokaryotic and eukaryotic cells differ in structure. [Photographs (left to right) by T. J. Beveridge/Visuals Unlimited/Getty Images (prokaryotes); W. Baumeister/Science Photo Library/Photo Researchers; G. Murti/Phototake; Biophoto Associates/Photo Researchers.]

2.1 Prokaryotic and Eukaryotic Cells Differ in a Number of Genetic Characteristics

Biologists traditionally classify all living organisms into two major groups, the *prokaryotes* and the *eukaryotes* (**Figure 2.1**). A **prokaryote** is a unicellular organism with a relatively simple cell structure. A **eukaryote** has a compartmentalized cell structure with components bounded by intracellular membranes; eukaryotes are either unicellular or multicellular.

Research indicates that a division of life into two major groups, the prokaryotes and eukaryotes, is not so simple. Although similar in cell structure, prokaryotes include at least two fundamentally distinct types of bacteria: the **eubacteria** (true bacteria) and the **archaea** (ancient bacteria). An examination of equivalent DNA sequences reveals that eubacteria and archaea are as distantly related to one another as they are to the eukaryotes. Although eubacteria and archaea are similar in cell structure, some genetic processes in archaea (such as transcription) are more similar to those in eukaryotes, and the archaea are actually closer evolutionarily to eukaryotes than to eubacteria. Thus, from an evolutionary perspective, there are three major groups of organisms: eubacteria, archaea, and eukaryotes. In this book, the prokaryotic–eukaryotic distinction will be made frequently, but important eubacterial–archaeal differences also will be noted.

From the perspective of genetics, a major difference between prokaryotic and eukaryotic cells is that a eukaryote has a nuclear envelope, which surrounds the genetic material to form a **nucleus** and separates the DNA from the other cellular contents. In prokaryotic cells, the genetic material is in close contact with other components of the cell—a property that has important consequences for the way in which genes are controlled.

Another fundamental difference between prokaryotes and eukaryotes lies in the packaging of their DNA. In eukaryotes, DNA is closely associated with a special class of proteins, the **histones**, to form tightly packed chromosomes. This complex of DNA and histone proteins is termed **chromatin**, which is the stuff of eukaryotic chromosomes (**Figure 2.2**). Histone proteins limit the accessibility of enzymes and other proteins that copy and read the DNA, but they enable the DNA to fit into the nucleus. Eukaryotic DNA must separate from the histones before the genetic information in the DNA can be accessed. Archaea also have some histone proteins that complex with DNA, but the structure of their chromatin is different from that found in eukaryotes. Eubacteria do not possess histones; so their DNA does not exist in the highly ordered, tightly packed arrangement found in eukaryotic cells (**Figure 2.3**). The copying and reading of DNA are therefore simpler processes in eubacteria.

Genes of prokaryotic cells are generally on a single, circular molecule of DNA—the chromosome of a prokaryotic cell. In eukaryotic cells, genes are located on multiple,

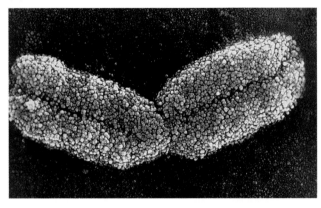

2.2 Eukaryotic chromosomes consist of DNA and histone proteins. (a) DNA wraps around the histone proteins to form chromatin, the material that makes up chromosomes. (b) A eukaryotic chromosome. [Part b: Biophoto Associates/Photo Researchers.]

usually linear DNA molecules (multiple chromosomes). Eukaryotic cells therefore require mechanisms that ensure that a copy of each chromosome is faithfully transmitted to each new cell. This generalization—a single, circular chromosome in prokaryotes and multiple, linear chromosomes in eukaryotes—is not always true. A few bacteria have more than one chromosome, and important bacterial genes are frequently found on other DNA molecules called *plasmids* (see Chapter 8). Furthermore, in some eukaryotes, a few genes are located on circular DNA molecules found in certain organelles.

2.3 Prokaryotic DNA (shown in red) is neither surrounded by a nuclear membrane nor complexed with histone proteins.
[A. B. Dowsett/Science Photo Library/Photo Researchers.]

CONCEPTS

Organisms are classified as prokaryotes or eukaryotes, and prokaryotes consist of archaea and eubacteria. A prokaryote is a unicellular organism that lacks a nucleus, its DNA is not complexed to histone proteins, and its genome is usually a single chromosome. Eukaryotes are either unicellular or multicellular, their cells possess a nucleus, their DNA is complexed to histone proteins, and their genomes consist of multiple chromosomes.

✔ CONCEPT CHECK 1

List several characteristics that eubacteria and archaea have in common and that distinguish them from eukaryotes.

Viruses are neither prokaryotic nor eukaryotic, because they do not possess a cellular structure. Viruses are actually simple structures composed of an outer protein coat surrounding nucleic acid (either DNA or RNA; **Figure 2.4**). Neither are viruses primitive forms of life: they can reproduce only within host cells, which means that they must have evolved after, rather than before, cells evolved. In addition, viruses are not an evolutionarily distinct group but are most closely related to their hosts: the genes of a plant virus are more similar to those in a plant cell than to those in animal viruses, which suggests that viruses evolved from their hosts, rather than from other viruses. The close relationship between the genes of virus and host makes viruses useful for studying the genetics of host organisms.

2.2 Cell Reproduction Requires the Copying of the Genetic Material, Separation of the Copies, and Cell Division

For any cell to reproduce successfully, three fundamental events must take place: (1) its genetic information must be copied, (2) the copies of genetic information must be separated from each other, and (3) the cell must divide. All cellular reproduction includes these three events, but the processes that lead to these events differ in prokaryotic and eukaryotic cells because of their structural differences.

Prokaryotic Cell Reproduction

When prokaryotic cells reproduce, the circular chromosome of the bacterium replicates and the cell divides in a process called binary fission (**Figure 2.5**). Replication usually begins at a specific place on the bacterial chromosome, called the origin of replication. In a process that is not fully understood, the origins of the two newly replicated chromosomes move away from each other and toward opposite ends of the cell. In at least some bacteria, proteins bind near the replication origins and anchor the new chromosomes to the plasma membrane at opposite ends of the cell. Finally, a new cell wall forms between the two chromosomes, producing two cells, each with an identical copy of the chromosome. Under optimal conditions, some bacterial cells divide every 20 minutes. At this rate, a single bacterial cell could produce a billion descendants in a mere 10 hours.

Eukaryotic Cell Reproduction

Like prokaryotic cell reproduction, eukaryotic cell reproduction requires the processes of DNA replication, copy separation, and division of the cytoplasm. However, the presence of multiple DNA molecules requires a more-complex mechanism to ensure that exactly one copy of each molecule ends up in each of the new cells.

Eukaryotic chromosomes are separated from the cytoplasm by the nuclear envelope. The nucleus was once thought to be a fluid-filled bag in which the chromosomes floated, but we now know that the nucleus has a highly organized internal scaffolding called the nuclear matrix. This matrix consists of a network of protein fibers that maintains precise spatial relations among the nuclear components and takes part in DNA replication, the expression of genes, and the modification of gene products before they leave the

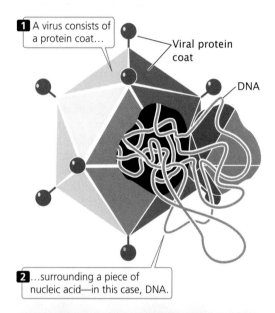

1 A virus consists of a protein coat…
Viral protein coat
DNA

2 …surrounding a piece of nucleic acid—in this case, DNA.

2.4 A virus is a simple replicative structure consisting of protein and nucleic acid. Adenoviruses are shown in the micrograph.
[Micrograph by Hans Gelderblom/Visuals Unlimited.]

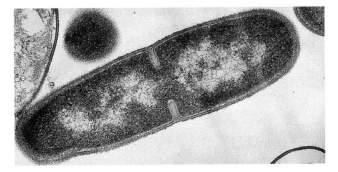

2.5 Prokaryotic cells reproduce by binary fission. (a) Process of binary fission. (b) Micrograph showing a bacterial cell undergoing binary fission. [Part b: Lee D. Simon/Photo Researchers.]

nucleus. We will now take a closer look at the structure of eukaryotic chromosomes.

Eukaryotic chromosomes Each eukaryotic species has a characteristic number of chromosomes per cell: potatoes have 48 chromosomes, fruit flies have 8, and humans have 46. There appears to be no special relation between the complexity of an organism and its number of chromosomes per cell.

In most eukaryotic cells, there are two sets of chromosomes. The presence of two sets is a consequence of sexual reproduction: one set is inherited from the male parent and the other from the female parent. Each chromosome in one set has a corresponding chromosome in the other set, together constituting a homologous pair (**Figure 2.6**). Human cells, for example, have 46 chromosomes, constituting 23 homologous pairs.

The two chromosomes of a **homologous pair** are usually alike in structure and size, and each carries genetic information for the same set of hereditary characteristics. (The sex chromosomes are an exception and will be discussed in Chapter 4.) For example, if a gene on a particular chromosome encodes a characteristic such as hair color, another copy of the gene (each copy is called an allele) at the same position on that chromosome's homolog *also* encodes hair color. However, these two alleles need not be identical: one might encode brown hair and the other might encode blond hair. Thus, most cells carry two sets of genetic information; these cells are **diploid**. But not all eukaryotic cells are diploid: reproductive cells (such as eggs, sperm, and spores) and even nonreproductive cells of some organisms may contain a single set of chromosomes. Cells with a single set of chromosomes are **haploid**. A haploid cell has only one copy of each gene.

CONCEPTS

Cells reproduce by copying and separating their genetic information and then dividing. Because eukaryotes possess multiple chromosomes, mechanisms exist to ensure that each new cell receives one copy of each chromosome. Most eukaryotic cells are diploid, and their two chromosome sets can be arranged in homologous pairs. Haploid cells contain a single set of chromosomes.

✔ **CONCEPT CHECK 2**

Diploid cells have

a. two chromosomes
b. two sets of chromosomes
c. one set of chromosomes
d. two pairs of homologous chromosomes

Chromosome structure The chromosomes of eukaryotic cells are larger and more complex than those found in prokaryotes, but each unreplicated chromosome nevertheless consists of a single molecule of DNA. Although linear, the DNA molecules in eukaryotic chromosomes are highly folded and condensed; if stretched out, some human chromosomes would be several centimeters long—thousands of

2.6 Diploid eukaryotic cells have two sets of chromosomes. (a) A set of chromosomes from a female human cell. Each pair of chromosomes is hybridized to a uniquely colored probe, giving it a distinct color. (b) The chromosomes are present in homologous pairs, which consist of chromosomes that are alike in size and structure and carry information for the same characteristics. [Part a: Courtesy of Dr. Thomas Ried and Dr. Evelin Schrock.]

times as long as the span of a typical nucleus. To package such a tremendous length of DNA into this small volume, each DNA molecule is coiled again and again and tightly packed around histone proteins, forming a rod-shaped chromosome. Most of the time, the chromosomes are thin and difficult to observe but, before cell division, they condense further into thick, readily observed structures; it is at this stage that chromosomes are usually studied.

A functional chromosome has three essential elements: a centromere, a pair of telomeres, and origins of replication. The centromere is the attachment point for spindle microtubules—the filaments responsible for moving chromosomes in cell division (**Figure 2.7**). The centromere appears as a constricted region. Before cell division, a multiprotein complex called the kinetochore assembles on the centromere; later, spindle microtubules attach to the kinetochore. Chromosomes lacking a centromere cannot be drawn into the newly formed nuclei; these chromosomes are lost, often with catastrophic consequences for the cell. On the basis of the location of the centromere, chromosomes are classified into four types: metacentric, submetacentric, acrocentric, and telocentric (**Figure 2.8**). One of the two arms of a chromosome (the short arm of a submetacentric or acrocentric chromosome) is designated by the letter p and the other arm is designated by q.

Telomeres are the natural ends, the tips, of a whole linear chromosome (see Figure 2.7). Just as plastic tips protect the ends of a shoelace, telomeres protect and stabilize the chromosome ends. If a chromosome breaks, producing new ends, the chromosome is degraded at the newly broken ends.

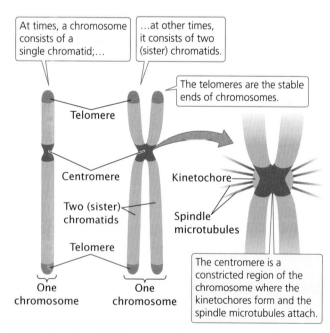

2.7 Each eukaryotic chromosome has a centromere and telomeres.

2.8 Eukaryotic chromosomes exist in four major types based on the position of the centromere.
[Micrograph by L. Lisco, D. W. Fawcett/Visuals Unlimited.]

Telomeres provide chromosome stability. Research shows that telomeres also participate in limiting cell division and may play important roles in aging and cancer.

Origins of replication are the sites where DNA synthesis begins; they are not easily observed by microscopy. In preparation for cell division, each chromosome replicates, making a copy of itself, as already mentioned. These two initially identical copies, called **sister chromatids**, are held together at the centromere (see Figure 2.7). Each sister chromatid consists of a single molecule of DNA.

CONCEPTS

Sister chromatids are copies of a chromosome held together at the centromere. Functional chromosomes contain centromeres, telomeres, and origins of replication. The kinetochore is the point of attachment for the spindle microtubules; telomeres are the stabilizing ends of a chromosome; origins of replication are sites where DNA synthesis begins.

✔ CONCEPT CHECK 3

What would be the result if a chromosome did not have a kinetochore?

The Cell Cycle and Mitosis

The **cell cycle** is the life story of a cell, the stages through which it passes from one division to the next (**Figure 2.9**). This process is critical to genetics because, through the cell cycle, the genetic instructions for all characteristics are passed from parent to daughter cells. A new cycle begins after a cell has divided and produced two new cells. Each new cell metabolizes, grows, and develops. At the end of its cycle, the cell divides to produce two cells, which can then undergo additional cell cycles. Progression through the cell cycle is regulated at key transition points called **checkpoints**.

The cell cycle consists of two major phases. The first is **interphase**, the period between cell divisions, in which the cell grows, develops, and functions. In interphase, critical events necessary for cell division also take place. The second major phase is the **M phase** (mitotic phase), the period of active cell division. The M phase includes **mitosis**, the process of nuclear division, and **cytokinesis**, or cytoplasmic division. Let's take a closer look at the details of interphase and the M phase.

Interphase Interphase is the extended period of growth and development between cell divisions. Although little activity can be observed with a light microscope, the cell is quite busy: DNA is being synthesized, RNA and proteins are being produced, and hundreds of biochemical reactions necessary for cellular functions are taking place. In addition to growth and development, interphase includes several checkpoints, which regulate the cell cycle by allowing or prohibiting the cell's division. These checkpoints, like the checkpoints in the M phase, ensure that all cellular components are present and in good working order before the cell proceeds to the next stage. Checkpoints are necessary to

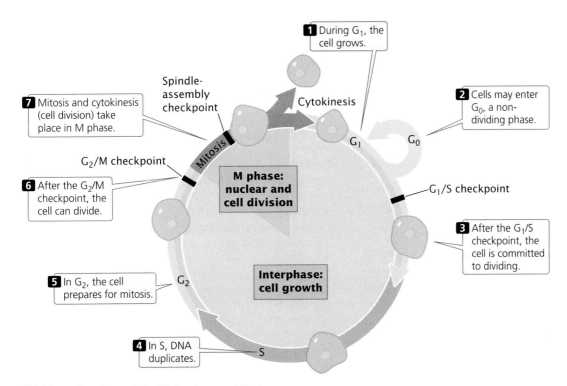

2.9 The cell cycle consists of interphase and M phase.

prevent cells with damaged or missing chromosomes from proliferating. Defects in checkpoints can lead to unregulated cell growth, as is seen in some cancers.

By convention, interphase is divided into three subphases: G_1, S, and G_2 (see Figure 2.9). Interphase begins with G_1 (for gap 1). In G_1, the cell grows, and proteins necessary for cell division are synthesized; this phase typically lasts several hours. Near the end of G_1, a critical point termed the G_1/S checkpoint holds the cell in G_1 until the cell has all of the enzymes necessary for the replication of DNA. After this checkpoint has been passed, the cell is committed to divide.

Before reaching the G_1/S checkpoint, cells may exit from the active cell cycle in response to regulatory signals and pass into a nondividing phase called G_0, which is a stable state during which cells usually maintain a constant size. They can remain in G_0 for an extended length of time, even indefinitely, or they can reenter G_1 and the active cell cycle. Many cells never enter G_0; rather, they cycle continuously.

After G_1, the cell enters the S phase (for DNA synthesis), in which each chromosome duplicates. Although the cell is committed to divide after the G_1/S checkpoint has been passed, DNA synthesis must take place before the cell can proceed to mitosis. If DNA synthesis is blocked (by drugs or by a mutation), the cell will not be able to undergo mitosis. Before the S phase, each chromosome is unreplicated; after the S phase, each chromosome is composed of two chromatids (see Figure 2.7).

After the S phase, the cell enters G_2 (gap 2). In this phase, several additional biochemical events necessary for cell division take place. The important G_2/M checkpoint is reached near the end of G_2. This checkpoint is passed only if the cell's DNA is undamaged. Damaged DNA can inhibit the activation of some proteins that are necessary for mitosis to take place. After the G_2/M checkpoint has been passed, the cell is ready to divide and enters the M phase. Although the length of interphase varies from cell type to cell type, a typical dividing mammalian cell spends about 10 hours in G_1, 9 hours in S, and 4 hours in G_2 (see Figure 2.9).

Throughout interphase, the chromosomes are in a relaxed, but by no means uncoiled, state, and individual chromosomes cannot be seen with a microscope. This condition changes dramatically when interphase draws to a close and the cell enters the M phase.

M phase The M phase is the part of the cell cycle in which the copies of the cell's chromosomes (sister chromatids) separate and the cell undergoes division. The separation of sister chromatids in the M phase is a critical process that results in a complete set of genetic information for each of the resulting cells. Biologists usually divide the M phase into six stages: the five stages of mitosis (prophase, prometaphase, metaphase, anaphase, and telophase), illustrated in **Figure 2.10**, and cytokinesis. It's important to keep in mind that the M phase is a continuous process, and its separation into these six stages is somewhat arbitrary.

Prophase. As a cell enters prophase, the chromosomes become visible under a light microscope. Because the chromosome was duplicated in the preceding S phase, each chromosome possesses two chromatids attached at the centromere. The mitotic spindle, an organized array of microtubules that move the chromosomes in mitosis, forms. In animal cells, the spindle grows out from a pair of centrosomes that migrate to opposite sides of the cell. Within each centrosome is a special organelle, the centriole, which also is composed of microtubules. Some plant cells do not have centrosomes or centrioles, but they do have mitotic spindles.

Prometaphase. Disintegration of the nuclear membrane marks the start of prometaphase. Spindle microtubules, which until now have been outside the nucleus, enter the nuclear region. The ends of certain microtubules make contact with the chromosomes. For each chromosome, a microtubule from one of the centrosomes anchors to the kinetochore of *one* of the sister chromatids; a microtubule from the opposite centrosome then attaches to the other sister chromatid, and so the chromosome is anchored to both of the centrosomes. The microtubules lengthen and shorten, pushing and pulling the chromosomes about. Some microtubules extend from each centrosome toward the center of the spindle but do not attach to a chromosome.

Metaphase. During metaphase, the chromosomes become arranged in a single plane, the metaphase plate, between the two centrosomes. The centrosomes, now at opposite ends of the cell with microtubules radiating outward and meeting in the middle of the cell, center at the spindle poles. A spindle-assembly checkpoint ensures that each chromosome is aligned on the metaphase plate and attached to spindle fibers from opposite poles.

The passage of a cell through the spindle-assembly checkpoint depends on tension generated at the kinetochore as the two conjoined chromatids are pulled in opposite directions by the spindle fibers. This tension is required for the cell to pass through the spindle-assembly checkpoint. If a microtubule attaches to one chromatid but not to the other, no tension is generated and the cell is unable to progress to the next stage of the cell cycle. The spindle-assembly checkpoint is able to detect even a single pair of chromosomes that are not properly attached to microtubules. The importance of this checkpoint is illustrated by cells that are defective in their spindle-assembly checkpoint; these cells often end up with abnormal numbers of chromosomes.

Anaphase. After the spindle-assembly checkpoint has been passed, the connection between sister chromatids breaks down and the sister chromatids separate. This chromatid separation marks the beginning of anaphase, during which the chromosomes move toward opposite spindle poles. The microtubules that connect the chromosomes to the spindle poles are composed of subunits of a protein

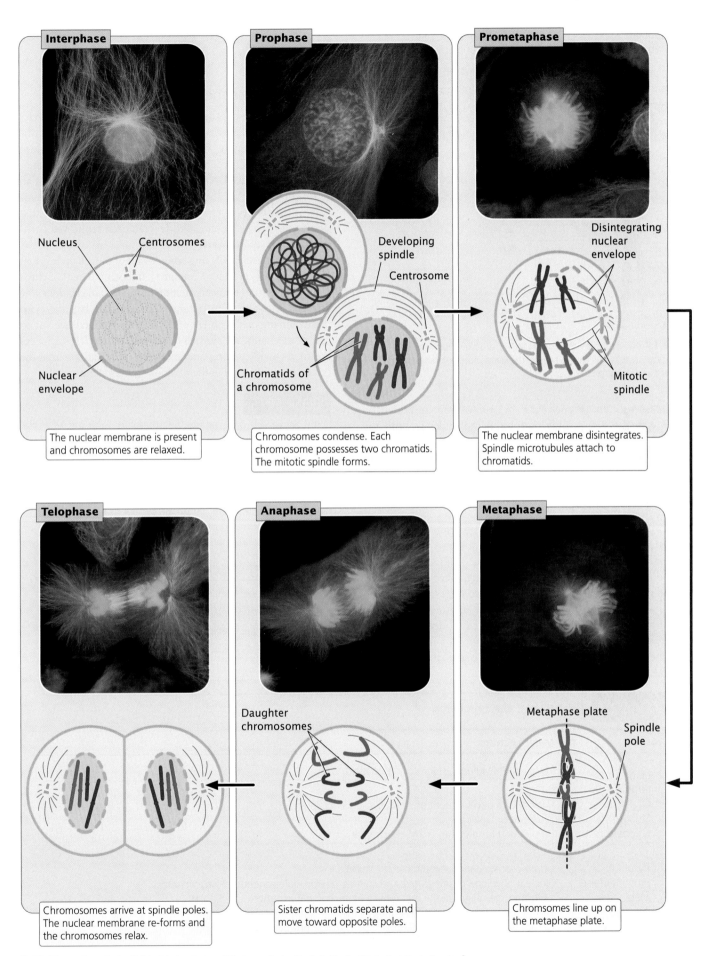

2.10 The cell cycle is divided into stages. [Photographs by Conly L. Rieder/Biological Photo Service.]

2.11 Microtubules are composed of tubulin subunits. Each microtubule has its plus (+) end at the kinetochore and its negative (−) end at the centrosome.

called tubulin (**Figure 2.11**). Chromosome movement is due to the disassembly of tubulin molecules at both the kinetochore end (called the + end) and the spindle end (called the − end) of the spindle fiber. Special proteins called molecular motors disassemble tubulin molecules from the spindle and generate forces that pull the chromosome toward the spindle pole.

Telophase. After the chromatids have separated, each is considered a separate chromosome. Telophase is marked by the arrival of the chromosomes at the spindle poles. The nuclear membrane re-forms around each set of chromosomes, producing two separate nuclei within the cell. The chromosomes relax and lengthen, once again disappearing from view. In many cells, division of the cytoplasm (cytokinesis) is simultaneous with telophase. The major features of the cell cycle are summarized in **Table 2.1**. **TRY PROBLEM 22** →

Genetic Consequences of the Cell Cycle

What are the genetically important results of the cell cycle? From a single cell, the cell cycle produces two cells that contain the same genetic instructions. The resulting daughter cells are genetically identical with each other and with their parent cell because DNA synthesis in the S phase creates an exact copy of each DNA molecule, giving rise to two genetically identical sister chromatids. Mitosis then ensures that one of the two sister chromatids from each replicated chromosome passes into each new cell.

Another genetically important result of the cell cycle is that each of the cells produced contains a full complement of chromosomes: there is no net reduction or increase in chromosome number. Each cell also contains approximately half the cytoplasm and organelle content of the original parental cell, but no precise mechanism analogous to mitosis ensures that organelles are evenly divided. Consequently, not all cells resulting from the cell cycle are identical in their cytoplasmic content.

CONCEPTS

The active cell cycle phases are interphase and the M phase. Interphase consists of G_1, S, and G_2. In G_1, the cell grows and prepares for cell division; in the S phase, DNA synthesis takes place; in G_2, other biochemical events necessary for cell division take place. Some cells enter a quiescent phase called G_0. The M phase includes mitosis and cytokinesis and is divided into prophase, prometaphase, metaphase, anaphase, and telophase. The cell cycle produces two genetically identical cells each of which possesses a full complement of chromosomes.

✔ CONCEPT CHECK 4

Which is the correct order of stages in the cell cycle?

a. G_1, S, prophase, metaphase, anaphase
b. S, G_1, prophase, metaphase, anaphase
c. Prophase, S, G_1, metaphase, anaphase
d. S, G_1, anaphase, prophase, metaphase

Table 2.1	Features of the cell cycle
Stage	Major Features
G_0 phase	Stable, nondividing period of variable length.
Interphase	
G_1 phase	Growth and development of the cell; G_1/S checkpoint.
S phase	Synthesis of DNA.
G_2 phase	Preparation for division; G_2/M checkpoint.
M phase	
Prophase	Chromosomes condense and mitotic spindle forms.
Prometaphase	Nuclear envelope disintegrates, and spindle microtubules anchor to kinetochores.
Metaphase	Chromosomes align on the metaphase plate; spindle-assembly checkpoint.
Anaphase	Sister chromatids separate, becoming individual chromosomes that migrate toward spindle poles.
Telophase	Chromosomes arrive at spindle poles, the nuclear envelope re-forms, and the condensed chromosomes relax.
Cytokinesis	Cytoplasm divides; cell wall forms in plant cells.

2.12 The number of chromosomes and the number of DNA molecules change in the course of the cell cycle. The number of chromosomes per cell equals the number of functional centromeres. The number of DNA molecules per cell equals the number of chromosomes when the chromosomes are unreplicated (no sister chromatids present) and twice the number of chromosomes when sister chromosomes *are* present.

CONNECTING CONCEPTS

Counting Chromosomes and DNA Molecules

The relations among chromosomes, chromatids, and DNA molecules frequently cause confusion. At certain times, chromosomes are unreplicated; at other times, each possesses two chromatids (see Figure 2.7). Chromosomes sometimes consist of a single DNA molecule; at other times, they consist of two DNA molecules. How can we keep track of the number of these structures in the cell cycle?

There are two simple rules for counting chromosomes and DNA molecules: (1) to determine the number of chromosomes, count the number of functional centromeres; (2) to determine the number of DNA molecules, first determine if sister chromatids are present. If sister chromatids are present, the chromosome has replicated and the number of DNA molecules is twice the number of chromosomes. If sister chromatids are not present, the chromosome has not replicated and the number of DNA molecules is the same as the number of chromosomes.

Let's examine a hypothetical cell as it passes through the cell cycle (**Figure 2.12**). At the beginning of G_1, this diploid cell has two complete sets of chromosomes, inherited from its parent cell. Each chromosome is unreplicated and consists of a single molecule of DNA, and so there are four DNA molecules in the cell during G_1. In the S phase, each DNA molecule is copied. The two resulting DNA molecules combine with histones and other proteins to form sister chromatids. Although the amount of DNA doubles in the S phase, the number of chromosomes remains the same because the sister chromatids are tethered together and share a single functional centromere. At the end of the S phase, this cell still contains four chromosomes, each with two sister chromatids; so $4 \times 2 = 8$ DNA molecules are present.

Through prophase, prometaphase, and metaphase, the cell has four chromosomes and eight DNA molecules. At anaphase, however, the sister chromatids separate. Each now has its own functional centromere, and so each is considered a separate chromosome. Until cytokinesis, the cell contains eight unreplicated chromosomes; thus, there are still eight DNA molecules present. After cytokinesis, the eight chromosomes (eight DNA molecules) are distributed equally between two cells; so each new cell contains four chromosomes and four DNA molecules, the number present at the beginning of the cell cycle.

In summary, the number of chromosomes increases only in anaphase, when the two chromatids of a chromosome separate and become distinct chromosomes. The number of chromosomes decreases only through cytokinesis. The number of DNA molecules increases only in the S phase and decreases only through cytokinesis. **TRY PROBLEM 23** →

2.3 Sexual Reproduction Produces Genetic Variation Through the Process of Meiosis

If all reproduction were accomplished through mitosis, life would be quite dull, because mitosis produces only genetically identical progeny. With only mitosis, you, your children, your parents, your brothers and sisters, your cousins, and many people you don't even know would be clones—copies of one another. Only the occasional mutation would introduce any genetic variability. All organisms reproduced in this way for the first 2 billion years of Earth's existence (and it is the way in which some organisms still reproduce today). Then, some 1.5 billion to 2 billion years ago, something remarkable evolved: cells that produce genetically variable offspring through sexual reproduction.

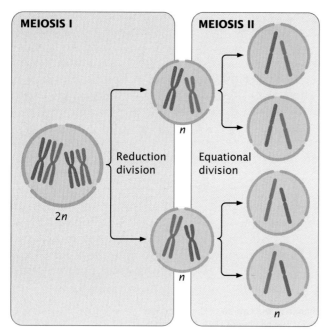

2.13 Meiosis includes two cell divisions. In this illustration, the original cell is $2n = 4$. After two meiotic divisions, each resulting cell is $1n = 2$.

Meiosis

The words *mitosis* and *meiosis* are sometimes confused. They sound a bit alike, and both refer to chromosome division and cytokinesis. But don't be deceived. The outcomes of mitosis and meiosis are radically different, and several unique events that have important genetic consequences take place only in meiosis.

How does meiosis differ from mitosis? Mitosis consists of a single nuclear division and is usually accompanied by a single cell division. Meiosis, on the other hand, consists of two divisions. After mitosis, chromosome number in newly formed cells is the same as that in the original cell, whereas meiosis causes chromosome number in the newly formed cells to be reduced by half. Finally, mitosis produces genetically identical cells, whereas meiosis produces genetically variable cells. Let's see how these differences arise.

Like mitosis, meiosis is preceded by an interphase stage that includes G_1, S, and G_2 phases. Meiosis consists of two distinct processes: meiosis I and meiosis II, each of which includes a cell division. The first division, which comes at the end of meiosis I, is termed the reduction division because the number of chromosomes per cell is reduced by half (**Figure 2.13**). The second division, which comes at the end of meiosis II, is sometimes termed the equational division. The events of meiosis II are similar to those of mitosis. However, meiosis II differs from mitosis in that chromosome number has already been halved in meiosis I, and the cell does not begin with the same number of chromosomes as it does in mitosis (see Figure 2.13).

Meiosis I During interphase, the chromosomes are relaxed and visible as diffuse chromatin. **Prophase I** is a lengthy stage, divided into five substages (**Figure 2.14**). In leptotene, the chromosomes contract and become visible. In zygotene, the chromosomes continue to condense; homologous chromosomes pair up and begin **synapsis**, a very close pairing association. Each homologous pair of

The evolution of sexual reproduction is one of the most significant events in the history of life. As will be discussed in Chapters 11 and 12, the pace of evolution depends on the amount of genetic variation present. By shuffling the genetic information from two parents, sexual reproduction greatly increases the amount of genetic variation and allows for accelerated evolution. Most of the tremendous diversity of life on Earth is a direct result of sexual reproduction.

Sexual reproduction consists of two processes. The first is **meiosis**, which leads to gametes in which chromosome number is reduced by half. The second process is **fertilization**, in which two haploid gametes fuse and restore chromosome number to its original diploid value.

2.14 Crossing over takes place in prophase I.

synapsed chromosomes consists of four chromatids called a **bivalent** or **tetrad**. In pachytene, the chromosomes become shorter and thicker, and a three-part synaptonemal complex develops between homologous chromosomes. The function of the synaptonemal complex is unclear, but the chromosomes of many cells deficient in this complex do not separate properly.

Crossing over takes place in prophase I, in which homologous chromosomes exchange genetic information. Crossing over generates genetic variation (see Sources of Genetic Variation in Meiosis later in this chapter) and is essential for the proper alignment and separation of homologous chromosomes. The centromeres of the paired chromosomes move apart in diplotene; the two homologs remain attached at each chiasma (plural, chiasmata), which is the result of crossing over. In diakinesis, chromosome condensation continues, and the chiasmata move toward the ends of the chromosomes as the strands slip apart; so the homologs remain paired only at the tips. Near the end of prophase I, the nuclear membrane breaks down and the spindle forms, setting the stage for metaphase I. The stages of meiosis are outlined in **Figure 2.15**.

Metaphase I is initiated when homologous pairs of chromosomes align along the metaphase plate (see Figure 2.15). A microtubule from one pole attaches to one chromosome of a homologous pair, and a microtubule from the other pole attaches to the other member of the pair. **Anaphase I** is marked by the separation of homologous chromosomes. The two chromosomes of a homologous pair are pulled toward opposite poles. Although the homologous chromosomes separate, the sister chromatids remain attached and travel together. In **telophase I**, the chromosomes arrive at the spindle poles and the cytoplasm divides.

Meiosis II The period between meiosis I and meiosis II is **interkinesis**, in which the nuclear membrane re-forms around the chromosomes clustered at each pole, the spindle breaks down, and the chromosomes relax. These cells then pass through **prophase II**, in which the events of interkinesis are reversed: the chromosomes recondense, the spindle re-forms, and the nuclear envelope once again breaks down. In interkinesis in some types of cells, the chromosomes remain condensed, and the spindle does not break down. These cells move directly from cytokinesis into **metaphase II**, which is similar to metaphase of mitosis: the individual chromosomes line up on the metaphase plate, with the sister chromatids facing opposite poles.

In **anaphase II**, the kinetochores of the sister chromatids separate and the chromatids are pulled to opposite poles. Each chromatid is now a distinct chromosome. In **telophase II**, the chromosomes arrive at the spindle poles, a nuclear envelope re-forms around the chromosomes, and the cytoplasm divides. The chromosomes relax and are no longer visible. The major events of meiosis are summarized in **Table 2.2**.

Table 2.2 Major events in each stage of meiosis

Stage	Major Events
Meiosis I	
Prophase I	Chromosomes condense, homologous chromosomes synapse, crossing over takes place, the nuclear envelope breaks down, and the mitotic spindle forms.
Metaphase I	Homologous pairs of chromosomes line up on the metaphase plate.
Anaphase I	The two chromosomes (each with two chromatids) of each homologous pair separate and move toward opposite poles.
Telophase I	Chromosomes arrive at the spindle poles.
Cytokinesis	The cytoplasm divides to produce two cells, each having half the original number of chromosomes.
Interkinesis	In some types of cells, the spindle breaks down, chromosomes relax, and a nuclear envelope re-forms, but no DNA synthesis takes place.
Meiosis II	
Prophase II*	Chromosomes condense, the spindle forms, and the nuclear envelope disintegrates.
Metaphase II	Individual chromosomes line up on the metaphase plate.
Anaphase II	Sister chromatids separate and move as individual chromosomes toward the spindle poles.
Telophase II	Chromosomes arrive at the spindle poles; the spindle breaks down and a nuclear envelope re-forms.
Cytokinesis	The cytoplasm divides.

*Only in cells in which the spindle has broken down, chromosomes have relaxed, and the nuclear envelope has re-formed in telophase I. Other types of cells proceed directly to metaphase II after cytokinesis.

CONCEPTS

Meiosis consists of two distinct processes: meiosis I and meiosis II. Meiosis I includes the reduction division, in which homologous chromosomes separate and chromosome number is reduced by half. In meiosis II (the equational division) chromatids separate.

✔ CONCEPT CHECK 5

Which of the following events takes place in metaphase I?

a. Crossing over
b. Chromosomes contract.
c. Homologous pairs of chromosomes line up on the metaphase plate.
d. Individual chromosomes line up on the metaphase plate.

2.15 Meiosis is divided into stages. [Photographs by C. A. Hasenkampf/Biological Photo Service.]

> www.whfreeman.com/pierce4e
> Examine meiosis and the consequences of its failure by viewing Animation 2.2.

Sources of Genetic Variation in Meiosis

What are the overall consequences of meiosis? First, meiosis comprises two divisions; so each original cell produces four cells (there are exceptions to this generalization, as, for example, in many female animals; see Figure 2.20b). Second, chromosome number is reduced by half; so cells produced by meiosis are haploid. Third, cells produced by meiosis are genetically different from one another and from the parental cell. Genetic differences among cells result from two processes that are unique to meiosis: crossing over and random separation of homologous chromosomes.

Crossing over Crossing over, which takes place in prophase I, refers to the exchange of genes between nonsister chromatids (chromatids from different homologous chromosomes). Evidence from yeast suggests that crossing over is initiated in zygotene, before the synaptonemal complex develops, and is not completed until near the end of prophase I (see Figure 2.14). In other organisms, recombination is initiated after the formation of the synaptonemal complex and, in yet others, there is no synaptonemal complex.

After crossing over has taken place, the sister chromatids may no longer be identical. Crossing over is the basis for

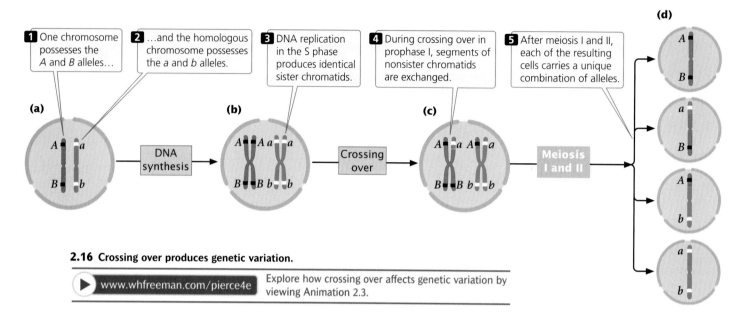

2.16 Crossing over produces genetic variation.

www.whfreeman.com/pierce4e — Explore how crossing over affects genetic variation by viewing Animation 2.3.

intrachromosomal **recombination**, creating new combinations of alleles on a chromatid. To see how crossing over produces genetic variation, consider two pairs of alleles, which we will abbreviate *Aa* and *Bb*. Assume that one chromosome possesses the *A* and *B* alleles and its homolog possesses the *a* and *b* alleles (**Figure 2.16a**). When DNA is replicated in the S phase, each chromosome duplicates, and so the resulting sister chromatids are identical (**Figure 2.16b**).

In the process of crossing over, there are breaks in the DNA strands and the breaks are repaired in such a way that segments of nonsister chromatids are exchanged (**Figure 2.16c**). The important thing here is that, after crossing over has taken place, the two sister chromatids are no longer identical: one chromatid has alleles *A* and *B*, whereas its sister chromatid (the chromatid that underwent crossing over) has alleles *a* and *B*. Likewise, one chromatid of the other chromosome has alleles *a* and *b*, and the other chromatid has alleles *A* and *b*. Each of the four chromatids now carries a unique combination of alleles: __A B__, __a B__, __A b__, and __a b__. Eventually, the two homologous chromosomes separate, each going into a different cell. In meiosis II, the two chromatids of each chromosome separate, and thus each of the four cells resulting from meiosis carries a different combination of alleles (**Figure 2.16d**).

Random Separation of Homologous Chromosomes

The second process of meiosis that contributes to genetic variation is the random distribution of chromosomes in anaphase I after their random alignment in metaphase I. To illustrate this process, consider a cell with three pairs of chromosomes, I, II, and III (**Figure 2.17a**). One chromosome of each pair is maternal in origin (I_m, II_m, and III_m); the other is paternal in origin (I_p, II_p, and III_p). The chromosome pairs line up in the center of the cell in metaphase I and, in anaphase I, the chromosomes of each homologous pair separate.

How each pair of homologs aligns and separates is random and independent of how other pairs of chromosomes align and separate (**Figure 2.17b**). By chance, all the maternal chromosomes might migrate to one side, with all the paternal chromosomes migrating to the other. After division, one cell would contain chromosomes I_m, II_m, and III_m, and the other, I_p, II_p, and III_p. Alternatively, the I_m, II_m, and III_p chromosomes might move to one side, and the I_p, II_p, and III_m chromosomes to the other. The different migrations would produce different combinations of chromosomes in the resulting cells (**Figure 2.17c**). There are four ways in which a diploid cell with three pairs of chromosomes can divide, producing a total of eight different combinations of chromosomes in the gametes. In general, the number of possible combinations is 2^n, where *n* equals the number of homologous pairs. As the number of chromosome pairs increases, the number of combinations quickly becomes very large. In humans, who have 23 pairs of chromosomes, there are 2^{23}, or 8,388,608, different combinations of chromosomes possible from the random separation of homologous chromosomes. The genetic consequences of this process, termed independent assortment, will be explored in more detail in Chapter 3.

In summary, crossing over shuffles alleles on the *same* chromosome into new combinations, whereas the random distribution of maternal and paternal chromosomes shuffles alleles on *different* chromosomes into new combinations. Together, these two processes are capable of producing tremendous amounts of genetic variation among the cells resulting from meiosis. **TRY PROBLEMS 30 AND 31**

CONCEPTS

The two mechanisms that produce genetic variation in meiosis are crossing over and the random distribution of maternal and paternal chromosomes.

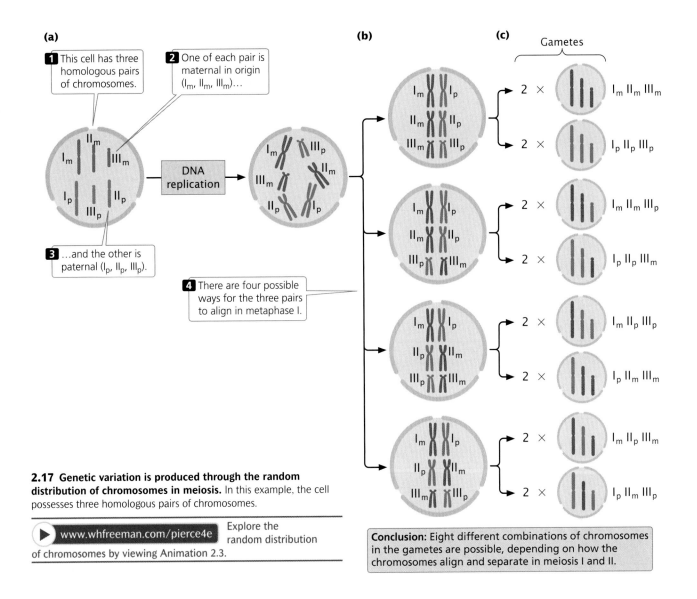

2.17 Genetic variation is produced through the random distribution of chromosomes in meiosis. In this example, the cell possesses three homologous pairs of chromosomes.

> www.whfreeman.com/pierce4e Explore the random distribution of chromosomes by viewing Animation 2.3.

Conclusion: Eight different combinations of chromosomes in the gametes are possible, depending on how the chromosomes align and separate in meiosis I and II.

CONNECTING CONCEPTS

Mitosis and Meiosis Compared

Now that we have examined the details of mitosis and meiosis, let's compare the two processes (**Figure 2.18** and **Table 2.3**). In both mitosis and meiosis, the chromosomes contract and become visible; both processes include the movement of chromosomes toward the spindle poles, and both are accompanied by cell division. Beyond these similarities, the processes are quite different.

Mitosis results in a single cell division and usually produces two daughter cells. Meiosis, in contrast, comprises two cell divisions and usually produces four cells. In diploid cells, homologous chromosomes are present before both meiosis and mitosis, but the pairing of homologs takes place only in meiosis.

Another difference is that, in meiosis, chromosome number is reduced by half as a consequence of the separation of homologous pairs of chromosomes in anaphase I, but no chromosome reduction takes place in mitosis. Furthermore, meiosis is characterized by two processes that produce genetic variation: crossing over (in prophase I) and the random distribution of maternal and paternal chromosomes (in anaphase I). There are normally no equivalent processes in mitosis.

Mitosis and meiosis also differ in the behavior of chromosomes in metaphase and anaphase. In metaphase I of meiosis, *homologous pairs* of chromosomes line up on the metaphase plate, whereas *individual chromosomes* line up on the metaphase plate in metaphase of mitosis (and in metaphase II of meiosis). In anaphase I of meiosis, *paired chromosomes* separate and migrate toward opposite spindle poles, each chromosome possessing two chromatids attached at the centromere. In contrast, in anaphase of mitosis (and in anaphase II of meiosis), *sister chromatids* separate, and each chromosome that moves toward a spindle pole is unreplicated.

TRY PROBLEMS 25 AND 26 →

The Separation of Sister Chromatids and Homologous Chromosomes

In recent years, some of the molecules required for the joining and separation of chromatids and homologous chromosomes have been identified. **Cohesin**, a protein that holds the chromatids together, is key to the behavior of chromosomes

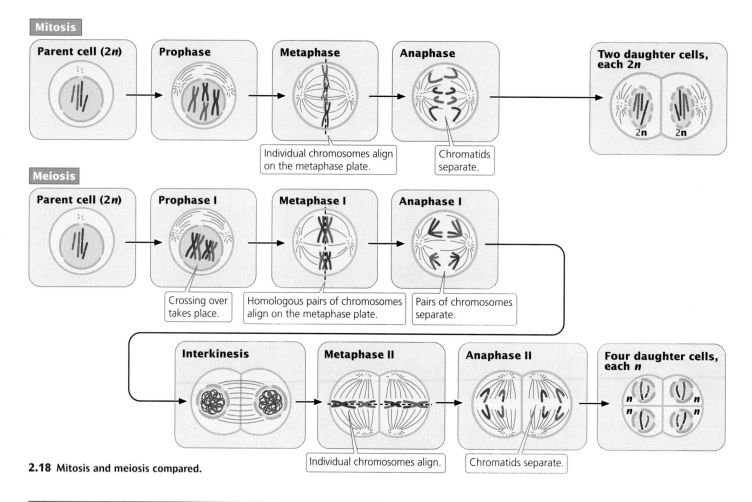

2.18 Mitosis and meiosis compared.

Table 2.3	Comparison of Mitosis, Meiosis I, and Meiosis II		
Event	Mitosis	Meiosis I	Meiosis II
Cell division	Yes	Yes	Yes
Chromosome reduction	No	Yes	No
Genetic variation produced	No	Yes	No
Crossing over	No	Yes	No
Random distribution of maternal and paternal chromosomes	No	Yes	No
Metaphase	Individual chromosomes line up	Homologous pairs line up	Individual chromosomes line up
Anaphase	Chromatids separate	Homologous chromosomes separate	Chromatids separate

in mitosis and meiosis (**Figure 2.19a**). The sister chromatids are held together by cohesin, which is established in the S phase and persists through G_2 and early mitosis. In anaphase of mitosis, cohesin along the entire length of the chromosome is broken down by an enzyme called separase, allowing the sister chromatids to separate.

As we have seen, mitosis and meiosis differ fundamentally in the behavior of chromosomes in anaphase (see Figure 2.18). Why do homologs separate in anaphase I of meiosis, whereas chromatids separate in anaphase of mitosis and anaphase II of meiosis? It is important to note that the forms of cohesin used in mitosis and meiosis differ. At the beginning of meiosis, the meiosis-specific cohesin is found along the entire length of a chromosome's arms (**Figure 2.19b**). The cohesin also acts on the chromosome arms of homologs at the chiasmata, tethering two homologs together at their ends.

In anaphase I, cohesin along the chromosome arms is broken, allowing the two homologs to separate. However, cohesin at the centromere is protected by a protein called shugoshin, which means "guardian spirit" in Japanese. Because of this protective action by shugoshin, the centromeric cohesin remains intact and prevents the separation of the two sister chromatids during anaphase I of meiosis. Shugoshin is subsequently degraded. At the end of meta-

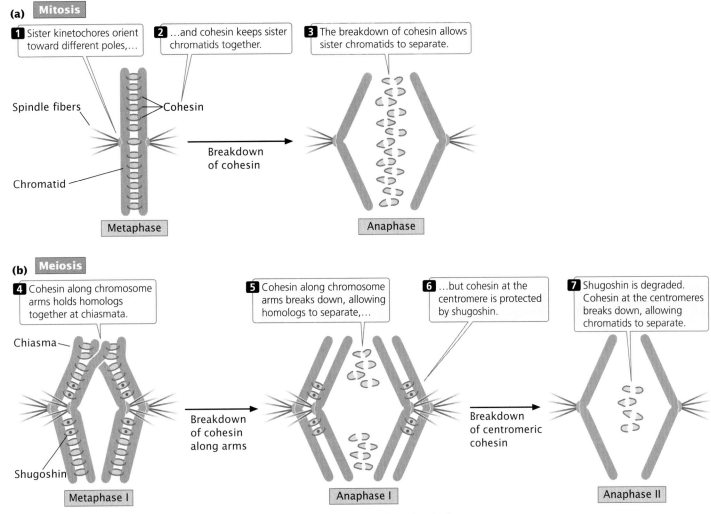

2.19 Cohesin controls the separation of chromatids and chromosomes in mitosis and meiosis.

phase II, the centromeric cohesin—no longer protected by shugoshin—breaks down, allowing the sister chromatids to separate in anaphase II, just as they do in mitosis (Figure 2.19b). **TRY PROBLEM 27**

CONCEPTS

Cohesin holds sister chromatids together during the early part of mitosis. In anaphase, cohesin breaks down, allowing sister chromatids to separate. In meiosis, cohesin is protected at the centromeres during anaphase I, and so homologous chromosomes, but not sister chromatids, separate in meiosis I. The breakdown of centromeric cohesin allows sister chromatids to separate in anaphase II of meiosis.

✔ **CONCEPT CHECK 6**

How does shugoshin affect sister chromatids in meiosis I and meiosis II?

Meiosis in the Life Cycles of Animals and Plants

The overall result of meiosis is four haploid cells that are genetically variable. Let's now see where meiosis fits into the life cycles of a multicellular animal and a multicellular plant.

Meiosis in animals The production of gametes in a male animal, a process called **spermatogenesis**, takes place in the testes. There, diploid primordial germ cells divide mitotically to produce diploid cells called **spermatogonia** (**Figure 2.20a**). Each spermatogonium can undergo repeated rounds of mitosis, giving rise to numerous additional spermatogonia. Alternatively, a spermatogonium can initiate meiosis and enter into prophase I. Now called a **primary spermatocyte**, the cell is still diploid because the homologous chromosomes have not yet separated. Each primary spermatocyte completes meiosis I, giving rise to two haploid **secondary spermatocytes** that then undergo meiosis II,

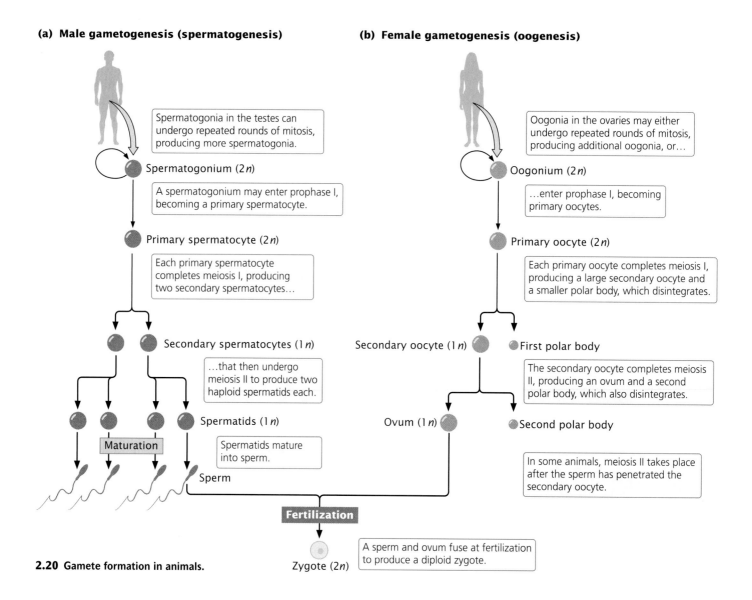

2.20 Gamete formation in animals.

with each producing two haploid **spermatids**. Thus, each primary spermatocyte produces a total of four haploid spermatids, which mature and develop into sperm.

The production of gametes in a female animal, a process called **oogenesis**, begins much as spermatogenesis does. Within the ovaries, diploid primordial germ cells divide mitotically to produce **oogonia** (**Figure 2.20b**). Like spermatogonia, oogonia can undergo repeated rounds of mitosis or they can enter into meiosis. When they enter prophase I, these still-diploid cells are called **primary oocytes**. Each primary oocyte completes meiosis I and divides.

At this point, the process of oogenesis begins to differ from that of spermatogenesis. In oogenesis, cytokinesis is unequal: most of the cytoplasm is allocated to one of the two haploid cells, the **secondary oocyte**. The smaller cell, which contains half of the chromosomes but only a small part of the cytoplasm, is called the **first polar body**; it may or may not divide further. The secondary oocyte completes meiosis II, and, again, cytokinesis is unequal—most of the cytoplasm passes into one of the cells. The larger cell, which acquires most of the cytoplasm, is the **ovum**, the mature female gamete. The smaller cell is the **second polar body**. Only the ovum is capable of being fertilized, and the polar bodies usually disintegrate. Oogenesis, then, produces a single mature gamete from each primary oocyte.

In mammals, oogenesis differs from spermatogenesis in another way. The formation of sperm takes place continuously in a male throughout his adult reproductive life. The formation of female gametes, however, is often a discontinuous process, and may take place over a period of years. Oogenesis begins before birth; at this time, oogonia initiate meiosis and give rise to primary oocytes. Meiosis is then arrested, stalled in prophase I. A female is born with primary oocytes arrested in prophase I. In humans, this period of suspended animation may last 30 or 40 years. Before ovulation, rising hormone levels stimulate one or more of the primary oocytes to recommence meiosis. The first division of meiosis is completed and a secondary oocyte is ovulated from the ovary. In humans and many other species, the second division of meiosis is then delayed until contact with the sperm. When the sperm penetrates the outer layer of the secondary oocyte, the second meiotic division takes place, the second polar body is extruded from the egg, and the nuclei of the sperm and newly formed ovum fuse, giving rise to the zygote.

CONCEPTS

In the testes, a diploid spermatogonium undergoes meiosis, producing a total of four haploid sperm cells. In the ovary, a diploid oogonium undergoes meiosis to produce a single large ovum and smaller polar bodies that normally disintegrate.

✔ CONCEPT CHECK 7

A secondary spermatocyte has 12 chromosomes. How many chromosomes will be found in the primary spermatocyte that gave rise to it?

a. 6
b. 12
c. 18
d. 24

Meiosis in plants Most plants have a complex life cycle that includes two distinct generations (stages): the diploid sporophyte and the haploid gametophyte. These two stages alternate; the sporophyte produces haploid spores through meiosis, and the gametophyte produces haploid gametes through mitosis (**Figure 2.21**). This type of life cycle is sometimes called alternation of generations. In this cycle, the immediate products of meiosis are called spores, not gametes; the spores undergo one or more mitotic divisions to produce gametes. Although the terms used for this process are somewhat different from those commonly used in regard to animals (and from some of those employed so far in this chapter), the processes in plants and animals are basically the same: in both, meiosis leads to a reduction in chromosome number, producing haploid cells.

In flowering plants, the sporophyte is the obvious, vegetative part of the plant; the gametophyte consists of only a few haploid cells within the sporophyte. The flower, which is part of the sporophyte, contains the reproductive structures. In some plants, both male and female reproductive structures are found in the same flower; in other plants, they exist in different flowers. In either case, the male part of the flower, the stamen, contains diploid reproductive cells called **microsporocytes**, each of which undergoes meiosis to produce four haploid **microspores** (**Figure 2.22a**). Each microspore divides mitotically, producing an immature pollen grain consisting of two haploid nuclei. One of these nuclei, called the tube nucleus, directs the growth of a pollen tube. The other, termed the generative nucleus, divides mitotically to produce two sperm cells. The pollen grain, with its two haploid nuclei, is the male gametophyte.

The female part of the flower, the ovary, contains diploid cells called **megasporocytes**, each of which undergoes meiosis to produce four haploid **megaspores** (**Figure 2.22b**), only one of which survives. The nucleus of the surviving megaspore divides mitotically three times, producing a total of eight haploid nuclei that make up the female gametophyte, the embryo sac. Division of the cytoplasm then produces separate cells, one of which becomes the egg.

When the plant flowers, the stamens open and release pollen grains. Pollen lands on a flower's stigma—a sticky platform that sits on top of a long stalk called the style. At

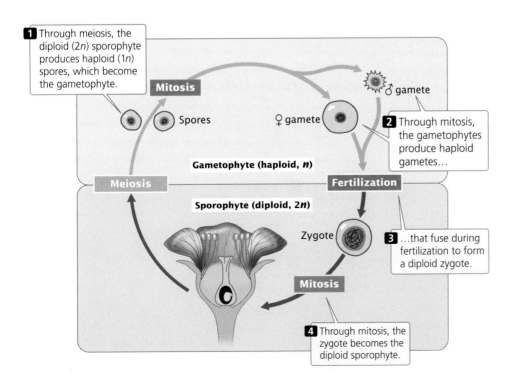

2.21 Plants alternate between diploid and haploid life stages (female, ♀; male, ♂).

2.22 Sexual reproduction in flowering plants.

the base of the style is the ovary. If a pollen grain germinates, it grows a tube down the style into the ovary. The two sperm cells pass down this tube and enter the embryo sac (**Figure 2.22c**). One of the sperm cells fertilizes the egg cell, producing a diploid zygote, which develops into an embryo. The other sperm cell fuses with two nuclei enclosed in a single cell, giving rise to a $3n$ (triploid) endosperm, which stores food that will be used later by the embryonic plant. These two fertilization events are termed double fertilization.

> **CONCEPTS**
>
> In the stamen of a flowering plant, meiosis produces haploid microspores that divide mitotically to produce haploid sperm in a pollen grain. Within the ovary, meiosis produces four haploid megaspores, only one of which divides mitotically three times to produce eight haploid nuclei. After pollination, one sperm fertilizes the egg cell, producing a diploid zygote; the other fuses with two nuclei to form the endosperm.

> ✔ **CONCEPT CHECK 8**
>
> Which structure is diploid?
> a. Microspore
> b. Egg
> c. Megaspore
> d. Microsporocyte

We have now examined the place of meiosis in the sexual cycle of two organisms, a typical multicellular animal and a flowering plant. These cycles are just two of the many variations found among eukaryotic organisms. Although the cellular events that produce reproductive cells in plants and animals differ in the number of cell divisions, the number of haploid gametes produced, and the relative size of the final products, the overall result is the same: meiosis gives rise to haploid, genetically variable cells that then fuse during fertilization to produce diploid progeny. **TRY PROBLEMS 33 AND 34**

CONCEPTS SUMMARY

- A prokaryotic cell possesses a simple structure, with no nuclear envelope and usually a single, circular chromosome. A eukaryotic cell possesses a more-complex structure, with a nucleus and multiple linear chromosomes consisting of DNA complexed to histone proteins.
- Cell reproduction requires the copying of the genetic material, separation of the copies, and cell division.
- In a prokaryotic cell, the single chromosome replicates, each copy moves toward opposite sides of the cell, and the cell divides. In eukaryotic cells, reproduction is more complex than in prokaryotic cells, requiring mitosis and meiosis to ensure that a complete set of genetic information is transferred to each new cell.
- In eukaryotic cells, chromosomes are typically found in homologous pairs. Each functional chromosome consists of a centromere, telomeres, and multiple origins of replication. After a chromosome has been copied, the two copies remain attached at the centromere, forming sister chromatids.
- The cell cycle consists of the stages through which a eukaryotic cell passes between cell divisions. It consists of (1) interphase, in which the cell grows and prepares for division, and (2) the M phase, in which nuclear and cell division take place. The M phase consists of (1) mitosis, the process of nuclear division, and (2) cytokinesis, the division of the cytoplasm.
- Progression through the cell cycle is controlled at checkpoints that regulate the cell cycle by allowing or prohibiting the cell to proceed to the next stage.
- Mitosis usually results in the production of two genetically identical cells.
- Sexual reproduction produces genetically variable progeny and allows for accelerated evolution. It includes meiosis, in which haploid sex cells are produced, and fertilization, the fusion of sex cells. Meiosis includes two cell divisions. In meiosis I, crossing over takes place and homologous chromosomes separate. In meiosis II, chromatids separate.
- The usual result of meiosis is the production of four haploid cells that are genetically variable. Genetic variation in meiosis is produced by crossing over and by the random distribution of maternal and paternal chromosomes.
- Cohesin holds sister chromatids together. In metaphase of mitosis and in metaphase II of meiosis, the breakdown of cohesin allows sister chromatids to separate. In meiosis I, centromeric cohesin remains intact and keeps sister chromatids together so that homologous chromosomes, but not sister chromatids, separate in anaphase I.
- In animals, a diploid spermatogonium undergoes meiosis to produce four haploid sperm cells. A diploid oogonium undergoes meiosis to produce one large haploid ovum and one or more smaller polar bodies.
- In plants, a diploid microsporocyte in the stamen undergoes meiosis to produce four pollen grains, each with two haploid sperm cells. In the ovary, a diploid megasporocyte undergoes meiosis to produce eight haploid nuclei, one of which forms the egg. During pollination, one sperm fertilizes the egg cell and the other fuses with two haploid nuclei to form a $3n$ endosperm.

IMPORTANT TERMS

prokaryote (p. 17)
eukaryote (p. 17)
eubacteria (p. 17)
archaea (p. 17)
nucleus (p. 17)
histone (p. 17)
chromatin (p. 17)
virus (p. 18)
homologous pair (p. 19)
diploid (p. 19)
haploid (p. 19)
telomere (p. 20)
origin of replication (p. 21)
sister chromatid (p. 21)
cell cycle (p. 21)
checkpoint (p. 21)
interphase (p. 21)
M phase (p. 21)
mitosis (p. 21)
cytokinesis (p. 21)

prophase (p. 22)
prometaphase (p. 22)
metaphase (p. 22)
anaphase (p. 22)
telophase (p. 24)
meiosis (p. 26)
fertilization (p. 26)
prophase I (p. 26)
synapsis (p. 26)
bivalent (p. 27)
tetrad (p. 27)
crossing over (p. 27)
metaphase I (p. 27)
anaphase I (p. 27)
telophase I (p. 27)
interkinesis (p. 27)
prophase II (p. 27)
metaphase II (p. 27)
anaphase II (p. 27)
telophase II (p. 27)

recombination (p. 30)
cohesin (p. 31)
spermatogenesis (p. 33)
spermatogonium (p. 33)
primary spermatocyte (p. 33)
secondary spermatocyte (p. 33)
spermatid (p. 34)
oogenesis (p. 34)
oogonium (p. 34)
primary oocyte (p. 34)
secondary oocyte (p. 34)
first polar body (p. 34)
ovum (p. 34)
second polar body (p. 34)
microsporocyte (p. 35)
microspore (p. 35)
megasporocyte (p. 35)
megaspore (p. 35)

ANSWERS TO CONCEPT CHECKS

1. Eubacteria and archaea are prokaryotes. They differ from eukaryotes in possessing no nucleus, a genome that usually consists of a single, circular chromosome, and a small amount of DNA.

2. b

3. The kinetochore is the point at which spindle microtubules attach to the chromosome. If the kinetochore were missing, spindle microtubules would not attach to the chromosome, the chromosome would not be drawn into the nucleus, and the resulting cells would be missing a chromosome.

4. a

5. c

6. During anaphase I, shugoshin protects cohesin at the centromeres from the action of separase; so cohesin remains intact and the sister chromatids remain together. Subsequently, shugoshin breaks down; so centromeric cohesin is cleaved in anaphase II and the chromatids separate.

7. d

8. d

WORKED PROBLEMS

1. A student examines a thin section of an onion-root tip and records the number of cells that are in each stage of the cell cycle. She observes 94 cells in interphase, 14 cells in prophase, 3 cells in prometaphase, 3 cells in metaphase, 5 cells in anaphase, and 1 cell in telophase. If the complete cell cycle in an onion-root tip requires 22 hours, what is the average duration of each stage in the cycle? Assume that all cells are in the active cell cycle (not G_0).

- **Solution**

This problem is solved in two steps. First, we calculate the proportions of cells in each stage of the cell cycle, which correspond to the amount of time that an average cell spends in each stage. For example, if cells spend 90% of their time in interphase, then, at any given moment, 90% of the cells will be in interphase. The second step is to convert the proportions into lengths of time, which is done by multiplying the proportions by the total time of the cell cycle (22 hours).

Step 1. Calculate the proportion of cells at each stage. The proportion of cells at each stage is equal to the number of cells found in that stage divided by the total number of cells examined:

Interphase	$94/120$	$= 0.783$
Prophase	$14/120$	$= 0.117$
Prometaphase	$3/120$	$= 0.025$
Metaphase	$3/120$	$= 0.025$
Anaphase	$5/120$	$= 0.042$
Telophase	$1/120$	$= 0.08$

We can check our calculations by making sure that the proportions sum to 1.0, which they do.

Step 2. Determine the average duration of each stage. To determine the average duration of each stage, multiply the proportion of cells in each stage by the time required for the entire cell cycle:

Interphase	0.783×22 hours =	17.23 hours
Prophase	0.117×22 hours =	2.57 hours
Prometaphase	0.025×22 hours =	0.55 hour
Metaphase	0.025×22 hours =	0.55 hour
Anaphase	0.042×22 hours =	0.92 hour
Telophase	0.008×22 hours =	0.18 hour

2. A cell in G_1 of interphase has 8 chromosomes. How many chromosomes and how many DNA molecules will be found per cell as this cell progresses through the following stages: G_2, metaphase of mitosis, anaphase of mitosis, after cytokinesis in mitosis, metaphase I of meiosis, metaphase II of meiosis, and after cytokinesis of meiosis II?

- **Solution**

Remember the rules about counting chromosomes and DNA molecules: (1) to determine the number of chromosomes, count the functional centromeres; (2) to determine the number of DNA molecules, determine whether sister chromatids exist. If sister chromatids are present, the number of DNA molecules is $2 \times$ the number of chromosomes. If the chromosomes are unreplicated (don't contain sister chromatids), the number of DNA molecules equals the number of chromosomes. Think carefully about when and how the numbers of chromosomes and DNA molecules change in the course of mitosis and meiosis.

The number of DNA molecules increases only in the S phase, when DNA replicates; the number of DNA molecules decreases only when the cell divides. Chromosome number increases only when sister chromatids separate in anaphase of mitosis and in anaphase II of meiosis (homologous chromosomes, not chromatids, separate in anaphase I of meiosis). Chromosome number, like the number of DNA molecules, is reduced only by cell division.

Let's now apply these principles to the problem. A cell in G_1 has 8 chromosomes, and sister chromatids are not present; so 8 DNA molecules are present in G_1. DNA replicates in the S phase and now each chromosome consists of two sister chromatids; so, in G_2, $2 \times 8 = 16$ DNA molecules are present per cell. However, the two copies of each DNA molecule remain attached at the centromere; so there are still only 8 chromosomes present. As the cell passes through prophase and metaphase of the cell cycle, the number of chromosomes and the number of DNA molecules remain the same; so, at metaphase, there are 16 DNA molecules and 8 chromosomes. In anaphase, the chromatids separate and each becomes an independent chromosome; at this point, the number of chromosomes increases from 8 to 16. This increase is temporary, lasting only until the cell divides in telophase or subsequent to it. The number of DNA molecules remains at 16 in anaphase. The number of DNA molecules and chromosomes per cell is reduced by cytokinesis after telophase, because the 16 chromosomes and DNA molecules are now distributed between two cells. Therefore, after cytokinesis, each cell has 8 DNA molecules and 8 chromosomes, the same numbers that were present at the beginning of the cell cycle.

Now, let's trace the numbers of DNA molecules and chromosomes through meiosis. At G_1, there are 8 chromosomes and 8 DNA molecules. The number of DNA molecules increases to 16 in the S phase, but the number of chromosomes remains at 8 (each chromosome has two chromatids). The cell therefore enters metaphase I having 16 DNA molecules and 8 chromosomes. In anaphase I of meiosis, homologous chromosomes separate, but the number of chromosomes remains at 8. After cytokinesis, the original 8 chromosomes are distributed between two cells; so the number of chromosomes per cell falls to 4 (each with two chromatids). The original 16 DNA molecules also are distributed between two cells; so the number of DNA molecules per cell is 8. There is no DNA synthesis in interkinesis, and each cell still maintains 4 chromosomes and 8 DNA molecules through metaphase II. In anaphase II, the two sister chromatids of each chromosome separate, temporarily raising the number of chromosomes per cell to 8, whereas the number of DNA molecules per cell remains at 8. After cytokinesis, the chromosomes and DNA molecules are again distributed between two cells, providing 4 chromosomes and 4 DNA molecules per cell. These results are summarized in the following table:

Stage	Number of chromosomes per cell	Number of DNA molecules per cell
G_1	8	8
G_2	8	16
Metaphase of mitosis	8	16
Anaphase of mitosis	16	16
After cytokinesis of mitosis	8	8
Metaphase I of meiosis	8	16
Metaphase II of meiosis	4	8
After cytokinesis of meiosis II	4	4

COMPREHENSION QUESTIONS

Section 2.1

*1. What are some genetic differences between prokaryotic and eukaryotic cells?

2. Why are the viruses that infect mammalian cells useful for studying the genetics of mammals?

Section 2.2

*3. List three fundamental events that must take place in cell reproduction.

4. Outline the process by which prokaryotic cells reproduce.

5. Name three essential structural elements of a functional eukaryotic chromosome and describe their functions.

*6. Sketch and identify four different types of chromosomes based on the position of the centromere.

7. List the stages of interphase and the major events that take place in each stage.

*8. List the stages of mitosis and the major events that take place in each stage.

9. Briefly describe how the chromosomes move toward the spindle poles during anaphase.

*10. What are the genetically important results of the cell cycle?

11. Why are the two cells produced by the cell cycle genetically identical?

12. What are checkpoints? List some of the important checkpoints in the cell cycle.

Section 2.3

13. What are the stages of meiosis and what major events take place in each stage?

*14. What are the major results of meiosis?

15. What two processes unique to meiosis are responsible for genetic variation? At what point in meiosis do these processes take place?

*16. List similarities and differences between mitosis and meiosis. Which differences do you think are most important and why?

17. Briefly explain why sister chromatids remain together in anaphase I but separate in anaphase II of meiosis.

18. Outline the process of spermatogenesis in animals. Outline the process of oogenesis in animals.

19. Outline the process by which male gametes are produced in plants. Outline the process of female gamete formation in plants.

APPLICATION QUESTIONS AND PROBLEMS

Introduction

*20. Answer the following questions regarding the Blind Men's Riddle, presented at the beginning of the chapter.
  a. What do the two socks of a pair represent in the cell cycle?
  b. In the riddle, each blind man buys his own pairs of socks, but the clerk places all pairs into one bag. Thus, there are two pairs of socks of each color in the bag (two black pairs, two blue pairs, two gray pairs, etc.). What do the two pairs (four socks in all) of each color represent?
  c. In the cell cycle, what is the thread that connects the two socks of a pair?
  d. In the cell cycle, what is the molecular knife that cuts the thread holding the two socks of a pair together?
  e. What in the riddle performs the same function as spindle fibers?
  f. What would happen if one man failed to grasp his sock of a particular pair and how does it relate to events in the cell cycle?

Section 2.2

21. A certain species has three pairs of chromosomes: an acrocentric pair, a metacentric pair, and a submetacentric pair. Draw a cell of this species as it would appear in metaphase of mitosis.

22. A biologist examines a series of cells and counts 160 cells in interphase, 20 cells in prophase, 6 cells in prometaphase, 2 cells in metaphase, 7 cells in anaphase, and 5 cells in telophase. If the complete cell cycle requires 24 hours, what is the average duration of the M phase in these cells? Of metaphase?

Section 2.3

*23. A cell in G_1 of interphase has 12 chromosomes. How many chromosomes and DNA molecules will be found per cell when this original cell progresses to the following stages?

  a. G_2 of interphase
  b. Metaphase I of meiosis
  c. Prophase of mitosis
  d. Anaphase I of meiosis
  e. Anaphase II of meiosis
  f. Prophase II of meiosis
  g. After cytokinesis following mitosis
  h. After cytokinesis following meiosis II

24. How are the events that take place in spermatogenesis and oogenesis similar? How are they different?

*25. All of the following cells, shown in various stages of mitosis and meiosis, come from the same rare species of plant. What is the diploid number of chromosomes in this plant? Give the names of each stage of mitosis or meiosis shown.

26. The amount of DNA per cell of a particular species is measured in cells found at various stages of meiosis, and the following amounts are obtained:

Amount of DNA per cell

_____ 3.7 pg   _____ 7.3 pg   _____ 14.6 pg

Match the amounts of DNA above with the corresponding stages of the cell cycle (*a* through *f*). You may use more than one stage for each amount of DNA.

Stage of meiosis
a. G_1
b. Prophase I
c. G_2
d. Following telophase II and cytokinesis
e. Anaphase I
f. Metaphase II

*27. How would each of the following events affect the outcome of mitosis or meiosis?
   a. Mitotic cohesin fails to form early in mitosis.
   b. Shugoshin is absent during meiosis.
   c. Shugoshin does not break down after anaphase I of meiosis.
   d. Separase is defective.

28. A cell in prophase II of meiosis contains 12 chromosomes. How many chromosomes would be present in a cell from the same organism if it were in prophase of mitosis? Prophase I of meiosis?

29. A cell has 8 chromosomes in G_1 of interphase. Draw a picture of this cell with its chromosomes at the following stages. Indicate how many DNA molecules are present at each stage.
   a. Metaphase of mitosis
   b. Anaphase of mitosis
   c. Anaphase II of meiosis
   d. Diplotene of meiosis I

*30. The fruit fly *Drosophila melanogaster* (left) has four pairs of chromosomes, whereas the house fly *Musca domestica* (right) has six pairs of chromosomes. Other things being equal, in which species would you expect to see more genetic variation among the progeny of a cross? Explain your answer.

[Herman Eisenbeiss/Photo Researchers; Materz Mali/iStockphoto.]

*31. A cell has two pairs of submetacentric chromosomes, which we will call chromosomes I_a, I_b, II_a, and II_b (chromosomes I_a and I_b are homologs, and chromosomes II_a and II_b are homologs). Allele M is located on the long arm of chromosome I_a, and allele m is located at the same position on chromosome I_b. Allele P is located on the short arm of chromosome I_a, and allele p is located at the same position on chromosome I_b. Allele R is located on chromosome II_a and allele r is located at the same position on chromosome II_b.
   a. Draw these chromosomes, identifying genes M, m, P, p, R, and r, as they might appear in metaphase I of meiosis. Assume that there is no crossing over.
   b. Taking into consideration the random separation of chromosomes in anaphase I, draw the chromosomes (with genes identified) present in all possible types of gametes that might result from this cell's undergoing meiosis. Assume that there is no crossing over.

32. A horse has 64 chromosomes and a donkey has 62 chromosomes. A cross between a female horse and a male donkey produces a mule, which is usually sterile. How many chromosomes does a mule have? Can you think of any reasons for the fact that most mules are sterile?

33. Normal somatic cells of horses have 64 chromosomes ($2n = 64$). How many chromosomes and DNA molecules will be present in the following types of horse cells?

[Tamara Didenko/iStockphoto.]

Cell type	Number of chromosomes	Number of DNA molecules
a. Spermatogonium	_____	_____
b. First polar body	_____	_____
c. Primary oocyte	_____	_____
d. Secondary spermatocyte	_____	_____

*34. A primary oocyte divides to give rise to a secondary oocyte and a first polar body. The secondary oocyte then divides to give rise to an ovum and a second polar body.
   a. Is the genetic information found in the first polar body identical with that found in the secondary oocyte? Explain your answer.
   b. Is the genetic information found in the second polar body identical with that in the ovum? Explain your answer.

CHALLENGE QUESTIONS

Section 2.3

35. From 80% to 90% of the most common human chromosome abnormalities arise because the chromosomes fail to divide properly in oogenesis. Can you think of a reason why failure of chromosome division might be more common in female gametogenesis than in male gametogenesis?

36. On average, what proportion of the genome in the following pairs of humans would be exactly the same if no crossing over took place? (For the purposes of this

question only, we will ignore the special case of the X and Y sex chromosomes and assume that all genes are located on nonsex chromosomes.)
  a. Father and child
  b. Mother and child
  c. Two full siblings (offspring that have the same two biological parents)
  d. Half siblings (offspring that have only one biological parent in common)
  e. Uncle and niece
  f. Grandparent and grandchild

*37. Female bees are diploid, and male bees are haploid. The haploid males produce sperm and can successfully mate with diploid females. Fertilized eggs develop into females and unfertilized eggs develop into males. How do you think the process of sperm production in male bees differs from sperm production in other animals?

3 Basic Principles of Heredity

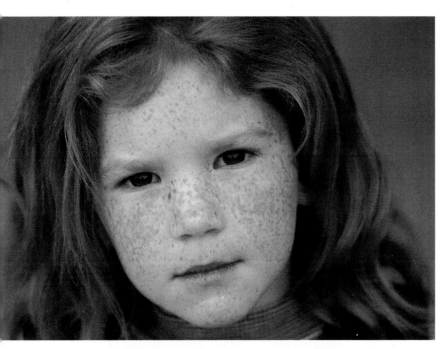

Red hair is caused by recessive mutations at the melanocortin 1 receptor gene. Reed Kaestner/Corbis.

THE GENETICS OF RED HAIR

Whether because of its exotic hue or its novelty, red hair has long been a subject of fascination for historians, poets, artists, and scientists. Historians made special note of the fact that Boudica, the Celtic queen who led a revolt against the Roman Empire, possessed a "great mass of red hair." Early Christian artists frequently portrayed Mary Magdalene as a striking red head (though there is no mention of her red hair in the Bible), and the famous artist Botticelli painted the goddess Venus as a red-haired beauty in his masterpiece *The Birth of Venus*. Queen Elizabeth I of England possessed curly red hair; during her reign, red hair was quite fashionable in London society.

The color of our hair is caused largely by a pigment called melanin that comes in two primary forms: eumelanin, which is black or brown, and pheomelanin, which is red or yellow. The color of a person's hair is determined by two factors: (1) the amount of melanin produced (more melanin causes darker hair; less melanin causes lighter hair) and (2) the relative amounts of eumelanin and pheomelanin (more eumelanin produces black or brown hair; more pheomelanin produces red or blond hair). The color of our hair is not just an academic curiosity; melanin protects against the harmful effects of sunlight, and people with red hair are usually fair skinned and particularly susceptible to skin cancer.

The inheritance of red hair has long been a subject of scientific debate. In 1909, Charles and Gertrude Davenport speculated on the inheritance of hair color in humans. Charles Davenport was an early enthusiast of genetics, particularly of inheritance in humans, and was the first director of the Biological Laboratory in Cold Spring Harbor, New York. He later became a leading proponent of eugenics, a movement—now discredited—that advocated improvement of the human race through genetics. The Davenports' study was based on family histories sent in by untrained amateurs and was methodologically flawed, but their results suggested that red hair is recessive to black and brown, meaning that a person must inherit two copies of a red-hair gene—one from each parent—to have red hair. Subsequent research contradicted this initial conclusion, suggesting that red hair is inherited instead as a dominant trait and that a person will have red hair even if possessing only a single red-hair gene. Controversy over whether red hair color is dominant or recessive or even dependent on combinations of several different genes continued for many years.

In 1993, scientists who were investigating a gene that affects the color of fur in mice discovered that the gene encodes the melanocortin-1 receptor. This receptor, when activated, increases the production of black eumelanin and decreases the production of red pheomelanin, resulting in black or brown fur. Shortly thereafter, the same melanocortin-1 receptor gene (*MC1R*) was located on human chromosome 16 and analyzed. When this gene is mutated in humans, red hair results. Most people with red hair carry two defective copies of the *MC1R* gene, which means that the trait is recessive (as originally proposed by the Davenports back in 1909). However, from 10% to 20% of red heads possess only a single mutant copy of *MC1R,* muddling the recessive interpretation of red hair (the people with a single mutant copy of the gene tend to have lighter red hair than those who harbor two mutant copies). The type and frequency of mutations at the *MC1R* gene vary widely among human populations, accounting for ethnic differences in the preponderance of red hair: among those of African and Asian descent, mutations for red hair are uncommon, whereas almost 40% of the people from the northern part of the United Kingdom carry at least one mutant copy of the gene for red hair.

Modern humans are not the only people with red hair. Analysis of DNA from ancient bones indicates that some Neanderthals also carried a mutation in the *MC1R* gene that almost certainly caused red hair, but the mutation is distinct from those seen in modern humans. This finding suggests that red hair arose independently in Neanderthals and modern humans.

This chapter is about the principles of heredity: how genes—such as the one for the melanocortin-1 receptor—are passed from generation to generation and how factors such as dominance influence that inheritance. The principles of heredity were first put forth by Gregor Mendel, and so we begin this chapter by examining Mendel's scientific achievements. We then turn to simple genetic crosses, those in which a single characteristic is examined. We will consider some techniques for predicting the outcome of genetic crosses and then turn to crosses in which two or more characteristics are examined. We will see how the principles applied to simple genetic crosses and the ratios of offspring that they produce serve as the key for understanding more-complicated crosses. The chapter ends with a discussion of statistical tests for analyzing crosses.

Throughout this chapter, a number of concepts are interwoven: Mendel's principles of segregation and independent assortment, probability, and the behavior of chromosomes. These concepts might at first appear to be unrelated, but they are actually different views of the same phenomenon, because the genes that undergo segregation and independent assortment are located on chromosomes. The principal aim of this chapter is to examine these different views and to clarify their relations.

3.1 Gregor Mendel Discovered the Basic Principles of Heredity

In 1909, when the Davenports speculated about the inheritance of red hair, the basic principles of heredity were just becoming widely known among biologists. Surprisingly, these principles had been discovered some 44 years earlier by Gregor Johann Mendel (1822–1884; **Figure 3.1**).

Mendel was born in what is now part of the Czech Republic. Although his parents were simple farmers with little money, he was able to achieve a sound education and was admitted to the Augustinian monastery in Brno in September 1843. After graduating from seminary, Mendel was ordained a priest and appointed to a teaching position in a local school. He excelled at teaching, and the abbot of the monastery recommended him for further study at

3.1 Gregor Johann Mendel, experimenting with peas, first discovered the principles of heredity. James King-Holmes/Photo Researchers.

the University of Vienna, which he attended from 1851 to 1853. There, Mendel enrolled in the newly opened Physics Institute and took courses in mathematics, chemistry, entomology, paleontology, botany, and plant physiology. It was probably there that Mendel acquired knowledge of the scientific method, which he later applied so successfully to his genetics experiments. After 2 years of study in Vienna, Mendel returned to Brno, where he taught school and began his experimental work with pea plants. He conducted breeding experiments from 1856 to 1863 and presented his results publicly at meetings of the Brno Natural Science Society in 1865. Mendel's paper from these lectures was published in 1866. In spite of widespread interest in heredity, the effect of his research on the scientific community was minimal. At the time, no one seemed to have noticed that Mendel had discovered the basic principles of inheritance.

In 1868, Mendel was elected abbot of his monastery, and increasing administrative duties brought an end to his teaching and eventually to his genetics experiments. He died at the age of 61 on January 6, 1884, unrecognized for his contribution to genetics.

The significance of Mendel's discovery was not recognized until 1900, when three botanists—Hugo de Vries, Erich von Tschermak, and Carl Correns—began independently conducting similar experiments with plants and arrived at conclusions similar to those of Mendel. Coming across Mendel's paper, they interpreted their results in accord with his principles and drew attention to his pioneering work.

Mendel's Success

Mendel's approach to the study of heredity was effective for several reasons. Foremost was his choice of experimental subject, the pea plant *Pisum sativum* (**Figure 3.2**), which offered clear advantages for genetic investigation. The plant is easy to cultivate, and Mendel had the monastery garden and greenhouse at his disposal. Compared with some other plants, peas grow relatively rapidly, completing an entire generation in a single growing season. By today's standards, one generation per year seems frightfully slow—fruit flies complete a generation in 2 weeks and bacteria in 20 minutes—but Mendel was under no pressure to publish quickly and was able to follow the inheritance of individual characteristics for several generations. Had he chosen to work on an organism with a longer generation time—horses, for example—he might never have discovered the basis of inheritance. Pea plants also produce many offspring—their seeds—which allowed Mendel to detect meaningful mathematical ratios in the traits that he observed in the progeny.

The large number of varieties of peas that were available to Mendel also was crucial, because these varieties differed in various traits and were genetically pure. Mendel was therefore able to begin with plants of variable, known genetic makeup.

Much of Mendel's success can be attributed to the seven characteristics that he chose for study (see Figure 3.2). He avoided characteristics that display a range of variation; instead, he focused his attention on those that exist in two easily differentiated forms, such as white versus gray seed coats, round versus wrinkled seeds, and inflated versus constricted pods.

Finally, Mendel was successful because he adopted an experimental approach and interpreted his results by using mathematics. Unlike many earlier investigators who just described the *results* of crosses, Mendel formulated *hypotheses* based on his initial observations and then conducted additional crosses to test his hypotheses. He kept careful records of the numbers of progeny possessing each type of trait and computed ratios of the different types. He was adept at seeing patterns in detail and was patient and thorough, conducting his experiments for 10 years before attempting to write up his results. TRY PROBLEM 13 →

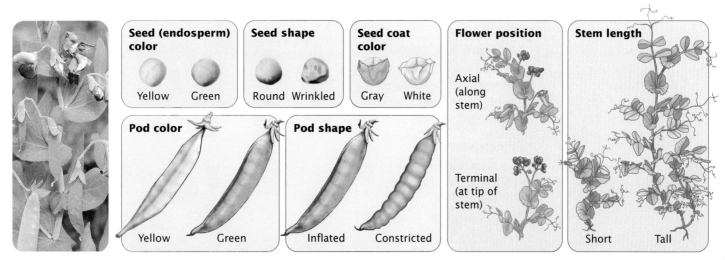

3.2 Mendel used the pea plant *Pisum sativum* in his studies of heredity. He examined seven characteristics that appeared in the seeds and in plants grown from the seeds. [Photograph by Wally Eberhart/Visuals Unlimited.]

CONCEPTS

Gregor Mendel put forth the basic principles of inheritance, publishing his findings in 1866. The significance of his work did not become widely known until 1900.

✔ CONCEPT CHECK 1

Which of the following factors did not contribute to Mendel's success in his study of heredity?
a. His use of the pea plant
b. His study of plant chromosomes
c. His adoption of an experimental approach
d. His use of mathematics

Genetic Terminology

Before we examine Mendel's crosses and the conclusions that he drew from them, a review of some terms commonly used in genetics will be helpful (**Table 3.1**). The term *gene* is a word that Mendel never knew. It was not coined until 1909, when Danish geneticist Wilhelm Johannsen first used it. The definition of a gene varies with the context of its use, and so its definition will change as we explore different aspects of heredity. For our present use in the context of genetic crosses, we will define a **gene** as an inherited factor that determines a characteristic.

Genes frequently come in different versions called **alleles** (**Figure 3.3**). In Mendel's crosses, seed shape was determined by a gene that exists as two different alleles: one allele encodes round seeds and the other encodes wrinkled seeds. All alleles for any particular gene will be found at a specific place on a chromosome called the **locus** for that gene. (The plural of locus is loci; it's bad form in genetics—and incorrect—to speak of locuses.) Thus, there is a specific place—a locus—on a chromosome in pea plants where the shape of seeds is determined. This locus might be occupied by an allele for round seeds or one for wrinkled seeds. We will use the term *allele* when referring to a specific version of a gene; we will use the term *gene* to refer more generally to any allele at a locus.

The **genotype** is the set of alleles that an individual organism possesses. A diploid organism with a genotype consisting of two identical alleles is **homozygous** for that locus. One that has a genotype consisting of two different alleles is **heterozygous** for the locus.

Another important term is **phenotype**, which is the manifestation or appearance of a characteristic. A phenotype can refer to any type of characteristic—physical, physiological, biochemical, or behavioral. Thus, the condition of having round seeds is a phenotype, a body weight of 50 kilograms (50 kg) is a phenotype, and having sickle-cell anemia is a phenotype. In this book, the term *characteristic* or *character* refers to a general feature such as eye color; the term *trait* or *phenotype* refers to specific manifestations of that feature, such as blue or brown eyes.

A given phenotype arises from a genotype that develops within a particular environment. The genotype determines the potential for development; it sets certain limits, or boundaries, on that development. How the phenotype develops within those limits is determined by the effects of other genes and of environmental factors, and the balance between these effects varies from character to character. For some characters, the differences between phenotypes are determined largely by differences in genotype. In Mendel's peas, for example, the genotype, not the environment, largely

Table 3.1	Summary of important genetic terms
Term	**Definition**
Gene	A genetic factor (region of DNA) that helps determine a characteristic
Allele	One of two or more alternative forms of a gene
Locus	Specific place on a chromosome occupied by an allele
Genotype	Set of alleles possessed by an individual organism
Heterozygote	An individual organism possessing two different alleles at a locus
Homozygote	An individual organism possessing two of the same alleles at a locus
Phenotype or trait	The appearance or manifestation of a character
Character or characteristic	An attribute or feature

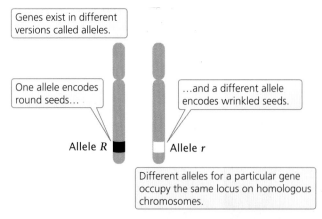

3.3 At each locus, a diploid organism possesses two alleles located on different homologous chromosomes. The alleles identified here refer to traits studied by Mendel.

determined the shape of the seeds. For other characters, environmental differences are more important. The height reached by an oak tree at maturity is a phenotype that is strongly influenced by environmental factors, such as the availability of water, sunlight, and nutrients. Nevertheless, the tree's genotype still imposes some limits on its height: an oak tree will never grow to be 300 meters (300 m) tall no matter how much sunlight, water, and fertilizer are provided. Thus, even the height of an oak tree is determined to some degree by genes. For many characteristics, both genes and environment are important in determining phenotypic differences.

An obvious but important concept is that only the alleles of the genotype are inherited. Although the phenotype is determined, at least to some extent, by genotype, organisms do not transmit their phenotypes to the next generation. The distinction between genotype and phenotype is one of the most important principles of modern genetics. The next section describes Mendel's careful observation of phenotypes through several generations of breeding experiments. These experiments allowed him to deduce not only the genotypes of the individual plants, but also the rules governing their inheritance.

CONCEPTS

Each phenotype results from a genotype developing within a specific environment. The alleles of the genotype, not the phenotype, are inherited.

✔ **CONCEPT CHECK 2**

Distinguish among the following terms: locus, allele, genotype, and phenotype.

3.2 Monohybrid Crosses Reveal the Principle of Segregation and the Concept of Dominance

Mendel started with 34 varieties of peas and spent 2 years selecting those varieties that he would use in his experiments. He verified that each variety was pure-breeding (homozygous for each of the traits that he chose to study) by growing the plants for two generations and confirming that all offspring were the same as their parents. He then carried out a number of crosses between the different varieties. Although peas are normally self-fertilizing (each plant crosses with itself), Mendel conducted crosses between different plants by opening the buds before the anthers (male sex organs) were fully developed, removing the anthers, and then dusting the stigma (female sex organs) with pollen from a different plant's anthers (**Figure 3.4**).

Mendel began by studying **monohybrid crosses**—those between parents that differed in a single characteristic. In one experiment, Mendel crossed a pure-breeding (homozygous) pea plant for round seeds with one that was pure-breeding for wrinkled seeds (see Figure 3.4). This first generation of a cross is the **P (parental) generation**.

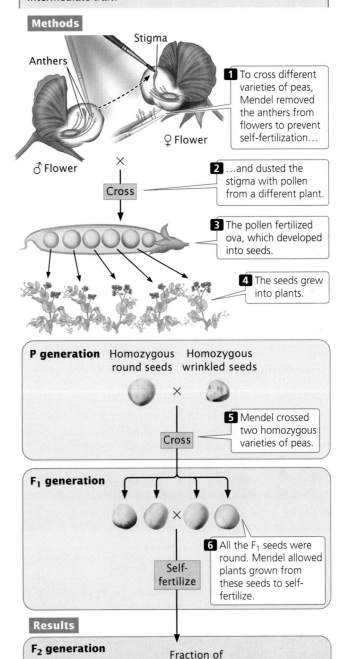

3.4 Mendel conducted monohybrid crosses.

After crossing the two varieties in the P generation, Mendel observed the offspring that resulted from the cross. In regard to seed characteristics, such as seed shape, the phenotype develops as soon as the seed matures, because the seed traits are determined by the newly formed embryo within the seed. For characters associated with the plant itself, such as stem length, the phenotype doesn't develop until the plant grows from the seed; for these characters, Mendel had to wait until the following spring, plant the seeds, and then observe the phenotypes on the plants that germinated.

The offspring from the parents in the P generation are the F_1 (**filial 1**) **generation**. When Mendel examined the F_1 generation of this cross, he found that they expressed only one of the phenotypes present in the parental generation: all the F_1 seeds were round. Mendel carried out 60 such crosses and always obtained this result. He also conducted **reciprocal crosses**: in one cross, pollen (the male gamete) was taken from a plant with round seeds and, in its reciprocal cross, pollen was taken from a plant with wrinkled seeds. Reciprocal crosses gave the same result: all the F_1 were round.

Mendel wasn't content with examining only the seeds arising from these monohybrid crosses. The following spring, he planted the F_1 seeds, cultivated the plants that germinated from them, and allowed the plants to self-fertilize, producing a second generation—the F_2 (**filial 2**) **generation**. Both of the traits from the P generation emerged in the F_2 generation; Mendel counted 5474 round seeds and 1850 wrinkled seeds in the F_2 (see Figure 3.4). He noticed that the number of the round and wrinkled seeds constituted approximately a 3 to 1 ratio; that is, about $3/4$ of the F_2 seeds were round and $1/4$ were wrinkled. Mendel conducted monohybrid crosses for all seven of the characteristics that he studied in pea plants and, in all of the crosses, he obtained the same result: all of the F_1 resembled only one of the two parents, but both parental traits emerged in the F_2 in an approximate ratio of 3 : 1.

What Monohybrid Crosses Reveal

First, Mendel reasoned that, although the F_1 plants display the phenotype of only one parent, they must inherit genetic factors from both parents because they transmit both phenotypes to the F_2 generation. The presence of both round and wrinkled seeds in the F_2 could be explained only if the F_1 plants possessed both round and wrinkled genetic factors that they had inherited from the P generation. He concluded that each plant must therefore possess two genetic factors encoding a character.

The genetic factors (now called alleles) that Mendel discovered are, by convention, designated with letters; the allele for round seeds is usually represented by R, and the allele for wrinkled seeds by r. The plants in the P generation of Mendel's cross possessed two identical alleles: RR in the round-seeded parent and rr in the wrinkled-seeded parent (**Figure 3.5a**).

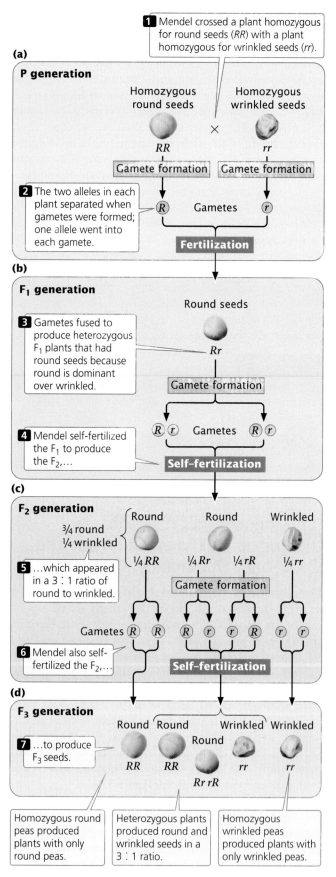

3.5 Mendel's monohybrid crosses revealed the principle of segregation and the concept of dominance.

The second conclusion that Mendel drew from his monohybrid crosses was that the two alleles in each plant separate when gametes are formed, and one allele goes into each gamete. When two gametes (one from each parent) fuse to produce a zygote, the allele from the male parent unites with the allele from the female parent to produce the genotype of the offspring. Thus, Mendel's F_1 plants inherited an R allele from the round-seeded plant and an r allele from the wrinkled-seeded plant (**Figure 3.5b**). However, only the trait encoded by the round allele (R) was *observed* in the F_1: all the F_1 progeny had round seeds. Those traits that appeared unchanged in the F_1 heterozygous offspring Mendel called **dominant**, and those traits that disappeared in the F_1 heterozygous offspring he called **recessive**. When dominant and recessive alleles are present together, the recessive allele is masked, or suppressed. The concept of dominance was the third important conclusion that Mendel derived from his monohybrid crosses.

Mendel's fourth conclusion was that the two alleles of an individual plant separate with equal probability into the gametes. When plants of the F_1 (with genotype Rr) produced gametes, half of the gametes received the R allele for round seeds and half received the r allele for wrinkled seeds. The gametes then paired randomly to produce the following genotypes in equal proportions among the F_2: RR, Rr, rR, rr (**Figure 3.5c**). Because round (R) is dominant over wrinkled (r), there were three round progeny in the F_2 (RR, Rr, rR) for every one wrinkled progeny (rr) in the F_2. This 3 : 1 ratio of round to wrinkled progeny that Mendel observed in the F_2 could be obtained only if the two alleles of a genotype separated into the gametes with equal probability.

The conclusions that Mendel developed about inheritance from his monohybrid crosses have been further developed and formalized into the principle of segregation and the concept of dominance. The **principle of segregation** (Mendel's first law, see **Table 3.2**) states that each individual diploid organism possesses two alleles for any particular characteristic. These two alleles segregate (separate) when gametes are formed, and one allele goes into each gamete. Furthermore, the two alleles segregate into gametes in equal proportions. The **concept of dominance** states that, when two different alleles are present in a genotype, only the trait encoded by one of them—the "dominant" allele—is observed in the phenotype.

Mendel confirmed these principles by allowing his F_2 plants to self-fertilize and produce an F_3 generation. He found that the plants grown from the wrinkled seeds—those displaying the recessive trait (rr)—produced an F_3 in which all plants produced wrinkled seeds. Because his wrinkled-seeded plants were homozygous for wrinkled alleles (rr), only wrinkled alleles could be passed on to their progeny (**Figure 3.5d**).

The plants grown from round seeds—the dominant trait—fell into two types (see Figure 3.5c). On self-fertilization, about $2/3$ of these plants produced both round and wrinkled seeds in the F_3 generation. These plants were heterozygous (Rr); so they produced $1/4$ RR (round), $1/2$ Rr (round), and $1/4$ rr (wrinkled) seeds, giving a 3 : 1 ratio of round to wrinkled in the F_3. About $1/3$ of the plants grown from round seeds were of the second type; they produced only the round-seeded trait in the F_3. These plants were homozygous for the round allele (RR) and could thus produce only round offspring in the F_3 generation. Mendel planted the seeds obtained in the F_3 and carried these plants through three more rounds of self-fertilization. In each generation, $2/3$ of the round-seeded plants produced round and wrinkled offspring, whereas $1/3$ produced only round offspring. These results are entirely consistent with the principle of segregation.

Table 3.2 Comparison of the principles of segregation and independent assortment

Principle	Observation	Stage of Meiosis
Segregation (Mendel's first law)	1. Each individual organism possesses two alleles encoding a trait.	Before meiosis
	2. Alleles separate when gametes are formed.	Anaphase I
	3. Alleles separate in equal proportions.	Anaphase I
Independent assortment (Mendel's second law)	Alleles at different loci separate independently.	Anaphase I

CONCEPTS

The principle of segregation states that each individual organism possesses two alleles that can encode a characteristic. These alleles segregate when gametes are formed, and one allele goes into each gamete. The concept of dominance states that, when the two alleles of a genotype are different, only the trait encoded by one of them—the "dominant" allele—is observed.

✔ **CONCEPT CHECK 3**

How did Mendel know that each of his pea plants carried two alleles encoding a characteristic?

CONNECTING CONCEPTS

Relating Genetic Crosses to Meiosis

We have now seen how the results of monohybrid crosses are explained by Mendel's principle of segregation. Many students find that they enjoy working genetic crosses but are frustrated by the abstract nature of the symbols. Perhaps you feel the same at this point. You may be asking, "What do these symbols really represent?

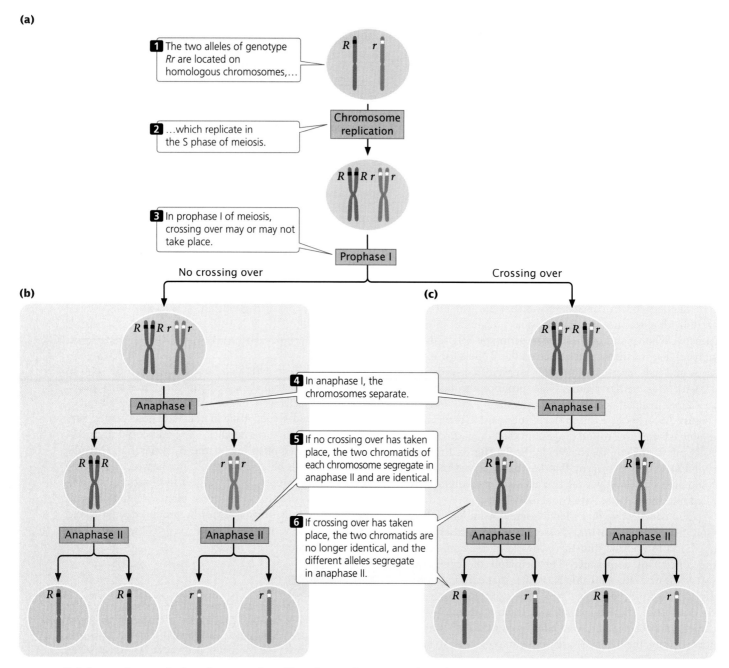

3.6 Segregation results from the separation of homologous chromosomes in meiosis.

What does the genotype *RR* mean in regard to the biology of the organism?" The answers to these questions lie in relating the abstract symbols of crosses to the structure and behavior of chromosomes, the repositories of genetic information (see Chapter 2).

In 1900, when Mendel's work was rediscovered and biologists began to apply his principles of heredity, the relation between genes and chromosomes was still unclear. The theory that genes are located on chromosomes (the **chromosome theory of heredity**) was developed in the early 1900s by Walter Sutton, then a graduate student at Columbia University. Through the careful study of meiosis in insects, Sutton documented the fact that each homologous pair of chromosomes consists of one maternal chromosome and one paternal chromosome. Showing that these pairs segregate independently into gametes in meiosis, he concluded that this process is the biological basis for Mendel's principles of heredity. German cytologist and embryologist Theodor Boveri came to similar conclusions at about the same time.

The symbols used in genetic crosses, such as *R* and *r*, are just shorthand notations for particular sequences of DNA in the chromosomes that encode particular phenotypes. The two alleles of a genotype are found on different but homologous chromosomes. One chromosome of each homologous pair is inherited from the mother and the other is inherited from the father. In the S phase of meiotic interphase, each chromosome replicates, producing two copies of each allele, one on each chromatid (**Figure 3.6a**). The homologous chromosomes segregate in anaphase I, thereby separating the two different alleles (**Figure 3.6b** and **c**). This chromosome segregation is the basis of the principle of segregation. In anaphase II of meiosis, the two chromatids of each replicated chromosome separate; so each gamete resulting from meiosis carries only a single allele at each locus, as Mendel's principle of segregation predicts.

If crossing over has taken place in prophase I of meiosis, then the two chromatids of each replicated chromosome are no longer identical, and the segregation of different alleles takes place at anaphase I and anaphase II (see Figure 3.6c). However, Mendel didn't know anything about chromosomes; he formulated his principles of heredity entirely on the basis of the results of the crosses that he carried out. Nevertheless, we should not forget that these principles work because they are based on the behavior of actual chromosomes in meiosis. **TRY PROBLEM 28 →**

Predicting the Outcomes of Genetic Crosses

One of Mendel's goals in conducting his experiments on pea plants was to develop a way to predict the outcome of crosses between plants with different phenotypes. In this section, you will first learn a simple, shorthand method for predicting outcomes of genetic crosses (the Punnett square), and then you will learn how to use probability to predict the results of crosses.

The Punnett square The Punnett square was developed by English geneticist Reginald C. Punnett in 1917. To illustrate the Punnett square, let's examine another cross carried out by Mendel. By crossing two varieties of peas that differed in height, Mendel established that tall (*T*) was dominant over short (*t*). He tested his theory concerning the inheritance of dominant traits by crossing an F_1 tall plant that was heterozygous (*Tt*) with the short homozygous parental variety (*tt*). This type of cross, between an F_1 genotype and either of the parental genotypes, is called a **backcross**.

To predict the types of offspring that result from this backcross, we first determine which gametes will be produced by each parent (**Figure 3.7a**). The principle of segregation tells us that the two alleles in each parent separate, and one allele passes to each gamete. All gametes from the homozygous *tt* short plant will receive a single short (*t*) allele. The tall plant in this cross is heterozygous (*Tt*); so 50% of its gametes will receive a tall allele (*T*) and the other 50% will receive a short allele (*t*).

A **Punnett square** is constructed by drawing a grid, putting the gametes produced by one parent along the upper edge and the gametes produced by the other parent down the left side (**Figure 3.7b**). Each cell (a block within the Punnett square) contains an allele from each of the corresponding gametes, generating the genotype of the progeny produced by fusion of those gametes. In the upper left-hand cell of the Punnett square in Figure 3.7b, a gamete containing *T* from the tall plant unites with a gamete containing *t* from the short plant, giving the genotype of the progeny (*Tt*). It is useful to write the phenotype expressed by each

3.7 The Punnett square can be used to determine the results of a genetic cross.

genotype; here the progeny will be tall, because the tall allele is dominant over the short allele. This process is repeated for all the cells in the Punnett square.

By simply counting, we can determine the types of progeny produced and their ratios. In Figure 3.7b, two cells contain tall (*Tt*) progeny and two cells contain short (*tt*) progeny; so the genotypic ratio expected for this cross is 2 *Tt* to 2 *tt* (a 1 : 1 ratio). Another way to express this result is to say that we expect $1/2$ of the progeny to have genotype *Tt* (and phenotype tall) and $1/2$ of the progeny to have genotype *tt* (and phenotype short). In this cross, the genotypic ratio and the phenotypic ratio are the same, but this outcome need not be the case. Try completing a Punnett square for the cross in which the F_1 round-seeded plants in Figure 3.5 undergo self-fertilization (you should obtain a phenotypic ratio of 3 round to 1 wrinkled and a genotypic ratio of 1 *RR* to 2 *Rr* to 1 *rr*).

CONCEPTS

The Punnett square is a shorthand method of predicting the genotypic and phenotypic ratios of progeny from a genetic cross.

✔ CONCEPT CHECK 4

If an F_1 plant depicted in Figure 3.5 is backcrossed to the parent with round seeds, what proportion of the progeny will have wrinkled seeds? (Use a Punnett square.)

a. 3/4 c. 1/4
b. 1/2 d. 0

Probability as a tool in genetics Another method for determining the outcome of a genetic cross is to use the rules of probability, as Mendel did with his crosses. **Probability** expresses the likelihood of the occurrence of a particular event. It is the number of times that a particular event occurs, divided by the number of all possible outcomes. For example, a deck of 52 cards contains only one king of hearts. The probability of drawing one card from the deck at random and obtaining the king of hearts is $1/52$, because there is only one card that is the king of hearts (one event) and there are 52 cards that can be drawn from the deck (52 possible outcomes). The probability of drawing a card and obtaining an ace is $4/52$, because there are four cards that are aces (four events) and 52 cards (possible outcomes). Probability can be expressed either as a fraction ($4/52$ in this case) or as a decimal number (0.077 in this case).

The probability of a particular event may be determined by knowing something about *how* the event occurs or *how often* it occurs. We know, for example, that the probability of rolling a six-sided die and getting a four is $1/6$, because the die has six sides and any one side is equally likely to end up on top. So, in this case, understanding the nature of the event—the shape of the thrown die—allows us to determine the probability. In other cases, we determine the probability of an event by making a large number of observations. When a weather forecaster says that there is a 40% chance of rain on a particular day, this probability was obtained by observing a large number of days with similar atmospheric conditions and finding that it rains on 40% of those days. In this case, the probability has been determined empirically (by observation).

The multiplication rule Two rules of probability are useful for predicting the ratios of offspring produced in genetic crosses. The first is the **multiplication rule**, which states that the probability of two or more independent events occurring together is calculated by multiplying their independent probabilities.

To illustrate the use of the multiplication rule, let's again consider the roll of a die. The probability of rolling one die and obtaining a four is $1/6$. To calculate the probability of rolling a die twice and obtaining 2 fours, we can apply the multiplication rule. The probability of obtaining a four on the first roll is $1/6$ and the probability of obtaining a four on the second roll is $1/6$; so the probability of rolling a four on both is $1/6 \times 1/6 = 1/36$ (**Figure 3.8a**). The key indicator for applying the multiplication rule is the word *and*; in the example just considered, we wanted to know the probability of obtaining a four on the first roll *and* a four on the second roll.

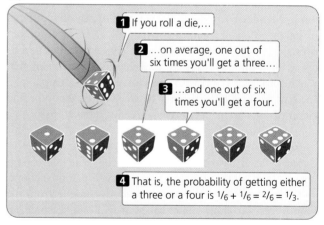

3.8 The multiplication and addition rules can be used to determine the probability of combinations of events.

For the multiplication rule to be valid, the events whose joint probability is being calculated must be independent—the outcome of one event must not influence the outcome of the other. For example, the number that comes up on one roll of the die has no influence on the number that comes up on the other roll; so these events are independent. However, if we wanted to know the probability of being hit on the head with a hammer and going to the hospital on the same day, we could not simply multiply the probability of being hit on the head with a hammer by the probability of going to the hospital. The multiplication rule cannot be applied here, because the two events are not independent—being hit on the head with a hammer certainly influences the probability of going to the hospital.

The addition rule The second rule of probability frequently used in genetics is the **addition rule**, which states that the probability of any one of two or more mutually exclusive events is calculated by adding the probabilities of these events. Let's look at this rule in concrete terms. To obtain the probability of throwing a die once and rolling *either* a three *or* a four, we would use the addition rule, adding the probability of obtaining a three ($1/6$) to the probability of obtaining a four (again, $1/6$), or $1/6 + 1/6 = 2/6 = 1/3$ (**Figure 3.8b**). The key indicators for applying the addition rule are the words *either* and *or*.

For the addition rule to be valid, the events whose probability is being calculated must be mutually exclusive, meaning that one event excludes the possibility of the occurrence of the other event. For example, you cannot throw a single die just once and obtain both a three and a four, because only one side of the die can be on top. These events are mutually exclusive.

CONCEPTS

The multiplication rule states that the probability of two or more independent events occurring together is calculated by multiplying their independent probabilities. The addition rule states that the probability that any one of two or more mutually exclusive events occurring is calculated by adding their probabilities.

✔ CONCEPT CHECK 5

If the probability of being blood-type A is $1/8$ and the probability of blood-type O is $1/2$, what is the probability of being either blood-type A or blood-type O?

a. $5/8$　　c. $1/8$
b. $1/2$　　d. $1/16$

The application of probability to genetic crosses The multiplication and addition rules of probability can be used in place of the Punnett square to predict the ratios of progeny expected from a genetic cross. Let's first consider a cross between two pea plants heterozygous for the locus that determines height, $Tt \times Tt$. Half of the gametes produced by each plant have a T allele, and the other half have a t allele; so the probability for each type of gamete is $1/2$.

The gametes from the two parents can combine in four different ways to produce offspring. Using the multiplication rule, we can determine the probability of each possible type. To calculate the probability of obtaining TT progeny, for example, we multiply the probability of receiving a T allele from the first parent ($1/2$) times the probability of receiving a T allele from the second parent ($1/2$). The multiplication rule should be used here because we need the probability of receiving a T allele from the first parent *and* a T allele from the second parent—two independent events. The four types of progeny from this cross and their associated probabilities are:

TT	(T gamete and T gamete)	$1/2 \times 1/2 = 1/4$	tall
Tt	(T gamete and t gamete)	$1/2 \times 1/2 = 1/4$	tall
tT	(t gamete and T gamete)	$1/2 \times 1/2 = 1/4$	tall
tt	(t gamete and t gamete)	$1/2 \times 1/2 = 1/4$	short

Notice that there are two ways for heterozygous progeny to be produced: a heterozygote can either receive a T allele from the first parent and a t allele from the second or receive a t allele from the first parent and a T allele from the second.

After determining the probabilities of obtaining each type of progeny, we can use the addition rule to determine the overall phenotypic ratios. Because of dominance, a tall plant can have genotype TT, Tt, or tT; so, using the addition rule, we find the probability of tall progeny to be $1/4 + 1/4 + 1/4 = 3/4$. Because only one genotype encodes short (tt), the probability of short progeny is simply $1/4$.

Two methods have now been introduced to solve genetic crosses: the Punnett square and the probability method. At this point, you may be asking, "Why bother with probability rules and calculations? The Punnett square is easier to understand and just as quick." For simple monohybrid crosses, the Punnett square is simpler than the probability method and is just as easy to use. However, for tackling more-complex crosses concerning genes at two or more loci, the probability method is both clearer and quicker than the Punnett square.

The binomial expansion and probability When probability is used, it is important to recognize that there may be several different ways in which a set of events can occur. Consider two parents who are both heterozygous for albinism, a recessive condition in humans that causes reduced pigmentation in the skin, hair, and eyes (**Figure 3.9**; see also the introduction to Chapter 1). When two parents heterozygous for albinism mate ($Aa \times Aa$), the probability of their having a child with albinism (aa) is $1/4$ and the probability of having a child with normal pigmentation (AA or Aa) is $3/4$. Suppose we want to know the probability of this couple having three children, all three with albinism. In this case, there is only one way in which they can have three children with albinism: their first child has albinism *and* their second child has albinism *and* their third child has albinism. Here, we simply apply the multiplication rule: $1/4 \times 1/4 \times 1/4 = 1/64$.

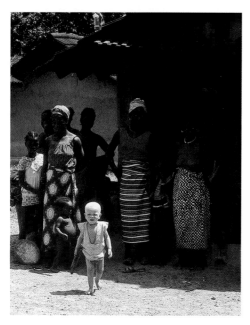

3.9 Albinism in human beings is usually inherited as a recessive trait. [Richard Dranitzke/SS/Photo Researchers.]

Suppose we now ask, What is the probability of this couple having three children, one with albinism and two with normal pigmentation? This situation is more complicated. The first child might have albinism, whereas the second and third are unaffected; the probability of this sequence of events is $\frac{1}{4} \times \frac{3}{4} \times \frac{3}{4} = \frac{9}{64}$. Alternatively, the first and third child might have normal pigmentation, whereas the second has albinism; the probability of this sequence is $\frac{3}{4} \times \frac{1}{4} \times \frac{3}{4} = \frac{9}{64}$. Finally, the first two children might have normal pigmentation and the third albinism; the probability of this sequence is $\frac{3}{4} \times \frac{3}{4} \times \frac{1}{4} = \frac{9}{64}$. Because *either* the first sequence *or* the second sequence *or* the third sequence produces one child with albinism and two with normal pigmentation, we apply the addition rule and add the probabilities: $\frac{9}{64} + \frac{9}{64} + \frac{9}{64} = \frac{27}{64}$.

If we want to know the probability of this couple having five children, two with albinism and three with normal pigmentation, figuring out *all* the different combinations of children and their probabilities becomes more difficult. This task is made easier if we apply the binomial expansion.

The binomial takes the form $(p + q)^n$, where p equals the probability of one event, q equals the probability of the alternative event, and n equals the number of times the event occurs. For figuring the probability of two out of five children with albinism:

p = the probability of a child having albinism ($\frac{1}{4}$)

q = the probability of a child having normal pigmentation ($\frac{3}{4}$)

The binomial for this situation is $(p + q)^5$ because there are five children in the family ($n = 5$). The expansion is:

$$(p + q)^5 = p^5 + 5p^4q + 10p^3q^2 + 10p^2q^3 + 5pq^4 + q^5$$

Each of the terms in the expansion provides the probability for one particular combination of traits in the children. The first term in the expansion (p^5) equals the probability of having five children all with albinism, because p is the probability of albinism. The second term ($5p^4q$) equals the probability of having four children with albinism and one with normal pigmentation, the third term ($10p^3q^2$) equals the probability of having three children with albinism and two with normal pigmentation, and so forth.

To obtain the probability of any combination of events, we insert the values of p and q; so the probability of having two out of five children with albinism is:

$$10p^2q^3 = 10(\tfrac{1}{4})^2(\tfrac{3}{4})^3 = \tfrac{270}{1024} = 0.26$$

We could easily figure out the probability of any desired combination of albinism and pigmentation among five children by using the other terms in the expansion.

How did we expand the binomial in this example? In general, the expansion of any binomial $(p + q)^n$ consists of a series of $n + 1$ terms. In the preceding example, $n = 5$; so there are $5 + 1 = 6$ terms: p^5, $5p^4q$, $10p^3q^2$, $10p^2q^3$, $5pq^4$, and q^5. To write out the terms, first figure out their exponents. The exponent of p in the first term always begins with the power to which the binomial is raised, or n. In our example, n equals 5, so our first term is p^5. The exponent of p decreases by one in each successive term; so the exponent of p is 4 in the second term (p^4), 3 in the third term (p^3), and so forth. The exponent of q is 0 (no q) in the first term and increases by 1 in each successive term, increasing from 0 to 5 in our example.

Next, determine the coefficient of each term. The coefficient of the first term is always 1; so, in our example, the first term is $1p^5$, or just p^5. The coefficient of the second term is always the same as the power to which the binomial is raised; in our example, this coefficient is 5 and the term is $5p^4q$. For the coefficient of the third term, look back at the preceding term; multiply the coefficient of the preceding term (5 in our example) by the exponent of p in that term (4) and then divide by the number of that term (second term, or 2). So the coefficient of the third term in our example is $(5 \times 4)/2 = 20/2 = 10$ and the term is $10p^3q^2$. Follow this procedure for each successive term.

Another way to determine the probability of any particular combination of events isn to use the following formula:

$$P = \frac{n!}{s!t!}p^sq^t$$

where P equals the overall probability of event X with probability p occurring s times and event Y with probability q occurring t times. For our albinism example, event X would be the occurrence of a child with albinism ($\frac{1}{4}$) and event Y would be the occurrence of a child with normal pigmentation ($\frac{3}{4}$); s would equal the number of children with albinism (2) and t would equal the number of children with normal pigmentation (3). The ! symbol stands for factorial, and it means the product of all the integers from n to 1. In this example, $n = 5$; so $n! = 5 \times 4 \times 3 \times 2 \times 1$. Applying this formula to obtain the probability of two out of five children having albinism, we obtain:

$$P = \frac{5!}{2!3!}(1/4)^2(3/4)^3$$

$$= \frac{5 \times 4 \times 3 \times 2 \times 1}{2 \times 1 \times 3 \times 2 \times 1}(1/4)^2(3/4)^3 = 0.26$$

This value is the same as that obtained with the binomial expansion. **TRY PROBLEMS 23, 24, AND 25** →

The Testcross

A useful tool for analyzing genetic crosses is the **testcross**, in which one individual of unknown genotype is crossed with another individual with a homozygous recessive genotype for the trait in question. Figure 3.7 illustrates a testcross (in this case, it is also a backcross). A testcross tests, or reveals, the genotype of the first individual.

Suppose you were given a tall pea plant with no information about its parents. Because tallness is a dominant trait in peas, your plant could be either homozygous (TT) or heterozygous (Tt), but you would not know which. You could determine its genotype by performing a testcross. If the plant were homozygous (TT), a testcross would produce all tall progeny ($TT \times tt \rightarrow$ all Tt); if the plant were heterozygous (Tt), half of the progeny would be tall and half would be short ($Tt \times tt \rightarrow \frac{1}{2} Tt$ and $\frac{1}{2} tt$). When a testcross is performed, any recessive allele in the unknown genotype is expressed in the progeny, because it will be paired with a recessive allele from the homozygous recessive parent. **TRY PROBLEMS 17 AND 19** →

CONCEPTS

The binomial expansion can be used to determine the probability of a particular set of events. A testcross is a cross between an individual with an unknown genotype and one with a homozygous recessive genotype. The outcome of the testcross can reveal the unknown genotype.

Genetic Symbols

As we have seen, genetic crosses are usually depicted with the use of symbols to designate the different alleles. Lowercase letters are traditionally used to designate recessive alleles, and uppercase letters are for dominant alleles. Two or three letters may be used for a single allele: the recessive allele for heart-shaped leaves in cucumbers is designated hl, and the recessive allele for abnormal sperm-head shape in mice is designated azh.

The common allele for a character—called the **wild type** because it is the allele usually found in the wild—is often symbolized by one or more letters and a plus sign ($+$). The letter or letters chosen are usually based on the mutant (unusual) phenotype. For example, the recessive allele for yellow eyes in the Oriental fruit fly is represented by ye, whereas the allele for wild-type eye color is represented by ye^+. At times, the letters for the wild-type allele are dropped and the allele is represented simply by a plus sign. In another way of distinguishing alleles, the first letter is lowercase if the mutant phenotype is recessive and it is uppercase if the mutant phenotype is dominant: for example, narrow leaflet (ln) in soybeans is recessive to broad leaflet (Ln). Superscripts and subscripts are sometimes added to distinguish between genes: Lfr_1 and Lfr_2 represent dominant mutant alleles at different loci that produce lacerate leaf margins in opium poppies; El^R represents an allele in goats that restricts the length of the ears.

A slash may be used to distinguish alleles present in an individual genotype. For example, the genotype of a goat that is heterozygous for restricted ears might be written El^+/El^R or simply $+/El^R$. If genotypes at more than one locus are presented together, a space separates the genotypes. For example, a goat heterozygous for a pair of alleles that produces restricted ears and heterozygous for another pair of alleles that produces goiter can be designated by $El^+/El^R\ G/g$.

CONNECTING CONCEPTS

Ratios in Simple Crosses

Now that we have had some experience with genetic crosses, let's review the ratios that appear in the progeny of simple crosses, in which a single locus is under consideration and one of the alleles is dominant over the other. Understanding these ratios and the parental genotypes that produce them will allow you to work simple genetic crosses quickly, without resorting to the Punnett square. Later, we will use these ratios to work more-complicated crosses entailing several loci.

There are only three phenotypic ratios to understand (**Table 3.3**). The 3 : 1 ratio arises in a simple genetic cross when both of the

Table 3.3 Phenotypic ratios for simple genetic crosses (crosses for a single locus) with dominance

Phenotypic Ratio	Genotypes of Parents	Genotypes of Progeny
3 : 1	$Aa \times Aa$	$\frac{3}{4} A\_ : \frac{1}{4} aa$
1 : 1	$Aa \times aa$	$\frac{1}{2} Aa : \frac{1}{2} aa$
Uniform progeny	$AA \times AA$	All AA
	$aa \times aa$	All aa
	$AA \times aa$	All Aa
	$AA \times Aa$	All $A\_$

Note: A line in a genotype, such as $A\_$, indicates that any allele is possible.

Table 3.4	Genotypic ratios for simple genetic crosses (crosses for a single locus)	
Genotypic Ratio	Genotypes of Parents	Genotypes of Progeny
1 : 2 : 1	$Aa \times Aa$	$\frac{1}{4} AA : \frac{1}{2} Aa : \frac{1}{4} aa$
1 : 1	$Aa \times aa$	$\frac{1}{2} Aa : \frac{1}{2} aa$
	$Aa \times AA$	$\frac{1}{2} Aa : \frac{1}{2} AA$
Uniform progeny	$AA \times AA$	All AA
	$aa \times aa$	All aa
	$AA \times aa$	All Aa

parents are heterozygous for a dominant trait ($Aa \times Aa$). The second phenotypic ratio is the 1 : 1 ratio, which results from the mating of a homozygous parent and a heterozygous parent. The homozygous parent in this cross must carry two recessive alleles ($Aa \times aa$) to obtain a 1 : 1 ratio, because a cross between a homozygous dominant parent and a heterozygous parent ($AA \times Aa$) produces offspring displaying only the dominant trait.

The third phenotypic ratio is not really a ratio: all the offspring have the same phenotype (uniform progeny). Several combinations of parents can produce this outcome (see Table 3.3). A cross between any two homozygous parents—either between two of the same homozygotes ($AA \times AA$ and $aa \times aa$) or between two different homozygotes ($AA \times aa$)—produces progeny all having the same phenotype. Progeny of a single phenotype also can result from a cross between a homozygous dominant parent and a heterozygote ($AA \times Aa$).

If we are interested in the ratios of genotypes instead of phenotypes, there are only three outcomes to remember (**Table 3.4**): the 1 : 2 : 1 ratio, produced by a cross between two heterozygotes; the 1 : 1 ratio, produced by a cross between a heterozygote and a homozygote; and the uniform progeny produced by a cross between two homozygotes. These simple phenotypic and genotypic ratios and the parental genotypes that produce them provide the key to understanding crosses for a single locus and, as you will see in the next section, for multiple loci.

3.3 Dihybrid Crosses Reveal the Principle of Independent Assortment

We will now extend Mendel's principle of segregation to more-complex crosses entailing alleles at multiple loci. Understanding the nature of these crosses will require an additional principle, the principle of independent assortment.

Dihybrid Crosses

In addition to his work on monohybrid crosses, Mendel crossed varieties of peas that differed in *two* characteristics—a **dihybrid cross**. For example, he had one homozygous variety of pea with seeds that were round and yellow; another homozygous variety with seeds that were wrinkled and green. When he crossed the two varieties, the seeds of all the F_1 progeny were round and yellow. He then self-fertilized the F_1 and obtained the following progeny in the F_2: 315 round, yellow seeds; 101 wrinkled, yellow seeds; 108 round, green seeds; and 32 wrinkled, green seeds. Mendel recognized that these traits appeared approximately in a 9 : 3 : 3 : 1 ratio; that is, $\frac{9}{16}$ of the progeny were round and yellow, $\frac{3}{16}$ were wrinkled and yellow, $\frac{3}{16}$ were round and green, and $\frac{1}{16}$ were wrinkled and green.

The Principle of Independent Assortment

Mendel carried out a number of dihybrid crosses for pairs of characteristics and always obtained a 9 : 3 : 3 : 1 ratio in the F_2. This ratio makes perfect sense in regard to segregation and dominance if we add a third principle, which Mendel recognized in his dihybrid crosses: the **principle of independent assortment** (Mendel's second law). This principle states that alleles at different loci separate independently of one another (see Table 3.2).

A common mistake is to think that the principle of segregation and the principle of independent assortment refer to two different processes. The principle of independent assortment is really an extension of the principle of segregation. The principle of segregation states that the two alleles of a locus separate when gametes are formed; the principle of independent assortment states that, when these two alleles separate, their separation is independent of the separation of alleles at *other* loci.

Let's see how the principle of independent assortment explains the results that Mendel obtained in his dihybrid cross. Each plant possesses two alleles encoding each characteristic, and so the parental plants must have had genotypes $RR\,YY$ and $rr\,yy$ (**Figure 3.10a**). The principle of segregation indicates that the alleles for each locus separate, and one allele for each locus passes to each gamete. The gametes produced by the round, yellow parent therefore contain alleles RY, whereas the gametes produced by the wrinkled, green parent contain alleles ry. These two types of gametes unite to produce the F_1, all with genotype $Rr\,Yy$. Because round is dominant over wrinkled and yellow is dominant over green, the phenotype of the F_1 will be round and yellow.

When Mendel self-fertilized the F_1 plants to produce the F_2, the alleles for each locus separated, with one allele going into each gamete. This event is where the principle of independent assortment becomes important. Each pair of alleles can separate in two ways: (1) R separates with Y, and r separates with y, to produce gametes RY and ry or (2) R separates with y, and r separates with Y, to produce gametes Ry and rY. The principle of independent assortment tells us that the alleles at each locus separate independently; thus, both kinds of separation occur equally and all four type of gametes (RY, ry, Ry, and rY) are produced in equal proportions (**Figure 3.10b**). When these four types of gametes are combined to produce the F_2 generation, the progeny con-

Basic Principles of Heredity

sist of $9/16$ round and yellow, $3/16$ wrinkled and yellow, $3/16$ round and green, and $1/16$ wrinkled and green, resulting in a 9 : 3 : 3 : 1 phenotypic ratio (**Figure 3.10c**).

Relating the Principle of Independent Assortment to Meiosis

An important qualification of the principle of independent assortment is that it applies to characters encoded by loci located on different chromosomes because, like the principle of segregation, it is based wholly on the behavior of chromosomes in meiosis. Each pair of homologous chromosomes separates independently of all other pairs in anaphase I of meiosis (see Figure 2.17); so genes located on different pairs of homologs will assort independently. Genes that happen to be located on the same chromosome will travel together during anaphase I of meiosis and will arrive at the same destination—within the same gamete (unless crossing over takes place). Genes located on the same chromosome therefore do not assort independently (unless they are located sufficiently far apart that crossing over takes place every meiotic division, as will be discussed fully in Chapter 7).

CONCEPTS

The principle of independent assortment states that genes encoding different characteristics separate independently of one another when gametes are formed, owing to the independent separation of homologous pairs of chromosomes in meiosis. Genes located close together on the same chromosome do not, however, assort independently.

✔ **CONCEPT CHECK 6**

How are the principles of segregation and independent assortment related and how are they different?

Applying Probability and the Branch Diagram to Dihybrid Crosses

When the genes at two loci separate independently, a dihybrid cross can be understood as two monohybrid crosses. Let's examine Mendel's dihybrid cross ($Rr\ Yy \times Rr\ Yy$) by considering each characteristic separately (**Figure 3.11a**). If we consider only the shape of the seeds, the cross was $Rr \times Rr$, which yields a 3 : 1 phenotypic ratio ($3/4$ round and $1/4$ wrinkled progeny, see Table 3.3). Next consider the other characteristic, the color of the seed. The cross was $Yy \times Yy$, which produces a 3 : 1 phenotypic ratio ($3/4$ yellow and $1/4$ green progeny).

We can now combine these monohybrid ratios by using the multiplication rule to obtain the proportion of progeny with different combinations of seed shape and color. The proportion of progeny with round and yellow seeds is $3/4$ (the probability of round) $\times$ $3/4$ (the probability of yellow) $= 9/16$. The proportion of progeny with round and

3.10 Mendel's dihybrid crosses revealed the principle of independent assortment.

 Set up your own crosses and explore Mendel's principles of heredity in Animation 3.1.

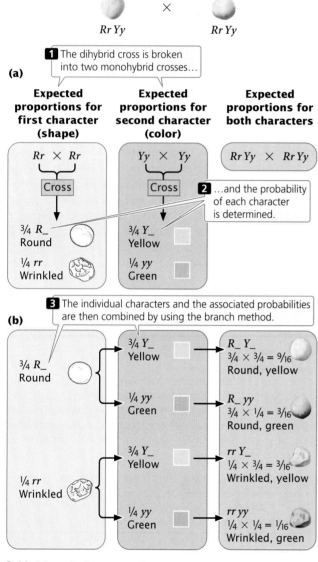

3.11 A branch diagram can be used to determine the phenotypes and expected proportions of offspring from a dihybrid cross ($Rr\ Yy \times Rr\ Yy$).

green seeds is $3/4 \times 1/4 = 3/16$; the proportion of progeny with wrinkled and yellow seeds is $1/4 \times 3/4 = 3/16$; and the proportion of progeny with wrinkled and green seeds is $1/4 \times 1/4 = 1/16$.

Branch diagrams are a convenient way of organizing all the combinations of characteristics (**Figure 3.11b**). In the first column, list the proportions of the phenotypes for one character (here, $3/4$ round and $1/4$ wrinkled). In the second column, list the proportions of the phenotypes for the second character ($3/4$ yellow and $1/4$ green) twice, next to each of the phenotypes in the first column: put $3/4$ yellow and $1/4$ green next to the round phenotype and again next to the wrinkled phenotype. Draw lines between the phenotypes in the first column and each of the phenotypes in the second column. Now follow each branch of the diagram, multiplying the probabilities for each trait along that branch. One branch leads from round to yellow, yielding round and yellow progeny. Another branch leads from round to green, yielding round and green progeny, and so forth. We calculate the probability of progeny with a particular combination of traits by using the multiplication rule: the probability of round ($3/4$) and yellow ($3/4$) seeds is $3/4 \times 3/4 = 9/16$. The advantage of the branch diagram is that it helps keep track of all the potential combinations of traits that may appear in the progeny. It can be used to determine phenotypic or genotypic ratios for any number of characteristics.

Using probability is much faster than using the Punnett square for crosses that include multiple loci. Genotypic and phenotypic ratios can be quickly worked out by combining, with the multiplication rule, the simple ratios in Tables 3.3 and 3.4. The probability method is particularly efficient if we need the probability of only a *particular* phenotype or genotype among the progeny of a cross. Suppose we needed to know the probability of obtaining the genotype $Rr\ yy$ in the F_2 of the dihybrid cross in Figure 3.10. The probability of obtaining the Rr genotype in a cross of $Rr \times Rr$ is $1/2$ and that of obtaining yy progeny in a cross of $Yy \times Yy$ is $1/4$ (see Table 3.4). Using the multiplication rule, we find the probability of $Rr\ yy$ to be $1/2 \times 1/4 = 1/8$.

To illustrate the advantage of the probability method, consider the cross $Aa\ Bb\ cc\ Dd\ Ee \times Aa\ Bb\ Cc\ dd\ Ee$. Suppose we wanted to know the probability of obtaining offspring with the genotype $aa\ bb\ cc\ dd\ ee$. If we used a Punnett square to determine this probability, we might be working on the solution for months. However, we can quickly figure the probability of obtaining this one genotype by breaking this cross into a series of single-locus crosses:

Progeny cross	Genotype	Probability
$Aa \times Aa$	aa	$1/4$
$Bb \times Bb$	bb	$1/4$
$cc \times Cc$	cc	$1/2$
$Dd \times dd$	dd	$1/2$
$Ee \times Ee$	ee	$1/4$

The probability of an offspring from this cross having genotype $aa\ bb\ cc\ dd\ ee$ is now easily obtained by using the multiplication rule: $1/4 \times 1/4 \times 1/2 \times 1/2 \times 1/4 = 1/256$. This calculation assumes that genes at these five loci all assort independently.

> **CONCEPTS**
>
> A cross including several characteristics can be worked by breaking the cross down into single-locus crosses and using the multiplication rule to determine the proportions of combinations of characteristics (provided that the genes assort independently).

3.12 A branch diagram can be used to determine the phenotypes and expected proportions of offspring from a dihybrid testcross (*Rr Yy* × *rr yy*).

The Dihybrid Testcross

Let's practice using the branch diagram by determining the types and proportions of phenotypes in a dihybrid testcross between the round and yellow F_1 plants (*Rr Yy*) obtained by Mendel in his dihybrid cross and the wrinkled and green plants (*rr yy*), as depicted in **Figure 3.12**. Break the cross down into a series of single-locus crosses. The cross *Rr* × *rr* yields $1/2$ round (*Rr*) progeny and $1/2$ wrinkled (*rr*) progeny. The cross *Yy* × *yy* yields $1/2$ yellow (*Yy*) progeny and $1/2$ green (*yy*) progeny. Using the multiplication rule, we find the proportion of round and yellow progeny to be $1/2$ (the probability of round) × $1/2$ (the probability of yellow) = $1/4$. Four combinations of traits with the following proportions appear in the offspring: $1/4$ *Rr Yy*, round yellow; $1/4$ *Rr yy*, round green; $1/4$ *rr Yy*, wrinkled yellow; and $1/4$ *rr yy*, wrinkled green.

Worked Problem

Not only are the principles of segregation and independent assortment important because they explain how heredity works, but they also provide the means for predicting the outcome of genetic crosses. This predictive power has made genetics a powerful tool in agriculture and other fields, and the ability to apply the principles of heredity is an important skill for all students of genetics. Practice with genetic problems is essential for mastering the basic principles of heredity; no amount of reading and memorization can substitute for the experience gained by deriving solutions to specific problems in genetics.

Students may have difficulty with genetics problems when they are unsure of where to begin or how to organize the problem and plan a solution. In genetics, every problem is different, and so no common series of steps can be applied to all genetics problems. Logic and common sense must be used to analyze a problem and arrive at a solution. Nevertheless, certain steps can facilitate the process, and solving the following problem will serve to illustrate these steps.

In mice, black coat color (*B*) is dominant over brown (*b*), and a solid pattern (*S*) is dominant over white spotted (*s*). Color and spotting are controlled by genes that assort independently. A homozygous black, spotted mouse is crossed with a homozygous brown, solid mouse. All the F_1 mice are black and solid. A testcross is then carried out by mating the F_1 mice with brown, spotted mice.

a. Give the genotypes of the parents and the F_1 mice.

b. Give the genotypes and phenotypes, along with their expected ratios, of the progeny expected from the testcross.

• Solution

Step 1. Determine the questions to be answered. What question or questions is the problem asking? Is it asking for genotypes, genotypic ratios, or phenotypic ratios? This problem asks you to provide the *genotypes* of the parents and the F_1, the *expected genotypes* and *phenotypes* of the progeny of the testcross, and their *expected proportions*.

Step 2. Write down the basic information given in the problem. This problem provides important information about the dominance relations of the characters and about the mice being crossed. Black is dominant over brown, and solid is dominant over white spotted. Furthermore, the genes for the two characters assort independently. In this problem, symbols are provided for the different alleles (*B* for black, *b* for brown, *S* for solid, and *s* for spotted); had these symbols not been provided, you would need to choose symbols to represent these alleles.

It is useful to record these symbols at the beginning of the solution:

B—black $\qquad$ S—solid
b—brown $\qquad$ s—white spotted

Next, write out the crosses given in the problem.

P $\qquad$ Homozygous $\quad \times \quad$ Homozygous
$\qquad\qquad$ black, spotted $\qquad\qquad$ brown, solid
$\qquad\qquad\qquad\qquad\qquad\downarrow$

F_1 $\qquad\qquad\qquad$ Black, solid

Testcross $\quad$ Black, solid $\quad \times \quad$ Brown, spotted

Step 3. Write down any genetic information that can be determined from the phenotypes alone. From the phenotypes and the statement that they are homozygous, you know that the P-generation mice must be $BB\ ss$ and $bb\ SS$. The F_1 mice are black and solid, both dominant traits, and so the F_1 mice must possess at least one black allele (B) and one solid allele (S). At this point, you cannot be certain about the other alleles; so represent the genotype of the F_1 as $B\_\ S\_$, where _ means that any allele is possible. The brown, spotted mice in the testcross must be $bb\ ss$, because both brown and spotted are recessive traits that will be expressed only if two recessive alleles are present. Record these genotypes on the crosses that you wrote out in step 2:

P $\qquad$ Homozygous $\quad \times \quad$ Homozygous
$\qquad\qquad$ black, spotted $\qquad\qquad$ brown, solid
$\qquad\qquad\quad BB\ ss \qquad\qquad\qquad\ bb\ SS$
$\qquad\qquad\qquad\qquad\qquad\downarrow$

F_1 $\qquad\qquad\qquad$ Black, solid
$\qquad\qquad\qquad\qquad\ B\_\ S\_$

Testcross $\quad$ Black, solid $\quad \times \quad$ Brown, spotted
$\qquad\qquad\quad B\_\ S\_ \qquad\qquad\qquad bb\ ss$

Step 4. Break the problem down into smaller parts. First, determine the genotype of the F_1. After this genotype has been determined, you can predict the results of the testcross and determine the genotypes and phenotypes of the progeny from the testcross. Second, because this cross includes two independently assorting loci, it can be conveniently broken down into two single-locus crosses: one for coat color and the other for spotting. Third, use a branch diagram to determine the proportion of progeny of the testcross with different combinations of the two traits.

Step 5. Work the different parts of the problem. Start by determining the genotype of the F_1 progeny. Mendel's first law indicates that the two alleles at a locus separate, one

going into each gamete. Thus, the gametes produced by the black, spotted parent contain $B\ s$ and the gametes produced by the brown, solid parent contain $b\ S$, which combine to produce F_1 progeny with the genotype $Bb\ Ss$:

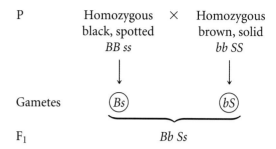

F_1 $\qquad\qquad\qquad\qquad Bb\ Ss$

Use the F_1 genotype to work the testcross ($Bb\ Ss \times bb\ ss$), breaking it into two single-locus crosses. First, consider the cross for coat color: $Bb \times bb$. Any cross between a heterozygote and a homozygous recessive genotype produces a 1 : 1 phenotypic ratio of progeny (see Table 3.3):

$$Bb \times bb$$
$$\downarrow$$
$$\tfrac{1}{2}\ Bb\ \text{black}$$
$$\tfrac{1}{2}\ bb\ \text{brown}$$

Next, do the cross for spotting: $Ss \times ss$. This cross also is between a heterozygote and a homozygous recessive genotype and will produce $\tfrac{1}{2}$ solid (Ss) and $\tfrac{1}{2}$ spotted (ss) progeny (see Table 3.3).

$$Ss \times ss$$
$$\downarrow$$
$$\tfrac{1}{2}\ Ss\ \text{solid}$$
$$\tfrac{1}{2}\ ss\ \text{spotted}$$

Finally, determine the proportions of progeny with combinations of these characters by using the branch diagram.

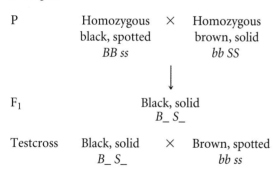

Step 6. Check all work. As a last step, reread the problem, checking to see if your answers are consistent with the information provided. You have used the genotypes $BB\ ss$ and

bb SS in the P generation. Do these genotypes encode the phenotypes given in the problem? Are the F$_1$ progeny phenotypes consistent with the genotypes that you assigned? The answers are consistent with the information.

> ➡ Now that we have stepped through a genetics problem together, try your hand at Problem 30 at the end of the chapter.

3.4 Observed Ratios of Progeny May Deviate from Expected Ratios by Chance

When two individual organisms of known genotype are crossed, we expect certain ratios of genotypes and phenotypes in the progeny; these expected ratios are based on the Mendelian principles of segregation, independent assortment, and dominance. The ratios of genotypes and phenotypes *actually* observed among the progeny, however, may deviate from these expectations.

For example, in German cockroaches, brown body color (*Y*) is dominant over yellow body color (*y*). If we cross a brown, heterozygous cockroach (*Yy*) with a yellow cockroach (*yy*), we expect a 1 : 1 ratio of brown (*Yy*) and yellow (*yy*) progeny. Among 40 progeny, we therefore expect to see 20 brown and 20 yellow offspring. However, the observed numbers might deviate from these expected values; we might in fact see 22 brown and 18 yellow progeny.

Chance plays a critical role in genetic crosses, just as it does in flipping a coin. When you flip a coin, you expect a 1 : 1 ratio—$\frac{1}{2}$ heads and $\frac{1}{2}$ tails. If you flip a coin 1000 times, the proportion of heads and tails obtained will probably be very close to that expected 1 : 1 ratio. However, if you flipped the coin 10 times, the ratio of heads to tails might be quite different from 1 : 1. You could easily get 6 heads and 4 tails, or 3 heads and 7 tails, just by chance. You might even get 10 heads and 0 tails. The same thing happens in genetic crosses. We may expect 20 brown and 20 yellow cockroaches, but 22 brown and 18 yellow progeny *could* arise as a result of chance.

The Goodness-of-Fit Chi-Square Test

If you expected a 1 : 1 ratio of brown and yellow cockroaches but the cross produced 22 brown and 18 yellow, you probably wouldn't be too surprised even though it wasn't a perfect 1 : 1 ratio. In this case, it seems reasonable to assume that chance produced the deviation between the expected and the observed results. But, if you observed 25 brown and 15 yellow, would the ratio still be 1 : 1? Something other than chance might have caused the deviation. Perhaps the inheritance of this character is more complicated than was assumed or perhaps some of the yellow progeny died before they were counted. Clearly, we need some means of evaluating how likely it is that chance is responsible for the deviation between the observed and the expected numbers.

To evaluate the role of chance in producing deviations between observed and expected values, a statistical test called the **goodness-of-fit chi-square test** is used. This test provides information about how well observed values fit expected values. Before we learn how to calculate the chi square, it is important to understand what this test does and does not indicate about a genetic cross.

The chi-square test cannot tell us whether a genetic cross has been correctly carried out, whether the results are correct, or whether we have chosen the correct genetic explanation for the results. What it does indicate is the *probability* that the difference between the observed and the expected values is due to chance. In other words, it indicates the likelihood that chance alone could produce the deviation between the expected and the observed values.

If we expected 20 brown and 20 yellow progeny from a genetic cross, the chi-square test gives the probability that we might observe 25 brown and 15 yellow progeny simply owing to chance deviations from the expected 20 : 20 ratio. This hypothesis, that chance alone is responsible for any deviations between observed and expected values, is sometimes called the null hypothesis. When the probability calculated from the chi-square test is high, we assume that chance alone produced the difference (the null hypothesis is true). When the probability is low, we assume that some factor other than chance—some significant factor—produced the deviation (the null hypothesis is false).

To use the goodness-of-fit chi-square test, we first determine the expected results. The chi-square test must always be applied to *numbers* of progeny, not to proportions or percentages. Let's consider a locus for coat color in domestic cats, for which black color (*B*) is dominant over gray (*b*). If we crossed two heterozygous black cats (*Bb* $\times$ *Bb*), we would expect a 3 : 1 ratio of black and gray kittens. A series of such crosses yields a total of 50 kittens—30 black and 20 gray. These numbers are our *observed* values. We can obtain the expected numbers by multiplying the *expected* proportions by the total number of observed progeny. In this case, the expected number of black kittens is $\frac{3}{4} \times 50 = 37.5$ and the expected number of gray kittens is $\frac{1}{4} \times 50 = 12.5$. The chi-square ($\chi^2$) value is calculated by using the following formula:

$$\chi^2 = \Sigma \frac{(\text{observed} - \text{expected})^2}{\text{expected}}$$

Table 3.5 Critical values of the χ^2 distribution

df	\multicolumn{9}{c}{P}
	0.995	0.975	0.9	0.5	0.1	0.05*	0.025	0.01	0.005
1	0.000	0.000	0.016	0.455	2.706	3.841	5.024	6.635	7.879
2	0.010	0.051	0.211	1.386	4.605	5.991	7.378	9.210	10.597
3	0.072	0.216	0.584	2.366	6.251	7.815	9.348	11.345	12.838
4	0.207	0.484	1.064	3.357	7.779	9.488	11.143	13.277	14.860
5	0.412	0.831	1.610	4.351	9.236	11.070	12.832	15.086	16.750
6	0.676	1.237	2.204	5.348	10.645	12.592	14.449	16.812	18.548
7	0.989	1.690	2.833	6.346	12.017	14.067	16.013	18.475	20.278
8	1.344	2.180	3.490	7.344	13.362	15.507	17.535	20.090	21.955
9	1.735	2.700	4.168	8.343	14.684	16.919	19.023	21.666	23.589
10	2.156	3.247	4.865	9.342	15.987	18.307	20.483	23.209	25.188
11	2.603	3.816	5.578	10.341	17.275	19.675	21.920	24.725	26.757
12	3.074	4.404	6.304	11.340	18.549	21.026	23.337	26.217	28.300
13	3.565	5.009	7.042	12.340	19.812	22.362	24.736	27.688	29.819
14	4.075	5.629	7.790	13.339	21.064	23.685	26.119	29.141	31.319
15	4.601	6.262	8.547	14.339	22.307	24.996	27.488	30.578	32.801

P, probability; df, degrees of freedom.
*Most scientists assume that, when $P < 0.05$, a significant difference exists between the observed and the expected values in a chi-square test.

where Σ means the sum. We calculate the sum of all the squared differences between observed and expected and divide by the expected values. To calculate the chi-square value for our black and gray kittens, we first subtract the number of *expected* black kittens from the number of *observed* black kittens ($30 - 37.5 = -7.5$) and square this value: $-7.5^2 = 56.25$. We then divide this result by the expected number of black kittens, $56.25/37.5 = 1.5$. We repeat the calculations on the number of expected gray kittens: $(20 - 12.5)^2/12.5 = 4.5$. To obtain the overall chi-square value, we sum the (observed - expected)^2/expected values: $1.5 + 4.5 = 6.0$.

The next step is to determine the probability associated with this calculated chi-square value, which is the probability that the deviation between the observed and the expected results could be due to chance. This step requires us to compare the calculated chi-square value (6.0) with theoretical values that have the same degrees of freedom in a chi-square table. The degrees of freedom represent the number of ways in which the expected classes are free to vary. For a goodness-of-fit chi-square test, the degrees of freedom are equal to $n - 1$, where n is the number of different expected phenotypes. In our example, there are two expected phenotypes (black and gray); so $n = 2$, and the degree of freedom equals $2 - 1 = 1$.

Now that we have our calculated chi-square value and have figured out the associated degrees of freedom, we are ready to obtain the probability from a chi-square table (**Table 3.5**). The degrees of freedom are given in the left-hand column of the table and the probabilities are given at the top; within the body of the table are chi-square values associated with these probabilities. First, find the row for the appropriate degrees of freedom; for our example with 1 degree of freedom, it is the first row of the table. Find where our calculated chi-square value (6.0) lies among the theoretical values in this row. The theoretical chi-square values increase from left to right and the probabilities decrease from left to right. Our chi-square value of 6.0 falls between the value of 5.024, associated with a probability of 0.025, and the value of 6.635, associated with a probability of 0.01.

Thus, the probability associated with our chi-square value is less than 0.025 and greater than 0.01. So there is less than a 2.5% probability that the deviation that we observed between the expected and the observed numbers of black and gray kittens could be due to chance.

Most scientists use the 0.05 probability level as their cutoff value: if the probability of chance being responsible for the deviation is greater than or equal to 0.05, they accept that chance may be responsible for the deviation between the observed and the expected values. When the probability is less than 0.05, scientists assume that chance is not responsible and a significant difference exists. The expression *significant difference* means that some factor other than chance is responsible for the observed values being different from the expected values. In regard to the kittens, perhaps one of the genotypes had a greater mortality rate before the progeny were counted or perhaps other genetic factors skewed the observed ratios.

In choosing 0.05 as the cutoff value, scientists have agreed to assume that chance is responsible for the deviations between observed and expected values unless there is strong evidence to the contrary. Bear in mind that, even if we obtain a probability of, say, 0.01, there is still a 1% probability that the deviation between the observed and the expected numbers is due to nothing more than chance. Calculation of the chi-square value is illustrated in **Figure 3.13**. TRY PROBLEM 35 →

CONCEPTS

Differences between observed and expected ratios can arise by chance. The goodness-of-fit chi-square test can be used to evaluate whether deviations between observed and expected numbers are likely to be due to chance or to some other significant factor.

✔ CONCEPT CHECK 7

A chi-square test comparing observed and expected progeny is carried out, and the probability associated with the calculated chi-square value is 0.72. What does this probability represent?

a. Probability that the correct results were obtained
b. Probability of obtaining the observed numbers
c. Probability that the difference between observed and expected numbers is significant
d. Probability that the difference between observed and expected numbers could be due to chance

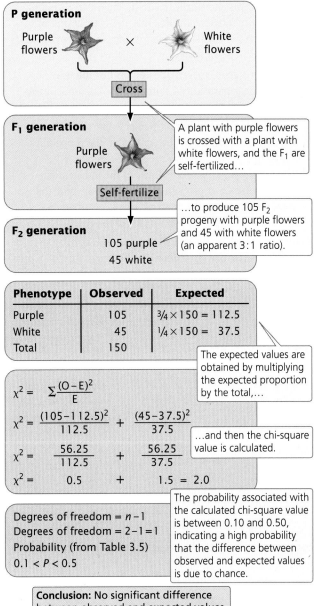

3.13 A chi-square test is used to determine the probability that the difference between observed and expected values is due to chance.

CONCEPTS SUMMARY

- Gregor Mendel discovered the principles of heredity. His success can be attributed to his choice of the pea plant as an experimental organism, the use of characters with a few easily distinguishable phenotypes, his experimental approach, the use of mathematics to interpret his results, and careful attention to detail.

- Genes are inherited factors that determine a characteristic. Alternative forms of a gene are called alleles. The alleles are located at a specific place, a locus, on a chromosome, and the set of genes that an individual organism possesses is its genotype. Phenotype is the manifestation or appearance of a characteristic and may refer to a physical, biochemical, or behavioral characteristic. Only the genotype—not the phenotype—is inherited.

- The principle of segregation states that an individual organism possesses two alleles encoding a trait and that these two alleles separate in equal proportions when gametes are formed.

- The concept of dominance indicates that, when two different alleles are present in a heterozygote, only the trait of one of them, the dominant allele, is observed in the phenotype. The other allele is said to be recessive.
- The two alleles of a genotype are located on homologous chromosomes. The separation of homologous chromosomes in anaphase I of meiosis brings about the segregation of alleles.
- Probability is the likelihood that a particular event will occur. The multiplication rule of probability states that the probability of two or more independent events occurring together is calculated by multiplying the probabilities of the independent events. The addition rule of probability states that the probability of any of two or more mutually exclusive events occurring is calculated by adding the probabilities of the events.
- The binomial expansion can be used to determine the probability of a particular combination of events.
- A testcross reveals the genotype (homozygote or heterozygote) of an individual organism having a dominant trait and consists of crossing that individual with one having the homozygous recessive genotype.
- The principle of independent assortment states that genes encoding different characters assort independently when gametes are formed. Independent assortment is based on the random separation of homologous pairs of chromosomes in anaphase I of meiosis; it takes place when genes encoding two characters are located on different pairs of chromosomes.
- Observed ratios of progeny from a genetic cross may deviate from the expected ratios owing to chance. The goodness-of-fit chi-square test can be used to determine the probability that a difference between observed and expected numbers is due to chance.

IMPORTANT TERMS

gene (p. 46)
allele (p. 46)
locus (p. 46)
genotype (p. 46)
homozygous (p. 46)
heterozygous (p. 46)
phenotype (p. 46)
monohybrid cross (p. 47)
P (parental) generation (p. 47)
F_1 (filial 1) generation (p. 48)

reciprocal crosses (p. 48)
F_2 (filial 2) generation (p. 48)
dominant (p. 49)
recessive (p. 49)
principle of segregation
 (Mendel's first law) (p. 49)
concept of dominance (p. 49)
chromosome theory of heredity (p. 50)
backcross (p. 51)
Punnett square (p. 51)

probability (p. 52)
multiplication rule (p. 52)
addition rule (p. 53)
testcross (p. 55)
wild type (p. 55)
dihybrid cross (p. 56)
principle of independent assortment
 (Mendel's second law) (p. 56)
goodness-of-fit chi-square test (p. 61)

ANSWERS TO CONCEPT CHECKS

1. b

2. A locus is a place on a chromosome where genetic information encoding a character is located. An allele is a copy of a gene that encodes a specific trait. A genotype is the set of alleles possessed by an individual organism, and a phenotype is the manifestation or appearance of a character.

3. Because the traits for both alleles appeared in the F_2 progeny

4. d

5. a

6. The principle of segregation and the principle of independent assortment both refer to the separation of alleles in anaphase I of meiosis. The principle of segregation says that these alleles separate, and the principle of independent assortment says that they separate independently of alleles at other loci.

7. d

WORKED PROBLEMS

1. Short hair (*S*) in rabbits is dominant over long hair (*s*). The following crosses are carried out, producing the progeny shown. Give all possible genotypes of the parents in each cross.

	Parents	Progeny
a.	short × short	4 short and 2 long
b.	short × short	8 short
c.	short × long	12 short
d.	short × long	3 short and 1 long
e.	long × long	2 long

- **Solution**

For this problem, it is useful to first gather as much information about the genotypes of the parents as possible on the basis of their phenotypes. We can then look at the types of progeny produced to provide the missing information. Notice that the problem asks for *all* possible genotypes of the parents.

a. short × short   4 short and 2 long

Because short hair is dominant over long hair, a rabbit having short hair could be either *SS* or *Ss*. The 2 long-haired offspring must be homozygous (*ss*) because long hair is recessive and will appear in the phenotype only when both alleles for long hair are present. Because each parent contributes one of the two alleles found in the progeny, each parent must be carrying the *s* allele and must therefore be *Ss*.

b. short × short   8 short

The short-haired parents could be *SS* or *Ss*. All 8 of the offspring are short (*S_*), and so at least one of the parents is likely to be homozygous (*SS*); if both parents were heterozygous, we would expect ¼ of the progeny to be long haired (*ss*), but we do not observe any long-haired progeny. The other parent could be homozygous (*SS*) or heterozygous (*Ss*); as long as one parent is homozygous, all the offspring will be short haired. It is theoretically possible, although unlikely, that both parents are heterozygous (*Ss* × *Ss*). If both were heterozygous, we would expect 2 of the 8 progeny to be long haired. Although no long-haired progeny are observed, it is possible that just by chance no long-haired rabbits would be produced among the 8 progeny of the cross.

c. short × long   12 short

The short-haired parent could be *SS* or *Ss*. The long-haired parent must be *ss*. If the short-haired parent were heterozygous (*Ss*), half of the offspring would be expected to be long haired, but we don't see any long-haired progeny. Therefore, this parent is most likely homozygous (*SS*). It is theoretically possible, although unlikely, that the parent is heterozygous and just by chance no long-haired progeny were produced.

d. short × long   3 short and 1 long

On the basis of its phenotype, the short-haired parent could be homozygous (*SS*) or heterozygous (*Ss*), but the presence of one long-haired offspring tells us that the short-haired parent must be heterozygous (*Ss*). The long-haired parent must be homozygous (*ss*).

e. long × long   2 long

Because long hair is recessive, both parents must be homozygous for a long-hair allele (*ss*).

2. In cats, black coat color is dominant over gray. A female black cat whose mother is gray mates with a gray male. If this female has a litter of six kittens, what is the probability that three will be black and three will be gray?

• **Solution**

Because black (*G*) is dominant over gray (*g*), a black cat may be homozygous (*GG*) or heterozygous (*Gg*). The black female in this problem must be heterozygous (*Gg*) because her mother is gray (*gg*) and she must inherit one of her mother's alleles. The gray male is homozygous (*gg*) because gray is recessive. Thus the cross is:

Gg × *gg*
Black female   Gray male
↓
½ *Gg* black
½ *gg* gray

We can use the binomial expansion to determine the probability of obtaining three black and three gray kittens in a litter of six. Let *p* equal the probability of a kitten being black and *q* equal the probability of a kitten being gray. The binomial is $(p + q)^6$, the expansion of which is:

$$(p + q)^6 = p^6 + 6p^5q + 15p^4q^2 + 20p^3q^3 + 15p^2q^4 + 6p^1q^5 + q^6$$

(See the section on The Binomial Expansion and Probability for an explanation of how to expand the binomial.) The probability of obtaining three black and three gray kittens in a litter of six is provided by the term $20p^3q^3$. The probabilities of *p* and *q* are both ½; so the overall probability is $20(½)^3(½)^3 = 20/64 = 5/16$.

3. The following genotypes are crossed:

Aa Bb Cc Dd × *Aa Bb Cc Dd*

Give the proportion of the progeny of this cross having each of the following genotypes: (**a**) *Aa Bb Cc Dd*, (**b**) *aa bb cc dd*, (**c**) *Aa Bb cc Dd*.

• **Solution**

This problem is easily worked if the cross is broken down into simple crosses and the multiplication rule is used to find the different combinations of genotypes:

Locus 1   *Aa* × *Aa* = ¼ *AA*, ½ *Aa*, ¼ *aa*
Locus 2   *Bb* × *Bb* = ¼ *BB*, ½ *Bb*, ¼ *bb*
Locus 3   *Cc* × *Cc* = ¼ *CC*, ½ *Cc*, ¼ *cc*
Locus 4   *Dd* × *Dd* = ¼ *DD*, ½ *Dd*, ¼ *dd*

To find the probability of any combination of genotypes, simply multiply the probabilities of the different genotypes:

a. *Aa Bb Cc Dd*   ½ (*Aa*) × ½ (*Bb*) × ½ (*Cc*) × ½ (*Dd*) = 1/16
b. *aa bb cc dd*   ¼ (*aa*) × ¼ (*bb*) × ¼ (*cc*) × ¼ (*dd*) = 1/256
c. *Aa Bb cc Dd*   ½ (*Aa*) × ½ (*Bb*) × ¼ (*cc*) × ½ (*Dd*) = 1/32

4. In corn, purple kernels are dominant over yellow kernels, and full kernels are dominant over shrunken kernels. A corn plant having purple and full kernels is crossed with a plant having yellow and shrunken kernels, and the following progeny are obtained:

purple, full	112
purple, shrunken	103
yellow, full	91
yellow, shrunken	94

What are the most likely genotypes of the parents and progeny? Test your genetic hypothesis with a chi-square test.

• **Solution**

The best way to begin this problem is by breaking the cross down into simple crosses for a single characteristic (seed color or seed shape):

P	purple × yellow	full × shrunken
F_1	112 + 103 = 215 purple	112 + 91 = 203 full
	91 + 94 = 185 yellow	103 + 94 = 197 shrunken

Purple × yellow produces approximately $\frac{1}{2}$ purple and $\frac{1}{2}$ yellow. A 1 : 1 ratio is usually caused by a cross between a heterozygote and a homozygote. Because purple is dominant, the purple parent must be heterozygous (*Pp*) and the yellow parent must be homozygous (*pp*). The purple progeny produced by this cross will be heterozygous (*Pp*) and the yellow progeny must be homozygous (*pp*).

Now let's examine the other character. Full × shrunken produces $\frac{1}{2}$ full and $\frac{1}{2}$ shrunken, or a 1 : 1 ratio, and so these progeny phenotypes also are produced by a cross between a heterozygote (*Ff*) and a homozygote (*ff*); the full-kernel progeny will be heterozygous (*Ff*) and the shrunken-kernel progeny will be homozygous (*ff*).

Now combine the two crosses and use the multiplication rule to obtain the overall genotypes and the proportions of each genotype:

P	Purple, full	×	Yellow, shrunken
	Pp Ff		*pp ff*
F_1	*Pp Ff* = $\frac{1}{2}$ purple	× $\frac{1}{2}$ full = $\frac{1}{4}$ purple, full	
	Pp ff = $\frac{1}{2}$ purple	× $\frac{1}{2}$ shrunken = $\frac{1}{4}$ purple, shrunken	
	pp Ff = $\frac{1}{2}$ yellow	× $\frac{1}{2}$ full = $\frac{1}{4}$ yellow, full	
	pp ff = $\frac{1}{2}$ yellow	× $\frac{1}{2}$ shrunken = $\frac{1}{4}$ yellow, shrunken	

Our genetic explanation predicts that, from this cross, we should see $\frac{1}{4}$ purple, full-kernel progeny; $\frac{1}{4}$ purple, shrunken-kernel progeny; $\frac{1}{4}$ yellow, full-kernel progeny; and $\frac{1}{4}$ yellow, shrunken-kernel progeny. A total of 400 progeny were produced; so $\frac{1}{4} \times 400 = 100$ of each phenotype are expected. These observed numbers do not fit the expected numbers exactly. Could the difference between what we observe and what we expect be due to chance? If the probability is high that chance alone is responsible for the difference between observed and expected, we will assume that the progeny have been produced in the 1 : 1 : 1 : 1 ratio predicted by the cross. If the probability that the difference between observed and expected is due to chance is low, the progeny are not really in the predicted ratio and some other, *significant* factor must be responsible for the deviation.

The observed and expected numbers are:

Phenotype	Observed	Expected
purple, full	112	$\frac{1}{4} \times 400 = 100$
purple, shrunken	103	$\frac{1}{4} \times 400 = 100$
yellow, full	91	$\frac{1}{4} \times 400 = 100$
yellow, shrunken	94	$\frac{1}{4} \times 400 = 100$

To determine the probability that the difference between observed and expected is due to chance, we calculate a chi-square value with the formula $\chi^2 = \Sigma[(\text{observed} - \text{expected})^2/\text{expected}]$:

$$\chi^2 = \frac{(112-100)^2}{100} + \frac{(103-100)^2}{100} + \frac{(91-100)^2}{100} + \frac{(94-100)^2}{100}$$

$$= \frac{12^2}{100} + \frac{3^2}{100} + \frac{9^2}{100} + \frac{6^2}{100}$$

$$= \frac{144}{100} + \frac{9}{100} + \frac{81}{100} + \frac{36}{100}$$

$$= 1.44 + 0.09 + 0.81 + 0.36 = 2.70$$

Now that we have the chi-square value, we must determine the probability that this chi-square value is due to chance. To obtain this probability, we first calculate the degrees of freedom, which for a goodness-of-fit chi-square test are $n - 1$, where n equals the number of expected phenotypic classes. In this case, there are four expected phenotypic classes; so the degrees of freedom equal $4 - 1 = 3$. We must now look up the chi-square value in a chi-square table (see Table 3.5). We select the row corresponding to 3 degrees of freedom and look along this row to find our calculated chi-square value. The calculated chi-square value of 2.7 lies between 2.366 (a probability of 0.5) and 6.251 (a probability of 0.1). The probability (*P*) associated with the calculated chi-square value is therefore $0.5 < P < 0.1$. It is the probability that the difference between what we observed and what we expected is due to chance, which in this case is high, and so chance is likely responsible for the deviation. We can conclude that the progeny *do* appear in the 1 : 1 : 1 : 1 ratio predicted by our genetic explanation.

COMPREHENSION QUESTIONS

Section 3.1

*1. Why was Mendel's approach to the study of heredity so successful?

2. What is the difference between genotype and phenotype?

Section 3.2

*3. What is the principle of segregation? Why is it important?

4. How are Mendel's principles different from the concept of blending inheritance discussed in Chapter 1?

5. What is the concept of dominance?

6. What are the addition and multiplication rules of probability and when should they be used?

7. Give the genotypic ratios that may appear among the progeny of simple crosses and the genotypes of the parents that may give rise to each ratio.

*8. What is the chromosome theory of heredity? Why was it important?

Section 3.3

*9. What is the principle of independent assortment? How is it related to the principle of segregation?

10. In which phases of mitosis and meiosis are the principles of segregation and independent assortment at work?

Section 3.4

11. How is the goodness-of-fit chi-square test used to analyze genetic crosses? What does the probability associated with a chi-square value indicate about the results of a cross?

APPLICATION QUESTIONS AND PROBLEMS

Introduction

12. The inheritance of red hair was discussed in the introduction to this chapter. At times in the past, red hair in humans was thought to be a recessive trait and, at other times, it was thought to be a dominant trait. What features of heritance is red hair expected to exhibit as a recessive trait? What features would it exhibit if it were a dominant trait?

Section 3.1

13. What characteristics of an organism would make it suitable for studies of the principles of inheritance? Can you name several organisms that have these characteristics?

Section 3.2

*14. In cucumbers, orange fruit color (R) is dominant over cream fruit color (r). A cucumber plant homozygous for orange fruits is crossed with a plant homozygous for cream fruits. The F_1 are intercrossed to produce the F_2.

  a. Give the genotypes and phenotypes of the parents, the F_1, and the F_2.

  b. Give the genotypes and phenotypes of the offspring of a backcross between the F_1 and the orange parent.

  c. Give the genotypes and phenotypes of a backcross between the F_1 and the cream parent.

15. J. W. McKay crossed a stock melon plant that produced tan seeds with a plant that produced red seeds and obtained the following results (J. W. McKay. 1936. *Journal of Heredity* 27:110–112).

Cross	F_1	F_2
tan ♀ × red ♂	13 tan seeds	93 tan, 24 red seeds

  a. Explain the inheritance of tan and red seeds in this plant.

  b. Assign symbols for the alleles in this cross and give genotypes for all the individual plants.

*16. White (w) coat color in guinea pigs is recessive to black (W). In 1909, W. E. Castle and J. C. Phillips transplanted an ovary from a black guinea pig into a white female whose ovaries had been removed. They then mated this white female with a white male. All the offspring from the mating were black in color (W. E. Castle and J. C. Phillips. 1909. *Science* 30:312–313).

[Wegner/ARCO/Nature Picture Library; Nigel Cattlin/Alamy.]

  a. Explain results of this cross.

  b. Give the genotype of the offspring of this cross.

  c. What, if anything, does this experiment indicate about the validity of the pangenesis and the germ-plasm theories discussed in Chapter 1?

*17. In cats, blood-type A results from an allele (I^A) that is dominant over an allele (i^B) that produces blood-type B. There is no O blood type. The blood types of male and female cats that were mated and the blood types of their kittens follow. Give the most likely genotypes for the parents of each litter.

	Male parent	Female parent	Kittens
a.	A	B	4 with type A, 3 with type B
b.	B	B	6 with type B
c.	B	A	8 with type A
d.	A	A	7 with type A, 2 with type B
e.	A	A	10 with type A
f.	A	B	4 with type A, 1 with type B

18. Joe has a white cat named Sam. When Joe crosses Sam with a black cat, he obtains $\frac{1}{2}$ white kittens and $\frac{1}{2}$ black kittens. When the black kittens are interbred, all the kittens that they produce are black. On the basis of these results, would you conclude that white or black coat color in cats is a recessive trait? Explain your reasoning.

19. In sheep, lustrous fleece results from an allele (*L*) that is dominant over an allele (*l*) for normal fleece. A ewe (adult female) with lustrous fleece is mated with a ram (adult male) with normal fleece. The ewe then gives birth to a single lamb with normal fleece. From this single offspring, is it possible to determine the genotypes of the two parents? If so, what are their genotypes? If not, why not?

[Jeffrey van daele/FeaturePics.]

*20. Alkaptonuria is a metabolic disorder in which affected persons produce black urine. Alkaptonuria results from an allele (*a*) that is recessive to the allele for normal metabolism (*A*). Sally has normal metabolism, but her brother has alkaptonuria. Sally's father has alkaptonuria, and her mother has normal metabolism.

  a. Give the genotypes of Sally, her mother, her father, and her brother.
  b. If Sally's parents have another child, what is the probability that this child will have alkaptonuria?
  c. If Sally marries a man with alkaptonuria, what is the probability that their first child will have alkaptonuria?

21. Suppose that you are raising Mongolian gerbils. You notice that some of your gerbils have white spots, whereas others have solid coats. What type of crosses could you carry out to determine whether white spots are due to a recessive or a dominant allele?

*22. Hairlessness in American rat terriers is recessive to the presence of hair. Suppose that you have a rat terrier with hair. How can you determine whether this dog is homozygous or heterozygous for the hairy trait?

23. What is the probability of rolling one six-sided die and obtaining the following numbers?
  a. 2
  b. 1 or 2
  c. An even number
  d. Any number but a 6

*24. What is the probability of rolling two six-sided dice and obtaining the following numbers?
  a. 2 and 3
  b. 6 and 6
  c. At least one 6
  d. Two of the same number (two 1s, or two 2s, or two 3s, etc.)
  e. An even number on both dice
  f. An even number on at least one die

*25. In a family of seven children, what is the probability of obtaining the following numbers of boys and girls?
  a. All boys
  b. All children of the same sex
  c. Six girls and one boy
  d. Four boys and three girls
  e. Four girls and three boys

26. Phenylketonuria (PKU) is a disease that results from a recessive gene. Two normal parents produce a child with PKU.
  a. What is the probability that a sperm from the father will contain the PKU allele?
  b. What is the probability that an egg from the mother will contain the PKU allele?
  c. What is the probability that their next child will have PKU?
  d. What is the probability that their next child will be heterozygous for the PKU gene?

*27. In German cockroaches, curved wing (*cv*) is recessive to normal wing (*cv*$^+$). A homozygous cockroach having normal wings is crossed with a homozygous cockroach having curved wings. The F$_1$ are intercrossed to produce the F$_2$. Assume that the pair of chromosomes containing the locus for wing shape is metacentric. Draw this pair of chromosomes as it would appear in the parents, the F$_1$, and each class of F$_2$ progeny at metaphase I of meiosis. Assume that no crossing over takes place. At each stage, label a location for the alleles for wing shape (*cv* and *cv*$^+$) on the chromosomes.

*28. In guinea pigs, the allele for black fur (*B*) is dominant over the allele for brown (*b*) fur. A black guinea pig is crossed with a brown guinea pig, producing five F$_1$ black guinea pigs and six F$_1$ brown guinea pigs.

a. How many copies of the black allele (*B*) will be present in each cell of an F_1 black guinea pig at the following stages: G_1, G_2, metaphase of mitosis, metaphase I of meiosis, metaphase II of meiosis, and after the second cytokinesis following meiosis? Assume that no crossing over takes place.

b. How many copies of the brown allele (*b*) will be present in each cell of an F_1 brown guinea pig at the same stages as those listed in part *a*? Assume that no crossing over takes place.

Section 3.3

29. In watermelons, bitter fruit (*B*) is dominant over sweet fruit (*b*), and yellow spots (*S*) are dominant over no spots (*s*). The genes for these two characteristics assort independently. A homozygous plant that has bitter fruit and yellow spots is crossed with a homozygous plant that has sweet fruit and no spots. The F_1 are intercrossed to produce the F_2.

    a. What are the phenotypic ratios in the F_2?
    b. If an F_1 plant is backcrossed with the bitter, yellow-spotted parent, what phenotypes and proportions are expected in the offspring?
    c. If an F_1 plant is backcrossed with the sweet, nonspotted parent, what phenotypes and proportions are expected in the offspring?

*30. In cats, curled ears result from an allele (*Cu*) that is dominant over an allele (*cu*) for normal ears. Black color results from an independently assorting allele (*G*) that is dominant over an allele for gray (*g*). A gray cat homozygous for curled ears is mated with a homozygous black cat with normal ears. All the F_1 cats are black and have curled ears.

[Biosphoto/J.-L. Klein & M.-L. Hubert/Peter Arnold.]

    a. If two of the F_1 cats mate, what phenotypes and proportions are expected in the F_2?
    b. An F_1 cat mates with a stray cat that is gray and possesses normal ears. What phenotypes and proportions of progeny are expected from this cross?

*31. The following two genotypes are crossed: *Aa Bb Cc dd Ee* × *Aa bb Cc Dd Ee*. What will the proportion of the following genotypes be among the progeny of this cross?

    a. *Aa Bb Cc Dd Ee*
    b. *Aa bb Cc dd ee*
    c. *aa bb cc dd ee*
    d. *AA BB CC DD EE*

32. In mice, an allele for apricot eyes (*a*) is recessive to an allele for brown eyes (*a*⁺). At an independently assorting locus, an allele for tan coat color (*t*) is recessive to an allele for black coat color (t^+). A mouse that is homozygous for brown eyes and black coat color is crossed with a mouse having apricot eyes and a tan coat. The resulting F_1 are intercrossed to produce the F_2. In a litter of eight F_2 mice, what is the probability that two will have apricot eyes and tan coats?

33. In cucumbers, dull fruit (*D*) is dominant over glossy fruit (*d*), orange fruit (*R*) is dominant over cream fruit (*r*), and bitter cotyledons (*B*) are dominant over nonbitter cotyledons (*b*). The three characters are encoded by genes located on different pairs of chromosomes. A plant homozygous for dull, orange fruit and bitter cotyledons is crossed with a plant that has glossy, cream fruit and nonbitter cotyledons. The F_1 are intercrossed to produce the F_2.

    a. Give the phenotypes and their expected proportions in the F_2.
    b. An F_1 plant is crossed with a plant that has glossy, cream fruit and nonbitter cotyledons. Give the phenotypes and expected proportions among the progeny of this cross.

*34. Alleles *A* and *a* are located on a pair of metacentric chromosomes. Alleles *B* and *b* are located on a pair of acrocentric chromosomes. A cross is made between individuals having the following genotypes: *Aa Bb* × *aa bb*.

    a. Draw the chromosomes as they would appear in each type of gamete produced by the individuals of this cross.
    b. For each type of progeny resulting from this cross, draw the chromosomes as they would appear in a cell at G_1, G_2, and metaphase of mitosis.

Section 3.4

*35. J. A. Moore investigated the inheritance of spotting patterns in leopard frogs (J. A. Moore. 1943. *Journal of Heredity* 34:3–7). The pipiens phenotype had the normal spots that give leopard frogs their name. In contrast, the burnsi phenotype lacked spots on its back. Moore carried out the following crosses, producing the progeny indicated.

Parent phenotypes	Progeny phenotypes
burnsi × burnsi	39 burnsi, 6 pipiens
burnsi × pipiens	23 burnsi, 3 pipiens
burnsi × pipiens	196 burnsi, 210 pipiens

    a. On the basis of these results, what is the most likely mode of inheritance of the burnsi phenotype?
    b. Give the most likely genotypes of the parent in each cross (use *B* for the burnsi allele and B^+ for pipiens allele)
    c. Use a chi-square test to evaluate the fit of the observed numbers of progeny to the number expected on the basis of your proposed genotypes.

36. In the 1800s, a man with dwarfism who lived in Utah produced a large number of descendants: 22 children, 49 grandchildren, and 250 great-grandchildren (see the illustration of a family pedigree on page 70), many of

whom also were dwarfs (F. F. Stephens. 1943. *Journal of Heredity* 34:229–235). The type of dwarfism found in this family is called Schmid-type metaphyseal chondrodysplasia, although it was originally thought to be achondroplastic dwarfism. Among the families of this kindred, dwarfism appeared only in members who had one parent with dwarfism. When one parent was a dwarf, the following numbers of children were produced.

Family in which one parent had dwarfism	Children with normal stature	Children with dwarfism
A	15	7
B	4	6
C	1	6
D	6	2
E	2	2
F	8	4
G	4	4
H	2	1
I	0	1
J	3	1
K	2	3
L	2	1
M	2	0
N	1	0
O	0	2
Total	52	40

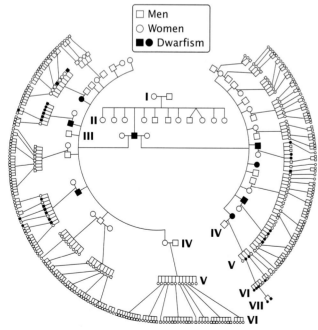

[Adapted from *The Journal of Heredity* 34:232.]

a. With the assumption that Schmid-type metaphyseal chondrodysplasia is rare, is this type of dwarfism inherited as a dominant or recessive trait? Explain your reasoning?

b. On the basis of your answer for part *a*, what is the expected ratio of dwarf and normal children in the families given in the table. Use a chi-square test to determine if the total number of children for these families (52 normal, 40 dwarfs) is significantly different from the number expected.

c. Use chi-square tests to determine if the number of children in family C (1 normal, 6 dwarf) and the number in family D (6 normal and 2 dwarf) are significantly different from the numbers expected on the basis of your proposed type of inheritance. How would you explain these deviations from the overall ratio expected?

37. Pink-eye and albino are two recessive traits found in the deer mouse *Peromyscus maniculatus*. In mice with pink-eye, the eye is devoid of color and appears pink from the blood vessels within it. Albino mice are completely lacking color both in their fur and in their eyes. F. H. Clark crossed pink-eyed mice with albino mice; the resulting F_1 had normal coloration in their fur and eyes. He then crossed these F_1 mice with mice that were pink eyed and albino and obtained the following mice. It is very hard to distinguish between mice that are albino and mice that are both pink-eye and albino, and so he combined these two phenotypes (F. H. Clark. 1936. *Journal of Heredity* 27: 259–260).

Phenotype	Number of progeny
wild-type fur, wild-type eye color	12
wild-type fur, pink-eye	62
albino ⎱ pink-eye ⎰	78
Total	152

a. Give the expected numbers of progeny with each phenotype if the genes for pink-eye and albino assort independently.

b. Use a chi-square test to determine if the observed numbers of progeny fit the number expected with independent assortment.

*38. In the California poppy, an allele for yellow flowers (*C*) is dominant over an allele for white flowers (*c*). At an independently assorting locus, an allele for entire petals (*F*) is dominant over an allele for fringed petals (*f*). A plant that is homozygous for yellow and entire petals is crossed with a plant that is white and fringed. A resulting F_1 plant is then crossed with a plant that is white and fringed, and the following progeny are produced: 54 yellow and entire; 58 yellow and fringed, 53 white and entire, and 10 white and fringed.

a. Use a chi-square test to compare the observed numbers with those expected for the cross.

b. What conclusion can you make from the results of the chi-square test?

c. Suggest an explanation for the results.

CHALLENGE QUESTIONS

Section 3.2

39. Dwarfism is a recessive trait in Hereford cattle. A rancher in western Texas discovers that several of the calves in his herd are dwarfs, and he wants to eliminate this undesirable trait from the herd as rapidly as possible. Suppose that the rancher hires you as a genetic consultant to advise him on how to breed the dwarfism trait out of the herd. What crosses would you advise the rancher to conduct to ensure that the allele causing dwarfism is eliminated from the herd?

*40. A geneticist discovers an obese mouse in his laboratory colony. He breeds this obese mouse with a normal mouse. All the F_1 mice from this cross are normal in size. When he interbreeds two F_1 mice, eight of the F_2 mice are normal in size and two are obese. The geneticist then intercrosses two of his obese mice, and he finds that all of the progeny from this cross are obese. These results lead the geneticist to conclude that obesity in mice results from a recessive allele.

    A second geneticist at a different university also discovers an obese mouse in her laboratory colony. She carries out the same crosses as those done by the first geneticist and obtains the same results. She also concludes that obesity in mice results from a recessive allele. One day the two geneticists meet at a genetics conference, learn of each other's experiments, and decide to exchange mice. They both find that, when they cross two obese mice from the different laboratories, all the offspring are normal; however, when they cross two obese mice from the same laboratory, all the offspring are obese. Explain their results.

41. Albinism is a recessive trait in humans (see the introduction to Chapter 1). A geneticist studies a series of families in which both parents are normal and at least one child has albinism. The geneticist reasons that both parents in these families must be heterozygotes and that albinism should appear in $1/4$ of the children of these families. To his surprise, the geneticist finds that the frequency of albinism among the children of these families is considerably greater than $1/4$. Can you think of an explanation for the higher-than-expected frequency of albinism among these families?

42. Two distinct phenotypes are found in the salamander *Plethodon cinereus*: a red form and a black form. Some biologists have speculated that the red phenotype is due to an autosomal allele that is dominant over an allele for black. Unfortunately, these salamanders will not mate in captivity; so the hypothesis that red is dominant over black has never been tested.

    One day a genetics student is hiking through the forest and finds 30 female salamanders, some red and some black, laying eggs. The student places each female and her eggs (from about 20 to 30 eggs per female) in separate plastic bags and takes them back to the lab. There, the student successfully raises the eggs until they hatch. After the eggs have hatched, the student records the phenotypes of the juvenile salamanders, along with the phenotypes of their mothers. Thus, the student has the phenotypes for 30 females and their progeny, but no information is available about the phenotypes of the fathers.

    [Phil Degginger/Alamy.]

    Explain how the student can determine whether red is dominant over black with this information on the phenotypes of the females and their offspring.

4 Sex Determination and Sex-Linked Characteristics

Sex in the platypus is determined by sex chromosomes. Females have 10 X chromosomes, whereas males have 5 X and 5 Y chromosomes. [Australian Geographic/Aflo Foto Agency/Photolibrary.]

THE STRANGE CASE OF PLATYPUS SEX

The platypus, *Ornithorhynchus anatinus*, is clearly one of life's strangest animals. It possesses a furry coat and, like a mammal, is warm blooded and produces milk to nourish its young, but it lacks teeth, has a bill, and lays eggs like a bird. The feet are webbed like those of a duck, and females have no nipples (offspring suck milk directly from the abdominal skin); males have spurs on their hind legs that deliver a deadly, snakelike venom. The platypus has such a hodgepodge of mammalian, avian, and reptilian traits that the first scientists to examine a platypus skin specimen thought it might be an elaborate hoax, produced by attaching parts culled from several different organisms. In spite of its strange appearance, the platypus is genetically a monotreme mammal, a side branch that diverged from the rest of mammals some 166 million years ago.

The platypus lives in eastern and southern Australia and on the island of Tasmania. An excellent swimmer, it spends most of its time in small rivers and streams foraging for worms, frogs, insect larvae, shrimp, and crayfish. Among other oddities, it locates its prey by using electroreception. The platypus genome was sequenced in 2008, providing a detailed view of the genetic makeup of this strange animal. It has a relatively small genome for a mammal, with 2.3 billion base pairs of DNA and about 18,500 protein-encoding genes. Almost 10% of its genes encode proteins that take part in odor and chemical reception. The platypus genome is a blend of mammalian, reptilian, and unique characteristics.

Platypus sex also is unusual. For most mammals, whether an individual organism is male or female is determined by sex chromosomes. Most females possess two X chromosomes, whereas males have a single X chromosome and a smaller sex chromosome called Y. This is the usual type of sex determination in mammals, but how sex is determined in the platypus remained a mystery for many years. The platypus possesses 52 chromosomes, and early geneticists observed a confusing mix of different chromosomes in male and female platypuses, including an unusual chainlike group of chromosomes in meiosis (**Figure 4.1**).

In 2004, Frank Grutzner and a group of other scientists created fluorescent paints to mark the platypus chromosomes so that they could follow the behavior of individual chromosomes in the course of meiosis. What they discovered was remarkable: platypuses

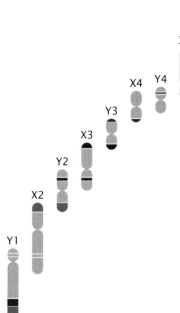

4.1 Sex chromosomes of the duckbill platypus. In meiosis, the sex chromosomes form chainlike structures. [Adapted from F. Veyrunes et al., *Genome Research* 18(6): 965–973, 2008. Copyright © 2008, Cold Spring Harbor Laboratory Press.]

possess ten sex chromosomes: female platypuses have ten X chromosomes, whereas male platypuses have five X chromosomes and five Y chromosomes. In meiosis, these sex chromosomes line up in a precise order, forming a long chain of sex chromosomes. In spite of what at first appears to be mass confusion, the platypus sex chromosomes pair and align with great precision so that each egg cell gets exactly five Xs; half the sperm get five Xs and the other half get five Ys. The mechanism that brings about this precise separation is not yet known. The complicated set of sex chromosomes in the platypus is just one example of the varied ways in which sex is determined and influences inheritance.

In Chapter 3, we studied Mendel's principles of segregation and independent assortment and saw how these principles explain much about the nature of inheritance. After Mendel's principles were rediscovered in 1900, biologists began to conduct genetic studies on a wide array of different organisms. As they applied Mendel's principles more widely, exceptions were observed, and it became necessary to devise extensions of his basic principles of heredity. In this chapter, we explore one of the major extensions of Mendel's principles: the inheritance of characteristics encoded by genes located on the sex chromosomes, which often differ in males and females (**Figure 4.2**). These characteristics and the genes that produce them are referred to as sex linked. To understand the inheritance of sex-linked characteristics, we must first know how sex is determined—why some members of a species are male and others are female. Sex determination is the focus of the first part of the chapter. The second part examines how characteristics encoded by genes on the sex chromosomes are inherited. In Chapter 5, we will explore some additional ways in which sex and inheritance interact.

As we consider sex determination and sex-linked characteristics, it will be helpful to think about two important principles. First, there are several different mechanisms of sex determination and, ultimately, the mechanism of sex determination controls the inheritance of sex-linked characteristics. Second, like other pairs of chromosomes, the X and Y sex chromosomes pair in the course of meiosis and segregate, but, throughout most of their length, they are not homologous (their gene sequences do not encode the same characteristics): most genes on the X chromosome are different from genes on the Y chromosome. Consequently, males and females do not possess the same number of alleles at sex-linked loci. This difference in the number of sex-linked alleles produces distinct patterns of inheritance in males and females. **TRY PROBLEM 14** →

4.1 Sex Is Determined by a Number of Different Mechanisms

Sexual reproduction is the formation of offspring that are genetically distinct from their parents; most often, two parents contribute genes to their offspring and the genes are assorted into new combinations through meiosis. Among most eukaryotes, sexual reproduction consists of two processes that lead to an alternation of haploid and diploid cells: meiosis produces haploid gametes (spores in plants), and fertilization produces diploid zygotes (**Figure 4.3**).

The term **sex** refers to sexual phenotype. Most organisms have only two sexual phenotypes: male and female. The fundamental difference between males and females is gamete size: males produce small gametes; females produce relatively larger gametes (**Figure 4.4**).

The mechanism by which sex is established is termed **sex determination**. We define the sex of an individual organism in reference to its phenotype. Sometimes an individual organism has chromosomes or genes that are normally associated with one sex but a morphology corresponding to the opposite sex. For instance, the cells of female humans normally have two X chromosomes, and the cells

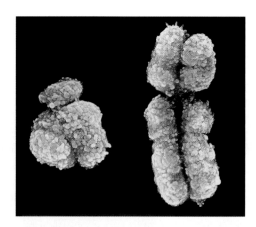

4.2 The male sex chromosome (Y, at the left) differs from the female sex chromosome (X, at the right) in size and shape.
[Biophoto Associates/Photo Researchers.]

4.3 In most eukaryotic organisms, sexual reproduction consists of an alternation of haploid (1*n*) and diploid (2*n*) cells.

of males have one X chromosome and one Y chromosome. A few rare persons have male anatomy, although their cells each contain two X chromosomes. Even though these people are genetically female, we refer to them as male because their sexual phenotype is male. (As we will see later in the chapter, these XX males usually have a small piece of the Y chromosome, which is attached to another chromosome.)

CONCEPTS

In sexual reproduction, parents contribute genes to produce an offspring that is genetically distinct from both parents. In most eukaryotes, sexual reproduction consists of meiosis, which produces haploid gametes (or spores), and fertilization, which produces a diploid zygote.

✔ CONCEPT CHECK 1

What process causes the genetic variation seen in offspring produced by sexual reproduction?

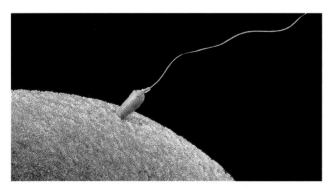

4.4 Male and female gametes (sperm and egg, respectively) differ in size. In this photograph, a human sperm (with flagellum) penetrates a human egg cell. [Francis Leroy, Biocosmos/Science Photo Library/Photo Researchers.]

There are many ways in which sex differences arise. In some species, both sexes are present in the same organism, a condition termed **hermaphroditism;** organisms that bear both male and female reproductive structures are said to be **monoecious** (meaning "one house"). Species in which the organism has either male or female reproductive structures are said to be **dioecious** (meaning "two houses"). Humans are dioecious. Among dioecious species, sex may be determined chromosomally, genetically, or environmentally.

Chromosomal Sex-Determining Systems

The chromosome theory of inheritance (discussed in Chapter 3) states that genes are located on chromosomes, which serve as vehicles for the segregation of genes in meiosis. Definitive proof of this theory was provided by the discovery that the sex of certain insects is determined by the presence or absence of particular chromosomes.

In 1891, Hermann Henking noticed a peculiar structure in the nuclei of cells from male insects. Understanding neither its function nor its relation to sex, he called this structure the X body. Later, Clarence E. McClung studied the X body in grasshoppers and recognized that it was a chromosome. McClung called it the accessory chromosome, but it eventually became known as the X chromosome, from Henking's original designation. McClung observed that the cells of female grasshoppers had one more chromosome than the number of chromosomes in the cells of male grasshoppers, and he concluded that accessory chromosomes played a role in sex determination. In 1905, Nettie Stevens and Edmund Wilson demonstrated that, in grasshoppers and other insects, the cells of females have two X chromosomes, whereas the cells of males have a single X. In some insects, they counted the same number of chromosomes in the cells of males and females but saw that one chromosome pair was different: two X chromosomes were found in female cells, whereas a single X chromosome plus a smaller chromosome, which they called Y, was found in male cells.

Stevens and Wilson also showed that the X and Y chromosomes separate into different cells in sperm formation; half of the sperm receive an X chromosome and the other half receive a Y. All egg cells produced by the female in meiosis receive one X chromosome. A sperm containing a Y chromosome unites with an X-bearing egg to produce an XY male, whereas a sperm containing an X chromosome unites with an X-bearing egg to produce an XX female. This distribution of X and Y chromosomes in sperm accounts for the 1 : 1 sex ratio observed in most dioecious organisms (**Figure 4.5**). Because sex is inherited like other genetically determined characteristics, Stevens and Wilson's discovery that sex is associated with the inheritance of a particular chromosome also demonstrated that genes are on chromosomes.

As Stevens and Wilson found for insects, sex in many organisms is determined by a pair of chromosomes, the **sex chromosomes**, which differ between males and females.

4.5 Inheritance of sex in organisms with X and Y chromosomes results in equal numbers of male and female offspring.

The nonsex chromosomes, which are the same for males and females, are called **autosomes**. We think of sex in these organisms as being determined by the presence of the sex chromosomes, but, in fact, the individual genes located on the sex chromosomes are usually responsible for the sexual phenotypes.

XX-XO sex determination The mechanism of sex determination in the grasshoppers studied by McClung is one of the simplest mechanisms of chromosomal sex determination and is called the XX-XO system. In this system, females have two X chromosomes (XX), and males possess a single X chromosome (XO). There is no O chromosome; the letter O signifies the absence of a sex chromosome.

In meiosis in females, the two X chromosomes pair and then separate, with one X chromosome entering each haploid egg. In males, the single X chromosome segregates in meiosis to half the sperm cells; the other half receive no sex chromosome. Because males produce two different types of gametes with respect to the sex chromosomes, they are said to be the **heterogametic sex**. Females, which produce gametes that are all the same with respect to the sex chromosomes, are the homogametic sex. In the XX-XO system, the sex of an individual organism is therefore determined by which type of male gamete fertilizes the egg. X-bearing sperm unite with X-bearing eggs to produce XX zygotes, which eventually develop as females. Sperm lacking an X chromosome unite with X-bearing eggs to produce XO zygotes, which develop into males.

XX-XY sex determination In many species, the cells of males and females have the same number of chromosomes, but the cells of females have two X chromosomes (XX) and the cells of males have a single X chromosome and a smaller sex chromosome, the Y chromosome (XY). In humans and many other organisms, the Y chromosome is acrocentric (**Figure 4.6**), not Y shaped as is commonly assumed. In this type of sex-determining system, the male is the heterogametic sex—half of his gametes have an X chromosome and half have a Y chromosome. The female is the homogametic sex—all her egg cells contain a single X chromosome. Many organisms, including some plants, insects, and reptiles, and all mammals (including humans), have the XX-XY sex-determining system. Other organisms have variations of the XX-XY system of sex determination, including the duck-billed platypus (discussed in the introduction to this chapter) in which females have five pairs of X chromosomes and males have five pairs of X and Y chromosomes.

Although the X and Y chromosomes are not generally homologous, they do pair and segregate into different cells in meiosis. They can pair because these chromosomes are homologous in small regions called the **pseudoautosomal regions** (see Figure 4.6), in which they carry the same genes. In humans, there are pseudoautosomal regions at both tips of the X and Y chromosomes.

ZZ-ZW sex determination In this system, the female is heterogametic and the male is homogametic. To prevent confusion with the XX-XY system, the sex chromosomes in this system are called Z and W, but the chromosomes do

4.6 The X and Y chromosomes in humans differ in size and genetic content. They are homologous only in the pseudoautosomal regions.

not resemble Zs and Ws. Females in this system are ZW; after meiosis, half of the eggs have a Z chromosome and the other half have a W chromosome. Males are ZZ; all sperm contain a single Z chromosome. The ZZ-ZW system is found in birds, snakes, butterflies, some amphibians, and some fishes.

> **CONCEPTS**
>
> The discovery that the presence or absence of particular chromosomes determines sex in insects provided evidence that genes are located on chromosomes. In XX-XO sex determination, the male is XO and heterogametic, and the female is XX and homogametic. In XX-XY sex determination, the male is XY and the female is XX; in this system, the male is heterogametic. In ZZ-ZW sex determination, the female is ZW and the male is ZZ; in this system, the female is the heterogametic sex.
>
> ✔ **CONCEPT CHECK 2**
>
> How does the heterogametic sex differ from the homogametic sex?
>
> a. The heterogametic sex is male; the homogametic sex is female.
> b. Gametes of the heterogametic sex have different sex chromosomes; gametes of homogametic sex have the same sex chromosome.
> c. Gametes of the heterogametic sex all contain a Y chromosome; gametes of the homogametic sex all contain an X chromosome.

Genic Sex Determination

In some plants, fungi, and protozoans, sex is genetically determined, but there are no obvious differences in the chromosomes of males and females: there are no sex chromosomes. These organisms have **genic sex determination**; genotypes at one or more loci determine the sex of an individual plant, fungus, or protozoan.

It is important to understand that, even in chromosomal sex-determining systems, sex is actually determined by individual genes. For example, in mammals, a gene (*SRY*, discussed later in this chapter) located on the Y chromosome determines the male phenotype. In both genic sex determination and chromosomal sex determination, sex is controlled by individual genes; the difference is that, with chromosomal sex determination, the chromosomes also look different in males and females.

Environmental Sex Determination

Genes have had a role in all of the examples of sex determination discussed thus far. However, in a number of organisms, sex is determined fully or in part by environmental factors.

A fascinating example of environmental sex determination is seen in the marine mollusk *Crepidula fornicata*, also known as the common slipper limpet (**Figure 4.7**). Slipper limpets live in stacks, one on top of another. Each limpet begins life as a swimming larva. The first larva to settle on a solid, unoccupied substrate develops into a female limpet. It then produces chemicals that attract other larvae, which settle on top of it. These larvae develop into males, which then serve as mates for the limpet below. After a period of time, the males on top develop into females and, in turn, attract additional larvae that settle on top of the stack, develop into males, and serve as mates for the limpets under them. Limpets can form stacks of a dozen or more animals; the uppermost animals are always male. This type of sexual development is called **sequential hermaphroditism**; each individual animal can be both male and female, although not at the same time. In *Crepidula fornicata*, sex is determined environmentally by the limpet's position in the stack.

Environmental factors are also important in determining sex in many reptiles. Although most snakes and lizards have sex chromosomes, the sexual phenotype of many turtles, crocodiles, and alligators is affected by temperature during embryonic development. In turtles, for example,

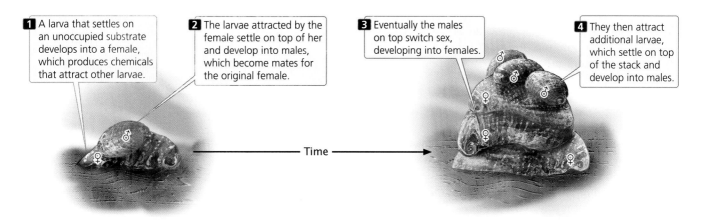

4.7 In *Crepidula fornicata*, the common slipper limpet, sex is determined by an environmental factor—the limpet's position in a stack of limpets.

Table 4.1 Some sex-determining systems

System	Mechanism		Heterogametic Sex
XX-XO	Females XX	Males X	Male
XX-XY	Females XX	Males XY	Male
ZZ-ZW	Females ZW	Males ZZ	Female
Genic sex determination	No distinct sex chromosomes Sex determined by genes on undifferentiated chromosomes		Varies
Environmental sex determination	Sex determined by environmental factors		None

warm temperatures produce females during certain times of the year, whereas cool temperatures produce males. In alligators, the reverse is true. Some of the different types of sex determination are summarized in **Table 4.1**.

Now that we have surveyed some of the different ways in which sex can be determined, we will examine in detail one mechanism (the XX-XY system). Sex determination is XX-XY in both fruit flies and humans but, as we will see, the way in which the X and Y chromosomes determine sex in these two organisms is quite different. **TRY PROBLEMS 4 AND 20**

CONCEPTS

In genic sex determination, sex is determined by genes at one or more loci, but there are no obvious differences in the chromosomes of males and females. In environmental sex determination, sex is determined fully or in part by environmental factors.

✔ CONCEPT CHECK 3

How do chromosomal, genic, and environmental sex determination differ?

Sex Determination in *Drosophila melanogaster*

The fruit fly *Drosophila melanogaster* has eight chromosomes: three pairs of autosomes and one pair of sex chromosomes. Thus, it has inherited one haploid set of autosomes and one sex chromosome from each parent. Normally, females have two X chromosomes and males have an X chromosome and a Y chromosome. However, the presence of the Y chromosome does not determine maleness in *Drosophila*; instead, each fly's sex is determined by a balance between genes on the autosomes and genes on the X chromosome. This type of sex determination is called the **genic balance system**. In this system, a number of different genes influence sexual development. The X chromosome contains genes with female-producing effects, whereas the autosomes contain genes with male-producing effects. Consequently, a fly's sex is determined by the **X : A ratio**, the number of X chromosomes divided by the number of haploid sets of autosomal chromosomes.

An X : A ratio of 1.0 produces a female fly; an X : A ratio of 0.5 produces a male. If the X : A ratio is less than 0.5, a male phenotype is produced, but the fly is weak and sterile—such flies are sometimes called metamales. An X : A ratio between 1.0 and 0.5 produces an intersex fly, with a mixture of male and female characteristics. If the X : A ratio is greater than 1.0, a female phenotype is produced, but this fly (called a metafemale) has serious developmental problems and many never complete development. **Table 4.2** presents some different chromosome complements in *Drosophila* and their associated sexual phenotypes. Normal females have two X chromosomes and two sets of autosomes (XX, AA), and so their X : A ratio is 1.0. Males, on the other hand, normally have a single X and two sets of autosomes (XY, AA), and so their X : A ratio is 0.5. Flies with XXY sex chromosomes and two sets of autosomes (an X : A ratio of 1.0) develop as fully fertile females, in spite of the presence of a Y chromosome. Flies with only a single X and two sets of autosomes (XO, AA, for an X : A ratio of 0.5) develop as males, although they are sterile. These observations confirm that the Y chromosome does not determine sex in *Drosophila*. **TRY PROBLEM 15**

Table 4.2 Chromosome complements and sexual phenotypes in *Drosophila*

Sex-Chromosome Complement	Haploid Sets of Autosomes	X : A Ratio	Sexual Phenotype
XX	AA	1.0	Female
XY	AA	0.5	Male
XO	AA	0.5	Male
XXY	AA	1.0	Female
XXX	AA	1.5	Metafemale
XXXY	AA	1.5	Metafemale
XX	AAA	0.67	Intersex
XO	AAA	0.33	Metamale
XXXX	AAA	1.3	Metafemale

CONCEPTS

The sexual phenotype of a fruit fly is determined by the ratio of the number of X chromosomes to the number of haploid sets of autosomal chromosomes (the X : A ratio).

✔ CONCEPT CHECK 4

What is the sexual phenotype of a fruit fly that has XXYYY sex chromosomes and two sets of autosomes?

a. Male c. Intersex
b. Female d. Metamale

Sex Determination in Humans

Humans, like *Drosophila*, have XX-XY sex determination, but, in humans, the presence of a gene (*SRY*) on the Y chromosome determines maleness. The phenotypes that result from abnormal numbers of sex chromosomes, which arise when the sex chromosomes do not segregate properly in meiosis or mitosis, illustrate the importance of the Y chromosome in human sex determination.

Turner syndrome Persons who have **Turner syndrome** are female and often have underdeveloped secondary sex characteristics. This syndrome is seen in 1 of 3000 female births. Affected women are frequently short and have a low hairline, a relatively broad chest, and folds of skin on the neck. Their intelligence is usually normal. Most women who have Turner syndrome are sterile. In 1959, Charles Ford used new techniques to study human chromosomes and discovered that cells from a 14-year-old girl with Turner syndrome had only a single X chromosome (**Figure 4.8**); this chromosome complement is usually referred to as XO.

There are no known cases in which a person is missing both X chromosomes, an indication that at least one X chromosome is necessary for human development. Presumably,

4.8 Persons with Turner syndrome have a single X chromosome in their cells. [Department of Clinical Cytogenetics, Addenbrookes Hospital/Science Photo Library/Photo Reseachers.]

4.9 Persons with Klinefelter syndrome have a Y chromosome and two or more X chromosomes in their cells. [Biophoto Associates/Science Source/Photo Researchers.]

embryos missing both Xs spontaneously abort in the early stages of development.

Klinefelter syndrome Persons who have **Klinefelter syndrome**, which occurs with a frequency of about 1 in 1000 male births, have cells with one or more Y chromosomes and multiple X chromosomes. The cells of most males having this condition are XXY (**Figure 4.9**), but the cells of a few Klinefelter males are XXXY, XXXXY, or XXYY. Men with this condition frequently have small testes and reduced facial and pubic hair. They are often taller than normal and sterile; most have normal intelligence.

Poly-X females In about 1 in 1000 female births, the infant's cells possess three X chromosomes, a condition often referred to as **triplo-X syndrome**. These persons have no distinctive features other than a tendency to be tall and thin. Although a few are sterile, many menstruate regularly and are fertile. The incidence of mental retardation among triple-X females is slightly greater than that in the general population, but most XXX females have normal intelligence. Much rarer are females whose cells contain four or five X chromosomes. These females usually have normal female anatomy but are mentally retarded and have a number of physical problems. The severity of mental retardation increases as the number of X chromosomes increases beyond three.

The role of sex chromosomes The phenotypes associated with sex-chromosome anomalies allow us to make several inferences about the role of sex chromosomes in human sex determination.

1. The X chromosome contains genetic information essential for both sexes; at least one copy of an X chromosome is required for human development.

2. The male-determining gene is located on the Y chromosome. A single copy of this chromosome, even in the presence of several X chromosomes, produces a male phenotype.

3. The absence of the Y chromosome results in a female phenotype.

4. Genes affecting fertility are located on the X and Y chromosomes. A female usually needs at least two copies of the X chromosome to be fertile.

5. Additional copies of the X chromosome may upset normal development in both males and females, producing physical and mental problems that increase as the number of extra X chromosomes increases.

The male-determining gene in humans The Y chromosome in humans and all other mammals is of paramount importance in producing a male phenotype. However, scientists discovered a few rare XX males whose cells apparently lack a Y chromosome. For many years, these males presented an enigma: How could a male phenotype exist without a Y chromosome? Close examination eventually revealed a small part of the Y chromosome attached to another chromosome. This finding indicates that it is not the entire Y chromosome that determines maleness in humans; rather, it is a gene on the Y chromosome.

Early in development, all humans possess undifferentiated gonads and both male and female reproductive ducts. Then, about 6 weeks after fertilization, a gene on the Y chromosome becomes active. By an unknown mechanism, this gene causes the neutral gonads to develop into testes, which begin to secrete two hormones: testosterone and Mullerian-inhibiting substance. Testosterone induces the development of male characteristics, and Mullerian-inhibiting substance causes the degeneration of the female reproductive ducts. In the absence of this male-determining gene, the neutral gonads become ovaries, and female features develop.

The male-determining gene in humans, called the **sex-determining region Y** (*SRY*) **gene,** was discovered in 1990 (**Figure 4.10**). This gene is found in XX males and is missing from XY females; it is also found on the Y chromosome of other mammals. Definitive proof that *SRY* is the male-determining gene came when scientists placed a copy of this gene into XX mice by means of genetic engineering. The XX mice that received this gene, although sterile, developed into anatomical males.

The *SRY* gene encodes a protein called a transcription factor that binds to DNA and stimulates the transcription of other genes that promote the differentiation of the testes. Although *SRY* is the primary determinant of maleness in humans, other genes (some X linked, others Y linked, and still others autosomal) also have roles in fertility and the development of sex differences.

CONCEPTS

The presence of the *SRY* gene on the Y chromosome causes a human embryo to develop as a male. In the absence of this gene, a human embryo develops as a female.

✔ **CONCEPT CHECK 5**

What is the phenotype of a person who has XXXY sex chromosomes?

a. Klinefelter syndrome
b. Turner syndrome
c. Poly-X female

Androgen-insensitivity syndrome Although the *SRY* gene is the primary determinant of sex in human embryos, several other genes influence sexual development, as illustrated by women with androgen-insensitivity syndrome. These persons have female external sexual characteristics and think of themselves as female. Indeed, most are unaware of their condition until they reach puberty and fail to menstruate. Examination by a gynecologist reveals that the vagina ends blindly and that the uterus, oviducts, and ovaries are absent. Inside the abdominal cavity, a pair of testes produce levels of testosterone normally seen in males. The cells of a woman with androgen-insensitivity syndrome contain an X and a Y chromosome.

How can a person be female in appearance when her cells contain a Y chromosome and she has testes that produce testosterone? The answer lies in the complex relation between genes and sex in humans. In a human embryo with a Y chromosome, the *SRY* gene causes the gonads to develop into testes, which produce testosterone. Testosterone stimulates embryonic tissues to develop male characteristics. But, for testosterone to have its effects, it must bind to an androgen receptor. This receptor is defective in females with androgen-insensitivity syndrome; consequently, their cells are insensitive to testosterone, and female characteristics develop. The gene for the androgen receptor is located on the X chromosome; so persons with this condition always inherit it from their mothers. (All XY persons inherit the X chromosome from their mothers.)

Androgen-insensitivity syndrome illustrates several important points about the influence of genes on a person's

4.10 The *SRY* gene is on the Y chromosome and causes the development of male characteristics.

sex. First, this condition demonstrates that human sexual development is a complex process, influenced not only by the *SRY* gene on the Y chromosome, but also by other genes found elsewhere. Second, it shows that most people carry genes for both male and female characteristics, as illustrated by the fact that those with androgen-insensitivity syndrome have the capacity to produce female characteristics, even though they have male chromosomes. Indeed, the genes for most male and female secondary sex characteristics are present not on the sex chromosomes but on autosomes. The key to maleness and femaleness lies not in the genes but in the control of their expression. TRY PROBLEM 16 →

4.2 Sex-Linked Characteristics Are Determined by Genes on the Sex Chromosomes

In Chapter 3, we learned several basic principles of heredity that Mendel discovered from his crosses among pea plants. A major extension of these Mendelian principles is the pattern of inheritance exhibited by **sex-linked characteristics**, characteristics determined by genes located on the sex chromosomes. Genes on the X chromosome determine **X-linked characteristics**; those on the Y chromosome determine **Y-linked characteristics**. Because the Y chromosome of many organisms contains little genetic information, most sex-linked characteristics are X linked. Males and females differ in their sex chromosomes; so the pattern of inheritance for sex-linked characteristics differs from that exhibited by genes located on autosomal chromosomes.

X-Linked White Eyes in *Drosophila*

The first person to explain sex-linked inheritance was American biologist Thomas Hunt Morgan (**Figure 4.11**). Morgan began his career as an embryologist, but the discovery of Mendel's principles inspired him to begin conducting genetic experiments, initially on mice and rats. In 1909, Morgan switched to *Drosophila melanogaster*; a year later, he discovered among the flies of his laboratory colony a single male that possessed white eyes, in stark contrast with the red eyes of normal fruit flies. This fly had a tremendous effect on the future of genetics and on Morgan's career as a biologist.

To investigate the inheritance of the white-eyed characteristic in fruit flies, Morgan systematically carried out a series of genetic crosses. First, he crossed pure-breeding, red-eyed females with his white-eyed male, producing F_1 progeny of which all had red eyes (**Figure 4.12a**). (In fact, Morgan found 3 white-eyed males among the 1237 progeny, but he assumed that the white eyes were due to new mutations.) Morgan's results from this initial cross were consistent with Mendel's principles: a cross between a homozygous dominant individual and a homozygous recessive individual produces heterozygous offspring exhibiting the dominant trait. His results suggested that white eyes are a simple recessive trait. However, when Morgan crossed the F_1 flies with one another, he found that all the female F_2 flies possessed red eyes but that half the male F_2 flies had red eyes and the other half had white eyes. This finding was clearly not the expected result for a simple recessive trait, which should appear in $1/4$ of both male and female F_2 offspring.

To explain this unexpected result, Morgan proposed that the locus affecting eye color is on the X chromosome (i.e., eye color is X linked). He recognized that the eye-color alleles are present only on the X chromosome; no homologous allele is present on the Y chromosome. Because the cells of females possess two X chromosomes, females can be homozygous or heterozygous for the eye-color alleles. The cells of males, on the other hand, possess only a single X chromosome and can carry only a single eye-color allele. Males therefore cannot be either homozygous or heterozygous but are said to be **hemizygous** for X-linked loci.

(a)

(b)

4.11 Thomas Hunt Morgan's work with *Drosophila* helped unravel many basic principles of genetics, including X-linked inheritance. (a) Morgan. (b) The Fly Room, where Morgan and his students conducted genetic research. [Part a: AP/Wide World Photos. Part b: American Philosophical Society.]

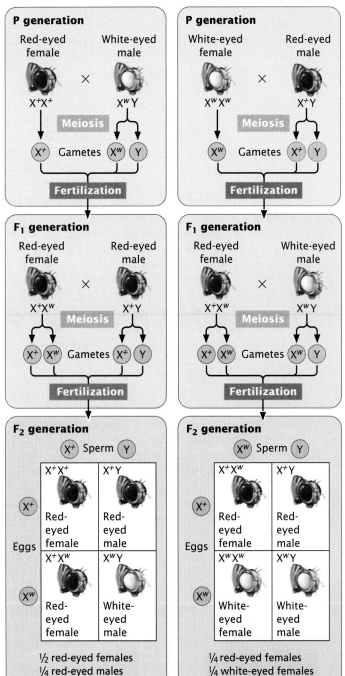

4.12 Morgan's X-linked crosses for white eyes in fruit flies. (a) Original and F₁ crosses. (b) Reciprocal crosses.

The results of Morgan's crosses are illustrated in Animation 4.1.

To verify his hypothesis that the white-eye trait is X linked, Morgan conducted additional crosses. He predicted that a cross between a white-eyed female and a red-eyed male would produce all red-eyed females and all white-eyed males (**Figure 4.12b**). When Morgan performed this cross, the results were exactly as predicted. Note that this cross is the reciprocal of the original cross and that the two reciprocal crosses produced different results in the F₁ and F₂ generations. Morgan also crossed the F₁ heterozygous females with their white-eyed father, the red-eyed F₂ females with white-eyed males, and white-eyed females with white-eyed males. In all of these crosses, the results were consistent with Morgan's conclusion that white eyes are an X-linked characteristic.

Nondisjunction and the Chromosome Theory of Inheritance

When Morgan crossed his original white-eyed male with homozygous red-eyed females, all 1237 of the progeny had red eyes, except for 3 white-eyed males. As already mentioned, Morgan attributed these white-eyed F₁ males to the occurrence of further random mutations. However, flies with these unexpected phenotypes continued to appear in his crosses. Although uncommon, they appeared far too often to be due to spontaneous mutation. Calvin Bridges, who was one of Morgan's students, set out to investigate the genetic basis of these exceptions.

Bridges found that the exceptions arose only in certain strains of white-eyed flies. When he crossed one of these exceptional white-eyed females with a red-eyed male, about 5% of the male offspring had red eyes and about 5% of the female offspring had white eyes. In this cross, the expected result is that every male fly should inherit its mother's X chromosome and should have the genotype $X^w Y$ and white eyes (see the F₁ progeny in Figure 4.12b). Every female fly should inherit a dominant red-eye allele on its father's X chromosome, along with a white-eye allele on its mother's X chromosome; thus, all the female progeny should be $X^+ X^w$ and have red eyes (see F₁ progeny in Figure 4.12b). The continual appearance of red-eyed males and white-eyed females in this cross was therefore unexpected.

Bridges's explanation To explain the appearance of red-eyed males and white-eyed females in his cross, Bridges hypothesized that the exceptional white-eyed females of this strain actually possessed two X chromosomes and a Y chromosome ($X^w X^w Y$). Sex in *Drosophila* is determined by the X : A ratio (see Table 4.2); for XXY females, the X : A ratio is 1.0, and so these flies developed as females in spite of possessing a Y chromosome.

About 90% of the time, the two X chromosomes of the $X^w X^w Y$ females separate from each other in anaphase I of meiosis, with an X and a Y chromosome going into one gamete and a single X going into another other gamete (**Figure 4.13**). When these gametes are fertilized by sperm from a normal red-eyed male, white-eyed males and red-eyed females are produced. About 10% of the time, the two X chromosomes in the females fail to separate in anaphase I of meiosis, a phenomenon known as **nondisjunction.** When nondisjunction of the Xs occurs, half of the eggs receive two copies of the X chromosome and the other half receive only a Y chromosome (see Figure 4.13). When these eggs are fertilized by sperm from a normal red-eyed male, four combinations of sex chromosomes are produced. An egg with two X chromosomes that is fertilized by an X-bearing sperm produces an $X^+ X^w X^w$ zygote, which usually dies. When an egg carrying two X chromosomes is fertilized by a Y-bearing sperm, the resulting zygote is $X^w X^w Y$, which develops into a white-eyed female. An egg with only a Y chromosome that is fertilized by an X-bearing sperm produces an $X^+ Y$ zygote, which develops into a normal red-eyed male. If the egg with only a Y chromosome is fertilized by a Y-bearing sperm, the resulting zygote has two Y chromosomes and no X chromosome and dies. Nondisjunction of the X chromosomes among $X^w X^w Y$ white-eyed females therefore produces a few white-eyed females and red-eyed males, which is exactly what Bridges found in his crosses.

Confirmation of Bridges's hypothesis Bridges's hypothesis predicted that the white-eyed females from these crosses would possess two X chromosomes and one Y chromosome and that the red-eyed males would possess a single X chromosome. To verify his hypothesis, Bridges examined the chromosomes of his flies and found precisely what he had predicted. The significance of Bridges's study is not that it explained the appearance of an occasional odd fly in his culture but that he was able link the inheritance of a specific gene (*w*) to the presence of a specific chromosome (X). This association between genotype and chromosomes gave unequivocal evidence that sex-linked genes are located on the X chromosome and confirmed the chromosome theory of inheritance. **TRY PROBLEM 21**

Experiment

Question: In a cross between a white-eyed female and a red-eyed male, why are a few white-eyed females and red-eyed males produced?

Methods

Hypothesis: White-eyed females and red-eyed males in F_1 result from nondisjunction in an XXY female.

F_1 flies	Chromosomes observed
Red-eyed females	½ XX, ½ XXY
White-eyed males	½ XY, ½ XYY
White-eyed females	XXY
Red-eyed males	XY

Conclusion: The white-eyed females and red-eyed males in the F_1 result from nondisjunction of the X chromosomes in an XXY female.

4.13 Bridges conducted experiments that proved that the gene for white eyes is located on the X chromosome.

 The results of Bridges's crosses are further explained in Animation 4.1.

> **CONCEPTS**
>
> By showing that the appearance of rare phenotypes is associated with the inheritance of particular chromosomes, Bridges proved that sex-linked genes are located on the X chromosome and that the chromosome theory of inheritance is correct.
>
> ✔ **CONCEPT CHECK 6**
>
> What was the genotype of the few F_1 red-eyed males obtained by Bridges when he crossed a white-eyed female with a red-eyed male?
>
> a. X^+
> b. $X^w X^+ Y$
> c. $X^+ Y$
> d. $X^+ X^+ Y$

X-Linked Color Blindness in Humans

To further examine X-linked inheritance, let's consider another X-linked characteristic: red–green color blindness in humans. Within the human eye, color is perceived in light-sensing cone cells that line the retina. Each cone cell contains one of three pigments capable of absorbing light of a particular wavelength; one absorbs blue light, a second absorbs red light, and a third absorbs green light. The human eye actually detects only three colors—red, green, and blue—but the brain mixes the signals from different cone cells to create the wide spectrum of colors that we perceive. Each of the three pigments is encoded by a separate locus; the locus for the blue pigment is found on chromosome 7, and those for the green and the red pigments lie close together on the X chromosome.

The most common types of human color blindness are caused by defects of the red and green pigments; we will refer to these conditions as red–green color blindness. Mutations that produce defective color vision are generally recessive and, because the genes encoding the red and the green pigments are located on the X chromosome, red–green color blindness is inherited as an X-linked recessive characteristic.

We will use the symbol X^c to represent an allele for red–green color blindness and the symbol X^+ to represent an allele for normal color vision. Females possess two X chromosomes; so there are three possible genotypes among females: X^+X^+ and X^+X^c, which produce normal vision, and X^cX^c, which produces color blindness. Males have only a single X chromosome and two possible genotypes: X^+Y, which produces normal vision, and X^cY which produces color blindness.

If a woman homozygous for normal color vision mates with a color-blind man (**Figure 4.14a**), all of the gametes produced by the woman will contain an allele for normal color vision. Half of the man's gametes will receive the X chromosome with the color-blind allele, and the other half will receive the Y chromosome, which carries no alleles affecting color vision. When an X^c-bearing sperm unites with the X^+-bearing egg, a heterozygous female with normal vision (X^+X^c) is produced. When a Y-bearing sperm unites with the X-bearing egg, a hemizygous male with normal vision (X^+Y) is produced.

In the reciprocal cross between a color-blind woman and a man with normal color vision (**Figure 4.14b**), the woman produces only X^c-bearing gametes. The man produces some gametes that contain the X chromosome and others that contain the Y chromosome. Males inherit the X chromosome from their mothers; because both of the mother's X chromosomes bear the X^c allele in this case, all the male offspring will be color blind. In contrast, females inherit an X chromosome from both parents; thus all the female offspring of this reciprocal cross will be heterozygous with normal vision. Females are color blind only when color-blind alleles have been inherited from both parents, whereas a color-blind male need inherit a color-blind allele from his mother only; for this reason, color blindness and most other rare X-linked recessive characteristics are more common in males.

In these crosses for color blindness, notice that an affected woman passes the X-linked recessive trait to her sons but not to her daughters, whereas an affected man passes the trait to his grandsons through his daughters but never to his sons. X-linked recessive characteristics may therefore appear to alternate

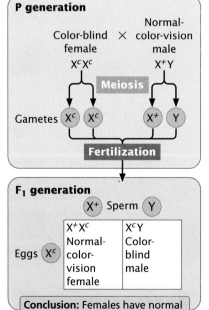

4.14 Red–green color blindness is inherited as an X-linked recessive trait in humans.

between the sexes, appearing in females one generation and in males the next generation.

Recall that the X and Y chromosomes pair in meiosis because they are homologous at the small pseudoautosomal regions. Genes in these regions of the X and Y chromosome are homologous, just like those on autosomes, and they exhibit autosomal patterns of inheritance rather than the sex-linked inheritance seen for most genes on the X and Y chromosomes.

Worked Problem

Now that we understand the pattern of X-linked inheritance, let's apply our knowledge to answer a specific question in regard to X-linked inheritance of color blindness in humans.

Betty has normal vision, but her mother is color blind. Bill is color blind. If Bill and Betty marry and have a child together, what is the probability that the child will be color blind?

• Solution

Because color blindness is an X-linked recessive characteristic, Betty's color-blind mother must be homozygous for the color-blind allele (X^cX^c). Females inherit one X chromosome from each of their parents; so Betty must have inherited a color-blind allele from her mother. Because Betty has normal color vision, she must have inherited an allele for normal vision (X^+) from her father; thus Betty is heterozygous (X^+X^c). Bill is color blind. Because males are hemizygous for X-linked alleles, he must be (X^cY). A mating between Betty and Bill is represented as:

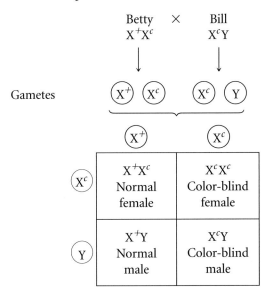

Thus, $1/4$ of the children are expected to be female with normal color vision, $1/4$ female with color blindness, $1/4$ male with normal color vision, and $1/4$ male with color blindness.

→ Get some additional practice with X-linked inheritance by working Problem 23 at the end of this chapter.

CONCEPTS

Characteristics determined by genes on the sex chromosomes are called sex-linked characteristics. Diploid females have two alleles at each X-linked locus, whereas diploid males possess a single allele at each X-linked locus. Females inherit X-linked alleles from both parents, but males inherit a single X-linked allele from their mothers.

✔ CONCEPT CHECK 7

Hemophilia (reduced blood clotting) is an X-linked recessive disease in humans. A woman with hemophilia mates with a man who exhibits normal blood clotting. What is the probability that their child will have hemophilia?

Symbols for X-Linked Genes

There are several different ways to record genotypes for X-linked traits. Sometimes the genotypes are recorded in the same fashion as they are for autosomal characteristics. In this case, the hemizygous males are simply given a single allele: the genotype of a female *Drosophila* with white eyes is *ww*, and the genotype of a white-eyed hemizygous male is *w*. Another method is to include the Y chromosome, designating it with a diagonal slash (/). With this method, the white-eyed female's genotype is still *ww* and the white-eyed male's genotype is *w/*. Perhaps the most useful method is to write the X and Y chromosomes in the genotype, designating the X-linked alleles with superscripts, as is done in this chapter. With this method, a white-eyed female is X^wX^w and a white-eyed male is X^wY. The use of Xs and Ys in the genotype has the advantage of reminding us that the genes are X linked and that the male must always have a single allele, inherited from the mother.

Z-Linked Characteristics

In organisms with ZZ-ZW sex determination, the males are the homogametic sex (ZZ) and carry two sex-linked (usually referred to as Z-linked) alleles; thus males may be homozygous or heterozygous. Females are the heterogametic sex (ZW) and possess only a single Z-linked allele. The inheritance of Z-linked characteristics is the same as that of X-linked characteristics, except that the pattern of inheritance in males and females is reversed.

An example of a Z-linked characteristic is the cameo phenotype in Indian blue peafowl *(Pavo cristatus)*. In these birds, the wild-type plumage is a glossy metallic blue. The female peafowl is ZW and the male is ZZ. Cameo plumage, which produces brown feathers, results from a Z-linked allele (Z^{ca}) that is recessive to the wild-type blue allele (Z^{Ca+}). If a blue-colored female ($Z^{Ca+}W$) is crossed with a cameo male ($Z^{ca}Z^{ca}$), all the F_1 females are cameo ($Z^{ca}W$) and all the F_1 males are blue ($Z^{Ca+}Z^{ca}$), as shown in **Figure 4.15**. When the F_1 are interbred, $1/4$ of the F_2 are blue males ($Z^{Ca+}Z^{ca}$), $1/4$ are blue females ($Z^{Ca+}W$), $1/4$ are cameo males ($Z^{ca}Z^{ca}$), and $1/4$ are cameo females ($Z^{ca}W$). The reciprocal cross of a cameo female with a homozygous blue male produces an F_1 generation in which all offspring are blue and an F_2 generation consisting of $1/2$ blue males ($Z^{Ca+}Z^{Ca+}$ and $Z^{Ca+}Z^{ca}$), $1/4$ blue females ($Z^{Ca+}W$), and $1/4$ cameo females ($Z^{ca}W$).

In organisms with ZZ-ZW sex determination, the female always inherits her W chromosome from her mother, and she inherits her Z chromosome, along with any Z-linked alleles, from her father. In this system, the male inherits Z chromosomes, along with any Z-linked alleles, from both his mother and his father. This pattern of inheritance is the reverse of that of X-linked alleles in organisms with XX-XY sex determination. TRY PROBLEM 32

Y-Linked Characteristics

Y-linked traits exhibit a distinct pattern of inheritance and are present only in males because only males possess a Y chromosome. All male offspring of a male with a Y-linked trait will display the trait, because every male inherits the Y chromosome from his father. TRY PROBLEM 42

Evolution of the Y chromosome Research on sex chromosomes has led to the conclusion the X and Y chromosomes in many organisms evolved from a pair of autosomes. The first step in this evolutionary process took place when one member of a pair of autosomes acquired a gene that determines maleness, such as the *SRY* gene found in humans today (**Figure 4.16**). This step took place in mammals about 250 millions years ago. Any individual organism with a copy of the chromosome containing this gene then became male. Additional mutations occurred on the proto-Y chromosome affecting traits that are beneficial only in males, such as the bright coloration used by male birds to attract females and the antlers used by a male elk in competition with other males. The genes that encode these types of traits are advantageous only if they are present in males. To prevent genes

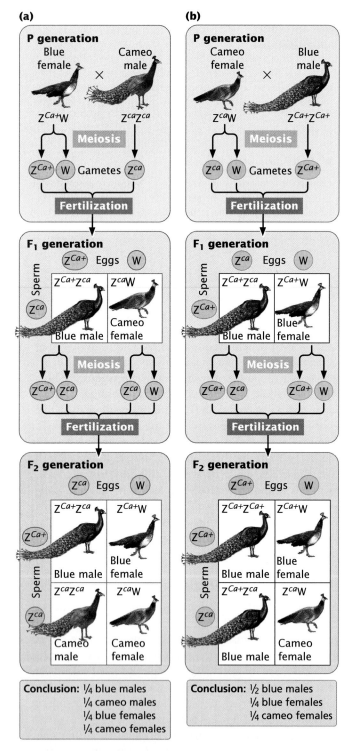

4.15 The cameo phenotype in Indian blue peafowl is inherited as a Z-linked recessive trait. (a) Blue female crossed with cameo male. (b) Reciprocal cross of cameo female crossed with homozygous blue male.

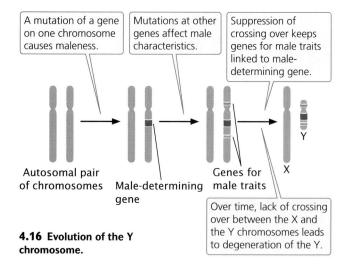

4.16 Evolution of the Y chromosome.

that encode male traits from appearing in females, crossing over was suppressed for most of the length of the X and Y chromosomes in meiosis. Crossing over can still take place between the two X chromosomes in females, but there is little crossing over between the X and the Y chromosomes, except for small pseudoautosomal regions in which the X and the Y chromosomes continue to pair and segregate in meiosis, as stated earlier.

For reasons that are beyond the discussion here, the lack of crossing over led to (and continues to lead to) an accumulation of mutations and the loss of genetic material from the Y chromosome (see Figure 4.16). Over millions of years, the Y chromosome slowly degenerated, losing DNA and genes until it became greatly reduced in size and contained little genetic information. This degeneration produced the Y chromosome found in males today. Indeed, the Y chromosomes of humans and many other organisms are usually small and contain little genetic information; therefore, few characteristics exhibit Y-linked inheritance. Some researchers have predicted that the human Y chromosome will continue to lose genetic information in the future and will completely disappear from the species in about 10 million years, a disheartening prospect for those of us with a Y chromosome. However, internal recombination within the Y chromosome (see next section) may prevent the ultimate demise of the human Y chromosome.

Characteristics of the human Y chromosome The genetic sequence of most of the human Y chromosome has now been determined as part of the Human Genome Project. This work reveals that about two-thirds of the Y chromosome consists of short DNA sequences that are repeated many times and contain no active genes. The other third consists of just a few genes. Only about 350 genes have been identified on the human Y chromosome, compared with thousands on most chromosomes, and only about half of those identified on the Y chromosome encode proteins. The function of most Y-linked genes is poorly understood; many appear to influence male sexual development and fertility. Some are expressed throughout the body, but many are expressed predominately or exclusively in the testes. Although the Y chromosome has relatively few genes, recent research in *Drosophila* suggests that the Y chromosome carries genetic elements that affect the expression of numerous genes on autosomal and X chromosomes.

A surprising feature revealed by sequencing is the presence of eight massive palindromic sequences on the Y chromosome. A palindrome is defined as a word, such as "rotator," or sentence that reads the same backward and forward. A palindromic sequence in DNA reads the same on both strands of the double helix, creating two nearly identical copies stretching out from a central point, such as:

$$\overbrace{\text{5'-TGGGAG...CTCCCA-3'}}^{\text{Arm 1}}$$
$$\underbrace{\text{3'-ACCCTC...GAGGGT-5'}}_{\text{Arm 2}}$$

Thus, a palindromic sequence in DNA appears twice, very much like the two copies of a DNA sequence that are found on two homologous chromosomes. Indeed, recombination takes place between the two palindromic sequences on the Y chromosome. As already mentioned, the X and the Y chromosomes are not homologous at almost all of their sequences, and most of the Y chromosome does not undergo crossing over with the X chromosome. This lack of interchromosomal recombination leads to an accumulation of deleterious mutations on the Y chromosome and the loss of genetic material. Evidence suggests that the two arms of the Y chromosome recombine with each other, which partly compensates for the absence of recombination between the X and the Y chromosomes. This internal recombination may help to maintain some sequences and functions of genes on the Y chromosome and prevent its total degeneration.

Although the palindromes afford opportunities for recombination, which helps prevent the decay of the Y chromosome over evolutionary time, they occasionally have harmful effects. Recent research has revealed that recombination between the palindromes can lead to rearrangements of the Y chromosome that cause anomalies of sexual development. In some cases, recombination between the palindromes leads to deletion of the *SRY* gene, producing an XY female. In other cases, recombination deletes other Y-chromosome genes that take part in sperm production. Sometimes, recombination produces a Y chromosome with two centromeres, which may break as the centromeres are pulled in opposite directions in mitosis. The broken Y chromosomes may be lost in mitosis, resulting in XO cells and Turner syndrome.

CONCEPTS

Y-linked characteristics exhibit a distinct pattern of inheritance: they are present only in males, and all male offspring of a male with a Y-linked trait inherit the trait. Palindromic sequences within the Y chromosome can undergo internal recombination, but such recombination may lead to chromosome anomalies.

✔ CONCEPT CHECK 8

What unusual feature of the Y chromosome allows some recombination among the genes found on it?

The use of Y-linked genetic markers DNA sequences in the Y chromosome undergo mutation with the passage of time and vary among individual males. These mutations create variations in DNA sequence—called genetic markers—that are passed from father to son like Y-linked traits and can be used to study male ancestry. Although the markers themselves do not encode any physical traits, they can be detected with the use of molecular methods. Much of the Y chromosome is nonfunctional; so mutations readily accumulate. Many of these mutations are unique; they arise only once and are passed down through the generations. Individual males possessing the same set of mutations are therefore related, and the distribution of these genetic markers on Y chromosomes provides clues about the genetic relationships of present-day people.

Y-linked markers have been used to study the offspring of Thomas Jefferson, principal author of the Declaration of Independence and third president of the United States. In 1802, Jefferson was accused by a political enemy of fathering a child by his slave Sally Hemings, but the evidence was circumstantial. Hemings, who worked in the Jefferson household and accompanied Jefferson on a trip to Paris, had five children. Jefferson was accused of fathering the first child, Tom, but rumors about the paternity of the other children circulated as well. Hemings's last child, Eston, bore a striking resemblance to Jefferson, and her fourth child, Madison, testified late in life that Jefferson was the father of all of Hemings's children. Ancestors of Hemings's children maintained that they were descendants of the Jefferson line, but some Jefferson descendants refused to recognize their claim.

To resolve this long-standing controversy, geneticists examined markers from the Y chromosomes of male-line descendants of Hemings's first son (Thomas Woodson), her last son (Eston Hemings), and a paternal uncle of Thomas Jefferson with whom Jefferson had Y chromosomes in common. (Descendants of Jefferson's uncle were used because Jefferson himself had no verified male descendants.) Geneticists determined that Jefferson possessed a rare and distinctive set of genetic markers on his Y chromosome. The same markers were also found on the Y chromosomes of the male-line descendants of Eston Hemings. The probability of such a match arising by chance is less than 1%. (The markers were not found on the Y chromosomes of the descendants of Thomas Woodson.) Together with the circumstantial historical evidence, these matching markers strongly suggest that Jefferson fathered Eston Hemings but not Thomas Woodson.

Y-chromosome sequences have also been used extensively to examine past patterns of human migration and the genetic relationships among different human populations.

CONNECTING CONCEPTS

Recognizing Sex-Linked Inheritance

What features should we look for to identify a trait as sex linked? A common misconception is that any genetic characteristic in which the phenotypes of males and females differ must be sex linked. In fact, the expression of many *autosomal* characteristics differs between males and females. The genes that encode these characteristics are the same in both sexes, but their expression is influenced by sex hormones. The different sex hormones of males and females cause the same genes to generate different phenotypes in males and females.

Another misconception is that any characteristic that is found more frequently in one sex is sex linked. A number of autosomal traits are expressed more commonly in one sex than in the other. These traits are said to be sex influenced. Some autosomal traits are expressed in only one sex; these traits are said to be sex limited. Both sex-influenced and sex-limited characteristics will be considered in more detail in Chapter 5.

Several features of sex-linked characteristics make them easy to recognize. Y-linked traits are found only in males, but this fact does not guarantee that a trait is Y linked, because some autosomal characteristics are expressed only in males. A Y-linked trait is unique, however, in that all the male offspring of an affected male will express the father's phenotype, and a Y-linked trait can be inherited only from the father's side of the family. Thus, a Y-linked trait can be inherited only from the paternal grandfather (the father's father), never from the maternal grandfather (the mother's father).

X-linked characteristics also exhibit a distinctive pattern of inheritance. X linkage is a possible explanation when reciprocal crosses give different results. If a characteristic is X-linked, a cross between an affected male and an unaffected female will not give the same results as a cross between an affected female and an unaffected male. For almost all autosomal characteristics, reciprocal crosses give the same result. We should not conclude, however, that, when the reciprocal crosses give different results, the characteristic is X linked. Other sex-associated forms of inheritance, considered in Chapter 5, also produce different results in reciprocal crosses. The key to recognizing X-linked inheritance is to remember that a male always inherits his X chromosome from his mother, not from his father. Thus, an X-linked characteristic is not passed directly from father to son; if a male clearly inherits a characteristic from his father—and the mother is not heterozygous—it cannot be X linked.

4.3 Dosage Compensation Equalizes the Amount of Protein Produced by X-Linked Genes in Males and Females

In species with XX-XY sex determination, the difference in the number of X chromosomes possessed by males and females presents a special problem in development. Because females have two copies of every X-linked gene and males have only one copy, the amount of gene product (protein) encoded by X-linked genes would differ in the two sexes: females would produce twice as much gene product as that produced by males. This difference could be highly detrimental because protein concentration plays a critical role in development. Animals overcome this potential problem through **dosage compensation,** which equalizes the amount of protein produced by X-linked genes in the two sexes. In fruit flies, dosage compensation is achieved by a doubling of the activity of the genes on the X chromosome of the male.

Table 4.3 Number of Barr bodies in human cells with different complements of sex chromosomes

Sex Chromosomes	Syndrome	Number of Barr Bodies
XX	None	1
XY	None	0
XO	Turner	0
XXY	Klinefelter	1
XXYY	Klinefelter	1
XXXY	Klinefelter	2
XXXXY	Klinefelter	3
XXX	Triplo-X	2
XXXX	Poly-X female	3
XXXXX	Poly-X female	4

In the worm *Caenorhabditis elegans*, it is achieved by a halving of the activity of genes on both of the X chromosomes in the female. Placental mammals use yet another mechanism of dosage compensation: genes on one of the X chromosomes in the female are inactivated, as described later in this section.

Lyon Hypothesis

In 1949, Murray Barr observed condensed, darkly staining bodies in the nuclei of cells from female cats (**Figure 4.17**); this darkly staining structure became known as a **Barr body**. Mary Lyon proposed in 1961 that the Barr body was an inactive X chromosome; her hypothesis has become known as the **Lyon hypothesis**. She suggested that, within each female cell, one of the two X chromosomes becomes inactive; which X chromosome is inactivated is random. If a cell contains more than two X chromosomes, all but one of them are inactivated. The number of Barr bodies present in human cells with different complements of sex chromosomes is shown in **Table 4.3**.

As a result of X inactivation, females are functionally hemizygous at the cellular level for X-linked genes. In females that are heterozygous at an X-linked locus, approximately 50% of the cells will express one allele and 50% will express the other allele; thus, in heterozygous females, proteins encoded by both alleles are produced, although not within the same cell. This functional hemizygosity means that cells in females are not identical with respect to the expression of the genes on the X chromosome; females are mosaics for the expression of X-linked genes.

Random X inactivation takes place early in development—in humans, within the first few weeks of development. After an X chromosome has become inactive in a cell, it remains inactive and is inactive in all somatic cells that descend from the cell. Thus, neighboring cells tend to have the same X chromosome inactivated, producing a patchy pattern (mosaic) for the expression of an X-linked characteristic in heterozygous females.

This patchy distribution can be seen in tortoiseshell and calico cats (**Figure 4.18**). Although many genes contribute to coat color and pattern in domestic cats, a single X-linked locus determines the presence of orange color. There are two possible alleles at this locus: X^+, which produces nonorange (usually black) fur, and X^o, which produces orange fur. Males are hemizygous and thus may be black (X^+Y) or orange (X^oY) but not black and orange. (Rare tortoiseshell males can arise from the presence of two X chromosomes, X^+X^oY.) Females may be black (X^+X^+), orange (X^oX^o), or tortoiseshell (X^+X^o), the tortoiseshell pattern arising from a patchy mixture of black and orange fur. Each orange patch is a clone of cells derived from an original cell in which the black allele is inactivated, and each black patch is a clone of cells derived from an original cell in which the orange allele is inactivated.

 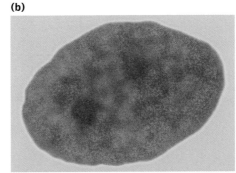

4.17 A Barr body is an inactivated X chromosome. (a) Female cell with a Barr body (indicated by arrow). (b) Male cell without a Barr body. [Chris Bjornberg/Photo Researchers.]

4.18 The patchy distribution of color on tortoiseshell cats results from the random inactivation of one X chromosome in females. [Chanan Photography.]

The Lyon hypothesis suggests that the presence of variable numbers of X chromosomes should not affect the phenotype in mammals, because any X chromosomes in excess of one X chromosome should be inactivated. However, persons with Turner syndrome (XO) differ from normal females, and those with Klinefelter syndrome (XXY) differ from normal males. How do these conditions arise in the face of dosage compensation?

These disorders probably arise because some X-linked genes escape inactivation. Indeed, the nature of X inactivation is more complex than originally envisioned. Studies of individual genes now reveal that only about 75% of X-linked human genes are permanently inactivated. About 15% completely escape X inactivation, meaning that these genes produce twice as much protein in females as they do in males. The remaining 10% are inactivated in some females but not in others. The reason for this variation among females in the activation of some X-linked genes is not known.

Mechanism of Random X Inactivation

Random inactivation of X chromosomes requires two steps. In the first step, the cell somehow assesses, or counts, how many X chromosomes are present. In the second step, one X chromosome is selected to become the active X chromosome and all others are silenced.

Although many details of X-chromosome inactivation remain unknown, several genes and sequences that participate in the process have been identified. Foremost among them is a gene called *Xist* (for X-inactivation-specific transcript). On the X chromosomes destined to become inactivated, the *Xist* gene is active, producing an RNA molecule that coats the X chromosome and inactivates the genes on it, probably by altering chromatin structure. On the X chromosome destined to become active, other genes repress the activity of *Xist* so that the *Xist* RNA never coats the X chromosome and genes on this chromosome remain active.

Dosage Imbalance Between X-Linked Genes and Autosomal Genes

X inactivation compensates for the difference in the number of X chromosomes in male and female mammals, but it creates another type of dosage problem. With X inactivation, males and females have a single active copy of each X-linked gene but possess two active copies of each autosomal gene, leading to different doses (amount) of proteins encoded by X-linked genes relative to proteins encoded by autosomal genes. Proteins often interact with one another and unbalanced dosage frequently leads to problems in development. Recent studies have shown that mammalian cells compensate for this imbalance by increasing the expression of (upregulating) genes on the single active X chromosome so that the total amount of protein produced by genes on the X chromosome is similar to that produced by genes on two copies of autosomal chromosomes. How genes on the single active X chromosome are upregulated is unknown at present. **TRY PROBLEM 41**

CONCEPTS

In mammals, dosage compensation ensures that the same amount of X-linked gene product will be produced in the cells of both males and females. All but one X chromosome are inactivated in each cell; which of the X chromosomes is inactivated is random and varies from cell to cell.

✔ **CONCEPT CHECK 9**

How many Barr bodies will a male with XXXYY chromosomes have in each of his cells? What are these Barr bodies?

CONCEPTS SUMMARY

- Sexual reproduction is the production of offspring that are genetically distinct from their parents. Most organisms have two sexual phenotypes—males and females. Males produce small gametes; females produce large gametes.
- The mechanism by which sex is specified is termed sex determination. Sex may be determined by differences in specific chromosomes, genotypes, or environment.
- The sex chromosomes of males and females differ in number and appearance. The homogametic sex produces gametes that are all identical with regard to sex chromosomes; the heterogametic sex produces gametes that differ in their sex-chromosome composition.
- In the XX-XO system of sex determination, females possess two X chromosomes, whereas males possess a single X chromosome. In the XX-XY system, females possess two X chromosomes, whereas males possess a single X chromosome and a single Y chromosome. In the ZZ-ZW system, males possess two Z chromosomes, whereas females possess a Z chromosome and a W chromosome.
- Some organisms have genic sex determination, in which genotypes at one or more loci determine the sex of an individual organism. Still others have environmental sex determination.
- In *Drosophila melanogaster*, sex is determined by a balance between genes on the X chromosomes and genes on the autosomes, the X : A ratio.
- In humans, sex is ultimately determined by the presence or absence of the *SRY* gene located on the Y chromosome.
- Sex-linked characteristics are determined by genes on the sex chromosomes; X-linked characteristics are encoded by genes on the X chromosome, and Y-linked characteristics are encoded by genes on the Y chromosome.
- A female inherits X-linked alleles from both parents; a male inherits X-linked alleles from his female parent only.
- The sex chromosomes evolved from autosomes. Crossing over between the X and the Y chromosomes has been suppressed, but palindromic sequences within the Y chromosome allow for internal recombination on the Y chromosome. This internal recombination sometimes leads to chromosome rearrangements that can adversely affect sexual development.
- Y-linked characteristics are found only in males and are passed from father to all sons.
- Dosage compensation equalizes the amount of protein produced by X-linked genes in males and females. In placental mammals, one of the two X chromosomes in females normally becomes inactivated. Which X chromosome is inactivated is random and varies from cell to cell. Some X-linked genes escape X inactivation, and other X-linked genes may be inactivated in some females but not in others. X inactivation is controlled by the *Xist* gene.

IMPORTANT TERMS

sex (p. 74)
sex determination (p. 74)
hermaphroditism (p. 75)
monoecious organism (p. 75)
dioecious organism (p. 75)
sex chromosome (p. 75)
autosome (p. 76)
heterogametic sex (p. 76)
homogametic sex (p. 76)
pseudoautosomal region (p. 76)
genic sex determination (p. 77)
sequential hermaphroditism (p. 77)
genic balance system (p. 78)
X : A ratio (p. 78)
Turner syndrome (p. 79)
Klinefelter syndrome (p. 79)
triplo-X syndrome (p. 79)
sex-determining region Y (*SRY*) gene (p. 80)
sex-linked characteristic (p. 81)
X-linked characteristic (p. 81)
Y-linked characteristic (p. 81)
hemizygosity (p. 81)
nondisjunction (p. 83)
dosage compensation (p. 88)
Barr body (p. 89)
Lyon hypothesis (p. 89)

ANSWERS TO CONCEPT CHECKS

1. Meiosis
2. b
3. In chromosomal sex determination, males and females have chromosomes that are distinguishable. In genic sex determination, sex is determined by genes, but the chromosomes of males and females are indistinguishable. In environmental sex determination, sex is determined fully or in part by environmental effects.
4. b
5. a
6. c
7. All male offspring will have hemophilia, and all female offspring will not have hemophilia; so the overall probability of hemophilia in the offspring is $1/2$.
8. Eight large palindromes that allow crossing over within the Y chromosome.
9. Two Barr bodies. A Barr body is an inactivated X chromosome.

WORKED PROBLEMS

1. A fruit fly has XXXYY sex chromosomes; all the autosomal chromosomes are normal. What sexual phenotype does this fly have?

• **Solution**

Sex in fruit flies is determined by the X : A ratio—the ratio of the number of X chromosomes to the number of haploid autosomal sets. An X : A ratio of 1.0 produces a female fly; an X : A ratio of 0.5 produces a male. If the X : A ratio is greater than 1.0, the fly is a metafemale; if it is less than 0.5, the fly is a metamale; if the X : A ratio is between 1.0 and 0.5, the fly is an intersex.

This fly has three X chromosomes and normal autosomes. Normal diploid flies have two autosomal sets of chromosomes; so the X : A ratio in this case is $3/2$, or 1.5. Thus, this fly is a metafemale.

2. Chickens, like all birds, have ZZ-ZW sex determination. The bar-feathered phenotype in chickens results from a Z-linked allele that is dominant over the allele for nonbar feathers. A barred female is crossed with a nonbarred male. The F_1 from this cross are intercrossed to produce the F_2. What are the phenotypes and their proportions in the F_1 and F_2 progeny?

• **Solution**

With the ZZ-ZW system of sex determination, females are the heterogametic sex, possessing a Z chromosome and a W chromosome; males are the homogametic sex, with two Z chromosomes. In this problem, the barred female is hemizygous for the bar phenotype (Z^BW). Because bar is dominant over nonbar, the nonbarred male must be homozygous for nonbar (Z^bZ^b). Crossing these two chickens, we obtain:

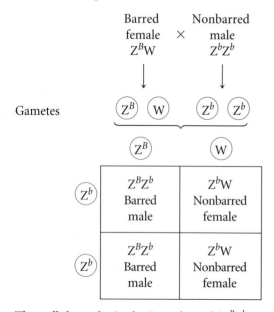

Thus, all the males in the F_1 are barred (Z^BZ^b), and all the females are nonbarred (Z^bW).

We now cross the F_1 to produce the F_2:

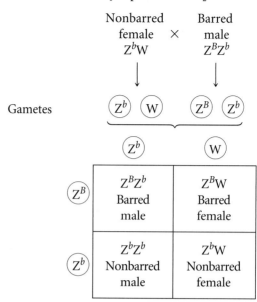

So, $1/4$ of the F_2 are barred males, $1/4$ are nonbarred males, $1/4$ are barred females, and $1/4$ are nonbarred females.

3. In *Drosophila melanogaster*, forked bristles are caused by an allele (X^f) that is X linked and recessive to an allele for normal bristles (X^+). Brown eyes are caused by an allele (b) that is autosomal and recessive to an allele for red eyes (b^+). A female fly that is homozygous for normal bristles and red eyes mates with a male fly that has forked bristles and brown eyes. The F_1 are intercrossed to produce the F_2. What will the phenotypes and proportions of the F_2 flies be from this cross?

• **Solution**

This problem is best worked by breaking the cross down into two separate crosses, one for the X-linked genes that determine the type of bristles and one for the autosomal genes that determine eye color.

Let's begin with the autosomal characteristics. A female fly that is homozygous for red eyes (b^+b^+) is crossed with a male with brown eyes. Because brown eyes are recessive, the male fly must be homozygous for the brown-eye allele (bb). All of the offspring of this cross will be heterozygous (b^+b) and will have red eyes:

P b^+b^+ × bb
            Red eyes   Brown eyes

Gametes   (b^+)   (b)

F_1 b^+b
            Red eyes

The F_1 are then intercrossed to produce the F_2. Whenever two individual organisms heterozygous for an autosomal recessive characteristic are crossed, $3/4$ of the offspring will have the dominant trait and $1/4$ will have the recessive trait; thus, $3/4$ of the F_2 flies will have red eyes and $1/4$ will have brown eyes:

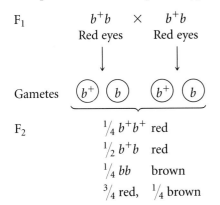

Next, we work out the results for the X-linked characteristic. A female that is homozygous for normal bristles (X^+X^+) is crossed with a male that has forked bristles (X^fY). The female F_1 from this cross are heterozygous (X^+X^f), receiving an X chromosome with a normal-bristle allele from their mother (X^+) and an X chromosome with a forked-bristle allele (X^f) from their father. The male F_1 are hemizygous (X^+Y), receiving an X chromosome with a normal-bristle allele from their mother (X^+) and a Y chromosome from their father:

P X^+X^+ × X^fY
    Normal bristles   Forked bristles

Gametes X^+ X^f Y

F_1 $1/2$ X^+X^f normal bristle
     $1/2$ X^+Y normal bristle

When these F_1 are intercrossed, $1/2$ of the F_2 will be normal-bristle females, $1/4$ will be normal-bristle males, and $1/4$ will be forked-bristle males:

F_1 X^+X^f × X^+Y

Gametes X^+ X^f X^+ Y

F_2:

	X^+	X^f
X^+	X^+X^+ Normal female	X^+X^f Normal female
Y	X^+Y Normal male	X^fY Forked-bristle male

$1/2$ normal female, $1/4$ normal male, $1/4$ forked-bristle male

To obtain the phenotypic ratio in the F_2, we now combine these two crosses by using the multiplication rule of probability and the branch diagram:

Eye color	Bristle and sex	F_2 phenotype	Probability
red ($3/4$)	normal female ($1/2$)	red normal female	$3/4 × 1/2 = 3/8 = 6/16$
	normal male ($1/4$)	red normal male	$3/4 × 1/4 = 3/16$
	forked-bristle male ($1/4$)	red forked-bristle male	$3/4 × 1/4 = 3/16$
brown ($1/4$)	normal female ($1/2$)	brown normal female	$1/4 × 1/2 = 1/8 = 2/16$
	normal male ($1/4$)	brown normal male	$1/4 × 1/4 = 1/16$
	forked-bristle male ($1/4$)	brown forked-bristle male	$1/4 × 1/4 = 1/16$

COMPREHENSION QUESTIONS

Section 4.1

*1. What is the most defining difference between the males and the females of any organism?

2. How do monoecious organisms differ from dioecious organisms?

3. Describe the XX-XO system of sex determination. In this system, which is the heterogametic sex and which is the homogametic sex?

4. How does sex determination in the XX-XY system differ from sex determination in the ZZ-ZW system?

*5. What is the pseudoautosomal region? How does the inheritance of genes in this region differ from the inheritance of other Y-linked characteristics?

6. What is meant by genic sex determination?

7. How does sex determination in *Drosophila* differ from sex determination in humans?

8. Give the typical sex-chromosome complement found in the cells of people with Turner syndrome, with Klinefelter syndrome, and with androgen-insensitivity syndrome. What is the sex-chromosome complement of triplo-X females?

Section 4.2

*9. What characteristics are exhibited by an X-linked trait?

10. Explain how Bridges's study of nondisjunction in *Drosophila* helped prove the chromosome theory of inheritance.

*11. What characteristics are exhibited by a Y-linked trait?

Section 4.3

12. Explain why tortoiseshell cats are almost always female and why they have a patchy distribution of orange and black fur.

13. What is a Barr body? How is it related to the Lyon hypothesis?

APPLICATION QUESTIONS AND PROBLEMS

Introduction

14. As described in the introduction to this chapter, platypuses possess 10 sex chromosomes. Females have five pairs of X chromosomes ($X_1X_1X_2X_2X_3X_3X_4X_4X_5X_5$) and males have five pairs of X and Y chromosomes ($X_1Y_1X_2Y_2X_3Y_3X_4Y_4X_5Y_5$). Do each of the XY chromosome pairs in males assort independently in meiosis? Why or why not?

Section 4.1

*15. What is the sexual phenotype of fruit flies having the following chromosomes?

	Sex chromosomes	Autosomal chromosomes
a.	XX	all normal
b.	XY	all normal
c.	XO	all normal
d.	XXY	all normal
e.	XYY	all normal
f.	XXYY	all normal
g.	XXX	all normal
h.	XX	four haploid sets
i.	XXX	four haploid sets
j.	XXX	three haploid sets
k.	X	three haploid sets
l.	XY	three haploid sets
m.	XX	three haploid sets

16. For each of the following sex-chromosome complements, what is the phenotypic sex of a person who has

a. XY with the *SRY* gene deleted?

b. XY with the *SRY* gene located on an autosomal chromosome?

c. XX with a copy of the *SRY* gene on an autosomal chromosome?

d. XO with a copy of the *SRY* gene on an autosomal chromosome?

e. XXY with the *SRY* gene deleted?

f. XXYY with one copy of the *SRY* gene deleted?

17. A normal female *Drosophila* produces abnormal eggs that contain all (a complete diploid set) of her chromosomes. She mates with a normal male *Drosophila* that produces normal sperm. What will the sex of the progeny from this cross be?

18. In certain salamanders, the sex of a genetic female can be altered, changing her into a functional male; these salamanders are called sex-reversed males. When a sex-reversed male is mated with a normal female, approximately $2/3$ of the offspring are female and $1/3$ are male. How is sex determined in these salamanders? Explain the results of this cross.

19. In some mites, males pass genes to their grandsons, but they never pass genes to their sons. Explain.

*20. In organisms with the ZZ-ZW sex-determining system, from which of the following possibilities can a female inherit her Z chromosome?

	Yes	No
Her mother's mother		
Her mother's father		
Her father's mother		
Her father's father		

Section 4.2

21. When Bridges crossed white-eyed females with red-eyed males, he obtained a few red-eyed males and white-eyed females (see Figure 4.13). What types of offspring would be produced if these red-eyed males and white-eyed females were crossed with each other?

*22. Joe has classic hemophilia, an X-linked recessive disease. Could Joe have inherited the gene for this disease from the following persons?

	Yes	No
a. His mother's mother		
b. His mother's father		
c. His father's mother		
d. His father's father		

*23. In *Drosophila*, yellow body is due to an X-linked gene that is recessive to the gene for gray body.

a. A homozygous gray female is crossed with a yellow male. The F_1 are intercrossed to produce F_2. Give the genotypes and phenotypes, along with the expected proportions, of the F_1 and F_2 progeny.

b. A yellow female is crossed with a gray male. The F_1 are intercrossed to produce the F_2. Give the genotypes and phenotypes, along with the expected proportions, of the F_1 and F_2 progeny.

c. A yellow female is crossed with a gray male. The F_1 females are backcrossed with gray males. Give the genotypes and phenotypes, along with the expected proportions, of the F_2 progeny.

d. If the F_2 flies in part *b* mate randomly, what are the expected phenotypic proportions of flies in the F_3?

[Courtesy Masa-Toshi Yamamoto, Drosophila Genetic Resource Center, Kyoto Institute of Technology.]

24. Coat color in cats is determined by genes at several different loci. At one locus on the X chromosome, one allele (X^+) encodes black fur; another allele (X^o) encodes orange fur. Females can be black (X^+X^+), orange (X^oX^o), or a mixture of orange and black called tortoiseshell (X^+X^o). Males are either black (X^+Y) or orange (X^oY). Bill has a female tortoiseshell cat named Patches. One night Patches escapes from Bill's house, spends the night out, and mates with a stray male. Patches later gives birth to the following kittens: one orange male, one black male, two tortoiseshell females, and one orange female. Give the genotypes of Patches, her kittens, and the stray male with which Patches mated.

*****25.** Red–green color blindness in humans is due to an X-linked recessive gene. Both John and Cathy have normal color vision. After 10 years of marriage to John, Cathy gave birth to a color-blind daughter. John filed for divorce, claiming that he is not the father of the child. Is John justified in his claim of nonpaternity? Explain why. If Cathy had given birth to a color-blind son, would John be justified in claiming nonpaternity?

26. Red–green color blindness in humans is due to an X-linked recessive gene. A woman whose father is color blind possesses one eye with normal color vision and one eye with color blindness.

a. Propose an explanation for this woman's vision pattern. Assume that no new mutations have spontaneously arisen.

b. Would it be possible for a man to have one eye with normal color vision and one eye with color blindness?

*****27.** Bob has XXY chromosomes (Klinefelter syndrome) and is color blind. His mother and father have normal color vision, but his maternal grandfather is color blind. Assume that Bob's chromosome abnormality arose from nondisjunction in meiosis. In which parent and in which meiotic division did nondisjunction take place? Explain your answer.

28. Xg is an antigen found on red blood cells. This antigen is caused by an X-linked allele (X^a) that is dominant over an allele for the absence of the antigen (X^-). The inheritance of these X-linked alleles was studied in children with chromosome abnormalities to determine where nondisjunction of the sex chromosomes took place. For each type of mating in parts *a* through *d*, indicate whether nondisjunction took place in the mother or in the father and, if possible, whether it took place in meiosis I or meiosis II.

a. $X^aY \times X^-X^- \rightarrow X^a$ (Turner syndrome)

b. $X^aY \times X^aX^- \rightarrow X^-$ (Turner syndrome)

c. $X^aY \times X^-X^- \rightarrow X^aX^-Y$ (Klinefelter syndrome)

d. $X^aY \times X^aX^- \rightarrow X^-X^-Y$ (Klinefelter syndrome)

29. The Talmud, an ancient book of Jewish civil and religious laws, states that, if a woman bears two sons who die of bleeding after circumcision (removal of the foreskin from the penis), any additional sons that she has should not be circumcised. (The bleeding is most likely due to the X-linked disorder hemophilia.) Furthermore, the Talmud states that the sons of her sisters must not be circumcised, whereas the sons of her brothers should be. Is this religious law consistent with sound genetic principles? Explain your answer.

30. Craniofrontonasal syndrome (CFNS) is a birth defect in which premature fusion of the cranial sutures leads to abnormal head shape, widely spaced eyes, nasal clefts, and various other skeletal abnormalities. George Feldman and his colleagues, looked at several families in which offspring had CFNS and recorded the results shown in the following table (G. J. Feldman. 1997. *Human Molecular Genetics* 6:1937–1941).

Family number	Parents		Offspring			
			Normal		CFNS	
	Father	Mother	Male	Female	Male	Female
1	normal	CFNS	1	0	2	1
5	normal	CFNS	0	2	1	2
6	normal	CFNS	0	0	1	2
8	normal	CFNS	1	1	1	0
10a	CFNS	normal	3	0	0	2
10b	normal	CFNS	1	1	2	0
12	CFNS	normal	0	0	0	1
13a	normal	CFNS	0	1	2	1
13b	CFNS	normal	0	0	0	2
7b	CFNS	normal	0	0	0	2

a. On the basis of these results, what is the most likely mode of inheritance for CFNS?

b. Give the most likely genotypes of the parents in family 1 and in family 10a.

*31. Miniature wings (X^m) in *Drosophila* result from an X-linked allele that is recessive to the allele for long wings (X^+). Give the genotypes of the parents in each of the following crosses.

	Male parent	Female parent	Male offspring	Female offspring
a.	long	long	231 long, 250 miniature	560 long
b.	miniature	long	610 long	632 long
c.	miniature	long	410 long, 417 miniature	412 long, 415 miniature
d.	long	miniature	753 miniature	761 long
e.	long	long	625 long	630 long

*32. In chickens, congenital baldness is due to a Z-linked recessive gene. A bald rooster is mated with a normal hen. The F_1 from this cross are interbred to produce the F_2. Give the genotypes and phenotypes, along with their expected proportions, among the F_1 and F_2 progeny.

33. Red–green color blindness is an X-linked recessive trait in humans. Polydactyly (extra fingers and toes) is an autosomal dominant trait. Martha has normal fingers and toes and normal color vision. Her mother is normal in all respects, but her father is color blind and polydactylous. Bill is color blind and polydactylous. His mother has normal color vision and normal fingers and toes. If Bill and Martha marry, what types and proportions of children can they produce?

34. A *Drosophila* mutation called *singed* (*s*) causes the bristles to be bent and misshapen. A mutation called *purple* (*p*) causes the fly's eyes to be purple in color instead of the normal red. Flies homozygous for *singed* and *purple* were crossed with flies that were homozygous for normal bristles and red eyes. The F_1 were intercrossed to produce the F_2, and the following results were obtained.

Cross 1

P   male, singed bristles, purple eyes × female, normal bristles, red eyes

F_1 420 female, normal bristles, red eyes
426 male, normal bristles, red eyes

F_2 337 female, normal bristles, red eyes
113 female, normal bristles, purple eyes
168 male, normal bristles, red eyes
170 male, singed bristles, red eyes
56 male, normal bristles, purple eyes
58 male, singed bristles, purple eyes

Cross 2

P   female, singed bristles, purple eyes × male, normal bristles, red eyes

F_1 504 female, normal bristles, red eyes
498 male, singed bristles, red eyes

F_2 227 female, normal bristles, red eyes
223 female, singed bristles, red eyes
225 male, normal bristles, red eyes
225 male, singed bristles, red eyes
78 female, normal bristles, purple eyes
76 female, singed bristles, purple eyes
74 male, normal bristles, purple eyes
72 male, singed bristles, purple eyes

a. What are the modes of inheritance of *singed* and *purple*? Explain your reasoning.

b. Give genotypes for the parents and offspring in the P, F_1, and F_2 generations of Cross 1 and Cross 2.

35. The following two genotypes are crossed: $Aa\ Bb\ Cc\ X^+X^r \times Aa\ BB\ cc\ X^+Y$, where *a*, *b*, and *c* represent alleles of autosomal genes and X^+ and X^r represent X-linked alleles in an organism with XX-XY sex determination. What is the probability of obtaining genotype $aa\ Bb\ Cc\ X^+X^+$ in the progeny?

*36. Miniature wings in *Drosophila melanogaster* are due to an X-linked gene (X^m) that is recessive to an allele for long wings (X^+). Sepia eyes are produced by an autosomal gene (*s*) that is recessive to an allele for red eyes (s^1).

a. A female fly that has miniature wings and sepia eyes is crossed with a male that has normal wings and is homozygous for red eyes. The F_1 flies are intercrossed to produce the F_2. Give the phenotypes, as well as their expected proportions, of the F_1 and F_2 flies.

b. A female fly that is homozygous for normal wings and has sepia eyes is crossed with a male that has miniature wings and is homozygous for red eyes. The F_1 flies are intercrossed to produce the F_2. Give the phenotypes, as well as their expected proportions, of the F_1 and F_2 flies.

37. Suppose that a recessive gene that produces a short tail in mice is located in the pseudoautosomal region. A short-tailed male mouse is mated with a female mouse that is homozygous for a normal tail. The F_1 mice from this cross are intercrossed to produce the F_2. Give the phenotypes, as well as their proportions, of the F_1 and F_2 mice?

*38. A color-blind woman and a man with normal vision have three sons and six daughters. All the sons are color blind. Five of the daughters have normal vision, but one of them is color blind. The color-blind daughter is 16 years old, is short for her age, and has not undergone puberty. Explain how this girl inherited her color blindness.

Section 4.3

*39. How many Barr bodies would you expect to see in a human cell containing the following chromosomes?

a. XX	d. XXY	g. XYY
b. XY	e. XXYY	h. XXX
c. XO	f. XXXY	i. XXXX

40. A woman with normal chromosomes mates with a man who also has normal chromosomes.

   a. Suppose that, in the course of oogenesis, the woman's sex chromosomes undergo nondisjunction in meiosis I; the man's chromosomes separate normally. Give all possible combinations of sex chromosomes that this couple's children might inherit and the number of Barr bodies that you would expect to see in each of the cells of each child.

   b. What chromosome combinations and numbers of Barr bodies would you expect to see if the chromosomes separate normally in oogenesis, but nondisjunction of the sex chromosomes takes place in meiosis I of spermatogenesis?

41. Anhidrotic ectodermal dysplasia is an X-linked recessive disorder in humans characterized by small teeth, no sweat glands, and sparse body hair. The trait is usually seen in men, but women who are heterozygous carriers of the trait often have irregular patches of skin with few or no sweat glands (see the illustration on the right).

   a. Explain why women who are heterozygous carriers of a recessive gene for anhidrotic ectodermal dysplasia have irregular patches of skin lacking sweat glands.

   b. Why does the distribution of the patches of skin lacking sweat glands differ among the females depicted in the illustration, even between the identical twins?

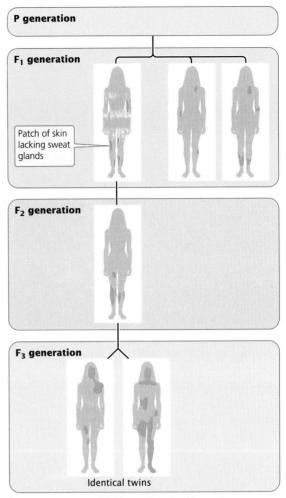

[After A. P. Mance and J. Mance, *Genetics: Human Aspects* (Sinauer, 1990), p. 133.]

CHALLENGE QUESTIONS

Section 4.2

42. A geneticist discovers a male mouse with greatly enlarged testes in his laboratory colony. He suspects that this trait results from a new mutation that is either Y linked or autosomal dominant. How could he determine whether the trait is autosomal dominant or Y linked?

Section 4.3

43. Female humans who are heterozygous for X-linked recessive genes sometimes exhibit mild expression of the trait. However, such mild expression of X-linked traits in females who are heterozygous for X-linked alleles is not seen in *Drosophila*. What might cause this difference in the expression of X-linked genes in female humans and female *Drosophila*? (Hint: In *Drosophila*, dosage compensation is accomplished by doubling the activity of genes on the X chromosome of males.)

44. Identical twins (also called monozygotic twins) are derived from a single egg fertilized by a single sperm, creating a zygote that later divides into two (see Chapter 6). Because identical twins originate from a single zygote, they are genetically identical.

    Caroline Loat and her colleagues examined 9 measures of social, behavioral, and cognitive ability in 1000 pairs of identical male twins and 1000 pairs of identical female twins (C. S. Loat et al. 2004. *Twin Research* 7:54–61). They found that, for three of the measures (prosocial behavior, peer problems, and verbal ability), the two male twins of a pair tended to be more alike in their scores than were two female twins of a pair. Propose a possible explanation for this observation. What might this observation indicate about the location of genes that influence prosocial behavior, peer problems, and verbal ability?

45. Occasionally, a mouse X chromosome is broken into two pieces and each piece becomes attached to a different autosomal chromosome. In this event, the genes on only one of the two pieces undergo X inactivation. What does this observation indicate about the mechanism of X-chromosome inactivation?

5 Extensions and Modifications of Basic Principles

Yellow coat color in mice is caused by a recessive lethal gene, producing distorted phenotypic ratios in the progeny of two yellow mice. William Castle and Clarence Little discovered the lethal nature of the yellow gene in 1910. [Reprinted with permission of Dr. Loan Phan and *In Vivo*, a publication of Columbia University Medical Center.]

CUÉNOT'S ODD YELLOW MICE

At the start of the twentieth century, Mendel's work on inheritance in pea plants became widely known (see Chapter 3), and a number of biologists set out to verify his conclusions by using other organisms to conduct crosses. One of these biologists was Lucien Cuénot, a French scientist working at the University of Nancy. Cuénot experimented with coat colors in mice and was among the first to show that Mendel's principles applied to animals.

Cuénot observed that the coat colors of his mice followed the same patterns of inheritance observed by Mendel in his pea plants. Cuénot found that, when he crossed pure-breeding gray mice with pure-breeding white mice, all of the F_1 progeny were gray, and interbreeding the F_1 produced a 3 : 1 ratio of gray and white mice in the F_2, as would be expected if gray were dominant over white. The results of Cuénot's breeding experiments perfectly fit Mendel's rules—with one exception. His crosses of yellow mice suggested that yellow coat color was dominant over gray, but he was never able to obtain true-breeding (homozygous) yellow mice. Whenever Cuénot crossed two yellow mice, he obtained yellow and gray mice in approximately a 3 : 1 ratio, suggesting that the yellow mice were heterozygous ($Yy \times Yy \rightarrow {}^3\!/_4\, Y\_$ and ${}^1\!/_4\, yy$). If yellow were indeed dominant over gray, some of the yellow progeny from this cross should have been homozygous for yellow (YY) and crossing two of these mice should have yielded all yellow offspring ($YY \times YY \rightarrow YY$). However, he never obtained all yellow progeny in his crosses.

Cuénot was puzzled by these results, which failed to conform to Mendel's predictions. He speculated that yellow gametes were incompatible with each other and would not fuse to form a zygote. Other biologists thought that additional factors might affect the inheritance of the yellow coat color, but the genetics of the yellow mice remained a mystery.

In 1910, William Ernest Castle and his student Clarence Little solved the mystery of Cuénot's unusual results. They carried out a large series of crosses between two yellow mice and showed that the progeny appeared, not in the 3 : 1 ratio that Cuénot thought he had observed but actually in a 2 : 1 ratio of yellow and nonyellow. Castle and Little recognized that the allele for yellow was lethal when homozygous (**Figure 5.1**), and thus all the yellow mice were heterozygous (Yy). A cross between two yellow heterozygous mice produces an

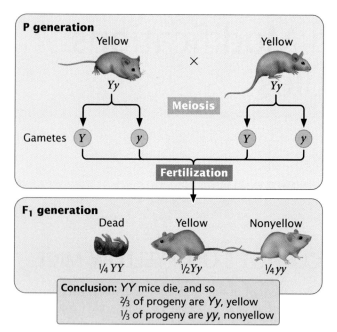

5.1 The 2 : 1 ratio produced by a cross between two yellow mice results from a lethal allele.

initial genotypic ratio of ¼ *YY*, ½ *Yy*, and ¼ *yy*, but the homozygous *YY* mice die early in development and do not appear among the progeny, resulting in a 2 : 1 ratio of *Yy* (yellow) to *yy* (nonyellow) in offspring. Indeed, Castle and Little found that crosses of yellow × yellow mice resulted in smaller litters compared with litters of yellow × nonyellow mice. Because only mice homozygous for the *Y* allele die, the *yellow* allele is a recessive lethal. The *yellow* allele in mice is unusual in that it acts as a *recessive* allele in its effect on development but acts as a *dominant* allele in its effect on coat color.

Cuénot went on to make a number of other important contributions to genetics. He was the first to propose that more than two alleles could exist at a single locus, and he described how genes at different loci could interact in the determination of coat color in mice (aspects of inheritance that we will consider in this chapter). He observed that some types of cancer in mice display a hereditary predisposition; he also proposed, far ahead of his time, that genes might encode enzymes. Unfortunately, Cuénot's work brought him little recognition in his lifetime and was not well received by other French biologists, many of them openly hostile to the idea of Mendelian genetics. Cuénot's studies were interrupted by World War I, when foreign troops occupied his town and he was forced to abandon his laboratory at the university. He later returned to find his stocks of mice destroyed, and he never again took up genetic investigations.

Like a number of other genetic phenomena, the lethal *yellow* gene discovered by Cuénot does not produce the ratios predicted by Mendel's principles of heredity. This lack of adherence to Mendel's rules doesn't mean that Mendel was wrong; rather, it demonstrates that Mendel's principles are not, by themselves, sufficient to explain the inheritance of all genetic characteristics. Our modern understanding of genetics has been greatly enriched by the discovery of a number of modifications and extensions of Mendel's basic principles, which are the focus of this chapter. **TRY PROBLEM 12**

5.1 Additional Factors at a Single Locus Can Affect the Results of Genetic Crosses

In Chapter 3, we learned that the principle of segregation and the principle of independent assortment allow us to predict the outcomes of genetic crosses. Here, we examine several additional factors acting at individual loci that can alter the phenotypic ratios predicted by Mendel's principles.

Types of Dominance

One of Mendel's important contributions to the study of heredity is the concept of dominance—the idea that an individual organism possesses two different alleles for a characteristic but the trait encoded by only one of the alleles is observed in the phenotype. With dominance, the heterozygote possesses the same phenotype as one of the homozygotes.

Mendel observed dominance in all of the traits that he chose to study extensively, but he was aware that not all characteristics exhibit dominance. He conducted some crosses concerning the length of time that pea plants take to flower. For example, when he crossed two homozygous varieties that differed in their flowering time by an average of 20 days, the length of time taken by the F_1 plants to flower was intermediate between those of the two parents. When the heterozygote has a phenotype intermediate between the phenotypes of the two homozygotes, the trait is said to display *incomplete dominance*.

Complete and incomplete dominance Dominance can be understood in regard to how the phenotype of the heterozygote relates to the phenotypes of the homozygotes. In the example presented in the top panel of **Figure 5.2**, flower color potentially ranges from red to white. One homozygous genotype, A^1A^1, produces red pigment, resulting in red flowers; another, A^2A^2, produces no pigment, resulting in white flowers. Where the heterozygote falls in the range of phenotypes determines the type of dominance. If the heterozygote (A^1A^2) produces the same amount of pigment as the A^1A^1 homozygote, resulting in red, then the A^1 allele displays **complete dominance** over the A^2 allele; that is, red is dominant over white. If, on the other hand, the heterozygote produces no pigment resulting in flowers with the same color as the A^2A^2 homozygote (white), then the A^2 allele is completely dominant, and white is dominant over red.

When the heterozygote falls in between the phenotypes of the two homozygotes, dominance is incomplete. With

5.2 The type of dominance exhibited by a trait depends on how the phenotype of the heterozygote relates to the phenotypes of the homozygotes.

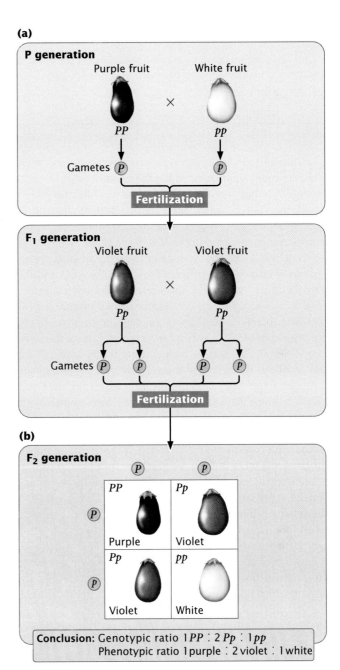

5.3 Fruit color in eggplant is inherited as an incompletely dominant trait.

incomplete dominance, the heterozygote need not be exactly intermediate (see the bottom panel of Figure 5.2) between the two homozygotes; it might be a slightly lighter shade of red or a slightly pink shade of white. As long as the heterozygote's phenotype can be differentiated and falls within the range of the two homozygotes, dominance is incomplete.

Incomplete dominance is also exhibited in the fruit color of eggplant. When a homozygous plant that produces purple fruit (PP) is crossed with a homozygous plant that produces white fruit (pp), all the heterozygous F_1 (Pp) produce violet fruit (**Figure 5.3a**). When the F_1 are crossed with each other, $1/4$ of the F_2 are purple (PP), $1/2$ are violet (Pp), and $1/4$ are white (pp), as shown in **Figure 5.3b**. Note that this 1 : 2 : 1 ratio is different from the 3 : 1 ratio that we would observe if eggplant fruit color exhibited complete dominance. Another example of incomplete dominance is feather color in chickens. A cross between a homozygous black chicken and a homozygous white chicken produces F_1 chickens that are gray. If these gray F_1 are intercrossed, they produce F_2 birds in a ratio of 1 black : 2 gray : 1 white.

We should now add the 1 : 2 : 1 ratio to those phenotypic ratios for simple crosses presented in Chapter 3 (see Table 3.3). A 1 : 2 : 1 phenotypic ratio arises in the progeny of a cross between two parents heterozygous for a character that exhibits incomplete dominance ($Aa \times Aa$). The genotypic ratio among these progeny also is 1 : 2 : 1. When a trait displays incomplete dominance, the genotypic ratios and phenotypic ratios of the offspring are the *same*, because each genotype has its own phenotype. The important thing to remember about dominance is that it affects the phenotype that genes produce but not the way in which genes are *inherited*.

CONCEPTS

Incomplete dominance is exhibited when the heterozygote has a phenotype intermediate between the phenotypes of the two homozygotes. When a trait exhibits incomplete dominance, a cross between two heterozygotes produces a 1 : 2 : 1 phenotypic ratio in the progeny.

✔ **CONCEPT CHECK 1**

If an F_1 eggplant in Figure 5.3 is used in a testcross, what proportion of the progeny from this cross will be white?

a. All the progeny   c. $1/4$
b. $1/2$ d. 0

Codominance Another type of interaction between alleles is **codominance**, in which the phenotype of the heterozygote is not intermediate between the phenotypes of the homozygotes; rather, the heterozygote simultaneously expresses the phenotypes of both homozygotes. An example of codominance is seen in the MN blood types.

The MN locus encodes one of the types of antigens on red blood cells. Unlike antigens foreign to the ABO and Rh blood groups (which also encode red-blood-cell antigens), foreign MN antigens do not elicit a strong immunological reaction; therefore, the MN blood types are not routinely considered in blood transfusions. At the MN locus, there are two alleles: the L^M allele, which encodes the M antigen; and the L^N allele, which encodes the N antigen. Homozygotes with genotype $L^M L^M$ express the M antigen on their red blood cells and have the M blood type. Homozygotes with genotype $L^N L^N$ express the N antigen and have the N blood type. Heterozygotes with genotype $L^M L^N$ exhibit codominance and express both the M and the N antigens; they have blood-type MN.

Some students might ask why the pink flowers illustrated in the bottom panel of Figure 5.2 exhibit incomplete dominance—that is, why is this outcome not an example of codominance? The flowers would exhibit codominance only if the heterozygote produced both red and white pigments, which then combined to produce a pink phenotype. However, in our example, the heterozygote produces only red pigment. The pink phenotype comes about because the amount of pigment produced by the heterozygote is less than the amount produced by the $A^1 A^1$ homozygote. So, here, the alleles clearly exhibit incomplete dominance, not codominance. The differences between dominance, incomplete dominance, and codominance are summarized in **Table 5.1**. **TRY PROBLEMS 14 AND 15** →

Dependency of type of dominance on level of phenotype observed Phenotypes can frequently be observed at several different levels, including the anatomical level, the physiological level, and the molecular level. The type of dominance exhibited by a character depends on the level of the phenotype examined. This dependency is seen in cystic fibrosis, a common genetic disorder found in Caucasians and usually considered to be a recessive disease. People who have cystic fibrosis produce large quantities of thick, sticky mucus, which plugs up the airways of the lungs and clogs the ducts leading from the pancreas to the intestine, causing frequent respiratory infections and digestive problems. Even

Table 5.1 Differences between dominance, incomplete dominance, and codominance

Type of Dominance	Definition
Dominance	Phenotype of the heterozygote is the same as the phenotype of one of the homozygotes.
Incomplete dominance	Phenotype of the heterozygote is intermediate (falls within the range) between the phenotypes of the two homozygotes.
Codominance	Phenotype of the heterozygote includes the phenotypes of both homozygotes.

with medical treatment, patients with cystic fibrosis suffer chronic, life-threatening medical problems.

The gene responsible for cystic fibrosis resides on the long arm of chromosome 7. It encodes a protein termed *cystic fibrosis transmembrane conductance regulator*, abbreviated CFTR, which acts as a gate in the cell membrane and regulates the movement of chloride ions into and out of the cell. Patients with cystic fibrosis have a mutated, dysfunctional form of CFTR that causes the channel to stay closed, and so chloride ions build up in the cell. This buildup causes the formation of thick mucus and produces the symptoms of the disease.

Most people have two copies of the normal allele for CFTR and produce only functional CFTR protein. Those with cystic fibrosis possess two copies of the mutated CFTR allele and produce only the defective CFTR protein. Heterozygotes, having one normal and one defective CFTR allele, produce both functional and defective CFTR protein. Thus, at the molecular level, the alleles for normal and defective CFTR are codominant, because both alleles are expressed in the heterozygote. However, because one functional allele produces enough functional CFTR protein to allow normal chloride ion transport, the heterozygote exhibits no adverse effects, and the mutated CFTR allele appears to be recessive at the physiological level. The type of dominance expressed by an allele, as illustrated in this example, is a function of the phenotypic aspect of the allele that is observed.

Characteristics of dominance Several important characteristics of dominance should be emphasized. First, dominance is a result of interactions between genes at the same locus; in other words, dominance is *allelic* interaction. Second, dominance does not alter the way in which the genes are inherited; it only influences the way in which they are expressed as a phenotype. The allelic interaction that characterizes dominance is therefore interaction between the *products* of the genes. Finally, dominance is frequently "in the eye of the beholder," meaning that the classification of dominance depends on the level at which the phenotype is examined. As seen for cystic fibrosis, an allele may exhibit codominance at one level and be recessive at another level.

> ### CONCEPTS
>
> Dominance entails interactions between genes at the same locus (allelic genes) and is an aspect of the phenotype; dominance does not affect the way in which genes are inherited. The type of dominance exhibited by a characteristic frequently depends on the level of the phenotype examined.
>
> ✔ **CONCEPT CHECK 2**
>
> How do complete dominance, incomplete dominance, and codominance differ?

Penetrance and Expressivity

In the genetic crosses presented thus far, we have considered only the interactions of alleles and have assumed that every individual organism having a particular genotype expresses the expected phenotype. We assumed, for example, that the genotype Rr always produces round seeds and that the genotype rr always produces wrinkled seeds. For some characters, however, such an assumption is incorrect: the genotype does not always produce the expected phenotype, a phenomenon termed **incomplete penetrance**.

Incomplete penetrance is seen in human polydactyly, the condition of having extra fingers and toes (**Figure 5.4**). There are several different forms of human polydactyly, but the trait is usually caused by a dominant allele. Occasionally, people possess the allele for polydactyly (as evidenced by the fact that their children inherit the polydactyly) but nevertheless have a normal number of fingers and toes. In these cases,

5.4 Human polydactyly (extra digits) exhibits incomplete penetrance and variable expressivity. [SPL/Photo Researchers.]

the gene for polydactyly is not fully penetrant. **Penetrance** is defined as the percentage of individual organisms having a particular genotype that express the expected phenotype. For example, if we examined 42 people having an allele for polydactyly and found that only 38 of them were polydactylous, the penetrance would be $38/42 = 0.90$ (90%).

A related concept is that of **expressivity**, the degree to which a character is expressed. In addition to incomplete penetrance, polydactyly exhibits variable expressivity. Some polydactylous persons possess extra fingers and toes that are fully functional, whereas others possess only a small tag of extra skin.

Incomplete penetrance and variable expressivity are due to the effects of other genes and to environmental factors that can alter or completely suppress the effect of a particular gene. For example, a gene may encode an enzyme that produces a particular phenotype only within a limited temperature range. At higher or lower temperatures, the enzyme does not function and the phenotype is not expressed; the allele encoding such an enzyme is therefore penetrant only within a particular temperature range. Many characters exhibit incomplete penetrance and variable expressivity; thus the mere presence of a gene does not guarantee its expression. **TRY PROBLEM 17 →**

> ### CONCEPTS
>
> Penetrance is the percentage of individuals having a particular genotype that express the associated phenotype. Expressivity is the degree to which a trait is expressed. Incomplete penetrance and variable expressivity result from the influence of other genes and environmental factors on the phenotype.
>
> ✔ **CONCEPT CHECK 3**
>
> How does incomplete dominance differ from incomplete penetrance?
>
> a. Incomplete dominance refers to alleles at the same locus; incomplete penetrance refers to alleles at different loci.
> b. Incomplete dominance ranges from 0% to 50%; incomplete penetrance ranges from 51% to 99%.
> c. In incomplete dominance, the heterozygote is intermediate to the homozygotes; in incomplete penetrance, heterozygotes express phenotypes of both homozygotes.
> d. In incomplete dominance, the heterozygote is intermediate to the homozygotes; in incomplete penetrance, some individuals do not express the expected phenotype.

Lethal Alleles

As described in the introduction to this chapter, Lucien Cuénot reported the first case of a lethal allele, the allele for yellow coat color in mice (see Figure 5.1). A **lethal allele** causes death at an early stage of development—often before birth—and so some genotypes may not appear among the progeny.

Another example of a lethal allele, originally described by Erwin Baur in 1907, is found in snapdragons. The *aurea* strain in these plants has yellow leaves. When two plants with yellow leaves are crossed, $2/3$ of the progeny have yellow leaves and $1/3$ have green leaves. When green is crossed with green, all the progeny have green leaves; however, when yellow is crossed with green, $1/2$ of the progeny have green leaves and $1/2$ have yellow leaves, confirming that all yellow-leaved snapdragons are heterozygous. A 2 : 1 ratio is almost always produced by a recessive lethal allele; so observing this ratio among the progeny of a cross between individuals with the same phenotype is a strong clue that one of the alleles is lethal.

In this example, like that of yellow coat color in mice, the lethal allele is recessive because it causes death only in homozygotes. Unlike its effect on *survival*, the effect of the allele on *color* is dominant; in both mice and snapdragons, a single copy of the allele in the heterozygote produces a yellow color. This example illustrates the point made earlier that the type of dominance depends on the aspect of the phenotype examined.

Many lethal alleles in nature are recessive, but lethal alleles can also be dominant; in this case, homozygotes and heterozygotes for the allele die. Truly dominant lethal alleles cannot be transmitted unless they are expressed after the onset of reproduction, as in Huntington disease. TRY PROBLEMS 18 AND 40

CONCEPTS

A lethal allele causes death, frequently at an early developmental stage, and so one or more genotypes are missing from the progeny of a cross. Lethal alleles modify the ratio of progeny resulting from a cross.

✔ CONCEPT CHECK 4

A cross between two green corn plants yields $2/3$ progeny that are green and $1/3$ progeny that are white. What is the genotype of the green progeny?

a. *WW* c. *ww*
b. *Ww* d. *W_* (*WW* and *Ww*)

Multiple Alleles

Most of the genetic systems that we have examined so far consist of two alleles. In Mendel's peas, for instance, one allele encoded round seeds and another encoded wrinkled seeds; in cats, one allele produced a black coat and another produced a gray coat. For some loci, more than two alleles are present within a group of organisms—the locus has **multiple alleles**. (Multiple alleles may also be referred to as an *allelic series*.) Although there may be more than two alleles present within a *group* of organisms, the genotype of each individual diploid organism still consists of only two alleles. The inheritance of characteristics encoded by multiple alleles is no different from the inheritance of characteristics encoded by two alleles, except that a greater variety of genotypes and phenotypes are possible.

Duck-feather patterns An example of multiple alleles is at a locus that determines the feather pattern of mallard ducks. One allele, M, produces the wild-type *mallard* pattern. A second allele, M^R, produces a different pattern called *restricted*, and a third allele, m^d, produces a pattern termed *dusky*. In this allelic series, restricted is dominant over mallard and dusky, and mallard is dominant over dusky: $M^R > M > m^d$. The six genotypes possible with these three alleles and their resulting phenotypes are:

Genotype	Phenotype
$M^R M^R$	restricted
$M^R M$	restricted
$M^R m^d$	restricted
MM	mallard
Mm^d	mallard
$m^d m^d$	dusky

In general, the number of genotypes possible will be $[n(n + 1)]/2$, where n equals the number of different alleles at a locus. Working crosses with multiple alleles is no different from working crosses with two alleles; Mendel's principle of segregation still holds, as shown in the cross between a restricted duck and a mallard duck (**Figure 5.5**). TRY PROBLEM 19

5.5 Mendel's principle of segregation applies to crosses with multiple alleles. In this example, three alleles determine the type of plumage in mallard ducks: M^R (restricted) $> M$ (mallard) $> m^d$ (dusky).

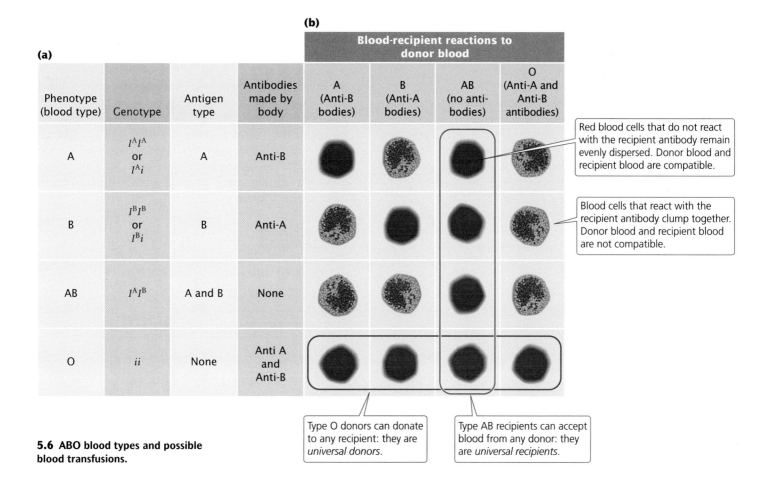

5.6 ABO blood types and possible blood transfusions.

The ABO blood group Another multiple-allele system is at the locus for the ABO blood group. This locus determines your ABO blood type and, like the MN locus, encodes antigens on red blood cells. The three common alleles for the ABO blood group locus are: I^A, which encodes the A antigen; I^B, which encodes the B antigen; and i, which encodes no antigen (O). We can represent the dominance relations among the ABO alleles as follows: $I^A > i$, $I^B > i$, $I^A = I^B$. Both the I^A and the I^B alleles are dominant over i and are codominant with each other; the AB phenotype is due to the presence of an I^A allele and an I^B allele, which results in the production of A and B antigens on red blood cells. A person with genotype ii produces neither antigen and has blood type O. The six common genotypes at this locus and their phenotypes are shown in **Figure 5.6a**.

Antibodies are produced against any foreign antigens (see Figure 5.6a). For instance, a person having blood-type A produces anti-B antibodies, because the B antigen is foreign. A person having blood-type B produces anti-A antibodies, and someone having blood-type AB produces neither anti-A nor anti-B antibodies, because neither A nor B antigen is foreign. A person having blood-type O possesses no A or B antigens; consequently, that person produces both anti-A antibodies and anti-B antibodies. The presence of antibodies against foreign ABO antigens means that successful blood transfusions are possible only between persons with certain compatible blood types (**Figure 5.6b**).

The inheritance of alleles at the ABO locus is illustrated by a paternity suit against the famous movie actor Charlie Chaplin. In 1941, Chaplin met a young actress named Joan Barry, with whom he had an affair. The affair ended in February 1942 but, 20 months later, Barry gave birth to a baby girl and claimed that Chaplin was the father. Barry then sued for child support. At this time, blood typing had just come into widespread use, and Chaplin's attorneys had Chaplin, Barry, and the child blood typed. Barry had blood-type A, her child had blood-type B, and Chaplin had blood-type O. Could Chaplin have been the father of Barry's child?

Your answer should be no. Joan Barry had blood-type A, which can be produced by either genotype I^AI^A or genotype I^Ai. Her baby possessed blood-type B, which can be produced by either genotype I^BI^B or genotype I^Bi. The baby could not have inherited the I^B allele from Barry (Barry could not carry an I^B allele if she were blood-type A); therefore the baby must have inherited the i allele from her.

Barry must have had genotype $I^A i$, and the baby must have had genotype $I^B i$. Because the baby girl inherited her i allele from Barry, she must have inherited the I^B allele from her father. Having blood-type O, produced only by genotype ii, Chaplin could not have been the father of Barry's child. Although blood types can be used to exclude the possibility of paternity (as in this case), they cannot prove that a person is the parent of a child, because many different people have the same blood type.

In the course of the trial to settle the paternity suit against Chaplin, three pathologists came to the witness stand and declared that it was genetically impossible for Chaplin to have fathered the child. Nevertheless, the jury ruled that Chaplin was the father and ordered him to pay child support and Barry's legal expenses. TRY PROBLEM 24 →

CONCEPTS

More than two alleles (multiple alleles) may be present within a group of individual organisms, although each individual diploid organism still has only two alleles at that locus.

✔ CONCEPT CHECK 5

How many genotypes are present at a locus with five alleles?

a. 30   c. 15
b. 27   d. 5

5.2 Gene Interaction Takes Place When Genes at Multiple Loci Determine a Single Phenotype

In the dihybrid crosses that we examined in Chapter 3, each locus had an independent effect on the phenotype. When Mendel crossed a homozygous round and yellow plant (*RR YY*) with a homozygous wrinkled and green plant (*rr yy*) and then self-fertilized the F_1, he obtained F_2 progeny in the following proportions:

9/16	R_ Y_	round, yellow
3/16	R_ yy	round, green
3/16	rr Y_	wrinkled, yellow
1/16	rr yy	wrinkled, green

In this example, the genes showed two kinds of independence. First, the genes at each locus are independent in their *assortment* in meiosis, which is what produces the 9 : 3 : 3 : 1 ratio of phenotypes in the progeny, in accord with Mendel's principle of independent assortment. Second, the genes are independent in their *phenotypic expression*, the *R* and *r* alleles affect only the shape of the seed and have no influence on the color of the seed; the *Y* and *y* alleles affect only color and have no influence on the shape of the seed.

Frequently, genes exhibit independent assortment but do not act independently in their phenotypic expression; instead, the effects of genes at one locus depend on the presence of genes at other loci. This type of interaction between the effects of genes at different loci (genes that are not allelic) is termed **gene interaction**. With gene interaction, the products of genes at different loci combine to produce new phenotypes that are not predictable from the single-locus effects alone. In our consideration of gene interaction, we will focus primarily on interaction between the effects of genes at two loci, although interactions among genes at three, four, or more loci are common.

CONCEPTS

In gene interaction, genes at different loci contribute to the determination of a single phenotypic characteristic.

✔ CONCEPT CHECK 6

How does gene interaction differ from dominance between alleles?

Gene Interaction That Produces Novel Phenotypes

Let's first examine gene interaction in which genes at two loci interact to produce a single characteristic. Fruit color in the pepper *Capsicum annuum* is determined in this way. Certain types of peppers produce fruits in one of four colors: red, peach, orange (sometimes called yellow), and cream (or white). If a homozygous plant with red peppers is crossed with a homozygous plant with cream peppers, all the F_1 plants have red peppers (**Figure 5.7a**). When the F_1 are crossed with one another, the F_2 are in a ratio of 9 red : 3 peach : 3 orange : 1 cream (**Figure 5.7b**). This dihybrid ratio (see Chapter 3) is produced by a cross between two plants that are both heterozygous for two loci ($Y^+y\, C^+c \times Y^+y\, C^+c$). In this example, the *Y* locus and the *C* locus interact to produce a single phenotype-the color of the pepper:

Genotype	Phenotype
$Y^+\_\ C^+\_$	red
$Y^+\_\ cc$	peach
$yy\ C^+\_$	orange
$yy\ cc$	cream

Color in peppers of *Capsicum annuum* results from the relative amounts of red and yellow carotenoids, compounds that are synthesized in a complex biochemical pathway. The *Y* locus encodes one enzyme (the first step in the pathway), and the *C* locus encodes a different enzyme (the last step in the pathway). When different loci influence different steps in a common biochemical pathway, gene interaction often arises because the product of one enzyme affects the substrate of another enzyme.

5.7 Gene interaction in which two loci determine a single characteristic, fruit color, in the pepper *Capsicum annuum*.

To illustrate how Mendel's rules of heredity can be used to understand the inheritance of characteristics determined by gene interaction, let's consider a testcross between an F_1 plant from the cross in Figure 5.7 ($Y^+y\ C^+c$) and a plant with cream peppers ($yy\ cc$). As outlined in Chapter 3 for independent loci, we can work this cross by breaking it down into two simple crosses. At the first locus, the heterozygote Y^+y is crossed with the homozygote yy; this cross produces $1/2\ Y^+y$ and $1/2\ yy$ progeny. Similarly, at the second locus, the heterozygous genotype C^+c is crossed with the homozygous genotype cc, producing $1/2\ C^+c$ and $1/2\ cc$ progeny. In accord with Mendel's principle of independent assortment, these single-locus ratios can be combined by using the multiplication rule: the probability of obtaining the genotype $Y^+y\ C^+c$ is the probability of Y^+y ($1/2$) multiplied by the probability of C^+c ($1/2$), or $1/4$. The probability of each progeny genotype resulting from the testcross is:

Progeny genotype	Probability at each locus		Overall probability	Phenotype
$Y^+y\ C^+c$	$1/2 \times 1/2$	=	$1/4$	red peppers
$Y^+y\ cc$	$1/2 \times 1/2$	=	$1/4$	peach peppers
$yy\ C^+c$	$1/2 \times 1/2$	=	$1/4$	orange peppers
$yy\ cc$	$1/2 \times 1/2$	=	$1/4$	cream peppers

When you work problems with gene interaction, it is especially important to determine the probabilities of single-locus genotypes and to multiply the probabilities of *genotypes*, not phenotypes, because the phenotypes cannot be determined without considering the effects of the genotypes at all the contributing loci. **TRY PROBLEM 25** →

Gene Interaction with Epistasis

Sometimes the effect of gene interaction is that one gene masks (hides) the effect of another gene at a different locus, a phenomenon known as **epistasis**. In the examples of genic interaction that we have examined, alleles at different loci interact to determine a single phenotype. In those examples, one allele did not *mask* the effect of an allele at another locus, meaning that there was no epistasis. Epistasis is similar to dominance, except that dominance entails the masking of genes at the *same* locus (allelic genes). In epistasis, the gene that does the masking is called an **epistatic gene**; the gene whose effect is masked is a **hypostatic gene**. Epistatic genes may be recessive or dominant in their effects.

Recessive epistasis Recessive epistasis is seen in the genes that determine coat color in Labrador retrievers. These dogs may be black, brown, or yellow; their different coat colors are determined by interactions between genes at two loci (although a number of other loci also help to determine coat color; see pp. 113–115). One locus determines the type of pigment produced by the skin cells: a dominant allele *B* encodes black pigment, whereas a recessive allele *b* encodes brown pigment. Alleles at a second locus affect the *deposition* of the pigment in the shaft of the hair; dominant allele *E* allows dark pigment (black or brown) to be deposited, whereas recessive allele *e* prevents the deposition of dark pigment, causing the hair to be yellow. The presence of genotype *ee* at the second locus therefore masks the expression of the black and brown alleles at the first locus. The genotypes that determine coat color and their phenotypes are:

Genotype	Phenotype
B_ E_	black
bb E_	brown (frequently called chocolate)
B_ ee	yellow
bb ee	yellow

If we cross a black Labrador homozygous for the dominant alleles (*BB EE*) with a yellow Labrador homozygous for the recessive alleles (*bb ee*) and then intercross the F_1, we obtain progeny in the F_2 in a 9 : 3 : 4 ratio:

P   BB EE   ×   bb ee
      Black   Yellow

F_1 Bb Ee
          Black

          Intercross

F_2 9/16 B_ E_ black
        3/16 bb E_ brown
        3/16 B_ ee yellow } 4/16 yellow
        1/16 bb ee yellow

Notice that yellow dogs can carry alleles for either black or brown pigment, but these alleles are not expressed in their coat color.

In this example of gene interaction, allele *e* is epistatic to *B* and *b*, because *e* masks the expression of the alleles for black and brown pigments, and alleles *B* and *b* are hypostatic to *e*. In this case, *e* is a recessive epistatic allele, because two copies of *e* must be present to mask the expression of the black and brown pigments. **TRY PROBLEM 29**

Another example of a recessive epistatic gene is the Bombay phenotype, which suppresses the expression of alleles at the ABO locus. As mentioned earlier in the chapter, the alleles at the ABO locus encode antigens on the red blood cells; the antigens consist of short chains of carbohydrates embedded in the membranes of red blood cells. The difference between the A and the B antigens is a function of chemical differences in the terminal sugar of the chain. The I^A and I^B alleles actually encode different enzymes, which add sugars designated A or B to the ends of the carbohydrate chains (**Figure 5.8**). The common substrate on which these enzymes act is a molecule called H. The enzyme encoded by the *i* allele apparently either adds no sugar to H or no functional enzyme is specified.

In most people, a dominant allele (*H*) at the H locus encodes an enzyme that makes H, but people with the Bombay phenotype are homozygous for a recessive mutation (*h*) that encodes a defective enzyme. The defective enzyme is incapable of making H and, because H is not produced, no ABO antigens are synthesized. Thus, the expression of the alleles at the ABO locus depends on the genotype at the H locus.

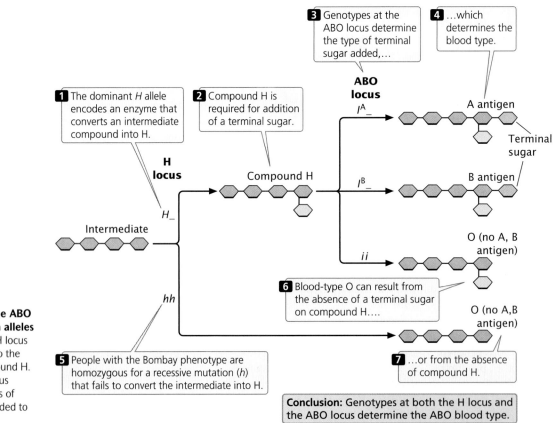

5.8 Expression of the ABO antigens depends on alleles at the H locus. The H locus encodes a precursor to the antigens called compound H. Alleles at the ABO locus determine which types of terminal sugars are added to compound H.

Genotype	H present	ABO phenotype
$H\_\ I^AI^A$, $H\_\ I^Ai$	Yes	A
$H\_\ I^BI^B$, $H\_\ I^Bi$	Yes	B
$H\_\ I^AI^B$	Yes	AB
$H\_\ ii$	Yes	O
$hh\ I^AI^A$, $hh\ I^Ai$, $hh\ I^BI^B$, $hh\ I^Bi$, $hh\ I^AI^B$, and $hh\ ii$	No	O

In this example, the alleles at the ABO locus are hypostatic to the recessive h allele.

The Bombay phenotype provides us with a good opportunity for considering how epistasis often arises when genes affect a series of steps in a biochemical pathway. The ABO antigens are produced in a multistep biochemical pathway (see Figure 5.8), which depends on enzymes that make H and on other enzymes that convert H into the A or B antigen. Note that blood-type O may arise in one of two ways: (1) from failure to add a terminal sugar to compound H (genotype $H\_\ ii$) or (2) from failure to produce compound H (genotype $hh\_$). Many cases of epistasis arise in this way. A gene (such as h) that has an effect on an early step in a biochemical pathway will be epistatic to genes (such as I^A and I^B) that affect subsequent steps, because the effects of the genes in a later step depend on the product of the earlier reaction.

Dominant epistasis In *recessive* epistasis, which we just considered, the presence of two recessive alleles (the homozygous genotype) inhibits the expression of an allele at a different locus. In *dominant* epistasis, only a single copy of an allele is required to inhibit the expression of the allele at a different locus.

Dominant epistasis is seen in the interaction of two loci that determine fruit color in summer squash, which is commonly found in one of three colors: yellow, white, or green. When a homozygous plant that produces white squash is crossed with a homozygous plant that produces green squash and the F_1 plants are crossed with each other, the following results are obtained:

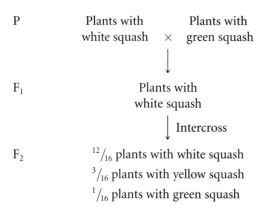

How can gene interaction explain these results?

In the F_2, $^{12}/_{16}$, or $^3/_4$, of the plants produce white squash and $^3/_{16} + ^1/_{16} = ^4/_{16} = ^1/_4$ of the plants produce squash having color. This outcome is the familiar 3 : 1 ratio produced by a cross between two heterozygotes, which suggests that a dominant allele at one locus inhibits the production of pigment, resulting in white progeny. If we use the symbol W to represent the dominant allele that inhibits pigment production, then genotype $W\_$ inhibits pigment production and produces white squash, whereas ww allows pigment and results in colored squash.

Among those ww F_2 plants with pigmented fruit, we observe $^3/_{16}$ yellow and $^1/_{16}$ green (a 3 : 1 ratio). In this outcome, a second locus determines the type of pigment produced in the squash, with yellow ($Y\_$) dominant over green (yy). This locus is expressed only in ww plants, which lack the dominant inhibitory allele W. We can assign the genotype $ww\ Y\_$ to plants that produce yellow squash and the genotype $ww\ yy$ to plants that produce green squash. The genotypes and their associated phenotypes are:

$W\_\ Y\_$	white squash
$W\_\ yy$	white squash
$ww\ Y\_$	yellow squash
$ww\ yy$	green squash

Allele W is epistatic to Y and y: it suppresses the expression of these pigment-producing genes. Allele W is a dominant epistatic allele because, in contrast with e in Labrador retriever coat color and with h in the Bombay phenotype, a single copy of the allele is sufficient to inhibit pigment production.

Yellow pigment in the squash is most likely produced in a two-step biochemical pathway (**Figure 5.9**). A colorless (white) compound (designated A in Figure 5.9) is converted by enzyme I into green compound B, which is then converted into compound C by enzyme II. Compound C is the yellow pigment in the fruit. Plants with the genotype ww produce enzyme I and may be green or yellow, depending on whether enzyme II is present. When allele Y is present at a second locus, enzyme II is produced and compound B is converted into compound C, producing a yellow fruit. When two copies of allele y, which does not encode a functional form of enzyme II, are present, squash remain green. The presence of W at the first locus inhibits the conversion of compound A into compound B; plants with genotype $W\_$ do not make compound B and their fruit remains white, regardless of which alleles are present at the second locus.

Duplicate recessive epistasis Finally, let's consider duplicate recessive epistasis, in which two recessive alleles at either of two loci are capable of suppressing a phenotype. This type of epistasis is illustrated by albinism in snails.

Albinism is the absence of pigment and is a common genetic trait in many plants and animals. Pigment is almost

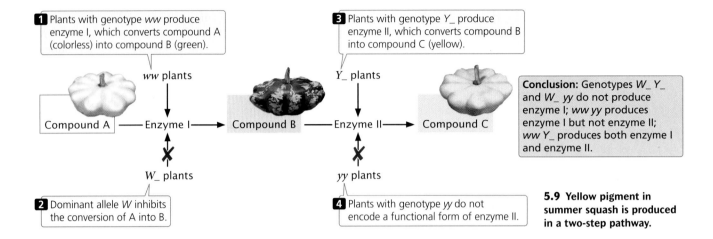

5.9 Yellow pigment in summer squash is produced in a two-step pathway.

always produced through a multistep biochemical pathway; thus, albinism may entail gene interaction. Robert T. Dillon and Amy R. Wethington found that albinism in the common freshwater snail *Physa heterostroha* can result from the presence of either of two recessive alleles at two different loci. Inseminated snails were collected from a natural population and placed in cups of water, where they laid eggs. Some of the eggs hatched into albino snails. When two albino snails were crossed, all of the F_1 were pigmented. When the F_1 were intercrossed, the F_2 consisted of $9/16$ pigmented snails and $7/16$ albino snails. How did this 9 : 7 ratio arise?

The 9 : 7 ratio seen in the F_2 snails can be understood as a modification of the 9 : 3 : 3 : 1 ratio obtained when two individuals heterozygous for two loci are crossed. The 9 : 7 ratio arises when dominant alleles at both loci ($A\_B\_$) produce pigmented snails; any other genotype produces albino snails:

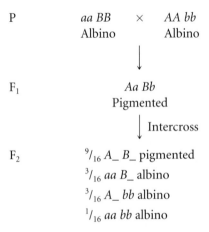

The 9 : 7 ratio in these snails is probably produced by a two-step pathway of pigment production (**Figure 5.10**). Pigment (compound C) is produced only after compound A has been converted into compound B by enzyme I and after compound B has been converted into compound C by enzyme II. At least one dominant allele A at the first locus is required to produce enzyme I; similarly, at least one dominant allele B at the second locus is required to produce enzyme II. Albinism arises from the absence of compound C, which may happen in one of three ways. First, two recessive alleles at the first locus (genotype $aa\ B\_$) may prevent the production of enzyme I, and so compound B is never produced. Second, two recessive alleles at the second locus (genotype A_bb) may prevent the production of enzyme II; in this case, compound B is never converted into compound C. Third, two recessive alleles may be present at both loci ($aa\ bb$), causing the absence of both enzyme I and enzyme II. In this example of gene interaction, a is epistatic to B, and b is epistatic to A; *both* are recessive epistatic alleles because the presence of two copies of either allele a or allele b is necessary to suppress pigment production. This example differs from the suppression of coat color in Labrador retrievers in that recessive alleles at either of two loci are capable of suppressing pigment production in the snails, whereas recessive alleles at a single locus suppress pigment expression in Labs.

CONCEPTS

Epistasis is the masking of the expression of one gene by another gene at a different locus. The epistatic gene does the masking; the hypostatic gene is masked. Epistatic genes can be dominant or recessive.

✔ CONCEPT CHECK 7

A number of all-white cats are crossed and they produce the following types of progeny: $12/16$ all-white, $3/16$ black, and $1/16$ gray. What is the genotype of the black progeny?

a. Aa
b. $Aa\ Bb$
c. $A\_B\_$
d. A_bb

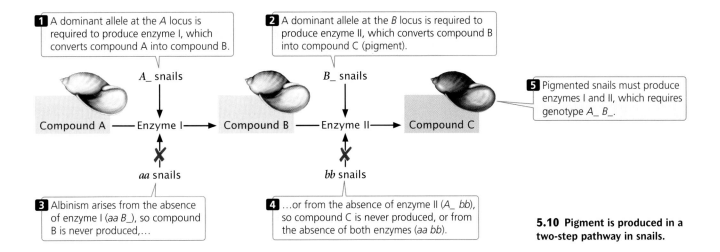

5.10 Pigment is produced in a two-step pathway in snails.

CONNECTING CONCEPTS

Interpreting Ratios Produced by Gene Interaction

A number of modified ratios that result from gene interaction are shown in **Table 5.2**. Each of these examples represents a modification of the basic 9 : 3 : 3 : 1 dihybrid ratio. In interpreting the genetic basis of modified ratios, we should keep several points in mind. First, the inheritance of the genes producing these characteristics is no different from the inheritance of genes encoding simple genetic characters. Mendel's principles of segregation and independent assortment still apply; each individual organism possesses two alleles at each locus, which separate in meiosis, and genes at the different loci assort independently. The only difference is in how the *products* of the genotypes interact to produce the phenotype. Thus, we cannot consider the expression of genes at each locus separately; instead, we must take into consideration how the genes at different loci interact.

A second point is that, in the examples that we have considered, the phenotypic proportions were always in sixteenths because, in all the crosses, pairs of alleles segregated at two independently assorting loci. The probability of inheriting one of the two alleles at a locus is $1/2$. Because there are two loci, each with two alleles, the probability of inheriting any particular combination of genes is $(1/2)^4 = 1/16$. For a trihybrid cross, the progeny proportions should be in sixty-fourths, because $(1/2)^6 = 1/64$. In general, the progeny proportions should be in fractions of $(1/2)^{2n}$, where n equals the number of loci with two alleles segregating in the cross.

Crosses rarely produce exactly 16 progeny; therefore, modifications of a dihybrid ratio are not always obvious. Modified dihybrid ratios are more easily seen if the number of individuals of each phenotype is expressed in sixteenths:

$$\frac{x}{16} = \frac{\text{number of progeny with a phenotype}}{\text{total number of progeny}}$$

Table 5.2 Modified dihybrid phenotypic ratios due to gene interaction

Ratio*	Genotype				Type of Interaction	Example
	A_ B_	A_ bb	aa B_	aa bb		
9 : 3 : 3 : 1	9	3	3	1	None	Seed shape and seed color in peas
9 : 3 : 4	9	3	4		Recessive epistasis	Coat color in Labrador retrievers
12 : 3 : 1	12		3	1	Dominant epistasis	Color in squash
9 : 7	9	7			Duplicate recessive epistasis	Albinism in snails
9 : 6 : 1	9	6		1	Duplicate interaction	—
15 : 1	15			1	Duplicate dominant epistasis	—
13 : 3	13		3		Dominant and recessive epistasis	—

*Each ratio is produced by a dihybrid cross (Aa Bb × Aa Bb). Shaded bars represent combinations of genotypes that give the same phenotype.

where $x/16$ equals the proportion of progeny with a particular phenotype. If we solve for x (the proportion of the particular phenotype in sixteenths), we have:

$$x = \frac{\text{number of progeny with a phenotype} \times 16}{\text{total number of progeny}}$$

For example, suppose that we cross two homozygotes, interbreed the F_1, and obtain 63 red, 21 brown, and 28 white F_2 individuals. Using the preceding formula, we find the phenotypic ratio in the F_2 to be: red = $(63 \times 16)/112 = 9$; brown = $(21 \times 16)/112 = 3$; and white = $(28 \times 16)/112 = 4$. The phenotypic ratio is 9 : 3 : 4.

A final point to consider is how to assign genotypes to the phenotypes in modified ratios that result from gene interaction. Don't try to *memorize* the genotypes associated with all the modified ratios in Table 5.2. Instead, practice relating modified ratios to known ratios, such as the 9 : 3 : 3 : 1 dihybrid ratio. Suppose that we obtain $^{15}/_{16}$ green progeny and $^{1}/_{16}$ white progeny in a cross between two plants. If we compare this 15 : 1 ratio with the standard 9 : 3 : 3 : 1 dihybrid ratio, we see that $^{9}/_{16} + ^{3}/_{16} + ^{3}/_{16}$ equals $^{15}/_{16}$. All the genotypes associated with these proportions in the dihybrid cross (A_ B_, A_ bb, and aa B_) must give the same phenotype, the green progeny. Genotype aa BB makes up $^{1}/_{16}$ of the progeny in a dihybrid cross, the white progeny in this cross.

In assigning genotypes to phenotypes in modified ratios, students sometimes become confused about which letters to assign to which phenotype. Suppose that we obtain the following phenotypic ratio: $^{9}/_{16}$ black : $^{3}/_{16}$ brown : $^{4}/_{16}$ white. Which genotype do we assign to the brown progeny, A_ bb or aa B_? Either answer is correct because the letters are just arbitrary symbols for the genetic information. The important thing to realize about this ratio is that the brown phenotype arises when two recessive alleles are present at one locus.

Worked Problem

A homozygous strain of yellow corn is crossed with a homozygous strain of purple corn. The F_1 are intercrossed, producing an ear of corn with 119 purple kernels and 89 yellow kernels (the progeny). What is the genotype of the yellow kernels?

• **Solution**

We should first consider whether the cross between yellow and purple strains might be a monohybrid cross for a simple dominant trait, which would produce a 3 : 1 ratio in the F_2 ($Aa \times Aa \rightarrow ^{3}/_{4} A\_$ and $^{1}/_{4}$ aa). Under this hypothesis, we would expect 156 purple progeny and 52 yellow progeny:

Phenotype	Genotype	Observed number	Expected number
purple	A_	119	$^{3}/_{4} \times 208 = 156$
yellow	aa	89	$^{1}/_{4} \times 208 = 52$
Total		208	

We see that the expected numbers do not closely fit the observed numbers. If we performed a chi-square test (see Chapter 3), we would obtain a calculated chi-square value of 35.08, which has a probability much less than 0.05, indicating that it is extremely unlikely that, when we expect a 3 : 1 ratio, we would obtain 119 purple progeny and 89 yellow progeny. Therefore, we can reject the hypothesis that these results were produced by a monohybrid cross.

Another possible hypothesis is that the observed F_2 progeny are in a 1 : 1 ratio. However, we learned in Chapter 3 that a 1 : 1 ratio is produced by a cross between a heterozygote and a homozygote ($Aa \times aa$) and, from the information given, the cross was not between a heterozygote and a homozygote, because both original parental strains were homozygous. Furthermore, a chi-square test comparing the observed numbers with an expected 1 : 1 ratio yields a calculated chi-square value of 4.32, which has a probability of less than 0.05.

Next, we should look to see if the results can be explained by a dihybrid cross ($Aa\ Bb \times Aa\ Bb$). A dihybrid cross results in phenotypic proportions that are in sixteenths. We can apply the formula given earlier in the chapter to determine the number of sixteenths for each phenotype:

$$x = \frac{\text{number of progeny with a phenotype} \times 16}{\text{total number of progeny}}$$

$$x_{(\text{purple})} = \frac{119 \times 16}{208} = 9.15$$

$$x_{(\text{yellow})} = \frac{89 \times 16}{208} = 6.85$$

Thus, purple and yellow appear in an approximate ratio of 9 : 7. We can test this hypothesis with a chi-square test:

Phenotype	Genotype	Observed number	Expected number
purple	?	119	$^{9}/_{16} \times 208 = 117$
yellow	?	89	$^{7}/_{16} \times 208 = 91$
Total		208	

$$x^2 = \sum \frac{(\text{observed} - \text{expected})^2}{\text{expected}}$$

$$= \frac{(119-117)^2}{117} + \frac{(89-91)^2}{91}$$

$$= 0.034 + 0.44 = 0.078$$

$$\text{Degree of freedom} = n - 1 = 2 - 1 = 1$$

$$P > 0.05$$

The probability associated with the chi-square value is greater than 0.05, indicating that there is a good fit between the observed results and a 9 : 7 ratio.

We now need to determine how a dihybrid cross can produce a 9 : 7 ratio and what genotypes correspond to the two phenotypes. A dihybrid cross without epistasis produces a 9 : 3 : 3 : 1 ratio:

$$Aa\ Bb \times Aa\ Bb$$
$$\downarrow$$
$$A\_\ B\_\ 9/16$$
$$A\_\ bb\ 3/16$$
$$aa\ B\_\ 3/16$$
$$aa\ bb\ 1/16$$

Because $9/16$ of the progeny from the corn cross are purple, purple must be produced by genotypes $A\_\ B\_$; in other words, individual kernels that have at least one dominant allele at the first locus and at least one dominant allele at the second locus are purple. The proportions of all the other genotypes ($A\_\ bb$, $aa\ B\_$, and $aa\ bb$) sum to $7/16$, which is the proportion of the progeny in the corn cross that are yellow, and so any individual kernel that does not have a dominant allele at both the first and the second locus is yellow.

→ Now test your understanding of epistasis by working Problem 26 at the end of the chapter.

Complementation: Determining Whether Mutations Are at the Same Locus or at Different Loci

How do we know whether different mutations that affect a characteristic occur at the same locus (are allelic) or at different loci? In fruit flies, for example, *white* is an X-linked recessive mutation that produces white eyes instead of the red eyes found in wild-type flies; *apricot* is an X-linked recessive mutation that produces light-orange-colored eyes. Do the *white* and *apricot* mutations occur at the same locus or at different loci? We can use the complementation test to answer this question.

To carry out a **complementation test** on recessive mutations, parents that are homozygous for different mutations are crossed, producing offspring that are heterozygous. If the mutations are allelic (occur at the same locus), then the heterozygous offspring have only mutant alleles (*a b*) and exhibit a mutant phenotype:

If, on the other hand, the mutations occur at different loci, each of the homozygous parents possesses wild-type genes at the other locus ($aa\ b^+b^+$ and $a^+a^+\ bb$); so the heterozygous offspring inherit a mutant allele and a wild-type allele at each locus. In this case, the mutations complement each other and the heterozygous offspring have the wild-type phenotype:

Complementation has occurred if an individual organism possessing two recessive mutations has a wild-type phenotype, indicating that the mutations are nonallelic genes. A lack of complementation occurs when two recessive mutations occur at the same locus, producing a mutant phenotype.

When the complementation test is applied to *white* and *apricot* mutations, all of the heterozygous offspring have light-colored eyes, demonstrating that white eyes and apricot eyes are produced by mutations that occur at the same locus and are allelic.

CONCEPTS

A complementation test is used to determine whether two mutations occur at the same locus (are allelic) or occur at different loci.

✔ CONCEPT CHECK 8

Brindle (tiger-striped appearance) is a recessive trait in bulldogs and in Chihuahuas. What types of crosses would you carry out to determine whether the *brindle* genes in bulldogs and in Chihuahuas are at the same locus?

The Complex Genetics of Coat Color in Dogs

The genetics of coat color in dogs is an excellent example of how complex interactions between genes may take part in the determination of a phenotype. Domestic dogs come in an amazing variety of shapes, sizes, and colors. For thousands of years, people have been breeding dogs for particular traits, producing the large number of types that we see today. Each breed of dog carries a selection of genes from the ancestral dog gene pool; these genes define the features of a particular breed. The genome of the domestic dog was completely sequenced in 2004, greatly facilitating the study of canine genetics.

An obvious difference between dogs is coat color. The genetics of coat color in dogs is quite complex; many genes participate, and there are numerous interactions between genes at different loci. We will consider four loci (in the list that follows) that are important in producing many of the noticeable differences in color and pattern among breeds of dogs. In interpreting the genetic basis of differences in the coat color of dogs, consider how the expression of a particular gene is modified by the effects of other genes. Keep in mind that additional loci not listed here can modify the colors produced by these four loci and that not all geneticists agree on the genetics of color variation in some breeds.

1. **Agouti (A) locus.** This locus has five common alleles that determine the depth and distribution of color in a dog's coat:

   A^s Solid black pigment.
   a^w Agouti, or wolflike gray. Hairs encoded by this allele have a "salt and pepper" appearance, produced by a band of yellow pigment on a black hair.
   a^y Yellow. The black pigment is markedly reduced; so the entire hair is yellow.
   a^s Saddle markings (dark color on the back, with extensive tan markings on the head and legs).
   a^t Bicolor (dark color over most of the body, with tan markings on the feet and eyebrows).

   Alleles A^s and a^y are generally dominant over the other alleles, but the dominance relations are complex and not yet completely understood.

2. **Black (B) locus.** This locus determines whether black pigment can be formed. The actual color of a dog's coat depends on the effects of genes at other loci (such as the A and E loci). Two alleles are common:

   B Allows black pigment to be produced.
   b Black pigment cannot be produced; pigmented dogs can be chocolate, liver, tan, or red.

   Allele B is dominant over allele b.

3. **Extension (E) locus.** Four alleles at this locus determine where the genotype at the A locus is expressed. For example, if a dog has the A^s allele (solid black) at the A locus, then black pigment will either be extended throughout the coat or be restricted to some areas, depending on the alleles present at the E locus. Areas where the A locus is not expressed may appear as yellow, red, or tan, depending on the presence of particular genes at other loci. When A^s is present at the A locus, the four alleles at the E locus have the following effects:

   E^m Black mask with a tan coat.
   E The A locus expressed throughout (solid black).
   e^{br} Brindle, in which black and yellow are in layers to give a tiger-striped appearance.
   e No black in the coat, but the nose and eyes may be black.

   The dominance relations among these alleles are poorly known.

4. **Spotting (S) locus.** Alleles at this locus determine whether white spots will be present. There are four common alleles:

   S No spots.
   s^i Irish spotting; numerous white spots.
   s^p Piebald spotting; various amounts of white.
   s^w Extreme white piebald; almost all white.

   Allele S is completely dominant over alleles s^i, s^p, and s^w; alleles s^i and s^p are dominant over allele s^w ($S > s^i$, $s^p > s^w$). The relation between s^i and s^p is poorly defined; indeed, they may not be separate alleles. Genes at other poorly known loci also modify spotting patterns.

To illustrate how genes at these loci interact in determining a dog's coat color, let's consider a few examples:

Labrador retriever Labrador retrievers (**Figure 5.11a**) may be black, brown, or yellow. Most are homozygous A^sA^s SS; thus, they vary only at the B and E loci. The A^s allele allows dark pigment to be expressed; whether a dog is black depends on which genes are present at the B and E loci. As discussed earlier in the chapter, all black Labradors must carry at least one B allele and one E allele ($B\_ E\_$). Brown dogs are homozygous bb and have at least one E allele ($bb E\_$). Yellow dogs

(a) (b) (c)

5.11 Coat color in dogs is determined by interactions between genes at a number of loci.
(a) Most Labrador retrievers are genotype A^sA^s SS, varying only at the B and E loci. (b) Most beagles are genotype a^sa^s BB s^ps^p. (c) Dalmatians are genotype A^sA^s EE s^ws^w, varying at the B locus, which makes the dogs black ($B\_$) or brown (bb). [Part a: Kent and Donna Dannen. Part b: Tara Darling. Part c: PhotoDisc.]

Table 5.3 Common genotypes in different breeds of dogs

Breed	Usual Homozygous Genes*	Other Genes Present Within the Breed
Basset hound	BB EE	a^y, a^t S, s^p, s^i
Beagle	$a^s a^s$ BB $s^p s^p$	E, e
English bulldog	BB	A^s, a^y, a^t E^m, E, e^{br} S, s^i, s^p, s^w
Chihuahua		A^s, a^y, a^s, a^t B, b E^m, E, e^{br}, e S, s^i, s^p, s^w
Collie	BB EE	a^y, a^t s^i, s^w
Dalmatian	$A^s A^s$ EE $s^w s^w$	B, b
Doberman	$a^t a^t$ EE SS	B, b
German shepherd	BB SS	a^y, a, a^s, a^t E^m, E, e
Golden retriever	$A^s A^s$ BB SS	E, e
Greyhound	BB	A^s, a^y E, e^{br}, e S, s^p, s^w, s^i
Irish setter	BB ee SS	A, a^t
Labrador retriever	$A^s A^s$ SS	B, b E, e
Poodle	SS	A^s, a^t B, b E, e
Rottweiler	$a^t a^t$ BB EE SS	
St. Bernard	$a^y a^y$ BB	E^m, E s^i, s^p, s^w

*Most dogs in the breed are homozygous for these genes; a few individual dogs may possess other alleles at these loci.

Source: Data from M. B. Willis, *Genetics of the Dog* (London: Witherby, 1989).

are a result of the presence of *ee* (*B_ ee* or *bb ee*). Labrador retrievers are homozygous for the *S* allele, which produces a solid color; the few white spots that appear in some dogs of this breed are due to other modifying genes.

Beagle Most beagles (**Figure 5.11b**) are homozygous $a^s a^s$ BB $s^p s^p$, although other alleles at these loci are occasionally present. The a^s allele produces the saddle markings—dark back and sides, with tan head and legs—that are characteristic of the breed. Allele *B* allows black to be produced, but its distribution is limited by the a^s allele. Most beagles are *E_*, but the genotype *ee* does occasionally arise, leading to a few all-tan beagles. White spotting in beagles is due to the s^p allele.

Dalmatian Dalmatians (**Figure 5.11c**) have an interesting genetic makeup. Most are homozygous $A^s A^s$ EE $s^w s^w$; so they vary only at the *B* locus. Notice that these dogs possess genotype $A^s A^s$ EE, which allows for a solid coat that would be black, if genotype *B_* were present, or brown (called liver), if genotype *bb* were present. However, the presence of the s^w allele produces a white coat, masking the expression of the solid color. The dog's color appears only in the pigmented spots, which are due to the presence of an allele at yet another locus that allows the color to penetrate in a limited number of spots.

Table 5.3 gives the common genotypes of other breeds of dogs. **TRY PROBLEM 33**

5.3 Sex Influences the Inheritance and Expression of Genes in a Variety of Ways

In Chapter 4, we considered characteristics encoded by genes located on the sex chromosomes (sex-linked traits) and how their inheritance differs from the inheritance of traits encoded by autosomal genes. X-linked traits, for example, are passed from father to daughter but never from father to son, and Y-linked traits are passed from father to all sons. Now, we will examine additional influences of sex, including the effect of the sex of an individual organism on the expression of genes on autosomal chromosomes, on characteristics determined by genes located in the cytoplasm, and on characteristics for which the genotype of only the maternal parent determines the phenotype of the offspring. Finally, we will look at situations in which the expression of genes on autosomal chromosomes is affected by the sex of the parent from whom the genes are inherited.

Sex-Influenced and Sex-Limited Characteristics

Sex-influenced characteristics are determined by autosomal genes and are inherited according to Mendel's principles, but they are expressed differently in males and females. In this case,

5.12 Genes that encode sex-influenced traits are inherited according to Mendel's principles but are expressed differently in males and females.

a particular trait is more readily expressed in one sex; in other words, the trait has higher penetrance in one of the sexes.

For example, the presence of a beard on some goats is determined by an autosomal gene (B^b) that is dominant in males and recessive in females. In males, a single allele is required for the expression of this trait: both the homozygote (B^bB^b) and the heterozygote (B^bB^+) have beards, whereas the B^+B^+ male is beardless.

Genotype	Males	Females
B^+B^+	beardless	beardless
B^+B^b	bearded	beardless
B^bB^b	bearded	bearded

In contrast, females require two alleles in order for this trait to be expressed: the homozygote B^bB^b has a beard, whereas the heterozygote (B^bB^+) and the other homozygote (B^+B^+) are beardless.

The key to understanding the expression of the bearded gene is to look at the heterozygote. In males (for which the presence of a beard is dominant), the heterozygous genotype produces a beard but, in females (for which the absence of a beard is dominant), the heterozygous genotype produces a goat without a beard.

Figure 5.12a illustrates a cross between a beardless male (B^+B^+) and a bearded female (B^bB^b). The alleles separate into gametes according to Mendel's principle of segregation, and all the F_1 are heterozygous (B^+B^b). Because the trait is dominant in males and recessive in females, all the F_1 males will be bearded and all the F_1 females will be beardless. When the F_1 are crossed with one another, $1/4$ of the F_2 progeny are B^bB^b, $1/2$ are B^bB^+, and $1/4$ are B^+B^+ (**Figure 5.12b**). Because male heterozygotes are bearded, $3/4$ of the males in the F_2 possess beards; because female heterozygotes are beardless, only $1/4$ of the females in the F_2 are bearded.

An extreme form of sex-influenced inheritance, a **sex-limited characteristic** is encoded by autosomal genes that are expressed in only one sex; the trait has zero penetrance in the other sex. In domestic chickens, some males display a plumage pattern called cock feathering (**Figure 5.13a**). Other males and all females display a pattern called hen feathering (**Figure 5.13b** and **c**). Cock feathering is an autosomal recessive trait that is sex-limited to males. Because the trait is autosomal, the genotypes of males and females are the same, but the phenotypes produced by these genotypes differ in males and females:

Genotype	Male phenotype	Female phenotype
HH	hen feathering	hen feathering
Hh	hen feathering	hen feathering
hh	cock feathering	hen feathering

5.13 A sex-limited characteristic is encoded by autosomal genes that are expressed in only one sex. An example is cock feathering in chickens, an autosomal recessive trait that is limited to males. (a) Cock-feathered male. (b) Hen-feathered female. (c) Hen-feathered male.
[Larry Lefever/Grant Heilman Photography.]

An example of a sex-limited characteristic in humans is male-limited precocious puberty. There are several types of precocious puberty in humans, most of which are not genetic. Male-limited precocious puberty, however, results from an autosomal dominant allele (*P*) that is expressed only in males; females with the gene are normal in phenotype. Males with precocious puberty undergo puberty at an early age, usually before the age of 4. At this time, the penis enlarges, the voice deepens, and pubic hair develops. There is no impairment of sexual function; affected males are fully fertile. Most are short as adults because the long bones stop growing after puberty.

Because the trait is rare, affected males are usually heterozygous (*Pp*). A male with precocious puberty who mates with a woman who has no family history of this condition will transmit the allele for precocious puberty to $1/2$ of their children (**Figure 5.14a**), but it will be expressed only in the sons. If one of the heterozygous daughters (*Pp*) mates with a male who has normal puberty (*pp*), $1/2$ of their sons will exhibit precocious puberty (**Figure 5.14b**). Thus a sex-limited characteristic can be inherited from either parent, although the trait appears in only one sex. TRY PROBLEM 35 →

CONCEPTS

Sex-influenced characteristics are encoded by autosomal genes that are more readily expressed in one sex. Sex-limited characteristics are encoded by autosomal genes whose expression is limited to one sex.

✔ CONCEPT CHECK 9

How do sex-influenced and sex-limited traits differ from sex-linked traits?

Cytoplasmic Inheritance

Mendel's principles of segregation and independent assortment are based on the assumption that genes are located on chromosomes in the nucleus of the cell. For most genetic characteristics, this assumption is valid, and Mendel's principles allow us to predict the types of offspring that will be produced in a genetic cross. However, not all the genetic

5.14 Sex-limited characteristics are inherited according to Mendel's principles. Precocious puberty is an autosomal dominant trait that is limited to males.

material of a cell is found in the nucleus; some characteristics are encoded by genes located in the cytoplasm. These characteristics exhibit **cytoplasmic inheritance**.

A few organelles, notably chloroplasts and mitochondria, contain DNA. The human mitochondrional genome contains about 15,000 nucleotides of DNA, encoding 37 genes. Compared with that of nuclear DNA, which contains some 3 billion nucleotides encoding some 20,000 to 25,000 genes, the size of the mitochondrial genome is very small; nevertheless, mitochondrial and chloroplast genes encode some important characteristics. Here, we will focus on patterns of cytoplasmic inheritance.

Cytoplasmic inheritance differs from the inheritance of characteristics encoded by nuclear genes in several important respects. A zygote inherits nuclear genes from both parents; but, typically, all its cytoplasmic organelles, and thus all its cytoplasmic genes, come from only one of the gametes, usually the egg. A sperm generally contributes only a set of nuclear genes from the male parent. In a few organisms, cytoplasmic genes are inherited from the male parent or from both parents; however, for most organisms, all the cytoplasm is inherited from the egg. In this case, cytoplasmically inherited traits are present in both males and females and are passed from mother to offspring, never from father to offspring. Reciprocal crosses, therefore, give different results when cytoplasmic genes encode a trait.

Cytoplasmically inherited characteristics frequently exhibit extensive phenotypic variation because no mechanism analogous to mitosis or meiosis ensures that cytoplasmic genes are evenly distributed in cell division. Thus, different cells and individual offspring will contain various proportions of cytoplasmic genes.

Consider mitochondrial genes. Most cells contain thousands of mitochondria, and each mitochondrion contains from 2 to 10 copies of mitochondrial DNA (mtDNA). Suppose that half of the mitochondria in a cell contain a normal wild-type copy of mtDNA and the other half contain a mutated copy (**Figure 5.15**). In cell division, the mitochondria segregate into progeny cells at random. Just by chance, one cell may receive mostly mutated mtDNA and another cell may receive mostly wild-type mtDNA. In this way, different progeny from the same mother and even cells within an individual offspring may vary in their phenotypes. Traits encoded by chloroplast DNA (cpDNA) are similarly variable. The characteristics that cytoplasmically inherited traits exhibit are summarized in **Table 5.4**.

Variegation in four-o'clocks In 1909, cytoplasmic inheritance was recognized by Carl Correns as an exception to Mendel's principles. Correns, one of the biologists who rediscovered Mendel's work, studied the inheritance of leaf variegation in the four-o'clock plant, *Mirabilis jalapa*. Correns found that the leaves and shoots of one variety of

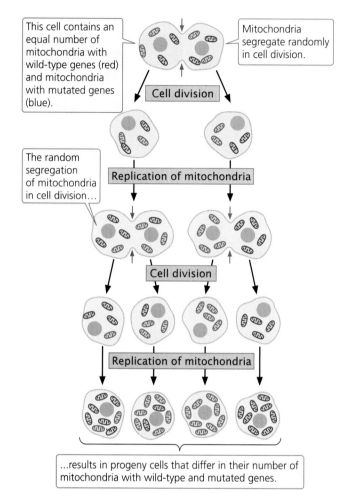

5.15 Cytoplasmically inherited characteristics frequently exhibit extensive phenotypic variation because cells and individual offspring contain various proportions of cytoplasmic genes. Mitochondria that have wild-type mtDNA are shown in red; those having mutant mtDNA are shown in blue.

four-o'clock were variegated, displaying a mixture of green and white splotches. He also noted that some branches of the variegated strain had all-green leaves; other branches had all-white leaves. Each branch produced flowers; so Correns was able to cross flowers from variegated, green, and

Table 5.4	Characteristics of cytoplasmically inherited traits.
1. Present in males and females.	
2. Usually inherited from one parent, usually the maternal parent.	
3. Reciprocal crosses give different results.	
4. Exhibit extensive phenotypic variation, even within a single family.	

white branches in all combinations (**Figure 5.16**). The seeds from green branches always gave rise to green progeny, no matter whether the pollen was from a green, white, or variegated branch. Similarly, flowers on white branches always produced white progeny. Flowers on the variegated branches gave rise to green, white, and variegated progeny, in no particular ratio.

Correns's crosses demonstrated cytoplasmic inheritance of variegation in the four-o'clocks. The phenotypes of the offspring were determined entirely by the maternal parent, never by the paternal parent (the source of the pollen). Furthermore, the production of all three phenotypes by flowers on variegated branches is consistent with cytoplasmic inheritance. Variegation in these plants is caused by a defective gene in the cpDNA, which results in a failure to produce the green pigment chlorophyll. Cells from green branches contain normal chloroplasts only, cells from white branches contain abnormal chloroplasts only, and cells from variegated branches contain a mixture of normal and abnormal chloroplasts. In the flowers from variegated branches, the random segregation of chloroplasts in the course of oogenesis produces some egg cells with normal cpDNA, which develop into green progeny; other egg cells with only abnormal cpDNA develop into white progeny; and, finally, still other egg cells with a mixture of normal and abnormal cpDNA develop into variegated progeny.

Mitochondrial diseases A number of human diseases (mostly rare) that exhibit cytoplasmic inheritance have been identified. These disorders arise from mutations in mtDNA, most of which occur in genes encoding components of the electron-transport chain, which generates most of the ATP (adenosine triphosphate) in aerobic cellular respiration. One such disease is Leber hereditary optic neuropathy (LHON). Patients who have this disorder experience rapid loss of vision in both eyes, resulting from the death of cells in the optic nerve. This loss of vision typically occurs in early adulthood (usually between the ages of 20 and 24), but it can occur any time after adolescence. There is much clinical variability in the severity of the disease, even within the same family. Leber hereditary optic neuropathy exhibits cytoplasmic inheritance: the trait is passed from mother to all children, sons and daughters alike.

Genetic Maternal Effect

A genetic phenomenon that is sometimes confused with cytoplasmic inheritance is **genetic maternal effect**, in which the phenotype of the offspring is determined by the genotype of the mother. In cytoplasmic inheritance, the genes for a characteristic are inherited from only one parent, usually the mother. In genetic maternal effect, the genes are inherited from both parents, but the offspring's phenotype is determined not by its own genotype but by the genotype of its mother.

Genetic maternal effect frequently arises when substances present in the cytoplasm of an egg (encoded by the mother's nuclear genes) are pivotal in early development. An excellent example is the shell coiling of the snail *Limnaea*

5.16 Crosses for leaf type in four-o'clocks illustrate cytoplasmic inheritance.

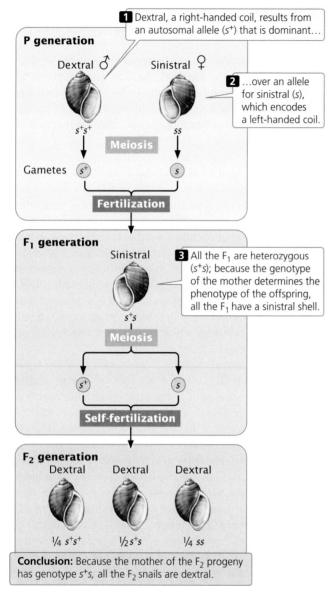

5.17 In genetic maternal effect, the genotype of the maternal parent determines the phenotype of the offspring. The shell coiling of a snail is a trait that exhibits genetic maternal effect.

peregra (**Figure 5.17**). In most snails of this species, the shell coils to the right, which is termed dextral coiling. However, some snails possess a left-coiling shell, exhibiting sinistral coiling. The direction of coiling is determined by a pair of alleles; the allele for dextral (s^+) is dominant over the allele for sinistral (s). However, the direction of coiling is determined not by that snail's own genotype, but by the genotype of its *mother*. The direction of coiling is affected by the way in which the cytoplasm divides soon after fertilization, which in turn is determined by a substance produced by the mother and passed to the offspring in the cytoplasm of the egg.

If a male homozygous for dextral alleles (s^+s^+) is crossed with a female homozygous for sinistral alleles (ss), all of the F_1 are heterozygous (s^+s) and have a sinistral shell because the genotype of the mother (ss) encodes sinistral coiling (see Figure 5.17). If these F_1 snails are self-fertilized, the genotypic ratio of the F_2 is 1 s^+s^+ : 2 s^+s : 1 ss.

Notice that that the phenotype of all the F_2 snails is dextral coiled, regardless of their genotypes. The F_2 offspring are dextral coiled because the genotype of their mother (s^+s) encodes a right-coiling shell and determines their phenotype. With genetic maternal effect, the phenotype of the progeny is not necessarily the same as the phenotype of the mother, because the progeny's phenotype is determined by the mother's *genotype*, not her phenotype. Neither the male parent's nor the offspring's own genotype has any role in the offspring's phenotype. However, a male does influence the phenotype of the F_2 generation: by contributing to the genotypes of his daughters, he affects the phenotypes of their offspring. Genes that exhibit genetic maternal effect are therefore transmitted through males to future generations. In contrast, genes that exhibit cytoplasmic inheritance are always transmitted through only one of the sexes (usually the female). **TRY PROBLEM 38** →

CONCEPTS

Characteristics exhibiting cytoplasmic inheritance are encoded by genes in the cytoplasm and are usually inherited from one parent, most commonly the mother. In genetic maternal effect, the genotype of the mother determines the phenotype of the offspring.

✔ CONCEPT CHECK 10

How might you determine whether a particular trait is due to cytoplasmic inheritance or to genetic maternal effect?

Genomic Imprinting

A basic tenet of Mendelian genetics is that the parental origin of a gene does not affect its expression and, therefore, reciprocal crosses give identical results. We have seen that there are some genetic characteristics—those encoded by X-linked genes and cytoplasmic genes—for which reciprocal crosses do not give the same results. In these cases, males and females do not contribute the same genetic material to the offspring. With regard to autosomal genes, males and females contribute the same number of genes, and paternal and maternal genes have long been assumed to have equal effects. However, the expression of some genes is significantly affected by their parental origin. This phenomenon, the differential expression of genetic material depending on whether it is inherited from the male or female parent, is called **genomic imprinting**.

A gene that exhibits genomic imprinting in both mice and humans is *Igf2*, which encodes a protein called insulin-like growth factor II (Igf2). Offspring inherit one *Igf2* allele from their mother and one from their father. The paternal copy of *Igf2* is actively expressed in the fetus and placenta, but the maternal copy is completely silent (**Figure 5.18**). Both male and female offspring possess *Igf2* genes; the key to whether the gene is expressed is the sex of the parent transmitting the gene. In the present example, the gene is expressed only when it is transmitted by a male parent. In other genomically imprinted traits, the trait is expressed only when the gene is transmitted by the female parent. In a way that is not completely understood, the paternal *Igf2* allele (but not the maternal allele) promotes placental and fetal growth; when the paternal copy of *Igf2* is deleted in mice, a small placenta and low-birth-weight offspring result.

Genomic imprinting has been implicated in several human disorders, including Prader–Willi and Angelman syndromes. Children with Prader–Willi syndrome have small hands and feet, short stature, poor sexual development, and mental retardation. These children are small at birth and suckle poorly; but, as toddlers, they develop voracious appetites and frequently become obese. Many persons with Prader–Willi syndrome are missing a small region on the long arm of chromosome 15. The deletion of this region is always inherited from the *father*. Thus, children with Prader–Willi syndrome lack a paternal copy of genes on the long arm of chromosome 15.

The deletion of this same region of chromosome 15 can also be inherited from the *mother*, but this inheritance results in a completely different set of symptoms, producing Angelman syndrome. Children with Angelman syndrome exhibit frequent laughter, uncontrolled muscle movement, a large mouth, and unusual seizures. They are missing a maternal copy of genes on the long arm of chromosome 15. For normal development to take place, copies of this region of chromosome 15 from both male and female parents are apparently required.

Many imprinted genes in mammals are associated with fetal growth. Imprinting has also been reported in plants, with differential expression of paternal and maternal genes in the endosperm, which, like the placenta in mammals, provides nutrients for the growth of the embryo. The mechanism of imprinting is still under investigation, but the methylation of DNA—the addition of methyl (CH_3) groups to DNA nucleotides—is essential to the process. In mammals, methylation is erased in the germ cells each generation and then reestablished in the course of gamete formation, with sperm and eggs undergoing different levels of methylation, which then causes the differential expression of male and female alleles in the offspring.

Imprinting and genetic conflict Why does genomic imprinting occur? One possible answer is the **genetic-conflict hypothesis**, which suggests that there are different and conflicting evolutionary pressures acting on maternal and paternal alleles for genes (such as *Igf2*) that affect fetal growth. From an evolutionary standpoint, paternal alleles that maximize the size of the offspring are favored, because birth weight is strongly associated with infant mortality and adult health. Thus, it is to the advantage of the male parent to pass on alleles that promote maximum fetal growth of their offspring. In contrast, maternal alleles that cause more-limited fetal growth are favored because committing too many of the female parent's nutrients to any one fetus may limit her ability to reproduce in the future and because giving birth to very large babies is difficult and risky for the mother. This hypothesis predicts that genomic imprinting will evolve: paternal copies of genes that affect fetal growth should be maximally expressed, whereas maternal copies of the same genes should be less actively expressed or even silent. Indeed, *Igf2* follows this pattern: the paternal allele is active and promotes growth; the maternal allele is silent and does not contribute to growth. Recent findings demonstrate that the paternal copy of *Igf2* promotes fetal growth by

5.18 Genomic imprinting of the *Igf2* gene in mice and humans affects fetal growth. (a) The paternal *Igf2* allele is active in the fetus and placenta, whereas the maternal allele is silent. (b) The human *Igf2* locus is on the short arm of chromosome 11; the locus in mice is on chromosome 7 [Courtesy of Dr. Thomas Ried and Dr. Evelin Schrock.]

Table 5.5 Sex influences on heredity

Genetic Phenomenon	Phenotype determined by
Sex-linked characteristic	Genes located on the sex chromosome
Sex-influenced characteristic	Genes on autosomal chromosomes that are more readily expressed in one sex
Sex-limited characteristic	Autosomal genes whose expression is limited to one sex
Genetic maternal effect	Nuclear genotype of the maternal parent
Cytoplasmic inheritance	Cytoplasmic genes, which are usually inherited entirely from only one parent
Genomic imprinting	Genes whose expression is affected by the sex of the transmitting parent

directing more maternal nutrients to the fetus through the placenta. Some of the different ways in which sex interacts with heredity are summarized in **Table 5.5**.

Epigenetics Genomic imprinting is just one form of a phenomenon known as **epigenetics**. Most traits are encoded by genetic information that resides in the sequence of nucleotide bases of DNA—the genetic code, which will be discussed in Chapter 15. However, some traits are caused by alterations to the DNA, such as DNA methylation, that affect the way in which the DNA sequences are expressed. These changes are often stable and heritable in the sense that they are passed from one cell to another.

In genomic imprinting, whether the gene passes through the egg or sperm determines how much methylation of the DNA takes place. The pattern of methylation on a gene is copied when the DNA is replicated and therefore remains on the gene as it is passed from cell to cell through mitosis. However, the pattern of methylation may be modified or removed when the DNA passes through a gamete, and so a gene methylated in sperm is unmethylated when it is eventually passed down to a daughter's egg. Ultimately, the amount of methylation determines whether the gene is expressed in the offspring.

These types of reversible changes to DNA that influence the expression of traits are termed epigenetic marks. The inactivation of one of the X chromosomes in female mammals (discussed in Chapter 4) is another type of epigenetic change.

A remarkable example of epigenetics is seen in honey bees. Queen bees and worker bees are both female, but there the resemblance ends. A queen is large and develops functional ovaries, whereas workers are small and sterile. The queen goes on a mating flight and spends her entire life reproducing, whereas workers spend all of their time collecting nectar and pollen, tending the queen, and raising her offspring. In spite of these significant differences in anatomy, physiology, and behavior, queens and workers are genetically the same; both develop from ordinary eggs. How they differ is in diet: worker bees produce and feed a few female larvae a special substance called royal jelly, which causes these larvae to develop as queens. Other larvae are fed ordinary bee food, and they develop as workers. This simple difference in diet greatly affects gene expression, causing different genes to be activated in queens and workers and resulting in a very different set of phenotypic traits. How royal jelly affects gene expression has long been a mystery, but recent research suggests that it changes an epigenetic mark.

Research conducted in 2008 by Ryszard Kucharski and his colleagues demonstrated that royal jelly silences the expression of a key gene called *Dnmt3*, whose product normally adds methyl groups to DNA. With *Dnmt3* shut down, bee DNA is less methylated and many genes that are normally silenced in workers are expressed, leading to the development of queen characteristics. Kucharski and his coworkers demonstrated the importance of DNA methylation in queen development by injecting into bee larvae small RNA molecules (small interfering RNAs, or siRNAs) that specifically inhibited the expression of *Dnmt3*. These larvae had lower levels of DNA methylation and many developed as queens with fully functional ovaries. This experiment demonstrates that royal jelly brings about epigenetic changes (less DNA methylation), which are transmitted through cell division and modify developmental pathways, eventually leading to a queen bee.

CONCEPTS

In genomic imprinting, the expression of a gene is influenced by the sex of the parent transmitting the gene to the offspring. Epigenetic marks are reversible changes in DNA that do not alter the base sequence but may affect how a gene is expressed.

✔ **CONCEPT CHECK 11**

What type of epigenetic mark is responsible for genomic imprinting?

5.4 Anticipation Is the Stronger or Earlier Expression of Traits in Succeeding Generations

Another genetic phenomenon that is not explained by Mendel's principles is **anticipation**, in which a genetic trait becomes more strongly expressed or is expressed at an ear-

5.19 The expression of the *vestigial* mutation in *Drosophila* is temperature dependent. When reared at temperatures below 29°C, flies homozygous for *vestigial* have greatly reduced wings; but, at temperatures above 31°C, the flies develop normal wings. [Data from M. H. Harnly, *Journal of Experimental Zoology* 56:363–379, 1936.]

lier age as it is passed from generation to generation. In the early 1900s, several physicians observed that many patients with moderate to severe myotonic dystrophy—an autosomal dominant muscle disorder—had ancestors who were only mildly affected by the disease. These observations led to the concept of anticipation. However, the concept quickly fell out of favor with geneticists because there was no obvious mechanism to explain it; traditional genetics held that genes are passed unaltered from parents to offspring. Geneticists tended to attribute anticipation to observational bias.

Research has now reestablished anticipation as a legitimate genetic phenomenon. The mutation causing myotonic dystrophy consists of an unstable region of DNA that can increase in size as the gene is passed from generation to generation. The age of onset and the severity of the disease are correlated with the size of the unstable region; an increase in the size of the region through generations produces anticipation. The phenomenon has now been implicated in a number of genetic diseases. We will examine these interesting types of mutations in more detail in Chapter 10.

> **CONCEPTS**
>
> Anticipation is the stronger or earlier expression of a genetic trait in succeeding generations. It is caused by an unstable region of DNA that increases in size from generation to generation.

5.5 The Expression of a Genotype May Be Affected by Environmental Effects

In Chapter 3, we learned that each phenotype is the result of a genotype developing within a specific environment; the genotype sets the potential for development, but how the phenotype actually develops within the limits imposed by the genotype depends on environmental effects. Stated another way, each genotype may produce several different phenotypes, depending on the environmental conditions in which development takes place. For example, a fruit fly homozygous for the vestigial mutation (*vg vg*) develops reduced wings when raised at a temperature below 29°C, but the same genotype develops much longer wings when raised at 31°C (**Figure 5.19**).

For most of the characteristics discussed so far, the effect of the environment on the phenotype has been slight. Mendel's peas with genotype *yy*, for example, developed green seeds regardless of the environment in which they were raised. Similarly, persons with genotype $I^A I^A$ have the A antigen on their red blood cells regardless of their diet, socioeconomic status, or family environment. For other phenotypes, however, environmental effects play a more important role.

Environmental Effects on the Phenotype

The phenotypic expression of some genotypes critically depends on the presence of a specific environment. For example, the *himalayan* allele in rabbits produces dark fur at the extremities of the body—on the nose, ears, and feet (**Figure 5.20**). The dark pigment develops, however, only when a rabbit is reared at a temperature of 25°C or lower; if a Himalayan rabbit is reared at 30°C, no dark patches develop. The expression of the *himalayan* allele is thus temperature dependent; an enzyme necessary for the production of dark pigment is inactivated at higher temperatures. The pigment is restricted to the nose, feet, and ears of a Himalayan rabbit because the animal's core body temperature is normally above 25°C and the enzyme is functional only in the cells of the relatively cool extremities. The *himalayan* allele is an example of a **temperature-sensitive allele**, an allele whose product is functional only at certain temperatures. Similarly, vestigial wings in *Drosophila melanogaster* is caused by a temperature-dependent mutation (see Figure 5.19).

Environmental factors also play an important role in the expression of a number of human genetic diseases. Phenylketonuria (PKU) is due to an autosomal recessive allele that causes mental retardation. The disorder arises from a defect in an enzyme that normally metabolizes the

Reared at 25°C or lower   Reared at temperatures above 30°C

5.20 The expression of *himalayan* depends on the temperature at which a rabbit is reared.

amino acid phenylalanine. When this enzyme is defective, phenylalanine is not metabolized, and its buildup causes brain damage in children. A simple environmental change, putting an affected child on a low-phenylalanine diet, prevents retardation.

These examples illustrate the point that genes and their products do not act in isolation; rather, they frequently interact with environmental factors. Occasionally, environmental factors alone can produce a phenotype that is the same as the phenotype produced by a genotype; this phenotype is called a **phenocopy**. In fruit flies, for example, the autosomal recessive mutation *eyeless* produces greatly reduced eyes. The eyeless phenotype can also be produced by exposing the larvae of normal flies to sodium metaborate.

CONCEPTS

The expression of many genes is modified by the environment. A phenocopy is a trait produced by environmental effects that mimics the phenotype produced by the genotype.

✔ **CONCEPT CHECK 12**

How can you determine whether a phenotype such as reduced eyes in fruit flies is due to a recessive mutation or is a phenocopy?

The Inheritance of Continuous Characteristics

So far, we've dealt primarily with characteristics that have only a few distinct phenotypes. In Mendel's peas, for example, the seeds were either smooth or wrinkled, yellow or green; the coats of dogs were black, brown, or yellow; blood types were of four distinct types, A, B, AB, or O. Such characteristics, which have a few easily distinguished phenotypes, are called **discontinuous characteristics**.

Not all characteristics exhibit discontinuous phenotypes. Human height is an example of such a characteristic; people do not come in just a few distinct heights but, rather, display a continuum of heights. Indeed, there are so many possible phenotypes of human height that we must use a measurement to describe a person's height. Characteristics that exhibit a continuous distribution of phenotypes are termed **continuous characteristics**. Because such characteristics have many possible phenotypes and must be described in quantitative terms, continuous characteristics are also called **quantitative characteristics**.

Continuous characteristics frequently arise because genes at many loci interact to produce the phenotypes. When a single locus with two alleles encodes a characteristic, there are three genotypes possible: *AA*, *Aa*, and *aa*. With two loci, each with two alleles, there are $3^2 = 9$ genotypes possible. The number of genotypes encoding a characteristic is 3^n, where *n* equals the number of loci, each with two alleles, that influence the characteristic. For example, when a characteristic is determined by eight loci, each with two alleles, there are $3^8 = 6561$ different genotypes possible for this characteristic. If each genotype produces a different phenotype, many phenotypes will be possible. The slight differences between the phenotypes will be indistinguishable, and the characteristic will appear continuous. Characteristics encoded by genes at many loci are called **polygenic characteristics**.

The converse of polygeny is **pleiotropy**, in which one gene affects multiple characteristics. Many genes exhibit pleiotropy. Phenylketonuria, mentioned earlier, results from a recessive allele; persons homozygous for this allele, if untreated, exhibit mental retardation, blue eyes, and light skin color. The lethal allele that causes yellow coat color in mice (discussed in the introduction to the chapter) also is pleiotropic. In addition to its lethality and effect on hair color, the gene causes a diabetes-like condition, obesity, and increased propensity to develop tumors.

Frequently, the phenotypes of continuous characteristics are also influenced by environmental factors. Each genotype is capable of producing a range of phenotypes. In this situation, the particular phenotype that results depends on both the genotype and the environmental conditions in which the genotype develops. For example, only three genotypes may encode a characteristic, but, because each genotype produces a range of phenotypes associated with different environments, the phenotype of the characteristic exhibits a continuous distribution. Many continuous characteristics are both polygenic and influenced by environmental factors; such characteristics are called **multifactorial characteristics** because many factors help determine the phenotype.

The inheritance of continuous characteristics may appear to be complex, but the alleles at each locus follow Mendel's principles and are inherited in the same way as alleles encoding simple, discontinuous characteristics. However, because many genes participate, because environmental factors influence the phenotype, and because the phenotypes do not sort out into a few distinct types, we cannot observe the distinct ratios that have allowed us to interpret the genetic basis of discontinuous characteristics. To analyze continuous characteristics, we must employ special statistical tools, as will be discussed in Chapter 11. TRY PROBLEMS 42 AND 43

CONCEPTS

Discontinuous characteristics exhibit a few distinct phenotypes; continuous characteristics exhibit a range of phenotypes. A continuous characteristic is frequently produced when genes at many loci and environmental factors combine to determine a phenotype.

✔ **CONCEPT CHECK 13**

What is the difference between polygeny and pleiotropy?

CONCEPTS SUMMARY

- Dominance always refers to genes at the same locus (allelic genes) and can be understood in regard to how the phenotype of the heterozygote relates to the phenotypes of the homozygotes.
- Dominance is complete when a heterozygote has the same phenotype as a homozygote, is incomplete when the heterozygote has a phenotype intermediate between those of two parental homozygotes, and is codominant when the heterozygote exhibits traits of both parental homozygotes.
- The type of dominance does not affect the inheritance of an allele; it does affect the phenotypic expression of the allele. The classification of dominance depends on the level of the phenotype examined.
- Penetrance is the percentage of individuals having a particular genotype that exhibit the expected phenotype. Expressivity is the degree to which a character is expressed.
- Lethal alleles cause the death of an individual organism possessing them, usually at an early stage of development, and may alter phenotypic ratios.
- Multiple alleles refer to the presence of more than two alleles at a locus within a group. Their presence increases the number of genotypes and phenotypes possible.
- Gene interaction refers to the interaction between genes at different loci to produce a single phenotype. An epistatic gene at one locus suppresses, or masks, the expression of hypostatic genes at other loci. Gene interaction frequently produces phenotypic ratios that are modifications of dihybrid ratios.
- Sex-influenced characteristics are encoded by autosomal genes that are expressed more readily in one sex. Sex-limited characteristics are encoded by autosomal genes expressed in only one sex.
- In cytoplasmic inheritance, the genes for the characteristic are found in the organelles and are usually inherited from a single (usually maternal) parent. Genetic maternal effect is present when an offspring inherits genes from both parents, but the nuclear genes of the mother determine the offspring's phenotype.
- Genomic imprinting refers to characteristics encoded by autosomal genes whose expression is affected by the sex of the parent transmitting the genes.
- Anticipation refers to a genetic trait that is more strongly expressed or is expressed at an earlier age in succeeding generations.
- Phenotypes are often modified by environmental effects. A phenocopy is a phenotype produced by an environmental effect that mimics a phenotype produced by a genotype.
- Continuous characteristics are those that exhibit a wide range of phenotypes; they are frequently produced by the combined effects of many genes and environmental effects.

IMPORTANT TERMS

complete dominance (p. 100)
incomplete dominance (p. 101)
codominance (p. 102)
incomplete penetrance (p. 103)
penetrance (p. 103)
expressivity (p. 103)
lethal allele (p. 103)
multiple alleles (p. 104)
gene interaction (p. 106)
epistasis (p. 107)

epistatic gene (p. 107)
hypostatic gene (p. 107)
complementation test (p. 113)
complementation (p. 113)
sex-influenced characteristic (p. 115)
sex-limited characteristic (p. 116)
cytoplasmic inheritance (p. 118)
genetic maternal effect (p. 119)
genomic imprinting (p. 120)
genetic-conflict hypothesis (p. 121)

epigenetics (p. 122)
anticipation (p. 122)
temperature-sensitive allele (p. 123)
phenocopy (p. 124)
discontinuous characteristic (p. 124)
continuous characteristic (p. 124)
quantitative characteristic (p. 124)
polygenic characteristic (p. 124)
pleiotropy (p. 124)
multifactorial characteristic (p. 124)

ANSWERS TO CONCEPT CHECKS

1. b
2. With complete dominance, the heterozygote expresses the same phenotype as that of one of the homozygotes. With incomplete dominance, the heterozygote has a phenotype that is intermediate between the two homozygotes. With codominance, the heterozygote has a phenotype that simultaneously expresses the phenotypes of both homozygotes.
3. d
4. b
5. c
6. Gene interaction is interaction between genes at different loci. Dominance is interaction between alleles at a single locus.
7. d
8. Cross a bulldog homozygous for *brindle* with a Chihuahua homozygous for *brindle*. If the two *brindle* genes are allelic, all the offspring will be brindle: $bb \times bb \rightarrow$ all bb (brindle). If, on the other hand, brindle in the two breeds is due to recessive genes at different loci, then none of the offspring will be brindle: $a^+a^+\ bb \times aa\ b^+b^+ \rightarrow a^+a\ b^+b$.
9. Both sex-influenced and sex-limited traits are encoded by autosomal genes whose expression is affected by the sex of the individual organism possessing the gene. Sex-linked traits are encoded by genes on the sex chromosomes.
10. Cytoplasmically inherited traits are encoded by genes in the cytoplasm, which is usually inherited only from the female parent. Therefore, a trait due to cytoplasmic inheritance will always be passed through females. Traits due to genetic maternal effect are encoded by autosomal genes and can therefore be passed through males, although any individual organism's trait is determined by the genotype of the maternal parent.
11. Methylation of DNA
12. Cross two eyeless flies, and cross an eyeless fly with a wild-type fly. Raise the offspring of both crosses in the same environment. If the trait is due to a recessive mutation, all the offspring of the two eyeless flies should be eyeless, whereas at least some of the offspring of the of the eyeless and wild-type flies should be wild type. If the trait is due to a phenocopy, there should be no differences in the progeny of the two crosses.
13. Polygeny refers to the influence of multiple genes on the expression of a single characteristic. Pleiotropy refers to the effect of a single gene on the expression of multiple characteristics.

WORKED PROBLEMS

1. The type of plumage found in mallard ducks is determined by three alleles at a single locus: M^R, which encodes restricted plumage; M, which encodes mallard plumage; and m^d, which encodes dusky plumage. The restricted phenotype is dominant over mallard and dusky; mallard is dominant over dusky ($M^R > M > m^d$). Give the expected phenotypes and proportions of offspring produced by the following crosses.

   a. $M^R M \times m^d m^d$
   b. $M^R m^d \times M m^d$
   c. $M^R m^d \times M^R M$
   d. $M^R M \times M m^d$

• **Solution**

We can determine the phenotypes and proportions of offspring by (1) determining the types of gametes produced by each parent and (2) combining the gametes of the two parents with the use of a Punnett square.

a.

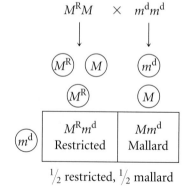

½ restricted, ½ mallard

b.

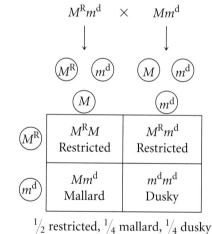

½ restricted, ¼ mallard, ¼ dusky

c.

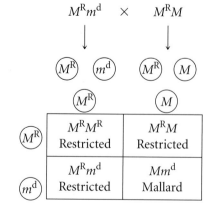

¾ restricted, ¼ mallard

d.

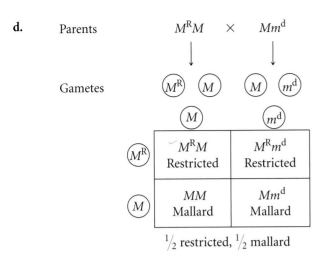

½ restricted, ½ mallard

2. A geneticist crosses two yellow mice with straight hair and obtains the following progeny:

½ yellow, straight
⅙ yellow, fuzzy
¼ gray, straight
1/12 gray, fuzzy

a. Provide a genetic explanation for the results and assign genotypes to the parents and progeny of this cross.
b. What additional crosses might be carried out to determine if your explanation is correct?

• Solution

a. This cross concerns two separate characteristics—color and type of hair; so we should begin by examining the results for each characteristic separately. First, let's look at the inheritance of color. Two yellow mice are crossed, producing ½ + ⅙ = 3/6 + 1/6 = 4/6 = ⅔ yellow mice and ¼ + 1/12 = 3/12 + 1/12 = 4/12 = ⅓ gray mice. We learned in this chapter that a 2 : 1 ratio is often produced when a recessive lethal gene is present:

$Yy \times Yy$

YY ¼ die
Yy ½ yellow, becomes ⅔
yy ¼ gray, becomes ⅓

Now, let's examine the inheritance of the hair type. Two mice with straight hair are crossed, producing ½ + ¼ = 2/4 + ¼ = ¾ mice with straight hair and ⅙ + 1/12 = 2/12 + 1/12 = 3/12 = ¼ mice with fuzzy hair. We learned in Chapter 3 that a 3 : 1 ratio is usually produced by a cross between two individuals heterozygous for a simple dominant allele:

$Ss \times Ss$

SS ¼ straight }
Ss ½ straight } ¾ straight
ss ¼ fuzzy

We can now combine both loci and assign genotypes to all the individual mice in the cross:

P $Yy\ Ss$ × $Yy\ Ss$
   Yellow, straight   Yellow, straight

Phenotype	Genotype	Probability at each locus	Combined probability
yellow, straight	$Yy\ S\_$	⅔ × ¾	= 6/12 = ½
yellow, fuzzy	$Yy\ ss$	⅔ × ¼	= 2/12 = ⅙
gray, straight	$yy\ S\_$	⅓ × ¾	= 3/12 = ¼
gray, fuzzy	$yy\ ss$	⅓ × ¼	= 1/12

b. We could carry out a number of different crosses to test our hypothesis that yellow is a recessive lethal and straight is dominant over fuzzy. For example, a cross between any two yellow mice should always produce ⅔ yellow and ⅓ gray offspring, and a cross between two gray mice should produce gray offspring only. A cross between two fuzzy mice should produce fuzzy offspring only.

3. In some sheep, the presence of horns is produced by an autosomal allele that is dominant in males and recessive in females. A horned female is crossed with a hornless male. One of the resulting F_1 females is crossed with a hornless male. What proportion of the male and female progeny from this cross will have horns?

• Solution

The presence of horns in these sheep is an example of a sex-influenced characteristic. Because the phenotypes associated with the genotypes differ for the two sexes, let's begin this problem by writing out the genotypes and phenotypes for each sex. We will let H represent the allele that encodes horns and H^+ represent the allele that encodes hornless. In males, the allele for horns is dominant over the allele for hornless, which means that males homozygous (HH) and heterozygous (H^+H) for this gene are horned. Only males homozygous for the recessive hornless allele (H^+H^+) are hornless. In females, the allele for horns is recessive, which means that only females homozygous for this allele (HH) are horned; females heterozygous (H^+H) and homozygous (H^+H^+) for the hornless allele are hornless. The following table summarizes genotypes and associated phenotypes:

Genotype	Male phenotype	Female phenotype
HH	horned	horned
HH^+	horned	hornless
H^+H^+	hornless	hornless

In the problem, a horned female is crossed with a hornless male. From the preceding table, we see that a horned female must be homozygous for the allele for horns (HH) and a hornless male must be homozygous for the allele for hornless (H^+H^+); so all the F_1 will be heterozygous; the F_1 males will be horned and the F_1 females will be hornless, as shown in the following diagram:

P H^+H^+ × HH

F_1 H^+H
Horned males and hornless females

A heterozygous hornless F_1 female (H^+H) is then crossed with a hornless male (H^+H^+):

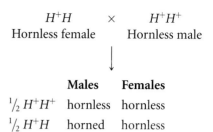

H^+H × H^+H^+
Hornless female   Hornless male

	Males	Females
½ H^+H^+	hornless	hornless
½ H^+H	horned	hornless

Therefore, ½ of the male progeny will be horned, but none of the female progeny will be horned.

COMPREHENSION QUESTIONS

Section 5.1

*1. How do incomplete dominance and codominance differ?

*2. What is incomplete penetrance and what causes it?

Section 5.2

3. What is gene interaction? What is the difference between an epistatic gene and a hypostatic gene?

4. What is a recessive epistatic gene?

*5. What is a complementation test and what is it used for?

Section 5.3

*6. What characteristics are exhibited by a cytoplasmically inherited trait?

7. What is genomic imprinting?

8. What is the difference between genetic maternal effect and genomic imprinting?

9. What is the difference between a sex-influenced gene and a gene that exhibits genomic imprinting?

Section 5.4

10. What characteristics do you expect to see in a trait that exhibits anticipation?

Section 5.5

*11. What are continuous characteristics and how do they arise?

APPLICATION QUESTIONS AND PROBLEMS

Introduction

12. As discussed in the introduction to this chapter, Cuénot studied the genetic basis of yellow coat color in mice. He carried out a number of crosses between two yellow mice and obtained what he thought was a 3 : 1 ratio of yellow to gray mice in the progeny. The following table gives Cuénot's actual results, along with the results of a much larger series of crosses carried out by Castle and Little (W. E. Castle and C. C. Little. 1910. *Science* 32:868–870).

Progeny Resulting from Crosses of Yellow × Yellow Mice

Investigators	Yellow progeny	Non-yellow progeny	Total progeny
Cuénot	263	100	363
Castle and Little	800	435	1235
Both combined	1063	535	1598

a. Using a chi-square test, determine whether Cuénot's results are significantly different from the 3 : 1 ratio that he thought he observed. Are they different from a 2 : 1 ratio?

b. Determine whether Castle and Little's results are significantly different from a 3 : 1 ratio. Are they different from a 2 : 1 ratio?

c. Combine the results of Castle and Cuénot and determine whether they are significantly different from a 3 : 1 ratio and a 2 : 1 ratio.

d. Offer an explanation for the different ratios that Cuénot and Castle obtained.

Sections 5.1 through 5.4

13. Match each of the following terms with its correct definition (parts *a* through *i*).

____ phenocopy   ____ genetic maternal effect
____ pleiotrophy   ____ genomic imprinting
____ polygenic trait ____ sex-influenced trait
____ penetrance   ____ anticipation
____ sex-limited trait

a. the percentage of individuals with a particular genotype that express the expected phenotype

b. a trait determined by an autosomal gene that is more easily expressed in one sex

c. a trait determined by an autosomal gene that is expressed in only one sex

d. a trait that is determined by an environmental effect and has the same phenotype as a genetically determined trait

e. a trait determined by genes at many loci

f. the expression of a trait is affected by the sex of the parent that transmits the gene to the offspring

g. the trait appears earlier or more severely in succeeding generations

h. a gene affects more than one phenotype

i. the genotype of the maternal parent influences the phenotype of the offspring

Section 5.1

*14. Palomino horses have a golden yellow coat, chestnut horses have a brown coat, and cremello horses have a coat that is almost white. A series of crosses between the three different types of horses produced the following offspring:

Cross	Offspring
palomino × palomino	13 palomino, 6 chestnut, 5 cremello
chestnut × chestnut	16 chestnut
cremello × cremello	13 cremello
palomino × chestnut	8 palomino, 9 chestnut
palomino × cremello	11 palomino, 11 cremello
chestnut × cremello	23 palomino

(a)

(b)

(c)

Coat color: (a) palomino; (b) chestnut; (c) cremello. [Part a: Keith J. Smith/Alamy; part b: jiri jura/iStock photo; part c: Caitlin Fagan.]

a. Explain the inheritance of the palomino, chestnut, and cremello phenotypes in horses.

b. Assign symbols for the alleles that determine these phenotypes, and list the genotypes of all parents and offspring given in the preceding table.

*15. The L^M and L^N alleles at the MN blood-group locus exhibit codominance. Give the expected genotypes and phenotypes and their ratios in progeny resulting from the following crosses.

a. $L^M L^M \times L^M L^N$
b. $L^N L^N \times L^N L^N$
c. $L^M L^N \times L^M L^N$
d. $L^M L^N \times L^N L^N$
e. $L^M L^M \times L^N L^N$

16. Assume that long ear lobes in humans are an autosomal dominant trait that exhibits 30% penetrance. A person who is heterozygous for long ear lobes mates with a person who is homozygous for normal ear lobes. What is the probability that their first child will have long ear lobes?

17. Club foot is one of the most common congenital skeletal abnormalities, with a worldwide incidence of about 1 in 1000 births. Both genetic and nongenetic factors are thought to be responsible for club foot. C. A. Gurnett et al. (2008. *American Journal of Human Genetics* 83:616–622) identified a family in which club foot was inherited as an autosomal dominant trait with reduced penetrance. They discovered a mutation in the *PITX1* gene that caused club foot in this family. Through DNA testing, they determined that 11 people in the family carried the *PITX1* mutation, but only 8 of these people had club foot. What is the penetrance of the *PITX1* mutation in this family?

*18. When a Chinese hamster with white spots is crossed with another hamster that has no spots, approximately $1/2$ of the offspring have white spots and $1/2$ have no spots. When two hamsters with white spots are crossed, $2/3$ of the offspring possess white spots and $1/3$ have no spots.

a. What is the genetic basis of white spotting in Chinese hamsters?

b. How might you go about producing Chinese hamsters that breed true for white spotting?

19. In the pearl-millet plant, color is determined by three alleles at a single locus: Rp^1 (red), Rp^2 (purple), and rp (green). Red is dominant over purple and green, and purple is dominant over green ($Rp^1 > Rp^2 > rp$). Give the expected phenotypes and ratios of offspring produced by the following crosses.

a. $Rp^1/Rp^2 \times Rp^1/rp$
b. $Rp^1/rp \times Rp^2/rp$
c. $Rp^1/Rp^2 \times Rp^1/Rp^2$
d. $Rp^2/rp \times rp/rp$
e. $rp/rp \times Rp^1/Rp^2$

20. If there are five alleles at a locus, how many genotypes may there be at this locus? How many different kinds of homozygotes will there be? How many genotypes and homozygotes may there be with eight alleles at a locus?

21. Turkeys have black, bronze, or black-bronze plumage. Examine the results of the following crosses:

Parents	Offspring
Cross 1: black and bronze	all black
Cross 2: black and black	3/4 black, 1/4 bronze
Cross 3: black-bronze and black-bronze	all black-bronze
Cross 4: black and bronze	1/2 black, 1/4 bronze, 1/4 black-bronze
Cross 5: bronze and black-bronze	1/2 bronze, 1/2 black-bronze
Cross 6: bronze and bronze	3/4 bronze, 1/4 black-bronze

Do you think these differences in plumage arise from incomplete dominance between two alleles at a single locus? If yes, support your conclusion by assigning symbols to each allele and providing genotypes for all turkeys in the crosses. If your answer is no, provide an alternative explanation and assign genotypes to all turkeys in the crosses.

22. In rabbits, an allelic series helps to determine coat color: C (full color), c^{ch} (chinchilla, gray color), c^{h} (Himalayan, white with black extremities), and c (albino, all-white). The C allele is dominant over all others, c^{ch} is dominant over c^{h} and c, c^{h} is dominant over c, and c is recessive to all the other alleles. This dominance hierarchy can be summarized as $C > c^{ch} > c^{h} > c$. The rabbits in the following list are crossed and produce the progeny shown. Give the genotypes of the parents for each cross:

Phenotypes of parents	Phenotypes of offspring
a. full color × albino	1/2 full color, 1/2 albino
b. Himalayan × albino	1/2 Himalayan, 1/2 albino
c. full color × albino	1/2 full color, 1/2 chinchilla
d. full color × Himalayan	1/2 full color, 1/4 Himalayan, 1/4 albino
e. full color × full color	3/4 full color, 1/4 albino

23. In this chapter, we considered Joan Barry's paternity suit against Charlie Chaplin and how, on the basis of blood types, Chaplin could not have been the father of her child.

a. What blood types are possible for the father of Barry's child?

b. If Chaplin had possessed one of these blood types, would that prove that he fathered Barry's child?

*24. A woman has blood-type A M. She has a child with blood-type AB MN. Which of the following blood types could *not* be that of the child's father? Explain your reasoning.

George	O	N
Tom	AB	MN
Bill	B	MN
Claude	A	N
Henry	AB	M

Section 5.2

*25. In chickens, comb shape is determined by alleles at two loci (R, r and P, p). A walnut comb is produced when at least one dominant allele R is present at one locus and at least one dominant allele P is present at a second locus (genotype $R\_ P\_$). A rose comb is produced when at least one dominant allele is present at the first locus and two recessive alleles are present at the second locus (genotype $R\_ pp$). A pea comb is produced when two recessive alleles are present at the first locus and at least one dominant allele is present at the second (genotype $rr P\_$). If two recessive alleles are present at the first and at the second locus ($rr\ pp$), a single comb is produced. Progeny with what types of combs and in what proportions will result from the following crosses?

a. $RR\ PP \times rr\ pp$
b. $Rr\ Pp \times rr\ pp$
c. $Rr\ Pp \times Rr\ Pp$
d. $Rr\ pp \times Rr\ pp$
e. $Rr\ pp \times rr\ Pp$
f. $Rr\ pp \times rr\ pp$

(a) (b)

(c) (d)

Comb shape: (a) walnut; (b) rose; (c) pea; (d) single. [Parts a and d: R. OSF Dowling/Animals Animals; part b: Robert Maier/Animals Animals; part c: George Godfrey/Animals Animals.]

*26. Tatuo Aida investigated the genetic basis of color variation in the Medaka (*Aplocheilus latipes*), a small fish found naturally in Japan (T. Aida. 1921. *Genetics* 6:554–573). Aida found that genes at two loci (*B, b* and *R, r*) determine the color of the fish: fish with a dominant allele at both loci (*B_R_*) are brown, fish with a dominant allele at the *B* locus only (*B_ rr*) are blue, fish with a dominant allele at the *R* locus only (*bb R_*) are red, and fish with recessive alleles at both loci (*bb rr*) are white. Aida crossed a homozygous brown fish with a homozygous white fish. He then backcrossed the F_1 with the homozygous white parent and obtained 228 brown fish, 230 blue fish, 237 red fish, and 222 white fish.

   a. Give the genotypes of the backcross progeny.

   b. Use a chi-square test to compare the observed numbers of backcross progeny with the number expected. What conclusion can you make from your chi-square results?

   c. What results would you expect for a cross between a homozygous red fish and a white fish?

   d. What results would you expect if you crossed a homozygous red fish with a homozygous blue fish and then backcrossed the F_1 with a homozygous red parental fish?

27. A variety of opium poppy (*Papaver somniferum* L.) having lacerate leaves was crossed with a variety that has normal leaves. All the F_1 had lacerate leaves. Two F_1 plants were interbred to produce the F_2. Of the F_2, 249 had lacerate leaves and 16 had normal leaves. Give genotypes for all the plants in the P, F_1, and F_2 generations. Explain how lacerate leaves are determined in the opium poppy.

28. E. W. Lindstrom crossed two corn plants with green seedlings and obtained the following progeny: 3583 green seedlings, 853 virescent-white seedlings, and 260 yellow seedlings (E. W. Lindstrom. 1921. *Genetics* 6:91–110).

   a. Give the genotypes for the green, virescent-white, and yellow progeny.

   b. Explain how color is determined in these seedlings.

   c. Is there epistasis among the genes that determine color in the corn seedlings? If so, which gene is epistatic and which is hypostatic.

*29. A dog breeder liked yellow and brown Labrador retrievers. In an attempt to produce yellow and brown puppies, he bought a yellow Labrador male and a brown Labrador female and mated them. Unfortunately, all the puppies produced in this cross were black. (See p. 107 for a discussion of the genetic basis of coat color in Labrador retrievers.)

   a. Explain this result.

   b. How might the breeder go about producing yellow and brown Labradors?

(a)   (b)

(c)

Coat color: (a) black; (b) brown; (c) yellow. [Part a: Cenorman/Dreamstime.com; parts b and c: Juniors Bildarchiv/Alamy.]

30. When a yellow female Labrador retriever was mated with a brown male, half of the puppies were brown and half were yellow. The same female, when mated with a different brown male, produced only brown offspring. Explain these results.

*31. A summer-squash plant that produces disc-shaped fruit is crossed with a summer-squash plant that produces long fruit. All the F_1 have disc-shaped fruit. When the F_1 are intercrossed, F_2 progeny are produced in the following ratio: $9/16$ disc-shaped fruit : $6/16$ spherical fruit : $1/16$ long fruit. Give the genotypes of the F_2 progeny.

32. Some sweet-pea plants have purple flowers and other plants have white flowers. A homozygous variety of pea that has purple flowers is crossed with a homozygous variety that has white flowers. All the F_1 have purple flowers. When these F_1 are self-fertilized, the F_2 appear in a ratio of $9/16$ purple to $7/16$ white.

   a. Give genotypes for the purple and white flowers in these crosses.

   b. Draw a hypothetical biochemical pathway to explain the production of purple and white flowers in sweet peas.

33. For the following questions, refer to pages 113–115 for a discussion of how coat color and pattern are determined in dogs.

   a. Why are Irish setters reddish in color?

   b. Can a poodle crossed with any other breed produce spotted puppies? Why or why not?

   c. If a St. Bernard is crossed with a Doberman, what will be the coat color of the offspring: solid, yellow, saddle, or bicolor?

d. If a Rottweiler is crossed with a Labrador retriever, what will be the coat color of the offspring: solid, yellow, saddle, or bicolor?

Section 5.3

34. Male-limited precocious puberty results from a rare, sex-limited autosomal allele (P) that is dominant over the allele for normal puberty (p) and is expressed only in males. Bill undergoes precocious puberty, but his brother Jack and his sister Beth underwent puberty at the usual time, between the ages of 10 and 14. Although Bill's mother and father underwent normal puberty, two of his maternal uncles (his mother's brothers) underwent precocious puberty. All of Bill's grandparents underwent normal puberty. Give the most likely genotypes for all the relatives mentioned in this family.

35. In some goats, the presence of horns is produced by an autosomal gene that is dominant in males and recessive in females. A horned female is crossed with a hornless male. The F_1 offspring are intercrossed to produce the F_2. What proportion of the F_2 females will have horns?

36. In goats, a beard is produced by an autosomal allele that is dominant in males and recessive in females. We'll use the symbol B^b for the beard allele and B^+ for the beardless allele. Another independently assorting autosomal allele that produces a black coat (W) is dominant over the allele for white coat (w). Give the phenotypes and their expected proportions for the following crosses.

a. $B^+B^b\ Ww$ male $\times\ B^+B^b\ Ww$ female
b. $B^+B^b\ Ww$ male $\times\ B^+B^b\ ww$ female
c. $B^+B^+\ Ww$ male $\times\ B^bB^b\ Ww$ female
d. $B^+B^b\ Ww$ male $\times\ B^bB^b\ ww$ female

37. J. K. Breitenbecher (1921. *Genetics* 6:65–86) investigated the genetic basis of color variation in the four-spotted cowpea weevil (*Bruchus quadrimaculatus*). The weevils were red, black, white, or tan. Breitenbecher found that four alleles (R, R^b, R^w, and r) at a single locus determine color. The alleles exhibit a dominance hierarchy, with red (R) dominant over all other alleles, black (R^b) dominant over white (R^w) and tan (r), white dominant over tan, and tan recessive to all others ($R > R^b > R^w > r$). The following genotypes encode each of the colors:

Genotype	Color
RR, RR^b, RR^w, Rr	red
R^bR^b, R^bR^w, R^br	black
R^wR^w, R^wr	white
rr	tan

Color variation in this species is sex-limited to females: males carry color genes but are always tan regardless of their genotype. For each of the following crosses carried out by Breitenbecher, give all possible genotypes of the parents.

Parents	Progeny
a. tan ♀ × tan ♂	78 reds ♀, 70 white ♀, 184 tan ♂
b. black ♀ × tan ♂	151 red ♀, 49 black ♀, 61 tan ♀, 249 tan ♂
c. white ♀ × tan ♂	32 red ♀, 31 tan ♂
d. black ♀ × tan ♂	3586 black ♀, 1282 tan ♀, 4791 tan ♂
e. white ♀ × tan ♂	594 white ♀, 189 tan ♀, 862 tan ♂
f. black ♀ × tan ♂	88 black ♀, 88 tan ♀, 186 tan ♂
g. tan ♀ × tan ♂	47 white ♀, 51 tan ♀, 100 tan ♂
h. red ♀ × tan ♂	1932 red ♀, 592 tan ♀, 2587 tan ♂
i. white ♀ × tan ♂	13 red ♀, 6 white ♀, 5 tan ♀, 19 tan ♂
j. red ♀ × tan ♂	190 red ♀, 196 black ♀, 311 tan ♂
k. black ♀ × tan ♂	1412 black ♀, 502 white ♀, 1766 tan ♂

38. Shell coiling of the snail *Limnaea peregra* results from a genetic maternal effect. An autosomal allele for a right-handed, or dextral, shell (s^+) is dominant over the allele for a left-handed, or sinistral, shell (s). A pet snail called Martha is sinistral and reproduces only as a female (the snails are hermaphroditic). Indicate which of the following statements are true and which are false. Explain your reasoning in each case.

a. Martha's genotype *must* be ss.
b. Martha's genotype *cannot* be s^+s^+.
c. All the offspring produced by Martha *must* be sinistral.
d. At least some of the offspring produced by Martha *must* be sinistral.
e. Martha's mother *must* have been sinistral.
f. All of Martha's brothers *must* be sinistral.

39. Hypospadias, a birth defect in male humans in which the urethra opens on the shaft instead of at the tip of the penis, results from an autosomal dominant gene in some families. Females who carry the gene show no effects. Is this birth defect an example of (a) an X-linked trait, (b) a Y-linked trait, (c) a sex-limited trait, (d) a sex-influenced trait, or (e) genetic maternal effect? Explain your answer.

40. In unicorns, two autosomal loci interact to determine the type of tail. One locus controls whether a tail is present at all; the allele for a tail (T) is dominant over the allele for tailless (t). If a unicorn has a tail, then alleles at a second locus determine whether the tail is curly or straight. Farmer Baldridge has two unicorns with curly tails. When he crosses these two unicorns, $1/2$ of the progeny have curly tails, $1/4$ have straight tails, and $1/4$ do not have a tail. Give the genotypes of the parents and progeny in Farmer Baldridge's cross. Explain how he obtained the 2 : 1 : 1 phenotypic ratio in his cross.

41. In 1983, a sheep farmer in Oklahoma noticed in his flock a ram that possessed increased muscle mass in his hindquarters. Many of the offspring of this ram possessed the same trait, which became known as the callipyge mutant (*callipyge* is Greek for "beautiful buttocks"). The mutation that caused the callipyge phenotype was eventually mapped to a position on the sheep chromosome 18.

    When the male callipyge offspring of the original mutant ram were crossed with normal females, they produced the following progeny: $1/4$ male callipyge, $1/4$ female callipyge, $1/4$ male normal, and $1/4$ female normal. When the female callipyge offspring of the original mutant ram were crossed with normal males, all of the offspring were normal. Analysis of the chromosomes of these offspring of callipyge females showed that half of them received a chromosome 18 with the allele encoding callipyge from their mother. Propose an explanation for the inheritance of the allele for callipyge. How might you test your explanation?

Section 5.4

42. Which of the following statements is an example of a phenocopy? Explain your reasoning.
    a. Phenylketonuria results from a recessive mutation that causes light skin as well as mental retardation.
    b. Human height is influenced by genes at many different loci.
    c. Dwarf plants and mottled leaves in tomatoes are caused by separate genes that are linked.
    d. Vestigial wings in *Drosophila* are produced by a recessive mutation. This trait is also produced by high temperature during development.
    e. Intelligence in humans is influenced by both genetic and environmental factors.

43. Long ears in some dogs are an autosomal dominant trait. Two dogs mate and produce a litter in which 75% of the puppies have long ears. Of the dogs with long ears in this litter, $1/3$ are known to be phenocopies. What are the most likely genotypes of the two parents of this litter?

CHALLENGE QUESTIONS

Section 5.1

44. Pigeons have long been the subject of genetic studies. Indeed, Charles Darwin bred pigeons in the hope of unraveling the principles of heredity but was unsuccessful. A series of genetic investigations in the early 1900s worked out the hereditary basis of color variation in these birds. W. R. Horlancher was interested in the genetic basis of kiteness, a color pattern that consists of a mixture of red and black stippling of the feathers. Horlancher knew from earlier experiments that black feather color was dominant over red. He carried out the following crosses to investigate the genetic relation of kiteness to black and red feather color (W. R. Horlancher. 1930. *Genetics* 15:312–346).

Cross	Offspring
kitey × kitey	16 kitey, 5 black, 3 red
kitey × black	6 kitey, 7 black
red × kitey	18 red, 9 kitey, 6 black

    a. On the basis of these results and the assumption that black is dominant over red, propose an hypothesis to explain the inheritance of kitey, black, and red feather color in pigeons.
    a. For each of the preceding crosses, test your hypothesis by using a chi-square test.

Section 5.3

45. Suppose that you are tending a mouse colony at a genetics research institute and one day you discover a mouse with twisted ears. You breed this mouse with twisted ears and find that the trait is inherited. Both male and female mice have twisted ears, but, when you cross a twisted-eared male with a normal-eared female, you obtain results that differ from those obtained when you cross a twisted-eared female with normal-eared male: the reciprocal crosses give different results. Describe how you would go about determining whether this trait results from a sex-linked gene, a sex-influenced gene, a genetic maternal effect, a cytoplasmically inherited gene, or genomic imprinting. What crosses would you conduct and what results would be expected with these different types of inheritance?

6 Pedigree Analysis, Applications, and Genetic Testing

Megan, a 5-year-old with Hutchinson–Gilford progeria syndrome. Today, researchers seek to understand aging from the study of children with this rare autosomal dominant disorder that causes premature aging. [Courtesy of John Hurley for the Progeria Research Foundation (progeriaresearch.org).]

HUTCHINSON–GILFORD SYNDROME AND THE SECRET OF AGING

Spanish explorer Juan Ponce de León searched in vain for the fabled fountain of youth, the healing waters of which were reported to prevent old age. In the course of his explorations, Ponce de León discovered Florida, but he never located the source of perpetual youth. Like Ponce de León, scientists today seek to forestall aging, not in the waters of a mythical spring, but rather in understanding the molecular changes associated with the aging process. Curiously, their search has led to a small group of special children, who experience, not perpetual youth, but alarmingly premature old age.

Such children were first noticed by an English physician, Jonathan Hutchinson, who, in 1886, published a report of a young patient only 6 years old with the appearance of an 80-year-old man; the patient was bald with aged facial features and stiffness of joints. A few years later, another physician, Hastings Gilford, observed an additional case of the disorder, giving it the name progeria, derived from the Greek word for "prematurely old."

This exceedingly rare disease, known today as Hutchinson–Gilford progeria syndrome (HGPS), affects about 50 children worldwide and is characterized by the same features described by Hutchinson and Gilford more than 100 years ago. Children with HGPS appear healthy at birth but, by 2 years of age, usually begin to show features of accelerated aging, including cessation of growth, hair loss, aged skin, stiffness of joints, heart disease, and osteoporosis. By the age of 13, many have died, usually from coronary artery disease. HGPS is clearly genetic in origin and is usually inherited as an autosomal dominant trait.

In 1998, Leslie Gordon and Scott Berns's 2-year-old child Sam was diagnosed with HGPS. Gordon and Berns—both physicians—begin to seek information about the disease but, to their dismay, virtually none existed. Shocked, Gordon and Berns established the Progeria Research Foundation, and Gordon focused her career on directing the foundation and conducting research on HGPS. The efforts of the foundation stimulated a number of scientists to study HGPS and, in 2003, the gene that causes the disease was identified. Gordon was a part of the research team that made this important discovery.

6.1 Hutchinson–Gilford progeria syndrome is caused by mutations in the *lamin A* gene, which encodes the lamin A nuclear protein. A structural model of the lamin A protein.

HGPS results from mutations in a gene called *lamin A* (*LMNA*), which is located on human chromosome 1. The protein normally encoded by *LMNA* helps to form an internal scaffold that supports the nuclear membrane of the cell (**Figure 6.1**). The specific mutations that cause HGPS produce an abnormal, truncated form of the lamin A protein, which disrupts the nuclear membrane and, in some unknown way, results in premature aging.

The finding that mutations in *LMNA* cause HGPS hints that the architecture of the nucleus may play a role in normal aging. Although the relation of lamin A protein to normal aging in humans is still unclear, research on patients with HGPS may lead to a better understanding of how we all change with age. The fountain of youth may yet be found, perhaps in the cells of children with HGPS.

Hutchinson–Gilford progeria syndrome is just one of a large number of human diseases that are currently the focus of intensive genetic research. In this chapter, we consider human genetic characteristics and examine three important techniques used by geneticists to investigate these characteristics: pedigrees, twin studies, and adoption studies. At the end of the chapter, we will see how the information garnered with these techniques can be used in genetic counseling and prenatal diagnosis.

Keep in mind as you read this chapter that many important characteristics are influenced by both genes and environment, and separating these factors in humans is always difficult. Studies of twins and adopted persons are designed to distinguish the effects of genes and environment, but such studies are based on assumptions that may be difficult to meet for some human characteristics, particularly behavioral ones. Therefore, it's always prudent to interpret the results of such studies with caution. TRY PROBLEM 19

6.1 The Study of Genetics in Humans Is Constrained by Special Features of Human Biology and Culture

Humans are the best and the worst of all organisms for genetic study. On the one hand, we know more about human anatomy, physiology, and biochemistry than we know about most other organisms, and so many well-characterized traits are available for study. Families often keep detailed records about their members extending back many generations. Additionally, a number of important human diseases have a genetic component, and so the incentive for understanding human inheritance is tremendous. On the other hand, the study of human genetic characteristics presents some major obstacles.

First, controlled matings are not possible. With other organisms, geneticists carry out specific crosses to test their hypotheses about inheritance. We have seen, for example, how the testcross provides a convenient way to determine whether an individual organism having a dominant trait is homozygous or heterozygous. Unfortunately (for the geneticist at least), matings between humans are usually determined by romance, family expectations, or—occasionally—accident rather than by the requirements of a geneticist.

Another obstacle is that humans have a long generation time. Human reproductive age is not normally reached until 10 to 14 years after birth, and most people do not reproduce until they are 18 years of age or older; thus, generation time in humans is usually about 20 years. This long generation time means that, even if geneticists could control human crosses, they would have to wait on average 40 years just to observe the F_2 progeny. In contrast, generation time in *Drosophila* is 2 weeks; in bacteria, it's a mere 20 minutes.

Finally, human family size is generally small. Observation of even the simple genetic ratios that we learned in Chapter 3 would require a substantial number of progeny in each family. When parents produce only 2 children, the detection of a 3 : 1 ratio is impossible. Even an extremely large family of 10 to 15 children would not permit the recognition of a dihybrid 9 : 3 : 3 : 1 ratio.

Although these special constraints make genetic studies of humans more complex, understanding human heredity is tremendously important. So geneticists have been forced to develop techniques that are uniquely suited to human biology and culture. TRY PROBLEM 20

> **CONCEPTS**
>
> Although the principles of heredity are the same in humans and in other organisms, the study of human inheritance is constrained by the inability to control genetic crosses, the long generation time, and the small number of offspring.

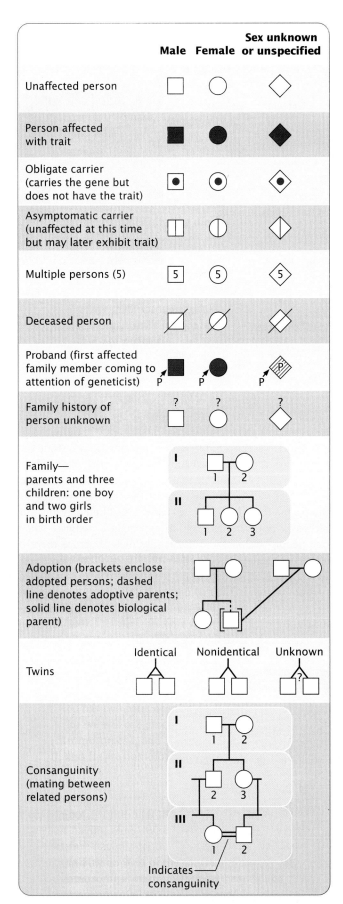

6.2 Standard symbols are used in pedigrees.

6.2 Geneticists Often Use Pedigrees to Study the Inheritance of Characteristics in Humans

An important technique used by geneticists to study human inheritance is the analysis of pedigrees. A **pedigree** is a pictorial representation of a family history, essentially a family tree that outlines the inheritance of one or more characteristics. When a particular characteristic or disease is observed in a person, a geneticist often studies the family of this affected person by drawing a pedigree.

Symbols Used in Pedigrees

The symbols commonly used in pedigrees are summarized in **Figure 6.2**. Males in a pedigree are represented by squares, females by circles. A horizontal line drawn between two symbols representing a man and a woman indicates a mating; children are connected to their parents by vertical lines extending downward from the parents. The pedigree shown in **Figure 6.3a** illustrates a family with Waardenburg syndrome, an autosomal dominant type of deafness that may be accompanied by fair skin, a white forelock, and visual problems (**Figure 6.3b**). Persons who exhibit the trait of interest are represented by filled circles and squares; in the pedigree of Figure 6.3a, the filled symbols represent members of the family who have Waardenburg syndrome. Unaffected members are represented by open circles and squares. The person from whom the pedigree is initiated is called the **proband** and is usually designated by an arrow (IV-2 in Figure 6.3a).

Let's look closely at Figure 6.3 and consider some additional features of a pedigree. Each generation in a pedigree is identified by a Roman numeral; within each generation, family members are assigned Arabic numerals, and children in each family are listed in birth order from left to right. Person II-4, a man with Waardenburg syndrome, mated with II-5, an unaffected woman, and they produced five children. The oldest of their children is III-8, a male with Waardenburg syndrome, and the youngest is III-14, an unaffected female. **TRY PROBLEM 21 →**

Analysis of Pedigrees

The limited number of offspring in most human families means that clear Mendelian ratios in a single pedigree are usually impossible to discern. Pedigree analysis requires a certain amount of genetic sleuthing, based on recognizing patterns associated with different modes of inheritance. For example, autosomal dominant traits should appear with equal frequency in both sexes and should not skip generations, provided that the trait is fully penetrant (see p. 103 in Chapter 5) and not sex influenced (see pp. 115–117 in Chapter 5).

Certain patterns may exclude the possibility of a particular mode of inheritance. For instance, a son inherits his X chromosome from his mother. If we observe that a trait

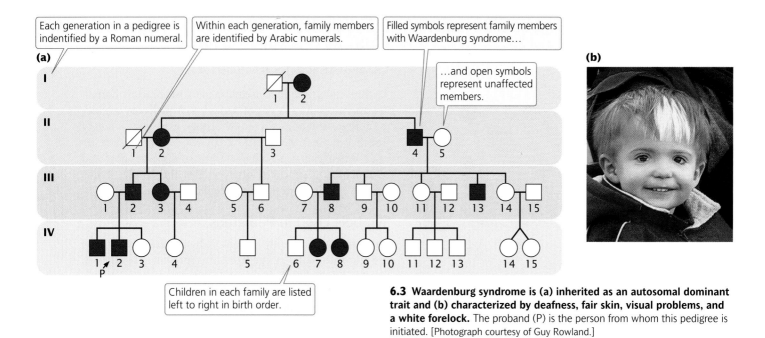

6.3 Waardenburg syndrome is (a) inherited as an autosomal dominant trait and (b) characterized by deafness, fair skin, visual problems, and a white forelock. The proband (P) is the person from whom this pedigree is initiated. [Photograph courtesy of Guy Rowland.]

is passed from father to son, we can exclude the possibility of X-linked inheritance. In the following sections, the traits discussed are assumed to be fully penetrant and rare.

Autosomal Recessive Traits

Autosomal recessive traits normally appear with equal frequency in both sexes (unless penetrance differs in males and females) and appear only when a person inherits two alleles for the trait, one from each parent. If the trait is uncommon, most parents of affected offspring are heterozygous and unaffected; consequently, the trait seems to skip generations (**Figure 6.4**). Frequently, a recessive allele may be passed for a number of generations without the trait appearing in a pedigree. Whenever both parents are heterozygous, approximately one-fourth of the offspring are expected to express the trait, but this ratio will not be obvious unless the family is large. In the rare event that both parents are affected by an autosomal recessive trait, all the offspring will be affected.

When a recessive trait is rare, persons from outside the family are usually homozygous for the normal allele. Thus, when an affected person mates with someone outside the family ($aa \times AA$), usually none of the children will display the trait, although all will be carriers (i.e., heterozygous). A recessive trait is more likely to appear in a pedigree when two people within the same family mate, because there is a greater chance of both parents carrying the same recessive allele. Mating between closely related people is called **consanguinity**. In the pedigree shown in Figure 6.4, persons III-3 and III-4 are first cousins, and both are heterozygous for the recessive allele; when they mate, $1/4$ of their children are expected to have the recessive trait.

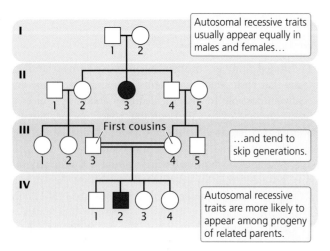

6.4 Autosomal recessive traits normally appear with equal frequency in both sexes and seem to skip generations.

CONCEPTS

Autosomal recessive traits appear with equal frequency in males and females. Affected children are commonly born to unaffected parents who are carriers of the gene for the trait, and the trait tends to skip generations. Recessive traits appear more frequently among the offspring of consanguine matings.

✔ CONCEPT CHECK 1

Autosomal recessive traits often appear in pedigrees in which there have been consanguine matings, because these traits

a. tend to skip generations.
b. appear only when both parents carry a copy of the gene for the trait, which is more likely when the parents are related.
c. usually arise in children born to parents who are unaffected.
d. appear equally in males and females.

A number of human metabolic diseases are inherited as autosomal recessive traits. One of them is Tay–Sachs disease. Children with Tay–Sachs disease appear healthy at birth but become listless and weak at about 6 months of age. Gradually, their physical and neurological conditions worsen, leading to blindness, deafness, and, eventually, death at 2 to 3 years of age. The disease results from the accumulation of a lipid called G_{M2} ganglioside in the brain. A normal component of brain cells, G_{M2} ganglioside is usually broken down by an enzyme called hexosaminidase A, but children with Tay–Sachs disease lack this enzyme. Excessive G_{M2} ganglioside accumulates in the brain, causing swelling and, ultimately, neurological symptoms. Heterozygotes have only one normal copy of the allele encoding hexosaminidase A and produce only about half the normal amount of the enzyme. However, this amount is enough to ensure that G_{M2} ganglioside is broken down normally, and heterozygotes are usually healthy.

Autosomal Dominant Traits

Autosomal dominant traits appear in both sexes with equal frequency, and both sexes are capable of transmitting these traits to their offspring. Every person with a dominant trait must have inherited the allele from at least one parent; autosomal dominant traits therefore do not skip generations (**Figure 6.5**). Exceptions to this rule arise when people acquire the trait as a result of a new mutation or when the trait has reduced penetrance.

If an autosomal dominant allele is rare, most people displaying the trait are heterozygous. When one parent is affected and heterozygous and the other parent is unaffected, approximately $1/2$ of the offspring will be affected. If both parents have the trait and are heterozygous, approximately $3/4$ of the children will be affected. Unaffected people do not transmit the trait to their descendants, provided that the trait is fully penetrant. In Figure 6.5, we see that none of the descendants of II-4 (who is unaffected) have the trait.

> **CONCEPTS**
>
> Autosomal dominant traits appear in both sexes with equal frequency. An affected person has an affected parent (unless the person carries new mutations), and the trait does not skip generations. Unaffected persons do not transmit the trait.
>
> ✔ **CONCEPT CHECK 2**
> When might you see an autosomal dominant trait skip generations?

A trait that is usually considered to be autosomal dominant is familial hypercholesterolemia, an inherited disease in which blood cholesterol is greatly elevated owing to a defect in cholesterol transport. Cholesterol is transported throughout the body in small soluble particles called lipoproteins (**Figure 6.6**). A principal lipoprotein in the transport of cholesterol is low-density lipoprotein (LDL). When an LDL molecule reaches a cell, it attaches to an LDL receptor, which

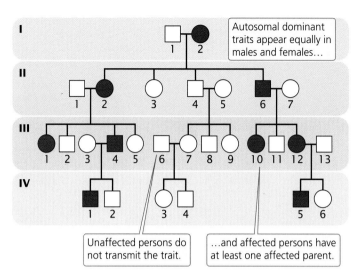

6.5 Autosomal dominant traits normally appear with equal frequency in both sexes and do not skip generations.

then moves the LDL through the cell membrane into the cytoplasm, where it is broken down and its cholesterol is released for use by the cell.

Familial hypercholesterolemia is due to a defect in the gene that normally encodes the LDL receptor. The disease is usually considered an autosomal dominant disorder because heterozygotes are deficient in LDL receptors and have elevated blood levels of cholesterol, leading to increased risk of coronary artery disease. Persons heterozygous for familial hypercholesterolemia have blood LDL levels that are twice normal and usually have heart attacks by the age of 35.

Very rarely, a person inherits two defective alleles for the LDL receptor. Such persons don't make any functional LDL receptors; their blood cholesterol levels are more than six times normal, and they may suffer a heart attack as early as age 2 and almost inevitably by age 20. Because homozygotes are more severely affected than heterozygotes, familial hypercholesterolemia is said to be incompletely dominant. However, homozygotes are rarely seen, and the common heterozygous form of the disease appears as a simple dominant trait in most pedigrees.

X-Linked Recessive Traits

X-linked recessive traits have a distinctive pattern of inheritance (**Figure 6.7**). First, these traits appear more frequently in males than in females because males need inherit only a single copy of the allele to display the trait, whereas females must inherit two copies of the allele, one from each parent, to be affected. Second, because a male inherits his X chromosome from his mother, affected males are usually born to unaffected mothers who carry an allele for the trait. Because the trait is passed from unaffected female to affected male to unaffected female, it tends to skip generations (see Figure 6.7). When a woman is heterozygous, approximately $1/2$ of her sons will be affected and $1/2$ of her

6.6 Low-density lipoprotein (LDL) particles transport cholesterol. The LDL receptor moves LDL from the bloodstream through the cell membrane into the cytoplasm.

daughters will be unaffected carriers. For example, we know that females I-2, II-2, and III-7 in Figure 6.7 are carriers because they transmit the trait to approximately half of their sons.

A third important characteristic of X-linked recessive traits is that they are not passed from father to son, because a son inherits his father's Y chromosome, not his X. In Figure 6.7, there is no case in which both a father and his son are affected. All daughters of an affected man, however, will be carriers (if their mother is homozygous for the normal allele). When a woman displays an X-linked recessive trait, she must be homozygous for the trait, and all of her sons also will display the trait.

CONCEPTS

Rare X-linked recessive traits appear more often in males than in females and are not passed from father to son. Affected sons are usually born to unaffected mothers who are carriers of the gene for the trait; thus X-linked recessive traits tend to skip generations.

✔ CONCEPT CHECK 3

How can you distinguish between an autosomal recessive trait with higher penetrance in males and an X-linked recessive trait?

An example of an X-linked recessive trait in humans is hemophilia A, also called classic hemophilia. Hemophila results from the absence of a protein necessary for blood to clot. The complex process of blood clotting consists of a cascade of reactions that includes more than 13 different factors. For this reason, there are several types of clotting disorders, each due to a glitch in a different step of the clotting pathway.

Hemophilia A results from abnormal or missing factor VIII, one of the proteins in the clotting cascade. The gene for factor VIII is located on the tip of the long arm of the X chromosome; so hemophilia A is an X-linked recessive disorder. People with hemophilia A bleed excessively; even small cuts and bruises can be life threatening. Spontaneous bleeding occurs in joints such as elbows, knees, and ankles, producing pain, swelling, and erosion of the bone. Fortunately, bleeding in people with hemophilia A can now be controlled by administering concentrated doses of factor VIII. The inheritance of hemophilia A is illustrated by the family of Queen Victoria of England (**Figure 6.8**).

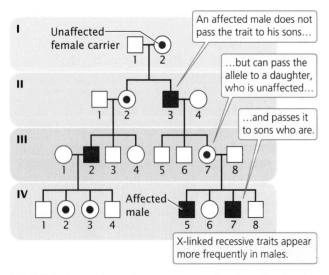

6.7 X-linked recessive traits appear more often in males than in females and are not passed from father to son.

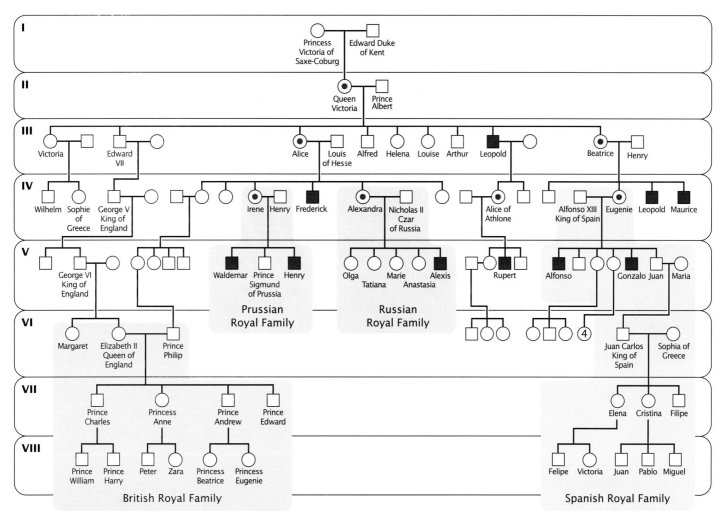

6.8 Classic hemophilia is inherited as an X-linked recessive trait. This pedigree is of hemophilia in the royal families of Europe.

X-Linked Dominant Traits

X-linked dominant traits appear in males and females, although they often affect more females than males. Each person with an X-linked dominant trait must have an affected parent (unless the person possesses a new mutation or the trait has reduced penetrance). X-linked dominant traits do not skip generations (**Figure 6.9**); affected men pass the trait on to all their daughters and none of their sons, as is seen in the children of I-1 in Figure 6.9. In contrast, affected women (if heterozygous) pass the trait on to about $1/2$ of their sons and about $1/2$ of their daughters, as seen in the children of III-6 in the pedigree. As with X-linked recessive traits, a male inherits an X-linked dominant trait only from his mother; the trait is not passed from father to son. This fact is what distinguishes X-linked dominant inheritance from autosomal dominant inheritance, in which a male can inherit the trait from his father. A female, on the other hand, inherits an X chromosome from both her mother and her father; so females can receive an X-linked trait from either parent.

6.9 X-linked dominant traits affect both males and females. An affected male must have an affected mother.

> **CONCEPTS**
>
> X-linked dominant traits affect both males and females. Affected males must have affected mothers (unless the males possess a new mutation), and they pass the trait on to all their daughters.
>
> ✔ **CONCEPT CHECK 4**
>
> A male affected with an X-linked dominant trait will have what proportion of offspring affected with the trait?
>
> a. ½ sons and ½ daughters   c. All daughters and no sons
> b. All sons and no daughters   d. ¾ daughters and ¼ sons

An example of an X-linked dominant trait in humans is hypophosphatemia, or familial vitamin-D-resistant rickets. People with this trait have features that superficially resemble those produced by rickets: bone deformities, stiff spines and joints, bowed legs, and mild growth deficiencies. This disorder, however, is resistant to treatment with vitamin D, which normally cures rickets. X-linked hypophosphatemia results from the defective transport of phosphate, especially in cells of the kidneys. People with this disorder excrete large amounts of phosphate in their urine, resulting in low levels of phosphate in the blood and reduced deposition of minerals in the bone. As is common with X-linked dominant traits, males with hypophosphatemia are often more severely affected than females.

Y-Linked Traits

Y-linked traits exhibit a specific, easily recognized pattern of inheritance. Only males are affected, and the trait is passed from father to son. If a man is affected, all his male offspring also should be affected, as is the case for I-1, II-4, II-6, III-6, and III-10 of the pedigree in **Figure 6.10**. Y-linked traits do not skip generations. As mentioned in Chapter 4,

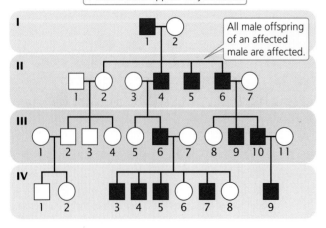

6.10 Y-linked traits appear only in males and are passed from a father to all his sons.

little genetic information is found on the human Y chromosome. Maleness is one of the few traits in humans that has been shown to be Y linked.

> **CONCEPTS**
>
> Y-linked traits appear only in males and are passed from a father to all his sons.
>
> ✔ **CONCEPT CHECK 5**
>
> What features of a pedigree would distinguish between a Y-linked trait and a trait that is rare, autosomal dominant, and sex-limited to males?

The major characteristics of autosomal recessive, autosomal dominant, X-linked recessive, X-linked dominant, and Y-linked traits are summarized in **Table 6.1**. **TRY PROBLEMS 23 AND 24 →**

Worked Problem

The following pedigree represents the inheritance of a rare disorder in an extended family. What is the most likely mode of inheritance for this disease? (Assume that the trait is fully penetrant.)

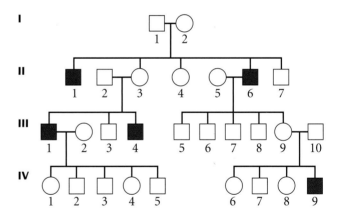

• Solution

To answer this question, we should consider each mode of inheritance and determine which, if any, we can eliminate. The trait appears only in males, and so autosomal dominant and autosomal recessive modes of inheritance are unlikely because traits with these modes appear equally in males and females. Additionally, autosomal dominance can be eliminated because some affected persons do not have an affected parent.

The trait is observed only among males in this pedigree, which might suggest Y-linked inheritance. However, for a Y-linked trait, affected men should pass the trait on to all their sons, which is not the case here; II-6 is an affected man

Table 6.1	Pedigree characteristics of autosomal recessive, autosomal dominant, X-linked recessive, X-linked dominant, and Y-linked traits

Autosomal recessive trait
1. Usually appears in both sexes with equal frequency.
2. Tends to skip generations.
3. Affected offspring are usually born to unaffected parents.
4. When both parents are heterozygous, approximately one-fourth of the offspring will be affected.
5. Appears more frequently among the children of consanguine marriages.

Autosomal dominant trait
1. Usually appears in both sexes with equal frequency.
2. Both sexes transmit the trait to their offspring.
3. Does not skip generations.
4. Affected offspring must have an affected parent unless they possess a new mutation.
5. When one parent is affected (heterozygous) and the other parent is unaffected, approximately half of the offspring will be affected.
6. Unaffected parents do not transmit the trait.

X-linked recessive trait
1. Usually more males than females are affected.
2. Affected sons are usually born to unaffected mothers; thus, the trait skips generations.
3. Approximately half of a carrier (heterozygous) mother's sons are affected.
4. Never passed from father to son.
5. All daughters of affected fathers are carriers.

X-linked dominant trait
1. Both males and females are usually affected; often more females than males are affected.
2. Does not skip generations. Affected sons must have an affected mother; affected daughters must have either an affected mother or an affected father.
3. Affected fathers will pass the trait on to all their daughters.
4. Affected mothers (if heterozygous) will pass the trait on to half of their sons and half of their daughters.

Y-linked trait
1. Only males are affected.
2. Passed from father to all sons.
3. Does not skip generations.

who has four unaffected male offspring. We can eliminate Y-linked inheritance.

X-linked dominance can be eliminated because affected men should pass an X-linked dominant trait on to all of their female offspring, and II-6 has an unaffected daughter (III-9).

X-linked recessive traits often appear more commonly in males, and affected males are usually born to unaffected female carriers; the pedigree shows this pattern of inheritance. For an X-linked trait, about half the sons of a heterozygous carrier mother should be affected. II-3 and III-9 are suspected carriers, and about half of their male children (three of five) are affected. Another important characteristic of an X-linked recessive trait is that it is not passed from father to son. We observe no father-to-son transmission in this pedigree. X-linked recessive is therefore the most likely mode of inheritance.

> For additional practice, try to determine the mode of inheritance for the pedigrees in Problem 25 at the end of the chapter.

6.3 Studying Twins and Adoptions Can Help Assess the Importance of Genes and Environment

Twins and adoptions provide natural experiments for separating the effects of genes and environmental factors in determining differences in traits. These two techniques have been widely used in genetic studies.

Types of Twins

Twins are of two types: **dizygotic** (nonidentical) **twins** arise when two separate eggs are fertilized by two different sperm, producing genetically distinct zygotes; **monozygotic** (identical) **twins** result when a single egg, fertilized by a single sperm, splits early in development into two separate embryos.

Because monozygotic twins result from a single egg and sperm (a single, "mono," zygote), they're genetically identical (except for rare somatic mutations), having 100% of their genes in common (**Figure 6.11a**). Dizygotic twins (**Figure 6.11b**), on the other hand, have on average only 50% of their genes in common, which is the same percentage that any pair of siblings has in common. Like other siblings, dizygotic twins may be of the same sex or of different sexes. The only difference between dizygotic twins and other siblings is that dizygotic twins are the same age and shared the same uterine environment. Dizygotic twinning often runs in families and the tendancy to produce dizygotic twins is influenced by heredity, but there appears to be little genetic tendency for producing monozygotic twins.

6.11 Monozygotic twins (a) are identical; dizygotic twins (b) are nonidentical. [Part a: Joe Carini/Index Stock Imagery/PictureQuest. Part b: Courtesy of Randi Rossignol.]

CONCEPTS

Dizygotic twins develop from two eggs fertilized by two separate sperm; they have, on average, 50% of their genes in common. Monozygotic twins develop from a single egg, fertilized by a single sperm, that splits into two embryos; they have 100% percent of their genes in common.

✔ CONCEPT CHECK 6

Why are monozygotic twins genetically identical, whereas dizygotic twins have only $\frac{1}{2}$ of their genes in common on average?

a. Monozygotic twins tend to look more similar.
b. Monozygotic twins are always the same sex.
c. Dizygotic twins arise more frequently with older mothers.
d. Monozygotic twins develop from a single embryo, whereas dizygotic twins develop from two embryos.

Concordance in Twins

Comparisons of dizygotic and monozygotic twins can be used to assess the importance of genetic and environmental factors in producing differences in a characteristic. This assessment is often made by calculating the concordance for a trait. If both members of a twin pair have a trait, the twins are said to be *concordant*; if only one member of the pair has the trait, the twins are said to be *discordant*. **Concordance** is the percentage of twin pairs that are concordant for a trait. Because identical twins have 100% of their genes in common and dizygotic twins have on average only 50% in common, genetically influenced traits should exhibit higher concordance in monozygotic twins. For instance, when one member of a monozygotic twin pair has epilepsy (**Table 6.2**), the other twin of the pair has epilepsy about 59% of the time; so the monozygotic concordance for epilepsy is 59%. However, when a dizygotic twin has epilepsy, the other twin has epilepsy only 19% of the time (19% dizygotic concordance). The higher concordance in the monozygotic twins suggests that genes influence epilepsy, a finding supported by the results of other family studies of this disease. Concordance values for several additional human traits and diseases are listed in Table 6.2.

The hallmark of a genetic influence on a particular trait is higher concordance in monozygotic twins compared with concordance in dizygotic twins. High concordance in monozygotic twins by itself does not signal a genetic influence. Twins normally share the same environment—they are raised in the same home, have the same friends, attend the same school—and so high concordance may be due to common genes or to common environment. If the high concordance is due to environmental factors, then dizygotic twins, who also share the same environment, should have just as high a concordance as that of monozygotic twins. When genes influence the trait, however, monozygotic twin pairs should exhibit higher concordance than that of dizygotic twin pairs, because monozygotic twins have a greater percentage of genes in common. It is important to note that

Table 6.2 Concordance of monozygotic and dizygotic twins for several traits

	Concordance (%)	
Trait	Monozygotic	Dizygotic
1. Heart attack (males)	39	26
2. Heart attack (females)	44	14
3. Bronchial asthma	47	24
4. Cancer (all sites)	12	15
5. Epilepsy	59	19
6. Rheumatoid arthritis	32	6
7. Multiple sclerosis	28	5

Sources: (1 and 2) B. Havald and M. Hauge, U.S. Public Health Service Publication 1103 (1963), pp. 61–67; (3, 4, and 5) B. Havald and M. Hauge, *Genetics and the Epidemiology of Chronic Diseases* (U.S. Department of Health, Education, and Welfare, 1965); (6) J. S. Lawrence, *Annals of Rheumatic Diseases* 26:357–379, 1970; (7) G. C. Ebers et al., *American Journal of Human Genetics* 36:495, 1984.

any discordance among monozygotic twins is usually due to environmental factors, because monozygotic twins are genetically identical.

The use of twins in genetic research rests on the important assumption that, when concordance for monozygotic twins is greater than that for dizygotic twins, it is because monozygotic twins are more similar in their genes and not because they have experienced a more similar environment. The degree of environmental similarity between monozygotic twins and dizygotic twins is assumed to be the same. This assumption may not always be correct, particularly for human behaviors. Because they look alike, identical twins may be treated more similarly by parents, teachers, and peers than are nonidentical twins. Evidence of this similar treatment is seen in the past tendency of parents to dress identical twins alike. In spite of this potential complication, twin studies have played a pivotal role in the study of human genetics. TRY PROBLEM 31

A Twin Study of Asthma

To illustrate the use of twins in genetic research, let's consider a study of asthma. Asthma is characterized by constriction of the airways and the secretion of mucus into the air passages, causing coughing, labored breathing, and wheezing (**Figure 6.12**). Severe cases can be life threatening. Asthma is a major health problem in industrialized countries and appears to be on the rise. The incidence of childhood asthma varies widely, the highest rates (from 17% to 30%) are in the United Kingdom, Australia, and New Zealand.

A number of environmental stimuli are known to precipitate asthma attacks, including dust, pollen, air pollution, respiratory infections, exercise, cold air, and emotional stress. Allergies frequently accompany asthma, suggesting that asthma is a disorder of the immune system, but the precise relation between immune function and asthma is poorly understood. Numerous studies have shown that genetic factors are important in asthma.

A genetic study of childhood asthma was conducted as a part of the Twins Early Development Study in England, an ongoing research project that studies more than 15,000 twins born in the United Kingdom between 1994 and 1996. These twins have been assessed for language, cognitive development, behavioral problems, and academic achievement at ages 7 and 9, and the genetic and environmental contributions to a number of their traits were examined. In the asthma study, researchers looked at a sample of 4910 twins at age 4. Parents of the twins were asked whether either of their twins had been prescribed medication to control asthma; those children receiving asthma medication were considered to have asthma.

Concordances of the monozygotic and dizygotic twins for asthma are shown in **Table 6.3**. The concordance value for the monozygotic twins (65%) was significantly higher than that for the dizygotic twins (37%), and the researchers concluded that, among the 4-year-olds included in the study, asthma was strongly influenced by genetic factors. The fact that even monozygotic twins were discordant 35% of the time indicates that environmental factors also play a role in asthma.

Table 6.3 Concordance values for asthma among monozygotic and dizygotic twins from the Twins Early Development Study

Group	Number of Pairs	Concordance (%)
Monozygotic	1658	65
Dizygotic	3252	37

CONCEPTS

Higher concordance for monozygotic twins compared with that for dizygotic twins indicates that genetic factors play a role in determining differences in a trait. Low concordance for monozygotic twins indicates that environmental factors play a significant role.

✔ CONCEPT CHECK 7

A trait exhibits 100% concordance for both monozygotic and dizygotic twins. What conclusion can you draw about the role of genetic factors in determining differences in the trait?

a. Genetic factors are extremely important.
b. Genetic factors are somewhat important.
c. Genetic factors are unimportant.
d. Both genetic and environmental factors are important.

6.12 Twin studies show that asthma, characterized by constriction of the airways, is caused by a combination of genetic and environmental factors. [Custom Medical Stock/Photolibrary.]

Experiment

Question: Is body-mass index (BMI) influenced by genetic factors?

Methods

Compare the body-mass index of adopted children with those of their adoptive and biological parents.

Results

Biological parents: Overweight biological parents tend to have overweight children.

Adoptive parents: There is no consistent association between the weight of children and that of their adoptive parents.

Conclusion: Genetic factors influence body-mass index.

6.13 Adoption studies demonstrate that obesity has a genetic influence. [Redrawn with the permission of the *New England Journal of Medicine* 314:195, 1986.]

Adoption Studies

Another technique used by geneticists to analyze human inheritance is the study of adopted people. This approach is one of the most powerful for distinguishing the effects of genes and environment on characteristics.

For a variety of reasons, many children each year are separated from their biological parents soon after birth and adopted by adults with whom they have no genetic relationship. These adopted persons have no more genes in common with their adoptive parents than do two randomly chosen persons; however, they do share an environment with their adoptive parents. In contrast, the adopted persons have 50% of their genes in common with each of their biological parents but do not share the same environment with them. If adopted persons and their adoptive parents show similarities in a characteristic, these similarities can be attributed to environmental factors. If, on the other hand, adopted persons and their biological parents show similarities, these similarities are likely to be due to genetic factors. Comparisons of adopted persons with their adoptive parents and with their biological parents can therefore help to define the roles of genetic and environmental factors in the determination of human variation. For example, adoption studies were instrumental in showing that schizophrenia has a genetic basis. Adoption studies have also shown that obesity, as measured by body-mass index, is at least partly influenced by genetics (**Figure 6.13**).

Adoption studies assume that the environments of biological and adoptive families are independent (i.e., not more alike than would be expected by chance). This assumption may not always be correct, because adoption agencies carefully choose adoptive parents and may select a family that resembles the biological family. Thus, some of the similarity between adopted persons and their biological parents may be due to these similar environments and not due to common genetic factors. In addition, offspring and the biological mother share the same environment during prenatal development. **TRY PROBLEM 33**

CONCEPTS

Similarities between adopted persons and their genetically unrelated adoptive parents indicate that environmental factors affect a particular characteristic; similarities between adopted persons and their biological parents indicate that genetic factors influence the characteristic.

✓ CONCEPT CHECK 8

What assumptions underlie the use of adoption studies in genetics?
a. Adoptees have no contact with their biological parents after birth.
b. The foster parents and biological parents are not related.
c. The environments of biological and adopted parents are independent.
d. All of the above.

6.4 Genetic Counseling and Genetic Testing Provide Information to Those Concerned about Genetic Diseases and Traits

Our knowledge of human genetic diseases and disorders has expanded rapidly in recent years. The *Online Mendelian Inheritance in Man* now lists more than 19,000 human genetic diseases, disorders, genes, and traits that have a simple genetic basis. Research has provided a great deal of information about the inheritance, chromosomal location, biochemical basis, and symptoms of many of these genetic traits, diseases, and disorders. This information is often useful to people who have a genetic condition.

Genetic Counseling

Genetic counseling is a field that provides information to patients and others who are concerned about hereditary conditions. It is an educational process that helps patients

Table 6.4	Common reasons for seeking genetic counseling

1. A person knows of a genetic disease in the family.
2. A couple has given birth to a child with a genetic disease, birth defect, or chromosomal abnormality.
3. A couple has a child who is mentally retarded or has a close relative who is mentally retarded.
4. An older woman becomes pregnant or wants to become pregnant. There is disagreement about the age at which a prospective mother who has no other risk factor should seek genetic counseling; many experts suggest that any prospective mother age 35 or older should seek genetic counseling.
5. Husband and wife are closely related (e.g., first cousins).
6. A couple experiences difficulties achieving a successful pregnancy.
7. A pregnant woman is concerned about exposure to an environmental substance (drug, chemical, or virus) that causes birth defects.
8. A couple needs assistance in interpreting the results of a prenatal or other test.
9. Both prospective parents are known carriers for a recessive genetic disease.

and family members deal with many aspects of a genetic condition including a diagnosis, information about symptoms and treatment, and information about the mode of inheritance. Genetic counseling also helps a patient and the family cope with the psychological and physical stress that may be associated with the disorder. Clearly, all of these considerations cannot be handled by a single person; so most genetic counseling is done by a team that can include counselors, physicians, medical geneticists, and laboratory personnel. **Table 6.4** lists some common reasons for seeking genetic counseling.

Genetic counseling usually begins with a diagnosis of the condition. On the bases of a physical examination, biochemical tests, DNA testing, chromosome analysis, family history, and other information, a physician determines the cause of the condition. An accurate diagnosis is critical, because treatment and the probability of passing the condition on may vary, depending on the diagnosis. For example, there are a number of different types of dwarfism, which may be caused by chromosome abnormalities, single-gene mutations, hormonal imbalances, or environmental factors. People who have dwarfism resulting from an autosomal dominant gene have a 50% chance of passing the condition on to their children, whereas people who have dwarfism caused by a rare recessive gene have a low likelihood of passing the trait on to their children.

When the nature of the condition is known, a genetic counselor meets with the patient and members of the patient's family and explains the diagnosis. A family pedigree may be constructed with the use of a computer program or by hand, and the probability of transmitting the condition to future generations can be calculated for different family members. The counselor helps the family interpret the genetic risks and explains various available reproductive options, including prenatal diagnosis, artificial insemination, and in vitro fertilization. Often family members have questions about genetic testing that may be available to help determine whether they carry a genetic mutation. The counselor helps them decide whether genetic testing is appropriate and which tests to apply. After the test results are in, the genetic counselor usually helps family members interpret the results.

A family's decision about future pregnancies frequently depends on the magnitude of the genetic risk, the severity and effects of the condition, the importance of having children, and religious and cultural views. Traditionally, genetic counselors have been trained to apply *nondirected* counseling, which means that they provide information and facilitate discussion but do not bring their own opinions and values into the discussion. The goal of nondirected counseling is for the family to reach its own decision on the basis of the best available information.

Because of the growing number of genetic tests and the complexity of assessing genetic risk, there is now some movement away from completely nondirected counseling. The goal is still to provide the family with information about all options and to reach the best decision for the family, but that goal may sometimes require the counselor to recommend certain options, much as a physician recommends the most appropriate medical treatments for his or her patient.

Genetic conditions are often perceived differently from other diseases and medical problems, because genetic conditions are intrinsic to the individual person and can be passed on to children. Such perceptions may produce feelings of guilt about past reproductive choices and intense personal dilemmas about future choices. Genetic counselors are trained to help patients and their families recognize and cope with these feelings.

Who does genetic counseling? In the United States, about 4500 health professionals are currently certified in genetics by the American Board of Medical Genetics or the American Board of Genetic Counseling. About half of them are specifically trained in genetic counseling, usually by completing a special 2-year masters program that provides education in both genetics and counseling. Most of the remainder are physicians and scientists certified in medical or clinical genetics. Because of the shortage of genetic counselors and medical geneticists, information about genetic testing and genetic risk is often conveyed by primary care physicians, nurses, and social workers. **TRY PROBLEMS 9 AND 10**

CONCEPTS

Genetic counseling is an educational process that provides patients and their families with information about a genetic condition, its medical implications, the probabilities that others in the family may have the disease, and reproductive options. It also helps patients and their families cope with the psychological and physical stress associated with a genetic condition.

Genetic Testing

The ultimate goal of genetic testing is to recognize the potential for a genetic condition at an early stage. In some cases, genetic testing allows people to make informed choices about reproduction. In other cases, genetic testing allows early intervention that may lessen or even prevent the development of the condition. For those who know that they are at risk for a genetic condition, genetic testing may help alleviate anxiety associated with the uncertainty of their situation. Genetic testing includes prenatal testing and postnatal testing.

Prenatal genetic tests are those that are conducted before birth and now include procedures for diagnosing several hundred genetic diseases and disorders (**Table 6.5**). The major purpose of prenatal tests is to provide families with the information that they need to make choices during pregnancies and, in some cases, to prepare for the birth of a child with a genetic condition. The Human Genome Project has accelerated the rate at which new genes are being isolated and new genetic tests are being developed. In spite of these advances, prenatal tests are still not available for many common genetic diseases, and no test can guarantee that a "perfect" child will be born. Several approaches to prenatal diagnosis are described in the following sections.

Ultrasonography Some genetic conditions can be detected through direct visualization of the fetus. Such visualization is most commonly done with the use of **ultrasonography**—usually referred to as ultrasound. In this technique, high-frequency sound is beamed into the uterus; when the sound waves encounter dense tissue, they bounce back and are transformed into a picture (**Figure 6.14**). The size of the fetus can be determined, as can genetic conditions such as neural-tube defects (defects in the development of the spinal column and the skull) and skeletal abnormalities.

Amniocentesis Most prenatal testing requires fetal tissue, which can be obtained in several ways. The most widely used method is **amniocentesis**, a procedure for obtaining a sample of amniotic fluid from a pregnant woman (**Figure 6.15**). Amniotic fluid—the substance that fills the amniotic sac and surrounds the developing fetus—contains fetal cells that can be used for genetic testing.

Amniocentesis is routinely performed as an outpatient procedure either with or without the use of a local anesthetic. First, ultrasonography is used to locate the position

Table 6.5 Examples of genetic diseases and disorders that can be detected prenatally and the techniques used in their detection

Disorder	Method of Detection
Chromosome abnormalities	Examination of a karyotype from cells obtained by amniocentesis or chorionic villus sampling
Cleft lip and palate	Ultrasound
Cystic fibrosis	DNA analysis of cells obtained by amniocentesis or chorionic villus sampling
Dwarfism	Ultrasound or X-ray; some forms can be detected by DNA analysis of cells obtained by amniocentesis or chorionic villus sampling
Hemophilia	Fetal blood sampling* or DNA analysis of cells obtained by amniocentesis or chorionic villus sampling
Lesch–Nyhan syndrome	Biochemical tests on cells obtained by amniocentesis or chorionic villus sampling
Neural-tube defects	Initial screening with maternal blood test, followed by biochemical tests on amniotic fluid obtained by amniocentesis or by the detection of birth defects with the use of ultrasound
Osteogenesis imperfecta (brittle bones)	Ultrasound or X-ray
Phenylketonuria	DNA analysis of cells obtained by amniocentesis or chorionic villus sampling
Sickle-cell anemia	Fetal blood sampling* or DNA analysis of cells obtained by amniocentesis or chorionic villus sampling
Tay–Sachs disease	Biochemical tests on cells obtained by amniocentesis or chorionic villus sampling

*A sample of fetal blood is obtained by inserting a needle into the umbilical cord.

6.14 Ultrasonography can be used to detect some genetic disorders in a fetus and to locate the fetus during amniocentesis and chorionic villus sampling. [Getty Images.]

these reasons, genetic information about the fetus may not be available until the 17th or 18th week of pregnancy. By this stage, abortion carries a risk of complications and is even more stressful for the parents. **Chorionic villus sampling** (CVS) can be performed earlier (between the 10th and 12th weeks of pregnancy) and collects a larger amount of fetal tissue, which eliminates the necessity of culturing the cells.

In CVS, a catheter—a soft plastic tube—is inserted into the vagina (**Figure 6.16**) and, with the use of ultrasonography for guidance, is pushed through the cervix into the uterus. The tip of the tube is placed into contact with the chorion, the outer layer of the placenta. Suction is then applied, and a small piece of the chorion is removed. Although the chorion is composed of fetal cells, it is a part of the placenta that is expelled from the uterus after birth; so the tissue that is removed is not actually from the fetus. This tissue contains millions of actively dividing cells that can be used directly in many genetic tests. Chorionic villus sampling has a somewhat higher risk of complication than that of amniocentesis; the results of several studies suggest that this procedure may increase the incidence of limb defects in the fetus when performed earlier than 10 weeks of gestation.

Fetal cells obtained by amniocentesis or by CVS can be used to prepare a **karyotype**, which is a picture of a complete set of metaphase chromosomes. Karyotypes can be studied for chromosome abnormalities (see Chapter 9). Biochemical analyses can be conducted on fetal cells to determine the presence of particular metabolic products of genes. For genetic diseases in which the DNA sequence of the causative gene has been determined, the DNA sequence (DNA testing; see Chapter 10) can be examined for defective alleles.

of the fetus in the uterus. Next, a long, sterile needle is inserted through the abdominal wall into the amniotic sac, and a small amount of amniotic fluid is withdrawn through the needle. Fetal cells are separated from the amniotic fluid and placed in a culture medium that stimulates them to grow and divide. Genetic tests are then performed on the cultured cells. Complications with amniocentesis (mostly miscarriage) are uncommon, arising in only about 1 in 400 procedures.

Chorionic villus sampling A major disadvantage of amniocentesis is that it is routinely performed at about the 15th to 18th week of a pregnancy (although some obstetricians successfully perform amniocentesis earlier). The cells obtained by amniocentesis must then be cultured before genetic tests can be performed, requiring yet more time. For

Maternal blood screening tests Increased risk of some genetic conditions can be detected by examining levels of certain substances in the blood of the mother (referred to as a **maternal blood screening test**). However, these tests do not determine the presence of a genetic problem; rather,

6.15 Amniocentesis is a procedure for obtaining fetal cells for genetic testing.

6.16 Chorionic villus sampling (CVS) is another procedure for obtaining fetal cells for genetic testing.

they simply indicate that the fetus is at increased risk and hence are referred to as *screening* tests. When increased risk is detected, follow-up tests (additional blood-screening tests, ultrasound, amniocentesis, or all three) are usually conducted.

One substance examined in maternal screening tests is α-fetoprotein, a protein that is normally produced by the fetus during development and is present in fetal blood, amniotic fluid, and the mother's blood during pregnancy. The level of α-fetoprotein is significantly higher than normal when the fetus has a neural-tube defect or one of several other disorders. Some chromosome abnormalities produce lower-than-normal levels of α-fetoprotein. Measuring the amount of α-fetoprotein in the mother's blood gives an indication of these conditions.

The American Collge of Obstetricians and Gynecologists recommends that physicians offer all pregnant women maternal blood screening tests. One typical test, carried out between 11 and 13 weeks of pregnancy, measures human chorionic gonadotropin (hCG, a pregnancy hormone) and a substance called pregnancy-associated plasma protein A (PAPP-A). When the fetus has certain chromosomal defects, the level of PAPP-A tends to be low and the level of hCG tends to be high. The risk of a chromosomal abnormality is calculated on the basis of the levels of hCG and PAPP-A in the mother's blood, along with the results of utrasound tests. Another test, referred to as the quad screen, is offered between 15 and 20 weeks of pregnancy; it measures the levels of four substances: α-fetoprotein, hCG, estriol, and inhibin. The risk of chromosomal abnormalities and certain other birth defects is calculated on the basis of the combined levels of the four substances plus the mother's age, weight, ethnic background, and number of fetuses. The quad screen successfully detects Down syndrome (due to three copies of chromosome 21) 81% of the time.

Noninvasive fetal diagnosis Prenatal tests that utilize only maternal blood are highly desirable because they are noninvasive and pose no risk to the fetus. In addition to maternal blood screening tests, which measure chemical substances produced by the fetus or placenta, some newer procedures diagnose genetic conditions by examining fetal DNA obtained from fetal cells in maternal blood. During pregnancy, a few fetal cells are released into the mother's circulatory system, where they mix and circulate with her blood. Recent advances have made it possible to detect and separate fetal cells from maternal blood cells (a procedure called **fetal cell sorting**) with the use of lasers and automated cell-sorting machines. The fetal cells obtained can be cultured for chromosome analysis or used as a source of fetal DNA for molecular testing (see Chapter 10). Maternal blood also contains low levels of free-floating fetal DNA, which also can be tested for mutations. Although still experimental, noninvasive fetal diagnosis has now been used to detect Down syndrome and diseases such as cystic fibrosis and thalassemia (a blood disorder). A current limitation is that the procedure is not as accurate as amniocentesis or CVS, because enough fetal cells for genetic testing cannot always be obtained. Additionally, only mutated genes from the father can be detected, because there is no way to completely separate maternal DNA from the fetal cells. Thus, if the mother carries a copy of the mutation, determining whether the fetus also carries the gene is impossible.

Preimplantation genetic diagnosis Prenatal genetic tests provide today's prospective parents with increasing amounts of information about the health of their future children. New reproductive technologies provide couples with options for using this information. One of these technologies is in vitro fertilization. In this procedure, hormones are used to induce ovulation. The ovulated eggs are surgically removed from the surface of the ovary, placed in a laboratory dish, and fertilized with sperm. The resulting embryo is then implanted in the uterus. Thousands of babies resulting from in vitro fertilization have now been born.

Genetic testing can be combined with in vitro fertilization to allow the implantation of embryos that are free of a specific genetic defect. Called **preimplantation genetic diagnosis** (PGD), this technique enables people who carry a genetic defect to avoid producing a child with the disorder. For example, if a woman is a carrier of an X-linked recessive disease, approximately half of her sons are expected to have the disease. Through in vitro fertilization and preimplantation testing, an embryo without the disorder can be selected for implantation in her uterus.

The procedure begins with the production of several single-celled embryos through in vitro fertilization. The embryos are allowed to divide several times until they reach the 8- or 16-cell stage. At this point, one cell is removed from each embryo and tested for the genetic abnormality. Removing a single cell at this early stage does not harm the embryo. After determination of which embryos are free of the disorder, a healthy embryo is selected and implanted in the woman's uterus.

Preimplantation genetic diagnosis requires the ability to conduct a genetic test on a single cell. Such testing is possible with the use of the polymerase chain reaction through which minute quantities of DNA can be amplified (replicated) quickly. After amplification of the cell's DNA, the DNA sequence is examined. Preimplantation genetic diagnosis has resulted in the birth of thousands of healthy children. Its use raises a number of ethical concerns, because it can be used as a means of selecting for or against genetic traits that have nothing to do with medical concerns. For example, it can potentially be used to select for a child with genes for a certain eye color or genes for increased height.

Newborn screening Testing for genetic disorders in newborn infants is called **newborn screening**. All states in the United States and many other countries require by law that newborn infants be tested for some genetic diseases and conditions. In 2006, the American College of Medical Genetics recommended mandatory screening for 29 conditions (**Table 6.6**), and many states have now adopted this list for newborn testing These genetic conditions were chosen because early identification can lead to effective treatment. For example, as mentioned in Chapter 5, phenylketonuria is an autosomal recessive disease that, if not treated at an early age, can result in mental retardation. But early intervention, through the administration of a modified diet, prevents retardation.

Table 6.6 Genetic conditions recommended for mandatory screening by the American College of Medical Genetics

Medium-chain acyl-CoA dehydrogenase deficiency
Congenital hypothyroidism
Phenylketonuria
Biotinidase deficiency
Sickle-cell anemia (Hb SS disease)
Congenital adrenal hyperplasia (21-hydroxylase deficiency)
Isovaleric acidemia
Very long chain acyl-CoA dehydrogenase deficiency
Maple syrup (urine) disease
Classical galactosemia
Hb S β-thalassemia
Hb S C disease
Long-chain L-3-hydroxyacyl-CoA dehydrogenase deficiency
Glutaric acidemia type I
3-Hydroxy-3-methyl glutaric aciduria
Trifunctional protein deficiency
Multiple carboxylase deficiency
Methylmalonic acidemia (mutase deficiency)
Homocystinuria (due to cystathionine β-synthase deficiency)
3-Methylcrotonyl-CoA carboxylase deficiency
Hearing loss
Methylmalonic acidemia
Propionic acidemia
Carnitine uptake defect
β-Ketothiolase deficiency
Citrullinemia
Argininosuccinic acidemia
Tyrosinemia type I
Cystic fibrosis

Most newborn testing is done by collecting a drop of an infant's blood on a card soon after birth. The card is then shipped to a laboratory for genetic analysis. These cards, which have been in use for more than 40 years, are often stored for many years—sometimes indefinitely—after their initial use in genetic testing. Recently, blood samples from archived cards have been used for research studies, such as examining genetic mutations that lead to childhood leukemia and birth defects and looking at the relation between genes and environmental toxins. These studies and other proposed studies have raised ethical concerns about the long-term storage of these blood samples and their use—in some cases, without consent—in some research studies. For example, in 2009, five families filed a federal lawsuit against the Texas Department of State Health Services, claiming that the agency violated privacy rights by collecting blood samples from Texas children and storing them for future research purposes, without the parents' knowledge or consent. This action prompted passage of a new Texas state law requiring that parents be informed at the time of a child's birth about potential uses of the samples; it also gives parents the right to withdraw their consent in the future and have the sample destroyed.

Presymptomatic testing In addition to testing for genetic diseases in fetuses and newborns, testing healthy adults for genes that might predispose them to a genetic condition in the future is now possible. This type of testing is known as **presymptomatic genetic testing**. For example, presymptomatic testing is available for members of families that have an autosomal dominant form of breast cancer. In this case, early identification of the disease-causing allele allows for closer surveillance and the early detection of tumors. Presymptomatic testing is also available for some genetic diseases for which no treatment is available, such as Huntington disease, an autosomal dominant disease that leads to slow physical and mental deterioration in middle age. Presymptomatic testing for untreatable conditions raises a number of social and ethical questions (see Chapter 10).

Heterozygote screening Another form of genetic testing in adults is **heterozygote screening**. In this type of screening, members of a population are tested to identify heterozygous carriers of recessive disease-causing alleles—people who are healthy but have the potential to produce children with a particular disease.

Testing for Tay–Sachs disease is a successful example of heterozygote screening. In the general population of North America, the frequency of Tay–Sachs disease is only about 1 person in 360,000. Among Ashkenazi Jews (descendants of Jewish people who settled in eastern and central Europe), the frequency is 100 times as great. A simple blood test is used to identify Ashkenazi Jews who carry the allele for Tay–Sachs disease. If a man and woman are both heterozygotes, approximately one in four of their children is expected to have Tay–Sachs disease. Couples identified as heterozygous carriers may use that information in deciding whether to have children. A prenatal test also is available for determining if the fetus of an at-risk couple will have Tay–Sachs disease. Screening programs have led to a significant decline in the number of children of Ashkenazi ancestry born with Tay–Sachs disease (now fewer than 10 children per year in the United States). **TRY PROBLEM 12**

CONCEPTS

Genetic testing is used to screen newborns for genetic diseases, detect persons who are heterozygous for recessive diseases, detect disease-causing alleles in those who have not yet developed symptoms of the disease, and detect defective alleles in unborn babies. Preimplantation genetic diagnosis combined with in vitro fertilization allows for the selection of embryos that are free from specific genetic diseases.

✔ **CONCEPT CHECK 9**

How does preimplantation genetic diagnosis differ from prenatal genetic testing?

Interpreting Genetic Tests

Today, more than 1200 genetic tests are clinically available and several hundred more are available through research studies. Information from the Human Genome Project and research examining the association of genetic markers and disease will greatly increase the number and complexity of genetic tests available in the future. Many of these tests will be for complex multifactorial diseases that are influenced by both genetics and environment, such as coronary artery disease, diabetes, asthma, some types of cancer, and depression.

Interpreting the results of genetic tests is often complicated by several factors. First, some genetic diseases are caused by numerous different mutations. For example, more than 1000 different mutations at a single locus can cause cystic fibrosis, an autosomal recessive disease in which chloride ion transport is defective (see p. 102 in Chapter 5). Genetic tests typically screen for only the most common mutations; uncommon and rare mutations are not detected. Therefore, a negative result does not mean that a genetic defect is absent; it indicates only that the person does not have a common mutation. When family members have the disease, their DNA can be examined to determine the nature of the mutation and other family members can then be screened for the same mutation, but this option is not possible if affected family members are unavailable or unwilling to be tested.

A second problem lies in interpreting the results of genetic tests. For a classic genetic disease such as Tay–Sachs disease, the inheritance of two copies of the gene virtually ensures that a person will have the disease. However, this is not the case for many genetic diseases in which penetrance is incomplete and environmental factors play a role. For these conditions, carrying a disease-predisposing mutation only elevates a person's risk of acquiring the disease. The risk associated with a particular mutation is a statistical estimate, based on the average effect of the mutation on many people. In this case, the calculated risk may provide little useful information to a specific person.

> **CONCEPTS**
>
> Interpreting genetic tests is complicated by the presence of multiple causative mutations, incomplete penetrance, and the influence of environmental factors.

Direct-to-Consumer Genetic Testing

An increasing number of genetic tests are now being offered to anyone interested in investigating his or her own hereditary conditions, without requiring a health-care provider. These **direct-to-consumer genetic tests** are available for testing a large and growing array of genetic conditions in adults and children, everything from single-gene disorders such as cystic fibrosis to multifactorial conditions such as obesity, cardiovascular disease, athletic performance, and predisposition to nicotine addiction. Direct-to-consumer tests are also available for paternity testing and for determining ancestry.

Many direct-to-consumer genetic tests are advertised and ordered through the Internet. After a person orders a test, the company sends a kit for collecting a sample of DNA (usually cells from inside the cheek or a spot of blood). The person collects the sample and sends it back to the company, which performs the test and sends the results to the person. Geneticists, public health officials, and consumer advocates have raised a number of concerns about direct-to-consumer genetic testing, including concerns that some tests are offered without appropriate information and genetic counseling and that consumers are often not equipped to interpret the results. Other concerns focus on the accuracy of some tests, the confidentiality of the results, and whether indications of risk provided by the test are even useful.

Advocates of direct-to-consumer genetic tests contend that the tests provide greater access to testing and enhanced confidentiality. Many states do not regulate direct-to-consumer testing and, currently, there is little federal oversight. **TRY PROBLEM 15**

Genetic Discrimination and Privacy

With the development of many new genetic tests, concerns have been raised about privacy regarding genetic information and the potential for genetic discrimination. Research shows that many people at risk for genetic diseases avoid genetic testing because they fear that the results would make it difficult for them to obtain health insurance or that the information might adversely affect their employability. Some of those who do seek genetic testing pay for it themselves and use aliases to prevent the results from becoming part of their health records. Fears about genetic discrimination have been reinforced by past practices. In the 1970s, some African Americans were forced to undergo genetic testing for sickle-cell anemia (an autosomal recessive disorder) and those who were identified as carriers faced employment discrimination, in spite of the fact that carriers are healthy.

In response to these concerns, the U. S. Congress passed the **Genetic Information Nondiscrimination Act**, commonly called GINA, in 2008. This law prohibits health insurers from using genetic information to make decisions about health-insurance coverage and rates. It also prevents employers from using genetic information in employment decisions and prohibits health insurers and employers from asking or requiring a person to take a genetic test. Results of genetic testing receive some degree of protection by other federal regulations that cover the uses and disclosure of individual health information. However, GINA covers health insurance and employment only; it does not apply to life, disability, and long-term care insurance. **TRY PROBLEM 16**

> **CONCEPTS**
>
> The growing number of genetic tests and their complexity have raised several concerns, including those about direct-to-consumer tests, genetic discrimination, and privacy regarding test results.

6.5 Comparison of Human and Chimpanzee Genomes Is Helping to Reveal Genes That Make Humans Unique

Now that we've examined the inheritance of human traits and how some of these conditions can be detected with genetic tests, let's see what our genes can tell us about our place in evolution. Chimpanzees are our closest living relatives, yet we differ from chimps anatomically and in a huge number of behavioral, social, and intellectual skills. The perceived large degree of difference between chimpanzees and humans is manifested by the placement of these two species in entirely different primate families (humans in Hominidae and chimpanzees in Pongidae).

In spite of the large phenotypic gulf between humans and chimpanzees, findings in early DNA-hybridization studies suggested that they differ in only about 1% of their DNA sequences. Recent sequencing of human and chimpanzee genomes has provided more-precise estimates of the genetic differences that separate these species: about 1% of the two genomes differ in base sequences and about 3% differ in regard to deletions and insertions. Thus, more than 95% of the DNA of humans and chimpanzees is identical. But, clearly, humans are not chimpanzees. What, then, are the genetic differences that make us distinctly human? What are the genes that give us our human qualities?

Geneticists are now identifying genes that contribute to human uniqueness and potentially played an important role in the evolution of modern humans. In some cases, these genes have been identified through the study of mutations that cause abnormalities in our human traits, such as brain size or language ability. One set of genes that potentially contribute to human uniqueness regulates brain size. Mutations at six loci, known as *microcephalin 1* through *microcephalin 6* (*MCPH1–MCPH6*), cause microcephaly, a condition in which the brain is severely reduced in overall size but brain structure is not affected. The observation that mutations in *microcephalin* drastically affect brain size has led to the suggestion that, in the course of human evolution, selection for alleles encoding large brains at one or more of these genes might have led to enlarged brain size in humans. Geneticists have studied variation in the *microcephalin* genes of different primates and have come to the conclusion that there was strong selection in the recent past for the sequences of the *microcephalin* genes that are currently found in humans.

Another candidate gene for helping to make humans unique is called *FOXP2*, which is required for human speech. Mutations in the *FOXP2* gene often cause speech and language disabilities. Evolutionary studies of sequence variation in the *FOXP2* gene of mice, humans, and other primates reveal evidence of past selection and suggest that humans acquired their version of the gene (which enables human speech) no more than 200,000 years ago, about the time at which modern humans emerged. In 2007, scientists extracted DNA from Neanderthal fossils and found that these now-extinct humans possessed the same version of the *FOXP2* gene seen in modern humans, perhaps giving them the ability to talk.

Other genes that may contribute to human uniqueness have been identified on the basis of their rapid evolution in the human lineage. For example, the *HAR1* gene evolved slowly in other vertebrates but has undergone rapid evolution since humans diverged from chimpanzees. This gene is expressed in brain cells and some evidence suggests that it may be important in the development of the cerebral cortex, an area of the brain that is greatly enlarged in humans.

CONCEPTS SUMMARY

- Constraints on the genetic study of human traits include the inability to conduct controlled crosses, long generation time, small family size, and the difficulty of separating genetic and environmental influences. Pedigrees are often used to study the inheritance of traits in humans.

- Autosomal recessive traits typically appear with equal frequency in both sexes and tend to skip generations. When both parents are heterozygous for a particular autosomal recessive trait, approximately one-fourth of their offspring will have the trait. Recessive traits are more likely to appear in families with consanguinity (mating between closely related persons).

- Autosomal dominant traits usually appear equally in both sexes and do not skip generations. When one parent is affected and heterozygous for an autosomal dominant trait, approximately half of the offspring will have the trait. Unaffected people do not normally transmit an autosomal dominant trait to their offspring.

- X-linked recessive traits appear more frequently in males than in females. When a woman is a heterozygous carrier for an X-linked recessive trait and a man is unaffected, approximately half of their sons will have the trait and half of their daughters will be unaffected carriers. X-linked traits are not passed on from father to son.

- X-linked dominant traits appear in males and females but more frequently in females. They do not skip generations. Affected men pass an X-linked dominant trait to all of their daughters but none of their sons. Heterozygous women pass the trait to half of their sons and half of their daughters.

- Y-linked traits appear only in males and are passed from father to all sons.

- A trait's higher concordance in monozygotic than in dizygotic twins indicates a genetic influence on the trait; less than 100% concordance in monozygotic twins indicates environmental influences on the trait.

- Similarities between adopted children and their biological parents indicate the importance of genetic factors in the expression of a trait; similarities between adopted children and their genetically unrelated adoptive parents indicate the influence of environmental factors.

- Genetic counseling provides information and support to people concerned about hereditary conditions in their families.

- Genetic testing includes prenatal diagnosis, screening for disease-causing alleles in newborns, the detection of people heterozygous for recessive alleles, and presymptomatic testing for the presence of a disease-causing allele in at-risk people.

- The interpretation of genetic tests may be complicated by the presence of numerous causitive mutations, incomplete penetrance, and the influence of environmental factors.

- The recent availability of direct-to-consumer genetic tests has raised concerns about the adequacey of information provided, the absence of genetic counseling, acccuracy, privacy, and the practical uses of some tests.

- Genetic testing has raised concerns about genetic descrimination and privacy regarding test results. The Genetic Information Nondiscrimination Act prohibits the use of genetic information in deciding health insurability and employment.

- Genetic research has identified a number of genes that may contribute to human uniqueness, including genes that influence brain size and speech.

IMPORTANT TERMS

pedigree (p. 137)
proband (p. 137)
consanguinity (p. 138)
dizygotic twins (p. 143)
monozygotic twins (p. 143)
concordance (p. 144)
genetic counseling (p. 146)
ultrasonography (p. 148)

amniocentesis (p. 148)
chorionic villus sampling (CVS) (p. 149)
karyotype (p. 149)
maternal blood screening test (p. 149)
fetal cell sorting (p. 150)
preimplantation genetic diagnosis (PGD) (p. 151)
newborn screening (p. 151)

presymptomatic genetic testing (p. 152)
heterozygote screening (p. 152)
direct-to-consumer genetic test (p. 153)
Genetic Information Nondiscrimination Act (GINA) (p. 153)

ANSWERS TO CONCEPT CHECKS

1. b

2. It might skip generations when a new mutation arises or the trait has reduced penetrance.

3. If X-linked recessive, the trait will not be passed from father to son.

4. c

5. If the trait were Y linked, an affected male would pass it on to all his sons, whereas, if the trait were autosomal and sex-limited, affected heterozygous males would pass it on to only half of their sons on average.

6. d

7. c

8. d

9. Preimplantation genetic diagnosis determines the presence of disease-causing genes in an embryo at an early stage, before it is implanted in the uterus and initiates pregnancy. Prenatal genetic diagnosis determines the presence of disease-causing genes or chromosomes in a developing fetus.

WORKED PROBLEMS

1. Joanna has "short fingers" (brachydactyly). She has two older brothers who are identical twins; both have short fingers. Joanna's two younger sisters have normal fingers. Joanna's mother has normal fingers, and her father has short fingers. Joanna's paternal grandmother (her father's mother) has short fingers; her paternal grandfather (her father's father), who is now deceased, had normal fingers. Both of Joanna's maternal grandparents (her mother's parents) have normal fingers. Joanna marries Tom, who has normal fingers; they adopt a son named Bill who has normal fingers. Bill's biological parents both have normal fingers. After adopting Bill, Joanna and Tom produce two children: an older daughter with short fingers and a younger son with normal fingers.

   a. Using standard symbols and labels, draw a pedigree illustrating the inheritance of short fingers in Joanna's family.

   b. What is the most likely mode of inheritance for short fingers in this family?

   c. If Joanna and Tom have another biological child, what is the probability (based on your answer to part *b*) that this child will have short fingers?

• Solution

   a. In the pedigree for the family, identify persons with the trait (short fingers) by filled circles (females) and filled squares (males). Connect Joanna's identical twin brothers to the line above by drawing diagonal lines that have a horizontal line between them. Enclose the adopted child of Joanna and Tom in brackets; connect him to his biological parents by drawing a diagonal line and to his adopted parents by a dashed line.

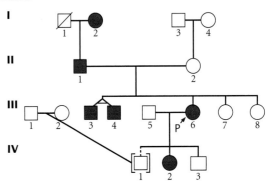

   b. The most likely mode of inheritance for short fingers in this family is autosomal dominant. The trait appears equally in males and females and does not skip generations. When one parent has the trait, it appears in approximately half of that parent's sons and daughters, although the number of children in the families is small. We can eliminate Y-linked inheritance because the trait is found in females as well as males. If short fingers were X-linked recessive, females with the trait would be expected to pass the trait to all their sons, but Joanna (III-6), who has short fingers, produced a son with normal fingers. For X-linked dominant traits, affected men should pass the trait to all their daughters; because male II-1 has short fingers and produced two daughters without short fingers (III-7 and III-8), we know that the trait cannot be

X-linked dominant. The trait is unlikely to be autosomal recessive, because it does not skip generations and approximately half the children of affected parents have the trait.

c. If having short fingers is autosomal dominant, Tom must be homozygous (*bb*) because he has normal fingers. Joanna must be heterozygous (*Bb*) because she and Tom have produced both short- and normal-fingered offspring. In a cross between a heterozygote and homozygote, half the progeny are expected to be heterozygous and the other half homozygous ($Bb \times bb \rightarrow \frac{1}{2} Bb, \frac{1}{2} bb$); so the probability that Joanna's and Tom's next biological child will have short fingers is $\frac{1}{2}$.

2. Concordance values for a series of traits were measured in monozygotic twins and dizygotic twins; the results are shown in the following table. For each trait, indicate whether the rates of concordance suggest genetic influences, environmental influences, or both. Explain your reasoning.

	Characteristic	Concordance (%) Monozygotic	Dizygotic
a.	ABO blood type	100	65
b.	Diabetes	85	36
c.	Coffee drinking	80	80
d.	Smoking	75	42
e.	Schizophrenia	53	16

• **Solution**

a. The concordance for ABO blood type in the monozygotic twins is 100%. This high concordance in monozygotic twins does not, by itself, indicate a genetic basis for the trait. An important indicator of a genetic influence on the trait is lower concordance in dizygotic twins. Because concordance for ABO blood type is substantially lower in the dizygotic twins, we would be safe in concluding that genes play a role in determining differences in ABO blood types.

b. The concordance for diabetes is substantially higher in monozygotic twins than in dizygotic twins; therefore, we can conclude that genetic factors play some role in susceptibility to diabetes. The fact that monozygotic twins show a concordance less than 100% suggests that environmental factors also play a role.

c. Both monozygotic and dizygotic twins exhibit the same high concordance for coffee drinking; so we can conclude that there is little genetic influence on coffee drinking. The fact that monozygotic twins show a concordance less than 100% suggests that environmental factors play a role.

d. The concordance for smoking is lower in dizygotic twins than in monozygotic twins; so genetic factors appear to influence the tendency to smoke. The fact that monozygotic twins show a concordance less than 100% suggests that environmental factors also play a role.

e. Monozygotic twins exhibit substantially higher concordance for schizophrenia than do dizygotic twins; so we can conclude that genetic factors influence this psychiatric disorder. Because the concordance of monozygotic twins is substantially less than 100%, we can also conclude that environmental factors play a role in the disorder as well.

COMPREHENSION QUESTIONS

Section 6.1

*1. What three factors complicate the task of studying the inheritance of human characteristics?

Section 6.2

2. Who is the proband in a pedigree? Is the proband always found in the last generation of the pedigree? Why or why not?

*3. For each of the following modes of inheritance, describe the features that will be exhibited in a pedigree in which the trait is present: autosomal recessive, autosomal dominant, X-linked recessive, X-linked dominant, and Y-linked inheritance.

4. How does the pedigree of an autosomal recessive trait differ from the pedigree of an X-linked recessive trait?

5. Other than the fact that a Y-linked trait appears only in males, how does the pedigree of a Y-linked trait differ from the pedigree of an autosomal dominant trait?

Section 6.3

*6. What are the two types of twins and how do they arise?

7. Explain how a comparison of concordance in monozygotic and dizygotic twins can be used to determine the extent to which the expression of a trait is influenced by genes or by environmental factors.

8. How are adoption studies used to separate the effects of genes and environment in the study of human characteristics?

Section 6.4

*9. What is genetic counseling?

10. Give at least four different reasons for seeking genetic counseling.

11. Briefly define newborn screening, heterozygote screening, presymptomatic testing, and prenatal diagnosis.

12. Compare the advantages and disadvantages of aminocentesis versus chorionic villus sampling for prenatal diagnosis.

13. What is preimplantation genetic diagnosis?

14. How does heterozygote screening differ from presymptomatic genetic testing?

15. What are direct-to-consumer genetic tests? What are some of the concerns about these tests?

16. What activities does the Genetic Information Nondiscrimination Act prohibit?
17. How might genetic testing lead to genetic discrimination?

Section 6.5

18. Briefly describe some of the recently discovered genes that contribute to human uniqueness and their possible importance in human evolution.

APPLICATION QUESTIONS AND PROBLEMS

Introduction

19. Hutchinson–Gilford progeria syndrome (HGPS) is caused by an autosomal dominant gene. Sam, the child discussed in the introduction to this chapter, has HGPS, but both of his parents are unaffected. Later in the chapter, we learned that autosomal dominant traits do not normally skip generations and that affected children have at least one affected parent. Why does HGPS not display these characteristics of an autosomal dominant trait?

Section 6.1

20. If humans have characteristics that make them unsuitable for genetic analysis, such as long generation time, small family size, and uncontrolled crosses, why do geneticists study humans? Give several reasons why humans have been the focus of so much genetic study.

Section 6.2

*21. Joe is color blind. Both his mother and his father have normal vision, but his mother's father (Joe's maternal grandfather) is color blind. All Joe's other grandparents have normal color vision. Joe has three sisters—Patty, Betsy, and Lora—all with normal color vision. Joe's oldest sister, Patty, is married to a man with normal color vision; they have two children, a 9-year-old color-blind boy and a 4-year-old girl with normal color vision.

   a. Using standard symbols and labels, draw a pedigree of Joe's family.
   b. What is the most likely mode of inheritance for color blindness in Joe's family?
   c. If Joe marries a woman who has no family history of color blindness, what is the probability that their first child will be a color-blind boy?
   d. If Joe marries a woman who is a carrier of the color-blind allele, what is the probability that their first child will be a color-blind boy?
   e. If Patty and her husband have another child, what is the probability that the child will be a color-blind boy?

22. Many studies have suggested a strong genetic predisposition to migraine headaches, but the mode of inheritance is not clear. L. Russo and colleagues examined migraine headaches in several families, two of which are shown below (L. Russo et al. 2005. *American Journal of Human Genetics* 76:327–333). What is the most likely mode of inheritance for migraine headaches in these families? Explain your reasoning.

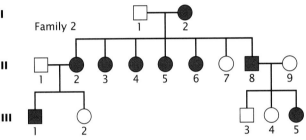

23. Dent disease is a rare disorder of the kidney, in which reabsorption of filtered solutes is impaired and there is progressive renal failure. R. R. Hoopes and colleagues studied mutations associated with Dent disease in the following family (R. R. Hoopes et al. 2005. *American Journal of Human Genetics* 76:260–267).

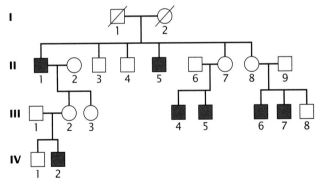

   a. On the basis of this pedigree, what is the most likely mode of inheritance for the disease? Explain your reasoning.
   b. From your answer to part *a*, give the most likely genotypes for all persons in the pedigree.

24. A man with a specific unusual genetic trait marries an unaffected woman and they have four children. Pedigrees of this family are shown in parts *a* through *e*, but the presence or absence of the trait in the children is not indicated. For each type of inheritance, indicate how many children of each sex are expected to express the trait by

filling in the appropriate circles and squares. Assume that the trait is rare and fully penetrant.

a. Autosomal recessive trait

b. Autosomal dominant trait

c. X-linked recessive trait

d. X-linked dominant trait

e. Y-linked trait

*25. For each of the following pedigrees, give the most likely mode of inheritance, assuming that the trait is rare. Carefully explain your reasoning.

a.

b.

c.

d.

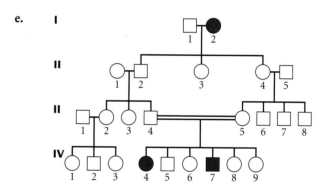

e.

26. The trait represented in the following pedigree is expressed only in the males of the family. Is the trait Y linked? Why or why not? If you believe that the trait is not Y linked, propose an alternate explanation for its inheritance.

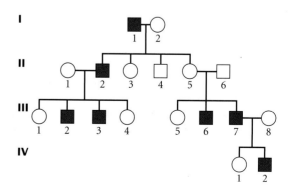

*27. The following pedigree illustrates the inheritance of Nance–Horan syndrome, a rare genetic condition in which affected persons have cataracts and abnormally shaped teeth.

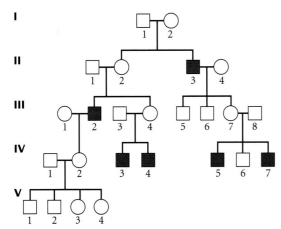

[Pedigree adapted from D. Stambolian, R. A. Lewis, K. Buetow, A. Bond, and R. Nussbaum. 1990. *American Journal of Human Genetics* 47:15.]

  a. On the basis of this pedigree, what do you think is the most likely mode of inheritance for Nance–Horan syndrome?
  b. If couple III-7 and III-8 have another child, what is the probability that the child will have Nance–Horan syndrome?
  c. If III-2 and III-7 were to mate, what is the probability that one of their children would have Nance–Horan syndrome?

28. The following pedigree illustrates the inheritance of ringed hair, a condition in which each hair is differentiated into light and dark zones. What mode or modes of inheritance are possible for the ringed-hair trait in this family?

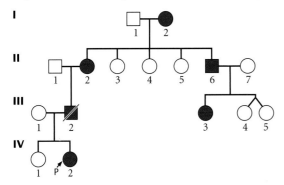

[Pedigree adapted from L. M. Ashley and R. S. Jacques. 1950. *Journal of Heredity* 41:83.]

29. Ectrodactyly is a rare condition in which the fingers are absent and the hand is split. This condition is usually inherited as an autosomal dominant trait. Ademar Freire-Maia reported the appearance of ectrodactyly in a family in São Paulo, Brazil, whose pedigree is shown here. Is this pedigree consistent with autosomal dominant inheritance? If not, what mode of inheritance is most likely? Explain your reasoning.

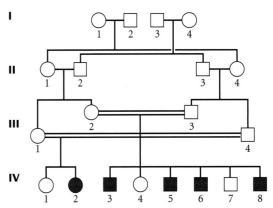

[Pedigree adapted from A. Freire-Maia. 1971. *Journal of Heredity* 62:53.]

30. The complete absence of one or more teeth (tooth agenesis) is a common trait in humans—indeed, more than 20% of humans lack one or more of their third molars. However, more-severe absence of teeth, defined as missing six or more teeth, is less common and frequently an inherited condition. L. Lammi and colleagues examined tooth agenesis in the Finnish family shown in the pedigree on the next page (L. Lammi. 2004. *American Journal of Human Genetics* 74:1043–1050).

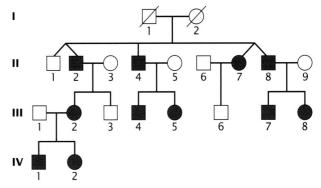

[Pedigree adapted from L. Lammi. 2004. *American Journal of Human Genetics* 74:1043–1050.]

  a. What is the most likely mode of inheritance for tooth agenesis in this family? Explain your reasoning.
  b. Are the two sets of twins in this family monozygotic or dizygotic twins? What is the basis of your answer?
  c. If IV-2 married a man who has a full set of teeth, what is the probability that their child would have tooth agenesis?
  d. If III-2 and III-7 married and had a child, what is the probability that their child would have tooth agenesis?

Section 6.4

*31. A geneticist studies a series of characteristics in monozygotic twins and dizygotic twins, obtaining the concordances listed on page 160. For each characteristic, indicate whether the rates of concordance suggest genetic influences, environmental influences, or both. Explain your reasoning.

	Concordance (%)	
Characteristic	Monozygotic	Dizygotic
Migraine headaches	60	30
Eye color	100	40
Measles	90	90
Clubfoot	30	10
High blood pressure	70	40
Handedness	70	70
Tuberculosis	5	5

32. M. T. Tsuang and colleagues studied drug dependence in male twin pairs (M. T. Tsuang et al. 1996. *American Journal of Medical Genetics* 67:473–477). They found that 4 out of 30 monozygotic twins were concordant for dependence on opioid drugs, whereas 1 out of 34 dizygotic twins were concordant for the same trait. Calculate the concordance rates for opioid dependence in these monozygotic and dizygotic twins. On the basis of these data, what conclusion can you make concerning the roles of genetic and environmental factors in opioid dependence?

Section 6.5

33. In a study of schizophrenia (a mental disorder including disorganization of thought and withdrawal from reality), researchers looked at the prevalence of the disorder in the biological and adoptive parents of people who were adopted as children; they found the following results:

	Prevalence of schizophrenia	
Adopted Persons	Biological parents	Adoptive parents
With schizophrenia	12	2
Without schizophrenia	6	4

[Source: S. S. Kety et al. 1978. in *The Nature of Schizophrenia: New Approaches to Research and Treatment*, L. C. Wynne, R. L. Cromwell, and S. Matthysse, Eds. New York: Wiley, 1978, pp. 25–37.]

What can you conclude from these results concerning the role of genetics in schizophrenia? Explain your reasoning.

Section 6.3

34. What, if any, ethical issues might arise from the widespread use of noninvasive fetal diagnosis, which can be carried out much earlier than amniocentesis or chorionic villus sampling?

CHALLENGE QUESTIONS

Section 6.1

35. Many genetic studies, particularly those of recessive traits, have focused on small isolated human populations, such as those on islands. Suggest one or more advantages that isolated populations might have for the study of recessive traits.

Section 6.2

36. Draw a pedigree that represents an autosomal dominant trait, sex-limited to males, and that excludes the possibility that the trait is Y linked.

37. A. C. Stevenson and E. A. Cheeseman studied deafness in a family in Northern Ireland and recorded the following pedigree (A. C. Stevenson and E. A. Cheeseman. 1956. *Annals of Human Genetics* 20:177–231).

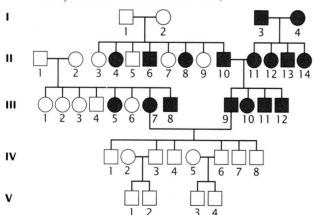

[Pedigree adapted from A. C. Stevenson and E. A. Cheeseman. 1956. *Annals of Human Genetics* 20:177–231.]

a. If you consider only generations I through III, what is the most likely mode of inheritance for this type of deafness?

b. Provide a possible explanation for the cross between III-7 and III-9 and the results for generations IV through V.

Section 6.3

38. Dizygotic twinning often runs in families and its frequency varies among ethnic groups, whereas monozygotic twinning rarely runs in families and its frequency is quite constant among ethnic groups. These observations have been interpreted as evidence for a genetic basis for variation in dizygotic twinning but for little genetic basis for variation in monozygotic twinning. Can you suggest a possible reason for these differences in the genetic tendencies of dizygotic and monozygotic twinning?

7 Linkage, Recombination, and Eukaryotic Gene Mapping

Pattern baldness is a hereditary trait. Recent research demonstrated that a gene for pattern baldness is linked to genetic markers located on the X chromosome, leading to the discovery that pattern baldness is influenced by variation in the androgen-receptor gene. The trait is seen in three generations of the Adams family: (a) John Adams (1735–1826), the second president of the United States, was father to (b) John Quincy Adams (1767–1848), sixth U.S. president, who was father to (c) Charles Francis Adams (1807–1886). [Part a: National Museum of American Art, Washington, D.C./Art Resource, New York. Part b: National Portrait Gallery, Washington, D.C./Art Resource, New York. Part c: Bettmann/Corbis.]

LINKED GENES AND BALD HEADS

For many, baldness is the curse of manhood. Twenty-five percent of men begin balding by age 30 and almost half are bald to some degree by age 50. In the United States, baldness affects more than 40 million men, who spend hundreds of millions of dollars each year on hair-loss treatment. Baldness is not just a matter of vanity: bald males are more likely to suffer from heart disease, high blood pressure, and prostate cancer.

Baldness can arise for a number of different reasons, including illness, injury, drugs, and heredity. The most-common type of baldness seen in men is pattern baldness—technically known as androgenic alopecia—in which hair is lost prematurely from the front and top of the head. More than 95% of hair loss in men is pattern baldness. Although pattern baldness is also seen in women, it is usually expressed weakly as mild thinning of the hair. The trait is clearly stimulated by male sex hormones (androgens), as evidenced by the observation that males castrated at an early age rarely become bald (but this is not recommended as a preventive treatment for baldness).

A strong hereditary influence on pattern baldness has long been recognized, but the exact mode of inheritance has been controversial. An early study suggested that it was autosomal dominant in males and recessive in females, an example of a sex influenced trait (see Chapter 4). Other evidence and common folklore suggested that a man inherits baldness from his mother's side of the family, exhibiting X-linked inheritance.

In 2005, geneticist Axel Hillmer and his colleagues set out to locate the gene that causes pattern baldness. They suspected that the gene for pattern baldness might be located on the X chromosome, but they had no idea where on the X chromosome the gene might reside. To identify the location of the gene, they conducted a linkage analysis study, in which they looked for an association between the inheritance of pattern baldness and the inheritance of genetic variants known to be located on the X chromosome. The genetic variants used in the study were single-nucleotide polymorphisms (SNPs, pronounced "snips"), which are positions in the genome where individuals vary in a single nucleotide. The geneticists studied the inheritance of pattern baldness and SNPs in 95 families in which at least two brothers developed pattern baldness at an early age.

What Hillmer and his colleagues found was that pattern baldness and SNPS from the X chromosome were not inherited independently, as predicted by Mendel's principle of independent assortment. Instead, they tended to be inherited together, which occurs when genes are physically linked on the same chromosome and segregate together in meiosis.

As we will learn in this chapter, linkage between genes is broken down by recombination (crossing over), and the amount of recombination between genes is directly related to the distance between them. In 1911, Thomas Hunt Morgan and his student Alfred Sturtevant demonstrated in fruit flies that genes can be mapped by determining the rates of recombination between the genes. Using this method for the families with pattern baldness, Hillmer and his colleagues demonstrated that the gene for pattern baldness is closely linked to SNPs located at position p12–22, a region that includes the androgen-receptor gene, on the X chromosome. The androgen receptor gene encodes a protein that binds male sex hormones. Given the clear involvement of male hormones in the development of pattern baldness, the androgen-receptor gene seemed a likely candidate for causing pattern baldness. Further analysis revealed that certain alleles of the androgen-receptor gene were closely associated with the inheritance of pattern baldness, and the androgen-receptor gene is almost certainly responsible for much of the differences in pattern baldness seen in the families examined. Additional studies conducted in 2008 found that genes on chromosomes 3 and 20 also appear to contribute to the expression of pattern baldness. **TRY PROBLEM 13**

This chapter explores the inheritance of genes located on the same chromosome. These linked genes do not strictly obey Mendel's principle of independent assortment; rather, they tend to be inherited together. This tendency requires a new approach to understanding their inheritance and predicting the types of offspring produced. A critical piece of information necessary for predicting the results of these crosses is the arrangement of the genes on the chromosomes; thus, it will be necessary to think about the relation between genes and chromosomes. A key to understanding the inheritance of linked genes is to make the conceptual connection between the genotypes in a cross and the behavior of chromosomes in meiosis.

We will begin our exploration of linkage by first comparing the inheritance of two linked genes with the inheritance of two genes that assort independently. We will then examine how crossing over, or recombination, breaks up linked genes. This knowledge of linkage and recombination will be used for predicting the results of genetic crosses in which genes are linked and for mapping genes. Later in the chapter, we will focus on physical methods of determining the chromosomal locations of genes. The final section examines variation in rates of recombination.

7.1 Linked Genes Do Not Assort Independently

Chapter 3 introduced Mendel's principles of segregation and independent assortment. Let's take a moment to review these two important concepts. The principle of segregation states that each individual diploid organism possesses two alleles at a locus that separate in meiosis, with one allele going into each gamete. The principle of independent assortment provides additional information about the process of segregation: it tells us that, in the process of separation, the two alleles at a locus act independently of alleles at other loci.

The independent separation of alleles results in *recombination*, the sorting of alleles into new combinations. Consider a cross between individuals homozygous for two different pairs of alleles: *AA BB* × *aa bb*. The first parent, *AA BB*, produces gametes with the alleles *A B*, and the second parent, *aa bb*, produces gametes with the alleles *a b*, resulting in F_1 progeny with genotype *Aa Bb* (**Figure 7.1**).

7.1 Recombination is the sorting of alleles into new combinations.

Recombination means that, when one of the F_1 progeny reproduces, the combination of alleles in its gametes may differ from the combinations in the gametes from its parents. In other words, the F_1 may produce gametes with alleles *A b* or *a B* in addition to gametes with *A B* or *a b*.

Mendel derived his principles of segregation and independent assortment by observing the progeny of genetic crosses, but he had no idea of what biological processes produced these phenomena. In 1903, Walter Sutton proposed a biological basis for Mendel's principles, called the chromosome theory of heredity (see Chapter 3). This theory holds that genes are found on chromosomes. Let's restate Mendel's two principles in relation to the chromosome theory of heredity. The principle of segregation states that a diploid organism possesses two alleles for a trait, each of which is located at the same position, or locus, on each of the two homologous chromosomes. These chromosomes segregate in meiosis, with each gamete receiving one homolog. The principle of independent assortment states that, in meiosis, each pair of homologous chromosomes assorts independently of other homologous pairs. With this new perspective, it is easy to see that the number of chromosomes in most organisms is limited and that there are certain to be more genes than chromosomes; so some genes must be present on the same chromosome and should not assort independently. Genes located close together on the same chromosome are called **linked genes** and belong to the same **linkage group**. Linked genes travel together in meiosis, eventually arriving at the same destination (the same gamete), and are not expected to assort independently.

All of the characteristics examined by Mendel in peas did display independent assortment and, after the rediscovery of Mendel's work, the first genetic characteristics studied in other organisms also seemed to assort independently. How could genes be carried on a limited number of chromosomes and yet assort independently?

The apparent inconsistency between the principle of independent assortment and the chromosome theory of heredity soon disappeared as biologists began finding genetic characteristics that did not assort independently. One of the first cases was reported in sweet peas by William Bateson, Edith Rebecca Saunders, and Reginald C. Punnett in 1905. They crossed a homozygous strain of peas having purple flowers and long pollen grains with a homozygous strain having red flowers and round pollen grains. All the F_1 had purple flowers and long pollen grains, indicating that purple was dominant over red and long was dominant over round. When they intercrossed the F_1, the resulting F_2 progeny did not appear in the 9 : 3 : 3 : 1 ratio expected with independent assortment (**Figure 7.2**). An excess of F_2 plants had purple flowers and long pollen or red flowers and round pollen (the parental phenotypes). Although Bateson, Saunders, and Punnett were unable to explain these results, we now know that the two loci that they examined lie close together on the same chromosome and therefore do not assort independently.

7.2 Nonindependent assortment of flower color and pollen shape in sweet peas.

7.2 Linked Genes Segregate Together and Crossing Over Produces Recombination Between Them

Genes that are close together on the same chromosome usually segregate as a unit and are therefore inherited together. However, genes occasionally switch from one homologous chromosome to the other through the process of crossing over (see Chapter 2), as illustrated

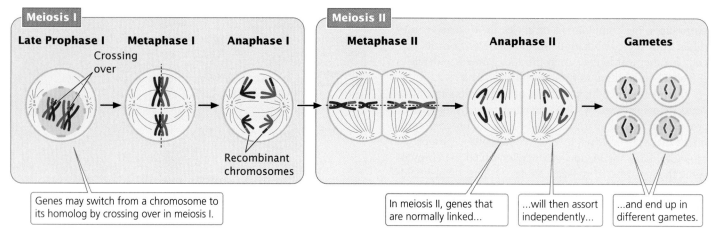

7.3 Crossing over takes place in meiosis and is responsible for recombination.

in **Figure 7.3**. Crossing over results in recombination; it breaks up the associations of genes that are close together on the same chromosome. Linkage and crossing over can be seen as processes that have opposite effects: linkage keeps particular genes together, and crossing over mixes them up. In Chapter 5, we considered a number of exceptions and extensions to Mendel's principles of heredity. The concept of linked genes adds a further complication to interpretations of the results of genetic crosses. However, with an understanding of how linkage affects heredity, we can analyze crosses for linked genes and successfully predict the types of progeny that will be produced.

Notation for Crosses with Linkage

In analyzing crosses with linked genes, we must know not only the genotypes of the individuals crossed, but also the arrangement of the genes on the chromosomes. To keep track of this arrangement, we introduce a new system of notation for presenting crosses with linked genes. Consider a cross between an individual homozygous for dominant alleles at two linked loci and another individual homozygous for recessive alleles at those loci ($AA\ BB \times aa\ bb$). For linked genes, it's necessary to write out the specific alleles as they are arranged on each of the homologous chromosomes:

$$\frac{A \quad B}{A \quad B} \times \frac{a \quad b}{a \quad b}$$

In this notation, each line represents one of the two homologous chromosomes. Inheriting one chromosome from each parent, the F_1 progeny will have the following genotype:

$$\frac{A \quad B}{a \quad b}$$

Here, the importance of designating the alleles on each chromosome is clear. One chromosome has the two dominant alleles A and B, whereas the homologous chromosome has the two recessive alleles a and b. The notation can be simplified by drawing only a single line, with the understanding that genes located on the same side of the line lie on the same chromosome:

$$\frac{A \quad B}{a \quad b}$$

This notation can be simplified further by separating the alleles on each chromosome with a slash: AB/ab.

Remember that the two alleles at a locus are always located on different homologous chromosomes and therefore must lie on opposite sides of the line. Consequently, we would *never* write the genotypes as

$$\frac{A \quad a}{B \quad b}$$

because the alleles A and a can *never* be on the same chromosome. It is also important to always keep the same order of the genes on both sides of the line; thus, we should *never* write

$$\frac{A \quad B}{b \quad a}$$

because it would imply that alleles A and b are allelic (at the same locus).

Complete Linkage Compared with Independent Assortment

We will first consider what happens to genes that exhibit complete linkage, meaning that they are located very close together on the same chromosome and do not exhibit crossing over. Genes are rarely completely linked but, by assuming that no crossing over occurs, we can see the effect of linkage more clearly. We will then consider what happens when genes assort independently. Finally, we will consider the results obtained if the genes are linked but exhibit some crossing over.

A testcross reveals the effects of linkage. For example, if a heterozygous individual is test-crossed with a homozygous recessive individual ($Aa\ Bb \times aa\ bb$), the alleles that are present in the gametes contributed by the heterozygous parent will be expressed in the phenotype of the offspring,

7.4 A testcross reveals the effects of linkage. Results of a testcross for two loci in tomatoes that determine leaf type and plant height.

because the homozygous parent could not contribute dominant alleles that might mask them. Consequently, traits that appear in the progeny reveal which alleles were transmitted by the heterozygous parent.

Consider a pair of linked genes in tomato plants. One of the genes affects the type of leaf: an allele for mottled leaves (m) is recessive to an allele that produces normal leaves (M). Nearby on the same chromosome the other gene determines the height of the plant: an allele for dwarf (d) is recessive to an allele for tall (D).

Testing for linkage can be done with a testcross, which requires a plant heterozygous for both characteristics. A geneticist might produce this heterozygous plant by crossing a variety of tomato that is homozygous for normal leaves and tall height with a variety that is homozygous for mottled leaves and dwarf height:

$$\text{P} \quad \frac{M \quad D}{M \quad D} \times \frac{m \quad d}{m \quad d}$$

$$\text{F}_1 \quad \frac{M \quad D}{m \quad d}$$

The geneticist would then use these F_1 heterozygotes in a testcross, crossing them with plants homozygous for mottled leaves and dwarf height:

$$\frac{M \quad D}{m \quad d} \times \frac{m \quad d}{m \quad d}$$

The results of this testcross are diagrammed in **Figure 7.4a**. The heterozygote produces two types of gametes: some

165

with the $\underline{\quad M \quad D \quad}$ chromosome and others with the $\underline{\quad m \quad d \quad}$ chromosome. Because no crossing over occurs, these gametes are the only types produced by the heterozygote. Notice that these gametes contain only combinations of alleles that were present in the original parents: either the allele for normal leaves together with the allele for tall height (M and D) or the allele for mottled leaves together with the allele for dwarf height (m and d). Gametes that contain only original combinations of alleles present in the parents are **nonrecombinant gametes**, or *parental* gametes.

The homozygous parent in the testcross produces only one type of gamete; it contains chromosome $\underline{\quad m \quad d \quad}$ and pairs with one of the two gametes generated by the heterozygous parent (see Figure 7.4a). Two types of progeny result: half have normal leaves and are tall:

$$\frac{M \quad D}{m \quad d}$$

and half have mottled leaves and are dwarf:

$$\frac{m \quad d}{m \quad d}$$

These progeny display the original combinations of traits present in the P generation and are **nonrecombinant progeny**, or *parental* progeny. No new combinations of the two traits, such as normal leaves with dwarf or mottled leaves with tall, appear in the offspring, because the genes affecting the two traits are completely linked and are inherited together. New combinations of traits could arise only if the physical connection between M and D or between m and d were broken.

These results are distinctly different from the results that are expected when genes assort independently (**Figure 7.4b**). If the M and D loci assorted independently, the heterozygous plant ($Mm\ Dd$) would produce four types of gametes: two nonrecombinant gametes containing the original combinations of alleles ($M\ D$ and $m\ d$) and two gametes containing new combinations of alleles ($M\ d$ and $m\ D$). Gametes with new combinations of alleles are called **recombinant gametes**. With independent assortment, nonrecombinant and recombinant gametes are produced in equal proportions. These four types of gametes join with the single type of gamete produced by the homozygous parent of the testcross to produce four kinds of progeny in equal proportions (see Figure 7.4b). The progeny with new combinations of traits formed from recombinant gametes are termed **recombinant progeny**.

> **CONCEPTS**
>
> A testcross in which one of the plants is heterozygous for two completely linked genes yields two types of progeny, each type displaying one of the original combinations of traits present in the P generation. Independent assortment, in contrast, produces four types of progeny in a 1 : 1 : 1 : 1 ratio—two types of recombinant progeny and two types of nonrecombinant progeny in equal proportions.

Crossing Over with Linked Genes

Usually, there is some crossing over between genes that lie on the same chromosome, producing new combinations of traits. Genes that exhibit crossing over are incompletely linked. Let's see how it takes place.

Theory The effect of crossing over on the inheritance of two linked genes is shown in **Figure 7.5**. Crossing over, which takes place in prophase I of meiosis, is the exchange of genetic material between nonsister chromatids (see Figures 2.16 and 2.18). After a single crossover has taken place, the two chromatids that did not participate in crossing over are unchanged; gametes that receive these chromatids are nonrecombinants. The other two chromatids, which did participate in crossing over, now contain new combinations of alleles; gametes that receive these chromatids are recombinants. For each meiosis in which a single crossover takes place, then, two nonrecombinant gametes and two recombinant gametes will be produced. This result is the same as that produced by independent assortment (see Figure 7.4b); so, when crossing over between two loci takes place in every meiosis, it is impossible to determine whether the genes are on the same chromosome and crossing over took place or whether the genes are on different chromosomes.

For closely linked genes, crossing over does not take place in every meiosis. In meioses in which there is no crossing over, only nonrecombinant gametes are produced. In meioses in which there is a single crossover, half the gametes are recombinants and half are nonrecombinants (because a single crossover affects only two of the four chromatids); so the total percentage of recombinant gametes is always half the percentage of meioses in which crossing over takes place. Even if crossing over between two genes takes place in every meiosis, only 50% of the resulting gametes will be recombinants. Thus, the frequency of recombinant gametes is always half the frequency of crossing over, and the maximum proportion of recombinant gametes is 50%.

> **CONCEPTS**
>
> Linkage between genes causes them to be inherited together and reduces recombination; crossing over breaks up the associations of such genes. In a testcross for two linked genes, each crossover produces two recombinant gametes and two nonrecombinants. The frequency of recombinant gametes is half the frequency of crossing over, and the maximum frequency of recombinant gametes is 50%.
>
> ✔ **CONCEPT CHECK 1**
>
> For single crossovers, the frequency of recombinant gametes is half the frequency of crossing over because
>
> a. a test cross between a homozygote and heterozygote produces $\frac{1}{2}$ heterozygous and $\frac{1}{2}$ homozygous progeny.

(a) No crossing over

1 Homologous chromosomes pair in prophase I.
2 If no crossing over takes place,...
3 ...all resulting chromosomes in gametes have original allele combinations and are nonrecombinants.

(b) Crossing over

1 A crossover may take place in prophase I.
2 In this case, half of the resulting gametes will have unchanged chromosomes (nonrecombinants)...
3 ...and half will have recombinant chromosomes.

Nonrecombinant
Recombinant
Recombinant
Nonrecombinant

7.5 A single crossover produces half nonrecombinant gametes and half recombinant gametes.

b. the frequency of recombination is always 50%.
c. each crossover takes place between only two of the four chromatids of a homologous pair.
d. crossovers occur in about 50% of meioses.

Application Let's apply what we have learned about linkage and recombination to a cross between tomato plants that differ in the genes that encode leaf type and plant height. Assume now that these genes are linked and that some crossing over takes place between them. Suppose a geneticist carried out the testcross described earlier:

$$\frac{M \quad D}{m \quad d} \times \frac{m \quad d}{m \quad d}$$

When crossing over takes place in the genes for leaf type and height, two of the four gametes produced are recombinants. When there is no crossing over, all four resulting gametes are nonrecombinants. Thus, over all meioses, the majority of gametes will be nonrecombinants (**Figure 7.6a**). These gametes then unite with gametes produced by the homozygous recessive parent, which contain only the recessive alleles, resulting in mostly nonrecombinant progeny and a few recombinant progeny (**Figure 7.6b**). In this cross, we see that 55 of the testcross progeny have normal leaves and are tall and 53 have mottled leaves and are dwarf. These plants are the nonrecombinant progeny, containing the original combinations of traits that were present in the parents. Of the 123 progeny, 15 have new combinations of traits that were not seen in the parents: 8 are normal leaved and dwarf, and 7 are mottle leaved and tall. These plants are the recombinant progeny.

The results of a cross such as the one illustrated in Figure 7.6 reveal several things. A testcross for two independently assorting genes is expected to produce a 1 : 1 : 1 : 1 phenotypic ratio in the progeny. The progeny of this cross clearly do not exhibit such a ratio; so we might suspect that the genes are not assorting independently. When linked genes undergo some crossing over, the result is mostly nonrecombinant progeny and fewer recombinant progeny. This result is what we observe among the progeny of the testcross illustrated in Figure 7.6; so we conclude that the two genes show evidence of linkage with some crossing over.

Calculating Recombination Frequency

The percentage of recombinant progeny produced in a cross is called the **recombination frequency**, which is calculated as follows:

$$\frac{\text{recombinant}}{\text{frequency}} = \frac{\text{number of recombinant progeny}}{\text{total number of progeny}} \times 100\%$$

In the testcross shown in Figure 7.6, 15 progeny exhibit new combinations of traits; so the recombination frequency is:

$$\frac{8+7}{55+53+8+7} \times 100\% = \frac{15}{123} \times 100\% = 12.2\%$$

Thus, 12.2% of the progeny exhibit new combinations of traits resulting from crossing over. The recombination frequency can also be expressed as a decimal fraction (0.122).

TRY PROBLEM 15 →

7.6 Crossing over between linked genes produces nonrecombinant and recombinant offspring. In this testcross, genes are linked and there is some crossing over.

Coupling and Repulsion

In crosses for linked genes, the arrangement of alleles on the homologous chromosomes is critical in determining the outcome of the cross. For example, consider the inheritance of two genes in the Australian blowfly, *Lucilia cuprina*. In this species, one locus determines the color of the thorax: a purple thorax (p) is recessive to the normal green thorax (p^+). A second locus determines the color of the puparium: a black puparium (b) is recessive to the normal brown puparium (b^+). The loci for thorax color and puparium color are located close together on the chromosome. Suppose we test-cross a fly that is heterozygous at both loci with a fly that is homozygous recessive at both. Because these genes are linked, there are two possible arrangements on the chromosomes of the heterozygous progeny fly. The dominant alleles for green thorax (p^+) and brown puparium (b^+) might reside on one chromosome of the homologous pair, and the recessive alleles for purple thorax (p) and black puparium (b) might reside on the other homologous chromosome:

$$\frac{p^+ \quad b^+}{p \quad b}$$

This arrangement, in which wild-type alleles are found on one chromosome and mutant alleles are found on the other chromosome, is referred to as the **coupling**, or **cis**, **configuration**. Alternatively, one chromosome might bear the alleles for green thorax (p^+) and black puparium (b), and the other chromosome carries the alleles for purple thorax (p) and brown puparium (b^+):

$$\frac{p^+ \quad b}{p \quad b^+}$$

This arrangement, in which each chromosome contains one wild-type and one mutant allele, is called the **repulsion**, or **trans**, **configuration**. Whether the alleles in the heterozygous parent are in coupling or repulsion determines which phenotypes will be most common among the progeny of a testcross.

When the alleles are in the coupling configuration, the most numerous progeny types are those with a green thorax and brown puparium and those with a purple thorax and black puparium (**Figure 7.7a**); but, when the alleles of the heterozygous parent are in repulsion, the most numerous progeny types are those with a green thorax and black puparium and those with a purple thorax and brown puparium (**Figure 7.7b**). Notice that the genotypes of the parents in Figure 7.7a and b are the same ($p^+p\ b^+b \times pp\ bb$) and that the dramatic difference in the phenotypic ratios of the progeny in the two crosses results entirely from the configuration—coupling or repulsion—of the chromosomes. Knowledge of the arrangement of the alleles on the chromosomes is essential to accurately predict the outcome of crosses in which genes are linked.

(a) Alleles in coupling configuration

Testcross: Green thorax, brown puparium ($p^+\ b^+ / p\ b$) × Purple thorax, black puparium ($p\ b / p\ b$)

Gamete formation:
- Nonrecombinant gametes: $p^+\ b^+$, $p\ b$
- Recombinant gametes: $p^+\ b$, $p\ b^+$

Other parent gamete: $p\ b$

Fertilization produces progeny:

Green thorax, brown puparium	Purple thorax, black puparium	Green thorax, black puparium	Purple thorax, brown puparium
$p^+\ b^+ / p\ b$	$p\ b / p\ b$	$p^+\ b / p\ b$	$p\ b^+ / p\ b$
40	40	10	10
Nonrecombinant progeny		Recombinant progeny	

(b) Alleles in repulsion configuration

Testcross: Green thorax, brown puparium ($p^+\ b / p\ b^+$) × Purple thorax, black puparium ($p\ b / p\ b$)

Gamete formation:
- Nonrecombinant gametes: $p^+\ b$, $p\ b^+$
- Recombinant gametes: $p^+\ b^+$, $p\ b$

Other parent gamete: $p\ b$

Fertilization produces progeny:

Green thorax, black puparium	Purple thorax, brown puparium	Green thorax, brown puparium	Purple thorax, black puparium
$p^+\ b / p\ b$	$p\ b^+ / p\ b$	$p^+\ b^+ / p\ b$	$p\ b / p\ b$
40	40	10	10
Nonrecombinant progeny		Recombinant progeny	

Conclusion: The phenotypes of the offspring are the same, but their numbers differ, depending on whether alleles are in coupling or in repulsion.

7.7 The arrangement (coupling or repulsion) of linked genes on a chromosome affects the results of a testcross. Linked loci in the Australian blowfly, *Lucilia cuprina*, determine the color of the thorax and that of the puparium.

CONCEPTS

In a cross, the arrangement of linked alleles on the chromosomes is critical for determining the outcome. When two wild-type alleles are on one homologous chromosome and two mutant alleles are on the other, they are in the coupling configuration; when each chromosome contains one wild-type allele and one mutant allele, the alleles are in repulsion.

✔ CONCEPT CHECK 2

The following testcross produces the progeny shown: *Aa Bb* × *aa bb* → 10 *Aa Bb*, 40 *Aa bb*, 40 *aa Bb*, 10 *aa bb*. Were the genes in the *Aa Bb* parent in coupling or in repulsion?

CONNECTING CONCEPTS

Relating Independent Assortment, Linkage, and Crossing Over

We have now considered three situations concerning genes at different loci. First, the genes may be located on different chromosomes; in this case, they exhibit independent assortment and combine randomly when gametes are formed. An individual heterozygous at two loci (*Aa Bb*) produces four types of gametes (*A B*, *a b*, *A b*, and *a B*) in equal proportions: two types of nonrecombinants and two types of recombinants. In a testcross, these gametes will result in four types of progeny in equal proportions (**Table 7.1**).

Second, the genes may be completely linked—meaning that they are on the same chromosome and lie so close together that crossing over between them is rare. In this case, the genes do not recombine. An individual heterozygous for two closely linked genes in the coupling configuration

$$\frac{A \quad B}{a \quad b}$$

produces only the nonrecombinant gametes containing alleles *A B* or *a b*; the alleles do not assort into new combinations such as *A b* or *a B*. In a testcross, completely linked genes will produce only two types of progeny, both nonrecombinants, in equal proportions (see Table 7.1).

The third situation, incomplete linkage, is intermediate between the two extremes of independent assortment and complete linkage. Here, the genes are physically linked on the same

Table 7.1	Results of a testcross ($Aa\ Bb \times aa\ bb$) with complete linkage, independent assortment, and linkage with some crossing over

Situation	Progeny of Test Cross	
Independent assortment	Aa Bb (nonrecombinant)	25%
	aa bb (nonrecombinant)	25%
	Aa bb (recombinant)	25%
	aa Bb (recombinant)	25%
Complete linkage (genes in coupling)	Aa Bb (nonrecombinant)	50%
	aa bb (nonrecombinant)	50%
Linkage with some crossing over (genes in coupling)	Aa Bb (nonrecombinant)	more than 50%
	aa bb (nonrecombinant)	
	Aa bb (recombinant)	less than 50%
	aa Bb (recombinant)	

chromosome, which prevents independent assortment. However, occasional crossovers break up the linkage and allow the genes to recombine. With incomplete linkage, an individual heterozygous at two loci produces four types of gametes—two types of recombinants and two types of nonrecombinants—but the nonrecombinants are produced more frequently than the recombinants because crossing over does not take place in every meiosis. In the testcross, these gametes result in four types of progeny, with the nonrecombinants more frequent than the recombinant (see Table 7.1).

Earlier in the chapter, the term recombination was defined as the sorting of alleles into new combinations. We've now considered two types of recombination that differ in the mechanism that generates these new combinations of alleles. **Interchromosomal recombination** takes place between genes located on *different* chromosomes. It arises from independent assortment—the random segregation of chromosomes in anaphase I of meiosis—and is the kind of recombination that Mendel discovered while studying dihybrid crosses. A second type of recombination, **intrachromosomal recombination**, takes place between genes located on the *same* chromosome. This recombination arises from crossing over—the exchange of genetic material in prophase I of meiosis. Both types of recombination produce new allele combinations in the gametes; so they cannot be distinguished by examining the types of gametes produced. Nevertheless, they can often be distinguished by the *frequencies* of types of gametes: interchromosomal recombination produces 50% nonrecombinant gametes and 50% recombinant gametes, whereas intrachromosomal recombination frequently produces less than 50% recombinant gametes. However, when the genes are very far apart on the same chromosome, they assort independently, as if they were on different chromosomes. In this case, intrachromosomal recombination also produces 50% recombinant gametes. Intrachromosomal recombination of genes that lie far apart on the same chromosome and interchromosomal recombination are genetically indistinguishable.

Evidence for the Physical Basis of Recombination

Walter Sutton's chromosome theory of inheritance, which stated that genes are physically located on chromosomes, was supported by Nettie Stevens and Edmund Wilson's discovery that sex was associated with a specific chromosome in insects (p. 75 in Chapter 4) and Calvin Bridges's demonstration that nondisjunction of X chromosomes was related to the inheritance of eye color in *Drosophila* (pp. 82–83). Further evidence for the chromosome theory of heredity came in 1931, when Harriet Creighton and Barbara McClintock (**Figure 7.8**) obtained evidence that intrachromosomal recombination was the result of physical exchange between chromosomes. Creighton and McClintock discovered a strain of corn that had an abnormal chromosome 9, containing a densely staining knob at one end and a small piece of another chromosome attached to the other end. This aberrant chromosome allowed them to visually distinguish the two members of a homologous pair.

They studied the inheritance of two traits in corn determined by genes on chromosome 9. At one locus, a dominant allele (*C*) produced colored kernels, whereas a recessive allele (*c*) produced colorless kernels. At a second, linked locus, a dominant allele (*Wx*) produced starchy kernels, whereas a recessive allele (*wx*) produced waxy kernels. Creighton and McClintock obtained a plant that was heterozygous at both loci in repulsion, with the alleles for colored and waxy on

7.8 Barbara McClintock (left) and Harriet Creighton (right) provided evidence that genes are located on chromosomes.
[Karl Maramorosch/Courtesy of Cold Spring Harbor Laboratory Archives.]

the aberrant chromosome and the alleles for colorless and starchy on the normal chromosome:

They crossed this heterozygous plant with a plant that was homozygous for colorless and heterozygous for waxy (with both chromosomes normal):

$$\frac{C \quad wx}{c \quad Wx} \times \frac{c \quad Wx}{c \quad wx}$$

This cross will produce different combinations of traits in the progeny, but the only way that colorless and waxy progeny can arise is through crossing over in the doubly heterozygous parent:

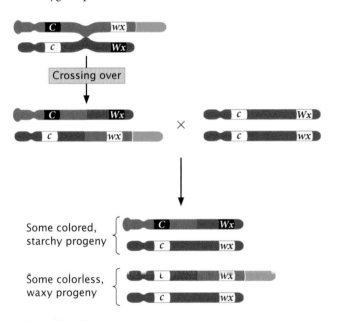

Note: Not all progeny genotypes are shown.

Notice that, if crossing over entails physical exchange between the chromosomes, then the colorless, waxy progeny resulting from recombination should have a chromosome with an extra piece but not a knob. Furthermore, some of the colored, starchy progeny should possess a knob but not the extra piece. This outcome is precisely what Creighton and McClintock observed, confirming the chromosomal theory of inheritance. Curt Stern provided a similar demonstration by using chromosomal markers in *Drosophila* at about the same time.

Predicting the Outcomes of Crosses with Linked Genes

Knowing the arrangement of alleles on a chromosome allows us to predict the types of progeny that will result from a cross entailing linked genes and to determine which of these types will be the most numerous. Determining the *proportions* of the types of offspring requires an additional piece of information—the recombination frequency. The recombination frequency provides us with information about how often the alleles in the gametes appear in new combinations and therefore allows us to predict the proportions of offspring phenotypes that will result from a specific cross with linked genes.

In cucumbers, smooth fruit (t) is recessive to warty fruit (T) and glossy fruit (d) is recessive to dull fruit (D). Geneticists have determined that these two genes exhibit a recombination frequency of 16%. Suppose we cross a plant homozygous for warty and dull fruit with a plant homozygous for smooth and glossy fruit and then carry out a testcross by using the F$_1$:

$$\frac{T \quad D}{t \quad d} \times \frac{t \quad d}{t \quad d}$$

What types and proportions of progeny will result from this testcross?

Four types of gametes will be produced by the heterozygous parent, as shown in **Figure 7.9**: two types of nonrecombinant gametes ($\underline{T \quad D}$ and $\underline{t \quad d}$) and two types of recombinant gametes ($\underline{T \quad d}$ and $\underline{t \quad D}$). The recombination frequency tells us that 16% of the gametes produced by the heterozygous parent will be recombinants. Because there are two types of recombinant gametes, each should arise with a frequency of 16%/2 = 8%. This frequency can also be represented as a probability of 0.08. All the other gametes will be nonrecombinants; so they should arise with a frequency of 100% − 16% = 84%. Because there are two types of nonrecombinant gametes, each should arise with a frequency of 84%/2 = 42% (or 0.42). The other parent in the testcross is homozygous and therefore produces only a single type of gamete ($\underline{t \quad d}$) with a frequency of 100% (or 1.00).

Four types of progeny result from the testcross (see Figure 7.9). The expected proportion of each type can be determined by using the multiplication rule, multiplying together the probability of each gamete. Testcross progeny with warty and dull fruit

$$\frac{T \quad D}{t \quad d}$$

appear with a frequency of 0.42 (the probability of inheriting a gamete with chromosome $\underline{T \quad D}$ from the heterozygous parent) × 1.00 (the probability of inheriting a gamete

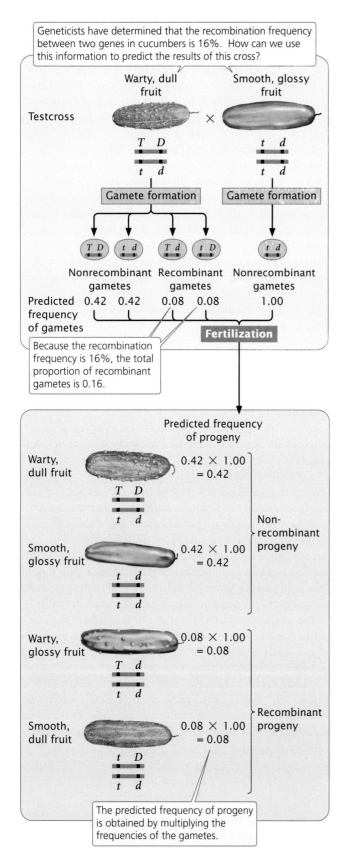

7.9 The recombination frequency allows a prediction of the proportions of offspring expected for a cross entailing linked genes.

with chromosome _t d_ from the recessive parent) = 0.42. The proportions of the other types of F_2 progeny can be calculated in a similar manner (see Figure 7.9). This method can be used for predicting the outcome of any cross with linked genes for which the recombination frequency is known. **TRY PROBLEM 17 →**

Testing for Independent Assortment

In some crosses, the genes are obviously linked because there are clearly more nonrecombinant progeny than recombinant progeny. In other crosses, the difference between independent assortment and linkage is not so obvious. For example, suppose we did a testcross for two pairs of genes, such as _Aa Bb_ × _aa bb_, and observed the following numbers of progeny: 54 _Aa Bb_, 56 _aa bb_, 42 _Aa bb_, and 48 _aa Bb_. Is this outcome the 1 : 1 : 1 : 1 ratio we would expect if _A_ and _B_ assorted independently? Not exactly, but it's pretty close. Perhaps these genes assorted independently and chance produced the slight deviations between the observed numbers and the expected 1 : 1 : 1 : 1 ratio. Alternatively, the genes might be linked, with considerable crossing over taking place between them, and so the number of nonrecombinants is only slightly greater than the number of recombinants. How do we distinguish between the role of chance and the role of linkage in producing deviations from the results expected with independent assortment?

We encountered a similar problem in crosses in which genes were unlinked—the problem of distinguishing between deviations due to chance and those due to other factors. We addressed this problem (in Chapter 3) with the goodness-of-fit chi-square test, which helps us evaluate the likelihood that chance alone is responsible for deviations between the numbers of progeny that we observed and the numbers that we expected by applying the principles of inheritance. Here, we are interested in a different question: Is the inheritance of alleles at one locus independent of the inheritance of alleles at a second locus? If the answer to this question is yes, then the genes are assorting independently; if the answer is no, then the genes are probably linked.

A possible way to test for independent assortment is to calculate the expected probability of each progeny type, assuming independent assortment, and then use the goodness-of-fit chi-square test to evaluate whether the observed numbers deviate significantly from the expected numbers. With independent assortment, we expect $1/4$ of each phenotype: $1/4$ _Aa Bb_, $1/4$ _aa bb_, $1/4$ _Aa bb_, and $1/4$ _aa Bb_. This expected probability of each genotype is based on the multiplication rule of probability, which we considered in Chapter 3. For example, if the probability of _Aa_ is $1/2$ and the probability of _Bb_ is $1/2$, then the probability of _Aa Bb_ is $1/2 \times 1/2 = 1/4$. In this calculation, we are making two assumptions: (1) the probability of each single-locus genotype is $1/2$, and (2) genotypes at the two loci are inherited independently ($1/2 \times 1/2 = 1/4$).

One problem with this approach is that a significant chi-square value can result from a violation of either assumption. If the genes are linked, then the inheritance of genotypes at the two loci are not independent (assumption 2), and we will get a significant deviation between observed and expected numbers. But we can also get a significant deviation if the probability of each single-locus genotype is not $1/2$ (assumption 1), even when the genotypes are assorting independently. We may obtain a significant deviation, for example, if individuals with one genotype have a lower probability of surviving or the penetrance of a genotype is not 100%. We could test both assumptions by conducting a series of chi-square tests, first testing the inheritance of genotypes at each locus separately (assumption 1) and then testing for independent assortment (assumption 2). However, a faster method is to test for independence in genotypes with a *chi-square test of independence*.

The chi-square test of independence The chi-square test of independence allows us to evaluate whether the segregation of alleles at one locus is independent of the segregation of alleles at another locus, without making any assumption about the probability of single-locus genotypes. To illustrate this analysis, we will examine results from a cross between German cockroaches, in which yellow body (y) is recessive to brown body (y^+) and curved wings (cv) are recessive to straight wings (cv^+). A testcross ($y^+y\ cv^+cv \times yy\ cvcv$) produced the progeny shown in **Figure 7.10a**.

To carry out the chi-square test of independence, we first construct a table of the observed numbers of progeny, somewhat like a Punnett square, except, here, we put the genotypes that result from the segregation of alleles at one locus along the top and the genotypes that result from the segregation of alleles at the other locus along the side (**Figure 7.10b**). Next, we compute the total for each row, the total for each column, and the grand total (the sum of all row totals or the sum of all column totals, which should be the same). These totals will be used to compute the expected values for the chi-square test of independence.

Now, we compute the values expected if the segregation of alleles at the y locus is independent of the segregation of alleles at the cv locus. If the segregation of alleles at each locus is independent, then the proportion of progeny with y^+y and yy genotypes should be the same for cockroaches with genotype cv^+cv and for cockroaches with genotype $cvcv$. The converse is also true; the proportions of progeny with cv^+cv and

7.10 A chi-square test of independence can be used to determine if genes at two loci are assorting independently.

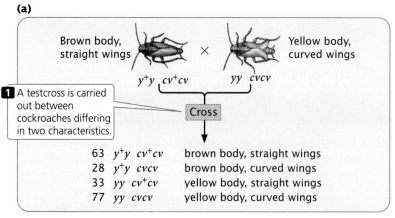

Conclusion: The genes for body color and type of wing are not assorting independently and must be linked.

cvcv genotypes should be the same for cockroaches with genotype y^+y and for cockroaches with genotype *yy*. With the assumption that the alleles at the two loci segregate independently, the expected number for each cell of the table can be computed by using the following formula:

$$\text{expected number} = \frac{\text{row total} \times \text{column total}}{\text{grand total}}$$

For the cell of the table corresponding to genotype $y^+y\ cv^+cv$ (the upper-left-hand cell of the table in Figure 7.10b) the expected number is:

$$\frac{96\ (\text{row total}) \times 91\ (\text{column total})}{201\ (\text{grand total})} = \frac{8736}{201} = 43.46$$

With the use of this method, the expected numbers for each cell are given in **Figure 7.10c**.

We now calculate a chi-square value by using the same formula that we used for the goodness-of-fit chi-square test in Chapter 3:

$$x^2 = \sum \frac{(\text{observed} - \text{expected})^2}{\text{expected}}$$

Recall that $\sum$ means "sum," and that we are adding together the (observed − expected)^2/expected value for each type of progeny. With the observed and expected numbers of cockroaches from the testcross, the calculated chi-square value is 30.73 (**Figure 7.10d**).

To determine the probability associated with this chi-square value, we need the degrees of freedom. Recall from Chapter 3 that the degrees of freedom are the number of ways in which the observed classes are free to vary from the expected values. In general, for the chi-square test of independence, the degrees of freedom equal the number of rows in the table minus 1 multiplied by the number of columns in the table minus 1 (**Figure 7.10e**), or

$$df = (\text{number of rows} - 1) \times (\text{number of columns} - 1)$$

In our example, there were two rows and two columns, and so the degrees of freedom are:

$$df = (2 - 1) \times (2 - 1) = 1 \times 1 = 1$$

Therefore, our calculated chi-square value is 30.73, with 1 degree of freedom. We can use Table 3.5 to find the associated probability. Looking at Table 3.5, we find that our calculated chi-square value is larger than the largest chi-square value given for 1 degree of freedom, which has a probability of 0.005. Thus, our calculated chi-square value has a probability less than 0.005. This very small probability indicates that the genotypes are not in the proportions that we would expect if independent assortment were taking place. Our conclusion, then, is that these genes are not assorting independently and must be linked. As is the case for the goodness-of-fit chi-square test, geneticists generally consider that any chi-square value for the test of independence with a probability less than 0.05 is significantly different from the expected values and is therefore evidence that the genes are not assorting independently. **TRY PROBLEM 16 →**

Gene Mapping with Recombination Frequencies

Thomas Hunt Morgan and his students developed the idea that physical distances between genes on a chromosome are related to the rates of recombination. They hypothesized that crossover events occur more or less at random up and down the chromosome and that two genes that lie far apart are more likely to undergo a crossover than are two genes that lie close together. They proposed that recombination frequencies could provide a convenient way to determine the order of genes along a chromosome and would give estimates of the relative distances between the genes. Chromosome maps calculated by using the genetic phenomenon of recombination are called **genetic maps**. In contrast, chromosome maps calculated by using physical distances along the chromosome (often expressed as numbers of base pairs) are called **physical maps**.

Distances on genetic maps are measured in **map units** (abbreviated m.u.); one map unit equals 1% recombination. Map units are also called **centiMorgans** (cM), in honor of Thomas Hunt Morgan; 100 centiMorgans equals 1 **Morgan**. Genetic distances measured with recombination rates are approximately additive: if the distance from gene *A* to gene *B* is 5 m.u., the distance from gene *B* to gene *C* is 10 m.u., and the distance from gene *A* to gene *C* is 15 m.u., then gene *B* must be located between genes *A* and *C*. On the basis of the map distances just given, we can draw a simple genetic map for genes *A*, *B*, and *C*, as shown here:

We could just as plausibly draw this map with *C* on the left and *A* on the right:

Both maps are correct and equivalent because, with information about the relative positions of only three genes, the most that we can determine is which gene lies in the middle. If we obtained distances to an additional gene, then we could position *A* and *C* relative to that gene. An additional gene *D*, examined through genetic crosses, might yield the following recombination frequencies:

Gene pair	Recombination frequency (%)
A and *D*	8
B and *D*	13
C and *D*	23

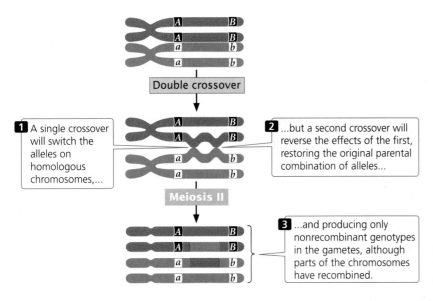

7.11 A two-strand double crossover between two linked genes produces only nonrecombinant gametes.

Notice that C and D exhibit the greatest amount of recombination; therefore, C and D must be farthest apart, with genes A and B between them. Using the recombination frequencies and remembering that 1 m.u. = 1% recombination, we can now add D to our map:

By doing a series of crosses between pairs of genes, we can construct genetic maps showing the linkage arrangements of a number of genes.

Two points should be emphasized about constructing chromosome maps from recombination frequencies. First, recall that we cannot distinguish between genes on different chromosomes and genes located far apart on the same chromosome. If genes exhibit 50% recombination, the most that can be said about them is that they belong to different groups of linked genes (different linkage groups), either on different chromosomes or far apart on the same chromosome.

The second point is that a testcross for two genes that are far apart on the same chromosome tends to underestimate the true physical distance, because the cross does not reveal double crossovers that might take place between the two genes (**Figure 7.11**). A double crossover arises when two separate crossover events take place between two loci. (For now, we will consider only double crossovers that occur between two of the four chromatids of a homologus pair—a two-strand double crossover. Double crossovers that take place among three and four chromatids will be considered later.) Whereas a single crossover produces combinations of alleles that were not present on the original parental chromosomes, a second crossover between the same two genes reverses the effects of the first, thus restoring the original parental combination of alleles (see Figure 7.11). We therefore cannot distinguish between the progeny produced by two-strand double crossovers and the progeny produced when there is no crossing over at all. As we shall see in the next section, we can detect double crossovers if we examine a third gene that lies between the two crossovers. Because double crossovers between two genes go undetected, map distances will be underestimated whenever double crossovers take place. Double crossovers are more frequent between genes that are far apart; therefore genetic maps based on short distances are usually more accurate than those based on longer distances.

CONCEPTS

A genetic map provides the order of the genes on a chromosome and the approximate distances from one gene to another based on recombination frequencies. In genetic maps, 1% recombination equals 1 map unit, or 1 centiMorgan. Double crossovers between two genes go undetected; so map distances between distant genes tend to underestimate genetic distances.

✔ **CONCEPT CHECK 3**

How does a genetic map differ from a physical map?

Constructing a Genetic Map with the Use of Two-Point Testcrosses

Genetic maps can be constructed by conducting a series of testcrosses. In each testcross, one of the parents is heterozygous for a different pair of genes, and recombination frequencies are calculated between pairs of genes. A testcross between two genes is called a **two-point testcross** or a two-point cross for short. Suppose that we carried out a series of two-point crosses for four genes, *a*, *b*, *c*, and *d*, and obtained the following recombination frequencies:

Gene loci in testcross	Recombination frequency (%)
a and *b*	50
a and *c*	50
a and *d*	50
b and *c*	20
b and *d*	10
c and *d*	28

We can begin constructing a genetic map for these genes by considering the recombination frequencies for each pair of genes. The recombination frequency between *a* and *b* is 50%, which is the recombination frequency expected with independent assortment. Therefore, genes *a* and *b* may either be on different chromosomes or be very far apart on the same chromosome; so we will place them in different linkage groups with the understanding that they may or may not be on the same chromosome:

Linkage group 1

a

Linkage group 2

b

The recombination frequency between *a* and *c* is 50%, indicating that they, too, are in different linkage groups. The recombination frequency between *b* and *c* is 20%; so these genes are linked and separated by 20 map units:

Linkage group 1

a

Linkage group 2

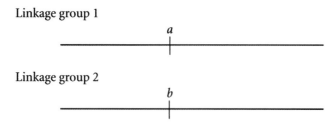

The recombination frequency between *a* and *d* is 50%, indicating that these genes belong to different linkage groups, whereas genes *b* and *d* are linked, with a recombination frequency of 10%. To decide whether gene *d* is 10 m.u. to the left or to the right of gene *b*, we must consult the *c*-to-*d* distance. If gene *d* is 10 m.u. to the left of gene *b*, then the distance between *d* and *c* should be approximately the sum of the distance between b and c and between c and d: 20 m.u. + 10 m.u. = 30 m.u. If, on the other hand, gene *d* lies to the right of gene *b*, then the distance between gene *d* and gene *c* will be much shorter, approximately 20 m.u. − 10 m.u. = 10 m.u. The summed distances will be only approximate because any double crossovers between the two genes will be missed and the map distance will be underestimated.

By examining the recombination frequency between *c* and *d*, we can distinguish between these two possibilities. The recombination frequency between *c* and *d* is 28%; so gene *d* must lie to the left of gene *b*. Notice that the sum of the recombination between *d* and *b* (10%) and between *b* and *c* (20%) is greater than the recombination between *d* and *c* (28%). As already discussed, this discrepancy arises because double crossovers between the two outer genes go undetected, causing an underestimation of the true map distance. The genetic map of these genes is now complete:

Linkage group 1

a

Linkage group 2

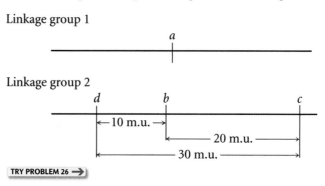

TRY PROBLEM 26

7.3 A Three-Point Testcross Can Be Used to Map Three Linked Genes

Genetic maps can be constructed from a series of testcrosses for pairs of genes, but this approach is not particularly efficient, because numerous two-point crosses must be carried out to establish the order of the genes and because double crossovers are missed. A more efficient mapping technique is a testcross for three genes—a **three-point testcross**, or three-point cross. With a three-point cross, the order of the three genes can be established in a single set of progeny and some double crossovers can usually be detected, providing more-accurate map distances.

Consider what happens when crossing over takes place among three hypothetical linked genes. **Figure 7.12** illustrates a pair of homologous chromosomes of an individual that is heterozygous at three loci (*Aa Bb Cc*). Notice that the genes are in the coupling configuration; that is, all the dominant alleles are on one chromosome (*A B C*) and all the recessive alleles are on the other chromosome (*a b c*). Three types of crossover events can take place between these three genes: two types of single crossovers (see Figure 7.12a and b) and a double crossover (see Figure 7.12c). In each type of crossover, two of the resulting chromosomes are recombinants and two are non-recombinants.

Notice that, in the recombinant chromosomes resulting from the double crossover, the outer two alleles are the

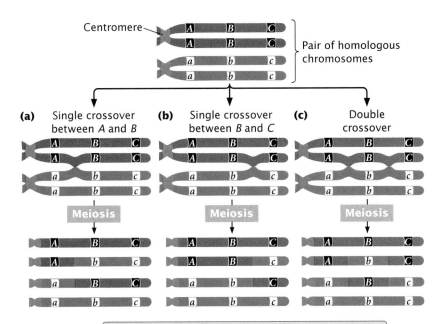

7.12 Three types of crossovers can take place among three linked loci.

same as in the nonrecombinants, but the middle allele is different. This result provides us with an important clue about the order of the genes. In progeny that result from a double crossover, only the middle allele should differ from the alleles present in the nonrecombinant progeny.

Constructing a Genetic Map with the Three-Point Testcross

To examine gene mapping with a three-point testcross, we will consider three recessive mutations in the fruit fly *Drosophila melanogaster*. In this species, scarlet eyes (st) are recessive to wild-type red eyes (st^+), ebony body color (e) is recessive to wild-type gray body color (e^+), and spineless (ss)—that is, the presence of small bristles—is recessive to wild-type normal bristles (ss^+). The loci encoding these three characteristics are linked and located on chromosome 3.

We will refer to these three loci as st, e, and ss, but keep in mind that either the recessive alleles (st, e, and ss) or the dominant alleles (st^+, e^+, and ss^+) may be present at each locus. So, when we say that there are 10 m.u. between st and ss, we mean that there are 10 m.u. between the loci at which mutations st and ss occur; we could just as easily say that there are 10 m.u. between st^+ and ss^+.

To map these genes, we need to determine their order on the chromosome and the genetic distances between them. First, we must set up a three-point testcross, a cross between a fly heterozygous at all three loci and a fly homozygous for recessive alleles at all three loci. To produce flies heterozygous for all three loci, we might cross a stock of flies that are homozygous for wild-type alleles at all three loci with flies that are homozygous for recessive alleles at all three loci:

$$P \quad \frac{st^+ \quad e^+ \quad ss^+}{st^+ \quad e^+ \quad ss^+} \times \frac{st \quad e \quad ss}{st \quad e \quad ss}$$

$$F_1 \quad \frac{st^+ \quad e^+ \quad ss^+}{st \quad e \quad ss}$$

The order of the genes has been arbitrarily assigned because, at this point, we do not know which is the middle gene. Additionally, the alleles in these heterozygotes are in coupling configuration (because all the wild-type dominant alleles were inherited from one parent and all the recessive mutations from the other parent), although the testcross can also be done with alleles in repulsion.

In the three-point testcross, we cross the F_1 heterozygotes with flies that are homozygous for all three recessive mutations. In many organisms, it makes no difference whether the heterozygous parent in the testcross is male or female (provided that the genes are autosomal) but, in *Drosophila*, no crossing over takes place in males. Because crossing over in the heterozygous parent is essential for determining recombination frequencies, the heterozygous flies in our testcross must be female. So we mate female F_1 flies that are heterozygous for all three traits with male flies that are homozygous for all the recessive traits:

$$\frac{st^+ \quad e^+ \quad ss^+}{st \quad e \quad ss} \text{ Female} \times \frac{st \quad e \quad ss}{st \quad e \quad ss} \text{ Male}$$

7.13 The results of a three-point testcross can be used to map linked genes. In this three-point testcross of *Drosophila melanogaster*, the recessive mutations scarlet eyes (*st*), ebony body color (*e*), and spineless bristles (*ss*) are at three linked loci. The order of the loci has been designated arbitrarily, as has the sex of the progeny flies.

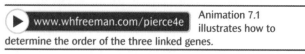 Animation 7.1 illustrates how to determine the order of the three linked genes.

The progeny produced from this cross are listed in **Figure 7.13**. For each locus, two classes of progeny are produced: progeny that are heterozygous, displaying the dominant trait, and progeny that are homozygous, display-

ing the recessive trait. With two classes of progeny possible for each of the three loci, there will be $2^3 = 8$ classes of phenotypes possible in the progeny. In this example, all eight phenotypic classes are present but, in some three-point crosses, one or more of the phenotypes may be missing if the number of progeny is limited. Nevertheless, the absence of a particular class can provide important information about which combination of traits is least frequent and, ultimately, about the order of the genes, as we will see.

To map the genes, we need information about where and how often crossing over has taken place. In the homozygous recessive parent, the two alleles at each locus are the same, and so crossing over will have no effect on the types of gametes produced; with or without crossing over, all gametes from this parent have a chromosome with three recessive alleles (st e ss). In contrast, the heterozygous parent has different alleles on its two chromosomes, and so crossing over can be detected. The information that we need for mapping, therefore, comes entirely from the gametes produced by the heterozygous parent. Because chromosomes contributed by the homozygous parent carry only recessive alleles, whatever alleles are present on the chromosome contributed by the heterozygous parent will be expressed in the progeny.

As a shortcut, we often do not write out the complete genotypes of the testcross progeny, listing instead only the alleles expressed in the phenotype, which are the alleles inherited from the heterozygous parent. This convention is used in the discussion that follows.

> **CONCEPTS**
>
> To map genes, information about the location and number of crossovers in the gametes that produced the progeny of a cross is needed. An efficient way to obtain this information is to use a three-point testcross, in which an individual heterozygous at three linked loci is crossed with an individual that is homozygous recessive at the three loci.
>
> ✔ **CONCEPT CHECK 4**
>
> Write the genotypes of all recombinant and nonrecombinant progeny expected from the following three-point cross:
>
> $$\frac{m^+ \quad p^+ \quad s^+}{m \quad p \quad s} \times \frac{m \quad p \quad s}{m \quad p \quad s}$$

Determining the gene order The first task in mapping the genes is to determine their order on the chromosome. In Figure 7.13, we arbitrarily listed the loci in the order *st*, *e*, *ss*, but we had no way of knowing which of the three loci was between the other two. We can now identify the middle locus by examining the double-crossover progeny.

First, determine which progeny are the nonrecombinants; they will be the two most-numerous classes of progeny. (Even if crossing over takes place in every meiosis, the nonrecombinants will constitute at least 50% of the progeny.) Among the progeny of the testcross in Figure 7.13, the most numerous are those with all three dominant traits (st^+ e^+ ss^+) and those with all three recessive traits (st e ss).

Next, identify the double-crossover progeny. These progeny should always have the two least-numerous phenotypes, because the probability of a double crossover is always less than the probability of a single crossover. The least-common progeny among those listed in Figure 7.13 are progeny with spineless bristles (st^+ e^+ ss) and progeny with scarlet eyes and ebony body (st e ss^+); so they are the double-crossover progeny.

Three orders of genes on the chromosome are possible: the eye-color locus could be in the middle (e st ss), the body-color locus could be in the middle (st e ss), or the bristle locus could be in the middle (st ss e). To determine which gene is in the middle, we can draw the chromosomes of the heterozygous parent with all three possible gene orders and then see if a double crossover produces the combination of genes observed in the double-crossover progeny. The three possible gene orders and the types of progeny produced by their double crossovers are:

Original chromosomes   **Chromosomes after crossing over**

1. e^+ st^+ ss^+ / e st ss → (crossover) → e^+ st ss^+ / e st^+ ss

2. st^+ e^+ ss^+ / st e ss → (crossover) → st^+ e ss^+ / st e^+ ss

3. st^+ ss^+ e^+ / st ss e → (crossover) → st^+ ss e^+ / st ss^+ e

The only gene order that produces chromosomes with the set of alleles observed in the least-numerous progeny or double crossovers (st^+ e^+ ss and st e ss^+ in Figure 7.13) is the one in which the ss locus for bristles lies in the middle (gene-order 3). Therefore, this order (st ss e) must be the correct sequence of genes on the chromosome.

With a little practice, we can quickly determine which locus is in the middle without writing out all the gene orders. The phenotypes of the progeny are expressions of the alleles inherited from the heterozygous parent. Recall that, when we looked at the results of double crossovers (see Figure 7.12), only the alleles at the middle locus differed from the nonrecombinants. If we compare the nonrecombinant progeny with double-crossover progeny, they should differ only in alleles of the middle locus (**Table 7.2**).

Let's compare the alleles in the double-crossover progeny st^+ e^+ ss with those in the nonrecombinant progeny st^+ e^+ ss^+. We see that both have an allele

Table 7.2 Steps in determining gene order in a three-point cross

1. Identify the nonrecombinant progeny (two most-numerous phenotypes).
2. Indentify the double-crossover progeny (two least-numerous phenotypes).
3. Compare the phenotype of double-crossover progeny with the phenotype of nonrecombinant progeny. They should be alike in two characteristics and differ in one.
4. The characteristic that differs between the double crossover and the nonrecombinant progeny is encoded by the middle gene.

for red eyes (st^+) and both have an allele for gray body (e^+), but the nonrecombinants have an allele for normal bristles (ss^+), whereas the double crossovers have an allele for spineless bristles (ss). Because the bristle locus is the only one that differs, it must lie in the middle. We would obtain the same results if we compared the other class of double-crossover progeny (st e ss^+) with other nonrecombinant progeny (st e ss). Again, the only locus that differs is the one for bristles. Don't forget that the nonrecombinants and the double crossovers should differ at only one locus; if they differ at two loci, the wrong classes of progeny are being compared.

CONCEPTS

To determine the middle locus in a three-point cross, compare the double-crossover progeny with the nonrecombinant progeny. The double crossovers will be the two least-common classes of phenotypes; the nonrecombinants will be the two most-common classes of phenotypes. The double-crossover progeny should have the same alleles as the nonrecombinant types at two loci and different alleles at the locus in the middle.

✔ **CONCEPT CHECK 5**

A three-point test cross is carried out between three linked genes. The resulting nonrecombinant progeny are s^+ r^+ c^+ and s r c and the double-crossover progeny are s r c^+ and s^+ r^+ c. Which is the middle locus?

Determining the locations of crossovers When we know the correct order of the loci on the chromosome, we should rewrite the phenotypes of the testcross progeny in Figure 7.13 with the alleles in the correct order so that we can determine where crossovers have taken place (**Figure 7.14**).

Among the eight classes of progeny, we have already identified two classes as nonrecombinants (st^+ ss^+ e^+ and st ss e) and two classes as double crossovers (st^+ ss e^+ and st ss^+ e). The other

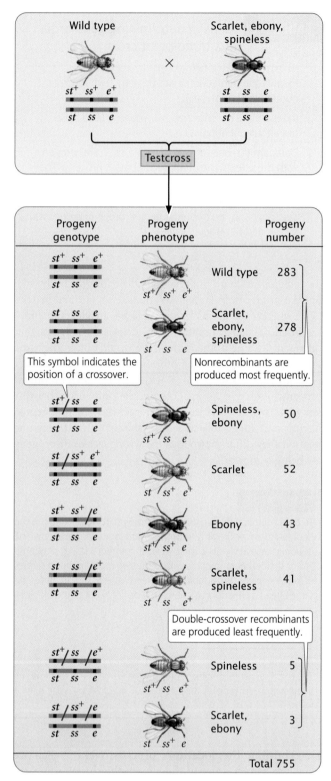

7.14 Writing the results of a three-point testcross with the loci in the correct order allows the locations of crossovers to be determined. These results are from the testcross illustrated in Figure 7.13, with the loci shown in the correct order. The location of a crossover is indicated by a slash (/). The sex of the progeny flies has been designated arbitrarily.

four classes include progeny that resulted from a chromosome that underwent a single crossover: two underwent single crossovers between st and ss, and two underwent single crossovers between ss and e.

To determine where the crossovers took place in these progeny, compare the alleles found in the single-crossover progeny with those found in the nonrecombinants, just as we did for the double crossovers. For example, consider progeny with chromosome $\underline{st^+ \quad ss \quad e}$. The first allele ($st^+$) came from the nonrecombinant chromosome $\underline{st^+ \quad ss^+ \quad e^+}$ and the other two alleles (ss and e) must have come from the other nonrecombinant chromosome $\underline{st \quad ss \quad e}$ through crossing over:

$$\frac{st^+ \quad ss^+ \quad e^+}{st \quad ss \quad e} \rightarrow \frac{st^+ \quad ss^+ \quad e^+}{st \quad ss \quad e} \rightarrow \frac{st^+ \quad ss \quad e}{st \quad ss^+ \quad e^+}$$

This same crossover also produces the $\underline{st \quad ss^+ \quad e^+}$ progeny.

This method can also be used to determine the location of crossing over in the other two types of single-crossover progeny. Crossing over between ss and e produces $\underline{st^+ \quad ss^+ \quad e}$ and $\underline{st \quad ss \quad e^+}$ chromosomes:

$$\frac{st^+ \quad ss^+ \quad e^+}{st \quad ss \quad e} \rightarrow \frac{st^+ \quad ss^+ \quad e^+}{st \quad ss \quad e} \rightarrow \frac{st^+ \quad ss^+ \quad e}{st \quad ss \quad e^+}$$

We now know the locations of all the crossovers; their locations are marked with a slash in Figure 7.14.

Calculating the recombination frequencies Next, we can determine the map distances, which are based on the frequencies of recombination. We calculate recombination frequency by adding up all of the recombinant progeny, dividing this number by the total number of progeny from the cross, and multiplying the number obtained by 100%. To determine the map distances accurately, we must include all crossovers (both single and double) that take place between two genes.

Recombinant progeny that possess a chromosome that underwent crossing over between the eye-color locus (st) and the bristle locus (ss) include the single crossovers ($\underline{st^+ \; / \; ss \quad e}$ and $\underline{st \; / \; ss^+ \quad e^+}$) and the two double crossovers ($\underline{st^+ \; / \; ss \; / \; e^+}$ and $\underline{st \; / \; ss^+ \; / \; e}$); see Figure 7.14. There are a total of 755 progeny; so the recombination frequency between ss and st is:

st–ss recombination frequency =
$$\frac{(50 + 52 + 5 + 3)}{755} \times 100\% = 14.6\%$$

The distance between the st and ss loci can be expressed as 14.6 m.u.

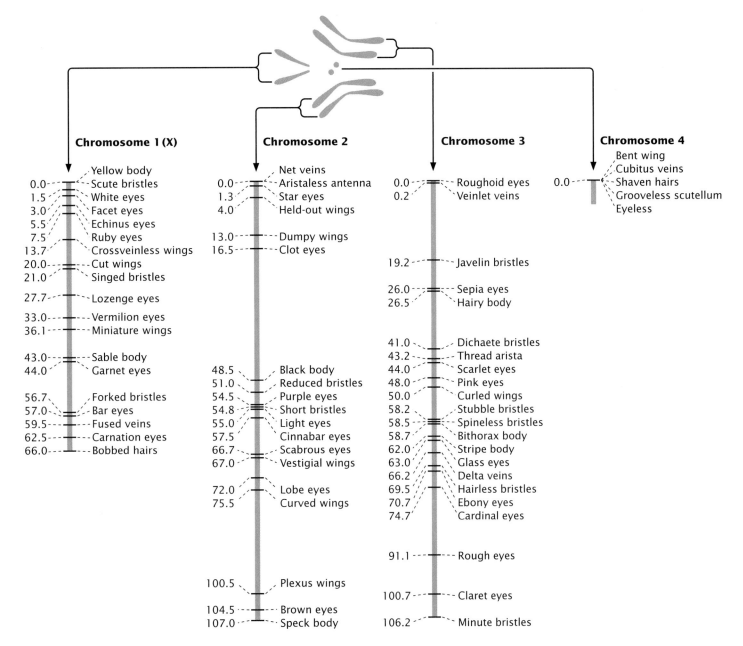

7.15 *Drosophila melanogaster* has four linkage groups corresponding to its four pairs of chromosomes. Distances between genes within a linkage group are in map units.

The map distance between the bristle locus (*ss*) and the body locus (*e*) is determined in the same manner. The recombinant progeny that possess a crossover between *ss* and *e* are the single crossovers $\underline{st^+ \quad ss^+ \ / \ e}$ and $\underline{st \quad ss \ / \ e^+}$ and the double crossovers $\underline{st \ / \ ss \ / \ e^+}$ and $\underline{st \ / \ ss^+ \ / \ e}$. The recombination frequency is:

ss–*e* recombination frequency =
$$\frac{(43 + 41 + 5 + 3)}{755} \times 100\% = 12.2\%$$

Thus, the map distance between *ss* and *e* is 12.2 m.u.

Finally, calculate the map distance between the outer two loci, *st* and *e*. This map distance can be obtained by summing the map distances between *st* and *ss* and between *ss* and *e* (14.6 m.u. + 12.2 m.u. = 26.8 m.u.). We can now use the map distances to draw a map of the three genes on the chromosome:

A genetic map of *D. melanogaster* is illustrated in **Figure 7.15**.

Interference and the coefficient of coincidence Map distances give us information not only about the distances that separate genes, but also about the proportions of recombinant and nonrecombinant gametes that will be produced in a cross. For example, knowing that genes *st* and *ss* on the third chromosome of *D. melanogaster* are separated by a distance of 14.6 m.u. tells us that 14.6% of the gametes produced by a fly heterozygous at these two loci will be recombinants. Similarly, 12.2% of the gametes from a fly heterozygous for *ss* and *e* will be recombinants.

Theoretically, we should be able to calculate the proportion of double-recombinant gametes by using the multiplication rule of probability (see Chapter 3), which states that the probability of two independent events occurring together is calculated by multiplying the probabilities of the independent events. Applying this principle, we should find that the proportion (probability) of gametes with double crossovers between *st* and *e* is equal to the probability of recombination between *st* and *ss* multiplied by the probability of recombination between *ss* and *e*, or $0.146 \times 0.122 = 0.0178$. Multiplying this probability by the total number of progeny gives us the *expected* number of double-crossover progeny from the cross: $0.0178 \times 755 = 13.4$. Only 8 double crossovers—considerably fewer than the 13 expected—were observed in the progeny of the cross (see Figure 7.14).

This phenomenon is common in eukaryotic organisms. The calculation assumes that each crossover event is independent and that the occurrence of one crossover does not influence the occurrence of another. But crossovers are frequently *not* independent events: the occurrence of one crossover tends to inhibit additional crossovers in the same region of the chromosome, and so double crossovers are less frequent than expected.

The degree to which one crossover interferes with additional crossovers in the same region is termed the **interference**. To calculate the interference, we first determine the **coefficient of coincidence**, which is the ratio of observed double crossovers to expected double crossovers:

$$\text{coefficient of coincidence} = \frac{\text{number of observed double crossovers}}{\text{number of expected double crossovers}}$$

For the loci that we mapped on the third chromosome of *D. melanogaster* (see Figure 7.14), we find that the

$$\text{coefficient of coincidence} = \frac{5+3}{0.146 \times 0.122 \times 755} = \frac{8}{13.4} = 0.6$$

which indicates that we are actually observing only 60% of the double crossovers that we expected on the basis of the single-crossover frequencies. The interference is calculated as

$$\text{interference} = 1 - \text{coefficient of coincidence}$$

So the interference for our three-point cross is:

$$\text{interference} = 1 - 0.6 = 0.4$$

This value of interference tells us that 40% of the double-crossover progeny expected will not be observed, because of interference. When interference is complete and no double-crossover progeny are observed, the coefficient of coincidence is 0 and the interference is 1.

Sometimes a crossover *increases* the probability of another crossover taking place nearby and we see *more* double-crossover progeny than expected. In this case, the coefficient of coincidence is greater than 1 and the interference is negative. **TRY PROBLEM 28**

CONCEPTS

The coefficient of coincidence equals the number of double crossovers observed divided by the number of double crossovers expected on the basis of the single-crossover frequencies. The interference equals 1 − the coefficient of coincidence; it indicates the degree to which one crossover interferes with additional crossovers.

✔ CONCEPT CHECK 6

In analyzing the results of a three-point testcross, a student determines that the interference is −0.23. What does this negative interference value indicate?

a. Fewer double crossovers took place than expected on the basis of single-crossover frequencies.
b. More double crossovers took place than expected on the basis of single-crossover frequencies.
c. Fewer single crossovers took place than expected.
d. A crossover in one region interferes with additional crossovers in the same region.

CONNECTING CONCEPTS

Stepping Through the Three-Point Cross

We have now examined the three-point cross in considerable detail and have seen how the information derived from the cross can be used to map a series of three linked genes. Let's briefly review the steps required to map genes from a three-point cross.

1. **Write out the phenotypes and numbers of progeny produced in the three-point cross.** The progeny phenotypes will be easier to interpret if you use allelic symbols for the traits (such as $st^+\ e^+\ ss$).

2. **Write out the genotypes of the original parents used to produce the triply heterozygous individual** in the testcross and, if known, the arrangement (coupling or repulsion) of the alleles on their chromosomes.

3. **Determine which phenotypic classes among the progeny are the nonrecombinants and which are the double crossovers.** The nonrecombinants will be the two most-

common phenotypes; double crossovers will be the two least-common phenotypes.

4. **Determine which locus lies in the middle.** Compare the alleles present in the double crossovers with those present in the nonrecombinants; each class of double crossovers should be like one of the nonrecombinants for two loci and should differ for one locus. The locus that differs is the middle one.

5. **Rewrite the phenotypes with the genes in correct order.**

6. **Determine where crossovers must have taken place to give rise to the progeny phenotypes.** To do so, compare each phenotype with the phenotype of the nonrecombinant progeny.

7. **Determine the recombination frequencies.** Add the numbers of the progeny that possess a chromosome with a crossover between a pair of loci. Add the double crossovers to this number. Divide this sum by the total number of progeny from the cross, and multiply by 100%; the result is the recombination frequency between the loci, which is the same as the map distance.

8. **Draw a map of the three loci.** Indicate which locus lies in the middle, and indicate the distances between them.

9. **Determine the coefficient of coincidence and the interference.** The coefficient of coincidence is the number of observed double-crossover progeny divided by the number of expected double-crossover progeny. The expected number can be obtained by multiplying the product of the two single-recombination probabilities by the total number of progeny in the cross. **TRY PROBLEM 31 →**

Worked Problem

In *D. melanogaster*, cherub wings (*ch*), black body (*b*), and cinnabar eyes (*cn*) result from recessive alleles that are all located on chromosome 2. A homozygous wild-type fly was mated with a cherub, black, and cinnabar fly, and the resulting F_1 females were test-crossed with cherub, black, and cinnabar males. The following progeny were produced from the testcross:

ch	b^+	cn	105
ch^+	b^+	cn^+	750
ch^+	b	cn	40
ch^+	b^+	cn	4
ch	b	cn	753
ch	b^+	cn^+	41
ch^+	b	cn^+	102
ch	b	cn^+	5
Total			1800

a. Determine the linear order of the genes on the chromosome (which gene is in the middle?).

b. Calculate the recombinant distances between the three loci.

c. Determine the coefficient of coincidence and the interference for these three loci.

• **Solution**

a. We can represent the crosses in this problem as follows:

$$P \quad \frac{ch^+ \quad b^+ \quad cn^+}{ch^+ \quad b^+ \quad cn^+} \times \frac{ch \quad b \quad cn}{ch \quad b \quad cn}$$

$$F_1 \quad \frac{ch^+ \quad b^+ \quad cn^+}{ch \quad b \quad cn}$$

$$\text{Testcross} \quad \frac{ch^+ \quad b^+ \quad cn^+}{ch \quad b \quad cn} \times \frac{ch \quad b \quad cn}{ch \quad b \quad cn}$$

Note that we do not know, at this point, the order of the genes; we have arbitrarily put *b* in the middle.

The next step is to determine which of the testcross progeny are nonrecombinants and which are double crossovers. The nonrecombinants should be the most-frequent phenotype; so they must be the progeny with phenotypes encoded by $\underline{ch^+ \; b^+ \; cn^+}$ and $\underline{ch \; b \; cn}$. These genotypes are consistent with the genotypes of the parents, given earlier. The double crossovers are the least-frequent phenotypes and are encoded by $\underline{ch^+ \; b^+ \; cn}$ and $\underline{ch \; b \; cn^+}$.

We can determine the gene order by comparing the alleles present in the double crossovers with those present in the nonrecombinants. The double-crossover progeny should be like one of the nonrecombinants at two loci and unlike it at one locus; the allele that differs should be in the middle. Compare the double-crossover progeny $\underline{ch \; b \; cn^+}$ with the nonrecombinant $\underline{ch \; b \; cn}$. Both have cherub wings (*ch*) and black body (*b*), but the double-crossover progeny have wild-type eyes (cn^+), whereas the nonrecombinants have cinnabar eyes (*cn*). The locus that determines cinnabar eyes must be in the middle.

b. To calculate the recombination frequencies among the genes, we first write the phenotypes of the progeny with the genes encoding them in the correct order. We have already identified the nonrecombinant and double-crossover progeny; so the other four progeny types must have resulted from single crossovers. To determine *where* single crossovers took place, we compare the alleles found in the single-crossover progeny with those in the nonrecombinants. Crossing over must have taken place where the alleles switch from those found in one nonrecombinant to those found in the other nonrecombinant. The locations of the crossovers are indicated with a slash:

ch		cn	/	b^+	105	single crossover
ch^+		cn^+		b^+	750	nonrecombinant
ch^+	/	cn		b	40	single crossover
ch^+	/	cn	/	b^+	4	double crossover
ch		cn		b	753	nonrecombinant
ch	/	cn^+		b^+	41	single crossover
ch^+		cn^+	/	b	102	single crossover
ch	/	cn^+	/	b	5	double crossover
Total					1800	

Next, we determine the recombination frequencies and draw a genetic map:

ch-cn recombination frequency =
$$\frac{40+4+41+5}{1800} \times 100\% = 5\%$$

cn-b recombination frequency =
$$\frac{105+4+102+5}{1800} \times 100\% = 12\%$$

ch-b map distance = 5% + 12% = 17%

c. The coefficient of coincidence is the number of observed double crossovers divided by the number of expected double crossovers. The number of expected double crossovers is obtained by multiplying the probability of a crossover between *ch* and *cn* (0.05) × the probability of a crossover between *cn* and *b* (0.12) × the total number of progeny in the cross (1800):

$$\text{coefficient of coincidence} = \frac{4+5}{0.05 \times 0.12 \times 1800} = 0.83$$

Finally, the interference is equal to 1 − the coefficient of coincidence:
$$\text{interference} = 1 - 0.83 = 0.17$$

→ To increase your skill with three-point crosses, try working Problem 29 at the end of this chapter.

Effect of Multiple Crossovers

So far, we have examined the effects of double crossovers taking place between only two of the four chromatids (strands) of a homologous pair. These crossovers are called two-strand crossovers. Double crossovers including three and even four of the chromatids of a homologous pair also may take place (**Figure 7.16**). If we examine only the alleles at loci on either side of both crossover events, two-strand double crossovers result in no new combinations of alleles, and no recombinant gametes are produced (see Figure 7.16). Three-strand double crossovers result in two of the four gametes being recombinant, and four-strand double crossovers result in all four gametes being recombinant. Thus, two-strand double crossovers produce 0% recombination, three-strand double crossovers produce 50% recombination, and four-strand double crossovers produce 100% recombination. The overall result is that all types of double crossovers, taken together, produce an average of 50% recombinant progeny.

As we have seen, two-strand double crossover cause alleles on either side of the crossovers to remain the same and produce no recombinant progeny. Three-strand and four-strand crossovers produce recombinant progeny, but these progeny are the same types produced by single crossovers. Consequently, some multiple crossovers go unde-

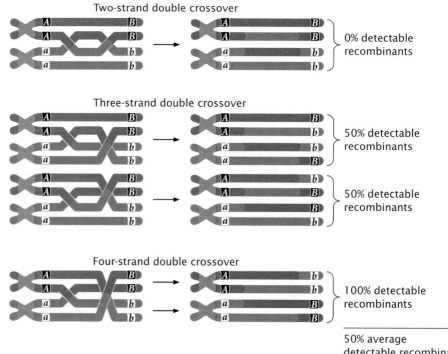

7.16 Results of two-, three-, and four-strand double crossovers on recombination between two genes.

7.17 Percent recombination underestimates the true physical distance between genes at higher map distances.

tected when the progeny of a genetic cross are observed. Therefore, map distances based on recombination rates will underestimate the true physical distances between genes, because some multiple crossovers are not detected among the progeny of a cross. When genes are very close together, multiple crossovers are unlikely, and the distances based on recombination rates accurately correspond to the physical distances on the chromosome. But, as the distance between genes increases, more multiple crossovers are likely, and the discrepancy between genetic distances (based on recombination rates) and physical distances increases. To correct for this discrepancy, geneticists have developed mathematical **mapping functions**, which relate recombination frequencies to actual physical distances between genes (**Figure 7.17**). Most of these functions are based on the Poisson distribution, which predicts the probability of multiple rare events. With the use of such mapping functions, map distances based on recombination rates can be more accurately estimated.

Mapping Human Genes

Efforts in mapping human genes are hampered by the inability to perform desired crosses and the small number of progeny in most human families. Geneticists are restricted to analyses of pedigrees, which are often incomplete and provide limited information. Nevertheless, a large number of human traits have been successfully mapped with the use of pedigree data to analyze linkage. Because the number of progeny from any one mating is usually small, data from several families and pedigrees are usually combined to test for independent assortment. The methods used in these types of analysis are beyond the scope of this book, but an example will illustrate how linkage can be detected from pedigree data.

One of the first documented demonstrations of linkage in humans was between the locus for nail–patella syndrome and the locus that determines the ABO blood types. Nail–patella syndrome is an autosomal dominant disorder characterized by abnormal fingernails and absent or rudimentary kneecaps. The ABO blood types are determined by an autosomal locus with multiple alleles (see Chapter 5). Linkage between the genes encoding these traits was established in families in which both traits segregate. Part of one such family is illustrated in **Figure 7.18**.

Nail–patella syndrome is rare, and so we can assume that people having this trait are heterozygous (Nn); unaffected people are homozygous (nn). The ABO genotypes can be inferred from the phenotypes and the types of offspring produced. Person I-2 in Figure 7.18, for example, has blood-type B, which has two possible genotypes: $I^B I^B$ or $I^B i$ (see Figure 5.6). Because some of her offspring are blood-type O (genotype ii) and must have therefore inherited an i allele from each parent, female I-2 must have genotype $I^B i$. Similarly, the presence of blood-type O offspring

7.18 Linkage between ABO blood types and nail–patella syndrome was established by examining families in whom both traits segregate. The pedigree shown here is for one such family. The ABO blood type is indicated in each circle or square. The genotype, inferred from phenotype, is given below each circle or square.

in generation II indicates that male I-1, with blood-type A, also must carry an i allele and therefore has genotype $I^A i$. The parents of this family are:

$$I^A i \, Nn \times I^B i \, nn$$

From generation II, we can see that the genes for nail–patella syndrome and the blood types do not appear to assort independently. All children in generation II with nail–patella syndrome have either blood-type B or blood-type O; all those with blood-type A have normal nails and kneecaps. If the genes encoding nail–patella syndrome and the ABO blood types assorted independently, we would expect that some children in generation II would have blood-type A and nail–patella syndrome, inheriting both the I^A and N alleles from their father. This outcome indicates that the arrangements of the alleles on the chromosomes of the crossed parents are:

$$\frac{I^A \quad n}{i \quad N} \times \frac{I^B \quad n}{i \quad n}$$

The pedigree indicates that there is no recombination among the offspring (generation II) of these parents, but there are two instances of recombination among the persons in generation III. Persons II-1 and II-2 have the following genotypes:

$$\frac{i^B \quad n}{i \quad N} \times \frac{I^A \quad n}{i \quad n}$$

Their child III-2 has blood-type A and does not have nail–patella syndrome; so he must have genotype

$$\frac{I^A \quad n}{i \quad n}$$

and must have inherited both the i and the n alleles from his father. These alleles are on different chromosomes in the father; so crossing over must have taken place. Crossing over also must have taken place to produce child III-3.

In the pedigree of Figure 7.18, 13 children are from matings in which the genes encoding nail–patella syndrome and ABO blood types segregate; 2 of them are recombinants. On this basis, we might assume that the loci for nail–patella syndrome and ABO blood types are linked, with a recombination frequency of $2/13 = 0.154$. However, it is possible that the genes *are* assorting independently and that the small number of children just makes it seem as though the genes are linked. To determine the probability that genes are actually linked, geneticists often calculate **lod** (logarithm of odds) scores.

To obtain a lod score, we calculate both the probability of obtaining the observed results with the assumption that the genes are linked with a specified degree of recombination and the probability of obtaining the observed results with the assumption of independent assortment. We then determine the ratio of these two probabilities, and the logarithm of this ratio is the lod score. Suppose that the probability of obtaining a particular set of observations with the assumption of linkage and a certain recombination frequency is 0.1 and that the probability of obtaining the same observations with the assumption of independent assortment is 0.0001. The ratio of these two probabilities is $0.1/0.0001 = 1000$, the logarithm of which (the lod score) is 3. Thus, linkage with the specified recombination is 1000 times as likely as independent assortment to produce what was observed. A lod score of 3 or higher is usually considered convincing evidence for linkage. **TRY PROBLEM 33**

Mapping with Molecular Markers

For many years, gene mapping was limited in most organisms by the availability of **genetic markers**—that is, variable genes with easily observable phenotypes for which inheritance could be studied. Traditional genetic markers include genes that encode easily observable characteristics such as flower color, seed shape, blood types, and biochemical differences. The paucity of these types of characteristics in many organisms limited mapping efforts.

In the 1980s, new molecular techniques made it possible to examine variations in DNA itself, providing an almost unlimited number of genetic markers that can be used for creating genetic maps and studying linkage relations. The earliest of these molecular markers consisted of restriction fragment length polymorphisms (RFLPs), which are variations in DNA sequence detected by cutting the DNA with restriction enzymes. Later, methods were developed for detecting variable numbers of short DNA sequences repeated in tandem, called microsatellites. More recently, DNA sequencing allows the direct detection of individual variations in the DNA nucleotides. All of these methods have expanded the availability of genetic markers and greatly facilitated the creation of genetic maps.

Gene mapping with molecular markers is done essentially in the same manner as mapping performed with traditional phenotypic markers: the cosegregation of two or more markers is studied, and map distances are based on the rates of recombination between markers.

Locating Genes with Genomewide Association Studies

The traditional approach to mapping genes, which we have learned in this chapter, is to examine progeny phenotypes in genetic crosses or among individuals in a pedigree, looking for associations between the inheritance of particular phenotype and the inheritance of alleles at other loci. This type of gene mapping is called **linkage analysis**, because it is based on the detection of physical linkage between genes, as measured by the rate of recombination, in progeny from a cross. Linkage analysis has been a powerful tool in the genetic analysis of many different types of organisms.

Another alternative approach to mapping genes is to conduct **genomewide association studies**, looking for nonrandom associations between the presence of a trait and alleles at many different loci scattered across the genome. Unlike linkage analysis, this approach does not trace the inheritance of genetic markers and a trait in a genetic cross or family. Rather, it looks for associations between traits and particular suites of alleles in a *population*.

Imagine that we are interested in finding genes that contribute to bipolar disease, a psychiatric illness characterized by severe depression and mania. When a mutation that predisposes a person to bipolar disease first arises in a population, it will occur on a particular chromosome and be associated with a specific set of alleles on that chromosome (**Figure 7.19**). A specific set of linked alleles, such as this, is called a **haplotype**, and the nonrandom association between alleles in a haplotype is called **linkage disequilibrium**. Because of the physical linkage between the bipolar mutation and the other alleles of the haplotype, bipolar illness and the haplotype will tend to be inherited together. Crossing over, however, breaks up the association between the alleles of the haplotype (see Figure 7.19), reducing the linkage disequilibrium between them. How long the linkage disequilibrium persists depends on the amount of recombination between alleles at different loci. When the loci are far apart, linkage disequilibrium breaks down quickly; when the loci are close together, crossing over is less common and linkage disequilibrium will persist longer. The important point is that linkage disequilibrium—the nonrandom association between alleles—provides information about the distance between genes. A strong association between a trait such as bipolar illness and a set of linked genetic markers indicates that one or more genes contributing to bipolar illness are likely to be near the genetic markers.

In recent years, geneticists have mapped millions of genetic variants called **single-nucleotide polymorphisms** (SNPs), which are positions in the genome where people vary in a single nucleotide base. It is now possible to quickly and inexpensively genotype people for hundreds of thousands or millions of SNPs. This genotyping has provided genetic markers needed for conducting genomewide association studies, in which SNP haplotypes of people with a particular disease, such as bipolar illness, are compared with the haplotypes of healthy persons. Nonrandom associations between SNPs and the disease suggest that one or more genes that contribute to the disease are closely linked to the SNPs. Genomewide association studies do not usually locate specific genes, but rather associate the inheritance of a trait or disease with a specific chromosomal region. After such an association has been established, geneticists can examine the chromosomal region for genes that might be responsible for the trait. Genomewide association studies have been instrumental in the discovery of genes or chromosomal regions that affect a number of genetic diseases and important human traits, including

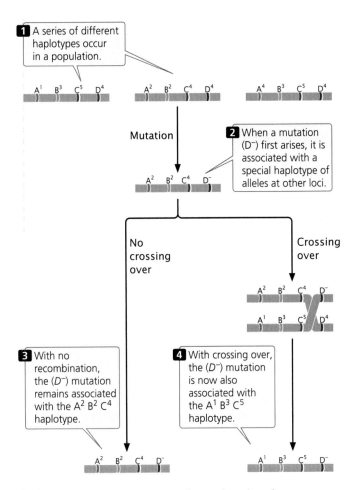

7.19 Genomewide association studies are based on the nonrandom association of a mutation (D^-) that produces a trait and closely linked genes that constitute a haplotype.

bipolar disease, height, skin pigmentation, eye color, body weight, coronary artery disease, blood-lipid concentrations, diabetes, heart attacks, bone density, and glaucoma, among others.

> **CONCEPTS**
>
> The development of molecular techniques for examining variation in DNA sequences has provided a large number of genetic markers that can be used to create genetic maps and study linkage relations. Genomewide association studies examine the nonrandom association of genetic markers and phenotypes to locate genes that contribute to the expression of traits.

7.4 Physical-Mapping Methods Are Used to Determine the Physical Positions of Genes on Particular Chromosomes

Genetic maps reveal the relative positions of genes on a chromosome on the basis of frequencies of recombination, but they do not provide information that allow us to place groups of linked genes on particular chromosomes. Furthermore, the units of a genetic map do not always precisely correspond to physical distances on the chromosome, because a number of factors other than physical distances between genes (such as the type and sex of the organism) can influence recombination. Because of these limitations, physical-mapping methods that do not rely on recombination frequencies have been developed.

Deletion Mapping

One method for determining the chromosomal location of a gene is **deletion mapping**. Special staining methods have been developed that reveal characteristic banding patterns on the chromosomes (see Chapter 9). The absence of one or more of the bands that are normally on a chromosome reveals the presence of a chromosome deletion, a mutation in which a part of a chromosome is missing. Genes can be assigned to regions of chromosomes by studying the association between a gene's phenotype or product and particular chromosome deletions.

In deletion mapping, an individual that is homozygous for a recessive mutation in the gene of interest is crossed with an individual that is heterozygous for a deletion (**Figure 7.20**). If the gene of interest is in the region of the chromosome represented by the deletion (the red part of the chromosomes in Figure 7.20), approximately half of the progeny will display the mutant phenotype (see Figure 7.20a). If the gene is not within the deleted region, all of the progeny will be wild type (see Figure 7.20b).

Deletion mapping has been used to reveal the chromosomal locations of a number of human genes. For example, Duchenne muscular dystrophy is a disease that causes progressive weakening and degeneration of the muscles. From its X-linked pattern of inheritance, the mutated allele causing this disorder was known to be on the X chromosome, but its precise location was uncertain. Examination of a number of patients having Duchenne muscular dystrophy, who also possessed small deletions, allowed researchers to position the gene on a small segment of the short arm of the X chromosome. TRY PROBLEM 34

Somatic-Cell Hybridization

Another method used for positioning genes on chromosomes is **somatic-cell hybridization**, which requires the fusion of different types of cells. Most mature somatic (non-sex) cells can undergo only a limited number of divisions and therefore cannot be grown continuously. However, cells that have been altered by viruses or derived from tumors that have lost the normal constraints on cell division will divide indefinitely; this type of cell can be cultured in the laboratory to produce a **cell line**.

Cells from two different cell lines can be fused by treating them with polyethylene glycol or other agents that alter their plasma membranes. After fusion, the cell possesses

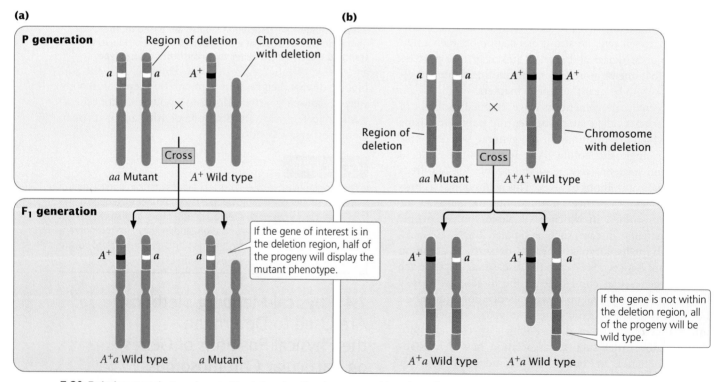

7.20 Deletion mapping can be used to determine the chromosomal location of a gene. An individual homozygous for a recessive mutation in the gene of interest (*aa*) is crossed with an individual heterozygous for a deletion.

7.21 Somatic-cell hybridization can be used to determine which chromosome contains a gene of interest.

two nuclei and is called a **heterokaryon**. The two nuclei of a heterokaryon eventually also fuse, generating a hybrid cell that contains chromosomes from both cell lines. If human and mouse cells are mixed in the presence of polyethylene glycol, fusion results in human–mouse somatic-cell hybrids (**Figure 7.21**). The hybrid cells tend to lose chromosomes as they divide and, for reasons that are not understood, chromosomes from one of the species are lost preferentially. In human–mouse somatic-cell hybrids, the human chromosomes tend to be lost, whereas the mouse chromosomes are retained. Eventually, the chromosome number stabilizes when all but a few of the human chromosomes have been lost. Chromosome loss is random and differs among cell lines. The presence of these "extra" human chromosomes in the mouse genome makes it possible to assign human genes to specific chromosomes.

To map genes by using somatic-cell hybridization requires a panel of different hybrid cell lines. Each cell line is examined microscopically and the human chromosomes that it contains are identified. The cell lines of the panel are chosen so that they differ in the human chromosomes that they have retained. For example, one cell line might possess human chromosomes 2, 4, 7, and 8, whereas another might possess chromosomes 4, 19, and 20. Each cell line in the panel is examined for evidence of a particular human gene. The human gene can be detected by looking either for the for the gene itself or for the protein that it produces. Correlation of the presence of the gene with the presence of specific human chromosomes often allows the gene to be assigned to the correct chromosome. For example, if a gene is detected in both of the aforementioned cell lines, the gene must be on chromosome 4, because it is the only human chromosome common to both cell lines (**Figure 7.22**).

Sometimes somatic-cell hybridization can be used to position a gene on a specific part of a chromosome. Some

Cell line	Gene product present	Human chromosomes present																					
		1	2	3	4	5	6	7	8	9	10	11	12	13	14	15	16	17	18	19	20	21	22	X
A	+		+		+			+	+															
B	+	+	+		+						+	+	+	+	+									
C	−															+		+	+			+		
D	+		+		+			+	+	+														
E	−													+							+			
F	+				+															+	+			

7.22 Somatic-cell hybridization is used to assign a gene to a particular human chromosome. A panel of six cell lines, each line containing a different subset of human chromosomes, is examined for the presence of the gene product (such as an enzyme). Four of the cell lines (A, B, D, and F) have the gene product. The only chromosome common to all four of these cell lines is chromosome 4, indicating that the gene is located on this chromosome.

7.23 Genes can be localized to a specific part of a chromosome by using somatic-cell hybridization.

hybrid cell lines carry a human chromosome with a chromosome mutation such as a deletion or a translocation. If the gene is present in a cell line with the intact chromosome but missing from a line with a chromosome deletion, the gene must be located in the deleted region (**Figure 7.23**). Similarly, if a gene is usually absent from a chromosome but consistently appears whenever a translocation (a piece of another chromosome that has broken off and attached itself to the chromosome in question) is present, it must be present on the translocated part of the chromosome.

Worked Problem

A panel of cell lines was created from human–mouse somatic-cell fusions. Each line was examined for the presence of human chromosomes and for the production of human haptoglobin (a protein). The following results were obtained:

Cell line	Human haptoglobin	Human chromosomes						
		1	2	3	14	15	16	21
A	−	+	−	+	−	+	−	−
B	+	+	−	+	−	−	+	−
C	+	+	−	−	−	+	+	−
D	−	+	+	−	−	+	−	−

On the basis of these results, which human chromosome carries the gene for haptoglobin?

• **Solution**

First, identify the cell lines that are positive for the protein (human haptoglobin) and determine the chromosomes that they have in common. Lines B and C produce human haptoglobin; the only chromosomes that they have in common are chromosomes 1 and 16. Next, examine all the cell lines that possess either chromosomes 1 and 16 and determine whether they produce haptoglobin. Chromosome 1 is found in cell lines A, B, C, and D. If the gene for human haptoglobin were found on chromosome 1, human haptoglobin would be present in all of these cell lines. However, lines A and D do not produce human haptoglobin; so the gene cannot be on chromosome 1. Chromosome 16 is found only in cell lines B and C, and only these lines produce human haptoglobin; so the gene for human haptoglobin lies on chromosome 16.

→ For more practice with somatic-cell hybridizations, work Problem 35 at the end of this chapter.

Physical Chromosome Mapping Through Molecular Analysis

So far, we have explored methods to indirectly determine the chromosomal location of a gene by deletion mapping or by looking for gene products. Researchers now have the information and technology to actually see where a gene lies. In situ hybridization is a method for determining the chromosomal location of a particular gene through molecular analysis. This method requires the creation of a probe for the gene, which is a single-stranded DNA complement to the gene of interest. The probe is radioactive or fluoresces under ultraviolet light so that it can be visualized. The probe binds to the DNA sequence of the gene on the chromosome. The presence of radioactivity or fluorescence from the bound probe reveals the location of the gene on a particular chromosome (**Figure 7.24**).

In addition to allowing us to see where a gene is located on a chromosome, modern laboratory techniques now allow researchers to identify the precise location of a gene at the nucleotide level. For example, with DNA sequencing, physical distances between genes can be determined in numbers of base pairs.

7.24 In situ hybridization is another technique for determining the chromosomal location of a gene. The red fluorescence is produced by a probe for sequences on chromosome 9; the green fluorescence is produced by a probe for sequences on chromosome 22. [Courtesy of Applied Imaging Corp.]

> **CONCEPTS**
>
> Physical-mapping methods determine the physical locations of genes on chromosomes and include deletion mapping, somatic-cell hybridization, in situ hybridization, and direct DNA sequencing.

7.5 Recombination Rates Exhibit Extensive Variation

In recent years, geneticists have studied variation in rates of recombination and found that levels of recombination vary widely—among species, among and along chromosomes of a single species, and even between males and females of the same species. For example, about twice as much recombination takes place in humans as in the mouse and the rat. Within the human genome, recombination varies among chromosomes, with chromosomes 21 and 22 having the highest rates and chromosomes 2 and 4 having the lowest rates. Researchers have also detected differences between male and female humans in rates of recombination: the autosomal chromosomes of females undergo about 50% more recombination than do the autosomal chromosomes of males.

Geneticists have found numerous recombination *hotspots*, where recombination is at least 10 times as high as the average elsewhere in the genome. The human genome may contain an estimated 25,000 to 50,000 such recombination hotspots. Approximately 60% of all crossovers take place in hotspots. For humans, recombination hotspots tend to be found near but not within active genes. Recombination hotspots have been detected in the genomes of other organisms as well. In comparing recombination hotspots in the genomes of humans and chimpanzees (our closest living relative), geneticists have determined that, although the DNA sequences of humans and chimpanzees are extremely similar, their recombination hotspots are at entirely different places.

> **CONCEPTS**
>
> Rates of recombination vary among species, among and along chromosomes, and even between males and females.

CONCEPTS SUMMARY

- Linked genes do not assort independently. In a testcross for two completely linked genes (no crossing over), only nonrecombinant progeny are produced. When two genes assort independently, recombinant progeny and nonrecombinant progeny are produced in equal proportions. When two genes are linked with some crossing over between them, more nonrecombinant progeny than recombinant progeny are produced.
- Recombination frequency is calculated by summing the number of recombinant progeny, dividing by the total number of progeny produced in the cross, and multiplying by 100%. The recombination frequency is half the frequency of crossing over, and the maximum frequency of recombinant gametes is 50%.
- Coupling and repulsion refer to the arrangement of alleles on a chromosome. Whether genes are in coupling configuration or in repulsion determines which combination of phenotypes will be most frequent in the progeny of a testcross.
- Interchromosomal recombination takes place among genes located on different chromosomes through the random segregation of chromosomes in meiosis. Intrachromosomal recombination takes place among genes located on the same chromosome through crossing over.
- A chi-square test of independence can be used to determine if genes are linked.
- Recombination rates can be used to determine the relative order of genes and distances between them on a chromosome. One percent recombination equals one map unit. Maps based on recombination rates are called genetic maps; maps based on physical distances are called physical maps.
- Genetic maps can be constructed by examining recombination rates from a series of two-point crosses or by examining the progeny of a three-point testcross.
- Some multiple crossovers go undetected; thus, genetic maps based on recombination rates underestimate the true physical distances between genes.
- Human genes can be mapped by examining the cosegregation of traits in pedigrees.
- A lod score is the logarithm of the ratio of the probability of obtaining the observed progeny with the assumption of linkage to the probability of obtaining the observed progeny with the assumption of independent assortment. A lod score of 3 or higher is usually considered evidence for linkage.
- Molecular techniques that allow the detection of variable differences in DNA sequence have greatly facilitated gene mapping.
- Genomewide association studies locate genes that affect particular traits by examining the nonrandom association of a trait with genetic markers from across the genome.
- Nucleotide sequencing is another method of physically mapping genes.
- Rates of recombination vary widely, differing among species, among and along chromosomes within a single species, and even between males and females of the same species.

IMPORTANT TERMS

linked genes (p. 163)
linkage group (p. 163)
nonrecombinant (parental) gamete (p. 166)
nonrecombinant (parental) progeny (p. 166)
recombinant gamete (p. 166)
recombinant progeny (p. 166)
recombination frequency (p. 167)
coupling (cis) configuration (p. 168)
repulsion (trans) configuration (p. 168)
interchromosomal recombination (p. 170)

intrachromosomal recombination (p. 170)
genetic map (p. 174)
physical map (p. 174)
map unit (m.u.) (p. 174)
centiMorgan (p. 174)
Morgan (p. 174)
two-point testcross (p. 175)
three-point testcross (p. 176)
interference (p. 182)
coefficient of coincidence (p. 182)
mapping function (p. 185)
lod (logarithm of odds) score (p. 186)

genetic marker (p. 186)
linkage analysis (p. 186)
genomewide association studies (p. 187)
haplotype (p. 187)
linkage disequilibrium (p. 187)
single-nucleotide polymorphism (SNP) (p. 187)
deletion mapping (p. 188)
somatic-cell hybridization (p. 188)
cell line (p. 188)
heterokaryon (p. 189)

ANSWERS TO CONCEPT CHECKS

1. c
2. Repulsion
3. Genetic maps are based on rates of recombination; physical maps are based on physical distances.
4. $\frac{m^+\ p^+\ s^+}{m\ p\ s}\ \frac{m^+\ p\ s}{m\ p\ s}\ \frac{m\ p^+\ s^+}{m\ p\ s}\ \frac{m^+\ p^+s}{m\ p\ s}\ \frac{m\ p\ s^+}{m\ p\ s}\ \frac{m^+\ p\ s^+}{m\ p\ s}\ \frac{m\ p^+\ s}{m\ p\ s}\ \frac{m\ p\ s}{m\ p\ s}$
5. The c locus
6. b

WORKED PROBLEMS

1. In guinea pigs, white coat (w) is recessive to black coat (W) and wavy hair (v) is recessive to straight hair (V). A breeder crosses a guinea pig that is homozygous for white coat and wavy hair with a guinea pig that is black with straight hair. The F_1 are then crossed with guinea pigs having white coats and wavy hair in a series of testcrosses. The following progeny are produced from these testcrosses:

black, straight	30
black, wavy	10
white, straight	12
white, wavy	31
Total	83

a. Are the genes that determine coat color and hair type assorting independently? Carry out chi-square tests to test your hypothesis.

b. If the genes are not assorting independently, what is the recombination frequency between them?

- **Solution**

a. Assuming independent assortment, outline the crosses conducted by the breeder:

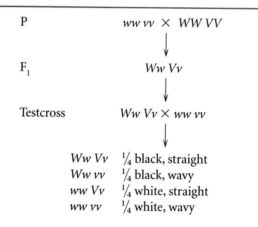

Because a total of 83 progeny were produced in the testcrosses, we expect $1/4 \times 83 = 20.75$ of each. The observed numbers of progeny from the testcross (30, 10, 12, 31) do not appear to fit the expected numbers (20.75, 20.75, 20.75, 20.75) well; so independent assortment may not have taken place.

To test the hypothesis, carry out a chi-square test of independence. Construct a table, with the genotypes of the first locus along the top and the genotypes of the second locus along the side. Compute the totals for the rows and columns and the grand total.

	Ww	ww	Row totals
Vv	30	12	42
vv	10	31	41
Column totals	40	43	83 ← Grand total

The expected value for each cell of the table is calculated with the following formula:

$$\text{expected number} = \frac{\text{row total} \times \text{column total}}{\text{grand total}}$$

Using this formula, we find the expected values (given in parentheses) to be:

	Ww	ww	Row totals
Vv	30 (20.24)	12 (21.76)	42
vv	10 (19.76)	31 (21.24)	41
Column totals	40	43	83 ← Grand total

Using these observed and expected numbers, we find the calculated chi-square value to be:

$$\chi^2 = \sum \frac{(\text{observed} - \text{expected})^2}{\text{expected}}$$

$$= \frac{(30-20.24)^2}{20.24} + \frac{(10-19.76)^2}{19.76} + \frac{(12-21.76)^2}{21.76} + \frac{(31-21.24)^2}{21.24}$$

$$= 4.71 + 4.82 + 4.38 + 4.48 = 18.39$$

The degrees of freedom for the chi-square test of independence are df = (number of rows − 1) × (number of columns − 1). There are two rows and two columns, and so the degrees of freedom are:

$$df = (2-1) \times (2-1) = 1 \times 1 = 1$$

In Table 3.5, the probability associated with a chi-square value of 18.39 and 1 degree of freedom is less than 0.005, indicating that chance is very unlikely to be responsible for the differences between the observed numbers and the numbers expected with independent assortment. The genes for coat color and hair type have therefore not assorted independently.

b. To determine the recombination frequencies, identify the recombinant progeny. Using the notation for linked genes, write the crosses:

P $\frac{w\quad v}{w\quad v}$ × $\frac{W\quad V}{W\quad V}$

↓

F_1 $\frac{W\quad V}{w\quad v}$

Testcross $\frac{W\quad V}{w\quad v}$ × $\frac{w\quad v}{w\quad v}$

↓

$\frac{W\quad V}{w\quad v}$ 30 black, straight (nonrecombinant progeny)

$\frac{w\quad v}{w\quad v}$ 31 white, wavy (nonrecombinant progeny)

$\frac{W\quad v}{w\quad v}$ 10 black, wavy (recombinant progeny)

$\frac{w\quad V}{w\quad v}$ 12 white, straight (recombinant progeny)

The recombination frequency is:

$$\frac{\text{number of recombinant progeny}}{\text{total number progeny}} \times 100\%$$

or

recombinant frequency =

$$\frac{10+12}{30+31+10+12} \times 100\% = \frac{22}{83} \times 100\% = 26.5$$

2. A series of two-point crosses were carried out among seven loci ($a, b, c, d, e, f,$ and g), producing the following recombination frequencies. Using these recombination frequencies, map the seven loci, showing their linkage groups, the order of the loci in each linkage group, and the distances between the loci of each group:

Loci	Recombination frequency (%)	Loci	Recombination frequency (%)
a and b	10	c and d	50
a and c	50	c and e	8
a and d	14	c and f	50
a and e	50	c and g	12
a and f	50	d and e	50
a and g	50	d and f	50
b and c	50	d and g	50
b and d	4	e and f	50
b and e	50	e and g	18
b and f	50	f and g	50
b and g	50		

• Solution

To work this problem, remember that 1% recombination equals 1 map unit and a recombination frequency of 50% means that genes at the two loci are assorting independently (located in different linkage groups).

The recombination frequency between *a* and *b* is 10%; so these two loci are in the same linkage group, approximately 10 m.u. apart.

The recombination frequency between *a* and *c* is 50%; so *c* must lie in a second linkage group.

The recombination frequency between *a* and *d* is 14%; so *d* is located in linkage group 1. Is locus *d* 14 m.u. to the right or to the left of gene *a*? If *d* is 14 m.u. to the left of *a*, then the *b*-to-*d* distance should be 10 m.u. + 14 m.u. = 24 m.u. On the other hand, if *d* is to the right of *a*, then the distance between *b* and *d* should be 14 m.u. − 10 m.u. = 4 m.u. The *b*–*d* recombination frequency is 4%; so *d* is 14 m.u. to the right of *a*. The updated map is:

The recombination frequencies between each of loci *a*, *b*, and *d*, and locus *e* are all 50%; so *e* is not in linkage group 1 with *a*, *b*, and *d*. The recombination frequency between *e* and *c* is 8 m.u.; so *e* is in linkage group 2:

There is 50% recombination between *f* and all the other genes; so *f* must belong to a third linkage group:

Linkage group 2

Linkage group 3

Finally, position locus *g* with respect to the other genes. The recombination frequencies between *g* and loci *a*, *b*, and *d* are all 50%; so *g* is not in linkage group 1. The recombination frequency between *g* and *c* is 12 m.u.; so *g* is a part of linkage group 2. To determine whether *g* is 12 m.u. to the right or left of *c*, consult the *g*-*e* recombination frequency. Because this recombination frequency is 18%, *g* must lie to the left of *c*:

Linkage group 3

f

Note that the *g*-to-*e* distance (18 m.u.) is shorter than the sum of the *g*-to-*c* (12 m.u.) and *c*-to-*e* distances (8 m.u.) because of undetectable double crossovers between *g* and *e*.

3. Ebony body color (*e*), rough eyes (*ro*), and brevis bristles (*bv*) are three recessive mutations that occur in fruit flies. The loci for these mutations have been mapped and are separated by the following map distances:

The interference between these genes is 0.4.

A fly with ebony body, rough eyes, and brevis bristles is crossed with a fly that is homozygous for the wild-type traits. The resulting F$_1$ females are test-crossed with males that have ebony body, rough eyes, and brevis bristles; 1800 progeny are produced. Give the expected numbers of phenotypes in the progeny of the testcross.

• **Solution**

The crosses are:

P $\quad \dfrac{e^+ \quad ro^+ \quad bv^+}{e^+ \quad ro^+ \quad bv^+} \times \dfrac{e \quad ro \quad bv}{e \quad ro \quad bv}$

$\downarrow$

$F_1 \quad \dfrac{e^+ \quad ro^+ \quad bv^+}{e \quad ro \quad bv}$

$\downarrow$

Testcross $\quad \dfrac{e^+ \quad ro^+ \quad bv^+}{e \quad ro \quad bv} \times \dfrac{e \quad ro \quad bv}{e \quad ro \quad bv}$

In this case, we know that *ro* is the middle locus because the genes have been mapped. Eight classes of progeny will be produced from this cross:

e^+	ro^+	bv^+	nonrecombinant
e	ro	bv	nonrecombinant
e^+ /	ro	bv	single crossover between *e* and *ro*
e /	ro^+	bv^+	single crossover between *e* and *ro*
e^+	ro^+ /	bv	single crossover between *ro* and *bv*
e	ro /	bv^+	single crossover between *ro* and *bv*
e^+ /	ro /	bv^+	double crossover
e /	ro^+ /	bv	double crossover

To determine the numbers of each type, use the map distances, starting with the double crossovers. The expected number of double crossovers is equal to the product of the single-crossover probabilities:

expected number of double crossovers = 0.20 × 0.12 × 1800
$\qquad\qquad\qquad\qquad\qquad\qquad$ = 43.2

However, there is some interference; so the observed number of double crossovers will be less than the expected. The interference is 1 − coefficient of coincidence; so the coefficient of coincidence is:

coefficient of coincidence = 1 − interference

The interference is given as 0.4; so the coefficient of coincidence equals 1 − 0.4 = 0.6. Recall that the coefficient of coincidence is:

coefficient of coincidence =
$\qquad\dfrac{\text{number of observed double crossovers}}{\text{number of expected double crossovers}}$

Rearranging this equation, we obtain:

number of observed double crossovers =
coefficient of coincidence × number of expected double crossovers

number of observed double crossovers = 0.6 × 43.2 = 26

A total of 26 double crossovers should be observed. Because there are two classes of double crossovers (e^+ / ro / bv^+ and e / ro^+ / bv), we expect to observe 13 of each class.

Next, we determine the number of single-crossover progeny. The genetic map indicates that the distance between *e* and *ro* is 20 m.u.; so 360 progeny (20% of 1800) are expected to have resulted from recombination between these two loci. Some of them will be single-crossover progeny and some will be double-crossover progeny. We have already determined that the number of double-crossover progeny is 26; so the number of progeny resulting from a single crossover between *e* and *ro* is 360 − 26 = 334, which will be divided equally between the two single-crossover phenotypes (e / ro^+ / bv^+ and e^+ / ro bv).

The distance between *ro* and *bv* is 12 m.u.; so the number of progeny resulting from recombination between these two genes is 0.12 × 1800 = 216. Again, some of these recombinants will be single-crossover progeny and some will be double-crossover progeny. To determine the number of progeny resulting from a single crossover, subtract the double crossovers: 216 − 26 = 190. These single-crossover progeny will be divided between the two single-crossover phenotypes (e^+ ro^+ / bv and e ro / bv^+); so there will be $^{190}/_2$ = 95 of each of these phenotypes. The remaining progeny will be nonrecombinants, and they can be obtained by subtraction: 1800 − 26 − 334 − 190 = 1250; there are two nonrecombinants (e^+ ro^+ bv^+ and e ro bv); so there will be $^{1250}/_2$ = 625 of each. The numbers of the various phenotypes are listed here:

e^+	ro^+	bv^+	625	nonrecombinant
e	ro	bv	625	nonrecombinant
e^+ /	ro	bv	167	single crossover between *e* and *ro*
e /	ro^+	bv^+	167	single crossover between *e* and *ro*
e^+	ro^+ /	bv	95	single crossover between *ro* and *bv*
e	ro /	bv^+	95	single crossover between *ro* and *bv*
e^+ /	ro /	bv^+	13	double crossover
e /	ro^+ /	bv	13	double crossover
Total			1800	

4. The locations of six deletions have been mapped to the *Drosophila* chromosome as shown in the following diagram.

```
                    Chromosome
                    _____

Deletion 1  ———————
   Deletion 2  ———————
      Deletion 3  ———————
         Deletion 4  ———————
                       Deletion 5  ———————
            Deletion 6  ———————
```

Recessive mutations *a*, *b*, *c*, *d*, *e*, *f*, and *g* are known to be located in the same regions as the deletions, but the order of the mutations on the chromosome is not known. When flies homozygous for the recessive mutations are crossed with flies homozygous for the deletions, the following results are obtained, where the letter "m" represents a mutant phenotype and a plus sign (+) represents the wild type. On the basis of these data, determine the relative order of the seven mutant genes on the chromosome:

Deletion	Mutations						
	a	b	c	d	e	f	g
1	+	m	m	m	+	+	+
2	+	+	m	m	+	+	+
3	+	+	+	m	m	+	+
4	m	+	+	m	m	+	+
5	m	+	+	+	+	m	m
6	m	+	+	m	m	m	+

- **Solution**

The offspring of the cross will be heterozygous, possessing one chromosome with the deletion and wild-type alleles and its homolog without the deletion and recessive mutant alleles. For loci within the deleted region, only the recessive mutations will be present in the offspring, which will exhibit the mutant phenotype. The presence of a mutant trait in the offspring therefore indicates that the locus for that trait is within the region covered by the deletion. We can map the genes by examining the expression of the recessive mutations in the flies with different deletions.

Mutation *a* is expressed in flies with deletions 4, 5, and 6 but not in flies with other deletions; so *a* must be in the area that is unique to deletions 4, 5, and 6:

Mutation *b* is expressed only when deletion 1 is present; so *b* must be located in the region of the chromosome covered by deletion 1 and none of the other deletions:

Using this procedure, we can map the remaining mutations. For each mutation, we look for the areas of overlap among deletions that express the mutations and exclude any areas of overlap that are covered by other deletions that do not express the mutation:

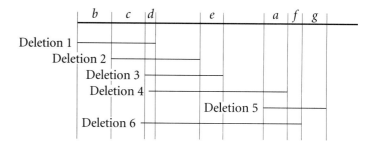

COMPREHENSION QUESTIONS

Section 7.1

*1. What does the term recombination mean? What are two causes of recombination?

Section 7.2

*2. In a testcross for two genes, what types of gametes are produced with (a) complete linkage, (b) independent assortment, and (c) incomplete linkage?

3. What effect does crossing over have on linkage?

4. Why is the frequency of recombinant gametes always half the frequency of crossing over?

*5. What is the difference between genes in coupling configuration and genes in repulsion? What effect does the arrangement of linked genes (whether they are in coupling configuration or in repulsion) have on the results of a cross?

6. How would you test to see if two genes are linked?

7. What is the difference between a genetic map and a physical map?

*8. Why do calculated recombination frequencies between pairs of loci that are located far apart underestimate the true genetic distances between loci?

Section 7.3

9. Explain how to determine, using the numbers of progeny from a three-point cross, which of three linked loci is the middle locus.

*10. What does the interference tell us about the effect of one crossover on another?

Section 7.4

11. What is a lod score and how is it calculated?

12. List some of the methods for physically mapping genes and explain how they are used to position genes on chromosomes.

APPLICATION QUESTIONS AND PROBLEMS

Introduction

13. The introduction to this chapter described the search for genes that determine pattern baldness in humans. In 1916, Dorothy Osborn suggested that pattern baldness is a sex-influenced trait (see Chapter 5) that is dominant in males and recessive in females. More research suggested that pattern baldness is an X-linked recessive trait. Would you expect to see independent assortment between genetic markers on the X chromosome and pattern baldness if (a) pattern baldness is sex-influenced and (b) if pattern baldness is X-linked recessive? Explain your answer.

Section 7.2

14. In the snail *Cepaea nemoralis,* an autosomal allele causing a banded shell (B^B) is recessive to the allele for an unbanded shell (B^O). Genes at a different locus determine the background color of the shell; here, yellow (C^Y) is recessive to brown (C^{Bw}). A banded, yellow snail is crossed with a homozygous brown, unbanded snail. The F_1 are then crossed with banded, yellow snails (a testcross).

   a. What will the results of the testcross be if the loci that control banding and color are linked with no crossing over?

   b. What will the results of the testcross be if the loci assort independently?

   c. What will the results of the testcross be if the loci are linked and 20 m.u. apart?

   [Otto Hahn/Photolibrary.]

15. In silkmoths (*Bombyx mori*), red eyes (*re*) and white-banded wing (*wb*) are encoded by two mutant alleles that are recessive to those that produce wild-type traits (*re*$^+$ and *wb*$^+$); these two genes are on the same chromosome. A moth homozygous for red eyes and white-banded wings is crossed with a moth homozygous for the wild-type traits. The F_1 have normal eyes and normal wings. The F_1 are crossed with moths that have red eyes and white-banded wings in a testcross. The progeny of this testcross are:

wild-type eyes, wild-type wings	418
red eyes, wild-type wings	19
wild-type eyes, white-banded wings	16
red eyes, white-banded wings	426

   a. What phenotypic proportions would be expected if the genes for red eyes and for white-banded wings were located on different chromosomes?

   b. What is the percent recombination between the genes for red eyes and those for white-banded wings?

*16. A geneticist discovers a new mutation in *Drosophila melanogaster* that causes the flies to shake and quiver. She calls this mutation spastic (*sps*) and determines that spastic is due to an autosomal recessive gene. She wants to determine if the gene encoding spastic is linked to the recessive gene for vestigial wings (*vg*). She crosses a fly homozygous for spastic and vestigial traits with a fly homozygous for the wild-type traits and then uses the resulting F_1 females in a testcross. She obtains the following flies from this testcross.

vg^+	sps^+	230
vg	sps	224
vg	sps^+	97
vg^+	sps	99
Total		650

   Are the genes that cause vestigial wings and the spastic mutation linked? Do a chi-square test of independence to determine whether the genes have assorted independently.

17. In cucumbers, heart-shaped leaves (*hl*) are recessive to normal leaves (*Hl*) and having numerous fruit spines (*ns*) is recessive to having few fruit spines (*Ns*). The genes for leaf shape and for number of spines are located on the same chromosome; findings from mapping experiments indicate that they are 32.6 m.u. apart. A cucumber plant having heart-shaped leaves and numerous spines is

crossed with a plant that is homozygous for normal leaves and few spines. The F₁ are crossed with plants that have heart-shaped leaves and numerous spines. What phenotypes and proportions are expected in the progeny of this cross?

*18. In tomatoes, tall (D) is dominant over dwarf (d) and smooth fruit (P) is dominant over pubescent fruit (p), which is covered with fine hairs. A farmer has two tall and smooth tomato plants, which we will call plant A and plant B. The farmer crosses plants A and B with the same dwarf and pubescent plant and obtains the following numbers of progeny:

	Progeny of	
	Plant A	Plant B
Dd Pp	122	2
Dd pp	6	82
dd Pp	4	82
dd pp	124	4

a. What are the genotypes of plant A and plant B?

b. Are the loci that determine height of the plant and pubescence linked? If so, what is the percent recombination between them?

c. Explain why different proportions of progeny are produced when plant A and plant B are crossed with the same dwarf pubescent plant.

19. Alleles A and a are at a locus on the same chromosome as is a locus with alleles B and b. $Aa\ Bb$ is crossed with $aa\ bb$ and the following progeny are produced:

Aa Bb	5
Aa bb	45
aa Bb	45
aa bb	5

What conclusion can be made about the arrangement of the genes on the chromosome in the $Aa\ Bb$ parent?

20. Daniel McDonald and Nancy Peer determined that eyespot (a clear spot in the center of the eye) in flour beetles is caused by an X-linked gene (es) that is recessive to the allele for the absence of eyespot (es^+). They conducted a series of crosses to determine the distance between the gene for eyespot and a dominant X-linked gene for striped (St), which causes white stripes on females and acts as a recessive lethal (is lethal when homozygous in females or hemizygous in males). The following cross was carried out (D. J. McDonald and N. J. Peer. 1961. *Journal of Heredity* 52:261–264).

♀ $\dfrac{es^+\quad St}{es\quad St^+}$ × $\dfrac{es\quad St^+}{Y}$ ♂

↓

$\dfrac{es^+\quad St}{es\quad St^+}$	1630
$\dfrac{es\quad St^+}{es\quad St^+}$	1665
$\dfrac{es\quad St}{es\quad St^+}$	935
$\dfrac{es^+\quad St^+}{es\quad St^+}$	1005
$\dfrac{es\quad St^+}{Y}$	1661
$\dfrac{es^+\quad St^+}{Y}$	1024

a. Which progeny are the recombinants and which progeny are the nonrecombinants?

b. Calculate the recombination frequency between es and St.

c. Are some potential genotypes missing among the progeny of the cross? If so, which ones and why?

*21. Recombination rates between three loci in corn are shown here.

Loci	Recombination rate
R and W_2	17%
R and L_2	35%
W_2 and L_2	18%

What is the order of the genes on the chromosome?

22. In tomatoes, dwarf (d) is recessive to tall (D) and opaque (light-green) leaves (op) are recessive to green leaves (Op). The loci that determine height and leaf color are linked and separated by a distance of 7 m.u. For each of the following crosses, determine the phenotypes and proportions of progeny produced.

a. $\dfrac{D\quad Op}{d\quad op} \times \dfrac{d\quad op}{d\quad op}$

b. $\dfrac{D\quad op}{d\quad Op} \times \dfrac{d\quad op}{d\quad op}$

c. $\dfrac{D\quad Op}{d\quad op} \times \dfrac{D\quad Op}{d\quad op}$

d. $\dfrac{D\quad op}{d\quad Op} \times \dfrac{D\quad op}{d\quad Op}$

23. In German cockroaches, bulging eyes (bu) are recessive to normal eyes (bu^+) and curved wings (cv) are recessive to straight wings (cv^+). Both traits are encoded by autosomal genes that are linked. A cockroach has genotype $bu^+bu\ cv^+cv$, and the genes are in repulsion. Which of the following sets of genes will be found in the most-common gametes produced by this cockroach?

[W. Willner/Photolibrary.]

a. $bu^+\ cv^+$
b. $bu\ cv$
c. $bu^+\ bu$
d. $cv^+\ cv$
e. $bu\ cv^+$

Explain your answer.

*24. In *Drosophila melanogaster*, ebony body (e) and rough eyes (ro) are encoded by autosomal recessive genes found on chromosome 3; they are separated by 20 m.u. The gene that encodes forked bristles (f) is X-linked recessive and assorts independently of e and ro. Give the phenotypes of progeny and their expected proportions when a female of each of the following genotypes is test-crossed with a male.

a. $\dfrac{e^+\quad ro^+}{e\quad ro}\ \ \dfrac{f^+}{f}$

b. $\dfrac{e^+\quad ro}{e\quad ro^+}\ \ \dfrac{f^+}{f}$

25. Honeybees have haplodiploid sex determination: females are diploid, developing from fertilized eggs, whereas males are haploid, developing from unfertilized eggs. Otto Mackensen studied linkage relations among eight mutations in honeybees (O. Mackensen. 1958. *Journal of Heredity* 49:99–102). The following table gives the results of two of MacKensen's crosses including three recessive mutations: cd (cordovan body color), h (hairless), and ch (chartreuse eye color).

Queen genotype	Phenotypes of drone (male) progeny
$\dfrac{cd\quad h^+}{cd^+\quad h}$	294 cordovan, 236 hairless, 262 cordovan and hairless, 289 wild .type
$\dfrac{h\quad ch^+}{h^+\quad ch}$	3131 hairless, 3064 chartreuse, 96 chartreuse and hairless, 132 wild type

a. Only the genotype of the queen is given. Why is the genotype of the male parent not needed for mapping these genes? Would the genotype of the male parent be required if we examined female progeny instead of male progeny?

b. Determine the nonrecombinant and recombinant progeny for each cross and calculate the map distances between cd, h, and ch. Draw a linkage map illustrating the linkage arrangements among these three genes.

*26. A series of two-point crosses were carried out among seven loci (a, b, c, d, e, f, and g), producing the following recombination frequencies. Map the seven loci, showing their linkage groups, the order of the loci in each linkage group, and the distances between the loci of each group.

Loci	Percent recombination	Loci	Percent recombination
a and b	50	c and d	50
a and c	50	c and e	26
a and d	12	c and f	50
a and e	50	c and g	50
a and f	50	d and e	50
a and g	4	d and f	50
b and c	10	d and g	8
b and d	50	e and f	50
b and e	18	e and g	50
b and f	50	f and g	50
b and g	50		

27. R. W. Allard and W. M. Clement determined recombination rates for a series of genes in lima beans (R. W. Allard and W. M. Clement. 1959. *Journal of Heredity* 50:63–67). The following table lists paired recombination rates for eight of the loci (D, Wl, R, S, L_1, Ms, C, and G) that they mapped. On the basis of these data, draw a series of genetic maps for the different linkage groups of the genes, indicating the distances between the genes. Keep in mind that these rates are estimates of the true recombination rates and that some error is associated with each estimate. An asterisk beside a recombination frequency indicates that the recombination frequency is significantly different from 50%.

Recombination Rates (%) among Seven Loci in Lima Beans

	Wl	R	S	L_1	Ms	C	G
D	2.1*	39.3*	52.4	48.1	53.1	51.4	49.8
Wl		38.0*	47.3	47.7	48.8	50.3	50.4
R			51.9	52.7	54.6	49.3	52.6
S				26.9*	54.9	52.0	48.0
L_1					48.2	45.3	50.4
Ms						14.7*	43.1
C							52.0

*Significantly different from 50%.

Section 7.3

28. Raymond Popp studied linkage among genes for pink eye (p), shaker-1 (sh-1, which causes circling behavior, head tossing, and deafness), and hemoglobin (Hb) in mice (R. A. Popp. 1962. *Journal of Heredity* 53:73–80). He performed a series of testcrosses, in which mice

heterozygous for pink eye, shaker-1, and hemoglobin 1 and 2 were crossed with mice that were homozygous for pink eye, shaker-1, and hemoglobin 2.

$$\frac{P\ Sh\text{-}1\ Hb^1}{p\ sh\text{-}1\ Hb^2} \times \frac{p\ sh\text{-}1\ Hb^2}{p\ sh\text{-}1\ Hb^2}$$

The following progeny were produced.

Progeny genotype	Number
$\frac{p\ sh\text{-}1\ Hb^2}{p\ sh\text{-}1\ Hb^2}$	274
$\frac{P\ Sh\text{-}1\ Hb^1}{p\ sh\text{-}1\ Hb^2}$	320
$\frac{P\ sh\text{-}1\ Hb^2}{p\ sh\text{-}1\ Hb^2}$	57
$\frac{p\ Sh\text{-}1\ Hb^1}{p\ sh\text{-}1\ Hb^2}$	45
$\frac{P\ Sh\text{-}1\ Hb^2}{p\ sh\text{-}1\ Hb^2}$	6
$\frac{p\ sh\text{-}1\ Hb^1}{p\ sh\text{-}1\ Hb^2}$	5
$\frac{p\ Sh\text{-}1\ Hb^2}{p\ sh\text{-}1\ Hb^2}$	0
$\frac{P\ sh\text{-}1\ Hb^1}{p\ sh\text{-}1\ Hb^2}$	1
Total	708

a. Determine the order of these genes on the chromosome.
b. Calculate the map distances between the genes.
c. Determine the coefficient of coincidence and the interference among these genes.

*29. Waxy endosperm (wx), shrunken endosperm (sh), and yellow seedling (v) are encoded by three recessive genes in corn that are linked on chromosome 5. A corn plant homozygous for all three recessive alleles is crossed with a plant homozygous for all the dominant alleles. The resulting F_1 are then crossed with a plant homozygous for the recessive alleles in a three-point testcross. The progeny of the testcross are:

wx	sh	V	87
Wx	Sh	v	94
Wx	Sh	V	3,479
wx	sh	v	3,478
Wx	sh	V	1,515
wx	Sh	v	1,531
wx	Sh	V	292
Wx	sh	v	280
Total			10,756

a. Determine the order of these genes on the chromosome.
b. Calculate the map distances between the genes.
c. Determine the coefficient of coincidence and the interference among these genes.

30. Priscilla Lane and Margaret Green studied the linkage relations of three genes affecting coat color in mice: mahogany (mg), agouti (a), and ragged (Rg). They carried out a series of three-point crosses, mating mice that were heterozygous at all three loci with mice that were homozygous for the recessive alleles at these loci (P W. Lane and M. C. Green. 1960. *Journal of Heredity* 51:228–230). The following table lists the results of the testcrosses.

Phenotype			Number
a	Rg	+	1
+	+	mg	1
a	+	+	15
+	Rg	mg	9
+	+	+	16
a	Rg	mg	36
a	+	mg	76
+	Rg	+	69
Total			213

Note: + represents a wild-type allele.

a. Determine the order of the loci that encode mahogany, agouti, and ragged on the chromosome, the map distances between them, and the interference and coefficient of coincidence for these genes.
b. Draw a picture of the two chromosomes in the triply heterozygous mice used in the testcrosses, indicating which of the alleles are present on each chromosome.

31. Fine spines (s), smooth fruit (tu), and uniform fruit color (u) are three recessive traits in cucumbers, the genes of which are linked on the same chromosome. A cucumber plant heterozygous for all three traits is used in a testcross, and the following progeny are produced from this testcross:

S	U	Tu	2
s	u	Tu	70
S	u	Tu	21
s	u	tu	4
S	U	tu	82
s	U	tu	21
s	U	Tu	13
S	u	tu	17
Total			230

a. Determine the order of these genes on the chromosome.
b. Calculate the map distances between the genes.
c. Determine the coefficient of coincidence and the interference among these genes.

d. List the genes found on each chromosome in the parents used in the testcross.

*32. In *Drosophila melanogaster*, black body (*b*) is recessive to gray body (*b*⁺), purple eyes (*pr*) are recessive to red eyes (*pr*⁺), and vestigial wings (*vg*) are recessive to normal wings (*vg*⁺). The loci encoding these traits are linked, with the following map distances:

The interference among these genes is 0.5. A fly with a black body, purple eyes, and vestigial wings is crossed with a fly homozygous for a gray body, red eyes, and normal wings. The female progeny are then crossed with males that have a black body, purple eyes, and vestigial wings. If 1000 progeny are produced from this testcross, what will be the phenotypes and proportions of the progeny?

33. A group of geneticists are interested in identifying genes that may play a role in susceptibility to asthma. They study the inheritance of genetic markers in a series of families that have two or more asthmatic children. They find an association between the presence or absence of asthma and a genetic marker on the short arm of chromosome 20 and calculate a lod score of 2 for this association. What does this lod score indicate about genes that may influence asthma?

Section 7.4

*34. The locations of six deletions have been mapped to the *Drosophila* chromosome, as shown in the following deletion map. Recessive mutations *a*, *b*, *c*, *d*, *e*, and *f* are known to be located in the same region as the deletions, but the order of the mutations on the chromosome is not known.

```
                Chromosome
                ─────────────────────────────────
Deletion 1  ──────────
    Deletion 2  ────
        Deletion 3  ──────────────
          Deletion 4  ────────
              Deletion 5  ──────
                  Deletion 6  ──────────
```

When flies homozygous for the recessive mutations are crossed with flies homozygous for the deletions, the following results are obtained, in which "m" represents a mutant phenotype and a plus sign (+) represents the wild type. On the basis of these data, determine the relative order of the seven mutant genes on the chromosome:

Deletion	Mutations					
	a	b	c	d	e	f
1	m	+	m	+	+	m
2	m	+	+	+	+	+
3	+	m	m	m	m	+
4	+	+	m	m	m	+
5	+	+	+	m	m	+
6	+	m	+	m	+	+

35. A panel of cell lines was created from human–mouse somatic-cell fusions. Each line was examined for the presence of human chromosomes and for the production of an enzyme. The following results were obtained:

Cell line	Enzyme	Human chromosomes										
		1	2	3	4	5	6	7	8	9	10	17	22
A	−	+	−	−	−	+	−	−	−	−	−	+	−
B	+	+	+	−	−	−	−	−	−	+	−	+	+
C	−	+	−	−	−	+	−	−	−	−	−	−	+
D	−	−	−	−	+	−	−	−	−	−	−	−	−
E	+	+	−	−	−	−	−	−	−	+	−	+	−

On the basis of these results, which chromosome has the gene that encodes the enzyme?

*36. A panel of cell lines was created from human–mouse somatic-cell fusions. Each line was examined for the presence of human chromosomes and for the production of three enzymes. The following results were obtained.

Cell line	Enzyme			Human chromosomes								
	1	2	3	4	8	9	12	15	16	17	22	X
A	+	−	+	−	−	+	−	+	+	−	−	+
B	+	−	−	−	−	+	−	−	+	+	−	−
C	−	+	+	+	−	−	−	−	−	+	−	+
D	−	+	+	+	+	−	−	−	+	−	−	+

On the basis of these results, give the chromosome location of enzyme 1, enzyme 2, and enzyme 3.

CHALLENGE QUESTION

Section 7.5

37. Transferrin is a blood protein that is encoded by the transferrin locus (*Trf*). In house mice, the two alleles at this locus (*Trf*ᵃ and *Trf*ᵇ) are codominant and encode three types of transferrin:

Genotype	Phenotype
*Trf*ᵃ/*Trf*ᵃ	Trf-a
*Trf*ᵃ/*Trf*ᵇ	Trf-ab
*Trf*ᵇ/*Trf*ᵇ	Trf-b

The dilution locus, found on the same chromosome, determines whether the color of a mouse is diluted or full; an allele for dilution (d) is recessive to an allele for full color (d^+):

Genotype	Phenotype
d^+d^+	d^+ (full color)
d^+d	d^+ (full color)
dd	d (dilution)

Donald Shreffler conducted a series of crosses to determine the map distance between the tranferrin locus and the dilution locus (D. C. Shreffler. 1963 *Journal of Heredity* 54:127–129). The following table presents a series of crosses carried out by Shreffler and the progeny resulting from these crosses.

Cross	♂	♀	d^+ Trf-ab	d^+ Trf-b	d Trf-ab	d Trf-b	Total
1	$\frac{d^+\ Trf^a}{d\ Trf^b}$	$\frac{d\ Trf^b}{d\ Trf^b}$	32	3	6	21	62
2	$\frac{d\ Trf^b}{d\ Trf^b}$	$\frac{d^+\ Trf^a}{d\ Trf^b}$	16	0	2	20	38
3	$\frac{d^+\ Trf^a}{d\ Trf^b}$	$\frac{d\ Trf^b}{d\ Trf^b}$	35	9	4	30	78
4	$\frac{d\ Trf^b}{d\ Trf^b}$	$\frac{d^+\ Trf^a}{d\ Trf^b}$	21	3	2	19	45
5	$\frac{d^+\ Trf^b}{d\ Trf^a}$	$\frac{d\ Trf^b}{d\ Trf^b}$	8	29	22	5	64
6	$\frac{d\ Trf^b}{d\ Trf^b}$	$\frac{d^+\ Trf^b}{d\ Trf^a}$	4	14	11	0	29

a. Calculate the recombinant frequency between the *Trf* and the *d* loci by using the pooled data from all the crosses.

b. Which crosses represent recombination in male gamete formation and which crosses represent recombination in female gamete formation?

c. On the basis of your answer to part *b*, calculate the frequency of recombination among male parents and female parents separately.

d. Are the rates of recombination in males and females the same? If not, what might produce the difference?

A mouse with the dilution trait. [L. Montoliu et al., *Color Genes*. (European Society for Pigment Cell Research, 2010).]

8 Bacterial and Viral Genetic Systems

Bacteria account for most of life's diversity and exist in almost every conceivable environment, including inhospitable habitats such as the highly saline Dead Sea. [PhotoStock-Israel/Alamy.]

LIFE IN A BACTERIAL WORLD

Humans like to think that that they rule the world but, compared with bacteria, we are clearly in a minor position. Bacteria first evolved some 2.5 billion years ago, a billion years before the first eukaryotes appeared. Today, bacteria are found in every conceivable environment, including boiling springs, highly saline lakes, and beneath more than 2 miles of ice in Antarctica. They are found at the top of Mt. Everest and at the bottoms of the deepest oceans. They are also present on and in *us*—in alarming numbers! Within the average human gut, there are about 10 trillion bacteria or more, ten times the total number of cells in the entire human body. No one knows how many bacteria populate the world, but an analysis conducted by scientists in 1998 estimated that the total number of living bacteria on Earth exceeded 5 million trillion (5×10^{30}).

Not only are bacteria numerically vast, they also constitute the majority of life's diversity. The total number of described species of bacteria is only about 6800, compared with about 1.4 million plants, animals, fungi, and single-celled eukaryotes. But the number of described species falls far short of the true microbial diversity. Species of bacteria are typically described only after they are cultivated and studied in the laboratory, and microbiologists now recognize that only a small fraction of all bacterial species are amenable to laboratory culture. For many years, it was impossible to identify and study most bacteria because they could not be grown in the laboratory. Then, in the 1970s, molecular techniques for analyzing DNA became available and opened up a whole new vista on microbial diversity. These techniques revealed several important facts about bacteria. First, many of the relations among bacteria that microbiologists had worked out on the basis of physical and biochemical traits turned out to be incorrect. Bacteria once thought to be related were in fact genetically quite different. Second, molecular analysis showed that one whole group of microbes—now called the archaea—were as different from other bacteria (the eubacteria) as they are from eukaryotes. Third, molecular analysis revealed that the number of different types of bacteria is astounding.

In 2007, Luiz Roesch and his colleagues set out to determine exactly how many types of bacteria exist in a gram of soil. They obtained soil samples from four locations: Brazil, Florida, Illinois, and Canada. From the soil samples, they extracted and purified bacterial DNA. From this DNA, they determined the sequences of a gene present in all bacteria, the 16S rRNA gene. Each different species of bacteria has a unique 16S rRNA gene sequence, and so they could determine how many species of bacteria there were in each soil sample by counting the number of different DNA sequences.

Roesch's results were amazing. The number of different eubacterial species in each gram of soil ranged from 26,140 for samples from Brazil to 53,533 for Canadian samples. Many unusual bacteria were detected that appeared dissimilar to all previously described groups of bacteria. Another interesting finding was that soil from agricultural fields harbored considerably fewer species than did soil from forests.

This study and others demonstrate that bacterial diversity far exceeds that of multicellular organisms and, undoubtedly, numerous groups of bacteria have yet to be discovered. Like it or not, we truly live in a bacterial world.

In this chapter, we examine some of the genetic properties of bacteria and viruses and the mechanisms by which they exchange and recombine their genes. Since the 1940s, the genetic systems of bacteria and viruses have contributed to the discovery of many important concepts in genetics. The study of molecular genetics initially focused almost entirely on their genes; today, bacteria and viruses are still essential tools for probing the nature of genes in more-complex organisms, in part because they possess a number of characteristics that make them suitable for genetic studies (**Table 8.1**).

The genetic systems of bacteria and viruses are also studied because these organisms play important roles in human society. Bacteria are found naturally in the mouth, gut, and on the skin, where they play essential roles in human ecology. They have been harnessed to produce a number of economically important substances, and they are of immense medical significance, causing many human diseases. In this chapter, we focus on several unique aspects of bacterial and viral genetic systems. Important processes of gene transfer and recombination will be described, and we will see how these processes can be used to map bacterial and viral genes. **TRY PROBLEM 16**

Table 8.1	Advantages of using bacteria and viruses for genetic studies
1.	Reproduction is rapid.
2.	Many progeny are produced.
3.	The haploid genome allows all mutations to be expressed directly.
4.	Asexual reproduction simplifies the isolation of genetically pure strains.
5.	Growth in the laboratory is easy and requires little space.
6.	Genomes are small.
7.	Techniques are available for isolating and manipulating their genes.
8.	They have medical importance.
9.	They can be genetically engineered to produce substances of commercial value.

8.1 Genetic Analysis of Bacteria Requires Special Methods

Heredity in bacteria is fundamentally similar to heredity in more-complex organisms, but the bacterial haploid genome and the small size of bacteria (which makes observation of their phenotypes difficult) require different approaches and methods.

Bacterial Diversity

Prokaryotes are unicellular organisms that lack nuclear membranes and membrane-bounded cell organelles. For many years, biologists considered all prokaryotes to be related, but DNA sequence information now provides convincing evidence that prokaryotes are divided into at least two distinct groups: the archaea and the eubacteria. The archaea are a group of diverse prokaryotes that are frequently found in extreme environments, such as hot springs and at the bottoms of oceans. The eubacteria are the remaining prokaryotes, including most of the familiar bacteria. Though superficially similar in their cell structure, eubacteria and archaea are distinct in their genetic makeup, and are as different from one another as each is to eukaryotic organisms. In fact, the archaea are *more* similar to eukaryotes than to eubacteria in a number of molecular features and genetic processes. In this book, the term bacteria is generally used to refer to eubacteria.

Bacteria are extremely diverse and come in a variety of shapes and sizes. Some are rod shaped, whereas others are spherical. Most are much smaller than eukaryotic cells, but at least one species isolated from the gut of fish is almost 1 mm long and can be seen with the naked eye. Some bacteria are photosynthetic. Others produce stalks and spores, superficially resembling fungi.

Bacteria have long been considered simple organisms that lack much of the cellular complexity of eukaryotes. However, recent evidence points to a number of similarities and parallels in bacterial and eukaryotic structure. For example, a bacterial protein termed FtsZ, which plays an integral part in cell division, is structurally similar to eukaryotic tubulin proteins, which are subunits of microtubules and help segregate chromosomes in mitosis and meiosis, as described in Chapter 2. Like eukaryotes, bacteria have

8.1 Bacteria can be grown (a) in liquid medium or (b) on solid medium.

proteins that help condense DNA. Other bacterial proteins function much as cytoskeletal proteins do in eukaryotes, helping to give bacterial cells shape and structure. And, although bacteria don't go through mitosis and meiosis, replication of the bacterial chromosome precedes binary fission, and there are bacterial processes that ensure that one copy of the chromosome is allocated to each cell.

Techniques for the Study of Bacteria

The culture and study of bacteria require special techniques. Microbiologists have defined the nutritional needs of a number of bacteria and developed culture media for growing them in the laboratory. Culture media typically contain a carbon source, essential elements such as nitrogen and phosphorus, certain vitamins, and other required ions and nutrients. Wild-type, or **prototrophic**, bacteria can use these simple ingredients to synthesize all the compounds that they need for growth and reproduction. A medium that contains only the nutrients required by prototrophic bacteria is termed **minimal medium**. Mutant strains called **auxotrophs** lack one or more enzymes necessary for synthesizing essential molecules and will grow only on medium supplemented with these essential molecules. For example, auxotrophic strains that are unable to synthesize the amino acid leucine will not grow on minimal medium but *will* grow on medium to which leucine has been added. **Complete medium** contains all the substances, such as the amino acid leucine, required by bacteria for growth and reproduction.

Cultures of bacteria are often grown in test tubes that contain sterile liquid medium (**Figure 8.1a**). A few bacteria are added to a tube, and they grow and divide until all the nutrients are used up or—more commonly—until the concentration of their waste products becomes toxic. Bacteria are also grown on petri plates (**Figure 8.1b**). Growth medium suspended in agar is poured into the bottom half of the petri plate, providing a solid, gel-like base for bacterial growth. In a process called plating, a dilute solution of bacteria is spread over the surface of an agar-filled petri plate. As each bacterium grows and divides, it gives rise to a visible clump of genetically identical cells (a **colony**). Genetically pure strains of the bacteria can be isolated by collecting bacteria from a single colony and transferring them to a new test tube or petri plate. The chief advantage of this method is that it allows one to isolate and count bacteria, which individually are too small to see without a microscope.

Because individual bacteria are too small to be seen directly, it is often easier to study phenotypes that affect the appearance of the colony (**Figure 8.2**) or can be detected

8.2 Bacteria have a variety of phenotypes. (a) *Serratia marcescens* with color variation. (b) *Bacillus cereus*. [Part a: Dr. E. Bottone/Peter Arnold. Part b: Biophoto Associates/Photo Researchers.]

by simple chemical tests. Auxotrophs are commonly studied phenotypes. Suppose we want to detect auxotrophs that cannot synthesize leucine (leu^- mutants). We first spread the bacteria on a petri plate containing medium that includes leucine; both prototrophs that have the leu^+ allele and auxotrophs that have leu^- allele will grow on it (**Figure 8.3**). Next, using a technique called replica plating, we transfer a few cells from each of the colonies on the original plate to two new replica plates: one plate contains medium to which leucine has been added; the other plate contains a medium lacking leucine. A medium that lacks an essential nutrient, such as the medium lacking leucine, is called a selective medium. The leu^+ bacteria will grow on both media, but the leu^- mutants will grow only on the selective medium supplemented by leucine, because they cannot synthesize their own leucine. Any colony that grows on medium that contains leucine but not on medium that lacks leucine consists of leu^- bacteria. The auxotrophs that grow on the supplemented medium can then be cultured for further study.

The Bacterial Genome

Most bacterial genomes consist of a circular chromosome that contains a single DNA molecule several million base pairs (bp) in length (**Figure 8.4**). For example, the *E. coli* genome has approximately 4.6 million base pairs of DNA. However, some bacteria contain multiple chromosomes. For example, *Vibrio cholerae*, which causes cholera, has two circular chromosomes, and *Rhizobium meliloti* has three chromosomes. There are even a few bacteria that have linear chromosomes. Many bacterial chromosomes are organized efficiently. For example, more than 90% of the DNA in *E. coli* encodes proteins. In contrast, only about 1% of human DNA encodes proteins.

Plasmids

In addition to having a chromosome, many bacteria possess **plasmids**—small, circular DNA molecules. Some plasmids are present in many copies per cell, whereas others are present in only one or two copies. In general, plasmids carry genes that are not essential to bacterial function but that may play an important role in the life cycle and growth of their bacterial hosts. There are many different types of plasmids; *E. coli* alone is estimated to have more than 270 different naturally occurring plasmids. Some plasmids promote mating between bacteria; others contain genes that kill other bacteria. Of great importance, plasmids are used extensively in genetic engineering, and some of them play a role in the spread of antibiotic resistance among bacteria.

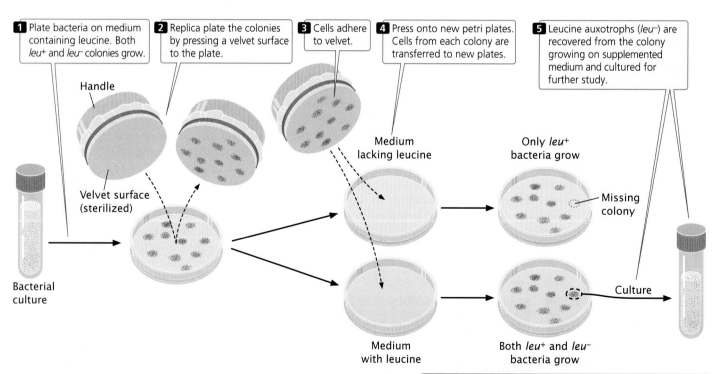

8.3 Mutant bacterial strains can be isolated on the basis of their nutritional requirements.

Conclusion: A colony that grows only on the supplemented medium has a mutation in a gene that encodes the synthesis of an essential nutrient.

Most plasmids are circular and several thousand base pairs in length, although plasmids consisting of several hundred thousand base pairs also have been found. Each plasmid possesses an origin of replication, a specific DNA sequence where DNA replication is initiated (see Chapter 2). The origin allows a plasmid to replicate independently of the bacterial chromosome (**Figure 8.5**). **Episomes** are plasmids that are capable of freely replicating and able to integrate into the bacterial chromosomes. The **F** (fertility) **factor** of *E. coli* (**Figure 8.6**) is an episome that controls mating and gene exchange between *E. coli* cells, as will be discussed in the next section.

8.4 Most bacterial cells possess a single, circular chromosome, shown here emerging from a ruptured bacterial cell. Many bacteria contain plasmids—small, circular molecules of DNA. [Drs. H. Potter and D. Dressler/Visuals Unlimited.]

CONCEPTS

Bacteria can be studied in the laboratory by growing them on defined liquid or solid medium. A typical bacterial genome consists of a single circular chromosome that contains several million base pairs. Some bacterial genes may be present on plasmids, which are small, circular DNA molecules that replicate independently of the bacterial chromosome.

✔ CONCEPT CHECK 1

Which is true of plasmids?

a. They are composed of RNA.
b. They normally exist outside of bacterial cells.
c. They possess only a single strand of DNA.
d. They contain an origin of replication.

8.5 A plasmid replicates independently of its bacterial chromosome. Replication begins at the origin of replication (*ori*) and continues around the circle. In this diagram, replication is taking place in both directions; in some plasmids, replication is in one direction only.

8.6 The F factor, a circular episome of *E. coli*, contains a number of genes that regulate transfer into the bacterial cell, replication, and insertion into the bacterial chromosome. Replication is initiated at *ori*. Insertion sequences *IS3* and *IS2* control insertion into the bacterial chromosome and excision from it.

8.2 Bacteria Exchange Genes Through Conjugation, Transformation, and Transduction

Bacteria exchange genetic material by three different mechanisms, all entailing some type of DNA transfer and recombination between the transferred DNA and the bacterial chromosome.

1. **Conjugation** takes place when genetic material passes directly from one bacterium to another (**Figure 8.7a**). In conjugation, two bacteria lie close together and a connection forms between them. A plasmid or a part of the bacterial chromosome passes from one cell (the donor) to the other (the recipient). Subsequent to conjugation, crossing over may take place between homologous sequences in the transferred DNA and the chromosome of the recipient cell. In conjugation, DNA is transferred only from donor to recipient, with no reciprocal exchange of genetic material.

2. **Transformation** takes place when a bacterium takes up DNA from the medium in which it is growing (**Figure 8.7b**). After transformation, recombination may take place between the introduced genes and those of the bacterial chromosome.

3. **Transduction** takes place when bacterial viruses (bacteriophages) carry DNA from one bacterium to another (**Figure 8.7c**). Inside the bacterium, the newly introduced DNA may undergo recombination with the bacterial chromosome.

Not all bacterial species exhibit all three types of genetic transfer. Conjugation takes place more frequently in some species than in others. Transformation takes place to a limited extent in many species of bacteria, but laboratory techniques increase the rate of DNA uptake. Most bacteriophages have a limited host range; so transduction is normally between bacteria of the same or closely related species only.

These processes of genetic exchange in bacteria differ from diploid eukaryotic sexual reproduction in two important ways. First, DNA exchange and reproduction are not coupled in bacteria. Second, donated genetic material that is not recombined into the host DNA is usually degraded, and so the recipient cell remains haploid. Each type of genetic transfer can be used to map genes, as will be discussed in the following sections.

CONCEPTS

DNA may be transferred between bacterial cells through conjugation, transformation, or transduction. Each type of genetic transfer consists of a one-way movement of genetic information to the recipient cell, sometimes followed by recombination. These processes are not connected to cellular reproduction in bacteria.

✔ CONCEPT CHECK 2

Which process of DNA transfer in bacteria requires a virus?
a. Conjugation
b. Transduction
c. Transformation
d. All of the above

Conjugation

In 1946, Joshua Lederberg and Edward Tatum demonstrated that bacteria can transfer and recombine genetic information, paving the way for the use of bacteria in genetic studies. In the course of their research, Lederberg and Tatum studied auxotrophic strains of *E. coli*. The Y10 strain required the amino acids threonine (and was genotypically *thr*$^-$) and leucine (*leu*$^-$) and the vitamin thiamine (*thi*$^-$) for growth but did not require the vitamin biotin (*bio*$^+$) or the amino acids phenylalanine (*phe*$^+$) and cysteine (*cys*$^+$); the genotype of this strain can be written as *thr*$^-$ *leu*$^-$ *thi*$^-$ *bio*$^+$ *phe*$^+$ *cys*$^+$. The Y24 strain had the opposite set of alleles: it required biotin, phenylalanine, and cysteine in its medium, but it did not require threonine, leucine, or thiamine; its genotype was *thr*$^+$ *leu*$^+$ *thi*$^+$ *bio*$^-$ *phe*$^-$ *cys*$^-$. In one experiment, Lederberg and Tatum mixed Y10 and Y24

8.7 Conjugation, transformation, and transduction are three processes of gene transfer in bacteria. For the transferred DNA to be stably inherited, all three processes require the transferred DNA to undergo recombination with the bacterial chromosome.

bacteria together and plated them on minimal medium (**Figure 8.8**). Each strain was also plated separately on minimal medium.

Alone, neither Y10 nor Y24 grew on minimal medium. Strain Y10 was unable to grow, because it required threonine, leucine, and thiamine, which were absent in the minimal medium; strain Y24 was unable to grow, because it required biotin, phenylalanine, and cysteine, which also were absent from the minimal medium. When Lederberg and Tatum mixed the two strains, however, a few colonies did grow on the minimal medium. These prototrophic bacteria must have had genotype $thr^+\ leu^+\ thi^+\ bio^+\ phe^+\ cys^+$. Where had they come from?

If mutations were responsible for the prototrophic colonies, then some colonies should also have grown on the plates containing Y10 or Y24 alone, but no bacteria grew on these plates. Multiple simultaneous mutations ($thr^- \rightarrow thr^+$, $leu^- \rightarrow leu^+$, and $thi^- \rightarrow thi^+$ in strain Y10 or $bio^- \rightarrow bio^+$, $phe^- \rightarrow phe^+$, and $cys^- \rightarrow cys^+$ in strain Y24) would have been required for either strain to become prototrophic by mutation, which was very improbable. Lederberg and Tatum concluded that some type of genetic transfer and recombination had taken place:

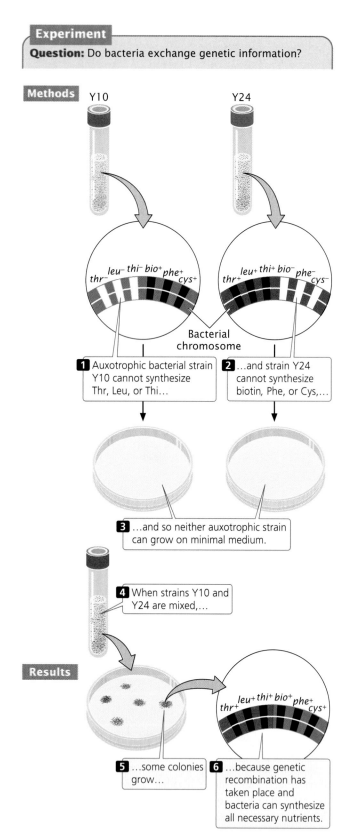

8.8 Lederberg and Tatum's experiment demonstrated that bacteria undergo genetic exchange.

What they did not know was *how* it had taken place.

To study this problem, Bernard Davis constructed a U-shaped tube (**Figure 8.9**) that was divided into two compart-

8.9 Davis's U-tube experiment.

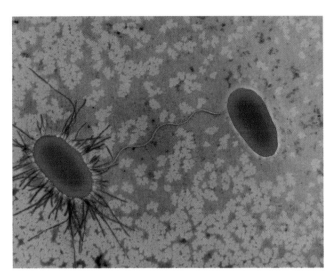

8.10 A sex pilus connects F⁺ and F⁻ cells during bacterial conjugation. [Dr. Dennis Kunkel/Phototake.]

ments by a filter having fine pores. This filter allowed liquid medium to pass from one side of the tube to the other, but the pores of the filter were too small to allow the passage of bacteria. Two auxotrophic strains of bacteria were placed on opposite sides of the filter, and suction was applied alternately to the ends of the U-tube, causing the medium to flow back and forth between the two compartments. Despite hours of incubation in the U-tube, bacteria plated out on minimal medium did not grow; there had been no genetic exchange between the strains. The exchange of bacterial genes clearly required direct contact, or conjugation, between the bacterial cells.

F⁺ and F⁻ cells In most bacteria, conjugation depends on a fertility (F) factor that is present in the donor cell and absent in the recipient cell. Cells that contain F are referred to as F⁺, and cells lacking F are F⁻.

The F factor contains an origin of replication and a number of genes required for conjugation (see Figure 8.6). For example, some of these genes encode sex **pili** (singular, pilus), slender extensions of the cell membrane. A cell containing F produces the sex pili, one of which makes contact with a receptor on an F⁻ cell (**Figure 8.10**) and pulls the two cells together. DNA is then transferred from the F⁺ cell to the F⁻ cell. Conjugation can take place only between a cell that possesses F and a cell that lacks F.

In most cases, the only genes transferred during conjugation between an F⁺ and F⁻ cell are those on the F factor (**Figure 8.11a** and **b**). Transfer is initiated when one of the DNA strands on the F factor is nicked at an origin (*oriT*). One end of the nicked DNA separates from the circle and passes into the recipient cell (**Figure 8.11c**). Replication takes place on the nicked strand, proceeding around the circular plasmid in the F⁺ cell and replacing the transferred strand (**Figure 8.11d**). Because the plasmid in the F⁺ cell is always nicked at the *oriT* site, this site always enters the recipient cell first, followed by the rest of the plasmid. Thus, the transfer of genetic material has a defined direction. Inside the recipient cell, the single strand replicates, producing a circular, double-stranded copy of the F plasmid (**Figure 8.11e**). If the entire F factor is transferred to the recipient F⁻ cell, that cell becomes an F⁺ cell.

Hfr cells Conjugation transfers genetic material in the F plasmid from F⁺ to F⁻ cells but does not account for the transfer of chromosomal genes observed by Lederberg and Tatum. In Hfr (high-frequency) strains, the F factor is integrated into the bacterial chromosome (**Figure 8.12**). Hfr cells behave as F⁺ cells, forming sex pili and undergoing conjugation with F⁻ cells.

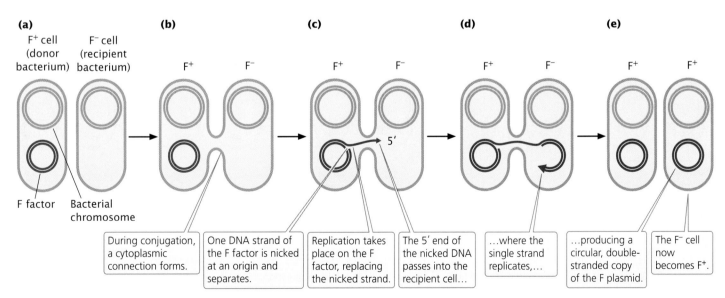

8.11 The F factor is transferred during conjugation between an F⁺ and F⁻ cell.

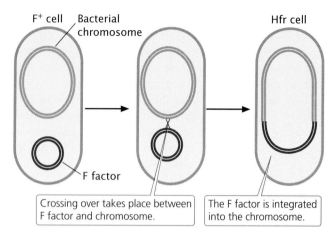

8.12 The F factor is integrated into the bacterial chromosome in an Hfr cell.

In conjugation between Hfr and F⁻ cells (**Figure 8.13a**), the integrated F factor is nicked, and the end of the nicked strand moves into the F⁻ cell (**Figure 8.13b**), just as it does in conjugation between F⁺ and F⁻ cells. Because, in an Hfr cell, the F factor has been integrated into the bacterial chromosome, the chromosome follows it into the recipient cell. How much of the bacterial chromosome is transferred depends on the length of time that the two cells remain in conjugation.

Inside the recipient cell, the donor DNA strand replicates (**Figure 8.13c**), and crossing over between it and the original chromosome of the F⁻ cell (**Figure 8.13d**) may take place. This gene transfer between Hfr and F⁻ cells is how the recombinant prototrophic cells observed by Lederberg and Tatum were produced. After crossing over has taken place in the recipient cell, the donated chromosome is degraded and the recombinant recipient chromosome remains (**Figure 8.13e**), to be replicated and passed on to later generations by binary fission.

In a mating of Hfr × F⁻, the F⁻ cell almost never becomes F⁺ or Hfr, because the F factor is nicked in the middle in the initiation of strand transfer, placing part of F at the beginning and part at the end of the strand to be transferred. To become F⁺ or Hfr, the recipient cell must receive the entire F factor, requiring the entire bacterial chromosome to be transferred. This event happens rarely, because most conjugating cells break apart before the entire chromosome has been transferred.

The F plasmid in F⁺ cells integrates into the bacterial chromosome, causing an F⁺ cell to become Hfr, at a frequency of only about $1/10{,}000$. This low frequency accounts for the low rate of recombination observed by Lederberg and Tatum in their F⁺ cells. The F factor is excised from the bacterial chromosome at a similarly low rate, causing a few Hfr cells to become F⁺.

F′ cells When an F factor does excise from the bacterial chromosome, a small amount of the bacterial chromosome may be removed with it, and these chromosomal genes will then be carried with the F plasmid (**Figure 8.14**). Cells containing an F plasmid with some bacterial genes are called F prime (F′). For example, if an F factor integrates into a chromosome adjacent to the *lac* genes (genes that enable a cell to metabolize the sugar lactose), the F factor may pick up *lac* genes when it excises, becoming F′*lac*. F′ cells can conjugate with F⁻ cells because F′ cells possess the F plasmid with all the genetic information necessary for conjugation

8.13 Bacterial genes may be transferred from an Hfr cell to an F⁻ cell in conjugation. In an Hfr cell, the F factor has been integrated into the bacterial chromosome.

8.14 An Hfr cell may be converted into an F′ cell when the F factor excises from the bacterial chromosome and carries bacterial genes with it. Conjugation produces a partial diploid.

and gene transfer. Characteristics of different mating types of *E. coli* (cells with different types of F) are summarized in **Table 8.2**.

During conjugation between an F′*lac* cell and an F⁻ cell, the F plasmid is transferred to the F⁻ cell, which means that any genes on the F plasmid, including those from the bacterial chromosome, may be transferred to F⁻ recipient cells. This process is called sexduction. It produces partial diploids, or merozygotes, which are cells with two copies of some genes, one on the bacterial chromosome and one on the newly introduced F plasmid. The outcomes of conjugation between different mating types of *E. coli* are summarized in **Table 8.3**.

CONCEPTS

Conjugation in *E. coli* is controlled by an episome called the F factor. Cells containing F (F⁺ cells) are donors during gene transfer; cells lacking F (F⁻ cells) are recipients. Hfr cells possess F integrated into the bacterial chromosome; they donate DNA to F⁻ cells at a high frequency. F′ cells contain a copy of F with some bacterial genes.

✔ CONCEPT CHECK 3

Conjugation between an F⁺ and an F⁻ cell usually results in
a. two F⁺ cells. b. two F⁻ cells.
c. an F⁺ and an F⁻ cell. d. an Hfr cell and an F⁺ cell.

Table 8.2	Characteristics of *E. coli* cells with different types of F factor	
Type	F Factor Characteristics	Role in Conjugation
F⁺	Present as separate circular DNA	Donor
F⁻	Absent	Recipient
Hfr	Present, integrated into bacterial chromosome	High-frequency donor
F′	Present as separate circular DNA, carrying some bacterial genes	Donor

Table 8.3	Results of conjugation between cells with different F factors
Conjugating	Cell Types Present after Conjugation
F⁺ × F⁻	Two F⁺ cells (F⁻ cell becomes F⁺)
Hfr × F⁻	One Hfr cell and one F⁻ (no change)*
F′ × F⁻	Two F′ cells (F⁻ cell becomes F′)

*Rarely, the F⁻ cell becomes F⁺ in an Hfr × F⁻ conjugation if the entire chromosome is transferred during conjugation.

Mapping bacterial genes with interrupted conjugation

The transfer of DNA that takes place during conjugation between Hfr and F⁻ cells allows bacterial genes to be mapped. In conjugation, the chromosome of the Hfr cell is transferred to the F⁻ cell. Transfer of the entire *E. coli* chromosome requires about 100 minutes; if conjugation is interrupted before 100 minutes have elapsed, only part of the chromosome will have passed into the F⁻ cell and have had an opportunity to recombine with the recipient chromosome.

Chromosome transfer always begins within the integrated F factor and proceeds in a continuous direction; so genes are transferred according to their sequence on the chromosome. The time required for individual genes to be transferred indicates their relative positions on the chromosome. In most genetic maps, distances are expressed as percent recombination; but, in bacterial maps constructed with interrupted conjugation, the basic unit of distance is a minute.

Worked Problem

To illustrate the method of mapping genes with interrupted conjugation, let's look at a cross analyzed by François Jacob and Elie Wollman, who first developed this method of gene mapping (**Figure 8.15a**). They used donor Hfr cells that were sensitive to the antibiotic streptomycin (genotype str^s), resistant to sodium azide (azi^r) and infection by bacteriophage T1 (ton^r), prototrophic for threonine (thr^+) and leucine (leu^+), and able to break down lactose (lac^+) and galactose (gal^+). They used F$^-$ recipient cells that were resistant to streptomycin (str^r), sensitive to sodium azide (azi^s) and to infection by bacteriophage T1 (ton^s), auxotrophic for threonine (thr^-) and leucine (leu^-), and unable to break down lactose (lac^-) and galactose (gal^-). Thus, the genotypes of the donor and recipient cells were:

Donor Hfr cells: $str^s\ leu^+\ thr^+\ azi^r\ ton^r\ lac^+\ gal^+$

Recipient F$^-$ cells: $str^r\ leu^-\ thr^-\ azi^s\ ton^s\ lac^-\ gal^-$

The two strains were mixed in nutrient medium and allowed to conjugate. After a few minutes, the medium was diluted to prevent any new pairings. At regular intervals, a sample of cells was removed and agitated vigorously in a kitchen blender to halt all conjugation and DNA transfer. The cells from each sample were plated on a selective medium that contained streptomycin and lacked leucine and threonine. The original donor cells were streptomycin sensitive (str^s) and would not grow on this medium. The F$^-$ recipient cells were auxotrophic for leucine and threonine, and they also failed to grow on this medium. Only cells that underwent conjugation and received at least the leu^+ and thr^+ genes from the Hfr donors could grow on this medium. All $str^r\ leu^+\ thr^+$ cells were then tested for the presence of other genes that might have been transferred from the donor Hfr strain.

All of the cells that grew on this selective medium were $str^r\ leu^+\ thr^+$; so we know that these genes were transferred. The percentage of $str^r\ leu^+\ thr^+$ cells receiving specific alleles (azi^r, ton^r, leu^+, and gal^+) from the Hfr strain are plotted against the duration of conjugation (**Figure 8.15b**). What are the order in which the genes are transferred and the distances among them?

• Solution

The first donor gene to appear in all of the recipient cells (at about 9 minutes) was azi^r. Gene ton^r appeared next (after about 10 minutes), followed by lac^+ (at about 18 minutes) and by gal^+ (after 25 minutes). These transfer times indicate the order of gene transfer and the relative distances among the genes (see Figure 8.15b).

Conclusion: The transfer times indicate the order and relative distances between genes and can be used to construct a genetic map.

8.15 Jacob and Wollman used interrupted conjugation to map bacterial genes.

 Animation 8.1 demonstrates the process of mapping genes through interrupted conjugation.

Notice that the frequency of gene transfer from donor to recipient cells decreased for the more distant genes. For example, about 90% of the recipients received the azi^r allele, but only about 30% received the gal^+ allele. The lower percentage for gal^+ is due to the fact that some conjugating cells spontaneously broke apart before they were disrupted by the blender. The probability of spontaneous disruption increases with time; so fewer cells had an opportunity to receive genes that were transferred later.

> For additional practice mapping bacterial genes with interrupted conjugation, try Problem 21 at the end of the chapter.

Directional transfer and mapping Different Hfr strains of a given species of bacteria have the F factor integrated into the bacterial chromosome at different sites and in different orientations. Gene transfer always begins within F, and the orientation and position of F determine the direction and starting point of gene transfer. **Figure 8.16a** shows that, in strain Hfr1, F is integrated between *leu* and *azi*; the orientation of F at this site dictates that gene transfer will proceed in a counterclockwise direction around the circular chromosome. Genes from this strain will be transferred in the order of:

$$\longleftarrow leu-thr-thi-his-gal-lac-pro-azi$$

In strain Hfr5, F is integrated between the *thi* and the *his* genes (**Figure 8.16b**) and in the opposite orientation. Here gene transfer will proceed in a clockwise direction:

$$\longleftarrow thi-thr-leu-azi-pro-lac-gal-his$$

Although the starting point and direction of transfer may differ between two strains, the relative distance in time between any two pairs of genes is constant.

Notice that the order of gene transfer is not the same for different Hfr strains (**Figure 8.17a**). For example, *azi* is transferred just after *leu* in strain HfrH but long after *leu* in strain Hfr1. Aligning the sequences (**Figure 8.17b**) shows that the two genes on either side of *azi* are always the same: *leu* and *pro*. That they are the same makes sense when we recognize that the bacterial chromosome is circular and the starting point of transfer varies from strain to strain. These data provided the first evidence that the bacterial chromosome is circular (**Figure 8.17c**). TRY PROBLEM 20 →

CONCEPTS

Conjugation can be used to map bacterial genes by mixing Hfr and F⁻ cells of different genotypes and interrupting conjugation at regular intervals. The amount of time required for individual genes to be transferred from the Hfr to the F⁻ cells indicates the relative positions of the genes on the bacterial chromosome.

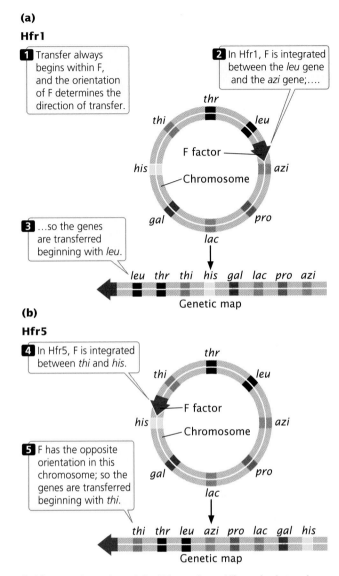

8.16 The orientation of the F factor in an Hfr strain determines the direction of gene transfer. Arrowheads indicate the origin and direction of transfer.

✔ CONCEPT CHECK 4

Interrupted conjugation was used to map three genes in *E. coli*. The donor genes first appeared in the recipient cells at the following times: *gal*, 10 minutes; *his*, 8 minutes; *pro*, 15 minutes. Which gene is in the middle?

Natural Gene Transfer and Antibiotic Resistance

Antibiotics are substances that kill bacteria. Their development and widespread use has greatly reduced the threat of infectious disease and saved countless lives. But many pathogenic bacteria have developed resistance to antibiotics,

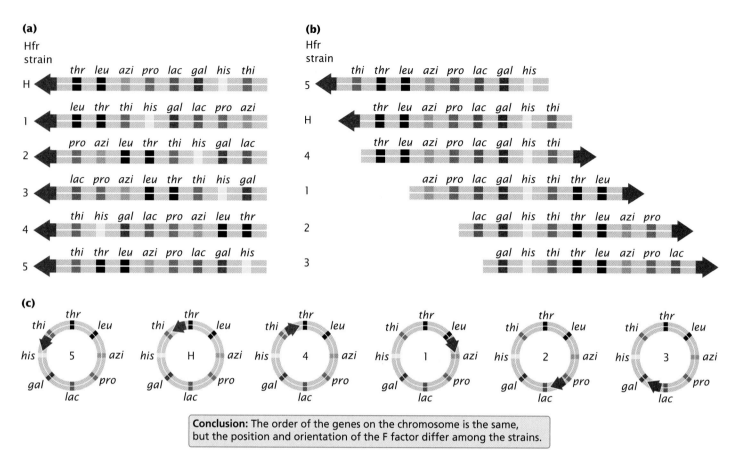

8.17 The order of gene transfer in a series of different Hfr strains indicates that the *E. coli* chromosome is circular.

Conclusion: The order of the genes on the chromosome is the same, but the position and orientation of the F factor differ among the strains.

particularly in environments where antibiotics are routinely used, such as hospitals, livestock operations, and fish farms. In these enviornments where antibiotics are continually present, the only bacteria to survive are those that possess resistance to antibiotics. No longer in competition with other bacteria, resistant bacteria multiply quickly and spread. In this way, the presence of antibiotics selects for resistant bacteria and reduces the effectiveness of antibiotic treatment for medically important infections.

Antibiotic resistance in bacteria frequently results from the action of genes located on *R plasmids*, small circular plasmids that can be transferred by conjugation. Drug-resistant R plasmids have evolved in the past 60 years (since the beginning of widespread use of antibiotics), and some of them convey resistance to several antibiotics simultaneously. Ironic but plausible sources of some of the resistance genes found in R plasmids are the microbes that produce antibiotics in the first place. R plasmids can spread easily throughout the environment, passing between related and unrelated bacteria in a variety of common situations. For example, research shows that plasmids carrying genes for resistance to multiple antibiotics were transferred from a cow udder infected with *E. coli* to a human strain of *E. coli* on a hand towel: a farmer wiping his hands after milking an infected cow might unwittingly transfer antibiotic resistance from bovine- to human-inhabiting microbes. Conjugation taking place in minced meat on a cutting board allowed R plasmids to be passed from porcine (pig) to human *E. coli*. The transfer of R plasmids also takes place in sewage, soil, lake water, and marine sediments. The fact that R plasmids can easily spread throughout the environment and pass between related and unrelated bacteria underscores both the importance of limiting antibiotic use to the treatment of medically important infections and the importance of hygiene in everyday life.

Transformation in Bacteria

A second way in which DNA can be transferred between bacteria is through transformation (see Figure 8.7b).

Transformation requires both the uptake of DNA from the surrounding medium and its incorporation into a bacterial chromosome or a plasmid. It may occur naturally when dead bacteria break up and release DNA fragments into the environment. In soil and marine environments, this means may be an important route of genetic exchange for some bacteria.

8.18 Genes can be transferred between bacteria through transformation.

Mechanism of transformation Cells that take up DNA through their membranes are said to be **competent**. Some species of bacteria take up DNA more easily than others: competence is influenced by growth stage, the concentration of available DNA in the environment, and other environmental factors. The DNA that a competent cell takes up need not be bacterial: virtually any type of DNA (bacterial or otherwise) can be taken up by competent cells under the appropriate conditions.

As a DNA fragment enters the cell in the course of transformation (**Figure 8.18**), one of the strands is broken up, whereas the other strand moves across the membrane and may pair with a homologous region and become integrated into the bacterial chromosome. This integration requires two crossover events, after which the remaining single-stranded DNA is degraded by bacterial enzymes.

Bacterial geneticists have developed techniques for increasing the frequency of transformation in the laboratory in order to introduce particular DNA fragments into cells. They have developed strains of bacteria that are more competent than wild-type cells. Treatment with calcium chloride, heat shock, or an electrical field makes bacterial membranes more porous and permeable to DNA, and the efficiency of transformation can also be increased by using high concentrations of DNA. These techniques enable researchers to transform bacteria such as *E. coli*, which are not naturally competent.

Gene mapping with transformation Transformation, like conjugation, is used to map bacterial genes, especially in those species that do not undergo conjugation or transduction (see Figure 8.7a and c). Transformation mapping requires two strains of bacteria that differ in several genetic traits; for example, the recipient strain might be $a^-\ b^-\ c^-$ (auxotrophic for three nutrients), and the donor cell prototrophic with alleles $a^+\ b^+\ c^+$ (**Figure 8.19**). DNA from

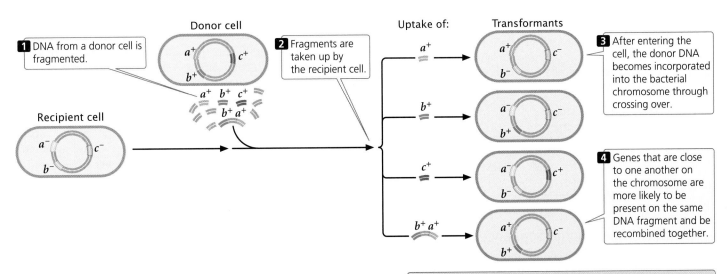

8.19 Transformation can be used to map bacterial genes.

the donor strain is isolated, purified, and fragmented. The recipient strain is treated to increase competency, and DNA from the donor strain is added to the medium. Fragments of the donor DNA enter the recipient cells and undergo recombination with homologous DNA sequences on the bacterial chromosome. Cells that receive genetic material through transformation are called **transformants**.

Genes can be mapped by observing the rate at which two or more genes are transferred to the host chromosome, or **cotransformed**, in transformation. When the donor DNA is fragmented before transformation, genes that are physically close on the chromosome are more likely to be present on the same DNA fragment and transferred together, as shown for genes a^+ and b^+ in Figure 8.19. Genes that are far apart are unlikely to be present on the same DNA fragment and will rarely be transferred together. Inside the cell, DNA becomes incorporated into the bacterial chromosome through recombination. If two genes are close together on the same fragment, any two crossovers are likely to take place on either side of the two genes, allowing both to become part of the recipient chromosome. If the two genes are far apart, there may be one crossover between them, allowing one gene but not the other to recombine with the bacterial chromosome. Thus, two genes are more likely to be transferred together when they are close together on the chromosome, and genes located far apart are rarely cotransformed. Therefore, the frequency of cotransformation can be used to map bacterial genes. If genes a and b are frequently cotransformed, and genes b and c are frequently cotransformed, but genes a and c are rarely cotransformed, then gene b must be between a and c—the gene order is $a\ b\ c$. TRY PROBLEMS 22 AND 23 →

CONCEPTS

Genes can be mapped in bacteria by taking advantage of transformation—the ability of cells to take up DNA from the environment and incorporate it into their chromosomes through crossing over. The relative rate at which pairs of genes are cotransformed indicates the distance between them: the higher the rate of cotransformation, the closer the genes are on the bacterial chromosome.

✔ CONCEPT CHECK 5

DNA from a bacterial strain with genotype $his^-\ leu^-\ thr^-$ is transformed with DNA from a strain that is $his^+\ leu^+\ thr^+$. A few $leu^+\ thr^+$ cells and a few $his^+\ thr^+$ cells are found, but no $his^+\ leu^+$ cells are observed. Which genes are farthest apart?

Bacterial Genome Sequences

Genetic maps serve as the foundation for more-detailed information provided by DNA sequencing, such as gene content and organization. Geneticists have now determined the complete nucleotide sequence of hundreds of bacterial genomes, and many additional microbial sequencing projects are underway.

Characteristics of bacterial genomes Most bacterial genomes contain from 1 million to 4 million base pairs of DNA, but a few are much smaller (e.g., 580,000 bp in *Mycoplasma genitalium*) and some are considerably larger (e.g., more than 7 million base pairs in *Mesorhizobium loti*). The small size of bacterial genomes relative to those found in multicellular eukaryotes, which often have billions of base pairs of DNA, is thought to be an adaptation for rapid cell division, because the rate of cell division is limited by the time required to replicate the DNA. On the other hand, a lack of mobility in most bacteria requires metabolic and environmental flexibility, and so genome size and content are likely to be a balance between the opposing evolutionary forces of gene loss to maintain rapid reproduction and gene acquisition to ensure flexibility.

The function of a substantial proportion of genes in all bacteria has not been determined. Certain genes, particularly those with related functions, tend to reside next to one another, but these clusters are in very different locations in different species, suggesting that bacterial genomes are constantly being reshuffled. Comparisons of the gene sequences of disease-causing and benign bacteria are helping to identify genes implicated in disease and may suggest new targets for antibiotics and other antimicrobial agents.

Horizontal Gene Transfer

The availability of genome sequences has provided evidence that many bacteria have acquired genetic information from other species of bacteria—and sometimes even from eukaryotic organisms—in a process called **horizontal gene transfer**. In most eukaryotes, genes are passed only among members of the same species through the process of reproduction (called vertical transmission); that is, genes are passed down from one generation to the next. In horizontal transfer, genes can be passed between individual members of different species by nonreproductive mechanisms, such as conjugation, transformation, and transduction. Evidence suggests that horizontal gene transfer has taken place repeatedly among bacteria. For example, as much as 17% of *E. coli*'s genome has been acquired from other bacteria through horizontal gene transfer. Of medical significance, some pathogenic bacteria have acquired, through horizontal gene transfer, the genes necessary for infection, whereas others have acquired genes that confer resistance to antibiotics.

Because of the widespread occurence of horizontal gene transfer, many bacterial chromosomes are a mixture of genes inherited through vertical transmission and genes acquired through horizontal transfer. This situation has caused some biologists to question whether the species concept is even appropriate for bacteria. A species is often defined as a group of organisms that are reproductively isolated from other

8.20 Viruses come in different structures and sizes.
(a) T4 bacteriophage (bright orange).
(b) Influenza A virus (green structures).
[(a) Biozentrum, University of Basel/Photo Researchers; (b) Eye of Science/Photo Researchers.]

groups, have a set of genes in common, and evolve together (see Chapter 13). Because of horizontal gene flow, the genes of one bacterial species are not isolated from the genes of other species, making the traditional species concept difficult to apply to bacteria. Horizontal gene flow also muddies the determination of the ancestral relationships among bacteria. The reconstruction of ancestral relationships is usually based on genetic similarities and differences: organisms that are genetically similar are assumed to have descended from a recent common ancestor, whereas organisms that are genetically distinct are assumed to be more distantly related. Through horizontal gene flow, however, even distantly related bacteria may have genes in common and thus appear to have descended from a recent common ancestor. The nature of species and how to classify bacteria are currently controversial topics within the field of microbiology.

8.3 Viruses Are Simple Replicating Systems Amenable to Genetic Analysis

All organisms—plants, animals, fungi, and bacteria—are infected by viruses. A **virus** is a simple replicating structure made up of nucleic acid surrounded by a protein coat (see Figure 2.4). Viruses come in a great variety of shapes and sizes (**Figure 8.20**). Some have DNA as their genetic material, whereas others have RNA; the nucleic acid may be double stranded or single stranded, linear or circular.

Viruses that infect bacteria (bacteriophages, or phages for short) have played a central role in genetic research since the late 1940s. They are ideal for many types of genetic research because they have small and easily manageable genomes, reproduce rapidly, and produce large numbers of progeny. Bacteriophages have two alternative life cycles: the *lytic* and the *lysogenic* cycles. In the lytic cycle, a phage attaches to a receptor on the bacterial cell wall and injects its DNA into the cell (**Figure 8.21**). Inside the host cell, the phage DNA is replicated, transcribed, and translated, producing more phage DNA and phage proteins. New phage particles are assembled from these components. The phages then produce an enzyme that breaks open the host cell, releasing the new phages. **Virulent phages** reproduce strictly through the lytic cycle and always kill their host cells.

Temperate phages can undergo either the lytic or the lysogenic cycle. The lysogenic cycle begins like the lytic cycle (see Figure 8.21) but, inside the cell, the phage DNA integrates into the bacterial chromosome, where it remains as an inactive **prophage**. The prophage is replicated along with the bacterial DNA and is passed on when the bacterium divides. Certain stimuli can cause the prophage to dissociate from the bacterial chromosome and enter into the lytic cycle, producing new phage particles and lysing the cell.

Techniques for the Study of Bacteriophages

Viruses reproduce only within host cells; so bacteriophages must be cultured in bacterial cells. To do so, phages and bacteria are mixed together and plated on solid medium on a petri plate. A high concentration of bacteria is used so that the colonies grow into one another and produce a continuous layer of bacteria, or "lawn," on the agar. An individual phage infects a single bacterial cell and goes through its lytic cycle. Many new phages are released from the lysed cell and infect additional cells; the cycle is then repeated. Because the bacteria grow on solid medium, the diffusion of the phages is restricted and only nearby cells are infected. After several rounds of phage reproduction, a clear patch of lysed cells, or **plaque**, appears on the plate (**Figure 8.22**). Each plaque represents a single phage that multiplied and lysed many cells. Plating a known volume of a dilute solution of phages on a bacterial lawn and counting the number of plaques that appear can be used to determine the original concentration of phage in the solution.

> **CONCEPTS**
>
> Viral genomes may be DNA or RNA, circular or linear, and double or single stranded. Bacteriophages are used in many types of genetic research.
>
> ✔ **CONCEPT CHECK 6**
>
> In which bacteriophage life cycle does the phage DNA become incorporated into the bacterial chromosome?
>
> a. Lytic
> b. Lysogenic
> c. Both lytic and lysogenic
> d. Neither lytic or lysogenic

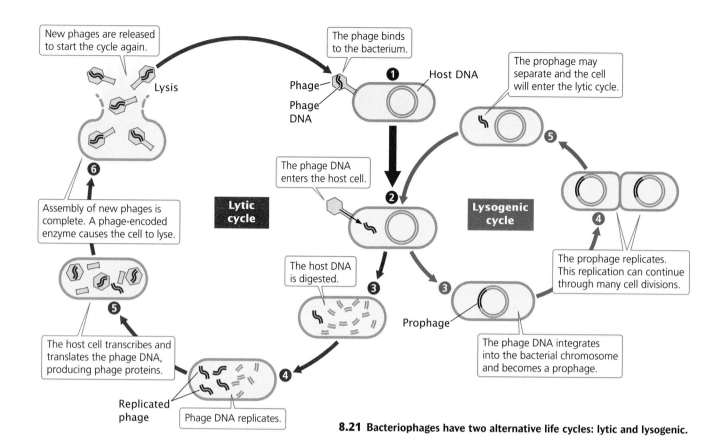

8.21 Bacteriophages have two alternative life cycles: lytic and lysogenic.

Transduction: Using Phages to Map Bacterial Genes

In the discussion of bacterial genetics, three mechanisms of gene transfer were identified: conjugation, transformation, and transduction (see Figure 8.7). Let's take a closer look at transduction, in which genes are transferred between bacteria by viruses. In **generalized transduction**, any gene may be transferred. In **specialized transduction**, only a few genes are transferred.

Generalized transduction Joshua Lederberg and Norton Zinder discovered generalized transduction in 1952. They were trying to produce recombination in the bacterium *Salmonella typhimurium* by conjugation. They mixed a strain of *S. typhimurium* that was $phe^+ \ trp^+ \ tyr^+ \ met^- \ his^-$ with a strain that was $phe^- \ trp^- \ tyr^- \ met^+ \ his^+$ (**Figure 8.23**) and plated them on minimal medium. A few prototrophic recombinants ($phe^+ \ trp^+ \ tyr^+ \ met^+ \ his^+$) appeared, suggesting that conjugation had taken place. However, when they tested the two strains in a U-shaped tube similar to the one used by Davis, some $phe^+ \ trp^+ \ tyr^+ \ met^+ \ his^+$ prototrophs were obtained on one side of the tube (compare Figure 8.23 with Figure 8.9). This apparatus separated the two strains by a filter with pores too small for the passage of bacteria; so how were genes being transferred between bacteria in the absence of conjugation? The results of subsequent studies revealed that the agent of transfer was a bacteriophage.

In the lytic cycle of phage reproduction, the bacterial chromosome is broken into random fragments (**Figure 8.24**). For some types of bacteriophage, a piece of the bacterial chromosome instead of phage DNA occasionally gets

8.22 Plaques are clear patches of lysed cells on a lawn of bacteria.
[Internationally Genetically Engineered Machine Competition, 2007.]

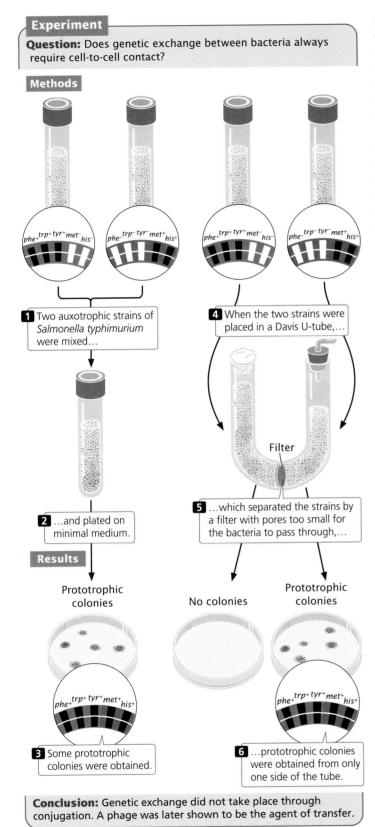

8.23 The Lederberg and Zinder experiment.

packaged into a phage coat; these phage particles are called **transducing phages**. The transducing phage infects a new cell, releasing the bacterial DNA, and the introduced genes may then become integrated into the bacterial chromosome by a double crossover. Bacterial genes can, by this process, be moved from one bacterial strain to another, producing recombinant bacteria called **transductants**.

Not all phages are capable of transduction, a rare event that requires (1) that the phage degrade the bacterial chromosome, (2) that the process of packaging DNA into the phage protein not be specific for phage DNA, and (3) that the bacterial genes transferred by the virus recombine with the chromosome in the recipient cell. The overall rate of transduction ranges from only about 1 in 100,000 to 1 in 1,000,000.

Because of the limited size of a phage particle, only about 1% of the bacterial chromosome can be transduced. Only genes located close together on the bacterial chromosome will be transferred together, or **cotransduced**. Because the chance of a cell being transduced by two separate phages is exceedingly small, any cotransduced genes are usually located close together on the bacterial chromosome. Thus, rates of cotransduction, like rates of cotransformation, give an indication of the physical distances between genes on a bacterial chromosome.

To map genes by using transduction, two bacterial strains with different alleles at several loci are used. The donor strain is infected with phages (**Figure 8.25**), which reproduce within the cell. When the phages have lysed the donor cells, a suspension of the progeny phages is mixed with a recipient strain of bacteria, which is then plated on several different kinds of media to determine the phenotypes of the transducing progeny phages.

TRY PROBLEM 29

CONCEPTS

In transduction, bacterial genes become packaged into a viral coat, are transferred to another bacterium by the virus, and become incorporated into the bacterial chromosome by crossing over. Bacterial genes can be mapped with the use of generalized transduction.

✔ CONCEPT CHECK 7

In gene mapping experiments using generalized transduction, bacterial genes that are cotransduced are

a. far apart on the bacterial chromosome.

b. on different bacterial chromosomes.

c. close together on the bacterial chromosome.

d. on a plasmid.

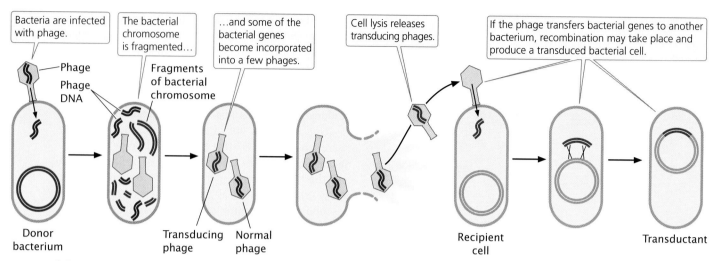

8.24 Genes can be transferred from one bacterium to another through generalized transduction.

Specialized transduction Like generalized transduction, specialized transduction requires gene transfer from one bacterium to another through phages, but, here, only genes near particular sites on the bacterial chromosome are transferred. This process requires lysogenic bacteriophages. The prophage may imperfectly excise from the bacterial chromosome, carrying with it a small part of the bacterial DNA adjacent to the site of prophage integration. A phage carrying this DNA will then inject it into another bacterial cell in the next round of infection. This process resembles the situation in F′ cells, in which the F plasmid carries genes from one bacterium into another (see Figure 8.14).

CONCEPTS

Specialized transduction transfers only those bacterial genes located near the site of prophage insertion.

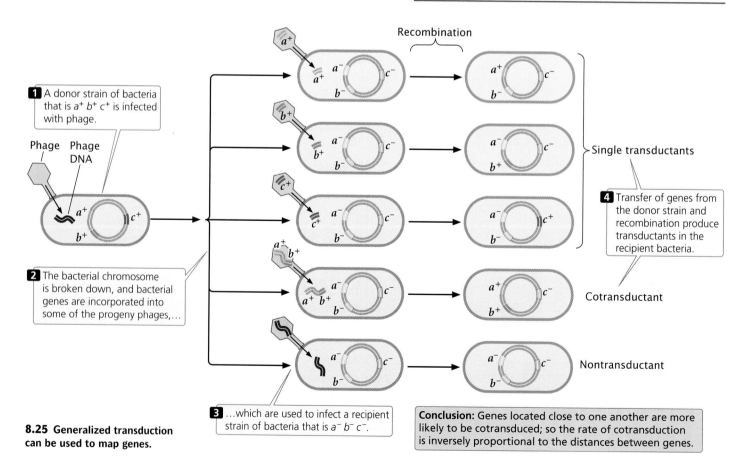

8.25 Generalized transduction can be used to map genes.

> **CONNECTING CONCEPTS**
>
> **Three Methods for Mapping Bacterial Genes**
>
> Three methods of mapping bacterial genes have now been outlined: (1) interrupted conjugation; (2) transformation; and (3) transduction. These methods have important similarities and differences.
>
> Mapping with interrupted conjugation is based on the time required for genes to be transferred from one bacterium to another by means of cell-to-cell contact. The key to this technique is that the bacterial chromosome itself is transferred, and the order of genes and the time required for their transfer provide information about the positions of the genes on the chromosome. In contrast with other mapping methods, the distance between genes is measured not in recombination frequencies but in units of time required for genes to be transferred. Here, the basic unit of conjugation mapping is a minute.
>
> In gene mapping with transformation, DNA from the donor strain is isolated, broken up, and mixed with the recipient strain. Some fragments pass into the recipient cells, where the transformed DNA may recombine with the bacterial chromosome. The unit of transfer here is a random fragment of the chromosome. Loci that are close together on the donor chromosome tend to be on the same DNA fragment; so the rates of cotransformation provide information about the relative positions of genes on the chromosome.
>
> Transduction mapping also relies on the transfer of genes between bacteria that differ in two or more traits, but, here, the vehicle of gene transfer is a bacteriophage. In a number of respects, transduction mapping is similar to transformation mapping. Small fragments of DNA are carried by the phage from donor to recipient bacteria, and the rates of cotransduction, like the rates of cotransformation, provide information about the relative distances between the genes.
>
> All of the methods use a common strategy for mapping bacterial genes. The movement of genes from donor to recipient is detected by using strains that differ in two or more traits, and the transfer of one gene relative to the transfer of others is examined. Additionally, all three methods rely on recombination between the transferred DNA and the bacterial chromosome. In mapping with interrupted conjugation, the relative order and timing of gene transfer provide the information necessary to map the genes; in transformation and transduction, the rate of cotransfer provides this information.
>
> In conclusion, the same basic strategies are used for mapping with interrupted conjugation, transformation, and transduction. The methods differ principally in their mechanisms of transfer: in conjugation mapping, DNA is transferred though contact between bacteria; in transformation, DNA is transferred as small naked fragments; and, in transduction, DNA is transferred by bacteriophages.

Gene Mapping in Phages

Mapping genes in the bacteriophages themselves depends on homolgous recombination between phage chromosomes and therefore requires the application of the same principles as those applied to mapping genes in eukaryotic organisms (see Chapter 7). Crosses are made between viruses that differ in two or more genes, and recombinant progeny phages are identified and counted. The proportion of recombinant progeny is then used to estimate the distances between the genes and their linear order on the chromosome.

In 1949, Alfred Hershey and Raquel Rotman examined rates of recombination in the T2 bacteriophage, which has single-stranded DNA. They studied recombination between genes in two strains that differed in plaque appearance and host range (the bacterial strains that the phages could infect). One strain was able to infect and lyse type B *E. coli* cells but not B/2 cells (making this strain of bacteria wild type with normal host range, or h^+) and produced an abnormal plaque that was large with distinct borders (r^-). The other strain was able to infect and lyse *both* B *and* B/2 cells (mutant host range, h^-) and produced wild-type plaques that were small with fuzzy borders (r^+).

Hershey and Rotman crossed the $h^+ \, r^-$ and $h^- \, r^+$ strains of T2 by infecting type B *E. coli* cells with a mixture of the two strains. They used a high concentration of phages so that most cells could be simultaneously infected by both strains (**Figure 8.26**). Within the bacteria cells, homologous recombination occasionally took place between the chromosomes of the different bacteriophage strains, producing $h^+ \, r^+$ and $h^- \, r^-$ chromosomes, which were then packaged into new phage particles. When the cells lysed, the recombinant phages were released, along with the nonrecombinant $h^+ \, r^-$ phages and $h^- \, r^+$ phages.

Hershey and Rotman diluted and plated the progeny phages on a bacterial lawn that consisted of a *mixture* of B and B/2 cells. Phages carrying the h^+ allele (which conferred the ability to infect only B cells) produced a cloudy plaque because the B/2 cells did not lyse. Phages carrying the h^- allele produced a clear plaque because all the cells within the plaque were lysed. The r^+ phages produced small plaques, whereas the r^- phages produced large plaques. The genotypes of these progeny phages could therefore be determined by the appearance of the plaque (see Figure 8.26 and **Table 8.4**).

In this type of phage cross, the recombination frequency (RF) between the two genes can be calculated by using the following formula:

$$RF = \frac{\text{recombinant plaques}}{\text{total plaques}}$$

In Hershey and Rotman's cross, the recombinant plaques were $h^+ \, r^+$ and $h^- \, r^-$; so the recombination frequency was

$$RF = \frac{(h^+ \, r^+) + (h^- \, r^-)}{\text{total plaques}}$$

Recombination frequencies can be used to determine the distances between genes and their order on the phage chromosome, just as recombination frequencies are used to map genes in eukaryotes. **TRY PROBLEMS 28 AND 32**

> **CONCEPTS**
>
> To map phage genes, bacterial cells are infected with viruses that differ in two or more genes. Recombinant plaques are counted, and rates of recombination are used to determine the linear order of the genes on the chromosome and the distances between them.

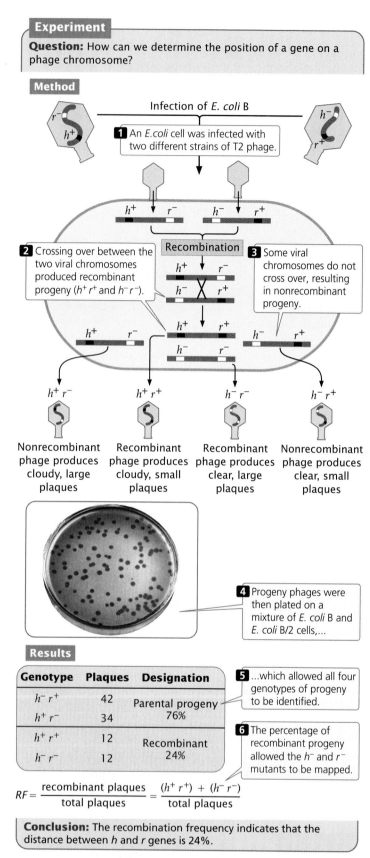

Table 8.4	Progeny phages produced from $h^- r^+ \times h^+ r^-$
Phenotype	Genotype
Clear and small	$h^- r^+$
Cloudy and large	$h^+ r^-$
Cloudy and small	$h^+ r^+$
Clear and large	$h^- r^-$

Fine-Structure Analysis of Bacteriophage Genes

In the 1950s and 1960s, Seymour Benzer conducted a series of experiments to examine the structure of a gene. Because no molecular techniques were available at the time for directly examining nucleotide sequences, Benzer was forced to infer gene structure from analyses of mutations and their effects. The results of his studies showed that different mutational sites *within* a single gene could be mapped (referred to as **intragenic mapping**) by using techniques similar to those described for mapping bacterial genes by transduction. Different sites within a single gene are very close together; so recombination between them takes place at a very low frequency. Because large numbers of progeny are required to detect these recombination events, Benzer used the bacteriophage T4, which reproduces rapidly and produces large numbers of progeny.

Benzer's mapping techniques Wild-type T4 phages normally produce small plaques with rough edges when grown on a lawn of *E. coli*. Certain mutants, called *r* for rapid lysis, produce larger plaques with sharply defined edges. Benzer isolated phages with a number of different *r* mutations, concentrating on one particular subgroup called *rII* mutants.

Wild-type T4 phages produce typical plaques (**Figure 8.27**) on *E. coli* strains B and K. In contrast, the *rII* mutants produce *r* plaques on strain B and do not form plaques at all on strain K. Benzer recognized the *r* mutants by their

8.27 T4 phage *rII* mutants produce distinct plaques when grown on *E. coli* B cells. (Left) Plaque produced by wild-type phage. (Right) Plaque produced by *rII* mutant. [Dr. D. P. Snustad, College of Biological Sciences, University of Minnesota.]

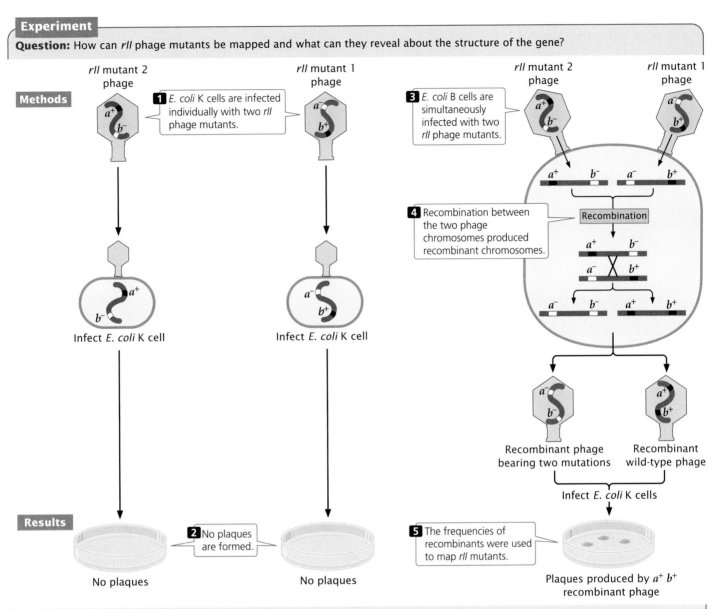

8.28 Benzer developed a procedure for mapping *rII* mutants. Two different *rII* mutants ($a^- b^+$ and $a^+ b^-$) are isolated on *E. coli* B cells. Only the $a^+ b^+$ recombinant can grow on *E. coli* K, allowing these recombinants to be identified.

distinctive plaques when grown on *E. coli* B. He then collected lysate from these plaques and used it to infect *E. coli* K. Phages that did not produce plaques on *E. coli* K were defined as the *rII* type.

Benzer collected thousands of *rII* mutations. He simultaneously infected bacterial cells with two different mutants and looked for recombinant progeny (**Figure 8.28**).

Consider two *rII* mutations, a^- and b^- (their wild-type alleles are a^+ and b^+). Benzer infected *E. coli* B cells with two different strains of phages, one $a^- b^+$ and the other $a^+ b^-$ (see Figure 8.28, step 3). Neither of these mutations is able to grow on *E. coli* K cells. While reproducing within the B cells, a few phages of the two strains recombined (see Figure 8.28, step 4). A single crossover produces two recombinant chromosomes; one with genotype $a^+ b^+$ and the other with genotype $a^- b^-$:

The resulting recombinant chromosomes, along with the nonrecombinant (parental) chromosomes, were incorporated into progeny phages, which were then used to infect *E. coli* K cells. The resulting plaques were examined to determine the genotype of the infecting phage and map the *rII* mutants (see Figure 8.28, step 5).

Neither of the *rII* mutants grew on *E. coli* K (see Figure 8.28, step 2), but wild-type phages grew; so progeny phages that produced plaques on *E. coli* K must have the recombinant genotype $a^+ b^+$. Each recombination event produces equal numbers of double mutants ($a^- b^-$) and wild-type chromosomes ($a^+ b^+$); so the number of recombinant progeny should be twice the number of wild-type plaques that appeared on *E. coli* K. The recombination frequency between the two *rII* mutants would be:

$$RF = \frac{2 \times \text{number of plaques on } E. coli \text{ K}}{\text{total number of plaques on } E. coli \text{ B}}$$

Because phages produce large numbers of progeny, Benzer was able to detect a single recombinant among billions of progeny phages. Recombination frequencies are proportional to physical distances along the chromosome (see p. 174 in Chapter 7), revealing the positions of the different mutations within the *rII* region of the phage chromosome. In this way, Benzer eventually mapped more than 2400 *rII* mutations, many corresponding to single base pairs in the viral DNA. His work provided the first molecular view of a gene.

Complementation experiments Benzer's mapping experiments demonstrated that some *rII* mutations were very closely linked. This finding raised the question whether they were at the same locus or at different loci. To determine whether different *rII* mutations belonged to different functional loci, Benzer used the complementation (cis–trans) test (see p. 113 in Chapter 5).

To carry out the complementation test in bacteriophage, Benzer infected cells of *E. coli* K with large numbers of two mutant strains of phage (**Figure 8.29**, step 1) so that cells would become doubly infected with both strains. Consider two *rII* mutations: r_{101}^- and r_{104}^-. Cells infected with both mutants

$$\frac{r_{101}^-}{r_{101}^+} \quad \frac{r_{104}^+}{r_{104}^-}$$

were effectively heterozygous for the phage genes, with the mutations in the trans configuration (see Figure 8.29, step 2). In the complementation testing, the phenotypes of progeny

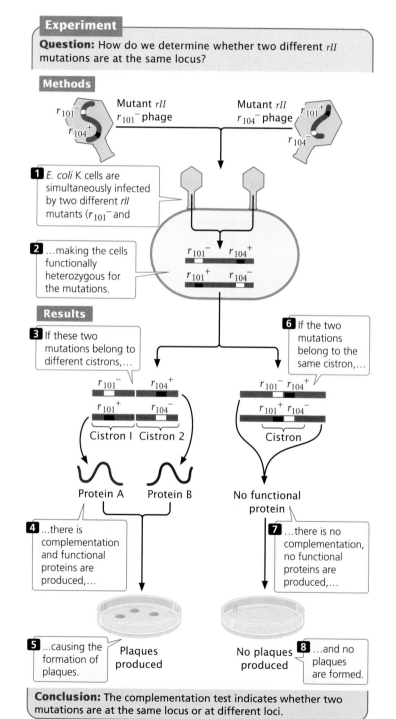

8.29 Complementation tests are used to determine whether different mutations are at the same functional gene.

phages were examined on the K strain, rather than the B strain as illustrated in Figure 8.28.

If the r_{101}^- and r_{104}^- mutations are at different functional loci that encode different proteins, then, in bacterial cells infected by both mutants, the wild-type sequences on the chromosome opposite each mutation will overcome the effects of the recessive mutations; the phages will produce normal plaques on E. coli K cells (Figure 8.29, steps 3, 4, and 5). If, on the other hand, the mutations are at the same locus, no functional protein is produced by either chromosome, and no plaques develop in the E. coli K cells (Figure 8.29, steps 6, 7, and 8). Thus, the absence of plaques indicates that the two mutations are at the same locus. Benzer coined the term *cistron* to designate a functional gene defined by the complementation test.

In the complementation test, the cis heterozygote is used as a control. Benzer simultaneously infected bacteria with wild-type phages (r_{101}^+ r_{104}^+) and with phages carrying both mutations (r_{101}^- r_{104}^-). This test produced cells that were heterozygous and in cis configuration for the phage genes:

$$\frac{r_{101}^+ \quad r_{104}^+}{r_{101}^- \quad r_{104}^-}$$

Regardless of whether the r_{101}^- and r_{104}^- mutations are in the same functional unit, these cells contain a copy of the wild-type phage chromosome (r_{101}^+ r_{104}^+) and will produce normal plaques in E. coli K.

Benzer carried out complementation testing on many pairs of rII mutants. He found that the rII region consists of two loci, designated *rIIA* and *rIIB*. Mutations belonging to the *rIIA* and *rIIB* groups complemented each other, but mutations in the *rIIA* group did not complement others in *rIIA*; nor did mutations in the *rIIB* group complement others in *rIIB*.

CONCEPTS

In a series of experiments with the bacteriophage T4, Seymour Benzer showed that recombination could take place within a single gene and created the first molecular map of a gene. Benzer used the complementation test to distinguish between functional genes (loci).

✔ CONCEPT CHECK 8

In complementation tests, Benzer simultaneously infected E. coli cells with two phages, each of which carried a different mutation. What conclusion did he make when the progeny phages produced normal plaques?

a. The mutations were at the same locus.
b. The mutations were at different loci.
c. The mutations were close together on the chromosome.
d. The genes were in the cis configuration.

At the time of Benzer's research, the relation between genes and DNA structure was unknown. A gene was defined as a functional unit of heredity that encoded a phenotype.

Many geneticists believed that genes were indivisible and that recombination could not take place within them. Benzer demonstrated that intragenic recombination did indeed take place (although at a very low rate) and gave geneticists their first glimpse at the structure of an individual gene.

TRY PROBLEMS 36 AND 37 →

RNA Viruses

Viral genomes may be encoded in either DNA or RNA, as stated earlier. RNA is the genetic material of some medically important human viruses, including those that cause influenza, common colds, polio, and AIDS. Almost all viruses that infect plants have RNA genomes. The medical and economic importance of RNA viruses has encouraged their study.

RNA viruses capable of integrating into the genomes of their hosts, much as temperate phages insert themselves into bacterial chromosomes, are called **retroviruses** (**Figure 8.30a**). Because the retroviral genome is RNA, whereas that of the host is DNA, a retrovirus must produce **reverse transcriptase**, an enzyme that synthesizes complementary DNA (cDNA) from either an RNA or a DNA template. A retrovirus uses reverse transcriptase to copy its RNA genome into a single-stranded DNA molecule, which it then copies to make double-stranded DNA. The DNA copy of the viral genome then integrates into the host chromosome. A viral genome incorporated into the host chromosome is called a **provirus**. The provirus is replicated by host enzymes when the host chromosome is duplicated (**Figure 8.30b**).

When conditions are appropriate, the provirus undergoes transcription to produce numerous copies of the original viral RNA genome. This RNA encodes viral proteins and serves as genomic RNA for new viral particles. As these viruses escape the cell, they collect patches of the cell membrane to use as their envelopes.

All known retroviral genomes have in common three genes: *gag*, *pol*, and *env*, each encoding a precursor protein that is cleaved into two or more functional proteins. The *gag* gene encodes proteins that make up the viral protein coat. The *pol* gene encodes reverse transcriptase and an enzyme called **integrase** that inserts the viral DNA into the host chromosome. The *env* gene encodes the glycoproteins that appear on the surface of the virus.

Some retroviruses contain **oncogenes** that may stimulate cell division and cause the formation of tumors. The first retrovirus to be isolated, the Rous sarcoma virus, was originally recognized by its ability to produce connective-tissue tumors (sarcomas) in chickens.

Human Immunodeficiency Virus and AIDS

The human immunodeficiency virus (HIV) causes acquired immune deficiency syndrome (AIDS), a disease that killed 2 million people worldwide in 2008 alone. AIDS was first recognized in 1982, when a number of homosexual males in the United States began to exhibit symptoms of a new immune-system-deficiency disease. In that year, Robert

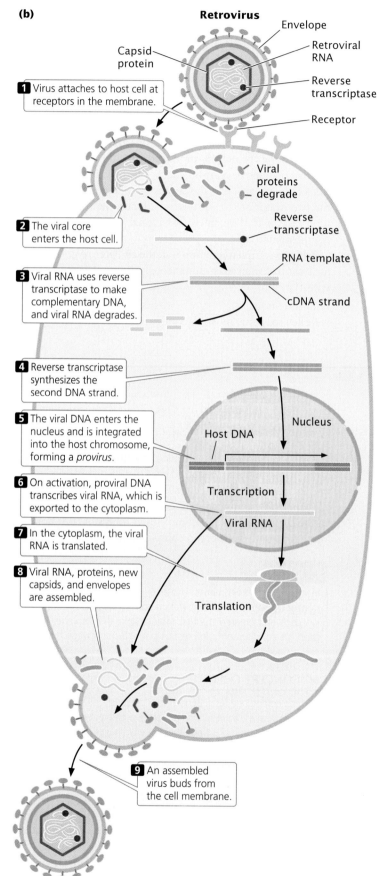

8.30 A retrovirus uses reverse transcription to incorporate its RNA into the host DNA. (a) Structure of a typical retrovirus. Two copies of the single-stranded RNA genome and the reverse transcriptase enzyme are shown enclosed within a protein capsid. The capsid is surrounded by a viral envelope that is studded with viral glycoproteins. (b) The retrovirus life cycle.

Gallo proposed that AIDS was caused by a retrovirus. Between 1983 and 1984, as the AIDS epidemic became widespread, the HIV retrovirus was isolated from AIDS patients. AIDS is now known to be caused by two different immunodeficiency viruses, HIV-1 and HIV-2, which together have infected more than 65 million people worldwide. Of those infected, 25 million have died. Most cases of AIDS are caused by HIV-1, which now has a global distribution; HIV-2 is found primarily in western Africa.

HIV illustrates the importance of genetic recombination in viral evolution. Studies of the DNA sequences of HIV and other retroviruses reveal that HIV-1 is closely related to the simian immunodeficiency virus found in chimpanzees (SIV_{cpz}). Many wild chimpanzees in Africa are infected with SIV_{cpz}, although it doesn't cause AIDS-like symptoms in chimps. SIV_{cpz} is itself a hybrid that resulted from recombination between a retrovirus found in the red-capped mangabey (a monkey) and a retrovirus found in the greater spot-nosed monkey (**Figure 8.31**). Apparently, one or more chimpanzees became infected with both viruses; recombination between the viruses produced SIV_{cpz}, which was then transmitted to humans through contact with infected chimpanzees. In humans, SIV_{cpz} underwent significant evolution to become HIV-1, which then spread throughout the world to produce the AIDS epidemic. Several independent transfers of SIV_{cpz} to humans gave rise to different strains of HIV-1. HIV-2 evolved from a different retrovirus, SIV_{sm}, found in sooty mangabeys.

HIV is transmitted by sexual contact between humans and through any type of blood-to-blood contact, such as that caused by the sharing of dirty needles by drug addicts.

8.31 HIV-1 evolved from a similar virus (SIV$_{cpz}$) found in chimpanzees and was transmitted to humans. SIV$_{cpz}$ arose from recombination taking place between retroviruses in red-capped mangabeys and greater spot-nosed monkeys.

Until screening tests could identify HIV-infected blood, transfusions and clotting factors used by hemophiliacs also were sources of infection.

HIV principally attacks a class of blood cells called helper T lymphocytes or, simply, helper T cells (**Figure 8.32**). HIV enters a helper T cell, undergoes reverse transcription, and integrates into the chromosome. The virus reproduces rapidly, destroying the T cell as new virus particles escape from the cell. Because helper T cells are central to immune function and are destroyed in the infection, AIDS patients have a diminished immune response; most AIDS patients die of secondary infections that develop because they have lost the ability to fight off pathogens.

The HIV genome is 9749 nucleotides long and carries *gag*, *pol*, *env*, and six other genes that regulate the life cycle of the virus. HIV's reverse transcriptase is very error prone, giving the virus a high mutation rate and allowing it to evolve rapidly, even within a single host. This rapid evolution makes the development of an effective vaccine against HIV particularly difficult. Genetic variation within the human population also affects the virus. To date, more than 10 loci in humans that affect HIV infection and the progression of AIDS have been identified.

8.32 HIV principally attacks T lymphocytes. Electron micrograph showing a T cell infected with HIV, visible as small circles with dark centers. [Courtesy of Dr. Hans Gelderblom.]

CONCEPTS

A retrovirus is an RNA virus that integrates into its host's chromosome by making a DNA copy of its RNA genome through the process of reverse transcription. Human immunodeficiency virus, the causative agent of AIDS, is a retrovirus. It evolved from related retroviruses found in other primates.

✔ **CONCEPT CHECK 9**

What enzyme is used by a retrovirus to make a DNA copy of its genome?

Influenza Virus

Influenza demonstrates how rapid changes in a pathogen can arise through recombination of its genetic material. Influenza, commonly called flu, is a respiratory disease caused by influenza viruses. In the United States, from 5% to 20% of the entire population is infected with influenza annually and, though most cases are mild, 36,000 people die from influenza-related causes each year. At certain times, particularly when new strains of influenza virus enter the human population, worldwide epidemics (called pandemics) take place; for example, the Spanish flu virus in 1918 killed an estimated 20 million to 100 million people worldwide.

Table 8.5	Strains of influenza virus responsible for major flu pandemics	
Year	Influenza Pandemic	Strain
1918	Spanish flu	H1N1
1957	Asian flu	H2N2
1968	Hong Kong flu	H3N2
2009	Swine flu	H1N1

Influenza viruses are RNA viruses that infect birds and mammals. They come in three main types: influenza A, influenza B, and influenza C. Most cases of the common flu are caused by influenza A and B. Influenza A is divided into subtypes on the basis of two proteins, hemagglutinin (HA) and neuraminidase (NA), found on the surface of the virus. The HA and NA proteins affect the ability of the virus to enter host cells and the host organism's immune response to infection. There are 16 types of HA and 9 types of NA, which can exist in a virus in different combinations. For example, common strains of influenza A circulating in humans today are H1N1 and H3N2 (**Table 8.5**), along with several strains of influenza B. Most of the different subtypes of influenza A are found in birds.

Although influenza is an RNA virus, it is not a retrovirus: its genome is not copied into DNA and incorporated into the host chromosome as is that of a retrovirus. The influenza viral genome consists of 7 or 8 pieces of RNA that are enclosed in a viral envelope. Each piece of RNA encodes one or two of the virus's proteins. The virus enters a host cell by attaching to specific receptors on the cell membrane. After the viral particle has entered the cell, the viral RNA is released, copied, and translated into viral proteins. Viral RNA molecules and viral proteins are then assembled into new viral particles, which exit the cell and infect additional cells.

One of the dangers of the influenza virus is that it evolves every rapid, with new strains appearing frequently. Influenza evolves in two ways. First, each strain continually changes through mutations arising in the viral RNA. The enzyme that copies the RNA is especially prone to making mistakes, and so new mutations are continually introduced into the viral genome. This type of continual change is call **antigenic drift**. Occasionally, major changes in the viral genome take place through **antigenic shift**, in which genetic material from different strains is combined in a process called reassortment. Reassortment takes place when a host is simultaneously infected with two different strains. The RNAs of both strains are replicated within the cell, and RNA segments from two different strains are incorporated into the same viral particle, creating a new strain. For example, in 2002, reassortment occurred between the H1N1 and H3N2 subtypes, creating a new H1N2 strain that contained the hemagluttinin from H1N1 and the neuraminidase from H3N2. New strains produced by antigenic shift are responsible for most pandemics, because no one has immunity to the radically different virus that is produced.

Most different strains of influenza A infect birds, but humans are not easily infected with bird influenza. The appearance of new strains in humans is thought to often arise from viruses that reassort in pigs, which can be infected by viruses from both humans and birds. In 2009, a new strain of H1N1 influenza (called swine flu) emerged in Mexico and quickly spread throughout the world. This virus arose from a series of reassortment events that combined gene sequences from human, bird, and pig influenza viruses to produce the new H1N1 virus (**Figure 8.33**).

CONCEPTS

Influenza is caused by RNA influenza viruses. New strains of influenza appear through antigenic shift, in which new viral genomes are created through the reassortment of RNA molecules of different strains.

8.33 New strains of influenza virus are created by reassortment of genetic material from different strains. A new H1N1 virus (swine flu virus) that appeared in 2009 contained genetic material from avian, swine, and human viruses.

SUMMARY

- Bacteria and viruses are well suited to genetic studies: they are small, have a small haploid genome, undergo rapid reproduction, and produce large numbers of progeny through asexual reproduction.
- The bacterial genome normally consists of a single, circular molecule of double-stranded DNA. Plasmids are small pieces of bacterial DNA that can replicate independently of the chromosome.
- DNA can be transferred between bacteria by conjugation, transformation, or transduction.
- Conjugation is the union of two bacterial cells and the transfer of genetic material between them. It is controlled by an episome called F. The rate at which individual genes are transferred during conjugation provides information about the order of the genes and the distances between them on the bacterial chromosome.
- Bacteria take up DNA from the environment through the process of transformation. Frequencies of the cotransformation of genes provide information about the physical distances between chromosomal genes.
- Complete DNA sequences of a number of bacterial species have been determined. This sequence information indicates that horizontal gene transfer—the movement of DNA between species—is common in bacteria.
- Viruses are replicating structures with DNA or RNA genomes that may be double stranded or single stranded and linear or circular.
- Bacterial genes become incorporated into phage coats and are transferred to other bacteria by phages through the process of transduction. Rates of cotransduction can be used to map bacterial genes.
- Phage genes can be mapped by infecting bacterial cells with two different phage strains and counting the number of recombinant plaques produced by the progeny phages.
- Benzer mapped a large number of mutations that occurred within the *rII* region of phage T4. The results of his complementation studies demonstrated that the *rII* region consists of two functional units that he called cistrons. He showed that intragenic recombination takes place.
- A number of viruses have RNA genomes. Retroviruses encode reverse transcriptase, an enzyme used to make a DNA copy of the viral genome, which then integrates into the host genome as a provirus. HIV is a retrovirus that is the causative agent for AIDS.
- Influenza is caused by RNA influenza viruses that evolve through small changes taking place through mutation (antigenic drift) and major changes taking place through reassortment of the genetic material from different strains.

IMPORTANT TERMS

prototrophic bacteria (p. 205)
minimal medium (p. 205)
auxotrophic bacteria (p. 205)
complete medium (p. 205)
colony (p. 205)
plasmid (p. 206)
episome (p. 207)
F factor (p. 207)
conjugation (p. 208)
transformation (p. 208)
transduction (p. 208)
pili (singular, pilus) (p. 211)

competent cell (p. 217)
transformant (p. 218)
cotransformation (p. 218)
horizontal gene transfer (p. 218)
virus (p. 219)
virulent phage (p. 219)
temperate phage (p. 219)
prophage (p. 219)
plaque (p. 219)
generalized transduction (p. 220)
specialized transduction (p. 222)
transducing phage (p. 221)

transductant (p. 221)
cotransduction (p. 221)
intragenic mapping (p. 224)
retrovirus (p. 227)
reverse transcriptase (p. 227)
provirus (p. 227)
integrase (p. 227)
oncogene (p. 227)
antigenic drift (p. 230)
antigenic shift (p. 230)

ANSWERS TO CONCEPT CHECKS

1. d
2. b
3. a
4. *gal*
5. *his* and *leu*
6. b
7. c
8. b
9. Reverse transcriptase

WORKED PROBLEMS

1. DNA from a strain of bacteria with genotype $a^+\ b^+\ c^+\ d^+\ e^+$ was isolated and used to transform a strain of bacteria that was $a^-\ b^-\ c^-\ d^-\ e^-$. The transformed cells were tested for the presence of donated genes. The following genes were cotransformed:

a^+ and d^+ $\quad$ c^+ and d^+
b^+ and e^+ $\quad$ c^+ and e^+

What is the order of genes *a*, *b*, *c*, *d*, and *e* on the bacterial chromosome?

- **Solution**

The rate at which genes are cotransformed is inversely proportional to the distance between them: genes that are close together are frequently cotransformed, whereas genes that are far apart are rarely cotransformed. In this transformation experiment, gene c^+ is cotransformed with both genes e^+ and d^+, but genes e^+ and d^+ are not cotransformed; therefore the *c* locus must be between the *d* and *e* loci:

Gene e^+ is also cotransformed with gene b^+; so the *e* and *b* loci must be located close together. Locus *b* could be on either side of locus *e*. To determine whether locus *b* is on the same side of *e* as locus *c*, we look to see whether genes b^+ and c^+ are cotransformed. They are not; so locus *b* must be on the opposite side of *e* from *c*:

Gene a^+ is cotransformed with gene d^+; so they must be located close together. If locus *a* were located on the same side of *d* as locus *c*, then genes a^+ and c^+ would be cotransformed. Because these genes display no cotransformation, locus *a* must be on the opposite side of locus *d*:

2. Consider three genes in *E. coli*: thr^+ (the ability to synthesize threonine), ara^+ (the ability to metabolize arabinose), and leu^+ (the ability to synthesize leucine). All three of these genes are close together on the *E. coli* chromosome. Phages are grown in a $thr^+\ ara^+\ leu^+$ strain of bacteria (the donor strain). The phage lysate is collected and used to infect a strain of bacteria that is $thr^-\ ara^-\ leu^-$. The recipient bacteria are then tested on medium lacking leucine. Bacteria that grow and form colonies on this medium (leu^+ transductants) are then replica plated onto medium lacking threonine and onto medium lacking arabinose to see which are thr^+ and which are ara^+.

Another group of recipient bacteria are tested on medium lacking threonine. Bacteria that grow and form colonies on this medium (thr^+ transductants) are then replica plated onto medium lacking leucine and onto medium lacking arabinose to see which are ara^+ and which are leu^+. Results from these experiments are as follows:

Selected marker	Cells with cotransduced genes (%)	
leu^+	3	thr^+
	76	ara^+
thr^+	3	leu^+
	0	ara^+

How are the loci arranged on the chromosome?

- **Solution**

Notice that, when we select for leu^+ (the top half of the table), most of the selected cells are also ara^+. This finding indicates that the *leu* and *ara* genes are located close together, because they are usually cotransduced. In contrast, thr^+ is only rarely cotransduced with leu^+, indicating that *leu* and *thr* are much farther apart. On the basis of these observations, we know that *leu* and *ara* are closer together than are *leu* and *thr*, but we don't yet know the order of three genes—whether *thr* is on the same side of *ara* as *leu* or on the opposite side, as shown here:

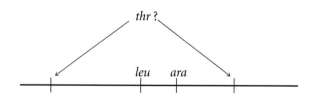

We can determine the position of *thr* with respect to the other two genes by looking at the cotransduction frequencies when thr^+ is selected (the bottom half of the preceding table). Notice that, although the cotransduction frequency for *thr* and *leu* also is 3%, no $thr^+\ ara^+$ cotransductants are observed. This finding indicates that *thr* is closer to *leu* than to *ara*, and therefore *thr* must be to the left of *leu*, as shown here:

COMPREHENSION QUESTIONS

Section 8.1

1. Explain how auxotrophic bacteria are isolated.

Section 8.2

2. Briefly explain the differences between F^+, F^-, Hfr, and F' cells.

*3. What types of matings are possible between F^+, F^-, Hfr, and F' cells? What outcomes do these matings produce? What is the role of F factor in conjugation?

*4. Explain how interrupted conjugation, transformation, and transduction can be used to map bacterial genes. How are these methods similar and how are they different?

5. What is horizontal gene transfer and how might it take place?

Section 8.2

*6. List some of the characteristics that make bacteria and viruses ideal organisms for many types of genetic studies.

7. What types of genomes do viruses have?

8. Briefly describe the differences between the lytic cycle of virulent phages and the lysogenic cycle of temperate phages.

9. Briefly explain how genes in phages are mapped.

*10. How does specialized transduction differ from generalized transduction?

*11. Briefly explain the method used by Benzer to determine whether two different mutations occurred at the same locus.

*12. Explain how a retrovirus, which has an RNA genome, is able to integrate its genetic material into that of a host having a DNA genome.

13. Briefly describe the genetic structure of a typical retrovirus.

14. What are the evolutionary origins of HIV-1 and HIV-2?

15. Most humans are not easily infected by avian influenza. How then do DNA sequences from avian influenza become incorporated into human influenza?

APPLICATION QUESTIONS AND PROBLEMS

Introduction

16. Suppose you want to compare the species of bacteria that exist in a polluted stream with the species that exist in an unpolluted stream. Traditionally, bacteria have been identified by growing them in the laboratory and comparing their physical and biochemical properties. You recognize that you will be unable to culture most of the bacteria that reside in the streams. How might you go about identifying the species in the two streams without culturing them in the lab?

Section 8.2

*17. John Smith is a pig farmer. For the past 5 years, Smith has been adding vitamins and low doses of antibiotics to his pig food; he says that these supplements enhance the growth of the pigs. Within the past year, however, several of his pigs died from infections of common bacteria, which failed to respond to large doses of antibiotics. Can you offer an explanation for the increased rate of mortality due to infection in Smith's pigs? What advice might you offer Smith to prevent this problem in the future?

18. Rarely, the conjugation of Hfr and F^- cells produces two Hfr cells. Explain how this event takes place.

19. Austin Taylor and Edward Adelberg isolated some new strains of Hfr cells that they then used to map several genes in *E. coli* by using interrupted conjugation (A. L. Taylor and E. A. Adelberg. 1960. *Genetics* 45:1233–1243). In one experiment, they mixed cells of Hfr strain AB-312, which were xyl^+ mtl^+ mal^+ met^+ and sensitive to phage T6, with F^- strain AB-531, which was xyl^- mtl^- mal^- met^- and resistant to phage T6. The cells were allowed to undergo conjugation. At regular intervals, the researchers removed a sample of cells and interrupted conjugation by killing the Hfr cells with phage T6. The F^- cells, which were resistant to phage T6, survived and were then tested for the presence of genes transferred from the Hfr strain. The results of this experiment are shown in the accompanying graph. On the basis of these data, give the order of the *xyl*, *mtl*, *mal*, and *met* genes on the bacterial chromosome and indicate the minimum distances between them.

*20. A series of Hfr strains that have genotype m^+ n^+ o^+ p^+ q^+ r^+ are mixed with an F^- strain that has genotype m^- n^- o^- p^- q^- r^-. Conjugation is interrupted at regular intervals and the order of the appearance of genes from the Hfr strain is determined in the recipient cells. The order of gene transfer for each Hfr strain is:

Hfr5	m^+ q^+ p^+ n^+ r^+ o^+
Hfr4	n^+ r^+ o^+ m^+ q^+ p^+
Hfr1	o^+ m^+ q^+ p^+ n^+ r^+
Hfr9	q^+ m^+ o^+ r^+ n^+ p^+

What is the order of genes on the circular bacterial chromosome? For each Hfr strain, give the location of the F factor in the chromosome and its polarity.

*21. Crosses of three different Hfr strains with separate samples of an F⁻ strain are carried out, and the following mapping data are provided from studies of interrupted conjugation:

Appearance of genes in F⁻ cells

Hfr1:	Genes	b^+	d^+	c^+	f^+	g^+
	Time	3	5	16	27	59
Hfr2:	Genes	e^+	f^+	c^+	d^+	b^+
	Time	6	24	35	46	48
Hfr3:	Genes	d^+	c^+	f^+	e^+	g^+
	Time	4	15	26	44	58

Construct a genetic map for these genes, indicating their order on the bacterial chromosome and the distances between them.

22. DNA from a strain of *Bacillus subtilis* with the genotype $trp^+\ tyr^+$ was used to transform a recipient strain with the genotype $trp^-\ tyr^-$. The following numbers of transformed cells were recovered:

Genotype	Number of transformed cells
$trp^+\ tyr^-$	154
$trp^-\ tyr^+$	312
$trp^+\ tyr^+$	354

What do these results suggest about the linkage of the *trp* and *tyr* genes?

23. DNA from a strain of *Bacillus subtilis* with genotype $a^+\ b^+\ c^+\ d^+\ e^+$ is used to transform a strain with genotype $a^-\ b^-\ c^-\ d^-\ e^-$. Pairs of genes are checked for cotransformation and the following results are obtained:

Pair of genes	Cotransformation	Pair of genes	Cotransformation
a^+ and b^+	no	b^+ and d^+	no
a^+ and c^+	no	b^+ and e^+	yes
a^+ and d^+	yes	c^+ and d^+	no
a^+ and e^+	yes	c^+ and e^+	yes
b^+ and c^+	yes	d^+ and e^+	no

On the basis of these results, what is the order of the genes on the bacterial chromosome?

24. DNA from a bacterial strain that is $his^+\ leu^+\ lac^+$ is used to transform a strain that is $his^-\ leu^-\ lac^-$. The following percentages of cells were transformed:

Donor strain	Recipient strain	Genotype of transformed cells	Percentage
$his^+\ leu^+\ lac^+$	$his^-\ leu^-\ lac^-$	$his^+\ leu^+\ lac^+$	0.02
		$his^+\ leu^+\ lac^-$	0.00
		$his^+\ leu^-\ lac^+$	2.00
		$his^+\ leu^-\ lac^-$	4.00
		$his^-\ leu^+\ lac^+$	0.10
		$his^-\ leu^+\ lac^-$	3.00
		$his^-\ leu^+\ lac^-$	1.50

a. What conclusions can you make about the order of these three genes on the chromosome?

b. Which two genes are closest?

25. Rollin Hotchkiss and Julius Marmur studied transformation in the bacterium *Streptococcus pneumoniae* (R. D. Hotchkiss and J. Marmur. 1954. *Proceedings of the National Academy of Sciences* 40:55–60). They examined four mutations in this bacterium: penicillin resistance (P), streptomycin resistance (S), sulfanilamide resistance (F), and the ability to utilize mannitol (M). They extracted DNA from strains of bacteria with different combinations of different mutations and used this DNA to transform wild-type bacterial cells ($P^+\ S^+\ F^+\ M^+$). The results from one of their transformation experiments are shown here.

Donor DNA	Recipient DNA	Transformants	Percentage of all cells
M S F	$M^+\ S^+\ F^+$	$M^+\ S\ F^+$	4.0
		$M^+\ S^+\ F$	4.0
		$M\ S^+\ F^+$	2.6
		$M\ S\ F^+$	0.41
		$M^+\ S\ F$	0.22
		$M\ S^+\ F$	0.0058
		$M\ S\ F$	0.0071

a. Hotchkiss and Marmur noted that the percentage of cotransformation was higher than would be expected on a random basis. For example, the results show that the 2.6% of the cells were transformed into M and 4% were transformed into S. If the M and S traits were inherited independently, the expected probability of cotransformation of M and S ($M\ S$) would be $0.026 \times 0.04 = 0.001$, or 0.1%. However, they observed 0.41% $M\ S$ cotransformants, four times more than they expected. What accounts for the relatively high frequency of cotransformation of the traits they observed?

b. On the basis of the results, what conclusion can you make about the order of the M, S, and F genes on the bacterial chromosome?

c. Why is the rate of cotransformation for all three genes ($M\ S\ F$) almost the same as the rate of cotransformation for $M\ F$ alone?

26. In the course of a study on the effects of the mechanical shearing of DNA, Eugene Nester, A. T. Ganesan, and Joshua Lederberg studied the transfer, by transformation, of sheared DNA from a wild-type strain of *Bacillus subtilis* ($his_2^+\ aro_3^+\ try_2^+\ aro_1^+\ tyr_1^+\ aro_2^+$) to strains of bacteria carrying a series of mutations (E. W. Nester, A. T. Ganesan, and J. Lederberg. 1963. *Proceedings of the National Academy of Sciences* 49:61–68). They reported the following rates of cotransformation between his_2^+ and the other genes (expressed as cotransfer rate).

Genes	Rate of cotransfer
his_2^+ and aro_3^+	0.015
his_2^+ and try_2^+	0.10
his_2^+ and aro_1^+	0.12
his_2^+ and tyr_1^+	0.23
his_2^+ and aro_2^+	0.05

On the basis of these data, which gene is farthest from his_2^+? Which gene is closest?

27. C. Anagnostopoulos and I. P. Crawford isolated and studied a series of mutations that affected several steps in the biochemical pathway leading to tryptophan in the bacterium *Bacillus subtilis* (C. Anagnostopoulos and I. P. Crawford. 1961. *Proceedings of the National Academy of Sciences* 47:378–390). Seven of the strains that they used in their study are listed here, along with the mutation found in that strain.

Strain	Mutation
T3	T^-
168	I^-
168PT	I^-
TI	I^-
TII	I^-
T8	A^-
H25	H^-

To map the genes for tryptophan synthesis, they carried out a series of transformation experiments on strains having different mutations and determined the percentage of recombinants among the transformed bacteria. Their results were as follows:

Recipient	Donor	Percent recombinants
T3	168PT	12.7
T3	T11	11.8
T3	T8	43.5
T3	H25	28.6
168	H25	44.9
TII	H25	41.4
TI	H25	31.3
T8	H25	67.4
H25	T3	19.0
H25	TII	26.3
H25	TI	13.4
H25	T8	45.0

On the basis of these two-point recombination frequencies, determine the order of the genes and the distances between them. Where more than one cross was completed for a pair of genes, average the recombination rates from the different crosses. Draw a map of the genes on the chromosome.

Section 8.3

28. Two mutations that affect plaque morphology in phages (a^- and b^-) have been isolated. Phages carrying both mutations ($a^-\ b^-$) are mixed with wild-type phages ($a^+\ b^+$) and added to a culture of bacterial cells. Subsequent to infection and lysis, samples of the phage lysate are collected and cultured on bacterial cells. The following numbers of plaques are observed:

Plaque phenotype	Number
$a^+\ b^+$	2043
$a^+\ b^-$	320
$a^-\ b^+$	357
$a^-\ b^-$	2134

What is the frequency of recombination between the *a* and *b* genes?

29. T. Miyake and M. Demerec examined proline-requiring mutations in the bacterium *Salmonella typhimurium* (T. Miyake and M. Demerec. 1960. *Genetics* 45:755–762). On the basis of complementation studies, they found four proline auxotrophs: *proA*, *proB*, *proC*, and *proD*. To determine whether *proA*, *proB*, *proC*, and *proD* loci were located close together on the bacterial chromosome, they conducted a transduction experiment. Bacterial strains that were $proC^+$ and had mutations at *proA*, *proB*, or *proD*, were used as donors. The donors were infected with bacteriophages, and progeny phages were allowed to infect recipient bacteria with genotype $proC^-\ proA^+\ proB^+\ proD^+$. The bacteria were then plated on a selective medium that allowed only $proC^+$ bacteria to grow. The following results were obtained:

S. typhimurium. [Kwangshin Kim/Photo Researchers.]

Donor genotype	Transductant genotype	Number
$proC^+\ proA^-\ proB^+\ proD^+$	$proC^+\ proA^+\ proB^+\ proD^+$	2765
	$proC^+\ proA^-\ proB^+\ proD^+$	3
$proC^+\ proA^+\ proB^-\ proD^+$	$proC^+\ proA^+\ proB^+\ proD^+$	1838
	$proC^+\ proA^+\ proB^-\ proD^+$	2
$proC^+\ proA^+\ proB^+\ proD^-$	$proC^+\ proA^+\ proB^+\ proD^+$	1166
	$proC^+\ proA^+\ proB^+\ proD^-$	0

a. Why are there no $proC^-$ genotypes among the transductants?

b. Which genotypes represent single transductants and which represent cotransductants?

c. Is there evidence that *proA*, *proB*, and *proD* are located close to *proC*? Explain your answer.

*30. A geneticist isolates two mutations in a bacteriophage. One mutation causes clear plaques (*c*), and the other produces minute plaques (*m*). Previous mapping experiments have established that the genes responsible for these two mutations are 8 m.u. apart. The geneticist mixes phages with genotype $c^+ m^+$ and genotype $c^- m^-$ and uses the mixture to infect bacterial cells. She collects the progeny phages and cultures a sample of them on plated bacteria. A total of 1000 plaques are observed. What numbers of the different types of plaques ($c^+ m^+, c^- m^-, c^+ m^-, c^- m^+$) should she expect to see?

31. The geneticist carries out the same experiment described in Problem 30, but this time she mixes phages with genotypes $c^+ m^-$ and $c^- m^+$. What results are expected from *this* cross?

*32. A geneticist isolates two bacteriophage *r* mutants (r_{13} and r_2) that cause rapid lysis. He carries out the following crosses and counts the number of plaques listed here:

Genotype of parental phage	Progeny	Number of plaques
$h^+ r_{13}^- \times h^- r_{13}^+$	$h^+ r_{13}^+$	1
	$h^- r_{13}^+$	104
	$h^+ r_{13}^-$	110
	$h^- r_{13}^-$	2
Total		216
$h^+ r_2^- \times h^- r_2^+$	$h^+ r_2^+$	6
	$h^- r_2^+$	86
	$h^+ r_2^-$	81
	$h^- r_2^-$	7
Total		180

a. Calculate the recombination frequencies between r_2 and *h* and between r_{13} and *h*.

b. Draw all possible linkage maps for these three genes.

*33. *E. coli* cells are simultaneously infected with two strains of phage λ. One strain has a mutant host range, is temperature sensitive, and produces clear plaques (genotype is *h st c*); another strain carries the wild-type alleles (genotype is $h^+ st^+ c^+$). Progeny phages are collected from the lysed cells and are plated on bacteria. The genotypes of the progeny phages are given here:

Progeny phage genotype	Number of plaques
$h^+ \; c^+ \; st^+$	321
$h \;\; c \;\; st$	338
$h^+ \; c \;\; st$	26
$h \;\; c^+ \; st^+$	30
$h^+ \; c \;\; st^+$	106
$h \;\; c^+ \; st$	110
$h^+ \; c^+ \; st$	5
$h \;\; c \;\; st^+$	6

a. Determine the order of the three genes on the phage chromosome.

b. Determine the map distances between the genes.

c. Determine the coefficient of coincidence and the interference (see p. 182 in Chapter 7).

34. A donor strain of bacteria with genes $a^+ b^+ c^+$ is infected with phages to map the donor chromosome with generalized transduction. The phage lysate from the bacterial cells is collected and used to infect a second strain of bacteria that are $a^- b^- c^-$. Bacteria with the a^+ gene are selected, and the percentage of cells with cotransduced b^+ and c^+ genes are recorded.

Donor	Recipient	Selected gene	Cells with cotransduced gene (%)
$a^+ b^+ c^+$	$a^- b^- c^-$	a^+	25 b^+
		a^+	3 c^+

Is gene *b* or gene *c* closer to gene *a*? Explain your reasoning.

35. A donor strain of bacteria with genotype $leu^+ gal^- pro^+$ is infected with phages. The phage lysate from the bacterial cells is collected and used to infect a second strain of bacteria that are $leu^- gal^+ pro^-$. The second strain is selected for leu^+, and the following cotransduction data are obtained:

Donor	Recipient	Selected gene	Cells with cotransduced gene (%)
$leu^+ gal^- pro^+$	$leu^- gal^+ pro^-$	leu^+	47 pro^+
		leu^+	26 gal^-

Which genes are closest, *leu* and *gal* or *leu* and *pro*?

36. A geneticist isolates two new mutations, called rII_x and rII_y, from the *rII* region of bacteriophage T4. *E. coli* B cells are simultaneously infected with phages carrying the rII_x mutation and with phages carrying the rII_y mutation. After the cells have lysed, samples of the phage lysate are collected. One sample is grown on *E. coli* K cells; a

second sample is grown on *E. coli* B cells. There are 8322 plaques on *E. coli* B and 3 plaques on *E. coli* K. What is the recombination frequency between these two mutations?

37. A geneticist is working with a new bacteriophage called phage Y3 that infects *E. coli*. He has isolated eight mutant phages that fail to produce plaques when grown on *E. coli* strain K. To determine whether these mutations occur at the same functional gene, he simultaneously infects *E. coli* K cells with paired combinations of the mutants and looks to see whether plaques are formed. He obtains the following results. (A plus sign means that plaques were formed on *E. coli* K; a minus sign means that no plaques were formed on *E. coli* K.)

Mutant	1	2	3	4	5	6	7	8
1								
2	+							
3	+	+						
4	+	−	+					
5	−	+	+	+				
6	−	+	+	+	−			
7	+	−	+	−	+	+		
8	−	+	+	+	−	−	+	

a. To how many functional genes (cistrons) do these mutations belong?

b. Which mutations belong to the same functional gene?

CHALLENGE QUESTIONS

Section 8.2

38. As a summer project, a microbiology student independently isolates two mutations in *E. coli* that are auxotrophic for glycine (gly^-). The student wants to know whether these two mutants are at the same functional unit. Outline a procedure that the student could use to determine whether these two gly^- mutations occur within the same functional unit.

39. A group of genetics students mix two auxotrophic strains of bacteria: one is leu^+ trp^+ his^- met^- and the other is leu^- trp^- his^+ met^+. After mixing the two strains, they plate the bacteria on minimal medium and observe a few prototrophic colonies (leu^+ trp^+ his^+ met^+). They assume that some gene transfer has taken place between the two strains. How can they determine whether the transfer of genes is due to conjugation, transduction, or transformation?

9 Chromosome Variation

Down syndrome is caused by the presence of three copies of one or more genes located on chromosome 21. [Stockbyte.]

TRISOMY 21 AND THE DOWN-SYNDROME CRITICAL REGION

In 1866, John Langdon Down, physician and medical superintendent of the Earlswood Asylum in Surrey, England, noticed a remarkable resemblance among a number of his mentally retarded patients: all of them possessed a broad, flat face, a thick tongue, a small nose, and oval-shaped eyes. Their features were so similar, in fact, that he felt that they might easily be mistaken as children of the same family. Down did not understand the cause of their retardation, but his original description faithfully records the physical characteristics of this most common genetic form of mental retardation. In his honor, the disorder is today known as Down syndrome.

As early as the 1930s, geneticists suggested that Down syndrome might be due to a chromosome abnormality, but not until 1959 did researchers firmly establish the cause of Down syndrome: most people with the disorder have three copies of chromosome 21, a condition known as trisomy 21. In a few rare cases, people having the disorder are trisomic for just specific parts of chromosome 21. By comparing the parts of chromosome 21 that these people have in common, geneticists have established that a specific segment—called the Down-syndrome critical region or DSCR—probably contains one or more genes responsible for the features of Down syndrome. Sequencing of the human genome has established that the DSCR consists of a little more than 5 million base pairs and only 33 genes.

Research by several groups in 2006 was a source of insight into the roles of two genes in the DSCR. Mice that have deficiencies in the regulatory pathway controlled by two DSCR genes, called *DSCR1* and *DYRK1A*, exhibit many of the features seen in Down syndrome. Mice genetically engineered to express *DYRK1A* and *DSCR1* at high levels have abnormal heart development, similar to congenital heart problems seen in many people with Down syndrome.

In spite of this exciting finding, the genetics of Down syndrome appears to be more complex than formerly thought. Geneticist Lisa Olson and her colleagues had conducted an earlier study on mice to determine whether genes within the DSCR are solely responsible for Down syndrome. Mouse breeders have developed several strains of mice that are trisomic for most of the genes found on human chromosome 21 (the equivalent mouse genes are found on mouse chromosome 16). These mice display many of the same anatomical features found in people with Down syndrome, as well as altered behavior, and they are considered an animal model for Down syndrome. Olson and her colleagues carefully created mice that were trisomic for genes in the DSCR but possessed the normal two copies of other genes found on human chromosome 21. If one or more of the genes in the DSCR are indeed responsible for Down syndrome, these mice should exhibit the features of Down syndrome. Surprisingly, the engineered mice exhibited none of the anatomical features of Down syndrome, demonstrating that three copies of genes in the DSCR are not the sole cause of the features of Down syndrome, at least in mice. Another study examined a human gene called *APP*, which lies outside of the DSCR. This gene appears to be responsible for at least some of the Alzheimer-like features observed in older Down-syndrome people.

Taken together, findings from these studies suggest that Down syndrome is not due to a single gene but is instead caused by complex interactions among multiple genes that are affected when an extra copy of chromosome 21 is present. Research on Down syndrome illustrates the principle that chromosome abnormalities often affect many genes that interact in complex ways.

Most species have a characteristic number of chromosomes, each with a distinct size and structure, and all the tissues of an organism (except for gametes) generally have the same set of chromosomes. Nevertheless, variations in chromosome number—such as the extra chromosome 21 that leads to Down syndrome—do periodically arise. Variations may also arise in chromosome structure: individual chromosomes may lose or gain parts and the order of genes within a chromosome may become altered. These variations in the number and structure of chromosomes are termed **chromosome mutations**, and they frequently play an important role in evolution.

We begin this chapter by briefly reviewing some basic concepts of chromosome structure, which we learned in Chapter 2. We then consider the different types of chromosome mutations, their definitions, features, phenotypic effects, and influence on evolution.

9.1 Chromosome Mutations Include Rearrangements, Aneuploids, and Polyploids

Before we consider the different types of chromosome mutations, their effects, and how they arise, we will review the basics of chromosome structure.

Chromosome Morphology

Each functional chromosome has a centromere, to which spindle fibers attach, and two telomeres, which stabilize the chromosome (see Figure 2.7). Chromosomes are classified into four basic types:

1. **Metacentric.** The centromere is located approximately in the middle, and so the chromosome has two arms of equal length.
2. **Submetacentric.** The centromere is displaced toward one end, creating a long arm and a short arm. (On human chromosomes, the short arm is designated by the letter p and the long arm by the letter q.)
3. **Acrocentric.** The centromere is near one end, producing a long arm and a knob, or satellite, at the other end.
4. **Telocentric.** The centromere is at or very near the end of the chromosome (see Figure 2.8).

The complete set of chromosomes possessed by an organism is called its *karyotype* and is usually presented as a picture of metaphase chromosomes lined up in descending order of their size (**Figure 9.1**). Karyotypes are prepared from actively dividing cells, such as white blood cells, bone-marrow cells, or cells from meristematic tissues of plants. After treatment with a chemical (such as colchicine) that prevents them from entering anaphase, the cells are chemically preserved, spread on a microscope slide, stained, and photographed. The photograph is then enlarged, and the individual chromosomes are cut out and arranged in a karyotype. For human chromosomes, karyotypes are often routinely prepared by automated machines, which scan a slide with a video camera attached to a microscope, looking for chromosome spreads. When a spread has been located, the camera

9.1 A human karyotype consists of 46 chromosomes. A karyotype for a male is shown here; a karyotype for a female would have two X chromosomes. [CNRI/Photo Researchers.]

9.2 Chromosome banding is revealed by special staining techniques. (a) G banding. (b) Q banding. (c) C banding. (d) R banding. [Part a: Leonard Lessin/Peter Arnold. Parts b and c: University of Washington Pathology Department. Part d: Dr. Ram Verma/Phototake.]

mosomes under ultraviolet light; variation in the brightness of Q bands results from differences in the relative amounts of cytosine–guanine (C–G) and adenine–thymine base pairs. Other techniques reveal C bands (**Figure 9.2c**), which are regions of DNA occupied by centromeric heterochromatin, and R bands (**Figure 9.2d**), which are rich in cytosine–guanine base pairs.

Types of Chromosome Mutations

Chromosome mutations can be grouped into three basic categories: chromosome rearrangements, aneuploids, and polyploids (**Figure 9.3**). Chromosome rearrangements alter the *structure* of chromosomes; for example, a piece of a chromosome might be duplicated, deleted, or inverted. In aneuploidy, the *number* of chromosomes is altered: one or more individual chromosomes are added or deleted. In polyploidy, one or more complete *sets* of chromosomes are added. Some organisms (such as yeast) possess a single chromosome set ($1n$) for most of their life cycles and are referred to as haploid, whereas others possess two chromosome sets and are referred to as diploid ($2n$). A polyploid is any organism that has more than two sets of chromosomes ($3n$, $4n$, $5n$, or more).

takes a picture of the chromosomes, the image is digitized, and the chromosomes are sorted and arranged electronically by a computer.

Preparation and staining techniques help to distinguish among chromosomes of similar size and shape. For instance, the staining of chromosomes with a special dye called Giemsa reveals G bands, which distinguish areas of DNA that are rich in adenine–thymine (A–T) base pairs (**Figure 9.2a**). Q bands (**Figure 9.2b**) are revealed by staining chromosomes with quinacrine mustard and viewing the chro-

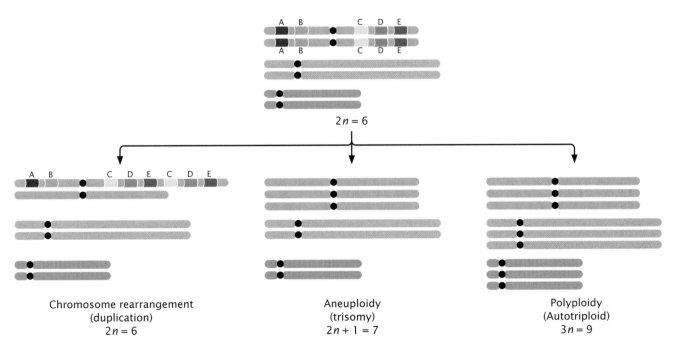

9.3 Chromosome mutations consist of chromosome rearrangements, aneuploids, and polyploids. Duplications, trisomy, and autotriploids are examples of each category of mutation.

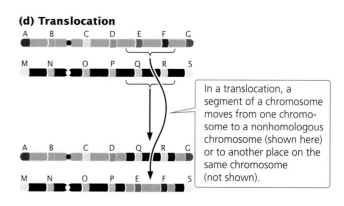

9.4 The four basic types of chromosome rearrangements are duplication, deletion, inversion, and translocation.

9.2 Chromosome Rearrangements Alter Chromosome Structure

Chromosome rearrangements are mutations that change the structures of individual chromosomes. The four basic types of rearrangements are duplications, deletions, inversions, and translocations (**Figure 9.4**).

Duplications

A **chromosome duplication** is a mutation in which part of the chromosome has been doubled (see Figure 9.4a). Consider a chromosome with segments AB•CDEFG, in which • represents the centromere. A duplication might include the EF segments, giving rise to a chromosome with segments AB•CDEFEFG. This type of duplication, in which the duplicated region is immediately adjacent to the original segment, is called a **tandem duplication**. If the duplicated segment is located some distance from the original segment, either on the same chromosome or on a different one, the chromosome rearrangement is called a **displaced duplication**. An example of a displaced duplication would be AB•CDEFGEF. A duplication can be either in the same orientation as that of the original sequence, as in the two preceding examples, or inverted: AB•CDEFFEG. When the duplication is inverted, it is called a **reverse duplication**.

Effects of chromosome duplications An individual homozygous for a duplication carries the duplication on both homologous chromosomes, and an individual heterozygous for a duplication has one normal chromosome and one chromosome with the duplication. In the heterozygotes (**Figure 9.5a**), problems arise in chromosome pairing at prophase I of meiosis, because the two chromosomes are not homologous throughout their length. The pairing and synapsis of homologous regions require that one or

9.5 In an individual heterozygous for a duplication, the duplicated chromosome loops out during pairing in prophase I.

both chromosomes loop and twist so that these regions are able to line up (**Figure 9.5b**). The appearance of this characteristic loop structure in meiosis is one way to detect duplications.

Duplications may have major effects on the phenotype. Among fruit flies (*Drosophila melanogaster*), for example, a fly having a *Bar* mutation has a reduced number of facets in the eye, making the eye smaller and bar shaped instead of oval (**Figure 9.6**). The *Bar* mutation results from a small duplication on the X chromosome that is inherited as an incompletely dominant, X-linked trait: heterozygous female flies have somewhat smaller eyes (the number of facets is reduced; see Figure 9.6b), whereas, in homozygous female and hemizygous male flies, the number of facets is greatly reduced (see Figure 9.6c). Occasionally, a fly carries three copies of the *Bar* duplication on its X chromosome; for flies carrying such mutations, which are termed *double Bar*, the number of facets is extremely reduced (see Figure 9.6d). The *Bar* mutation arises from unequal crossing over, a duplication-generating process (**Figure 9.7**; see also Figure 10.14).

Unbalanced gene dosage How does a chromosome duplication alter the phenotype? After all, gene sequences are not altered by duplications, and no genetic information is missing; the only change is the presence of additional copies of normal sequences. The answer to this question is not well understood, but the effects are most likely due to imbalances in the amounts of gene products (abnormal gene dosage). The amount of a particular protein synthesized by a cell is often directly related to the number of copies of its corresponding gene: an individual organism with three functional copies of a gene often produces 1.5 times as much of the protein encoded by that gene as that produced by an individual with two copies. Because developmental processes require the interaction of many proteins, they often depend critically on proper gene dosage. If the amount of one protein increases while the amounts

9.6 The Bar phenotype in *Drosophila melanogaster* results from an X-linked duplication. (a) Wild-type fruit flies have normal-size eyes. (b) Flies heterozygous and (c) homozygous for the *Bar* mutation have smaller, bar-shaped eyes. (d) Flies with *double Bar* have three copies of the duplication and much smaller bar-shaped eyes.

of others remain constant, problems can result (**Figure 9.8**). Although duplications can have severe consequences when the precise balance of a gene product is critical to cell function, duplications have arisen frequently throughout the evolution of many eukaryotic organisms and are a source of new genes that may provide novel functions. For example, humans have a series of genes that encode different globin chains, some of which function as an oxygen carrier during adult stages and others that function during embryonic and fetal development. All of these globin genes arose from an original ancestral gene that underwent a

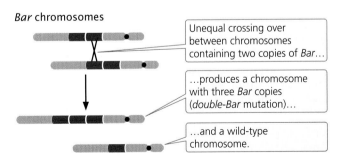

9.7 Unequal crossing over produces *Bar* and *double-Bar* mutations.

9.8 Unbalanced gene dosage leads to developmental abnormalities.

series of duplications. Human phenotypes associated with some duplications are summarized in **Table 9.1**.

Segmental duplications The human genome consists of numerous duplicated sequences called **segmental duplications**, which are defined as duplications greater than 1000 bp in length. In most segmental duplications, the two copies are found on the same chromosome (an intrachromosomal duplication) but, in others, the two copies are found on different chromosomes (an interchromosomal duplication). Many segmental duplications have been detected with molecular techniques that examine DNA sequences on a chromosome. These techniques reveal that about 4% of the human genome consists of segmental duplications. In the human genome, the average size of segmental duplications is 15,000 bp.

CONCEPTS

A chromosome duplication is a mutation that doubles part of a chromosome. In individuals heterozygous for a chromosome duplication, the duplicated region of the chromosome loops out when homologous chromosomes pair in prophase I of meiosis. Duplications often have major effects on the phenotype, possibly by altering gene dosage. Segmental duplications are common within the human genome.

✔ CONCEPT CHECK 1

Chromosome duplications often result in abnormal phenotypes because

a. developmental processes depend on the relative amounts of proteins encoded by different genes.
b. extra copies of the genes within the duplicated region do not pair in meiosis.
c. the chromosome is more likely to break when it loops in meiosis.
d. extra DNA must be replicated, which slows down cell division.

Deletions

A second type of chromosome rearrangement is a **chromosome deletion**, the loss of a chromosome segment (see Figure 9.4b). A chromosome with segments AB•CDEFG that undergoes a deletion of segment EF would generate the mutated chromosome AB•CDG.

A large deletion can be easily detected because the chromosome is noticeably shortened. In individuals heterozygous for deletions, the normal chromosome must loop during the pairing of homologs in prophase I of meiosis (**Figure 9.9**) to allow the homologous regions of the two chromosomes to align and undergo synapsis. This looping out generates a structure that looks very much like that seen for individuals heterozygous for duplications.

Effects of deletions The phenotypic consequences of a deletion depend on which genes are located in the deleted region. If the deletion includes the centromere, the chromosome will not segregate in meiosis or mitosis and will usually be lost. Many deletions are lethal in the homozygous state because all copies of any essential genes located in the deleted region are missing. Even individuals heterozygous for a deletion may have multiple defects for three reasons.

First, the heterozygous condition may produce imbalances in the amounts of gene products, similar to the imbalances produced by extra gene copies. Second, normally recessive mutations on the homologous chromosome lacking the deletion may be expressed when the wild-type

Table 9.1 Effects of some human chromosome rearrangements

Type of Rearrangement	Chromosome	Disorder	Symptoms
Duplication	4, short arm	—	Small head, short neck, low hairline, growth and mental retardation
Duplication	4, long arm	—	Small head, sloping forehead, hand abnormalities
Duplication	7, long arm	—	Delayed development, asymmetry of the head, fuzzy scalp, small nose, low-set ears
Duplication	9, short arm	—	Characteristic face, variable mental retardation, high and broad forehead, hand abnormalities
Deletion	5, short arm	*Cri-du-chat* syndrome	Small head, distinctive cry, widely spaced eyes, round face, mental retardation
Deletion	4, short arm	Wolf–Hirschhorn syndrome	Small head with high forehead, wide nose, cleft lip and palate, severe mental retardation
Deletion	4, long arm	—	Small head, from mild to moderate mental retardation, cleft lip and palate, hand and foot abnormalities
Deletion	7, long arm	Williams–Beuren syndrome	Facial features, heart defects, mental impairment
Deletion	15, long arm	Prader–Willi syndrome	Feeding difficulty at early age, but becoming obese after 1 year of age, from mild to moderate mental retardation
Deletion	18, short arm	—	Round face, large low-set ears, from mild to moderate mental retardation
Deletion	18, long arm	—	Distinctive mouth shape, small hands, small head, mental retardation

allele has been deleted (and is no longer present to mask the recessive allele's expression). The expression of a normally recessive mutation is referred to as **pseudodominance**, and it is an indication that one of the homologous chromosomes has a deletion. Third, some genes must be present in two copies for normal function. When a single copy of a gene is not sufficient to produce a wild-type phenotype, it is said to be a **haploinsufficient gene**. *Notch* is a series of X-linked wing mutations in *Drosophila* that often result from chromosome deletions. *Notch* deletions behave as

9.9 In an individual heterozygous for a deletion, the normal chromosome loops out during chromosome pairing in prophase I.

9.10 The Notch phenotype is produced by a chromosome deletion that includes the *Notch* gene. (Left) Normal wing veination. (Right) Wing veination produced by a *Notch* mutation. [Spyros Artavanis-Tsakonas, Kenji Matsuno, and Mark E. Fortini.]

dominant mutations: when heterozygous for a *Notch* deletion, a fly has wings that are notched at the tips and along the edges (**Figure 9.10**). The *Notch* mutation is therefore haploinsufficient. Females that are homozygous for a *Notch* deletion (or males that are hemizygous) die early in embryonic development. The *Notch* locus, which is deleted in *Notch* mutations, encodes a receptor that normally transmits signals received from outside the cell to the cell's interior and is important in fly development. The deletion acts as a recessive lethal because loss of all copies of the *Notch* gene prevents normal development.

Chromosome deletions in humans In humans, a deletion on the short arm of chromosome 5 is responsible for *cri-du-chat* syndrome. The name (French for "cry of the cat") derives from the peculiar, catlike cry of infants with this syndrome. A child who is heterozygous for this deletion has a small head, widely spaced eyes, and a round face and is mentally retarded. Deletion of part of the short arm of chromosome 4 results in another human disorder—Wolf–Hirschhorn syndrome, which is characterized by seizures and severe mental and growth retardation. A deletion of a tiny segment of chromosome 7 causes haploinsufficiency of the gene encoding elastin and a few other genes and leads to a condition known as Williams–Beuren syndrome, characterized by distinctive facial features, heart defects, high blood pressure, and cognitive impairments. Effects of deletions in human chromosomes are summarized in Table 9.1.

> **CONCEPTS**
>
> A chromosomal deletion is a mutation in which a part of a chromosome is lost. In individuals heterozygous for a deletion, the normal chromosome loops out during prophase I of meiosis. Deletions cause recessive genes on the homologous chromosome to be expressed and may cause imbalances in gene products.
>
> ✔ **CONCEPT CHECK 2**
>
> What is pseudodominance and how is it produced by a chromosome deletion?

Inversions

A third type of chromosome rearrangement is a **chromosome inversion**, in which a chromosome segment is inverted—turned 180 degrees (see Figure 9.4c). If a chromosome originally had segments AB•CDEFG, then chromosome AB•CFEDG represents an inversion that includes segments DEF. For an inversion to take place, the chromosome must break in two places. Inversions that do not include the centromere, such as AB•CFEDG, are termed **paracentric inversions** (*para* meaning "next to"), whereas inversions that include the centromere, such as ADC•BEFG, are termed **pericentric inversions** (*peri* meaning "around").

Inversion heterozygotes are common in many organisms, including a number of plants, some species of *Drosophila*, mosquitoes, and grasshoppers. Inversions may have played an important role in human evolution: G-banding patterns reveal that several human chromosomes differ from those of chimpanzees by only a pericentric inversion (**Figure 9.11**).

Effects of inversions Individual organisms with inversions have neither lost nor gained any genetic material; just the DNA sequence has been altered. Nevertheless, these mutations often have pronounced phenotypic effects. An inversion may break a gene into two parts, with one part moving to a new location and destroying the function of that gene. Even when the chromosome breaks are between genes, phenotypic effects may arise from the inverted gene order in an inversion. Many genes are regulated in a position-dependent manner; if their positions are altered by an inversion, they may be expressed at inappropriate times or in inappropriate tissues, an outcome referred to as a **position effect**.

Inversions in meiosis When an individual is homozygous for a particular inversion, no special problems arise in meiosis, and the two homologous chromosomes can pair and separate normally. When an individual is heterozygous for an inversion, however, the gene order of the two homologs differs, and the homologous sequences can align and pair only if the two chromosomes form an inversion loop (**Figure 9.12**).

Individuals heterozygous for inversions also exhibit reduced recombination among genes located in the inverted region. The frequency of crossing over within the inversion

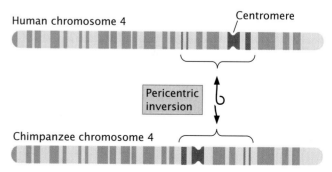

9.11 Chromosome 4 differs in humans and chimpanzees in a pericentric inversion.

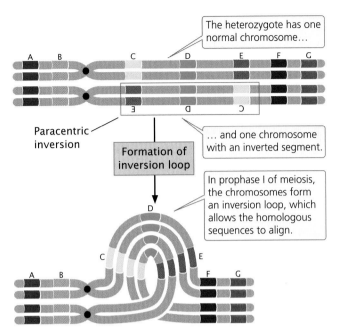

9.12 In an individual heterozygous for a paracentric inversion, the chromosomes form an inversion loop during pairing in prophase I.

is not actually diminished but, when crossing over does take place, the result is abnormal gametes that result in nonviable offspring, and thus no recombinant progeny are observed. Let's see why this result occurs.

Figure 9.13 illustrates the results of crossing over within a paracentric inversion. The individual is heterozygous for an inversion (see Figure 9.13a), with one wild-type, unmutated chromosome (AB•CDEFG) and one inverted chromosome (AB•EDCFG). In prophase I of meiosis, an inversion loop forms, allowing the homologous sequences to pair up (see Figure 9.13b). If a single crossover takes place in the inverted region (between segments C and D in Figure 9.13), an unusual structure results (see Figure 9.13c). The two outer chromatids, which did not participate in crossing over, contain original, nonrecombinant gene sequences. The two inner chromatids, which did cross over, are highly abnormal: each has two copies of some genes and no copies of others. Furthermore, one of the four chromatids now has two centromeres and is said to be a **dicentric chromatid**; the other lacks a centromere and is an **acentric chromatid**.

In anaphase I of meiosis, the centromeres are pulled toward opposite poles and the two homologous chromosomes separate. This action stretches the dicentric chromatid across the center of the nucleus, forming a structure called a **dicentric bridge** (see Figure 9.13d). Eventually, the dicentric bridge breaks, as the two centromeres are pulled farther apart. Spindle fibers do not attach to the acentric fragment, and so this fragment does not segregate into a nucleus in meiosis and is usually lost.

In the second division of meiosis, the sister chromatids separate and four gametes are produced (see Figure 9.13e). Two of the gametes contain the original, nonrecombinant chromosomes (AB•CDEFG and AB•EDCFG). The other

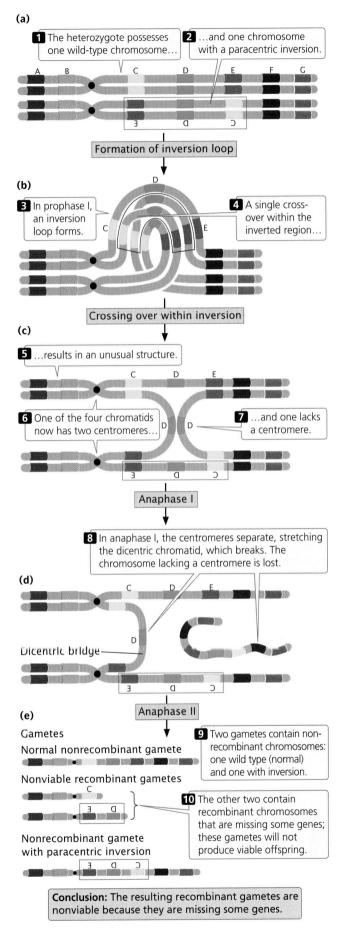

9.13 In a heterozygous individual, a single crossover within a paracentric inversion leads to abnormal gametes.

two gametes contain recombinant chromosomes that are missing some genes; these gametes will not produce viable offspring. Thus, no recombinant progeny result when crossing over takes place within a paracentric inversion. The key is to recognize that crossing over still takes place, but, when it does so, the resulting recombinant gametes are not viable; so no recombinant progeny are observed.

Recombination is also reduced within a pericentric inversion (**Figure 9.14**). No dicentric bridges or acentric fragments are produced, but the recombinant chromosomes have too many copies of some genes and no copies of others; so gametes that receive the recombinant chromosomes cannot produce viable progeny.

Figures 9.13 and 9.14 illustrate the results of single crossovers within inversions. Double crossovers in which both crossovers are on the same two strands (two-strand double crossovers) result in functional recombinant chromosomes. (Try drawing out the results of a double crossover.) Thus, even though the overall rate of recombination is reduced within an inversion, some viable recombinant progeny may still be produced through two-strand double crossovers. **TRY PROBLEM 26**

CONCEPTS

In an inversion, a segment of a chromosome is turned 180 degrees. Inversions cause breaks in some genes and may move others to new locations. In individuals heterozygous for a chromosome inversion, the homologous chromosomes form a loop in prophase I of meiosis. When crossing over takes place within the inverted region, nonviable gametes are usually produced, resulting in a depression in observed recombination frequencies.

✔ CONCEPT CHECK 3

A dicentric chromosome is produced when crossing over takes place in an individual heterozygous for which type of chromosome rearrangement?

a. Duplication   c. Paracentric inversion
b. Deletion   d. Pericentric inversion

Translocations

A **translocation** entails the movement of genetic material between nonhomologous chromosomes (see Figure 9.4d) or within the same chromosome. Translocation should not be confused with crossing over, in which there is an exchange of genetic material between *homologous* chromosomes.

In a **nonreciprocal translocation**, genetic material moves from one chromosome to another without any reciprocal exchange. Consider the following two nonhomologous chromosomes: AB•CDEFG and MN•OPQRS. If chromosome segment EF moves from the first chromosome to the second without any transfer of segments from the second chromosome to the first, a nonreciprocal translocation has taken place, producing chromosomes AB•CDG and MN•OPEFQRS. More commonly, there is a two-way

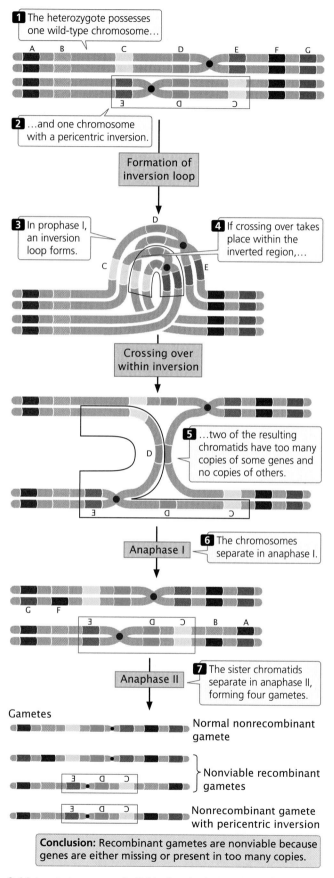

9.14 In a heterozygous individual, a single crossover within a pericentric inversion leads to abnormal gametes.

exchange of segments between the chromosomes, resulting in a **reciprocal translocation**. A reciprocal translocation between chromosomes AB•CDEFG and MN•OPQRS might give rise to chromosomes AB•CD<u>QR</u>G and MN•OP<u>EF</u>S.

Effects of translocations Translocations can affect a phenotype in several ways. First, they can physically link genes that were formerly located on different chromosomes. These new linkage relations may affect gene expression (a position effect): genes translocated to new locations may come under the control of different regulatory sequences or other genes that affect their expression.

Second, the chromosomal breaks that bring about translocations may take place within a gene and disrupt its function. Molecular geneticists have used these types of effects to map human genes. Neurofibromatosis is a genetic disease characterized by numerous fibrous tumors of the skin and nervous tissue; it results from an autosomal dominant mutation. Linkage studies first placed the locus that, when mutated, causes neurofibromatosis on chromosome 17, but the precise location of the locus was unknown. Geneticists later narrowed down the location when they identified two patients with neurofibromatosis who possessed a translocation affecting chromosome 17. These patients were assumed to have developed neurofibromatosis because one of the chromosome breaks that occurred in the translocation disrupted a particular gene, resulting in neurofibromatosis. DNA from the regions around the breaks was sequenced, eventually leading to the identification of the gene responsible for neurofibromatosis.

Deletions frequently accompany translocations. In a **Robertsonian translocation**, for example, the long arms of two acrocentric chromosomes become joined to a common centromere through a translocation, generating a metacentric chromosome with two long arms and another chromosome with two very short arms (**Figure 9.15**). The smaller chromosome often fails to segregate, leading to an overall reduction in chromosome number. As we will see, Robertsonian translocations are the cause of some cases of Down syndrome, a chromosome disorder discussed in the introduction to this chapter.

Translocations in meiosis The effects of a translocation on chromosome segregation in meiosis depend on the nature of the translocation. Let's consider what happens in an individual heterozygous for a reciprocal translocation. Suppose that the original chromosomes were AB•CDEFG and MN•OPQRS (designated N_1 and N_2 respectively, for normal chromosomes 1 and 2) and that a reciprocal translocation takes place, producing chromosomes AB•CD<u>QRS</u> and MN•OP<u>EFG</u> (designated T_1 and T_2, respectively, for translocated chromosomes 1 and 2). An individual heterozygous for this translocation would possess one normal copy of each chromosome and one translocated copy (**Figure 9.16a**). Each of these chromosomes contains segments that are homologous to two other chromosomes. When the homologous sequences pair in prophase I of meiosis, crosslike configurations consisting of all four chromosomes (**Figure 9.16b**) form.

Notice that N_1 and T_1 have homologous centromeres (in both chromosomes, the centromere is between segments B and C); similarly, N_2 and T_2 have homologous centromeres (between segments N and O). Normally, homologous centromeres separate and move toward opposite poles in anaphase I of meiosis. With a reciprocal translocation, the chromosomes may segregate in three different ways. In **alternate segregation** (**Figure 9.16c**), N_1 and N_2 move toward one pole and T_1 and T_2 move toward the opposite pole. In **adjacent-1 segregation**, N_1 and T_2 move toward one pole and T_1 and N_2 move toward the other pole. In both alternate and adjacent-1 segregation, homologous centromeres segregate toward opposite poles. **Adjacent-2 segregation**, in which N_1 and T_1 move toward one pole and T_2 and N_2 move toward the other, is rare because the two homologous chromosmes usually separate in meiosis.

The products of the three segregation patterns are illustrated in **Figure 9.16d**. As you can see, the gametes produced by alternate segregation possess one complete set of the chromosome segments. These gametes are therefore functional and can produce viable progeny. In contrast, gametes produced by adjacent-1 and adjacent-2 segregation are not viable, because some chromosome segments are present in two copies, whereas others are missing. Because adjacent-2 segregation is rare, most gametes are produced by alternate or adjacent-1 segregation. Therefore, approximately half of the gametes from an individual heterozygous for a reciprocal translocation are expected to be functional.

The role of translocations in evolution Translocations frequently play an important role in the evolution of karyotypes. Chimpanzees, gorillas, and orangutans all

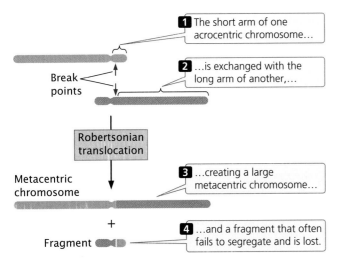

9.15 In a Robertsonian translocation, the short arm of one acrocentric chromosome is exchanged with the long arm of another.

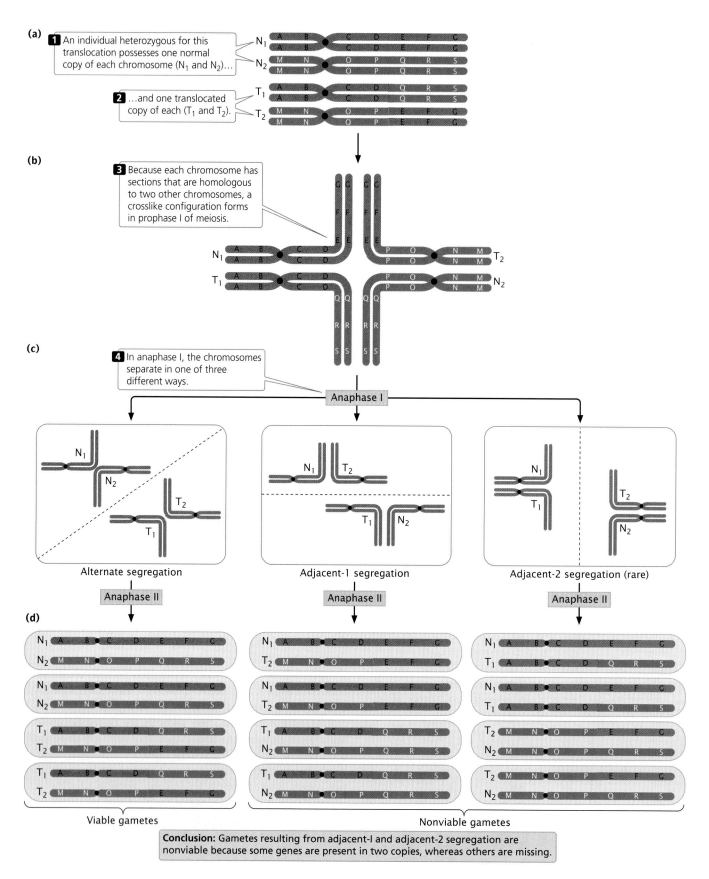

9.16 In an individual heterozygous for a reciprocal translocation, crosslike structures form in homologous pairing.

9.17 Human chromosome 2 contains a Robertsonian translocation that is not present in chimpanzees, gorillas, or orangutans. G-banding reveals that a Robertsonian translocation in a human ancestor switched the long and short arms of the two acrocentric chromosomes that are still found in the other three primates. This translocation created the large metacentric human chromosome 2.

have 48 chromosomes, whereas humans have 46. Human chromosome 2 is a large, metacentric chromosome with G-banding patterns that match those found on two different acrocentric chromosomes of the apes (**Figure 9.17**). Apparently, a Robertsonian translocation took place in a human ancestor, creating a large metacentric chromosome from the two long arms of the ancestral acrocentric chromosomes and a small chromosome consisting of the two short arms. The small chromosome was subsequently lost, leading to the reduced chromosome number in humans relative to that of the other apes. **TRY PROBLEM 27**

CONCEPTS

In translocations, parts of chromosomes move to other, nonhomologous chromosomes or to other regions of the same chromosome. Translocations can affect the phenotype by causing genes to move to new locations, where they come under the influence of new regulatory sequences, or by breaking genes and disrupting their function.

✔ CONCEPT CHECK 4

What is the outcome of a Robertsonian translocation?

a. Two acrocentric chromosomes
b. One metacentric chromosome and one chromosome with two very short arms
c. One metacentric and one acrocentric chromosome
d. Two metacentric chromosomes

Fragile Sites

Chromosomes of cells grown in culture sometimes develop constrictions or gaps at particular locations called **fragile sites** (Figure 9.18) because they are prone to breakage under certain conditions. More than 100 fragile sites have been identified on human chromosomes.

Fragile sites fall into two groups. *Common fragile sites* are present in all humans and are a normal feature of chromosomes. Common fragile sites are often the location of chromosome breakage and rearrangements in cancer cells, leading to chromosome deletions, translocations, and other chromosome rearrangements. *Rare fragile sites* are found in few people and are inherited as a Mendelian trait. Rare fragile sites are often associated with genetic disorders, such as mental retardation. Most of them consist of expanding nucleotide repeats, in which the number of repeats of a set of nucleotides is increased (see Chapter 10).

One of the most intensively studied rare fragile sites is located on the human X chromosome. This site is associated with mental retardation and is known as the **fragile-X syndrome**. Exhibiting X-linked inheritance and arising with a frequency of about 1 in 1250 male births, fragile-X syndrome has been shown to result from an increase in the number of repeats of a CGG trinucleotide.

Molecular studies of fragile sites have shown that many of these sites are more than 100,000 bp in length and include one or more genes. Fragile sites are often late in being replicated. At these places, the enzymes that replicate DNA may stall while unwinding of the DNA continues, leading to long stretches of DNA that are unwound and vulnerable to breakage. In spite of recent advances in our understanding of fragile sites, much about their nature remains poorly understood.

9.18 Fragile sites are chromosomal regions susceptible to breakage under certain conditions. Shown here is a fragile site on human chromosome X. [Courtesy of Dr. Christine Harrison.]

Copy-Number Variations

Chromosome rearrangements have traditionally been detected by examination of the chromosomes with a microscope. Visual examination identifies chromosome rearrangements on the basis of changes in the overall size of a chromosome, alteration of banding patterns revealed by chromosome staining, or the behavior of chromosomes in meiosis. Microscopy, however, can detect only large chromosome rearrangements, typically those that are at least 5 million bp in length.

With the completion of the Human Genome Project, detailed information about DNA sequences found on individual chromosomes has become available. Using this information, geneticists can now examine the number of copies of specific DNA sequences present in a cell and detect duplications, deletions, and other chromosome rearrangements that cannot be observed with microscopy alone. The work has been greatly facilitated by the availability of microarrays, which allow the simultaneous detection of hundreds of thousands of DNA sequences across the genome. Because these methods measure the number of copies of particular DNA sequences, the variations that they detect are called **copy-number variations** (CNVs). Copy-number variations include duplications and deletions that range in size from thousands of base pairs to several million base pairs. Many of these variants encompass at least one gene and may encompass several genes.

Recent studies of copy-number variation have revealed that submicroscopic chromosome duplications and deletions are quite common: research suggests that each person may possess as many as 1000 copy-number variations. Many of them probably have no observable phenotypic effects, but some copy-number variations have now been implicated in causing a number of diseases and disorders. For example, Janine Wagenstaller and her colleagues studied copy-number variation in 67 children with unexplained mental retardation and found that 11 (16%) of them had duplications and deletions. Copy-number variations have also been associated with osteoporosis, autism, schizophrenia, and a number of other diseases and disorders. TRY PROBLEM 19 →

> **CONCEPTS**
>
> Variations in the number of copies of particular DNA sequences (copy-number variations) are surprisingly common in the human genome.

9.3 Aneuploidy Is an Increase or Decrease in the Number of Individual Chromosomes

In addition to chromosome rearrangements, chromosome mutations include changes in the number of chromosomes. Variations in chromosome number can be classified into two basic types: **aneuploidy**, which is a change in the number of individual chromosomes, and **polyploidy**, which is a change in the number of chromosome sets.

Aneuploidy can arise in several ways. First, a chromosome may be lost in the course of mitosis or meiosis if, for example, its centromere is deleted. Loss of the centromere prevents the spindle fibers from attaching; so the chromosome fails to move to the spindle pole and does not become incorporated into a nucleus after cell division. Second, the small chromosome generated by a Robertsonian translocation may be lost in mitosis or meiosis. Third, aneuploids may arise through nondisjunction, the failure of homologous chromosomes or sister chromatids to separate in meiosis or mitosis (see p. 83 in Chapter 4). Nondisjunction leads to some gametes or cells that contain an extra chromosome and others that are missing a chromosome (**Figure 9.19**). TRY PROBLEM 28 →

Types of Aneuploidy

We will consider four types of common aneuploid conditions in diploid individuals: nullisomy, monosomy, trisomy, and tetrasomy.

1. **Nullisomy** is the loss of both members of a homologous pair of chromosomes. It is represented as $2n - 2$, where n refers to the haploid number of chromosomes. Thus, among humans, who normally possess $2n = 46$ chromosomes, a nullisomic zygote has 44 chromosomes.

2. **Monosomy** is the loss of a single chromosome, represented as $2n - 1$. A human monosomic zygote has 45 chromosomes.

3. **Trisomy** is the gain of a single chromosome, represented as $2n + 1$. A human trisomic zygote has 47 chromosomes. The gain of a chromosome means that there are three homologous copies of one chromosome. Most cases of Down syndrome, discussed in the introduction to the chapter, result from trisomy of chromosome 21.

4. **Tetrasomy** is the gain of two homologous chromosomes, represented as $2n + 2$. A human tetrasomic zygote has 48 chromosomes. Tetrasomy is not the gain of *any* two extra chromosomes, but rather the gain of two homologous chromosomes; so there will be four homologous copies of a particular chromosome.

More than one aneuploid mutation may occur in the same individual organism. An individual that has an extra copy of two different (nonhomologous) chromosomes is referred to as being double trisomic and represented as $2n + 1 + 1$. Similarly, a double monosomic has two fewer nonhomologous chromosomes ($2n - 1 - 1$), and a double tetrasomic has two extra pairs of homologous chromosomes ($2n + 2 + 2$).

Effects of Aneuploidy

One of the first aneuploids to be recognized was a fruit fly with a single X chromosome and no Y chromosome discov-

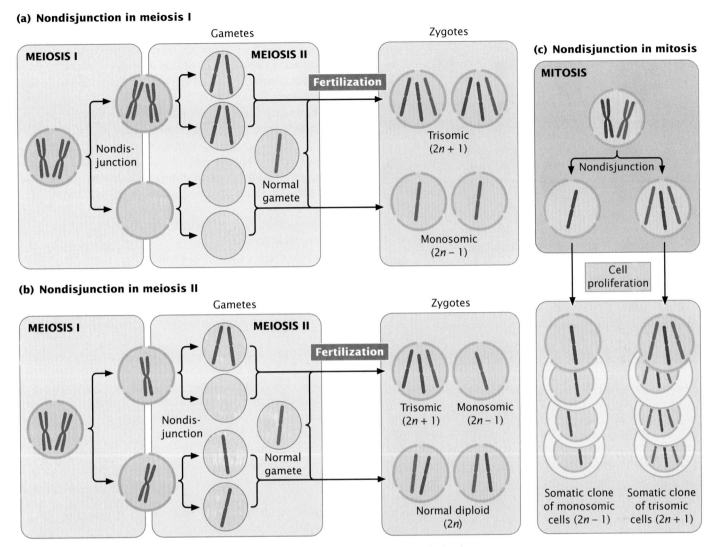

9.19 Aneuploids can be produced through nondisjunction in meiosis I, meiosis II, and mitosis. The gametes that result from meioses with nondisjunction combine with a gamete (with blue chromosome) that results from normal meiosis to produce the zygotes.

ered by Calvin Bridges in 1913 (see pp. 82–83 in Chapter 4). Another early study of aneuploidy focused on mutants in the Jimson weed, *Datura stramonium*. A. Francis Blakeslee began breeding this plant in 1913, and he observed that crosses with several Jimson mutants produced unusual ratios of progeny. For example, the *globe* mutant (producing a seedcase globular in shape) was dominant but was inherited primarily from the female parent. When plants having the *globe* mutation were self-fertilized, only 25% of the progeny had the globe phenotype. If the *globe* mutant were strictly dominant, Blakeslee should have seen 75% of the progeny with the trait (see Chapter 3), and so the 25% that he observed was unusual. Blakeslee isolated 12 different mutants (**Figure 9.20**) that exhibited peculiar patterns of inheritance. Eventually, John Belling demonstrated that these 12 mutants are in fact trisomics. *Datura stramonium* has 12 pairs of chromosomes ($2n = 24$), and each of the 12 mutants is trisomic for a different chromosome pair. The aneuploid nature of the mutants explained the unusual ratios that Blakeslee had observed in the progeny. Many of the extra chromosomes in the trisomics were lost in meiosis; so fewer than 50% of the gametes carried the extra chromosome, and the proportion of trisomics in the progeny was low. Furthermore, the pollen containing an extra chromosome was not as successful in fertilization, and trisomic zygotes were less viable.

Aneuploidy usually alters the phenotype drastically. In most animals and many plants, aneuploid mutations are lethal. Because aneuploidy affects the number of gene copies but not their nucleotide sequences, the effects of aneuploidy are most likely due to abnormal gene dosage. Aneuploidy alters the dosage for some, but not all, genes, disrupting the relative concentrations of gene products and often interfering with normal development.

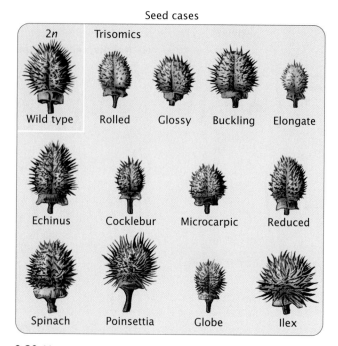

9.20 Mutant capsules in Jimson weed (*Datura stramonium*) result from different trisomies. Each type of capsule is a phenotype that is trisomic for a different chromosome.

A major exception to the relation between gene number and protein dosage pertains to genes on the mammalian X chromosome. In mammals, X-chromosome inactivation ensures that males (who have a single X chromosome) and females (who have two X chromosomes) receive the same functional dosage for X-linked genes (see pp. 88–90 in Chapter 4 for further discussion of X-chromosome inactivation). Extra X chromosomes in mammals are inactivated; so we might expect that aneuploidy of the sex chromosomes would be less detrimental in these animals. Indeed, it is the case for mice and humans, for whom aneuploids of the sex chromosomes are the most common form of aneuploidy seen in living organisms. Y-chromosome aneuploids are probably common because there is so little information in the Y-chromosome.

> **CONCEPTS**
>
> Aneuploidy, the loss or gain of one or more individual chromosomes, may arise from the loss of a chromosome subsequent to translocation or from nondisjunction in meiosis or mitosis. It disrupts gene dosage and often has severe phenotypic effects.
>
> ✔ **CONCEPT CHECK 5**
>
> A diploid organism has $2n = 36$ chromosomes. How many chromosomes will be found in a trisomic member of this species?

Aneuploidy in Humans

For unknown reasons, an incredibly high percentage of all human embryos that are conceived possess chromosome abnormalities. Findings from studies of women who are attempting pregnancy suggest that more than 30% of all conceptions spontaneously abort (miscarry), usually so early in development that the mother is not even aware of her pregnancy. Chromosome defects are present in at least 50% of spontaneously aborted human fetuses, with aneuploidy accounting for most of them. This rate of chromosome abnormality in humans is higher than in other organisms that have been studied; in mice, for example, aneuploidy is found in no more than 2% of fertilized eggs. Aneuploidy in humans usually produces such serious developmental problems that spontaneous abortion results. Only about 2% of all fetuses with a chromosome defect survive to birth.

Sex-chromosome aneuploids The most common aneuploidy seen in living humans has to do with the sex chromosomes. As is true of all mammals, aneuploidy of the human sex chromosomes is better tolerated than aneuploidy of autosomal chromosomes. Both Turner syndrome and Klinefelter syndrome (see Figures 4.8 and 4.9) result from aneuploidy of the sex chromosomes.

Autosomal aneuploids Autosomal aneuploids resulting in live births are less common than sex-chromosome aneuploids in humans, probably because there is no mechanism of dosage compensation for autosomal chromosomes. Most autosomal aneuploids are spontaneously aborted, with the exception of aneuploids of some of the small autosomes such as chromosome 21. Because these chromosomes are small and carry fewer genes, the presence of extra copies is less detrimental than it is for larger chromosomes.

Down syndrome The most common autosomal aneuploidy in humans is **trisomy 21**, also called **Down syndrome** (discussed in the introduction to the chapter). The incidence of Down syndrome in the United States is similar to that of the world, about 1 in 700 human births, although the incidence increases among children born to older mothers. Approximately 92% of those who have Down syndrome have three full copies of chromosome 21 (and therefore a total of 47 chromosomes), a condition termed **primary Down syndrome** (**Figure 9.21**). Primary Down syndrome usually arises from spontaneous nondisjunction in egg formation: about 75% of the nondisjunction events that cause Down syndrome are maternal in origin, most arising in meiosis I. Most children with Down syndrome are born to normal parents, and the failure of the chromosomes to divide has little hereditary tendency. A couple who has conceived one child with primary Down syndrome has only a slightly higher risk of conceiving a second child with Down syndrome (compared with other couples of similar age who have not had any Down-syndrome children). Similarly, the couple's relatives are not more likely to have a child with primary Down syndrome.

About 4% of people with Down syndrome are not trisomic for a complete chromosome 21. Instead, they have 46 chromosomes, but an extra copy of part of chromosome 21 is attached to another chromosome through a translocation.

9.21 Primary Down syndrome is caused by the presence of three copies of chromosome 21. [L. Willatt, East Anglian Regional Genetics Service/Science Photo Library/Photo Researchers.]

9.22 The translocation of chromosome 21 onto another chromosome results in familial Down syndrome. Here, the long arm of chromosome 21 is attached to chromosome 14. This karyotype is from a translocation carrier, who is phenotypically normal but is at increased risk for producing children with Down syndrome. [Copyright Centre for Genetics Education for and on behalf of the Crown in right of the State of New South Wales.]

This condition is termed **familial Down syndrome** because it has a tendency to run in families. The phenotypic characteristics of familial Down syndrome are the same as those of primary Down syndrome.

Familial Down syndrome arises in offspring whose parents are carriers of chromosomes that have undergone a Robertsonian translocation, most commonly between chromosome 21 and chromosome 14: the long arm of 21 and the short arm of 14 exchange places (**Figure 9.22**). This exchange produces a chromosome that includes the long arms of chromosomes 14 and 21, and a very small chromosome that consists of the short arms of chromosomes 21 and 14. The small chromosome is generally lost after several cell divisions. Although exchange between chromosomes 21 and 14 is the most common cause of familial Down syndrome, the condition can also be caused by translocations between 21 and other chromosomes such as 15.

Persons with the translocation, called **translocation carriers**, do not have Down syndrome. Although they possess only 45 chromosomes, their phenotypes are normal because they have two copies of the long arms of chromosomes 14 and 21, and apparently the short arms of these chromosomes (which are lost) carry no essential genetic information. Although translocation carriers are completely healthy, they have an increased chance of producing children with Down syndrome.

When a translocation carrier produces gametes, the translocation chromosome may segregate in three different ways. First, it may separate from the normal chromosomes 14 and 21 in anaphase I of meiosis (**Figure 9.23a**). In this type of segregation, half of the gametes will have the translocation chromosome and no other copies of chromosomes 21 and 14; the fusion of such a gamete with a normal gamete will give rise to a translocation carrier. The other half of the gametes produced by this first type of segregation will be normal, each with a single copy of chromosomes 21 and 14, and will result in normal offspring.

Alternatively, the translocation chromosome may separate from chromosome 14 and pass into the same cell with the normal chromosome 21 (**Figure 9.23b**). This type of segregation produces abnormal gametes only; half will have two functional copies of chromosome 21 (one normal and one attached to chromosome 14) and the other half will lack chromosome 21. If a gamete with the two functional copies of chromosome 21 fuses with a normal gamete carrying a single copy of chromosome 21, the resulting zygote will have familial Down syndrome. If a gamete lacking chromosome 21 fuses with a normal gamete, the resulting zygote will have monosomy 21 and will be spontaneously aborted.

In the third type of segregation, the translocation chromosome and the normal copy of chromosome 14 segregate together (**Figure 9.23c**). This pattern is presumably rare, because the two centromeres are both derived from chromosome 14 and usually separate from each other. All the gametes produced by this process are abnormal: half result in monosomy 14 and the other half result in trisomy 14. All are spontaneously aborted.

Thus, only three of the six types of gametes that can be produced by a translocation carrier will result in the birth of a baby and, theoretically, these gametes should arise with equal frequency. One-third of the offspring of a translocation carrier should be translocation carriers like their

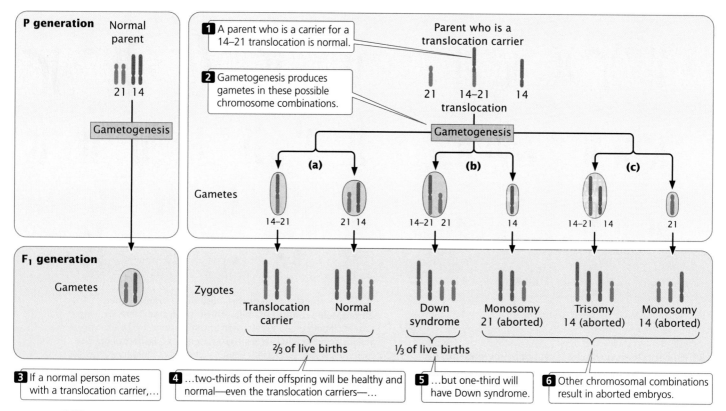

9.23 Translocation carriers are at increased risk for producing children with Down syndrome.

parent, one-third should have familial Down syndrome, and one-third should be normal. In reality, however, fewer than one-third of the children born to translocation carriers have Down syndrome, which suggests that some of the embryos with Down syndrome are spontaneously aborted.

TRY PROBLEM 31 →

Other human trisomies Few autosomal aneuploids besides trisomy 21 result in human live births. **Trisomy 18**, also known as **Edward syndrome**, arises with a frequency of approximately 1 in 8000 live births. Babies with Edward syndrome are severely retarded and have low-set ears, a short neck, deformed feet, clenched fingers, heart problems, and other disabilities. Few live for more than a year after birth. **Trisomy 13** has a frequency of about 1 in 15,000 live births and produces features that are collectively known as **Patau syndrome**. Characteristics of this condition include severe mental retardation, a small head, sloping forehead, small eyes, cleft lip and palate, extra fingers and toes, and numerous other problems. About half of children with trisomy 13 die within the first month of life, and 95% die by the age of 3. Rarer still is **trisomy 8**, which arises with a frequency ranging from about 1 in 25,000 to 1 in 50,000 live births. This aneuploid is characterized by mental retardation, contracted fingers and toes, low-set malformed ears, and a prominent forehead. Many who have this condition have normal life expectancy.

Aneuploidy and maternal age Most cases of Down syndrome and other types of aneuploidy in humans arise from maternal nondisjunction, and the frequency of aneuploidy increases with maternal age (**Figure 9.24**). Why maternal age is associated with nondisjunction is not known for certain. Female mammals are born with primary oocytes suspended in the diplotene substage of prophase I of meiosis. Just before ovulation, meiosis resumes and the first division is completed, producing a secondary oocyte. At this point, meiosis is suspended again and remains so until the secondary oocyte is penetrated by a sperm. The second meiotic division takes place immediately before the nuclei of egg and sperm unite to form a zygote.

Primary oocytes may remain suspended in diplotene for many years before ovulation takes place and meiosis recommences. Components of the spindle and other structures required for chromosome segregation may break down in the long arrest of meiosis, leading to more aneuploidy in children born to older mothers. According to this theory, no age effect is seen in males, because sperm are produced continuously after puberty with no long suspension of the meiotic divisions.

Aneuploidy and cancer Many tumor cells have extra or missing chromosomes or both; some types of tumors are consistently associated with specific chromosome mutations, including aneuploidy and chromosome rearrangements.

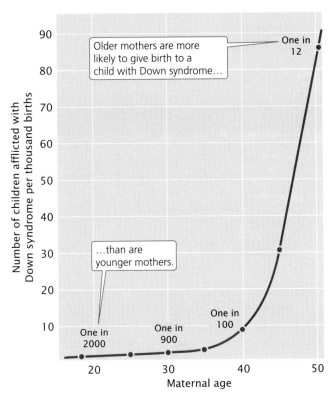

9.24 The incidence of primary Down syndrome and other aneuploids increases with maternal age.

> **CONCEPTS**
>
> In humans, sex-chromosome aneuploids are more common than autosomal aneuploids. X-chromosome inactivation prevents problems of gene dosage for X-linked genes. Down syndrome results from three functional copies of chromosome 21, either through trisomy (primary Down syndrome) or a Robertsonian translocation (familial Down syndrome).
>
> ✔ **CONCEPT CHECK 6**
>
> Briefly explain why, in humans and mammals, sex-chromosome aneuploids are more common than autosomal aneuploids?

Uniparental Disomy

Normally, the two chromosomes of a homologous pair are inherited from different parents—one from the father and one from the mother. The development of molecular techniques that facilitate the identification of specific DNA sequences has made the determination of the parental origins of chromosomes possible. Surprisingly, sometimes both chromosomes are inherited from the same parent, a condition termed **uniparental disomy**.

Many cases of uniparental disomy probably originate as a trisomy. Although most autosomal trisomies are lethal, a trisomic embryo can survive if one of the three chromosomes is lost early in development. If, just by chance, the two remaining chromosomes are both from the same parent, uniparental disomy results.

Uniparental disomy violates the rule that children affected with a recessive disorder appear only in families where both parents are carriers. For example, cystic fibrosis is an autosomal recessive disease; typically, both parents of an affected child are heterozygous for the cystic fibrosis mutation on chromosome 7. However, for a small proportion of people with cystic fibrosis, only one of the parents is heterozygous for the cystic fibrosis mutation. In these cases, cystic fibrosis is due to uniparental disomy: the person who has cystic fibrosis inherited from the heterozygous parent two copies of the chromosome 7 that carries the defective cystic fibrosis allele and no copy of the normal allele from the other parent.

Uniparental disomy has also been observed in Prader–Willi syndrome, a rare condition that arises when a paternal copy of a gene on chromosome 15 is missing. Although most cases of Prader–Willi syndrome result from a chromosome deletion that removes the paternal copy of the gene (see p. 121 in Chapter 5), from 20% to 30% arise when both copies of chromosome 15 are inherited from the mother and no copy is inherited from the father.

Mosaicism

Nondisjunction in a mitotic division may generate patches of cells in which every cell has a chromosome abnormality and other patches in which every cell has a normal karyotype. This type of nondisjunction leads to regions of tissue with different chromosome constitutions, a condition known as **mosaicism**. Growing evidence suggests that mosaicism is common.

Only about 50% of those diagnosed with Turner syndrome have the 45,X karyotype (presence of a single X chromosome) in all their cells; most others are mosaics, possessing some 45,X cells and some normal 46,XX cells. A few may even be mosaics for two or more types of abnormal karyotypes. The 45,X/46,XX mosaic usually arises when an X chromosome is lost soon after fertilization in an XX embryo.

Fruit flies that are XX/XO mosaics (O designates the absence of a homologous chromosome; XO means that the cell has a single X chromosome and no Y chromosome) develop a mixture of male and female traits, because the presence of two X chromosomes in fruit flies produces female traits and the presence of a single X chromosome produces male traits (**Figure 9.25**). In fruit flies, sex is determined independently in each cell in the course of development. Those cells that are XX express female traits; those that are XO express male traits. Such sexual mosaics are called **gynandromorphs**. Normally, X-linked recessive genes are masked in heterozygous females but, in XX/XO mosaics, any X-linked recessive genes present in the cells with a single X chromosome will be expressed.

> **CONCEPTS**
>
> In uniparental disomy, an individual organism has two copies of a chromosome from one parent and no copy from the other. Uniparental disomy may arise when a trisomic embryo loses one of the triplicate chromosomes early in development. In mosaicism, different cells within the same individual organism have different chromosome constitutions.

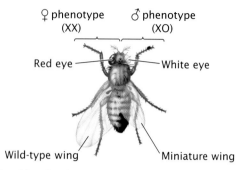

9.25 Mosaicism for the sex chromosomes produces a gynandromorph. This XX/XO gynandromorph fruit fly carries one wild-type X chromosome and one X chromosome with recessive alleles for white eyes and miniature wings. The left side of the fly has a normal female phenotype, because the cells are XX and the recessive alleles on one X chromosome are masked by the presence of wild-type alleles on the other. The right side of the fly has a male phenotype with white eyes and miniature wing, because the cells are missing the wild-type X chromosome (are XO), allowing the white and miniature alleles to be expressed.

9.4 Polyploidy Is the Presence of More Than Two Sets of Chromosomes

Most eukaryotic organisms are diploid ($2n$) for most of their life cycles, possessing two sets of chromosomes. Occasionally, whole sets of chromosomes fail to separate in meiosis or mitosis, leading to polyploidy, the presence of more than two genomic sets of chromosomes. Polyploids include *triploids* ($3n$), *tetraploids* ($4n$), *pentaploids* ($5n$), and even higher numbers of chromosome sets.

Polyploidy is common in plants and is a major mechanism by which new plant species have evolved. Approximately 40% of all flowering-plant species and from 70% to 80% of grasses are polyploids. They include a number of agriculturally important plants, such as wheat, oats, cotton, potatoes, and sugar cane. Polyploidy is less common in animals but is found in some invertebrates, fishes, salamanders, frogs, and lizards. No naturally occurring, viable polyploids are known in birds, but at least one polyploid mammal—a rat in Argentina—has been reported.

We will consider two major types of polyploidy: **autopolyploidy**, in which all chromosome sets are from a single species; and **allopolyploidy**, in which chromosome sets are from two or more species.

Autopolyploidy

Autopolyploidy is due to accidents of meiosis or mitosis that produce extra sets of chromosomes, all derived from a single species. Nondisjunction of all chromosomes in mitosis in an early $2n$ embryo, for example, doubles the chromosome number and produces an autotetraploid ($4n$), as depicted in **Figure 9.26a**. An autotriploid ($3n$) may arise when nondisjunction in meiosis produces a diploid gamete that then fuses with a normal haploid gamete to produce a triploid zygote (**Figure 9.26b**). Alternatively, triploids may arise from a cross between an autotetraploid that produces $2n$ gametes and a diploid that produces $1n$ gametes.

(a) Autopolyploidy through mitosis

(b) Autopolyploidy through meiosis

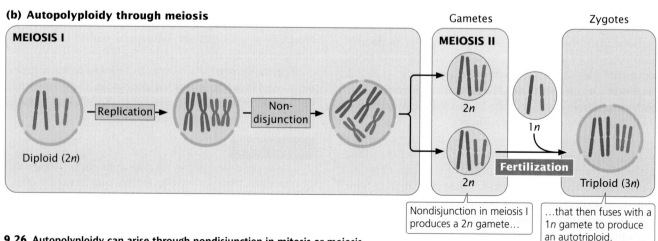

9.26 Autopolyploidy can arise through nondisjunction in mitosis or meiosis.

Because all the chromosome sets in autopolyploids are from the same species, they are homologous and attempt to align in prophase I of meiosis, which usually results in sterility. Consider meiosis in an autotriploid (**Figure 9.27**). In meiosis in a diploid cell, two chromosome homologs pair and align, but, in autotriploids, three homologs are present. One of the three homologs may fail to align with the other two, and this unaligned chromosome will segregate randomly (see Figure 9.27a). Which gamete gets the extra chromosome will be determined by chance and will differ for each homologous group of chromosomes. The resulting gametes will have two copies of some chromosomes and one copy of others. Even if all three chromosomes do align, two chromosomes must segregate to one gamete and one chromosome to the other (see Figure 9.27b). Occasionally, the presence of a third chromosome interferes with normal alignment, and all three chromosomes move to the same gamete (see Figure 9.27c).

No matter how the three homologous chromosomes align, their random segregation will create **unbalanced gametes**, with various numbers of chromosomes. A gamete produced by meiosis in such an autotriploid might receive, say, two copies of chromosome 1, one copy of chromosome 2, three copies of chromosome 3, and no copies of chromosome 4. When the unbalanced gamete fuses with a normal gamete (or with another unbalanced gamete), the resulting zygote has different numbers of the four types of chromosomes. This difference in number creates unbalanced gene dosage in the zygote, which is often lethal. For this reason, triploids do not usually produce viable offspring.

In even-numbered autopolyploids, such as autotetraploids, the homologous chromosomes can theoretically form pairs and divide equally. However, this event rarely happens; so these types of autotetraploids also produce unbalanced gametes.

The sterility that usually accompanies autopolyploidy has been exploited in agriculture. Wild diploid bananas ($2n = 22$), for example, produce seeds that are hard and inedible, but triploid bananas ($3n = 33$) are sterile and produce no seeds; they are the bananas sold commercially. Similarly, seedless triploid watermelons have been created and are now widely sold.

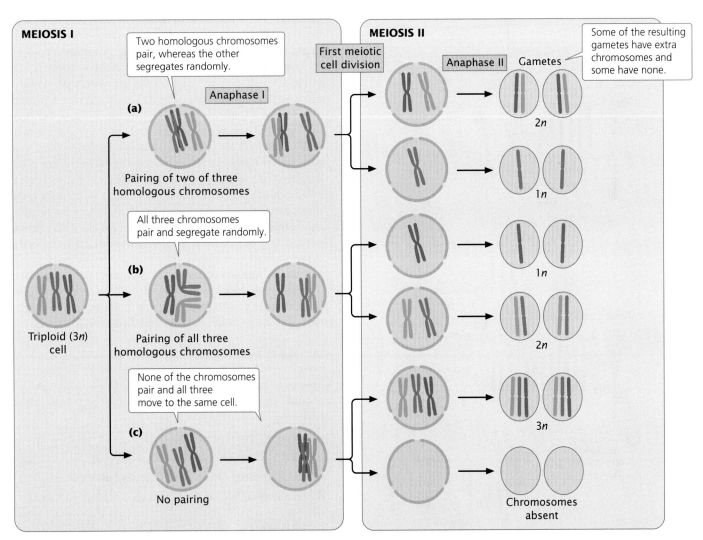

9.27 In meiosis of an autotriploid, homologous chromosomes can pair or not pair in three ways.
This example illustrates the pairing and segregation of a single homologous set of chromosomes.

Allopolyploidy

Allopolyploidy arises from hybridization between two species; the resulting polyploid carries chromosome sets derived from two or more species. **Figure 9.28** shows how allopolyploidy can arise from two species that are sufficiently related that hybridization takes place between them. Species 1 (AABBCC, $2n = 6$) produces haploid gametes with chromosomes ABC, and species 2 (GGHHII, $2n = 6$) produces gametes with chromosomes GHI. If gametes from species 1 and 2 fuse, a hybrid with six chromosomes (ABCGHI) is created. The hybrid has the same chromosome number as that of both diploid species; so the hybrid is considered diploid. However, because the hybrid chromosomes are not homologous, they will not pair and segregate properly in meiosis; so this hybrid is functionally haploid and sterile.

The sterile hybrid is unable to produce viable gametes through meiosis, but it may be able to perpetuate itself through mitosis (asexual reproduction). On rare occasions, nondisjunction takes place in a mitotic division, which leads to a doubling of chromosome number and an allotetraploid with chromosomes AABBCCGGHHII. This type of allopolyploid, consisting of two combined diploid genomes, is sometimes called an **amphidiploid**. Although the chromosome number has doubled compared with what was present in each of the parental species, the amphidiploid is functionally diploid: every chromosome has one and only one homologous partner, which is exactly what meiosis requires for proper segregation. The amphidiploid can now undergo normal meiosis to produce balanced gametes having six chromosomes.

George Karpechenko created polyploids experimentally in the 1920s. Today, as well as in the early twentieth century, cabbage (*Brassica oleracea*, $2n = 18$) and radishes (*Raphanus sativa*, $2n = 18$) are agriculturally important plants, but only the leaves of the cabbage and the roots of the radish are normally consumed. Karpechenko wanted to produce a plant that had cabbage leaves and radish roots so that no part of the plant would go to waste. Because both cabbage and radish possess 18 chromosomes, Karpechenko was able to successfully cross them, producing a hybrid with $2n = 18$, but, unfortunately, the hybrid was sterile. After several crosses, Karpechenko noticed that one of his hybrid plants produced a few seeds. When planted, these seeds grew into plants that were viable and fertile. Analysis of their chromosomes revealed that the plants were allotetraploids, with $2n = 36$ chromosomes. To Karpechencko's great disappointment, however, the new plants possessed the roots of a cabbage and the leaves of a radish.

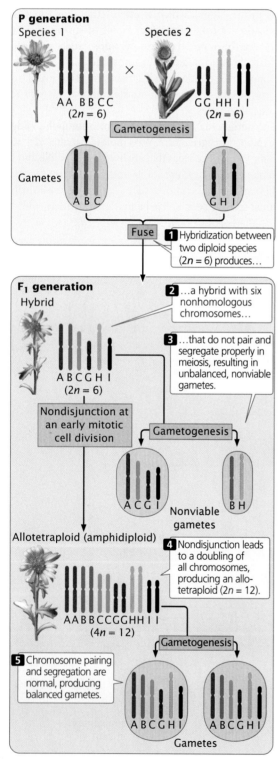

9.28 Most allopolyploids arise from hybridization between two species followed by chromosome doubling.

Worked Problem

Species I has $2n = 14$ and species II has $2n = 20$. Give all possible chromosome numbers that may be found in the following individuals.

a. An autotriploid of species I

b. An autotetraploid of species II

c. An allotriploid formed from species I and species II

d. An allotetraploid formed from species I and species II

- **Solution**

The haploid number of chromosomes (n) for species I is 7 and for species II is 10.

a. A triploid individual is 3n. A common mistake is to assume that 3n means three times as many chromosomes as in a normal individual, but remember that normal individuals are 2n. Because n for species I is 7 and all genomes of an autopolyploid are from the same species, $3n = 3 \times 7 = 21$.

b. An autotetraploid is 4n with all genomes from the same species. The n for species II is 10, so $4n = 4 \times 10 = 40$.

c. A triploid is 3n. By definition, an allopolyploid must have genomes from two different species. An allotriploid could have 1n from species I and 2n from species II or $(1 \times 7) + (2 \times 10) = 27$. Alternatively, it might have 2n from species I and 1n from species II, or $(2 \times 7) + (1 \times 10) = 24$. Thus, the number of chromosomes in an allotriploid could be 24 or 27.

d. A tetraploid is 4n. By definition, an allotetraploid must have genomes from at least two different species. An allotetraploid could have 3n from species I and 1n from species II or $(3 \times 7) + (1 \times 10) = 31$; or 2n from species I and 2n from species II or $(2 \times 7) + (2 \times 10) = 34$; or 1n from species I and 3n from species II or $(1 \times 7) + (3 \times 10) = 37$. Thus, the number of chromosomes could be 31, 34, or 37.

> For additional practice, try Problem 37 at the end of this chapter.

The Significance of Polyploidy

In many organisms, cell volume is correlated with nuclear volume, which, in turn, is determined by genome size. Thus, the increase in chromosome number in polyploidy is often associated with an increase in cell size, and many polyploids are physically larger than diploids. Breeders have used this effect to produce plants with larger leaves, flowers, fruits, and seeds. The hexaploid ($6n = 42$) genome of wheat probably contains chromosomes derived from three different wild species (**Figure 9.29**). As a result, the seeds of modern wheat are larger than those of its ancestors. Many other cultivated plants also are polyploid (**Table 9.2**).

Polyploidy is less common in animals than in plants for several reasons. As discussed, allopolyploids require hybridization between different species, which happens less frequently in animals than in plants. Animal behavior often prevents interbreeding among species, and the complexity of animal development causes most interspecific hybrids to be nonviable. Many of the polyploid animals that do arise are in groups that reproduce through parthenogenesis (a type of reproduction in which the animal develops from an unfertilized egg). Thus, asexual reproduction may facilitate the development of polyploids, perhaps because the perpetuation of hybrids through asexual reproduction provides greater opportunities for nondisjunction than does sexual reproduction. Only a few human polyploid babies have been reported, and most died within a few days of birth. Polyploidy—usually triploidy—is seen in about 10% of all spontaneously aborted human fetuses.

9.29 Modern bread wheat, *Triticum aestivum*, is a hexaploid with genes derived from three different species. Two diploid species, *T. monococcum* ($n = 14$) and probably *T. searsii* ($n = 14$), originally crossed to produce a diploid hybrid ($2n = 14$) that underwent chromosome doubling to create *T. turgidum* ($4n = 28$). A cross between *T. turgidum* and *T. tauschi* ($2n = 14$) produced a triploid hybrid ($3n = 21$) that then underwent chromosome doubling to produce *T. aestivum*, which is a hexaploid ($6n = 42$).

Table 9.2 Examples of polyploid crop plants

Plant	Type of Polyploidy	Ploidy	Chromosome Number
Potato	Autopolyploid	$4n$	48
Banana	Autopolyploid	$3n$	33
Peanut	Autopolyploid	$4n$	40
Sweet potato	Autopolyploid	$6n$	90
Tobacco	Allopolyploid	$4n$	48
Cotton	Allopolyploid	$4n$	52
Wheat	Allopolyploid	$6n$	42
Oats	Allopolyploid	$6n$	42
Sugar cane	Allopolyploid	$8n$	80
Strawberry	Allopolyploid	$8n$	56

Source: After F. C. Elliot, *Plant Breeding and Cytogenetics* (New York: McGraw-Hill, 1958).

CONCEPTS

Polyploidy is the presence of extra chromosome sets: autopolyploids possess extra chromosome sets from the same species; allopolyploids possess extra chromosome sets from two or more species. Problems in chromosome pairing and segregation often lead to sterility in autopolyploids, but many allopolyploids are fertile.

✔ CONCEPT CHECK 7

Species A has $2n = 16$ chromosomes and species B has $2n = 14$. How many chromosomes would be found in an allotriploid of these two species?

a. 21 or 24   c. 22 or 23
b. 42 or 48   d. 45

9.5 Chromosome Variation Plays an Important Role in Evolution

Chromosome variations are potentially important in evolution and, within a number of different groups of organisms, have clearly played a significant role in past evolution. Chromosome duplications provide one way in which new genes evolve. In many cases, existing copies of a gene are not free to vary, because they encode a product that is essential to development or function. However, after a chromosome undergoes duplication, extra copies of genes within the duplicated region are present. The original copy can provide the essential function, whereas an extra copy from the duplication is free to undergo mutation and change. Over evolutionary time, the extra copy may acquire enough mutations to assume a new function that benefits the organism.

Inversions also can play important evolutionary roles by suppressing recombination among a set of genes. As we have seen, crossing over within an inversion in an individual

Table 9.3 Different types of chromosome mutations

Chromosome Mutation	Definition
Chromosome rearrangement	Change in chromosome structure
Chromosome duplication	Duplication of a chromosome segment
Chromosome deletion	Deletion of a chromosome segment
Inversion	Chromosome segment inverted 180 degrees
Paracentric inversion	Inversion that does not include the centromere in the inverted region
Pericentric inversion	Inversion that includes the centromere in the inverted region
Translocation	Movement of a chromosome segment to a nonhomologous chromosome or to another region of the same chromosome
Nonreciprocal translocation	Movement of a chromosome segment to a nonhomologous chromosome or to another region of the same chromosome without reciprocal exchange
Reciprocal translocation	Exchange between segments of nonhomologous chromosomes or between regions of the same chromosome
Aneuploidy	Change in number of individual chromosomes
Nullisomy	Loss of both members of a homologous pair
Monosomy	Loss of one member of a homologous pair
Trisomy	Gain of one chromosome, resulting in three homologous chromosomes
Tetrasomy	Gain of two homologous chromosomes, resulting in four homologous chromosomes
Polyploidy	Addition of entire chromosome sets
Autopolyploidy	Polyploidy in which extra chromosome sets are derived from the same species
Allopolyploidy	Polyploidy in which extra chromosome sets are derived from two or more species

heterozygous for a pericentric or paracentric inversion leads to unbalanced gametes and no recombinant progeny. This suppression of recombination allows particular sets of co-adapted alleles that function well together to remain intact, unshuffled by recombination.

Polyploidy, particularly allopolyploidy, often gives rise to new species and has been particularly important in the evolution of flowering plants. Occasional genome doubling through polyploidy has been a major contributor to evolutionary success in animal groups. For example, *Saccharomyces cerevisiae* (yeast) is a tetraploid, having undergone whole-genome duplication about 100 million years ago. The vertebrate genome has duplicated twice, once in the common ancestor to jawed vertebrates and again in the ancestor of fishes. Certain groups of vertebrates, such as some frogs and some fishes, have undergone additional polyploidy. Cereal plants have undergone several genome-duplication events. Different types of chromosome mutations are summarized in **Table 9.3**.

CONCEPTS SUMMARY

- Three basic types of chromosome mutations are: (1) chromosome rearrangements, which are changes in the structures of chromosomes; (2) aneuploidy, which is an increase or decrease in chromosome number; and (3) polyploidy, which is the presence of extra chromosome sets.
- Chromosome rearrangements include duplications, deletions, inversions, and translocations.
- In individuals heterozygous for a duplication, the duplicated region will form a loop when homologous chromosomes pair in meiosis. Duplications often have pronounced effects on the phenotype owing to unbalanced gene dosage.
- Segmental duplications are common in the human genome.
- In individuals heterozygous for a deletion, one of the chromosomes will loop out during pairing in meiosis. Deletions may cause recessive alleles to be expressed.
- Pericentric inversions include the centromere; paracentric inversions do not. In individuals heterozygous for an inversion, the homologous chromosomes form inversion loops in meiosis, with reduced recombination taking place within the inverted region.
- In translocation heterozygotes, the chromosomes form crosslike structures in meiosis, and the segregation of chromosomes produces unbalanced gametes.
- Fragile sites are constrictions or gaps that appear at particular regions on the chromosomes of cells grown in culture and are prone to breakage under certain conditions.
- Copy-number variations (CNVs) are differences in the number of copies of DNA sequences and include duplications and deletions. These variants are common in the human genome; some are associated with diseases and disorders.
- Nullisomy is the loss of two homologous chromosomes; monosomy is the loss of one homologous chromosome; trisomy is the addition of one homologous chromosome; tetrasomy is the addition of two homologous chromosomes.
- Aneuploidy usually causes drastic phenotypic effects because it leads to unbalanced gene dosage.
- Primary Down syndrome is caused by the presence of three full copies of chromosome 21, whereas familial Down syndrome is caused by the presence of two normal copies of chromosome 21 and a third copy that is attached to another chromosome through a translocation.
- Uniparental disomy is the presence of two copies of a chromosome from one parent and no copy from the other. Mosaicism is caused by nondisjunction in an early mitotic division that leads to different chromosome constitutions in different cells of a single individual.
- All the chromosomes in an autopolyploid derive from one species; chromosomes in an allopolyploid come from two or more species.
- Chromosome variations have played an important role in the evolution of many groups of organisms.

IMPORTANT TERMS

chromosome mutation (p. 240)
metacentric chromosome (p. 240)
submetacentric chromosome (p. 240)
acrocentric chromosome (p. 240)
telocentric chromosome (p. 240)
chromosome rearrangement (p. 242)
chromosome duplication (p. 242)
tandem duplication (p. 242)
displaced duplication (p. 242)
reverse duplication (p. 242)
segmental duplication (p. 244)

chromosome deletion (p. 244)
pseudodominance (p. 245)
haploinsufficient gene (p. 245)
chromosome inversion (p. 246)
paracentric inversion (p. 246)
pericentric inversion (p. 246)
position effect (p. 246)
dicentric chromatid (p. 247)
acentric chromatid (p. 247)
dicentric bridge (p. 247)
translocation (p. 248)

nonreciprocal translocation (p. 248)
reciprocal translocation (p. 249)
Robertsonian translocation (p. 249)
alternate segregation (p. 249)
adjacent-1 segregation (p. 249)
adjacent-2 segregation (p. 249)
fragile site (p. 251)
fragile-X syndrome (p. 251)
copy-number variation (CNV) (p. 252)
aneuploidy (p. 252)
polyploidy (p. 252)

nullisomy (p. 252)
monosomy (p. 252)
trisomy (p. 252)
tetrasomy (p. 252)
Down syndrome (trisomy 21) (p. 254)
primary Down syndrome (p. 254)

familial Down syndrome (p. 255)
translocation carrier (p. 255)
Edward syndrome (trisomy 18) (p. 256)
Patau syndrome (trisomy 13) (p. 256)
trisomy 8 (p. 256)
uniparental disomy (p. 257)

mosaicism (p. 257)
gynandromorph (p. 257)
autopolyploidy (p. 258)
allopolyploidy (p. 258)
unbalanced gametes (p. 259)
amphidiploidy (p 260)

ANSWERS TO CONCEPT CHECKS

1. a

2. Pseudodominance is the expression of a recessive mutation. It is produced when the dominant wild-type allele in a heterozygous individual is absent due to a deletion on one chromosome.

3. c

4. b

5. 37

6. Dosage compensation prevents the expression of additional copies of X-linked genes in mammals, and there is little information in the Y chromosome; so extra copies of the X and Y chromosomes do not have major effects on development. In contrast, there is no mechanism of dosage compensation for autosomes, and so extra copies of autosomal genes are expressed, upsetting development and causing the spontaneous abortion of aneuploid embryos.

7. c

WORKED PROBLEMS

1. A chromosome has the following segments, where • represents the centromere.

    A B C D E • F G

    What types of chromosome mutations are required to change this chromosome into each of the following chromosomes? (In some cases, more than one chromosome mutation may be required.)

    a. A B E • F G
    b. A E D C B • F G
    c. A B A B C D E • F G
    d. A F • E D C B G
    e. A B C D E E D C • F G

• **Solution**

The types of chromosome mutations are identified by comparing the mutated chromosome with the original, wild-type chromosome.

a. The mutated chromosome (A B E • F G) is missing segment C D; so this mutation is a deletion.

b. The mutated chromosome (A E D C B • F G) has one and only one copy of all the gene segments, but segment B C D E has been inverted 180 degrees. Because the centromere has not changed location and is not in the inverted region, this chromosome mutation is a paracentric inversion.

c. The mutated chromosome (A B A B C D E • F G) is longer than normal, and we see that segment A B has been duplicated. This mutation is a tandem duplication.

d. The mutated chromosome (A F • E D C B G) is normal length, but the gene order and the location of the centromere have changed; this mutation is therefore a pericentric inversion of region (B C D E • F).

e. The mutated chromosome (A B C D E E D C • F G) contains a duplication (C D E) that is also inverted; so this chromosome has undergone a duplication and a paracentric inversion.

2. Species I is diploid ($2n = 4$) with chromosomes AABB; related species II is diploid ($2n = 6$) with chromosomes MMNNOO. Give the chromosomes that would be found in individuals with the following chromosome mutations.

    a. Autotriploidy in species I
    b. Allotetraploidy including species I and II
    c. Monosomy in species I
    d. Trisomy in species II for chromosome M
    e. Tetrasomy in species I for chromosome A
    f. Allotriploidy including species I and II
    g. Nullisomy in species II for chromosome N

• **Solution**

To work this problem, we should first determine the haploid genome complement for each species. For species I, $n = 2$ with chromosomes AB and, for species II, $n = 3$ with chromosomes MNO.

a. An autotriploid is $3n$, with all the chromosomes coming from a single species; so an autotriploid of species I would have chromosomes AAABBB ($3n = 6$).

b. An allotetraploid is $4n$, with the chromosomes coming from more than one species. An allotetraploid could consist of $2n$ from species I and $2n$ from species II, giving the allotetraploid

($4n = 2 + 2 + 3 + 3 = 10$) chromosomes AABBMMNNOO. An allotetraploid could also possess $3n$ from species I and $1n$ from species II ($4n = 2 + 2 + 2 + 3 = 9$; AAABBBMNO) or $1n$ from species I and $3n$ from species II ($4n = 2 + 3 + 3 + 3 = 11$; ABMMMNNNOOO).

c. A monosomic is missing a single chromosome; so a monosomic for species I would be $2n - 1 = 4 - 1 = 3$. The monosomy might include either of the two chromosome pairs, with chromosomes ABB or AAB.

d. Trisomy requires an extra chromosome; so a trisomic of species II for chromosome M would be $2n + 1 = 6 + 1 = 7$ (MMMNNOO).

e. A tetrasomic has two extra homologous chromosomes; so a tetrasomic of species I for chromosome A would be $2n + 2 = 4 + 2 = 6$ (AAAABB).

f. An allotriploid is $3n$ with the chromosomes coming from two different species; so an allotriploid could be $3n = 2 + 2 + 3 = 7$ (AABBMNO) or $3n = 2 + 3 + 3 = 8$ (ABMMNNOO).

g. A nullisomic is missing both chromosomes of a homologous pair; so a nullisomic of species II for chromosome N would be $2n - 2 = 6 - 2 = 4$ (MMOO).

COMPREHENSION QUESTIONS

Section 9.1

*1. List the different types of chromosome mutations and define each one.

Section 9.2

*2. Why do extra copies of genes sometimes cause drastic phenotypic effects?

3. Draw a pair of chromosomes as they would appear during synapsis in prophase I of meiosis in an individual heterozygous for a chromosome duplication.

4. What is haploinsufficiency?

*5. What is the difference between a paracentric and a pericentric inversion?

6. How do inversions cause phenotypic effects?

7. Explain, with the aid of a drawing, how a dicentric bridge is produced when crossing over takes place in an individual heterozygous for a paracentric inversion.

8. Explain why recombination is suppressed in individuals heterozygous for paracentric and pericentric inversions.

*9. How do translocations produce phenotypic effects?

10. Sketch the chromosome pairing and the different segregation patterns that can arise in an individual heterozygous for a reciprocal translocation.

11. What is a Robertsonian translocation?

Section 9.3

12. List four major types of aneuploidy.

*13. What is the difference between primary Down syndrome and familial Down syndrome? How does each type arise?

*14. What is uniparental disomy and how does it arise?

15. What is mosaicism and how does it arise?

Section 9.4

*16. What is the difference between autopolyploidy and allopolyploidy? How does each arise?

17. Explain why autopolyploids are usually sterile, whereas allopolyploids are often fertile.

APPLICATION QUESTIONS AND PROBLEMS

Section 9.1

*18. Which types of chromosome mutations
  a. increase the amount of genetic material in a particular chromosome?
  b. increase the amount of genetic material in all chromosomes?
  c. decrease the amount of genetic material in a particular chromosome?
  d. change the position of DNA sequences in a single chromosome without changing the amount of genetic material?
  e. move DNA from one chromosome to a nonhomologous chromosome?

Section 9.2

*19. A chromosome has the following segments, where • represents the centromere:

$$A\ B \cdot C\ D\ E\ F\ G$$

What types of chromosome mutations are required to change this chromosome into each of the following chromosomes? (In some cases, more than one chromosome mutation may be required.)

  a. $\underline{A\ B\ A\ B \cdot C\ D\ E\ F\ G}$
  b. $\underline{A\ B \cdot C\ D\ E\ A\ B\ F\ G}$
  c. $\underline{A\ B \cdot C\ F\ E\ D\ G}$
  d. $\underline{A \cdot C\ D\ E\ F\ G}$

e. A B · C D E
f. A B · E D C F G
g. C · B A D E F G
h. A B · C F E D F E D G
i. A B · C D E F C D F E G

20. A chromosome initially has the following segments:

    A B · C D E F G

    Draw the chromosome, identifying its segments, that would result from each of the following mutations.
    a. Tandem duplication of DEF
    b. Displaced duplication of DEF
    c. Deletion of FG
    d. Paracentric inversion that includes DEFG
    e. Pericentric inversion of BCDE

21. The following diagram represents two nonhomologous chromosomes:

    A B · C D E F G
    R S · T U V W X

    What type of chromosome mutation would produce each of the following chromosomes?
    a. A B · C D
       R S · T U V W X E F G
    b. A U V B · C D E F G
       R S · T W X
    c. A B · T U V F G
       R S · C D E W X
    d. A B · C W G
       R S · T U V D E F X

*22. The *Notch* mutation is a deletion on the X chromosome of *Drosophila melanogaster*. Female flies heterozygous for *Notch* have an indentation on the margins of their wings; *Notch* is lethal in the homozygous and hemizygous conditions. The *Notch* deletion covers the region of the X chromosome that contains the locus for white eyes, an X-linked recessive trait. Give the phenotypes and proportions of progeny produced in the following crosses.
   a. A red-eyed, Notch female is mated with a white-eyed male.
   b. A white-eyed, Notch female is mated with a red-eyed male.
   c. A white-eyed, Notch female is mated with a white-eyed male.

23. The green-nose fly normally has six chromosomes, two metacentric and four acrocentric. A geneticist examines the chromosomes of an odd-looking green-nose fly and discovers that it has only five chromosomes; three of them are metacentric and two are acrocentric. Explain how this change in chromosome number might have taken place.

*24. A wild-type chromosome has the following segments:

    A B C · D E F G H I

    An individual is heterozygous for the following chromosome mutations. For each mutation, sketch how the wild-type and mutated chromosomes would pair in prophase I of meiosis, showing all chromosome strands.
    a. A B C · D E F D E F G H I
    b. A B C · D H I
    c. A B C · D G F E H I
    d. A B E D · C F G H I

25. Draw the chromatids that would result from a two-strand double crossover between E and F in Problem 24.

26. As discussed in this chapter, crossing over within a pericentric inversion produces chromosomes that have extra copies of some genes and no copies of other genes. The fertilization of gametes containing such duplication or deficient chromosomes often results in children with syndromes characterized by developmental delay, mental retardation, abnormal development of organ systems, and early death. Maarit Jaarola and colleagues examined individual sperm cells of a male who was heterozygous for a pericentric inversion on chromosome 8 and determined that crossing over took place within the pericentric inversion in 26% of the meiotic divisions (M. Jaarola, R. H. Martin, and T. Ashley. 1998. *American Journal of Human Genetics* 63:218–224).

    Assume that you are a genetic counselor and that a couple seeks counseling from you. Both the man and the woman are phenotypically normal, but the woman is heterozygous for a pericentric inversion on chromosome 8. The man is karyotypically normal. What is the probability that this couple will produce a child with a debilitating syndrome as the result of crossing over within the pericentric inversion?

*27. An individual heterozygous for a reciprocal translocation possesses the following chromosomes:

    A B · C D E F G
    A B · C D V W X
    R S · T U E F G
    R S · T U V W X

   a. Draw the pairing arrangement of these chromosomes in prophase I of meiosis.
   b. Diagram the alternate, adjacent-1, and adjacent-2 segregation patterns in anaphase I of meiosis.
   c. Give the products that result from alternate, adjacent-1, and adjacent-2 segregation.

Section 9.3

28. Red–green color blindness is a human X-linked recessive disorder. A young man with a 47,XXY karyotype

(Klinefelter syndrome) is color blind. His 46,XY brother also is color blind. Both parents have normal color vision. Where did the nondisjunction that gave rise to the young man with Klinefelter syndrome take place?

29. Junctional epidermolysis bullosa (JEB) is a severe skin disorder that results in blisters over the entire body. The disorder is caused by autosomal recessive mutations at any one of three loci that help to encode laminin 5, a major component in the dermal–epidermal basement membrane. Leena Pulkkinen and colleagues described a male newborn who was born with JEB and died at 2 months of age (L. Pulkkinen et al. 1997. *American Journal of Human Genetics* 61:611–619); the child had healthy unrelated parents. Chromosome analysis revealed that the infant had 46 normal-appearing chromosomes. Analysis of DNA showed that his mother was heterozygous for a JEB-causing allele at the *LAMB3* locus, which is on chromosome 1. The father had two normal alleles at this locus. DNA fingerprinting demonstrated that the male assumed to be the father had, in fact, conceived the child.

   a. Assuming that no new mutations occurred in this family, explain the presence of an autosomal recessive disease in the child when the mother is heterozygous and the father is homozygous normal.
   b. How might you go about proving your explanation? Assume that a number of genetic markers are available for each chromosome.

30. Some people with Turner syndrome are 45,X/46,XY mosaics. Explain how this mosaicism could arise.

*31. Bill and Betty have had two children with Down syndrome. Bill's brother has Down syndrome and his sister has two children with Down syndrome. On the basis of these observations, indicate which of the following statements are most likely correct and which are most likely incorrect. Explain your reasoning.

   a. Bill has 47 chromosomes.
   b. Betty has 47 chromosomes.
   c. Bill and Betty's children each have 47 chromosomes.
   d. Bill's sister has 45 chromosomes.
   e. Bill has 46 chromosomes.
   f. Betty has 45 chromosomes.
   g. Bill's brother has 45 chromosomes.

*32. In mammals, sex-chromosome aneuploids are more common than autosomal aneuploids but, in fish, sex-chromosome aneuploids and autosomal aneuploids are found with equal frequency. Explain these differences in mammals and fish.

*33. A young couple is planning to have children. Knowing that there have been a substantial number of stillbirths, miscarriages, and fertility problems on the husband's side of the family, they see a genetic counselor. A chromosome analysis reveals that, whereas the woman has a normal karyotype, the man possesses only 45 chromosomes and is a carrier of a Robertsonian translocation between chromosomes 22 and 13.

   a. List all the different types of gametes that might be produced by the man.
   b. What types of zygotes will develop when each of gametes produced by the man fuses with a normal gamete produced by the woman?
   c. If trisomies and monosomies entailing chromosomes 13 and 22 are lethal, what proportion of the surviving offspring will be carriers of the translocation?

34. Using breeding techniques, Andrei Dyban and V. S. Baranov (*Cytogenetics of Mammalian Embryonic Development*. Oxford: Oxford University Press, Clarendon Press; New York: Oxford University Press, 1987) created mice that were trisomic for each of the different mouse chromosomes. They found that only mice with trisomy 19 developed. Mice trisomic for all other chromosomes died in the course of development. For some of these trisomics, they compared the length of development (number of days after conception before the embryo died) as a function of the size of the mouse chromosome that was present in three copies (see the adjoining graph). Summarize their findings as presented in this graph and provide a possible explanation for the results.

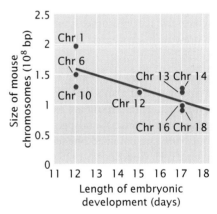

[E. Torres, B. R. Williams, and A. Amon. 2008. *Genetics* 179:737–746, Fig. 2B.]

Section 9.4

*35. Species I has $2n = 16$ chromosomes. How many chromosomes will be found per cell in each of the following mutants in this species?

   a. Monosomic
   b. Autotriploid
   c. Autotetraploid
   d. Trisomic
   e. Double monosomic
   f. Nullisomic
   g. Autopentaploid
   h. Tetrasomic

36. Species I is diploid ($2n = 8$) with chromosomes AABBCCDD; related species II is diploid ($2n = 8$) with chromosomes MMNNOOPP. What types of chromosome mutations do individual organisms with the following sets of chromosomes have?

a. AAABBCCDD
b. MMNNOOOOPP
c. AABBCDD
d. AAABBBCCCDDD
e. AAABBCCDDD
f. AABBDD
g. AABBCCDDMMNNOOPP
h. AABBCCDDMNOP

37. Species I has $2n = 8$ chromosomes and species II has $2n = 14$ chromosomes. What would the expected chromosome numbers be in individual organisms with the following chromosome mutations? Give all possible answers.

    a. Allotriploidy including species I and II
    b. Autotetraploidy in species II
    c. Trisomy in species I
    d. Monosomy in species II
    e. Tetrasomy in species I
    f. Allotetraploidy including species I and II

38. Consider a diploid cell that has $2n = 4$ chromosomes—one pair of metacentric chromosomes and one pair of acrocentric chromosomes. Suppose this cell undergoes nondisjunction giving rise to an autotriploid cell ($3n$). The triploid cell then undergoes meiosis. Draw the different types of gametes that may result from meiosis in the triploid cell, showing the chromosomes present in each type. To distinguish between the different metacentric and acrocentric chromosomes, use a different color to draw each metacentric chromosome; similarly, use a different color to draw each acrocentric chromosome. (Hint: See Figure 9.27).

39. *Nicotiana glutinosa* ($2n = 24$) and *N. tabacum* ($2n = 48$) are two closely related plants that can be intercrossed, but the F_1 hybrid plants that result are usually sterile. In 1925, Roy Clausen and Thomas Goodspeed crossed *N. glutinosa* and *N. tabacum* and obtained one fertile F_1 plant (R. E. Clausen and T. H. Goodspeed. 1925 *Genetics* 10:278–284). They were able to self-pollinate the flowers of this plant to produce an F_2 generation. Surprisingly, the F_2 plants were fully fertile and produced viable seed. When Clausen and Goodspeed examined the chromosomes of the F_2 plants, they observed 36 pairs of chromosomes in metaphase I and 36 individual chromosomes in metaphase II. Explain the origin of the F_2 plants obtained by Clausen and Goodspeed and the numbers of chromosomes observed.

40. What would be the chromosome number of progeny resulting from the following crosses in wheat (see Figure 9.29)? What type of polyploid (allotriploid, allotetraploid, etc.) would result from each cross?

    a. Einkorn wheat and emmer wheat
    b. Bread wheat and emmer wheat
    c. Einkorn wheat and bread wheat

41. Karl and Hally Sax crossed *Aegilops cylindrica* ($2n = 28$), a wild grass found in the Mediterranean region, with *Triticum vulgare* ($2n = 42$), a type of wheat (K. Sax and H. J. Sax. 1924. *Genetics* 9:454–464). The resulting F_1 plants from this cross had 35 chromosomes. Examination of metaphase I in the F_1 plants revealed the presence of 7 pairs of chromosomes (bivalents) and 21 unpaired chromosomes (univalents).

    a. If the unpaired chromosomes segregate randomly, what possible chromosome numbers will appear in the gametes of the F_1 plants?
    b. What does the appearance of the bivalents in the F_1 hybrids suggest about the origin of *Triticum vulgare* wheat?

Aegilops cylindrica, jointed goatgrass. [Sam Brinker, OMNR-NHIC, 2008. Canadian Food Inspection Agency.]

Triticum vulgare, wheat. [Michael Hieber/123RE.com.]

CHALLENGE QUESTIONS

Section 9.3

42. Red–green color blindness is a human X-linked recessive disorder. Jill has normal color vision, but her father is color blind. Jill marries Tom, who also has normal color vision. Jill and Tom have a daughter who has Turner syndrome and is color blind.

    a. How did the daughter inherit color blindness?
    b. Did the daughter inherit her X chromosome from Jill or from Tom?

43. Progeny of triploid tomato plants often contain parts of an extra chromosome, in addition to the normal complement of 24 chromosomes (J. W. Lesley and M. M. Lesley. 1929. *Genetics* 14:321–336). Mutants with a part of an extra chromosome are referred to as secondaries. James and Margaret Lesley observed that secondaries arise from triploid ($3n$), trisomic ($3n + 1$), and double trisomic ($3n + 1 + 1$) parents, but never from diploids ($2n$). Give one or more possible reasons that secondaries arise from parents that have unpaired chromosomes but not from parents that are normal diploids.

44. Mules result from a cross between a horse ($2n = 64$) and a donkey ($2n = 62$), have 63 chromosomes, and are almost always sterile. However, in the summer of 1985, a female mule named Krause who was pastured with a male donkey gave birth to a newborn foal (O. A. Ryder et al. 1985. *Journal of Heredity* 76:379–381). Blood tests established that the male foal, appropriately named Blue Moon, was the offspring of Krause and that Krause was indeed a mule. Both Blue Moon and Krause were fathered by the same donkey (see the illustration). The foal, like his mother, had 63 chromosomes—half of them horse chromosomes and the other half donkey chromosomes. Analyses of genetic markers showed that, remarkably, Blue Moon seemed to have inherited a complete set of horse chromosomes from his mother, instead of the random mixture of horse and donkey chromosomes that would be expected with normal meiosis. Thus, Blue Moon and Krause were not only mother and son, but also brother and sister.

   a. With the use of a diagram, show how, if Blue Moon inherited only horse chromosomes from his mother, Blue Moon and Krause are both mother and son as well as brother and sister.

   b. Although rare, additional cases of fertile mules giving births to offspring have been reported. In these cases, when a female mule mates with a male horse, the offspring is horselike in appearance but, when a female mule mates with a male donkey, the offspring is mulelike in appearance. Is this observation consistent with the idea that the offspring of fertile female mules inherit only a set of horse chromosomes from their mule mothers? Explain your reasoning.

   c. Can you suggest a possible mechanism for how the offspring of fertile female mules might pass on a complete set of horse chromosomes to their offspring?

Section 9.5

45. Humans and many other complex organisms are diploid, possessing two sets of genes, one inherited from the mother and one from the father. However, a number of eukaryotic organisms spend most of their life cycles in a haploid state. Many of these eukaryotes, such as *Neurospora* and yeast, still undergo meiosis and sexual reproduction, but most of the cells that make up the organism are haploid.

    Considering that haploid organisms are fully capable of sexual reproduction and generating genetic variation, why are most complex eukaryotes diploid? In other words, what might be the evolutionary advantage of existing in a diploid state instead of a haploid state? And why might a few organisms, such as *Neurospora* and yeast, exist as haploids?

10 Gene Mutations and DNA Repair

People with a mutation in the *tinman* gene, named after Tin Man in *The Wizard of Oz*, often have congenital heart defects. [Mary Evans Picture Library/Alamy.]

A FLY WITHOUT A HEART

The heart of a fruit fly is a simple organ, an open-ended tube that rhythmically contracts, pumping fluid—rather inefficiently—around the body of the fly. Although simple and inelegant, the fruit fly's heart is nevertheless essential, at least for a fruit fly. Remarkably, a few rare mutant fruit flies never develop a heart and die (not surprisingly) at an early embryonic stage. Geneticist Rolf Bodmer analyzed these mutants in the 1980s and made an important discovery—a gene that specifies the development of a heart. He named the gene *tinman*, after the character in *The Wizard of Oz* who also lacked a heart. Bodmer's research revealed that *tinman* encodes a transcription factor, which binds to DNA and turns on other genes that are essential for the normal development of a heart. In the mutant flies studied by Bodmer, this gene was lacking, the transcription factor was never produced, and the heart never developed. Findings from subsequent research revealed the existence of a human gene (called *Nkx2.5*) with a sequence similar to that of *tinman*, but the function of the human gene was unknown.

Then, in the 1990s, physicians Jonathan and Christine Seidman began studying people born with abnormal hearts, such as those with a hole in the septum that separates the chambers on the left and right sides of the heart. Such defects result in abnormal blood flow through the heart, causing the heart to work harder than normally and the mixing of oxygenated and deoxygenated blood. Congenital heart defects are not uncommon; they're found in about 1 of every 125 babies. Some of the defects heal on their own, but others require corrective surgery. Although surgery is often successful in reversing congenital heart problems, many of these patients begin to have irregular heartbeats in their 20s and 30s. The Seidmans and their colleagues found several families in which congenital heart defects and irregular heart beats were inherited together in an autosomal dominant fashion. Detailed molecular analysis of one of these families revealed that the gene responsible for the heart problems was located on chromosome 5, at a spot where the human *tinman* gene (*Nkx2.5*) had been previously mapped. All members of this family who inherited the heart defects also inherited a mutation in the *tinman* gene. Subsequent studies with additional patients found that many people with congenital heart defects have a mutation in the *tinman* gene. The human version of this gene, like its counterpart in flies, encodes a transcription factor that controls heart development. Despite tremendous differences in size, anatomy, and physiology, humans and flies use the same gene to make a heart.

The story of *tinman* illustrates the central importance of studying mutations: the analysis of mutants is often a source of key insights into important biological processes. This chapter focuses on gene mutations—how errors arise in genetic instructions and how those errors are studied. We begin with a brief examination of the different types of mutations, including their phenotypic effects, how they can be suppressed, and mutation rates. The next section explores how mutations spontaneously arise in the course of DNA replication and afterward, as well as how chemicals and radiation induce mutations. We then consider the analysis of mutations. Finally, we take a look at DNA repair and some of the diseases that arise when DNA repair is defective.

10.1 Mutations Are Inherited Alterations in the DNA Sequence

DNA is a highly stable molecule that is replicated with amazing accuracy, but changes in DNA structure and errors of replication do occur. A **mutation** is defined as an inherited change in genetic information; the descendants may be cells or organisms.

The Importance of Mutations

Mutations are both the sustainer of life and the cause of great suffering. On the one hand, mutation is the source of all genetic variation, the raw material of evolution. The ability of organisms to adapt to environmental change depends critically on the presence of genetic variation in natural populations, and genetic variation is produced—at least partly—by mutation. On the other hand, many mutations have detrimental effects, and mutation is the source of many diseases and disorders.

Much of the study of genetics focuses on how variants produced by mutation are inherited; genetic crosses are meaningless if all individual members of a species are identically homozygous for the same alleles. Much of Gregor Mendel's success in unraveling the principles of inheritance can be traced to his use of carefully selected variants of the garden pea; similarly, Thomas Hunt Morgan and his students discovered many basic principles of genetics by analyzing mutant fruit flies.

Mutations are also useful for probing fundamental biological processes. Finding or creating mutations that affect different components of a biological system and studying their effects can often lead to an understanding of the system. This method, referred to as genetic dissection, is analogous to figuring out how an automobile works by breaking different parts of a car and observing the effects; for example, smash the radiator and the engine overheats, revealing that the radiator cools the engine. Using mutations to disrupt function can likewise be a source of insight into biological processes. For example, geneticists have begun to unravel the molecular details of development by studying mutations, such as *tinman*, that interrupt various embryonic stages in *Drosophila*. Although this method of breaking "parts" to determine their function might seem like a crude approach to understanding a system, it is actually very powerful and has been used extensively in biochemistry, developmental biology, physiology, and behavioral science (but this method is *not* recommended for learning how your car works).

> **CONCEPTS**
>
> Mutations are heritable changes in DNA. They are essential to the study of genetics and are useful in many other biological fields.
>
> ✔ **CONCEPT CHECK 1**
>
> How are mutations used to help in understanding basic biological processes?

Categories of Mutations

In multicellular organisms, we can distinguish between two broad categories of mutations: somatic mutations and germ-line mutations. **Somatic mutations** arise in somatic tissues, which do not produce gametes (**Figure 10.1**). When

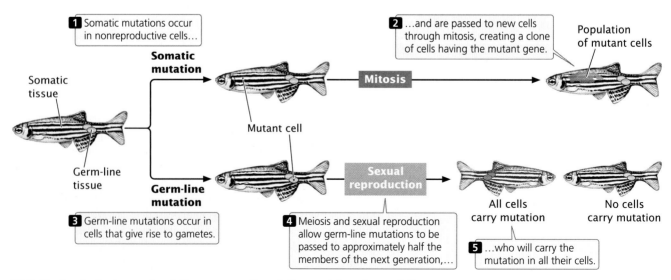

10.1 The two basic classes of mutations are somatic mutations and germ-line mutations.

10.2 Three basic types of gene mutations are base substitutions, insertions, and deletions.

a somatic cell with a mutation divides (mitosis), the mutation is passed on to the daughter cells, leading to a population of genetically identical cells (a clone). The earlier in development that a somatic mutation occurs, the larger the clone of cells within that individual organism that will contain the mutation.

Because of the huge number of cells present in a typical eukaryotic organism, somatic mutations are numerous. For example, there are about 10^{14} cells in the human body. Typically, a mutation arises once in every million cell divisions, and so hundreds of millions of somatic mutations must arise in each person. Many somatic mutations have no obvious effect on the phenotype of the organism, because the function of the mutant cell (even the cell itself) is replaced by that of normal cells. However, cells with a somatic mutation that stimulates cell division can increase in number and spread; this type of mutation can give rise to cells with a selective advantage and is the basis for cancers.

Germ-line mutations arise in cells that ultimately produce gametes. A germ-line mutation can be passed to future generations, producing individual organisms that carry the mutation in all their somatic and germ-line cells (see Figure 10.1). When we speak of mutations in multicellular organisms, we're usually talking about germ-line mutations.

Historically, mutations have been partitioned into those that affect a single gene, called *gene mutations*, and those that affect the number or structure of chromosomes, called *chromosome mutations*. This distinction arose because chromosome mutations could be observed directly, by looking at chromosomes with a microscope, whereas gene mutations could be detected only by observing their phenotypic effects. Now, DNA sequencing allows direct observation of gene mutations, and chromosome mutations are distinguished from gene mutations somewhat arbitrarily on the basis of the size of the DNA lesion. Nevertheless, it is practical to use the term *chromosome mutation* for a large-scale genetic alteration that affects chromosome structure or the number of chromosomes and to use the term **gene mutation** for a relatively small DNA lesion that affects a single gene. This chapter focuses on gene mutations; chromosome mutations were discussed in Chapter 9.

Types of Gene Mutations

There are a number of ways to classify gene mutations. Some classification schemes are based on the nature of the phenotypic effect, others are based on the causative agent of the mutation, and still others focus on the molecular nature of the defect. Here, we will categorize mutations primarily on the basis of their molecular nature, but we will also encounter some terms that relate the causes and the phenotypic effects of mutations.

Base substitutions The simplest type of gene mutation is a **base substitution**, the alteration of a single nucleotide in the DNA (**Figure 10.2a**). Base substitutions are of two types. In a **transition**, a purine is replaced by a different purine or, alternatively, a pyrimidine is replaced by a different pyrimidine (**Figure 10.3**). In a **transversion**, a purine is replaced by a pyrimidine or a pyrimidine is replaced by a purine. The number of possible transversions (see Figure 10.3) is twice the number of possible transitions, but transitions arise more frequently. TRY PROBLEM 18

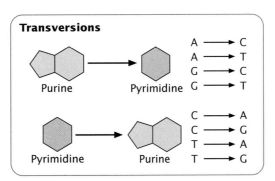

10.3 A transition is the substitution of a purine for a purine or of a pyrimidine for a pyrimidine; a transversion is the substitution of a pyrimidine for a purine or of a purine for a pyrimidine.

Insertions and deletions The second major class of gene mutations contains **insertions** and **deletions**—the addition or the removal, respectively, of one or more nucleotide pairs (collectively called indels, **Figure 10.2b** and **c**). Although base substitutions are often assumed to be the most common type of mutation, molecular analysis has revealed that insertions and deletions are often more frequent. Insertions and deletions within sequences that encode proteins may lead to **frameshift mutations**, changes in the reading frame of the gene. Frameshift mutations usually alter all amino acids encoded by nucleotides following the mutation, and so they generally have drastic effects on the phenotype. Not all insertions and deletions lead to frameshifts, however; insertions and deletions consisting of any multiple of three nucleotides will leave the reading frame intact, although the addition or removal of one or more amino acids may still affect the phenotype. Mutations not affecting the reading frame are called **in-frame insertions** and **deletions**, respectively.

10.4 The fragile-X chromosome is associated with a characteristic constriction (fragile site) on the long arm.
[Visuals Unlimited.]

CONCEPTS

Gene mutations consist of changes in a single gene and can be base substitutions (a single pair of nucleotides is altered) or insertions or deletions (nucleotides are added or removed). A base substitution can be a transition (substitution of like bases) or a transversion (substitution of unlike bases). Insertions and deletions often lead to a change in the reading frame of a gene.

✔ CONCEPT CHECK 2

Which of the following changes is a transition base substitution?

a. Adenine is replaced by thymine.
b. Cytosine is replaced by adenine.
c. Guanine is replaced by adenine.
d. Three nucleotide pairs are inserted into DNA.

Expanding nucleotide repeats Mutations in which the number of copies of a set of nucleotides increases in number are called **expanding nucleotide repeats**. This type of mutation was first observed in 1991 in a gene called *FMR-1*, which causes fragile-X syndrome, the most common hereditary cause of mental retardation. The disorder is so named because, in specially treated cells from persons having the condition, the tip of each long arm of the X chromosome is attached by only a slender thread (**Figure 10.4**). The normal *FMR-1* allele (not containing the mutation) has 60 or fewer copies of CGG but, in persons with fragile-X syndrome, the allele may harbor hundreds or even thousands of copies.

Expanding nucleotide repeats have been found in almost 30 human diseases, several of which are listed in **Table 10.1**. Most of these diseases are caused by the expansion of a set of three nucleotides (called a trinucleotide), most often CNG, where N can be any nucleotide. However,

Table 10.1 Examples of genetic diseases caused by expanding nucleotide repeats

Disease	Repeated Sequence	Number of Copies of Repeat	
		Normal Range	Disease Range
Spinal and bulbar muscular atrophy	CAG	11–33	40–62
Fragile-X syndrome	CGG	6–54	50–1500
Jacobsen syndrome	CGG	11	100–1000
Spinocerebellar ataxia (several types)	CAG	4–44	21–130
Autosomal dominant cerebellar ataxia	CAG	7–19	37–220
Myotonic dystrophy	CTG	5–37	44–3000
Huntington disease	CAG	9–37	37–121
Friedreich ataxia	GAA	6–29	200–900
Dentatorubral-pallidoluysian atrophy	CAG	7–25	49–75
Myoclonus epilepsy of the Unverricht–Lundborg type	CCCCGCCCCGCG	2–3	12–13

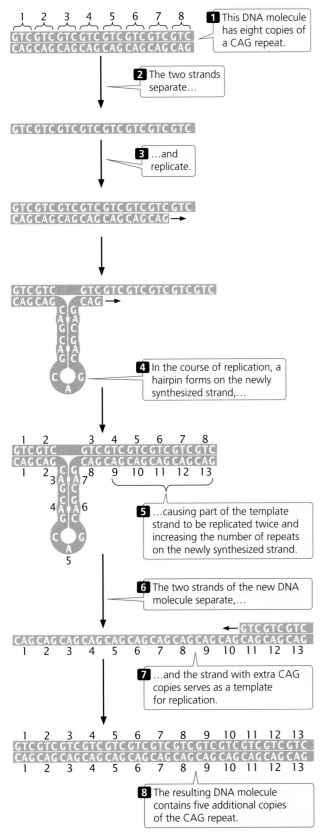

10.5 A model of how the number of copies of a nucleotide repeat may increase in replication.

www.whfreeman.com/pierce4e Try Animation 18.1 to see how copies of a nucleotide repeat can increase.

some diseases are caused by repeats of four, five, and even twelve nucleotides. The number of copies of the nucleotide repeat often correlates with the severity or age of onset of the disease. The number of copies of the repeat also corresponds to the instability of nucleotide repeats: when more repeats are present, the probability of expansion to even more repeats increases. This association between the number of copies of nucleotide repeats, the severity of the disease, and the probability of expansion leads to a phenomenon known as anticipation (see pp. 122–123 in Chapter 5), in which diseases caused by nucleotide-repeat expansions become more severe in each generation. Less commonly, the number of nucleotide repeats may decrease within a family. Expanding nucleotide repeats have also been observed in some microbes and plants.

Increases in the number of nucleotide repeats can produce disease symptoms in different ways. In several of the diseases (e.g., Huntington disease), the nucleotide expands within the coding part of a gene, producing a toxic protein that has extra glutamine residues (the amino acid encoded by CAG). In other diseases, the repeat is outside the coding region of a gene and affects its expression. In fragile-X syndrome, additional copies of the nucleotide repeat cause the DNA to become methylated, which turns off the transcription of an essential gene.

The mechanism that leads to the expansion of nucleotide repeats is not completely understood. Evidence suggests that expansion occurs in the course of DNA replication and appears to be related to the formation of hairpins and other special DNA structures that form in single-stranded DNA consisting of nucleotide repeats. Such structures may interfere with normal replication by causing strand slippage, misalignment of the sequences, or stalling of replication. One model of how repeat hairpins might result in repeat expansion is shown in **Figure 10.5**.

CONCEPTS

Expanding nucleotide repeats are regions of DNA that consist of repeated copies of sets of nucleotides. Increased numbers of nucleotide repeats are associated with several genetic diseases.

Phenotypic Effects of Mutations

Another way that mutations are classified is on the basis of their phenotypic effects. At the most general level, we can distinguish a mutation on the basis of its phenotype compared with the wild-type phenotype. A mutation that alters the wild-type phenotype is called a **forward mutation**, whereas a **reverse mutation** (a *reversion*) changes a mutant phenotype back into the wild type.

Geneticists use other terms to describe the effects of mutations on protein structure. A base substitution that results in a different amino acid in the protein is referred to as a **missense mutation** (**Figure 10.6a**). A **nonsense mutation**

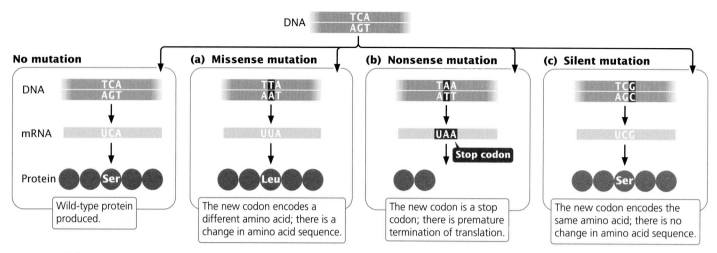

10.6 Base substitutions can cause (a) missense, (b) nonsense, and (c) silent mutations.

changes a sense codon (one that specifies an amino acid) into a nonsense codon (one that terminates translation), as shown in **Figure 10.6b**. If a nonsense mutation occurs early in the mRNA sequence, the protein will be greatly shortened and will usually be nonfunctional.

Because of the redundancy of the genetic code, some different codons specify the same amino acid. A **silent mutation** changes a codon to a synonymous codon that specifies the same amino acid (**Figure 10.6c**), altering the DNA sequence without changing the amino acid sequence of the protein. Not all silent mutations, however, are truly silent: some do have phenotypic effects. For example, silent mutations may have phenotypic effects when different tRNAs are used for different synonymous codons. Because some isoaccepting tRNAs are more abundant than others, which synonymous codon is used may affect the rate of protein synthesis. The rate of protein synthesis can influence the phenotype by affecting the amount of protein present in the cell and, in a few cases, the folding of the protein. Other silent mutations may alter sequences near the exon–intron junctions that affect splicing. Still other silent mutations may influence the folding of the mRNA, affecting its stability.

A **neutral mutation** is a missense mutation that alters the amino acid sequence of the protein but does not change its function. Neutral mutations occur when one amino acid is replaced by another that is chemically similar or when the affected amino acid has little influence on protein function. For example, neutral mutations occur in the genes that encode hemoglobin; although these mutations alter the amino acid sequence of hemoglobin, they do not affect its ability to transport oxygen.

Loss-of-function mutations cause the complete or partial absence of normal protein function. A loss-of-function mutation so alters the structure of the protein that the protein no longer works correctly or the mutation can occur in regulatory regions that affect the transcription, translation, or splicing of the protein. Loss-of-function mutations are frequently recessive, and an individual diploid organism must be homozygous for a loss-of-function mutation before the effects of the loss of the functional protein can be exhibited. The mutations that cause cystic fibrosis are loss-of-function mutations: these mutations produce a nonfunctional form of the cystic fibrosis transmembrane conductance regulator protein, which normally regulates the movement of chloride ions into and out of the cell (see Chapter 5).

In contrast, a **gain-of-function mutation** produces an entirely new trait or it causes a trait to appear in an inappropriate tissue or at an inappropriate time in development. For example, a mutation in a gene that encodes a receptor for a growth factor might cause the mutated receptor to stimulate growth all the time, even in the absence of the growth factor. Gain-of-function mutations are frequently dominant in their expression. Still other types of mutations are **conditional mutations**, which are expressed only under certain conditions. For example, some conditional mutations affect the phenotype only at elevated temperatures. Others are **lethal mutations**, causing premature death (Chapter 5).

TRY PROBLEM 21 →

Suppressor Mutations

A **suppressor mutation** is a genetic change that hides or suppresses the effect of another mutation. This type of mutation is distinct from a reverse mutation, in which the mutated site changes back into the original wild-type sequence (**Figure 10.7**). A suppressor mutation occurs at a site that is distinct from the site of the original mutation; thus, an individual with a suppressor mutation is a double mutant, possessing both the original mutation and the suppressor mutation but exhibiting the phenotype of an unmutated wild type. Like other mutations, suppressors arise randomly. Geneticists distinguish between two classes of suppressor mutations: intragenic and intergenic.

10.7 Relation of forward, reverse, and suppressor mutations.

Intragenic suppressor mutation An **intragenic suppressor mutation** is in the same gene as that containing the mutation being suppressed and may work in several ways. The suppressor may change a second nucleotide in the same codon altered by the original mutation, producing a codon that specifies the same amino acid as that specified by the original, unmutated codon (**Figure 10.8**). Intragenic suppressors may also work by suppressing a frameshift mutation. If the original mutation is a one-base deletion, then the addition of a single base elsewhere in the gene will restore the former reading frame. Consider the following nucleotide sequence in DNA and the amino acids that it encodes:

DNA	AAA	TCA	CTT	GGC	GTA	CAA
Amino acids	Phe	Ser	Glu	Pro	His	Val

Suppose a one-base deletion occurs in the first nucleotide of the second codon. This deletion shifts the reading frame by one nucleotide and alters all the amino acids that follow the mutation.

One-nucleotide deletion

AAA	⊼CAC	TTG	GCG	TAC	AA
Phe	Val	Asn	Arg	Met	

If a single nucleotide is added to the third codon (the suppressor mutation), the reading frame is restored, although two of the amino acids differ from those specified by the original sequence.

One-nucleotide insertion

AAA	CAC	TTT	GGC	GTA	CAA
Phe	Val	Lys	Pro	His	Val

Similarly, a mutation due to an insertion may be suppressed by a subsequent deletion in the same gene.

A third way in which an intragenic suppressor may work is by making compensatory changes in the protein. A first missense mutation may alter the folding of a polypeptide chain by changing the way in which amino acids in the protein interact with one another. A second missense mutation at a different site (the suppressor) may recreate the original folding pattern by restoring interactions between the amino acids.

Intergenic suppressors mutations An **intergenic suppressor mutation**, in contrast, occurs in a gene other than the one bearing the original mutation. These suppressors sometimes work by changing the way that the mRNA is translated. In the example illustrated in **Figure 10.9a**, the

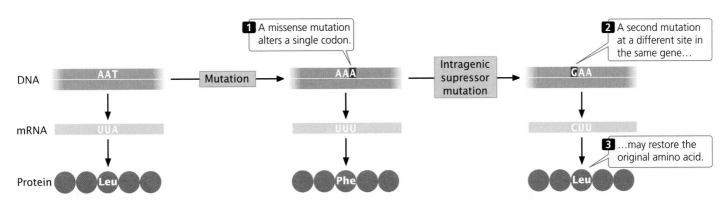

10.8 An intragenic suppressor mutation occurs in the gene containing the mutation being suppressed.

10.9 An intergenic suppressor mutation occurs in a gene other than the one bearing the original mutation. (a) The wild-type sequence produces a full-length, functional protein. (b) A base substitution at a site in the same gene produces a premature stop codon, resulting in a shortened, nonfunctional protein. (c) A base substitution at a site in another gene, which in this case encodes tRNA, alters the anticodon of tRNA^Tyr; tRNA^Tyr can pair with the stop codon produced by the original mutation, allowing tyrosine to be incorporated into the protein and translation to continue.

original DNA sequence is AAC (UUG in the mRNA) and specifies leucine. This sequence mutates to A*T*C (U*A*G in mRNA), a termination codon (**Figure 10.9b**). The ATC nonsense mutation could be suppressed by a second mutation in a different gene that encodes a tRNA; this second mutation would result in a codon capable of pairing with the UAG termination codon (**Figure 10.9c**). For example, the gene that encodes the tRNA for tyrosine (tRNA^Tyr), which has the anticodon AUA, might be mutated to have the anticodon AU*C*, which will then pair with the UAG stop codon. Instead of translation terminating at the UAG codon, tyrosine would be inserted into the protein and a full-length protein would be produced, although tyrosine would now substitute for leucine. The effect of this change would depend on the role of this amino acid in the overall structure of the protein, but the effect of the suppressor mutation is likely to be less detrimental than the effect of the nonsense mutation, which would halt translation prematurely.

Because cells in many organisms have multiple copies of tRNA genes, other unmutated copies of tRNA^Tyr would remain available to recognize tyrosine codons in the transcripts of the mutant gene in question and other genes being expressed concurrently. We might expect that the tRNAs that have undergone the suppressor mutation described would also suppress the normal termination codons at the ends of other coding sequences, resulting in the production of longer-than-normal proteins, but this event does not usually take place.

Intergenic suppressors can also work through genic interactions (see p. 106 in Chapter 5). Polypeptide chains that are produced by two genes may interact to produce a

Table 10.2 Characteristics of different types of mutations

Type of Mutation	Definition
Base substitution	Changes the base of a single DNA nucleotide
Transition	Base substitution in which a purine replaces a purine or a pyrimidine replaces a pyrimidine
Transversion	Base substitution in which a purine replaces a pyrimidine or a pyrimidine replaces a purine
Insertion	Addition of one or more nucleotides
Deletion	Deletion of one or more nucleotides
Frameshift mutation	Insertion or deletion that alters the reading frame of a gene
In-frame deletion or insertion	Deletion or insertion of a multiple of three nucleotides that does not alter the reading frame
Expanding nucleotide repeats	Repeated sequence of a set of nucleotides in which the number of copies of the sequence increases
Forward mutation	Changes the wild-type phenotype to a mutant phenotype
Reverse mutation	Changes a mutant phenotype back to the wild-type phenotype
Missense mutation	Changes a sense codon into a different sense codon, resulting in the incorporation of a different amino acid in the protein
Nonsense mutation	Changes a sense codon into a nonsense codon, causing premature termination of translation
Silent mutation	Changes a sense codon into a synonymous codon, leaving unchanged the amino acid sequence of the protein
Neutral mutation	Changes the amino acid sequence of a protein without altering its ability to function
Loss-of-function mutation	Causes a complete or partial loss of function
Gain-of-function mutation	Causes the appearance of a new trait or function or causes the appearance of a trait in inappropriate tissue or at an inappropriate time
Lethal mutation	Causes premature death
Suppressor mutation	Suppresses the effect of an earlier mutation at a different site
Intragenic suppressor mutation	Suppresses the effect of an earlier mutation within the same gene
Intergenic suppressor mutation	Suppresses the effect of an earlier mutation in another gene

functional protein. A mutation in one gene may alter the encoded polypeptide such that the interaction between the two polypeptides is destroyed, in which case a functional protein is not produced. A suppressor mutation in the second gene may produce a compensatory change in its polypeptide, therefore restoring the original interaction. Characteristics of some of the different types of mutations are summarized in **Table 10.2**. TRY PROBLEM 22 →

CONCEPTS

A suppressor mutation overrides the effect of an earlier mutation at a different site. An intragenic suppressor mutation occurs within the *same* gene as that containing the original mutation, whereas an intergenic suppressor mutation occurs in a *different* gene.

✔ **CONCEPT CHECK 3**

How is a suppressor mutation different from a reverse mutation?

Worked Problem

A gene encodes a protein with the following amino acid sequence:

Met-Arg-Cys-Ile-Lys-Arg

A mutation of a single nucleotide alters the amino acid sequence to:

Met-Asp-Ala-Tyr-Lys-Gly-Glu-Ala-Pro-Val

A second single-nucleotide mutation occurs in the same gene and suppresses the effects of the first mutation (an intragenic suppressor). With the original mutation and the intragenic suppressor present, the protein has the following amino acid sequence:

Met-Asp-Gly-Ile-Lys-Arg

What is the nature and location of the first mutation and of the intragenic suppressor mutation?

• **Solution**

The first mutation alters the reading frame, because all amino acids after Met are changed. Insertions and deletions affect the reading frame; the original mutation consists of a single-nucleotide insertion or deletion in the second codon. The intragenic suppressor restores the reading frame; the intragenic suppressor also is most likely a single-nucleotide insertion or deletion. If the first mutation is an insertion, the suppressor must be a deletion; if the first mutation is a deletion, then the suppressor must be an insertion. Notice that the protein produced by the suppressor still differs from the original protein at the second and third amino acids, but the second amino acid produced by the suppressor is the same as that in the protein produced by the original mutation. Thus, the suppressor mutation must have occurred in the third codon, because the suppressor does not alter the second amino acid.

> For more practice with analyzing mutations, try working Problem 23 at the end of the chapter.

Mutation Rates

The frequency with which a wild-type allele at a locus changes into a mutant allele is referred to as the **mutation rate** and is generally expressed as the number of mutations per biological unit, which may be mutations per cell division, per gamete, or per round of replication. For example, achondroplasia is a type of hereditary dwarfism in humans that results from a dominant mutation. On average, about four achondroplasia mutations arise in every 100,000 gametes, and so the mutation rate is $4/100{,}000$, or 0.00004 mutations per gamete. The mutation rate provides information about how often a mutation arises.

Factors affecting mutation rates Calculations of mutation rates are affected by three factors. First, they depend on the frequency with which a change takes place in DNA. A change in the DNA can arise from spontaneous molecular changes in DNA or it can be induced by chemical, biological, or physical agents in the environment.

The second factor influencing the mutation rate is the probability that, when a change takes place, it will be repaired. Most cells possess a number of mechanisms for repairing altered DNA, and so most alterations are corrected before they are replicated. If these repair systems are effective, mutation rates will be low; if they are faulty, mutation rates will be elevated. Some mutations increase the overall rate of mutation at other genes; these mutations usually occur in genes that encode components of the replication machinery or DNA-repair enzymes.

The third factor is the probability that a mutation will be recognized and recorded. When DNA is sequenced, all mutations are potentially detectable. In practice, however, mutations are usually detected by their phenotypic effects. Some mutations may appear to arise at a higher rate simply because they are easier to detect.

Variation in mutation rates Mutation rates vary among genes and species (**Table 10.3**), but we can draw several general conclusions about mutation rates. First, spontaneous mutation rates are low for all organisms studied. Typical mutation rates for bacterial genes range from about 1 to 100 mutations per 10 billion cells (from 1×10^{-8} to 1×10^{-10}). The mutation rates for most eukaryotic genes are a bit higher, from about 1 to 10 mutations per million gametes (from 1×10^{-5} to 1×10^{-6}). These higher values in eukaryotes may be due to the fact that the rates are calculated per gamete, and several cell divisions are required to produce a gamete, whereas mutation rates in prokaryotic cells are calculated *per cell division.*

The differences in mutation rates among species may be due to differing abilities to repair mutations, unequal exposures to mutagens, or biological differences in rates of spontaneously arising mutations. Even within a single species, spontaneous rates of mutation vary among genes. The reason for this variation is not entirely understood, but some regions of DNA are known to be more susceptible to mutation than others.

Several recent studies have measured mutation rates directly by sequencing genes of organisms before and after a number of generations. These new studies suggest that mutation rates are often higher than those previously measured on the basis of changes in phenotype. In one study, geneticists sequenced randomly chosen stretches of DNA in the nematode worm *Caenorhabditis elegans* and found about 2.1 mutations per genome per generation, which was 10 times as high as previous estimates based on phenotypic changes. The researchers found that about half of the mutations were insertions and deletions.

Adaptive mutation As will be discussed in Chapters 11 through 13, evolutionary change that brings about adaptation to new environments depends critically on the presence of genetic variation. New genetic variants arise primarily through mutation. For many years, genetic variation was assumed to arise randomly and at rates that are independent of the need for adaptation. However, some evidence now suggests that stressful environments—where adaptation may be necessary to survive—can induce more mutations in bacteria, a process that has been termed **adaptive mutation**. The idea of adaptive mutation has been intensely debated; critics counter that most mutations are expected to be deleterious, and so increased mutagenesis would likely be harmful most of the time.

Research findings have shown that mutation rates in bacteria collected from the wild do increase in stressful environments, such as those in which nutrients are limited. There

Table 10.3 Mutation rates of different genes in different organisms

Organism	Mutation	Rate	Unit
Bacteriophage T2	Lysis inhibition	1×10^{-8}	Per replication
	Host range	3×10^{-9}	
Escherichia coli	Lactose fermentation	2×10^{-7}	Per cell division
	Histidine requirement	2×10^{-8}	
Neurospora crassa	Inositol requirement	8×10^{-8}	Per asexual spore
	Adenine requirement	4×10^{-8}	
Corn	Kernel color	2.2×10^{-6}	Per gamete
Drosophila	Eye color	4×10^{-5}	Per gamete
	Allozymes	5.14×10^{-6}	
Mouse	Albino coat color	4.5×10^{-5}	Per gamete
	Dilution coat color	3×10^{-5}	
Human	Huntington disease	1×10^{-6}	Per gamete
	Achondroplasia	1×10^{-5}	
	Neurofibromatosis (Michigan)	1×10^{-4}	
	Hemophilia A (Finland)	3.2×10^{-5}	
	Duchenne muscular dystrophy (Wisconsin)	9.2×10^{-5}	

is considerable disagreement about whether (1) the increased mutagenesis is a genetic strategy selected to increase the probability of evolving beneficial traits in the stressful environment or (2) mutagenesis is merely an accidental consequence of error-prone DNA-repair mechanisms that are more active in stressful environments (see Section 10.4 on DNA repair).

CONCEPTS

Mutation rate is the frequency with which a specific mutation arises. Rates of mutations are generally low and are affected by environmental and genetic factors.

✔ **CONCEPT CHECK 4**

What three factors affect mutation rates?

10.2 Mutations Are Potentially Caused by a Number of Different Natural and Unnatural Factors

Mutations result from both internal and external factors. Those that occur under normal conditions are termed **spontaneous mutations**, whereas those that result from changes caused by environmental chemicals or radiation are **induced mutations**.

Spontaneous Replication Errors

Replication is amazingly accurate: less than one error in a billion nucleotides arises in the course of DNA synthesis. However, spontaneous replication errors do occasionally occur.

Tautomeric shifts The primary cause of spontaneous replication errors was formerly thought to be tautomeric shifts, in which the positions of protons in the DNA bases change. Purine and pyrimidine bases exist in different chemical forms called tautomers (**Figure 10.10a**). The two tautomeric forms of each base are in dynamic equilibrium, although one form is more common than the other. The standard Watson-and-Crick base pairings—adenine with thymine, and cytosine with guanine—are between the common forms of the bases, but, if the bases are in their rare tautomeric forms, other base pairings are possible (**Figure 10.10b**).

Watson and Crick proposed that tautomeric shifts might produce mutations, and, for many years, their proposal was the accepted model for spontaneous replication errors. However, there has never been convincing evidence that the rare tautomers are the cause of spontaneous mutations. Furthermore, research now shows little evidence of tautomers in DNA.

Mispairing due to other structures Mispairing can also occur through wobble, in which normal, protonated,

10.10 Purine and pyrimidine bases exist in different forms called tautomers. (a) A tautomeric shift takes place when a proton changes its position, resulting in a rare tautomeric form. (b) Standard and anomalous base-pairing arrangements that arise if bases are in the rare tautomeric forms.

and other forms of the bases are able to pair because of flexibility in the DNA helical structure (**Figure 10.11**). These structures have been detected in DNA molecules and are now thought to be responsible for many of the mispairings in replication.

Incorporation errors and replication errors When a mismatched base has been incorporated into a newly synthesized nucleotide chain, an **incorporated error** is said to have occurred. Suppose that, in replication, thymine (which normally pairs with adenine) mispairs with guanine through wobble (**Figure 10.12**). In the next round of replication, the two mismatched bases separate, and each serves as template for the synthesis of a new nucleotide strand. This time, thymine pairs with adenine, producing another copy of the original DNA sequence. On the other strand, however, the incorrectly incorporated guanine serves as the template and pairs with cytosine, producing a new DNA molecule that has an error—a C·G pair in place of the original T·A pair (a T·A→C·G base substitution). The original incorporated error leads to a **replication error**, which creates a permanent mutation, because all the base pairings are correct and there is no mechanism for repair systems to detect the error.

Causes of deletions and insertions Mutations due to small insertions and deletions also arise spontaneously in

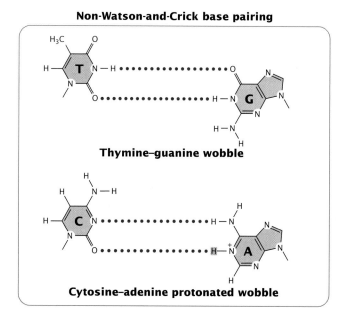

10.11 Nonstandard base pairings can occur as a result of the flexibility in DNA structure. Thymine and guanine can pair through wobble between normal bases. Cytosine and adenine can pair through wobble when adenine is protonated (has an extra hydrogen atom).

10.12 Wobble base pairing leads to a replicated error.

replication and crossing over. **Strand slippage** can occur when one nucleotide strand forms a small loop (**Figure 10.13**). If the looped-out nucleotides are on the newly synthesized strand, an insertion results. At the next round of replication, the insertion will be replicated and both strands will contain the insertion. If the looped-out nucleotides are on the template strand, then the newly replicated strand has a deletion, and this deletion will be perpetuated in subsequent rounds of replication.

Another process that produces insertions and deletions is unequal crossing over. In normal crossing over, the homologous sequences of the two DNA molecules align, and crossing over produces no net change in the number of nucleotides in either molecule. Misaligned pairing can cause **unequal crossing over**, which results in one DNA molecule with an insertion and the other with a deletion (**Figure 10.14**). Some DNA sequences are more likely than others to undergo strand slippage or unequal crossing over. Stretches of repeated sequences, such as nucleotide repeats or homopolymeric repeats (more than five repeats of the same base in a row), are prone to strand slippage. Duplicated or repetitive sequences may misalign during pairing, leading to unequal crossing over. Both strand slippage and unequal crossing over produce duplicated copies of sequences, which in turn promote further strand slippage and unequal crossing over. This chain of events may explain the phenomenon of anticipation often observed for expanding nucleotide repeats.

CONCEPTS

Spontaneous replication errors arise from altered base structures and from wobble base pairing. Small insertions and deletions can occur through strand slippage in replication and through unequal crossing over.

✔ **CONCEPT CHECK 5**

How does an incorporated error differ from a replicated error?

Spontaneous Chemical Changes

In addition to spontaneous mutations that arise in replication, mutations also result from spontaneous chemical changes in DNA. One such change is **depurination**, the loss of a purine base from a nucleotide. Depurination results when the covalent bond connecting the purine to the 1′-carbon atom of the deoxyribose sugar breaks (**Figure 10.15a**), producing an apurinic site, a nucleotide that lacks its purine base. An apurinic site cannot act as a template for a complementary base in replication. In the absence of base-pairing constraints, an

10.13 Insertions and deletions may result from strand slippage.

10.14 Unequal crossing over produces insertions and deletions.

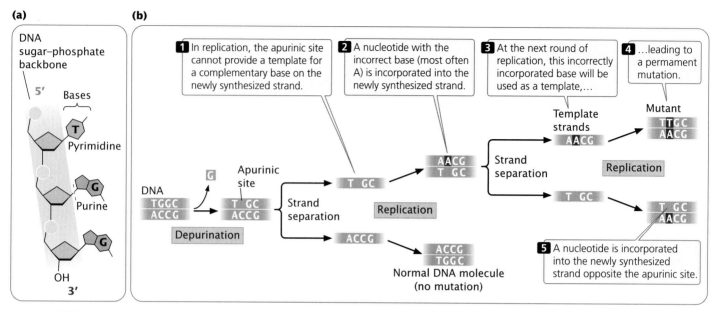

10.15 Depurination (the loss of a purine base from a nucleotide) produces an apurinic site.

incorrect nucleotide (most often adenine) is incorporated into the newly synthesized DNA strand opposite the apurinic site (**Figure 10.15b**), frequently leading to an incorporated error. The incorporated error is then transformed into a replication error at the next round of replication. Depurination is a common cause of spontaneous mutation; a mammalian cell in culture loses approximately 10,000 purines every day.

Another spontaneously occurring chemical change that takes place in DNA is **deamination**, the loss of an amino group (NH_2) from a base. Deamination can be spontaneous or be induced by mutagenic chemicals.

Deamination can alter the pairing properties of a base: the deamination of cytosine, for example, produces uracil (**Figure 10.16a**), which pairs with adenine in replication. After another round of replication, the adenine will pair with thymine, creating a T·A pair in place of the original C·G pair (C·G→U·A→T·A); this chemical change is a transition mutation. This type of mutation is usually repaired by enzymes that remove uracil whenever it is found in DNA. The ability to recognize the product of cytosine deamination may explain why thymine, not uracil, is found in DNA. In mammals, including humans, some cytosine bases in DNA are naturally methylated and exist in the form of 5-methylcytosine (5mC). When deaminated, 5mC becomes thymine (**Figure 10.16b**). Because thymine pairs with adenine in replication, the deamination of 5-methylcytosine changes an original C·G pair to T·A (C·G→5mC·G→T·G→T·A). Consequently, C·G→T·A transitions are frequent in mammalian cells, and 5mC sites are mutation hotspots in humans.

TRY PROBLEM 27 →

CONCEPTS

Some mutations arise from spontaneous alterations to DNA structure, such as depurination and deamination, which may alter the pairing properties of the bases and cause errors in subsequent rounds of replication.

Chemically Induced Mutations

Although many mutations arise spontaneously, a number of environmental agents are capable of damaging DNA, including certain chemicals and radiation. Any environmental agent

10.16 Deamination alters DNA bases.

10.17 5-Bromouracil (a base analog) resembles thymine, except that it has a bromine atom in place of a methyl group on the 5-carbon atom. Because of the similarity in their structures, 5-bromouracil may be incorporated into DNA in place of thymine. Like thymine, 5-bromouracil normally pairs with adenine but, when ionized, it may pair with guanine through wobble.

that significantly increases the rate of mutation above the spontaneous rate is called a **mutagen**.

The first discovery of a chemical mutagen was made by Charlotte Auerbach, who started her career in Berlin researching the development of mutants in *Drosophila*. Faced with increasing anti-Semitism in Nazi Germany, Auerbach emigrated to Britain. There she continued her research on *Drosophila* and collaborated with pharmacologist John Robson on the mutagenic effects of mustard gas, which had been used as a chemical weapon in World War I. The experimental conditions were crude. They heated liquid mustard gas over a Bunsen burner on the roof of the pharmacology building, and the flies were exposed to the gas in a large chamber. After developing serious burns on her hands from the gas, Auerbach let others carry out the exposures, and she analyzed the flies. Auerbach and Robson showed that mustard gas is indeed a powerful mutagen, reducing the viability of gametes and increasing the numbers of mutations seen in the offspring of exposed flies. Because the research was part of the secret war effort, publication of their findings was delayed until 1947.

Base analogs One class of chemical mutagens consists of **base analogs**, chemicals with structures similar to that of any of the four standard bases of DNA. DNA polymerases cannot distinguish these analogs from the standard bases; so, if base analogs are present during replication, they may be incorporated into newly synthesized DNA molecules. For example, 5-bromouracil (5BU) is an analog of thymine; it has the same structure as that of thymine except that it has a bromine (Br) atom on the 5-carbon atom instead of a methyl group (**Figure 10.17a**). Normally, 5-bromouracil pairs with adenine just as thymine does, but it occasionally mispairs with guanine (**Figure 10.17b**), leading to a transition (T·A→5BU·A→5BU·G→C·G), as shown in **Figure 10.18**. Through mispairing, 5-bromouracil can also be incorporated into a newly synthesized DNA strand opposite guanine. In the next round of replication 5-bromouracil pairs with adenine, leading to another transition (G·C→G·5BU→A·5BU→A·T).

Another mutagenic chemical is 2-aminopurine (2AP), which is a base analog of adenine. Normally, 2-aminopurine base pairs with thymine, but it may mispair with cytosine, causing a transition mutation (T·A→T·2AP→C·2AP→C·G). Alternatively, 2-aminopurine may be incorporated through mispairing into the newly synthesized DNA opposite cytosine and then later pair with thymine, leading to a C·G→C·2AP→T·2AP→T·A transition.

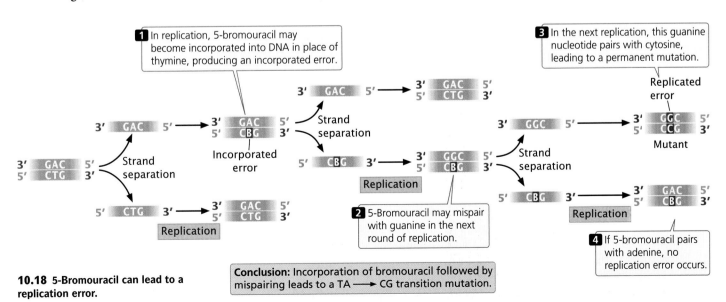

10.18 5-Bromouracil can lead to a replication error.

10.19 Chemicals may alter DNA bases.

	Original base	Mutagen	Modified base	Pairing partner	Type of mutation
(a)	Guanine	EMS Alkylation	O^6-Ethylguanine	Thymine	CG → TA
(b)	Cytosine	Nitrous acid (HNO_2) Deamination	Uracil	Adenine	CG → TA
(c)	Cytosine	Hydroxylamine (NH_2OH) Hydroxylation	Hydroxylaminocytosine	Adenine	CG → TA

Thus, both 5-bromouracil and 2-aminopurine can produce transition mutations. In the laboratory, mutations caused by base analogs can be reversed by treatment with the same analog or by treatment with a different analog.

Alkylating agents Alkylating agents are chemicals that donate alkyl groups, such as methyl (CH_3) and ethyl (CH_3–CH_2) groups, to nucleotide bases. For example, ethylmethylsulfonate (EMS) adds an ethyl group to guanine, producing O^6-ethylguanine, which pairs with thymine (**Figure 10.19a**). Thus, EMS produces C·G→T·A transitions. EMS is also capable of adding an ethyl group to thymine, producing 4-ethylthymine, which then pairs with guanine, leading to a T·A→C·G transition. Because EMS produces both C·G→T·A and T·A→C·G transitions, mutations produced by EMS can be reversed by additional treatment with EMS. Mustard gas is another alkylating agent.

Deamination In addition to its spontaneous occurrence (see Figure 10.16), deamination can be induced by some chemicals. For instance, nitrous acid deaminates cytosine, creating uracil, which in the next round of replication pairs with adenine (**Figure 10.19b**), producing a C·G→T·A transition mutation. Nitrous acid changes adenine into hypoxanthine, which pairs with cytosine, leading to a T·A→C·G transition. Nitrous acid also deaminates guanine, producing xanthine, which pairs with cytosine just as guanine does; however, xanthine can also pair with thymine, leading to a C·G→T·A transition. Nitrous acid produces exclusively transition mutations and, because both C·G→T·A and T·A→C·G transitions are produced, these mutations can be reversed with nitrous acid.

Hydroxylamine Hydroxylamine is a very specific base-modifying mutagen that adds a hydroxyl group to cytosine, converting it into hydroxylaminocytosine (**Figure 10.19c**). This conversion increases the frequency of a rare tautomer that pairs with adenine instead of guanine and leads to C·G→T·A transitions. Because hydroxylamine acts only on cytosine, it will not generate T·A→C·G transitions; thus, hydroxylamine will not reverse the mutations that it produces. **TRY PROBLEM 26**

Oxidative reactions Reactive forms of oxygen (including superoxide radicals, hydrogen peroxide, and hydroxyl radicals) are produced in the course of normal aerobic metabolism, as well as by radiation, ozone, peroxides, and certain drugs. These reactive forms of oxygen damage DNA and induce mutations by bringing about chemical changes in DNA. For example, oxidation converts guanine into 8-oxy-7,8-dihydrodeoxyguanine (**Figure 10.20**), which frequently mispairs with adenine instead of cytosine, causing a G·C→T·A transversion mutation.

Intercalating agents Proflavin, acridine orange, ethidium bromide, and dioxin are **intercalating agents** (**Figure 10.21a**), which produce mutations by sandwiching

10.20 Oxidative radicals convert guanine into 8-oxy-7, 8-dihydrodeoxyguanine.

themselves (intercalating) between adjacent bases in DNA, distorting the three-dimensional structure of the helix and causing single-nucleotide insertions and deletions in replication (**Figure 10.21b**). These insertions and deletions frequently produce frameshift mutations, and so the mutagenic effects of intercalating agents are often severe. Because intercalating agents generate both additions and deletions, they can reverse the effects of their own mutations.

CONCEPTS

Chemicals can produce mutations by a number of mechanisms. Base analogs are inserted into DNA and frequently pair with the wrong base. Alkylating agents, deaminating chemicals, hydroxylamine, and oxidative radicals change the structure of DNA bases, thereby altering their pairing properties. Intercalating agents wedge between the bases and cause single-base insertions and deletions in replication.

✔ CONCEPT CHECK 6

Base analogs are mutagenic because of which characteristic?
a. They produce changes in DNA polymerase that cause it to malfunction.
b. They distort the structure of DNA.
c. They are similar in structure to the normal bases.
d. They chemically modify the normal bases.

Radiation

In 1927, Hermann Muller demonstrated that mutations in fruit flies could be induced by X-rays. The results of subsequent studies showed that X-rays greatly increase mutation rates in all organisms. Because of their high energies, X-rays, gamma rays, and cosmic rays are all capable of penetrating tissues and damaging DNA. These forms of radiation, called ionizing radiation, dislodge electrons from the atoms that they encounter, changing stable molecules into free radicals and reactive ions, which then alter the structures of bases and break phosphodiester bonds in DNA. Ionizing radiation also frequently results in double-strand breaks in DNA. Attempts to repair these breaks can produce chromosome mutations (discussed in Chapter 9).

Ultraviolet (UV) light has less energy than that of ionizing radiation and does not eject electrons but is nevertheless highly mutagenic. Purine and pyrimidine bases readily absorb UV light, resulting in the formation of chemical bonds between adjacent pyrimidine molecules on the same strand of DNA and in the creation of **pyrimidine dimers** (**Figure 10.22a**). Pyrimidine dimers consisting of two thymine bases (called thymine dimers) are most frequent, but cytosine dimers and thymine–cytosine dimers also can form. Dimers distort the configuration of DNA (**Figure 10.22b**) and often block replication. Most pyrimidine dimers are immediately repaired by mechanisms discussed later in this chapter, but some escape repair and inhibit replication and transcription.

When pyrimidine dimers block replication, cell division is inhibited and the cell usually dies; for this reason, UV light kills bacteria and is an effective sterilizing agent. For a mutation—a hereditary error in the genetic instructions—to occur, the replication block must be overcome. Bacteria can sometimes circumvent replication blocks produced by pyrimidine dimers and other types of DNA damage by means of the **SOS system**. This system allows replication blocks to be overcome but, in the process, makes numerous mistakes and greatly increases the rate of mutation. Indeed, the very reason that replication can proceed in the presence of a block is that the enzymes in the SOS system do not strictly adhere to the base-pairing rules. The trade-off is that

10.21 Intercalating agents. Proflavin and acridine orange (a) insert themselves between adjacent bases in DNA, distorting the three-dimensional structure of the helix (b).

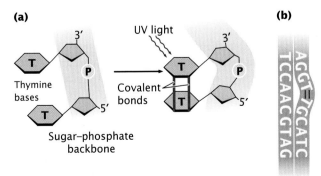

10.22 Pyrimidine dimers result from ultraviolet light.
(a) Formation of thymine dimer. (b) Distorted DNA.

replication may continue and the cell survives, but only by sacrificing the normal accuracy of DNA synthesis.

The SOS system is complex, including the products of at least 25 genes. A protein called RecA binds to the damaged DNA at the blocked replication fork and becomes activated. This activation promotes the binding of a protein called LexA, which is a repressor of the SOS system. The activated RecA complex induces LexA to undergo self-cleavage, destroying its repressive activity. This inactivation enables other SOS genes to be expressed, and the products of these genes allow replication of the damaged DNA to proceed. The SOS system allows bases to be inserted into a new DNA strand in the absence of bases on the template strand, but these insertions result in numerous errors in the base sequence.

> **CONCEPTS**
>
> Ionizing radiation such as X-rays and gamma rays damage DNA by dislodging electrons from atoms; these electrons then break phosphodiester bonds and alter the structure of bases. Ultraviolet light causes mutations primarily by producing pyrimidine dimers that disrupt replication and transcription. The SOS system enables bacteria to overcome replication blocks but introduces mistakes in replication.

10.3 Mutations Are the Focus of Intense Study by Geneticists

Because mutations often have detrimental effects, they are often studied by geneticists. These studies have included the development of tests to determine the mutagenic properties of chemical compounds and the investigation of human populations tragically exposed to high levels of radiation.

Detecting Mutations with the Ames Test

People in industrial societies are surrounded by a multitude of artificially produced chemicals: more than 50,000 different chemicals are in commercial and industrial use today, and from 500 to 1000 new chemicals are introduced each year. Some of these chemicals are potential carcinogens and may cause harm to humans. One method for testing the cancer-causing potential of chemicals is to administer them to laboratory animals (rats or mice) and compare the incidence of cancer in the treated animals with that of control animals. These tests are unfortunately time consuming and expensive. Furthermore, the ability of a substance to cause cancer in rodents is not always indicative of its effect on humans. After all, we aren't rats!

In 1974, Bruce Ames developed a simple test for evaluating the potential of chemicals to cause cancer. The **Ames test** is based on the principle that both cancer and mutations result from damage to DNA, and the results of experiments have demonstrated that 90% of known carcinogens are also mutagens. Ames proposed that mutagenesis in bacteria could serve as an indicator of carcinogenesis in humans.

The Ames test uses different auxotrophic strains of the bacterium *Salmonella typhimurium* that have defects in the lipopolysaccharide coat, which normally protects the bacteria from chemicals in the environment. Furthermore, the DNA-repair system in these strains has been inactivated, enhancing their susceptibility to mutagens.

The most recent version of the test (called Ames II) uses several auxotrophic strains that detect different types of base-pair substitutions. Other strains detect different types of frameshift mutations. Each strain carries a his^- mutation, which renders it unable to synthesize the amino acid histidine, and the bacteria are plated onto medium that lacks histidine (**Figure 10.23**). Only bacteria that have undergone a reverse mutation of the histidine gene ($his^- \rightarrow his^+$) are able to synthesize histidine and grow on the medium. Different dilutions of a chemical to be tested are added to plates inoculated with the bacteria, and the number of mutant bacterial colonies that appear on each plate is compared with the number that appear on control plates with no chemical (i.e., that arose through spontaneous mutation). Any chemical that significantly increases the number of colonies appearing on a treated plate is mutagenic and is probably also carcinogenic.

Some compounds are not active carcinogens but can be converted into cancer-causing compounds in the body. To make the Ames test sensitive for such *potential* carcinogens, a compound to be tested is first incubated in mammalian liver extract that contains metabolic enzymes.

The Ames test has been applied to thousands of chemicals and commercial products. An early demonstration of its usefulness was the discovery, in 1975, that many hair dyes sold in the United States contained compounds that were mutagenic to bacteria. These compounds were then removed from most hair dyes.

> **CONCEPTS**
>
> The Ames test uses his^- strains of bacteria to test chemicals for their ability to produce $his^- \rightarrow his^+$ mutations. Because mutagenic activity and carcinogenic potential are closely correlated, the Ames test is widely used to screen chemicals for their cancer-causing potential.

Radiation Exposure in Humans

People are routinely exposed to low levels of radiation from cosmic, medical, and environmental sources, but there have also been tragic events that produced exposures of much higher degree.

On August 6, 1945, a high-flying American airplane dropped a single atomic bomb on the city of Hiroshima, Japan. The explosion devastated an area of the city measuring 4.5 square miles, killed from 90,000 to 140,000 people,

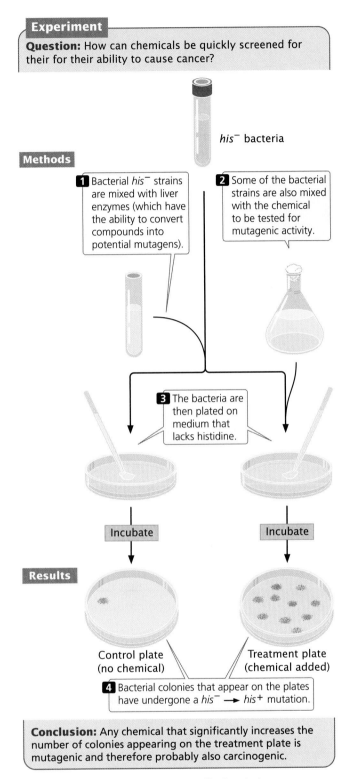

10.23 The Ames test is used to identify chemical mutagens.

After the war, a joint Japanese–U.S. effort was made to study the biological effects of radiation exposure on the survivors of the atomic blasts and their children. Somatic mutations were examined by studying radiation sickness and cancer among the survivors; germ-line mutations were assessed by looking at birth defects, chromosome abnormalities, and gene mutations in children born to people that had been exposed to radiation.

Geneticist James Neel and his colleagues examined almost 19,000 children of people who were within 2000 meters (1.2 miles) of the center of the atomic blast at Hiroshima or Nagasaki, along with a similar number of children whose parents did not receive radiation exposure. Radiation doses were estimated for a child's parents on the basis of careful assessment of the parents' location, posture, and position at the time of the blast. A blood sample was collected from each child, and gel electrophoresis was used to investigate amino acid substitutions in 28 proteins. When rare variants were detected, blood samples from the child's parents also were analyzed to establish whether the variant was inherited or a new mutation.

Of a total of 289,868 genes examined by Neel and his colleagues, only one mutation was found in the children of exposed parents; no mutations were found in the control group. From these findings, a mutation rate of 3.4×10^{-6} was estimated for the children whose parents were exposed to the blast, which is within the range of spontaneous mutation rates observed for other eukaryotes. Neel and his colleagues also examined the frequency of chromosome mutations, sex ratios of children born to exposed parents, and frequencies of chromosome aneuploidy. There was no evidence in any of these assays for increased mutations among the children of the people who were exposed to radiation from the atomic explosions, suggesting that germ-line mutations were not elevated.

Animal studies clearly show that radiation causes germ-line mutations; so why was there no apparent increase in germ-line mutations among the inhabitants of Hiroshima and Nagasaki? The exposed parents did exhibit an increased incidence of leukemia and other types of cancers; so somatic mutations were clearly induced. The answer to the question is not known, but the lack of germ-line mutations may be due to the fact that those persons who received the largest radiation doses died soon after the blasts.

The Techa River in southern Russia is another place where people have been tragically exposed to high levels of radiation. The Mayak nuclear facility produced plutonium for nuclear warheads in the early days of the Cold War. Between 1949 and 1956, this plant dumped some 76 million cubic meters of radioactive sludge into the Techa River. People downstream used the river for drinking water and crop irrigation; some received radiation doses 1700 times the annual amount considered safe by today's standards. Radiation in the area was further elevated by a series of nuclear accidents at the Mayak plant; the worst was an explosion of a radioactive liquid storage tank in 1957, which

and injured almost as many (**Figure 10.24**). Three days later, the United States dropped an atomic bomb on the city of Nagasaki, this time destroying an area measuring 1.5 square miles and killing between 60,000 and 80,000 people. Huge amounts of radiation were released during these explosions, and many people were exposed.

10.24 Hiroshima was destroyed by an atomic bomb on August 6, 1945. The atomic explosion produced many somatic mutations among the survivors. [Stanley Troutman/AP.]

showered radiation over a 27,000-square-kilometer (10,425-square-mile) area.

Although Soviet authorities suppressed information about the radiation problems along the Techa until the 1990s, Russian physicians led by Mira Kossenko quietly began studying cancer and other radiation-related illnesses among the inhabitants in the 1960s. They found that the overall incidence of cancer was elevated among people who lived on the banks of the Techa River.

Most data on radiation exposure in humans are from the intensive study of the survivors of the atomic bombing of Hiroshima and Nagasaki. However, the inhabitants of Hiroshima and Nagasaki were exposed in one intense burst of radiation, and these data may not be appropriate for understanding the effects of long-term low-dose radiation. Today, U.S. and Russian scientists are studying the people of the Techa River region in an attempt to better understand the effects of chronic radiation exposure on human populations.

10.4 A Number of Pathways Repair Changes in DNA

The integrity of DNA is under constant assault from radiation, chemical mutagens, and spontaneously arising changes. In spite of this onslaught of damaging agents, the rate of mutation remains remarkably low, thanks to the efficiency with which DNA is repaired.

There are a number of complex pathways for repairing DNA, but several general statements can be made about DNA repair. First, most DNA-repair mechanisms require two nucleotide strands of DNA because most replace whole nucleotides, and a template strand is needed to specify the base sequence.

A second general feature of DNA repair is redundancy, meaning that many types of DNA damage can be corrected by more than one pathway of repair. This redundancy testifies to the extreme importance of DNA repair to the survival of the cell: it ensures that almost all mistakes are corrected. If a mistake escapes one repair system, it's likely to be repaired by another system.

We will consider several general mechanisms of DNA repair: mismatch repair, direct repair, base-excision repair, nucleotide-excision repair, and repair of double-strand breaks (**Table 10.4**).

Table 10.4	Summary of common DNA repair mechanisms
Repair System	**Type of Damage Repaired**
Mismatch	Replication errors, including mispaired bases and strand slippage
Direct	Pyrimidine dimers; other specific types of alterations
Base excision	Abnormal bases, modified bases, and pyrimidine dimers
Nucleotide excision	DNA damage that distorts the double helix, including abnormal bases, modified bases, and pyrimidine dimers
Homologous recombination	Double-strand breaks
Nonhomologous end joining	Double-strand breaks

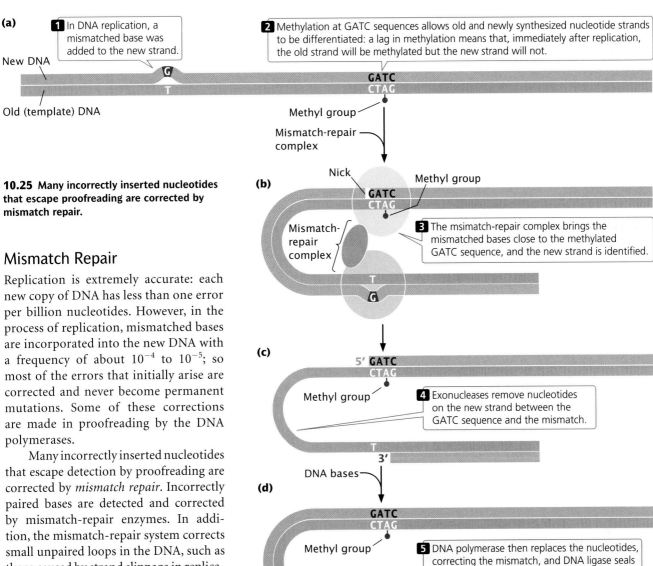

10.25 Many incorrectly inserted nucleotides that escape proofreading are corrected by mismatch repair.

Mismatch Repair

Replication is extremely accurate: each new copy of DNA has less than one error per billion nucleotides. However, in the process of replication, mismatched bases are incorporated into the new DNA with a frequency of about 10^{-4} to 10^{-5}; so most of the errors that initially arise are corrected and never become permanent mutations. Some of these corrections are made in proofreading by the DNA polymerases.

Many incorrectly inserted nucleotides that escape detection by proofreading are corrected by *mismatch repair*. Incorrectly paired bases are detected and corrected by mismatch-repair enzymes. In addition, the mismatch-repair system corrects small unpaired loops in the DNA, such as those caused by strand slippage in replication (see Figure 10.13). Some nucleotide repeats may form secondary structures on the unpaired strand (see Figure 10.5), allowing them to escape detection by the mismatch-repair system.

After the incorporation error has been recognized, mismatch-repair enzymes cut out a section of the newly synthesized strand and fill the gap with new nucleotides by using the original DNA strand as a template. For this strategy to work, mismatch repair must have some way of distinguishing between the old and the new strands of the DNA so that the incorporation error, but not part of the original strand, is removed.

The proteins that carry out mismatch repair in *E. coli* differentiate between old and new strands by the presence of methyl groups on special sequences of the old strand. After replication, adenine nucleotides in the sequence GATC are methylated. The process of methylation is delayed and so, immediately after replication, the old strand is methylated and the new strand is not (**Figure 10.25a**). The mismatch-repair complex brings an unmethylated GATC sequence in close proximity to the mismatched bases. It nicks the unmethylated strand at the GATC site (**Figure 10.25b**), and degrades the strand between the nick and the mismatched bases (**Figure 10.25c**). DNA polymerase and DNA ligase fill in the gap on the unmethylated strand with correctly paired nucleotides (**Figure 10.25d**).

Mismatch repair in eukaryotic cells is similar to that in *E. coli*, but how the old and new strands are recognized in eukaryotic cells is not known. In some eukaryotes, such as yeast and fruit flies, there is no detectable methylation of DNA, and yet mismatch repair still takes place. Humans who possess mutations in mismatch-repair genes often exhibit elevated somatic mutations and are frequently susceptible to colon cancer.

10.26 Direct repair changes nucleotides back into their original structures.

CONCEPTS

Mismatched bases and other DNA lesions are corrected by mismatch repair. Enzymes cut out a section of the newly synthesized strand of DNA and replace it with new nucleotides.

✔ CONCEPT CHECK 7

Mismatch repair in *E. coli* distinguishes between old and new strands of DNA on the basis of

a. differences in base composition of the two strands.
b. modification of histone proteins.
c. base analogs on the new strand.
d. methyl groups on the old strand.

Direct Repair

Direct repair does not replace altered nucleotides but, instead, changes them back into their original (correct) structures. One of the most well understood direct-repair mechanisms is the photoreactivation of UV-induced pyrimidine dimers. *E. coli* and some eukaryotic cells possess an enzyme called photolyase, which uses energy captured from light to break the covalent bonds that link the pyrimidines in a dimer.

Direct repair also corrects O^6-methylguanine, an alkylation product of guanine that pairs with adenine, producing $G \cdot C \rightarrow T \cdot A$ transversions. An enzyme called O^6-methylguanine-DNA methyltransferase removes the methyl group from O^6-methylguanine, restoring the base to guanine (**Figure 10.26**).

Base-Excision Repair

In **base-excision repair**, a modified base is first excised and then the entire nucleotide is replaced. The excision of modified bases is catalyzed by a set of enzymes called DNA glycosylases, each of which recognizes and removes a specific type of modified base by cleaving the bond that links that base to the 1′-carbon atom of deoxyribose sugar (**Figure 10.27a**). Uracil glycosylase, for example, recognizes and removes uracil produced by the deamination of cytosine. Other glycosylases recognize hypoxanthine, 3-**methyl**adenine, 7-methylguanine, and other modified bases.

10.27 Base-excision repair excises modified bases and then replaces one or more nucleotides.

After the base has been removed, an enzyme called AP (apurinic or apyrimidinic) endonuclease cuts the phosphodiester bond, and other enzymes remove the deoxyribose sugar (**Figure 10.27b**). DNA polymerase then adds one or more new nucleotides to the exposed 3′-OH group (**Figure 10.27c**), replacing a section of nucleotides on the damaged strand. The nick in the phosphodiester backbone is sealed by DNA ligase (**Figure 10.27d**), and the original intact sequence is restored (**Figure 10.27e**).

Bacteria use DNA polymerase I to replace excised nucleotides, but eukaryotes use DNA polymerase β, which has no proofreading ability and tends to make mistakes. On average, DNA polymerase β makes one mistake per 4000 nucleotides inserted. About 20,000 to 40,000 base modifications per day are repaired by base excision, and so DNA polymerase β may introduce as many as 10 mutations per day into the human genome. How are these errors corrected? Recent research results show that some AP endonucleases have the ability to proofread. When DNA polymerase β inserts a nucleotide with the wrong base into the DNA, DNA ligase cannot seal the nick in the sugar–phosphate backbone, because the 3′-OH and 5′-P groups of adjacent nucleotides are not in the correct orientation for ligase to connect them. In this case, AP endonuclease 1 detects the mispairing and uses its 3′→5′ exonuclease activity to excise the incorrectly paired base. DNA polymerase β then uses its polymerase activity to fill in the missing nucleotide. In this way, the fidelity of base-excision repair is maintained.

> ### CONCEPTS
>
> Direct-repair mechanisms change altered nucleotides back into their correct structures. In base-excision repair, glycosylase enzymes recognize and remove specific types of modified bases. The entire nucleotide is then removed and a section of the polynucleotide strand is replaced.
>
> ✔ **CONCEPT CHECK 8**
>
> How do direct-repair mechanisms differ from mismatch repair and base-excision repair?

Nucleotide-Excision Repair

Another repair pathway is **nucleotide-excision repair**, which removes bulky DNA lesions (such as pyrimidine dimers) that distort the double helix. Nucleotide-excision repair is quite versatile and can repair many different types of DNA damage. It is found in cells of all organisms from bacteria to humans and is among the most important of all repair mechanisms.

The process of nucleotide excision is complex; in humans, a large number of genes take part. First, a complex of enzymes scans DNA, looking for distortions of its

10.28 Nucleotide-excision repair removes bulky DNA lesions that distort the double helix.

three-dimensional configuration (**Figure 10.28a** and **b**). When a distortion is detected, additional enzymes separate the two nucleotide strands at the damaged region, and single-strand-binding proteins stabilize the separated strands (**Figure 10.28c**). Next, the sugar–phosphate backbone of the damaged strand is cleaved on both sides of the damage (**Figure 10.28d**). Part of the damaged strand is peeled away by helicase enzymes (**Figure 10.28e**), and the gap is filled in by DNA polymerase and sealed by DNA ligase (**Figure 10.28f**).

> **CONCEPTS**
>
> Nucleotide-excision repair removes and replaces many types of damaged DNA that distort the DNA structure. The two strands of DNA are separated, a section of the DNA containing the distortion is removed, DNA polymerase fills in the gap, and DNA ligase seals the filled-in gap.

CONNECTING CONCEPTS

The Basic Pathway of DNA Repair

We have now examined several different mechanisms of DNA repair. What do these methods have in common? How are they different? Most methods of DNA repair depend on the presence of two strands, because nucleotides in the damaged area are removed and replaced. Nucleotides are replaced in mismatch repair, base-excision repair, and nucleotide-excision repair but are not replaced by direct-repair mechanisms.

Repair mechanisms that include nucleotide removal utilize a common four-step pathway:

1. Detection: The damaged section of the DNA is recognized.
2. Excision: DNA-repair endonucleases nick the phosphodiester backbone on one or both sides of the DNA damage and one or more nucleotides are removed.
3. Polymerization: DNA polymerase adds nucleotides to the newly exposed 3′-OH group by using the other strand as a template and replacing damaged (and frequently some undamaged) nucleotides.
4. Ligation: DNA ligase seals the nicks in the sugar–phosphate backbone.

The primary differences in the mechanisms of mismatch, base-excision, and nucleotide-excision repair are in the details of detection and excision. In base-excision and mismatch repair, a single nick is made in the sugar–phosphate backbone on one side of the damage; in nucleotide-excision repair, nicks are made on both sides of the DNA lesion. In base-excision repair, DNA polymerase displaces the old nucleotides as it adds new nucleotides to the 3′ end of the nick; in mismatch repair, the old nucleotides are degraded; and, in nucleotide-excision repair, nucleotides are displaced by helicase enzymes. All three mechanisms use DNA polymerase and ligase to fill in the gap produced by the excision and removal of damaged nucleotides.

Repair of Double-Strand Breaks

A common type of DNA damage is a double-strand break, in which both strands of the DNA helix are broken. Double-strand breaks are caused by ionizing radiation, oxidative free radicals, and other DNA-damaging agents. These types of breaks are particularly detrimental to the cell because they stall DNA replication and may lead to chromosome rearrangements, such as deletions, duplications, inversions, and translocations. There are two major pathways for repairing double-strand breaks: homologous recombination and nonhomologous end joining.

Homologous recombination Homologous recombination repairs a broken DNA molecule by using the identical or nearly identical genetic information contained in another DNA molecule, usually a sister chromatid. DNA repair through homologous recombination uses the same mechanism employed in the process of homologous recombination that is responsible for crossing over. Homologous recombination begins with the removal of some nucleotides at the broken ends, followed by strand invasion, displacement, and replication. Many of the same enzymes that carry out crossing over are utilized in the repair of double-strand breaks by homologous recombination. Two enzymes that play a role in homologous recombination are BRCA1 and BRCA2. The genes for these proteins are frequently mutated in breast cancer cells.

Nonhomologous end joining Nonhomologous end joining repairs double-strand breaks without using a homologous template. This pathway is often used when the cell is in G_1 and a sister chromatid is not available for repair through homologous recombination. Nonhomologous end joining uses proteins that recognize the broken ends of DNA, bind to the ends, and then joins them together. Nonhomologous end joining is more error prone than homologous recombination and often leads to deletions, insertions, and translocations.

Translesion DNA Polymerases

As discussed in Chapter 12, the high-fidelity DNA polymerases that normally carry out replication operate at high speed and, like a high-speed train, require a smooth track—an undistorted template. Some mutations, such as pyrimidine dimers, produce distortions in the three-dimensional structure of the DNA helix, blocking replication by the high-speed polymerases. When distortions of the template are encountered, specialized translesion DNA polymerases take over replication and bypass the lesion.

The translesion polymerases are able to bypass bulky lesions but, in the process, often make errors. Thus, the translesion polymerases allow replication to proceed past bulky lesion but at the cost of introducing mutations into the sequence. Some of these mutations are corrected by DNA-repair systems, but others escape detection.

An example of a translesion DNA polymerase is polymerase η (eta), which bypasses pyrimidine dimers in eukaryotes.

Polymerase η inserts AA opposite a pyrimidine dimer. This strategy seems to be reasonable because about two-thirds of pyrimidine dimers are thymine dimers. However, the insertion of AA opposite a CT dimer results in a C·G→A·T transversion. Polymerase η therefore tends to introduce mutations into the DNA sequence.

> **CONCEPTS**
>
> Two major pathways exist for the repair of double-strand breaks in DNA: homologous recombination and nonhomologous end joining. Special translesion DNA polymerases allow replication to proceed past bulky distortions in the DNA but often introduce errors as they bypass the distorted region.

Genetic Diseases and Faulty DNA Repair

Several human diseases are connected to defects in DNA repair. These diseases are often associated with high incidences of specific cancers, because defects in DNA repair lead to increased rates of mutation.

Among the best studied of the human DNA-repair diseases is xeroderma pigmentosum (**Figure 10.29**), a rare autosomal recessive condition that includes abnormal skin pigmentation and acute sensitivity to sunlight. Persons who have this disease also have a strong predisposition to skin cancer, with an incidence ranging from 1000 to 2000 times that found in unaffected people.

Sunlight includes a strong UV component; so exposure to sunlight produces pyrimidine dimers in the DNA of skin cells. Although human cells lack photolyase (the enzyme that repairs pyrimidine dimers in bacteria), most pyrimidine dimers in humans can be corrected by nucleotide-excision repair (see Figure 10.28). However, the cells of most people with xeroderma pigmentosum are defective in nucleotide-excision repair, and many of their pyrimidine dimers go uncorrected and may lead to cancer.

Xeroderma pigmentosum can result from defects in several different genes. Some persons with xeroderma pigmentosum have mutations in a gene encoding the protein that recognizes and binds to damaged DNA; others have mutations in a gene encoding helicase. Still others have defects in the genes that play a role in cutting the damaged strand on the 5' or 3' sides of the pyrimidine dimer. Some persons have a slightly different form of the disease (xeroderma pigmentosum variant) owing to mutations in the gene encoding polymerase η, a translesion DNA polymerase that bypasses pyrimidine dimers.

Two other genetic diseases due to defects in nucleotide-excision repair are Cockayne syndrome and trichothiodystrophy (also known as brittle-hair syndrome). Persons who have either of these diseases do not have an increased risk of cancer but do exhibit multiple developmental and neurological problems. Both diseases result from mutations in some of the same genes that cause xeroderma pigmentosum. Several of the genes taking part in nucleotide-excision repair produce proteins that also play a role in recombination and the initiation of transcription. These other functions may account for the developmental symptoms seen in Cockayne syndrome and trichothiodystrophy.

Another genetic disease caused by faulty DNA repair is an inherited form of colon cancer called hereditary non-polyposis colon cancer (HNPCC). This cancer is one of the most common hereditary cancers, accounting for about 15% of colon cancers. Research findings indicate that HNPCC arises from mutations in the proteins that carry out mismatch repair (see Figure 10.25). Some additional genetic diseases associated with defective DNA repair are summarized in **Table 10.5**.   **TRY PROBLEM 32** →

10.29 Xeroderma pigmentosum results from defects in DNA repair. The disease is characterized by frecklelike spots on the skin (shown here) and a predisposition to skin cancer. [Ken Greer/Visuals Unlimited.]

> **CONCEPTS**
>
> Defects in DNA repair are the underlying cause of several genetic diseases. Many of these diseases are characterized by a predisposition to cancer.
>
> ✔ **CONCEPT CHECK 9**
>
> Why are defects in DNA repair often associated with increases in cancer?

Table 10.5 Genetic diseases associated with defects in DNA-repair systems

Disease	Symptoms	Genetic Defect
Xeroderma pigmentosum	Frecklelike spots on skin, sensitivity to sunlight, predisposition to skin cancer	Defects in nucleotide-excision repair
Cockayne syndrome	Dwarfism, sensitivity to sunlight, premature aging, deafness, mental retardation	Defects in nucleotide-excision repair
Trichothiodystrophy	Brittle hair, skin abnormalities, short stature, immature sexual development, characteristic facial features	Defects in nucleotide-excision repair
Hereditary nonpolyposis colon cancer	Predisposition to colon cancer	Defects in mismatch repair
Fanconi anemia	Increased skin pigmentation, abnormalities of skeleton, heart, and kidneys, predisposition to leukemia	Possibly defects in the repair of interstrand cross-links
Li-Fraumeni syndrome	Predisposition to cancer in many different tissues	Defects in DNA damage response
Werner syndrome	Premature aging, predisposition to cancer	Defect in homologous recombination

CONCEPTS SUMMARY

- Mutations are heritable changes in genetic information. Somatic mutations occur in somatic cells; germ-line mutations occur in cells that give rise to gametes.

- The simplest type of mutation is a base substitution, a change in a single base pair of DNA. Transitions are base substitutions in which purines are replaced by purines or pyrimidines are replaced by pyrimidines. Transversions are base substitutions in which a purine replaces a pyrimidine or a pyrimidine replaces a purine.

- Insertions are the addition of nucleotides, and deletions are the removal of nucleotides; these mutations often change the reading frame of the gene.

- Expanding nucleotide repeats are mutations in which the number of copies of a set of nucleotides increases with the passage of time; they are responsible for several human genetic diseases.

- A missense mutation alters the coding sequence so that one amino acid substitutes for another. A nonsense mutation changes a codon that specifies an amino acid into a termination codon. A silent mutation produces a synonymous codon that specifies the same amino acid as the original sequence, whereas a neutral mutation alters the amino acid sequence but does not change the functioning of the protein. A suppressor mutation reverses the effect of a previous mutation at a different site and may be intragenic (within the same gene as the original mutation) or intergenic (within a different gene).

- Mutation rate is the frequency with which a particular mutation arises in a population. Mutation rates are influenced by both genetic and environmental factors.

- Some mutations occur spontaneously. These mutations include the mispairing of bases in replication and spontaneous depurination and deamination.

- Insertions and deletions can arise from strand slippage in replication or from unequal crossing over.

- Base analogs can become incorporated into DNA in the course of replication and pair with the wrong base in subsequent replication events. Alkylating agents and hydroxylamine modify the chemical structure of bases and lead to mutations. Intercalating agents insert into the DNA molecule and cause single-nucleotide additions and deletions. Oxidative reactions alter the chemical structures of bases.

- Ionizing radiation is mutagenic, altering base structures and breaking phosphodiester bonds. Ultraviolet light produces pyrimidine dimers, which block replication. Bacteria use the SOS response to overcome replication blocks produced by pyrimidine dimers and other lesions in DNA.

- The Ames tests uses bacteria to assess the mutagenic potential of chemical substances.

- Most damage to DNA is corrected by DNA-repair mechanisms. These mechanisms include mismatch repair, direct repair, base-excision repair, nucleotide-excision repair, and other repair pathways. Most require two strands of DNA and exhibit some overlap in the types of damage repaired.

- Double-strand beaks are repaired by homologous recombination and nonhomologous end joining. Special translesion DNA polymerases allow replication to proceed past bulky distortions in the DNA.

- Defects in DNA repair are the underlying cause of several genetic diseases.

IMPORTANT TERMS

mutation (p. 272)
somatic mutation (p. 272)
germ-line mutation (p. 273)
gene mutation (p. 273)
base substitution (p. 273)
transition (p. 273)
transversion (p. 273)
insertion (p. 274)
deletion (p. 274)
frameshift mutation (p. 274)
in-frame insertion (p. 274)
in-frame deletion (p. 274)
expanding nucleotide repeat (p. 274)
forward mutation (p. 275)
reverse mutation (reversion) (p. 275)
missense mutation (p. 275)
nonsense mutation (p. 275)
silent mutation (p. 276)
neutral mutation (p. 276)
loss-of-function mutation (p. 276)
gain-of-function mutation (p. 276)
conditional mutation (p. 276)
lethal mutation (p. 276)
suppressor mutation (p. 276)
intragenic suppressor mutation (p. 277)
intergenic suppressor mutation (p. 277)
mutation rate (p. 280)
adaptive mutation (p. 280)
spontaneous mutation (p. 281)
induced mutation (p. 281)
incorporated error (p. 282)
replication error (p. 282)
strand slippage (p. 283)
unequal crossing over (p. 283)
depurination (p. 283)
deamination (p. 284)
mutagen (p. 285)
base analog (p. 285)
intercalating agent (p. 286)
pyrimidine dimer (p. 287)
SOS system (p. 287)
Ames test (p. 288)
direct repair (p. 292)
base-excision repair (p. 292)
nucleotide-excision repair (p. 293)

ANSWERS TO CONCEPT CHECKS

1. Studying mutations that disrupt normal processes often leads to the identification of genes that normally play a role in the process and can help in understanding the molecular details of a process.

2. c

3. A reverse mutation restores the original phenotype by changing the DNA sequence back to the wild-type sequence. A suppressor mutation restores the phenotype by causing an additional change in the DNA at a site that is different from that of the original mutation.

4. The frequency with which changes arise in DNA, how often these changes are repaired by DNA-repair mechanisms, and our ability to detect the mutation

5. An incorporated error is due to a change that takes place in DNA. This change may be corrected by a DNA-repair pathway. However, if the error has been replicated, it is permanent and cannot be detected by repair pathways.

6. c

7. d

8. Direct-repair mechanisms return an altered base to its correct structure without removing and replacing nucleotides. Mismatch repair and base-excision repair remove and replace nucleotides.

9. Changes in DNA structure do not undergo repair in people with defects in DNA-repair mechanisms. Consequently, increased numbers of mutations occur at all genes, including those that predispose to cancer. This observation indicates that cancer arises from mutations in DNA.

WORKED PROBLEMS

1. A codon that specifies the amino acid Asp undergoes a single-base substitution that yields a codon that specifies Ala. Refer to the genetic code in Figure 15.10 and give all possible DNA sequences for the original and the mutated codon. Is the mutation a transition or a transversion?

• **Solution**

There are two possible RNA codons for Asp: GAU and GAC. The DNA sequences that encode these codons will be complementary to the RNA codons: CTA and CTG. There are four possible RNA codons for Ala: GCU, GCC, GCA, and GCG, which correspond to DNA sequences CGA, CGG, CGT, and CGC. If we organize the original and the mutated sequences as shown in the following table, the type of mutations that may have occurred can be easily seen:

Possible original sequence for Asp	Possible mutated sequence for Ala
CTA	CGA
CTG	CGG
	CGT
	CGC

If the mutation is confined to a single-base substitution, then the only mutations possible are that CTA mutated to CGA or that CTG mutated to CGG. In both, there is a T→G transversion in the middle nucleotide of the codon.

2. The mutations produced by the following compounds are reversed by the substances shown. What conclusions can you make about the nature of the mutations originally produced by these compounds?

	Mutations produced by compound	Reversed by			
		5-Bromouracil	EMS	Hydroxylamine	Acridine orange
a.	1	Yes	Yes	No	No
b.	2	Yes	Yes	Some	No
c.	3	No	No	No	Yes
d.	4	Yes	Yes	Yes	Yes

- **Solution**

The ability of various compounds to produce reverse mutations reveals important information about the nature of the original mutation.

a. Mutations produced by compound 1 are reversed by 5-bromouracil, which produces both $A \cdot T \rightarrow G \cdot C$ and $G \cdot C \rightarrow A \cdot T$ transitions, which tells us that compound 1 produces single-base substitutions that may include the generation of either $A \cdot T$ or $G \cdot C$ pairs. The mutations produced by compound 1 are also reversed by EMS, which, like 5-bromouracil, produces both $A \cdot T \rightarrow G \cdot C$ and $G \cdot C \rightarrow A \cdot T$ transitions; so no additional information is provided here. Hydroxylamine does not reverse the mutations produced by compound 1. Because hydroxylamine produces only $C \cdot G \rightarrow T \cdot A$ transitions, we know that compound 1 does not generate $C \cdot G$ base pairs. Acridine orange, an intercalating agent that produces frameshift mutations, also does not reverse the mutations, revealing that compound 1 produces only single-base-pair substitutions, not insertions or deletions. In summary, compound 1 appears to cause single-base substitutions that generate $T \cdot A$ but not $G \cdot C$ base pairs.

b. Compound 2 generates mutations that are reversed by 5-bromouracil and EMS, indicating that it may produce $G \cdot C$ or $A \cdot T$ base pairs. Some of these mutations are reversed by hydroxylamine, which produces only $C \cdot G \rightarrow T \cdot A$ transitions, indicating that some of the mutations produced by compound 2 are $C \cdot G$ base pairs. None of the mutations are reversed by acridine orange; so compound 2 does not induce insertions or deletions. In summary, compound 2 produces single-base substitutions that generate both $G \cdot C$ and $A \cdot T$ base pairs.

c. Compound 3 produces mutations that are reversed only by acridine orange; so compound 3 appears to produce only insertions and deletions.

d. Compound 4 is reversed by 5 bromouracil, EMS, hydroxylamine, and acridine orange, indicating that this compound produces single-base substitutions, which include $G \cdot C$ and $A \cdot T$ base pairs, insertions, and deletions.

COMPREHENSION QUESTIONS

Section 10.1

*1. What is the difference between somatic mutations and germ-line mutations?

*2. What is the difference between a transition and a transversion? Which type of base substitution is usually more common?

*3. Briefly describe expanding nucleotide repeats. How do they account for the phenomenon of anticipation?

4. What is the difference between a missense mutation and a nonsense mutation? Between a silent mutation and a neutral mutation?

5. Briefly describe two different ways that intragenic suppressors may reverse the effects of mutations.

*6. How do intergenic suppressors work?

Section 10.2

*7. What causes errors in DNA replication?

8. How do insertions and deletions arise?

*9. How do base analogs lead to mutations?

10. How do alkylating agents, nitrous acid, and hydroxylamine produce mutations?

11. What types of mutations are produced by ionizing and UV radiation?

*12. What is the SOS system and how does it lead to an increase in mutations?

Section 10.3

13. What is the purpose of the Ames test? How are his^- bacteria used in this test?

Section 10.4

*14. List at least three different types of DNA repair and briefly explain how each is carried out.

15. What features do mismatch repair, base-excision repair, and nucleotide-excision repair have in common?

16. What are two major mechanisms for the repair of double-strand breaks? How do they differ?

APPLICATION QUESTIONS AND PROBLEMS

Section 10.1

*17. A codon that specifies the amino acid Gly undergoes a single-base substitution to become a nonsense mutation. In accord with the genetic code given in Figure 15.10, is this mutation a transition or a transversion? At which position of the codon does the mutation occur?

*18a. If a single transition occurs in a codon that specifies Phe, what amino acids can be specified by the mutated sequence?

b. If a single transversion occurs in a codon that specifies Phe, what amino acids can be specified by the mutated sequence?

c. If a single transition occurs in a codon that specifies Leu, what amino acids can be specified by the mutated sequence?

d. If a single transversion occurs in a codon that specifies Leu, what amino acids can be specified by the mutated sequence?

19. Hemoglobin is a complex protein that contains four polypeptide chains. The normal hemoglobin found in adults—called adult hemoglobin—consists of two alpha and two beta polypeptide chains, which are encoded by different loci. Sickle-cell hemoglobin, which causes sickle-cell anemia, arises from a mutation in the beta chain of adult hemoglobin. Adult hemoglobin and sickle-cell hemoglobin differ in a single amino acid: the sixth amino acid from one end in adult hemoglobin is glutamic acid, whereas sickle-cell hemoglobin has valine at this position. After consulting the genetic code provided in Figure 15.10, indicate the type and location of the mutation that gave rise to sickle-cell anemia.

*20. The following nucleotide sequence is found on the template strand of DNA. First, determine the amino acids of the protein encoded by this sequence by using the genetic code provided in Figure 15.10. Then, give the altered amino acid sequence of the protein that will be found in each of the following mutations:

Sequence of DNA template
3′–TAC TGG CCG TTA GTT GAT ATA ACT–5′
Nucleotide number 1   24

a. Mutant 1:   A transition at nucleotide 11
b. Mutant 2:   A transition at nucleotide 13
c. Mutant 3:   A one-nucleotide deletion at nucleotide 7
d. Mutant 4:   A T→A transversion at nucleotide 15
e. Mutant 5:   An addition of TGG after nucleotide 6
f. Mutant 6:   A transition at nucleotide 9

21. A polypeptide has the following amino acid sequence:

Met-Ser-Pro-Arg-Leu-Glu-Gly

The amino acid sequence of this polypeptide was determined in a series of mutants listed in parts *a* through *e*. For each mutant, indicate the type of mutation that occurred in the DNA (single-base substitution, insertion, deletion) and the phenotypic effect of the mutation (nonsense mutation, missense mutation, frameshift, etc.).

a. Mutant 1:   Met-Ser-Ser-Arg-Leu-Glu-Gly
b. Mutant 2:   Met-Ser-Pro
c. Mutant 3:   Met-Ser-Pro-Asp-Trp-Arg-Asp-Lys
d. Mutant 4:   Met-Ser-Pro-Glu-Gly
e. Mutant 5:   Met-Ser-Pro-Arg-Leu-Leu-Glu-Gly

*22. A gene encodes a protein with the following amino acid sequence:

Met-Trp-His-Arg-Ala-Ser-Phe

A mutation occurs in the gene. The mutant protein has the following amino acid sequence:

Met-Trp-His-Ser-Ala-Ser-Phe

An intragenic suppressor restores the amino acid sequence to that of the original protein:

Met-Trp-His-Arg-Ala-Ser-Phe

Give at least one example of base changes that could produce the original mutation and the intragenic suppressor. (Consult the genetic code in Figure 15.10.)

23. A gene encodes a protein with the following amino acid sequence:

Met-Lys-Ser-Pro-Ala-Thr-Pro

A nonsense mutation from a single-base-pair substitution occurs in this gene, resulting in a protein with the amino acid sequence Met-Lys. An intergenic suppressor mutation allows the gene to produce the full-length protein. With the original mutation and the intergenic suppressor present, the gene now produces a protein with the following amino acid sequence:

Met-Lys-Cys-Pro-Ala-Thr-Pro

Give the location and nature of the original mutation and of the intergenic suppressor.

24. XG syndrome is a rare genetic disease that is due to an autosomal dominant gene. A complete census of a small European country reveals that 77,536 babies were born in 2004, of whom 3 had XG syndrome. In the same year, this country had a population of 5,964,321 people, and there were 35 living persons with XG syndrome. What is the mutation rate of XG syndrome in this country?

Section 10.2

*25. Can nonsense mutations be reversed by hydroxylamine? Why or why not?

*26. The following nucleotide sequence is found in a short stretch of DNA:

5′–ATGT–3′
3′–TACA–5′

If this sequence is treated with hydroxylamine, what sequences will result after replication?

27. The following nucleotide sequence is found in a short stretch of DNA:

5′–AG–3′
3′–TC–5′

a. Give all the mutant sequences that may result from spontaneous depurination in this stretch of DNA.

b. Give all the mutant sequences that may result from spontaneous deamination in this stretch of DNA.

28. In many eukaryotic organisms, a significant proportion of cytosine bases are naturally methylated to 5-methylcytosine. Through evolutionary time, the proportion of AT base pairs in the DNA of these organisms increases. Can you suggest a possible mechanism for this increase?

Section 10.3

*29. A chemist synthesizes four new chemical compounds in the laboratory and names them PFI1, PFI2, PFI3, and PFI4. He gives the PFI compounds to a geneticist friend and asks her to determine their mutagenic potential. The geneticist finds that all four are highly mutagenic. She also tests the capacity of mutations produced by the PFI compounds to be reversed by other known mutagens and obtains the following results. What conclusions can you make about the nature of the mutations produced by these compounds?

Mutations produced by	Reversed by			
	2-Amino-purine	Nitrous acid	Hydroxyl-amine	Acridine orange
PFI1	Yes	Yes	Some	No
PFI2	No	No	No	No
PFI3	Yes	Yes	No	No
PFI4	No	No	No	Yes

30. Mary Alexander studied the effects of radiation on mutation rates in the sperm of *Drosophila melanogaster*. She irradiated *Drosophila* larvae with either 3000 roentgens (r) or 3975 r, collected the adult males that developed from irradiated larvae, mated them with unirradiated females, and then counted the number of mutant F$_1$ flies produced by each male. All mutant flies that appeared were used in subsequent crosses to determine if their mutant phenotypes were genetic. She obtained the following results (M. L. Alexander. 1954. *Genetics* 39:409–428):

Group	Number of offspring	Offspring with a genetic mutation
Control (0 r)	45,504	0
Irradiated (3000 r)	49,512	71
Irradiated (3975 r)	50,159	70

a. Calculate the mutation rates of the control group and the two groups of irradiated flies.

b. On the basis of these data, do you think radiation has any effect on mutation? Explain your answer.

31. A genetics instructor designs a laboratory experiment to study the effects of UV radiation on mutation in bacteria. In the experiment, the students expose bacteria plated on petri plates to UV light for different lengths of time, place the plates in an incubator for 48 hours, and then count the number of colonies that appear on each plate. The plates that have received more UV radiation should have more pyrimidine dimers, which block replication; thus, fewer colonies should appear on the plates exposed to UV light for longer periods of time. Before the students carry out the experiment, the instructor warns them that, while the bacteria are in the incubator, the students must not open the incubator door unless the room is darkened. Why should the bacteria not be exposed to light?

Section 10.4

*32. A plant breeder wants to isolate mutants in tomatoes that are defective in DNA repair. However, this breeder does not have the expertise or equipment to study enzymes in DNA-repair systems. How can the breeder identify tomato plants that are deficient in DNA repair? What are the traits to look for?

CHALLENGE QUESTIONS

Section 10.1

33. Robert Bost and Richard Cribbs studied a strain of *E. coli* (*araB14*) that possessed a nonsense mutation in the structural gene that encodes L-ribulokinase, an enzyme that allows the bacteria to metabolize the sugar arabinose (R. Bost and R. Cribbs. 1969. *Genetics* 62:1–8). From the *araB14* strain, they isolated some bacteria that possessed mutations that caused them to revert back to the wild type. Genetic analysis of these revertants showed that they possessed two different suppressor mutations. One suppressor mutation (*R1*) was linked to the original mutation in the L-ribulokinase and probably occurred at the same locus. By itself, this mutation allowed the production of L-ribulokinase, but the enzyme was not as effective in metabolizing arabinose as the enzyme encoded by the wild-type allele. The second suppressor mutation (*Su^B*) was not linked to the original mutation. In conjunction with the *R1* mutation, *Su^B* allowed the production of L-ribulokinase, but *Su^B* by itself was not able to suppress the original mutation.

a. On the basis of this information, are the *R1* and *Su^B* mutations intragenic suppressors or intergenic suppressors? Explain your reasoning.

b. Propose an explanation for how *R1* and *Su^B* restore the ability of *araB14* to metabolize arabinose and why *Su^B* is able to more fully restore the ability to metabolize arabinose.

34. Achondroplasia is an autosomal dominant disorder characterized by disproportionate short stature: the legs and arms are short compared with the head and trunk. The disorder is due to a base substitution in the gene, located on the short arm of chromosome 4, for fibroblast growth factor receptor 3 (FGFR3).

Although achondroplasia is clearly inherited as an autosomal dominant trait, more than 80% of the people who have achondroplasia are born to parents with normal stature. This high percentage indicates that most cases are caused by newly arising mutations; these cases (not inherited from an affected parent) are referred to as sporadic. Findings from molecular studies have demonstrated that sporadic cases of achondroplasia are almost always caused by mutations inherited from the father (paternal mutations). In addition, the occurrence of achondroplasia is higher among older fathers; indeed, approximately 50% of children with achondroplasia are born to fathers older than 35 years of age. There is no association with maternal age. The mutation rate for achondroplasia (about 4×10^{-5} mutations per gamete) is high compared with those for other genetic disorders. Explain why most spontaneous mutations for achondroplasia are paternal in origin and why the occurrence of achondroplasia is higher among older fathers.

A family of three who have achondroplasia. [Gail Burton/AP.]

35. Tay–Sachs disease is a severe, autosomal recessive genetic disease that produces deafness, blindness, seizures, and, eventually, death. The disease results from a defect in the *HEXA* gene, which encodes hexosaminidase A. This enzyme normally degrades G_{M2} gangliosides. In the absence of hexosaminidase A, G_{M2} gangliosides accumulate in the brain. The results of recent molecular studies showed that the most common mutation causing Tay–Sachs disease is a 4-bp insertion that produces a downstream premature stop codon. Results of further studies have revealed that the transcription of the *HEXA* gene is normal in people who have Tay–Sachs disease, but the *HEXA* mRNA is unstable. Propose a mechanism to account for how a premature stop codon could cause mRNA instability.

36. Mutations *ochre* and *amber* are two types of nonsense mutations. Before the genetic code was worked out, Sydney Brenner, Anthony O. Stretton, and Samuel Kaplan applied different types of mutagens to bacteriophages in an attempt to determine the bases present in the codons responsible for *amber* and *ochre* mutations. They knew that *ochre* and *amber* mutants were suppressed by different types of mutations, demonstrating that each is a different termination codon. They obtained the following results:

   (1) A single-base substitution could convert an *ochre* mutation into an *amber* mutation.
   (2) Hydroxylamine induced both *ochre* and *amber* mutations in wild-type phages.
   (3) 2-Aminopurine caused *ochre* to mutate to *amber*.
   (4) Hydroxylamine did not cause *ochre* to mutate to *amber*.

   These data do not allow the complete nucleotide sequence of the *amber* and *ochre* codons to be worked out, but they do provide some information about the bases found in the nonsense mutations.

   a. What conclusions about the bases found in the codons of *amber* and *ochre* mutations can be made from these observations?
   b. Of the three nonsense codons (UAA, UAG, UGA), which represents the *ochre* mutation?

Section 10.3

37. To determine whether radiation associated with the atomic bombings of Hiroshima and Nagasaki produced recessive germ-line mutations, scientists examined the sex ratio of the children of the survivors of the blasts. Can you explain why an increase in germ-line mutations might be expected to alter the sex ratio?

Section 10.4

38. Trichothiodystrophy is a human inherited disorder characterized by premature aging, including osteoporosis, osteosclerosis, early graying, infertility, and reduced life span. The results of recent studies showed that the mutation that causes this disorder occurs in a gene that encodes a DNA helicase. Propose a mechanism for how a mutation in a DNA helicase might cause premature aging. Be sure to relate the symptoms of the disorder to possible functions of the helicase enzyme.

11 QUANTITATIVE GENETICS

Methods of quantitative genetics coupled with molecular techniques have been used to identify a gene that determines oil content in corn. [Walter Bibikow/Getty Images.]

CORN OIL AND QUANTITATIVE GENETICS

In the year 2010, the world's population was estimated to be more than 6.8 billion people. The United Nations projects that the world population will surpass 7 billion in 2012 and reach 9 billion by 2050. Feeding those billions of additional people will be a major challenge for agriculture in next few decades. Crop plants will have to provide most of the calories and nutrients required for world's future population. Because of dwindling petroleum supplies and concerns about global warming, plants are also increasingly being utilized as sources of biofuels, placing additional demands on crop production. How will agriculture provide sufficient crop yields to supply the food and energy requirements of the future?

To help meet the need for increased crop production, plant breeders are using the latest techniques of genetics in their quest to develop higher-yielding, more-efficient crop plants. The power of this approach is demonstrated by research aimed at increasing the oil content of corn. The oil content of corn is inherited, but it is not a simple single-gene characteristic such as seed shape in peas. Numerous genes and environmental factors contribute to the oil content of corn, and the inheritance of oil content is more complex than that of the characteristics that we have studied so far. For traits such as oil content, several loci frequently interact and their expression is affected by environmental factors. Can the inheritance of a complex characteristic such as oil content be studied? Is it possible to predict the oil content of a plant on the basis of its breeding? The answers are yes—at least in part—but these questions cannot be addressed with the methods that we used for simple genetic characteristics. Instead, we must use statistical procedures that have been developed for analyzing complex characteristics. The genetic analysis of complex characteristics such as the oil content of corn is known as **quantitative genetics**.

In 2008, geneticists used a combination of quantitative genetics and molecular techniques to identify a key gene that controls oil content in corn. First, they conducted crosses between high-oil corn plants and low-oil plants to identify chromosomal regions that play an important role in determining oil production. Chromosome regions containing genes that influence a quantitative trait are termed **quantitative trait loci** (QTLs).

Through these crosses, the geneticists located a QTL that affected oil content on corn chromosome 6.

Fine-scale genetic mapping further narrowed the QTL down to a small region of 4.2 centiMorgans on chromosome 6. Researchers sequenced DNA from the region and found that it contained five genes, one of which was *DGAT1-2*, a gene known to produce an enzyme that catalyzes the final step in a pathway for triacyglycerol biosynthesis. DNA from the *DGAT1-2* gene in a high-oil-producing strain contained an insertion of a codon that added phenylalanine to the enzyme, an insertion that was missing from a low-oil strain. The researchers confirmed the effect of the additional phenylalanine codon on oil production by producing transgenic corn that contained the extra codon; oil production of these transgenic strains increased compared with that of transgenic strains without the extra codon. Interestingly, the extra phenylalanine codon is present in wild relatives of corn, suggesting that the codon was lost in the process of domestication or subsequent breeding of modern varieties.

This research suggests that the oil content of corn and other plants might be increased by genetically modifying their *DGAT1-2* genes to contain the extra codon for phenylalanine. Other studies that similarly combine quantitative and molecular analyses have recently led to the identification of genes that increase the vitamin A content of rice and increase sugar production by tomatoes.

This chapter is about the genetic analysis of complex characteristics such as oil content in corn. We begin by considering the differences between quantitative and qualitative characteristics and why the expression of some characteristics varies continuously. We'll see how quantitative characteristics are often influenced by many genes, each of which has a small effect on the phenotype. Next, we will examine statistical procedures for describing and analyzing quantitative characteristics. We will consider how much of phenotypic variation can be attributed to genetic and environmental influences and will conclude by looking at the effects of selection on quantitative characteristics.

11.1 Quantitative Characteristics Vary Continuously and Many Are Influenced by Alleles at Multiple Loci

Qualitative, or discontinuous, characteristics possess only a few distinct phenotypes (**Figure 11.1a**); these characteristics are the types studied by Mendel and have been the focus of our attention thus far. However, many characteristics vary continuously along a scale of measurement with many overlapping phenotypes (**Figure 11.1b**). They are referred to as *continuous characteristics*; they are also called *quantitative characteristics* because any individual's phenotype must be described with a quantitative measurement. Quantitative characteristics might include height, weight, and blood pressure in humans, growth rate in mice, seed weight in plants, and milk production in cattle.

Quantitative characteristics arise from two phenomena. First, many are polygenic: they are influenced by genes at many loci. If many loci take part, many genotypes are possible, each producing a slightly different phenotype. Second,

11.1 Discontinuous and continuous characteristics differ in the number of phenotypes exhibited.

Table 11.1 Hypothetical example of plant height determined by pairs of alleles at each of three loci

Plant Genotype	Doses of Hormone	Height (cm)
$A^-A^-\ B^-B^-\ C^-C^-$	0	10
$A^+A^-\ B^-B^-\ C^-C^-$	1	11
$A^-A^-\ B^+B^-\ C^-C^-$		
$A^-A^-\ B^-B^-\ C^-C^+$		
$A^+A^+\ B^-B^-\ C^-C^-$	2	12
$A^-A^-\ B^+B^+\ C^-C^-$		
$A^-A^-\ B^-B^-\ C^+C^+$		
$A^+A^-\ B^+B^-\ C^-C^-$		
$A^+A^-\ B^-B^-\ C^+C^-$		
$A^-A^-\ B^+B^-\ C^+C^-$		
$A^+A^+\ B^+B^-\ C^-C^-$	3	13
$A^+A^+\ B^-B^-\ C^+C^-$		
$A^+A^-\ B^+B^+\ C^-C^-$		
$A^-A^-\ B^+B^+\ C^+C^-$		
$A^+A^-\ B^-B^-\ C^+C^+$		
$A^-A^-\ B^-B^-\ C^+C^+$		
$A^+A^-\ B^+B^-\ C^+C^-$		
$A^+A^+\ B^+B^+\ C^-C^-$	4	14
$A^+A^+\ B^+B^-\ C^+C^-$		
$A^+A^-\ B^+B^+\ C^+C^-$		
$A^-A^-\ B^+B^+\ C^+C^+$		
$A^+A^+\ B^-B^-\ C^+C^+$		
$A^+A^-\ B^+B^-\ C^+C^+$		
$A^+A^+\ B^+B^+\ C^+C^-$	5	15
$A^+A^-\ B^+B^+\ C^+C^+$		
$A^+A^+\ B^+B^-\ C^+C^+$		
$A^+A^+\ B^+B^+\ C^+C^+$	6	16

Note: Each + allele contributes 1 cm in height above a baseline of 10 cm.

quantitative characteristics often arise when environmental factors affect the phenotype because environmental differences result in a single genotype producing a range of phenotypes. Most continuously varying characteristics are *both* polygenic *and* influenced by environmental factors, and these characteristics are said to be multifactorial.

The Relation Between Genotype and Phenotype

For many discontinuous characteristics, the relation between genotype and phenotype is straightforward. Each genotype produces a single phenotype, and most phenotypes are encoded by a single genotype. Dominance and epistasis may allow two or three genotypes to produce the same phenotype, but the relation remains simple. This simple relation between genotype and phenotype allowed Mendel to decipher the basic rules of inheritance from his crosses with pea plants; it also permits us both to predict the outcome of genetic crosses and to assign genotypes to individuals.

For quantitative characteristics, the relation between genotype and phenotype is often more complex. If the characteristic is polygenic, many different genotypes are possible, several of which may produce the same phenotype. For instance, consider a plant whose height is determined by three loci (A, B, and C), each of which has two alleles. Assume that one allele at each locus (A^+, B^+, and C^+) encodes a plant hormone that causes the plant to grow 1 cm above its baseline height of 10 cm. The other allele at each locus (A^-, B^-, and C^-) does not encode a plant hormone and thus does not contribute to additional height. If we consider only the two alleles at a single locus, 3 genotypes are possible (A^+A^+, A^+A^-, and A^-A^-). If all three loci are taken into account, there are a total of $3^3 = 27$ possible multilocus genotypes ($A^+A^+\ B^+B^+\ C^+C^+$, $A^+A^-\ B^+B^+\ C^+C^+$, etc.). Although there are 27 genotypes, they produce only seven phenotypes (10 cm, 11 cm, 12 cm, 13 cm, 14 cm, 15 cm, and 16 cm in height). Some of the genotypes produce the same phenotype (**Table 11.1**); for example, genotypes $A^+A^-\ B^-B^-\ C^-C^-$, $A^-A^-\ B^+B^-\ C^-C^-$, and $A^-A^-\ B^-B^-\ C^+C^-$ all have one gene that encodes a plant hormone. These genotypes produce one dose of the hormone and a plant that is 11 cm tall. Even in this simple example of only three loci, the relation between genotype and phenotype is quite complex. The more loci encoding a characteristic, the greater the complexity.

The influence of environment on a characteristic also can complicate the relation between genotype and phenotype. Because of environmental effects, the same genotype may produce a range of potential phenotypes. The phenotypic ranges of different genotypes may overlap, making it difficult to know whether individuals differ in phenotype because of genetic or environmental differences (**Figure 11.2**).

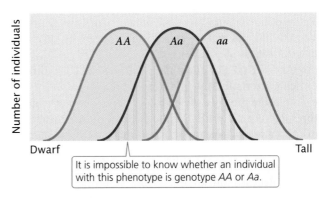

11.2 For a quantitative characteristic, each genotype may produce a range of possible phenotypes. In this hypothetical example, the phenotypes produced by genotypes AA, Aa, and aa overlap.

In summary, the simple relation between genotype and phenotype that exists for many qualitative (discontinuous) characteristics is absent for quantitative characteristics, and it is impossible to assign a genotype to an individual on the basis of its phenotype alone. The methods used for analyzing qualitative characteristics (examining the phenotypic ratios of progeny from a genetic cross) will not work for quantitative characteristics. Our goal remains the same: we wish to make predictions about the phenotypes of offspring produced in a genetic cross. We may also want to know how much of the variation in a characteristic results from genetic differences and how much results from environmental differences. To answer these questions, we must turn to statistical methods that allow us to make predictions about the inheritance of phenotypes in the absence of information about the underlying genotypes. TRY PROBLEM 16 →

Types of Quantitative Characteristics

Before we look more closely at polygenic characteristics and relevant statistical methods, we need to more clearly define what is meant by a quantitative characteristic. Thus far, we have considered only quantitative characteristics that vary continuously in a population. A *continuous characteristic* can theoretically assume any value between two extremes; the number of phenotypes is limited only by our ability to precisely measure the phenotype. Human height is a continuous characteristic because, within certain limits, people can theoretically have any height. Although the number of phenotypes possible with a continuous characteristic is infinite, we often group similar phenotypes together for convenience; we may say that two people are both 5 feet 11 inches tall, but careful measurement may show that one is slightly taller than the other.

Some characteristics are not continuous but are nevertheless considered quantitative because they are determined by multiple genetic and environmental factors. **Meristic characteristics**, for instance, are measured in whole numbers. An example is litter size: a female mouse may have 4, 5, or 6 pups but not 4.13 pups. A meristic characteristic has a limited number of distinct phenotypes, but the underlying determination of the characteristic may still be quantitative. These characteristics must therefore be analyzed with the same techniques that we use to study continuous quantitative characteristics.

Another type of quantitative characteristic is a **threshold characteristic**, which is simply present or absent. For example, the presence of some diseases can be considered a threshold characteristic. Although threshold characteristics exhibit only two phenotypes, they are considered quantitative because they, too, are determined by multiple genetic and environmental factors. The expression of the characteristic depends on an underlying susceptibility (usually referred to as liability or risk) that varies continuously. When the susceptibility is larger than a threshold value, a specific trait is expressed (**Figure 11.3**). Diseases are often threshold

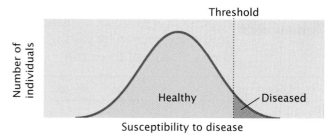

11.3 Threshold characteristics display only two possible phenotypes—the trait is either present or absent—but they are quantitative because the underlying susceptibility to the characteristic varies continuously. When the susceptibility exceeds a threshold value, the characteristic is expressed.

characteristics because many factors, both genetic and environmental, contribute to disease susceptibility. If enough of the susceptibility factors are present, the disease develops; otherwise, it is absent. Although we focus on the genetics of continuous characteristics in this chapter, the same principles apply to many meristic and threshold characteristics.

Just because a characteristic can be measured on a continuous scale does not mean that it exhibits quantitative variation. One of the characteristics studied by Mendel was height of the pea plant, which can be described by measuring the length of the plant's stem. However, Mendel's particular plants exhibited only two distinct phenotypes (some were tall and others short), and these differences were determined by alleles at a single locus. The differences that Mendel studied were therefore discontinuous in nature.

CONCEPTS

Characteristics whose phenotypes vary continuously are called quantitative characteristics. For most quantitative characteristics, the relation between genotype and phenotype is complex. Some characteristics whose phenotypes do not vary continuously also are considered quantitative because they are influenced by multiple genes and environmental factors.

Polygenic Inheritance

After the rediscovery of Mendel's work in 1900, questions soon arose about the inheritance of continuously varying characteristics. These characteristics had already been the focus of a group of biologists and statisticians, led by Francis Galton, who used statistical procedures to examine the inheritance of quantitative characteristics such as human height and intelligence. The results of these studies showed that quantitative characteristics are inherited, although the mechanism of inheritance was not yet known. Some biometricians argued that the inheritance of quantitative characteristics could not be explained by Mendelian principles, whereas others felt that Mendel's principles acting on numerous genes (polygenes) could adequately account for the inheritance of quantitative characteristics.

This conflict began to be resolved by the work of Wilhelm Johannsen, who showed that continuous variation in the weight of beans was influenced by both genetic and environmental factors. George Udny Yule, a mathematician, proposed in 1906 that several genes acting together could produce continuous characteristics. This hypothesis was later confirmed by Herman Nilsson-Ehle, working on wheat and tobacco, and by Edward East, working on corn. The argument was finally laid to rest in 1918, when Ronald Fisher demonstrated that the inheritance of quantitative characteristics could indeed be explained by the cumulative effects of many genes, each following Mendel's rules.

Kernel Color in Wheat

To illustrate how multiple genes acting on a characteristic can produce a continuous range of phenotypes, let us examine one of the first demonstrations of polygenic inheritance. Nilsson-Ehle studied kernel color in wheat and found that the intensity of red pigmentation was determined by three unlinked loci, each of which had two alleles.

Nilsson-Ehle's cross Nilsson-Ehle obtained several homozygous varieties of wheat that differed in color. Like Mendel, he performed crosses between these homozygous varieties and studied the ratios of phenotypes in the progeny. In one experiment, he crossed a variety of wheat that possessed white kernels with a variety that possessed purple (very dark red) kernels and obtained the following results:

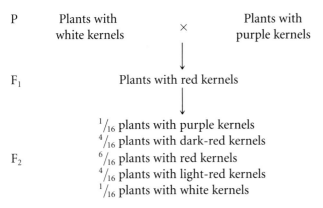

Interpretation of the cross Nilsson-Ehle interpreted this phenotypic ratio as the result of the segregation of alleles at two loci. (Although he found alleles at three loci that affect kernel color, the two varieties used in this cross differed at only two of the loci.) He proposed that there were two alleles at each locus: one that produced red pigment and another that produced no pigment. We'll designate the alleles that encoded pigment A^+ and B^+ and the alleles that encoded no pigment A^- and B^-. Nilsson-Ehle recognized that the effects of the genes were additive. Each gene seemed to contribute equally to color; so the overall phenotype could be determined by adding the effects of all the genes, as shown in the following table.

Genotype	Doses of pigment	Phenotype
$A^+A^+\ B^+B^+$	4	Purple
$A^+A^+\ B^+B^-$ $A^+A^-\ B^+B^+$	3	Dark red
$A^+A^+\ B^-B^-$ $A^-A^-\ B^+B^+$ $A^+A^-\ B^+B^-$	2	Red
$A^+A^-\ B^-B^-$ $A^-A^-\ B^+B^-$	1	Light red
$A^-A^-\ B^-B^-$	0	White

Notice that the purple and white phenotypes are each encoded by a single genotype, but other phenotypes may result from several different genotypes.

From these results, we see that five phenotypes are possible when alleles at two loci influence the phenotype and the effects of the genes are additive. When alleles at more than two loci influence the phenotype, more phenotypes are possible, and the color would appear to vary continuously between white and purple. If environmental factors had influenced the characteristic, individuals of the same genotype would vary somewhat in color, making it even more difficult to distinguish between discrete phenotypic classes. Luckily, environment played little role in determining kernel color in Nilsson-Ehle's crosses, and only a few loci encoded color; so Nilsson-Ehle was able to distinguish among the different phenotypic classes. This ability allowed him to see the Mendelian nature of the characteristic.

Let's now see how Mendel's principles explain the ratio obtained by Nilsson-Ehle in his F_2 progeny. Remember that Nilsson-Ehle crossed the homozygous purple variety ($A^+A^+\ B^+B^+$) with the homozygous white variety ($A^-A^-\ B^-B^-$), producing F_1 progeny that were heterozygous at both loci ($A^+A^-\ B^+B^-$). This is a dihybrid cross, like those that we worked in Chapter 3, except that both loci encode the same trait. All the F_1 plants possessed two pigment-producing alleles that allowed two doses of color to make red kernels. The types and proportions of progeny expected in the F_2 can be found by applying Mendel's principles of segregation and independent assortment.

Let's first examine the effects of each locus separately. At the first locus, two heterozygous F_1s are crossed ($A^+A^- \times A^+A^-$). As we learned in Chapter 3, when two heterozygotes are crossed, we expect progeny in the proportions $1/4\ A^+A^+$, $1/2\ A^+A^-$, and $1/4\ A^-A^-$. At the second locus, two heterozygotes also are crossed, and, again, we expect progeny in the proportions $1/4\ B^+B^+$, $1/2\ B^+B^-$, and $1/4\ B^-B^-$.

To obtain the probability of combinations of genes at both loci, we must use the multiplication rule of probability (see Chapter 3), which is based on Mendel's principle of independent assortment. The expected proportion of F_2 progeny with genotype $A^+A^+\ B^+B^+$ is the product of the probability of obtaining genotype A^+A^+ ($1/4$) and the probability of obtaining genotype B^+B^+ ($1/4$), or $1/4 \times 1/4 = 1/16$

(Figure 11.4). The probabilities of each of the phenotypes can then be obtained by adding the probabilities of all the genotypes that produce that phenotype. For example, the red phenotype is produced by three genotypes:

Genotype	Probability
$A^+A^+\ B^-B^-$	$1/16$
$A^-A^-\ B^+B^+$	$1/16$
$A^+A^-\ B^+B^-$	$1/4$

Thus, the overall probability of obtaining red kernels in the F_2 progeny is $1/16 + 1/16 + 1/4 = 6/16$. Figure 11.4 shows that the phenotypic ratio expected in the F_2 is $1/16$ purple, $4/16$ dark red, $6/16$ red, $4/16$ light red, and $1/16$ white. This phenotypic ratio is precisely what Nilsson-Ehle observed in his F_2 progeny, demonstrating that the inheritance of a continuously varying characteristic such as kernel color is indeed according to Mendel's basic principles.

Conclusions and implications Nilsson-Ehle's crosses demonstrated that the difference between the inheritance of genes influencing quantitative characteristics and the inheritance of genes influencing discontinuous characteristics is in the *number* of loci that determine the characteristic. When multiple loci affect a character, more genotypes are possible; so the relation between the genotype and the phenotype is less obvious. As the number of loci affecting a character increases, the number of phenotypic classes in the F_2 increases (**Figure 11.5**).

Several conditions of Nilsson-Ehle's crosses greatly simplified the polygenic inheritance of kernel color and made it possible for him to recognize the Mendelian nature of the characteristic. First, genes affecting color segregated at only two or three loci. If genes at many loci had been segregating, he would have had difficulty in distinguishing the phenotypic classes. Second, the genes affecting kernel color had strictly additive effects, making the relation between genotype and phenotype simple. Third, environment played almost no role in the phenotype; had environmental factors modified the phenotypes, distinguishing between the five phenotypic classes would have been difficult. Finally, the loci that Nilsson-Ehle studied were not linked; so the genes assorted independently. Nilsson-Ehle was fortunate: for many polygenic characteristics, these simplifying conditions are not present and Mendelian inheritance of these characteristics is not obvious. TRY PROBLEM 17 →

Determining Gene Number for a Polygenic Characteristic

The proportion of F_2 individuals that resemble one of the original parents can be used to estimate the number of genes affecting a polygenic trait. When two individuals homozygous for different alleles at a single locus are crossed ($A^1A^1 \times A^2A^2$) and the resulting F_1 are interbred ($A^1A^2 \times A^1A^2$), one-fourth of the F_2 should be homozygous like each of the original

11.4 Nilsson-Ehle demonstrated that kernel color in wheat is inherited according to Mendelian principles. The ratio of phenotypes in the F_2 can be determined by breaking the dihybrid cross into two simple single-locus crosses and combining the results by using the multiplication rule.

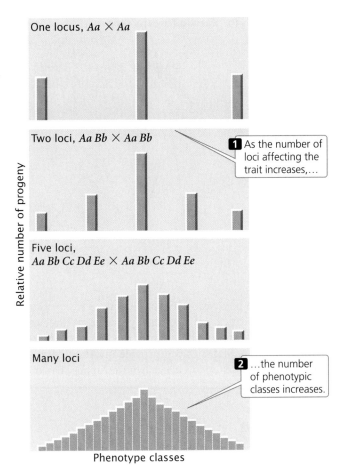

11.5 The results of crossing individuals heterozygous for different numbers of loci affecting a characteristic.

parents. If the original parents are homozygous for different alleles at two loci, as are those in Nilsson-Ehle's crosses, then $1/4 \times 1/4 = 1/16$ of the F_2 should resemble one of the original homozygous parents. Generally, $(1/4)^n$ will be the number of individuals in the F_2 progeny that should resemble each of the original homozygous parents, where n equals the number of loci with a segregating pair of alleles that affects the characteristic. This equation provides us with a possible means of determining the number of loci influencing a quantitative characteristic.

To illustrate the use of this equation, assume that we cross two different homozygous varieties of pea plants that differ in height by 16 cm, interbreed the F_1, and find that approximately $1/256$ of the F_2 are similar to one of the original homozygous parental varieties. This outcome would suggest that four loci with segregating pairs of alleles ($1/256 = (1/4)^4$) are responsible for the height difference between the two varieties. Because the two homozygous strains differ in height by 16 cm and there are four loci each of which has two alleles (eight alleles in all), each of the alleles contributes 16 cm/8 = 2 cm in height.

This method for determining the number of loci affecting phenotypic differences requires the use of homozygous strains, which may be difficult to obtain in some organisms.

It also assumes that all the genes influencing the characteristic have equal effects, that their effects are additive, and that the loci are unlinked. For many polygenic characteristics, these assumptions are not valid, and so this method of determining the number of genes affecting a characteristic has limited application. **TRY PROBLEM 19**

CONCEPTS

The principles that determine the inheritance of quantitative characteristics are the same as the principles that determine the inheritance of discontinuous characteristics, but more genes take part in the determination of quantitative characteristics.

✔ **CONCEPT CHECK 1**

Briefly explain how the number of genes influencing a polygenic trait can be determined?

11.2 Statistical Methods Are Required for Analyzing Quantitative Characteristics

Because quantitative characteristics are described by a measurement and are influenced by multiple factors, their inheritance must be analyzed statistically. This section will explain the basic concepts of statistics that are used to analyze quantitative characteristics.

Distributions

Understanding the genetic basis of any characteristic begins with a description of the numbers and kinds of phenotypes present in a group of individuals. Phenotypic variation in a group can be conveniently represented by a **frequency distribution**, which is a graph of the frequencies (numbers or proportions) of the different phenotypes (**Figure 11.6**). In a typical frequency distribution, the phenotypic classes

11.6 A frequency distribution is a graph that displays the number or proportion of different phenotypes. Phenotypic values are plotted on the horizontal axis, and the numbers (or proportions) of individuals in each class are plotted on the vertical axis.

11.7 Distributions of phenotypes can assume several different shapes.

are plotted on the horizontal (*x*) axis, and the numbers (or proportions) of individuals in each class are plotted on the vertical (*y*) axis. A frequency distribution is a concise method of summarizing all phenotypes of a quantitative characteristic.

Connecting the points of a frequency distribution with a line creates a curve that is characteristic of the distribution. Many quantitative characteristics exhibit a symmetrical (bell-shaped) curve called a **normal distribution** (**Figure 11.7a**). Normal distributions arise when a large number of independent factors contribute to a measurement, as is often the case in quantitative characteristics. Two other common types of distributions (skewed and bimodal) are illustrated in **Figure 11.7b and c**.

Samples and Populations

Biologists frequently need to describe the distribution of phenotypes exhibited by a group of individuals. We might want to describe the height of students at the University of Texas (UT), but there are more than 40,000 students at UT, and measuring every one of them would be impractical. Scientists are constantly confronted with this problem: the group of interest, called the **population**, is too large for a complete census. One solution is to measure a smaller collection of individuals, called a **sample**, and use measurements made on the sample to describe the population.

To provide an accurate description of the population, a good sample must have several characteristics. First, it must be representative of the whole population. If our sample consisted entirely of members of the UT basketball team, for instance, we would probably overestimate the true height of the students. One way to ensure that a sample is representative of the population is to select the members of the sample randomly. Second, the sample must be large enough that chance differences between individuals in the sample and the overall population do not distort the estimate of the population measurements. If we measured only three students at UT and just by chance all three were short, we would underestimate the true height of the student population. Statistics can provide information about how much confidence to expect from estimates based on random samples.

> **CONCEPTS**
>
> In statistics, the population is the group of interest; a sample is a subset of the population. The sample should be representative of the population and large enough to minimize chance differences between the population and the sample.
>
> ✔ **CONCEPT CHECK 2**
>
> A geneticist is interested in whether asthma is caused by a mutation in the *DS112* gene. The geneticist collects DNA from 120 people with asthma and 100 healthy people and sequences the DNA. She finds that 35 of the people with asthma and none of the healthy people have a mutation in the *DS112* gene. What is the population in this study?
>
> a. The 120 people with asthma
> b. The 100 healthy people
> c. The 35 people with a mutation in their gene
> d. All people with asthma

The Mean

The **mean**, also called the average, provides information about the center of the distribution. If we measured the heights of 10-year-old and 18-year-old boys and plotted a frequency distribution for each group, we would find that both distributions are normal, but the two distributions would be centered at different heights, and this difference would be indicated in their different means (**Figure 11.8**).

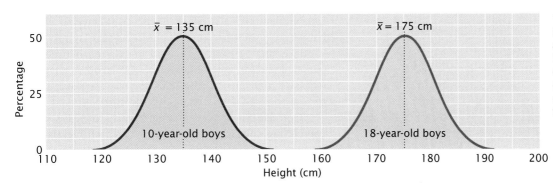

11.8 The mean provides information about the center of a distribution. Both distributions of heights of 10-year-old and 18-year-old boys are normal, but they have different locations along a continuum of height, which makes their means different.

Suppose we have five measurements of height in centimeters: 160, 161, 167, 164, and 165. If we represent a group of measurements as x_1, x_2, x_3, and so forth, then the mean ($\bar{x}$) is calculated by adding all the individual measurements and dividing by the total number of measurements in the sample (n):

$$\bar{x} = \frac{x_1 + x_2 + x_3 + \ldots + x_n}{n} \quad (11.1)$$

In our example, $x_1 = 160$, $x_2 = 161$, $x_3 = 167$, and so forth. The mean height ($\bar{x}$) equals:

$$\bar{x} = \frac{160 + 161 + 167 + 164 + 165}{5} = \frac{817}{5} = 163.4$$

A shorthand way to represent this formula is

$$\bar{x} = \frac{\sum x_i}{n} \quad (11.2)$$

or

$$\bar{x} = \frac{1}{n}\sum x_i \quad (11.3)$$

where the symbol $\sum$ means "the summation of" and x_i represents individual x values.

The Variance and Standard Deviation

A statistic that provides key information about a distribution is the **variance**, which indicates the variability of a group of measurements, or how spread out the distribution is. Distributions may have the same mean but different variances (**Figure 11.9**). The larger the variance, the greater the spread of measurements in a distribution about its mean.

The variance (s^2) is defined as the average squared deviation from the mean:

$$s^2 = \frac{\sum (x_i - \bar{x})^2}{n - 1} \quad (11.4)$$

To calculate the variance, we (1) subtract the mean from each measurement and square the value obtained, (2) add all the squared deviations, and (3) divide this sum by the number of original measurements minus 1.

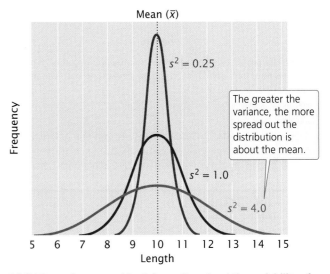

11.9 The variance provides information about the variability of a group of phenotypes. Shown here are three distributions with the same mean but different variances.

Another statistic that is closely related to the variance is the **standard deviation** (s), which is defined as the square root of the variance:

$$s = \sqrt{s^2} \quad (11.5)$$

Whereas the variance is expressed in units squared, the standard deviation is in the same units as the original measurements; so the standard deviation is often preferred for describing the variability of a measurement.

A normal distribution is symmetrical; so the mean and standard deviation are sufficient to describe its shape. The mean plus or minus one standard deviation ($\bar{x} \pm s$) includes approximately 66% of the measurements in a normal distribution; the mean plus or minus two standard deviations ($\bar{x} \pm 2s$) includes approximately 95% of the measurements, and the mean plus or minus three standard deviations ($\bar{x} \pm 3s$) includes approximately 99% of the measurements (**Figure 11.10**). Thus, only 1% of a normally distributed population lies outside the range of ($\bar{x} \pm 3s$).

TRY PROBLEM 22

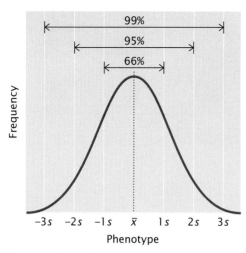

11.10 The proportions of a normal distribution occupied by plus or minus one, two, and three standard deviations from the mean.

> **CONCEPTS**
>
> The mean and variance describe a distribution of measurements: the mean provides information about the location of the center of a distribution, and the variance provides information about its variability.
>
> ✔ **CONCEPT CHECK 3**
>
> The measurements of a distribution with a higher _____ will be more spread out.
>
> a. mean c. standard deviation
> b. variance d. variance and standard deviation

Correlation

The mean and the variance can be used to describe an individual characteristic, but geneticists are frequently interested in more than one characteristic. Often, two or more characteristics vary together. For instance, both the number and the weight of eggs produced by hens are important to the poultry industry. These two characteristics are not independent of each other. There is an inverse relation between egg number and weight: hens that lay more eggs produce smaller eggs. This kind of relation between two characteristics is called a **correlation**. When two characteristics are correlated, a change in one characteristic is likely to be associated with a change in the other.

Correlations between characteristics are measured by a **correlation coefficient** (designated r), which measures the strength of their association. Consider two characteristics, such as human height (x) and arm length (y). To determine how these characteristics are correlated, we first obtain the covariance (cov) of x and y:

$$\text{cov}_{xy} = \frac{\sum(x_i - \bar{x})(y_i - \bar{y})}{n-1} \quad (11.6)$$

The covariance is computed by (1) taking an x value for an individual and subtracting it from the mean of x ($\bar{x}$); (2) taking the y value for the same individual and subtracting it from the mean of y ($\bar{y}$); (3) multiplying the results of these two subtractions; (4) adding the results for all the xy pairs; and (5) dividing this sum by $n - 1$ (where n equals the number of xy pairs).

The correlation coefficient (r) is obtained by dividing the covariance of x and y by the product of the standard deviations of x and y:

$$r = \frac{\text{cov}_{xy}}{s_x s_y} \quad (11.7)$$

A correlation coefficient can theoretically range from -1 to $+1$. A positive value indicates that there is a direct association between the variables (**Figure 11.11a**): as one variable increases, the other variable also tends to increase. A positive correlation exists for human height and weight: tall people tend to weigh more. A negative correlation coefficient indicates that there is an inverse relation between the two variables (**Figure 11.11b**): as one variable increases, the other tends to decrease (as is the case for egg number and egg weight in chickens).

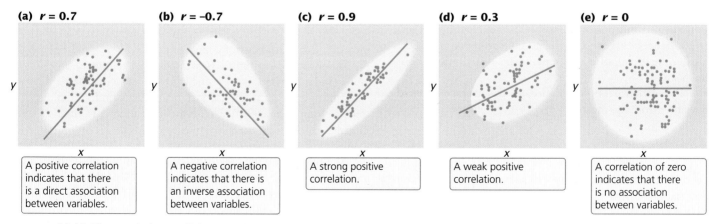

11.11 The correlation coefficient describes the relation between two or more variables.

The absolute value of the correlation coefficient (the size of the coefficient, ignoring its sign) provides information about the strength of association between the variables. A coefficient of −1 or +1 indicates a perfect correlation between the variables, meaning that a change in x is always accompanied by a proportional change in y. Correlation coefficients close to −1 or close to +1 indicate a strong association between the variables: a change in x is almost always associated with a proportional increase in y, as seen in **Figure 11.11c**. On the other hand, a correlation coefficient closer to 0 indicates a weak correlation: a change in x is associated with a change in y but not always (**Figure 11.11d**). A correlation of 0 indicates that there is no association between variables (**Figure 11.11e**).

A correlation coefficient can be computed for two variables measured for the same individual, such as height (x) and weight (y). A correlation coefficient can also be computed for a single variable measured for pairs of individuals. For example, we can calculate for fish the correlation between the number of vertebrae of a parent (x) and the number of vertebrae of its offspring (y), as shown in **Figure 11.12**. This approach is often used in quantitative genetics.

A correlation between two variables indicates only that the variables are associated; it does not imply a cause-and-effect relation. Correlation also does not mean that the values of two variables are the same; it means only that a change in one variable is associated with a proportional change in the other variable. For example, the x and y variables in the following list are almost perfectly correlated, with a correlation coefficient of 0.99.

	x value	y value
	12	123
	14	140
	10	110
	6	61
	3	32
Average:	9	90

A high correlation is found between these x and y variables; larger values of x are always associated with larger values of y. Note that the y values are about 10 times as large as the corresponding x values; so, although x and y are correlated, they are not identical. The distinction between correlation and identity becomes important when we consider the effects of heredity and environment on the correlation of characteristics. **TRY PROBLEM 25**

Regression

Correlation provides information only about the strength and direction of association between variables. However, we often want to know more than just whether two variables are associated; we want to be able to predict the value of one variable, given a value of the other.

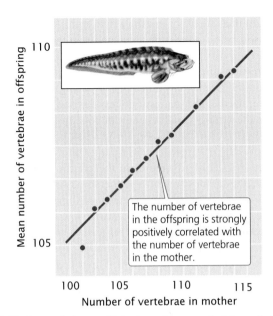

11.12 A correlation coefficient can be computed for a single variable measured for pairs of individuals. Here, the numbers of vertebrae in mothers and offspring of the fish *Zoarces viviparus* are compared.

A positive correlation exists between the body weight of parents and the body weight of their offspring; this correlation exists in part because genes influence body weight, and parents and children have genes in common. Because of this association between parental and offspring phenotypes, we can predict the weight of an individual on the basis of the weights of its parents. This type of statistical prediction is called **regression**. This technique plays an important role in quantitative genetics because it allows us to predict the characteristics of offspring from a given mating, even without knowledge of the genotypes that encode the characteristics.

Regression can be understood by plotting a series of x and y values. **Figure 11.13** illustrates the relation between

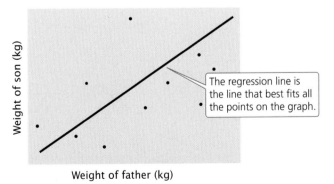

11.13 A regression line defines the relation between two variables. Illustrated here is a regression of the weights of fathers against the weights of sons. Each father–son pair is represented by a point on the graph: the x value of a point is the father's weight and the y value of the point is the son's weight.

the weight of a father (x) and the weight of his son (y). Each father–son pair is represented by a point on the graph. The overall relation between these two variables is depicted by the regression line, which is the line that best fits all the points on the graph (deviations of the points from the line are minimized). The regression line defines the relation between the x and y variables and can be represented by

$$y = a + bx \tag{11.8}$$

In Equation 11.8, x and y represent the x and y variables (in this case, the father's weight and the son's weight, respectively). The variable a is the y intercept of the line, which is the expected value of y when x is 0. Variable b is the slope of the regression line, also called the **regression coefficient**.

Trying to position a regression line by eye is not only very difficult but also inaccurate when there are many points scattered over a wide area. Fortunately, the regression coefficient and the y intercept can be obtained mathematically. The regression coefficient (b) can be computed from the covariance of x and y (cov_{xy}) and the variance of x (s) by

$$b = \frac{\text{cov}_{xy}}{s_x^2} \tag{11.9}$$

The regression coefficient indicates how much y increases, on average, per increase in x. Several regression lines with different regression coefficients are illustrated in **Figure 11.14**. Notice that as the regression coefficient increases, the slope of the regression line increases.

After the regression coefficient has been calculated, the y intercept can be calculated by substituting the regression coefficient and the mean values of x and y into the following equation:

$$a = \bar{y} - b\bar{x} \tag{11.10}$$

The regression equation ($y = a + bx$, Equation 24.8) can then be used to predict the value of any y given the value of x.

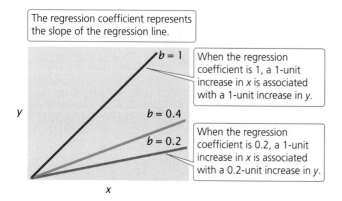

11.14 The regression coefficient, b, represents the change in y per unit change in x. Shown here are regression lines with different regression coefficients.

CONCEPTS

A correlation coefficient measures the strength of association between two variables. The sign (positive or negative) indicates the direction of the correlation; the absolute value measures the strength of the association. Regression is used to predict the value of one variable on the basis of the value of a correlated variable.

✔ **CONCEPT CHECK 4**

In Lubbock, Texas, rainfall and temperature exhibit a significant correlation of -0.7. Which conclusion is correct?

a. There is usually more rainfall when the temperature is high.
b. There is usually more rainfall when the temperature is low.
c. Rainfall is equally likely when the temperature is high or low.

Worked Problem

Body weights of 11 female fishes and the numbers of eggs that they produce are:

Weight (mg)	Eggs (thousands)
x	y
14	61
17	37
24	65
25	69
27	54
33	93
34	87
37	89
40	100
41	90
42	97

What are the correlation coefficient and the regression coefficient for body weight and egg number in these 11 fishes?

• **Solution**

The computations needed to answer this question are given in the table on the facing page. To calculate the correlation and regression coefficients, we first obtain the sum of all the x_i values ($\sum x_i$) and the sum of all the y_i values ($\sum y_i$); these sums are shown in the last row of the table. We can calculate the means of the two variables by dividing the sums by the number of measurements, which is 11:

$$\bar{x} = \frac{\sum x_i}{n} = \frac{334}{11} = 30.36$$

$$\bar{y} = \frac{\sum y_i}{n} = \frac{842}{11} = 76.55$$

After the means have been calculated, the deviations of each value from the mean are computed; these deviations are shown in columns B and E of the table.

A Weight (mg) x	B $x_i - \bar{x}$	C $(x_i - \bar{x})^2$	D Eggs (thousands) y	E $y_i - \bar{y}$	F $(y_i - \bar{y})^2$	G $(x_i - \bar{x})(y_i - \bar{y})$
14	−16.36	267.65	61	−15.55	241.80	254.40
17	−13.36	178.49	37	−39.55	1564.20	528.39
24	−6.36	40.45	65	−11.55	133.40	73.46
25	−5.36	28.73	69	−7.55	57.00	40.47
27	−3.36	11.29	54	−22.55	508.50	75.77
33	2.64	6.97	93	16.45	270.60	43.43
34	3.64	13.25	87	10.45	109.20	38.04
37	6.64	44.09	89	12.45	155.00	82.67
40	9.64	92.93	100	23.45	549.90	226.06
41	10.64	113.21	90	13.45	180.90	143.11
42	11.64	135.49	97	20.45	418.20	238.04
$\sum x_i = 334$		$\sum (x_i - \bar{x})^2 = 932.55$	$\sum y_i = 842$		$\sum (y_i - \bar{y})^2 = 4188.70$	$\sum (x_i - \bar{x})(y_i - \bar{y}) = 1743.84$

Source: R. R. Sokal and F. J. Rohlf, *Biometry*, 2d ed. (San Francisco: W. H. Freeman and Company, 1981).

The deviations are then squared (columns C and F) and summed (last row of columns C and F). Next, the products of the deviation of the x values and the deviation of the y values (($x_i - \bar{x}$) ($y_i - \bar{y}$)) are calculated; these products are shown in column G, and their sum is shown in the last row of column G.

To calculate the covariance, we use Equation 11.6:

$$\text{cov}_{xy} = \frac{\sum (x_i - \bar{x})(y_i - \bar{y})}{n-1} = \frac{1743.84}{10} = 174.38$$

To calculate the covariance and the regression requires the variances and standard deviations of x and y:

$$s_x^2 = \frac{\sum (x_i - \bar{x})^2}{n-1} = \frac{932.55}{10} = 93.26$$

$$s_x = \sqrt{s_x^2} = \sqrt{93.26} = 9.66$$

$$s_y^2 = \frac{\sum (y_i - \bar{y})^2}{n-1} = \frac{4188.70}{10} = 418.87$$

$$s_y = \sqrt{s_y^2} = \sqrt{418.87} = 20.47$$

We can now compute the correlation and regression coefficients as shown here.

Correlation coefficient:

$$r = \frac{\text{cov}_{xy}}{s_x s_y} = \frac{174.38}{9.66 \times 20.47} = 0.88$$

Regression coefficient:

$$b = \frac{\text{cov}_{xy}}{s_x^2} = \frac{174.38}{93.26} = 1.87$$

→ Practice your understanding of correlation and regression by working Problem 26 at the end of the chapter.

Applying Statistics to the Study of a Polygenic Characteristic

Edward East carried out one early statistical study of polygenic inheritance on the length of flowers in tobacco (*Nicotiana longiflora*). He obtained two varieties of tobacco that differed in flower length: one variety had a mean flower length of 40.5 mm, and the other had a mean flower length of 93.3 mm (**Figure 11.15**). These two varieties had been inbred for many generations and were homozygous at all loci contributing to flower length. Thus, there was no genetic variation in the original parental strains; the small differences in flower length within each strain were due to environmental effects on flower length.

When East crossed the two strains, he found that flower length in the F_1 was about halfway between that in the two parents (see Figure 11.15), as would be expected if the genes determining the differences in the two strains were additive in their effects. The variance of flower length in the F_1 was similar to that seen in the parents, because all the F_1 had the same genotype, as did each parental strain (the F_1 were all heterozygous at the genes that differed between the two parental varieties).

East then interbred the F_1 to produce F_2 progeny. The mean flower length of the F_2 was similar to that of the F_1, but the variance of the F_2 was much greater (see Figure 11.15).

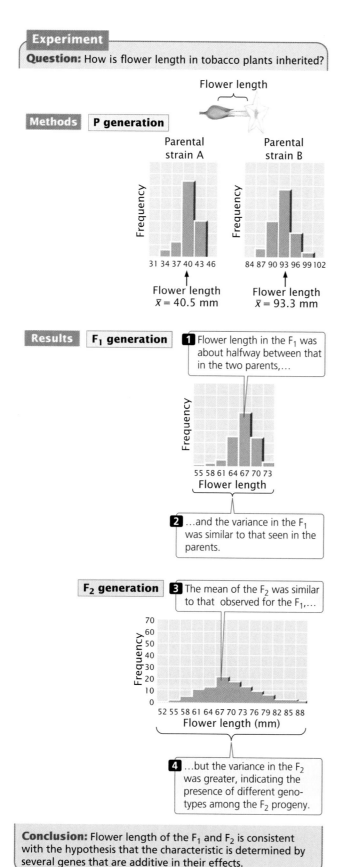

11.15 Edward East conducted an early statistical study of the inheritance of flower length in tobacco.

This greater variability indicates that not all of the F_2 progeny had the same genotype.

East selected some F_2 plants and interbred them to produce F_3 progeny. He found that flower length of the F_3 depended on flower length in the plants selected as their parents. This finding demonstrated that flower-length differences in the F_2 were partly genetic and were therefore passed to the next generation. None of the 444 F_2 plants raised by East exhibited flower lengths similar to those of the two parental strains. This result suggested that more than four loci with pairs of alleles affected flower length in his varieties, because four allelic pairs are expected to produce 1 of 256 progeny $[(1/4)^4 = 1/256]$ having one or the other of the original parental phenotypes.

11.3 Heritability Is Used to Estimate the Proportion of Variation in a Trait That Is Genetic

In addition to being polygenic, quantitative characteristics are frequently influenced by environmental factors. Knowing how much of the variation in a quantitative characteristic is due to genetic differences and how much is due to environmental differences is often useful. The proportion of the total phenotypic variation that is due to genetic differences is known as the **heritability**.

Consider a dairy farmer who owns several hundred milk cows. The farmer notices that some cows consistently produce more milk than others. The nature of these differences is important to the profitability of his dairy operation. If the differences in milk production are largely genetic in origin, then the farmer may be able to boost milk production by selectively breeding the cows that produce the most milk. On the other hand, if the differences are largely environmental in origin, selective breeding will have little effect on milk production, and the farmer might better boost milk production by adjusting the environmental factors associated with higher milk production. To determine the extent of genetic and environmental influences on variation in a characteristic, phenotypic variation in the characteristic must be partitioned into components attributable to different factors.

Phenotypic Variance

To determine how much of phenotypic differences in a population is due to genetic and environmental factors, we must first have some quantitative measure of the phenotype under consideration. Consider a population of wild plants that differ in size. We could collect a representative sample of plants from the population, weigh each plant in the sample, and calculate the mean and variance of plant weight. This **phenotypic variance** is represented by V_P.

Components of phenotypic variance First, some of the phenotypic variance may be due to differences in genotypes among individual members of the population. These differences are termed the **genetic variance** and are represented by V_G.

Second, some of the differences in phenotype may be due to environmental differences among the plants; these differences are termed the **environmental variance**, V_E. Environmental variance includes differences in environmental factors such as the amount of light or water that the plant receives; it also includes random differences in development that cannot be attributed to any specific factor. Any variation in phenotype that is not inherited is, by definition, a part of the environmental variance.

Third, **genetic–environmental interaction variance** (V_{GE}) arises when the effect of a gene depends on the specific environment in which it is found. An example is shown in **Figure 11.16**. In a dry environment, genotype AA produces a plant that averages 12 g in weight, and genotype aa produces a smaller plant that averages 10 g. In a wet environment, genotype aa produces the larger plant, averaging 24 g in weight, whereas genotype AA produces a plant that averages 20 g. In this example, there are clearly differences in the two environments: both genotypes produce heavier plants in the wet environment. There are also differences in the weights of the two genotypes, but the relative performances of the genotypes depend on whether the plants are grown in a wet or a dry environment. In this case, the influences on phenotype cannot be neatly allocated into genetic and environmental components, because the expression of the genotype depends on the environment in which the plant grows. The phenotypic variance must therefore include a component that accounts for the way in which genetic and environmental factors interact.

In summary, the total phenotypic variance can be apportioned into three components:

$$V_P = V_G + V_E + V_{GE} \qquad (11.11)$$

Components of genetic variance Genetic variance can be further subdivided into components consisting of different types of genetic effects. First, **additive genetic variance** (V_A) comprises the additive effects of genes on the phenotype, which can be summed to determine the overall effect on the phenotype. For example, suppose that, in a plant, allele A^1 contributes 2 g in weight and allele A^2 contributes 4 g. If the alleles are strictly additive, then the genotypes would have the following weights:

$$A^1A^1 = 2 + 2 = 4 \text{ g}$$
$$A^1A^2 = 2 + 4 = 6 \text{ g}$$
$$A^2A^2 = 4 + 4 = 8 \text{ g}$$

The genes that Nilsson-Ehle studied, which affect kernel color in wheat, were additive in this way. The additive genetic variance primarily determines the resemblance between parents and offspring. For example, if all of the phenotypic variance is due to additive genetic variance, then the phenotypes of the offspring will be exactly intermediate between those of the parents, but, if some genes have dominance, then offspring may be phenotypically different from both parents (i.e., $Aa \times Aa \rightarrow aa$ offspring).

Second, there is **dominance genetic variance** (V_D) when some genes have a dominance component. In this case, the alleles at a locus are not additive; rather, the effect of an allele depends on the identity of the other allele at that locus. For example, with a dominant allele (T), genotypes TT and Tt have the same phenotype. Here, we cannot simply add the effects of the alleles together, because the effect of the small t allele is masked by the presence of the large T allele. Instead, we must add a component (V_D) to the genetic variance to account for the way in which alleles interact.

Third, genes at different loci may interact in the same way that alleles at the same locus interact. When this genic interaction takes place, the effects of the genes are not additive. For example, Chapter 5 described how coat color in Labrador retrievers exhibits genic interaction; genotypes $BB\ ee$ and $bb\ ee$ both produce yellow dogs, because the effect of alleles at the B locus are masked when ee alleles are present at the E locus. With genic interaction, we must add

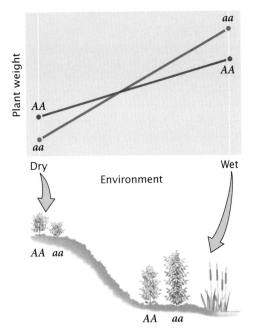

11.16 Genetic–environmental interaction variance is obtained when the effect of a gene depends on the specific environment in which it is found. In this example, the genotype affects plant weight, but the environmental conditions determine which genotype produces the heavier plant.

a third component, called **genic interaction variance** (V_I), to the genetic variance:

$$V_G = V_A + V_D + V_I \quad (11.12)$$

Summary equation We can now integrate these components into one equation to represent all the potential contributions to the phenotypic variance:

$$V_P = V_A + V_D + V_I + V_E + V_{GE} \quad (11.13)$$

This equation provides us with a model that describes the potential causes of differences that we observe among individual phenotypes. It's important to note that this model deals strictly with the observable *differences* (variance) in phenotypes among individual members of a population; it says nothing about the absolute value of the characteristic or about the underlying genotypes that produce these differences.

Types of Heritability

The model of phenotypic variance that we've just developed can be used to determine how much of the phenotypic variance in a characteristic is due to genetic differences. **Broad-sense heritability** (H^2) represents the proportion of phenotypic variance that is due to genetic variance and is calculated by dividing the genetic variance by the phenotypic variance:

$$\text{broad-sense heritability} = H^2 = \frac{V_G}{V_P} \quad (11.14)$$

The symbol H^2 represents broad-sense heritability because it is a measure of variance, which is in units squared.

Broad-sense heritability can potentially range from 0 to 1. A value of 0 indicates that none of the phenotypic variance results from differences in genotype and all of the differences in phenotype result from environmental variation. A value of 1 indicates that all of the phenotypic variance results from differences in genotypes. A heritability value between 0 and 1 indicates that both genetic and environmental factors influence the phenotypic variance.

Often, we are more interested in the proportion of the phenotypic variance that results from the additive genetic variance because, as mentioned earlier, the additive genetic variance primarily determines the resemblance between parents and offspring. **Narrow-sense heritability** (h^2) is equal to the additive genetic variance divided by the phenotypic variance:

$$\text{narrow-sense heritability} = h^2 = \frac{V_A}{V_P} \quad (11.15)$$

TRY PROBLEM 27 →

Calculating Heritability

Having considered the components that contribute to phenotypic variance and having developed a general concept of heritability, we can ask, How do we go about estimating these different components and calculating heritability? There are several ways to measure the heritability of a characteristic. They include eliminating one or more variance components, comparing the resemblance of parents and offspring, comparing the phenotypic variances of individuals with different degrees of relatedness, and measuring the response to selection. The mathematical theory that underlies these calculations of heritability is complex and beyond the scope of this book. Nevertheless, we can develop a general understanding of how heritability is measured.

Heritability by elimination of variance components One way of calculating the broad-sense heritability is to eliminate one of the variance components. We have seen that $V_P = V_G + V_E + V_{GE}$. If we eliminate all environmental variance ($V_E = 0$), then $V_{GE} = 0$ (because, if either V_G or V_E is zero, no genetic–environmental interaction can take place), and $V_P = V_G$. In theory, we might make V_E equal to 0 by ensuring that all individuals were raised in exactly the same environment but, in practice, it is virtually impossible. Instead, we could make V_G equal to 0 by raising genetically identical individuals, causing V_P to be equal to V_E. In a typical experiment, we might raise cloned or highly inbred, identically homozygous individuals in a defined environment and measure their phenotypic variance to estimate V_E. We could then raise a group of genetically variable individuals and measure their phenotypic variance (V_P). Using V_E calculated on the genetically identical individuals, we could obtain the genetic variance of the variable individuals by subtraction:

$$V_{G\text{(of genetically varying individuals)}} = V_{P\text{(of genetically varying individuals)}} - V_{E\text{(of genetically identical individuals)}} \quad (11.16)$$

The broad-sense heritability of the genetically variable individuals would then be calculated as follows:

$$H^2 = \frac{V_{G\text{ (of genetically varying individuals)}}}{V_{P\text{ (of genetically varying individuals)}}} \quad (11.17)$$

Sewall Wright used this method to estimate the heritability of white spotting in guinea pigs. He first measured the phenotypic variance for white spotting in a genetically variable population and found that $V_P = 573$. Then he inbred the guinea pigs for many generations so that they were essentially homozygous and genetically identical. When he measured their phenotypic variance in white spotting, he obtained V_P equal to 340. Because $V_G = 0$ in this group, their $V_P = V_E$. Wright assumed this value of environmental vari-

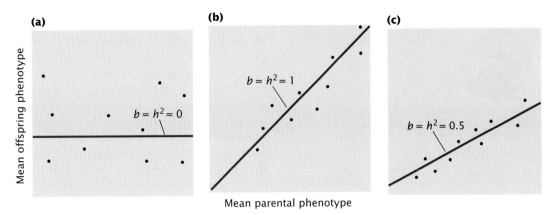

11.17 The narrow-sense heritability, h^2, equals the regression coefficient, b, in a regression of the mean phenotype of the offspring against the mean phenotype of the parents. (a) There is no relation between the parental phenotype and the offspring phenotype. (b) The offspring phenotype is the same as the parental phenotypes. (c) Both genes and environment contribute to the differences in phenotype.

ance for the original (genetically variable) population and estimated their genetic variance:

$$V_P - V_E = V_G$$
$$573 - 340 = 233$$

He then estimated the broad-sense heritability from the genetic and phenotypic variance:

$$H^2 = \frac{V_G}{V_P}$$

$$H^2 = \frac{233}{573} = 0.41$$

This value implies that 41% of the variation in spotting of guinea pigs in Wright's population was due to differences in genotype.

Estimating heritability by using this method assumes that the environmental variance of genetically identical individuals is the same as the environmental variance of the genetically variable individuals, which may not be true. Additionally, this approach can be applied only to organisms for which it is possible to create genetically identical individuals. **TRY PROBLEM 32**

Heritability by parent–offspring regression Another method for estimating heritability is to compare the phenotypes of parents and offspring. When genetic differences are responsible for phenotypic variance, offspring should resemble their parents more than they resemble unrelated individuals, because offspring and parents have some genes in common that help determine their phenotype. Correlation and regression can be used to analyze the association of phenotypes in different individuals.

To calculate the narrow-sense heritability in this way, we first measure the characteristic on a series of parents and offspring. The data are arranged into families, and the mean parental phenotype is plotted against the mean offspring phenotype (**Figure 11.17**). Each data point in the graph represents one family; the value on the x (horizontal) axis is the mean phenotypic value of the parents in a family, and the value on the y (vertical) axis is the mean phenotypic value of the offspring for the family.

Let's assume that there is no narrow-sense heritability for the characteristic ($h^2 = 0$), meaning that genetic differences do not contribute to the phenotypic differences among individuals. In this case, offspring will be no more similar to their parents than they are to unrelated individuals, and the data points will be scattered randomly, generating a regression coefficient of zero (see Figure 11.17a). Next, let's assume that all of the phenotypic differences are due to additive genetic differences ($h^2 = 1.0$). In this case, the mean phenotype of the offspring will be equal to the mean phenotype of the parents, and the regression coefficient will be 1 (see Figure 11.17b). If genes and environment both contribute to the differences in phenotype, both heritability and the regression coefficient will lie between 0 and 1 (see Figure 11.17c). The regression coefficient therefore provides information about the magnitude of the heritability.

A complex mathematical proof (which we will not go into here) demonstrates that, in a regression of the mean phenotype of the offspring against the mean phenotype of the parents, narrow-sense heritability (h^2) equals the regression coefficient (b):

$$h^2 = b_{\text{(regression of mean offspring against mean of both parents)}} \quad (11.18)$$

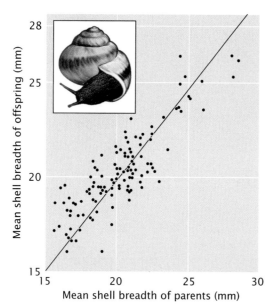

11.18 The heritability of shell breadth in snails can be determined by regression of the phenotype of offspring against the mean phenotype of the parents. The regression coefficient, which equals the heritability, is 0.70. [From L. M. Cook, *Evolution* 19:86–94, 1965.]

An example of calculating heritability by regression of the phenotypes of parents and offspring is illustrated in **Figure 11.18**.

Sometimes, only the phenotype of one parent is known. In a regression of the mean offspring phenotype against the phenotype of only one parent, the narrow-sense heritability equals twice the regression coefficient:

$$h^2 = 2b_{\text{(regression of mean offspring against mean of one parent)}} \quad (11.19)$$

With only one parent, the heritability is twice the regression coefficient because only half the genes of the offspring come from one parent; thus, we must double the regression coefficient to obtain the full heritability.

Heritability and degrees of relatedness A third method for calculating heritability is to compare the phenotypes of individuals having different degrees of relatedness. This method is based on the concept that the more closely related two individuals are, the more genes they have in common.

Monozygotic (identical) twins have 100% of their genes in common, whereas dizygotic (nonidentical) twins have, on average, 50% of their genes in common. If genes are important in determining variability in a characteristic, then monozygotic twins should be more similar in a particular characteristic than dizygotic twins. By using correlation to compare the phenotypes of monozygotic and dizygotic twins, we can estimate broad-sense heritability. A rough estimate of the broad-sense heritability can be obtained by taking twice the difference of the correlation coefficients for a quantitative characteristic in monozygotic and dizygotic twins:

$$H^2 = 2(r_{\text{MZ}} - r_{\text{DZ}}) \quad (11.20)$$

where r_{MZ} equals the correlation coefficient among monozygotic twins and r_{DZ} equals the correlation coefficient among dizygotic twins. For example, suppose we found the correlation of height among the two members of monozygotic twin pairs (r_{MZ}) to be 0.9 and the correlation of height among the two members of dizygotic twins (r_{DZ}) to be 0.5. The broad-sense heritability for height would be $H^2 = 2(0.9 - 0.5) = 2(0.4) = 0.8$. This calculation assumes that the two individuals of a monozygotic twin pair experience environments that are no more similar to each other than those experienced by the two individuals of a dizygotic twin pair. This assumption is often not met when twins have been reared together.

Narrow-sense heritability also can be estimated by comparing the phenotypic variances for a characteristic in full sibs (siblings who have both parents in common as well as 50% of their genes on the average) and half sibs (who have only one parent in common and thus 25% of their genes on the average).

All estimates of heritability depend on the assumption that the environments of related individuals are not more similar than those of unrelated individuals. This assumption is difficult to meet in human studies, because related people are usually reared together. Heritability estimates for humans should therefore always be viewed with caution.

TRY PROBLEM 35

CONCEPTS

Broad-sense heritability is the proportion of phenotypic variance that is due to genetic variance. Narrow-sense heritability is the proportion of phenotypic variance that is due to additive genetic variance. Heritability can be measured by eliminating one of the variance components, by analyzing parent–offspring regression, or by comparing individuals having different degrees of relatedness.

✔ CONCEPT CHECK 5

If the environmental variance (V_E) increases and all other variance components remain the same, what will the effect be?

a. Broad-sense heritability will decrease.
b. Broad-sense heritability will increase.
c. Narrow-sense heritability will increase.
d. Broad-sense heritability will increase, but narrow-sense heritability will decrease.

The Limitations of Heritability

Knowledge of heritability has great practical value because it allows us to statistically predict the phenotypes of offspring on the basis of their parent's phenotype. It also provides useful information about how characteristics will respond to selection (see Section 11.4). In spite of its importance, heritability is frequently misunderstood. Heritability does not provide information about an individual's genes or the environmental factors that control the development of a characteristic, and it says nothing about the nature of differences between groups. This section outlines some limitations and common misconceptions concerning broad- and narrow-sense heritability.

Heritability does not indicate the degree to which a characteristic is genetically determined Heritability is the proportion of the phenotypic variance that is due to genetic variance; it says nothing about the degree to which genes determine a characteristic. Heritability indicates only the degree to which genes determine *variation* in a characteristic. The determination of a characteristic and the determination of variation in a characteristic are two very different things.

Consider polydactyly (the presence of extra digits) in rabbits, which can be caused either by environmental factors or by a dominant gene. Suppose we have a group of rabbits all homozygous for a gene that produces the usual numbers of digits. None of the rabbits in this group carries a gene for polydactyly, but a few of the rabbits are polydactylous because of environmental factors. Broad-sense heritability for polydactyly in this group is zero, because there is no genetic variation for polydactyly; all of the variation is due to environmental factors. However, it would be incorrect for us to conclude that genes play no role in determining the number of digits in rabbits. Indeed, we know that there are specific genes that can produce extra digits. Heritability indicates nothing about whether genes control the development of a characteristic; it provides information only about causes of the variation in a characteristic within a defined group.

An individual does not have heritability Broad- and narrow-sense heritabilities are statistical values based on the genetic and phenotypic variances found in a *group* of individuals. Heritability cannot be calculated for an individual, and heritability has no meaning for a specific individual. Suppose we calculate the narrow-sense heritability of adult body weight for the students in a biology class and obtain a value of 0.6. We could conclude that 60% of the variation in adult body weight among the students in this class is determined by additive genetic variation. We should not, however, conclude that 60% of any particular student's body weight is due to additive genes.

There is no universal heritability for a characteristic The value of heritability for a characteristic is specific for a given population in a given environment. Recall that broad-sense heritability is genetic variance divided by phenotypic variance. Genetic variance depends on which genes are present, which often differs between populations. In the example of polydactyly in rabbits, there were no genes for polydactyly in the group; so the heritability of the characteristic was zero. A different group of rabbits might contain many genes for polydactyly, and the heritability of the characteristic might then be high.

Environmental differences also may affect heritability because V_P is composed of both genetic and environmental variance. When the environmental differences that affect a characteristic differ between two groups, the heritabilities for the two groups also often differ.

Because heritability is specific to a defined population in a given environment, it is important not to extrapolate heritabilities from one population to another. For example, human height is determined by environmental factors (such as nutrition and health) and by genes. If we measured the heritability of height in a developed country, we might obtain a value of 0.8, indicating that the variation in height in this population is largely genetic. This population has a high heritability because most people have adequate nutrition and health care (V_E is low); so most of the phenotypic variation in height is genetically determined. It would be incorrect for us to assume that height has a high heritability in all human populations. In developing countries, there may be more variation in a range of environmental factors; some people may enjoy good nutrition and health, whereas others may have a diet deficient in protein and suffer from diseases that affect stature. If we measured the heritability of height in such a country, we would undoubtedly obtain a lower value than we observed in the developed country because there is more environmental variation and the genetic variance in height constitutes a smaller proportion of the phenotypic variation, making the heritability lower. The important point to remember is that heritability must be calculated separately for each population and each environment.

Even when heritability is high, environmental factors may influence a characteristic High heritability does not mean that environmental factors cannot influence the expression of a characteristic. High heritability indicates only that the environmental variation to which the population is *currently* exposed is not responsible for variation in the characteristic. Let's look again at human height. In most developed countries, heritability of human height is high, indicating that genetic differences are responsible for most of the variation in height. It would be wrong for us to conclude that human height cannot be changed by alteration of the environment. Indeed, height decreased in several European cities during World War II owing to hunger and disease, and height can be increased dramatically by the administration of growth hormone to children. The absence of environmental variation in a characteristic does not mean that the characteristic will not respond to environmental change.

Heritabilities indicate nothing about the nature of population differences in a characteristic A common misconception about heritability is that it provides information about population differences in a characteristic.

Heritability is specific for a given population in a given environment, and so it cannot be used to draw conclusions about why populations differ in a characteristic.

Suppose we measured heritability for human height in two groups. One group is from a small town in a developed country, where everyone consumes a high-protein diet. Because there is little variation in the environmental factors that affect human height and there is some genetic variation, the heritability of height in this group is high. The second group comprises the inhabitants of a single village in a developing country. The consumption of protein by these people is only 25% of that consumed by those in the first group; so their average adult height is several centimeters less than that in the developed country. Again, there is little variation in the environmental factors that determine height in this group, because everyone in the village eats the same types of food and is exposed to the same diseases. Because there is little environmental variation and there is some genetic variation, the heritability of height in this group also is high.

Thus, the heritability of height in both groups is high, and the average height in the two groups is considerably different. We might be tempted to conclude that the difference in height between the two groups is genetically based—that the people in the developed country are genetically taller than the people in the developing country. This conclusion is obviously wrong, however, because the differences in height are due largely to diet—an environmental factor. Heritability provides no information about the causes of differences between populations.

These limitations of heritability have often been ignored, particularly in arguments about possible social implications of genetic differences between humans. Soon after Mendel's principles of heredity were rediscovered, some geneticists began to claim that many human behavioral characteristics are determined entirely by genes. This claim led to debates about whether characteristics such as human intelligence are determined by genes or environment. Many of the early claims of genetically based human behavior were based on poor research; unfortunately, the results of these studies were often accepted at face value and led to a number of eugenic laws that discriminated against certain groups of people. Today, geneticists recognize that many behavioral characteristics are influenced by a complex interaction of genes and environment, and separating genetic effects from those of the environment is very difficult.

The results of a number of modern studies indicate that human intelligence as measured by IQ and other intelligence tests has a moderately high heritability (usually from 0.4 to 0.8). On the basis of this observation, some people have argued that intelligence is innate and that enhanced educational opportunities cannot boost intelligence. This argument is based on the misconception that, when heritability is high, changing the environment will not alter the characteristic. In addition, because heritabilities of intelligence range from 0.4 to 0.8, a considerable amount of the variance in intelligence originates from environmental differences.

Another argument based on a misconception about heritability is that ethnic differences in measures of intelligence are genetically based. Because the results of some genetic studies show that IQ has moderately high heritability and because other studies find differences in the average IQ of ethnic groups, some people have suggested that ethnic differences in IQ are genetically based. As in the example of the effects of diet on human height, heritability provides no information about causes of differences among groups; it indicates only the degree to which phenotypic variance within a single group is genetically based. High heritability for a characteristic does not mean that phenotypic differences between ethnic groups are genetic. We should also remember that separating genetic and environmental effects in humans is very difficult; so heritability estimates themselves may be unreliable.

TRY PROBLEM 34

CONCEPTS

Heritability provides information only about the degree to which *variation* in a characteristic is genetically determined. There is no universal heritability for a characteristic; heritability is specific for a given population in a specific environment. Environmental factors can potentially affect characteristics with high heritability, and heritability says nothing about the nature of population differences in a characteristic.

✔ CONCEPT CHECK 6

Suppose that you just learned that the narrow-sense heritability of blood pressure measured among a group of African Americans in Detroit, Michigan, is 0.40. What does this heritability tell us about genetic and environmental contributions to blood pressure?

Locating Genes That Affect Quantitative Characteristics

The statistical methods described for use in analyzing quantitative characteristics can be used both to make predictions about the average phenotype expected in offspring and to estimate the overall contribution of genes to variation in the characteristic. These methods do not, however, allow us to identify and determine the influence of individual genes that affect quantitative characteristics. As discussed in the introduction to this chapter, chromosome regions with genes that control polygenic characteristics are referred to as quantitative trait loci. Although quantitative genetics has made important contributions to basic biology and to plant and animal breeding, the past inability to identify QTLs and measure their individual effects severely limited the application of quantitative genetic methods.

of specific genes that influence the quantitative trait. The introduction to the chapter describes how this approach was used to identify a major gene that affects oil production in corn. QTL mapping has been used to detect genes affecting a variety of characteristics in plant and animal species (**Table 11.2**).

Genomewide association studies The traditional method of identifying QTLs is to carry out crosses between varieties that differ in a quantitative trait and then genotype numerous progeny for many markers. Although effective, this method is slow and labor intensive.

An alternative technique for identifying genes that affect quantitative traits is to conduct genomewide association studies, which were introduced in Chapter 7. Unlike traditional linkage analysis, which examines the association of a trait and gene markers among the *progeny of a cross*, genomewide association studies look for associations between traits and genetic markers in a *biological*

11.19 The availability of molecular markers makes the mapping of QTLs possible in many organisms. QTL mapping is used to identify genes that determine muscle mass in pigs. [USDA.]

Mapping QTLs In recent years, numerous genetic markers have been identified and mapped with the use of molecular techniques, making it possible to identify QTLs by linkage analysis. The underlying idea is simple: if the inheritance of a genetic marker is associated consistently with the inheritance of a particular characteristic (such as increased height), then that marker must be linked to a QTL that affects height. The key is to have enough genetic markers so that QTLs can be detected throughout the genome. With the introduction of restriction fragment length polymorphisms, microsatellite variations, and single-nucleotide polymorphisms (SNPs), variable markers are now available for mapping QTLs in a number of different organisms (**Figure 11.19**).

A common procedure for mapping QTLs is to cross two homozygous strains that differ in alleles at many loci. The resulting F_1 progeny are then intercrossed or backcrossed to allow the genes to recombine through independent assortment and crossing over. Genes on different chromosomes and genes that are far apart on the same chromosome will recombine freely; genes that are closely linked will be inherited together. The offspring are measured for one or more quantitative characteristics; at the same time, they are genotyped for numerous genetic markers that span the genome. Any correlation between the inheritance of a particular marker allele and a quantitative phenotype indicates that a QTL is linked to that marker. If enough markers are used, the detection of all the QTLs affecting a characteristic is theoretically possible. It is important to recognize that a QTL is not a gene; rather, it is a map location for a chromosome region that is associated with that trait. After a QTL has been identified, it can be studied for the presence

Table 11.2 Quantitative characteristics for which QTLs have been detected

Organism	Quantitative Characteristic	Number of QTLs Detected
Tomato	Soluble solids	7
	Fruit mass	13
	Fruit pH	9
	Growth	5
	Leaflet shape	9
	Height	9
Corn	Height	11
	Leaf length	7
	Tiller number	1
	Glume hardness	5
	Grain yield	18
	Number of ears	9
	Thermotolerance	6
Common bean	Number of nodules	4
Mung bean	Seed weight	4
Cow pea	Seed weight	2
Wheat	Preharvest sprout	4
Pig	Growth	2
	Length of small intestine	1
	Average back fat	1
	Abdominal fat	1
Mouse	Epilepsy	2
Rat	Hypertension	2

Source: After S. D. Tanksley, Mapping polygenes, *Annual Review of Genetics* 27:218, 1993.

population, a group of interbreeding individuals. The presence of an association between genetic markers and a trait indicate that the genetic markers are closely linked to one or more genes that affect variation in the trait. Genomewide association studies have been facilitated by the identification of single-nucleotide polymorphisms, which are positions in the genome where individual organisms vary in a single nucleotide base. Many SNPs have been identified in some organisms through genomic sequencing. Individual organisms can often be quickly and inexpensively genotyped for numerous SNPs, providing the genetic markers necessary to conduct genomewide association studies.

Genomewide association studies have been widely used to locate genes that affect quantitative traits in humans, including disease susceptibility, obesity, intelligence, and height. A number of quantitative traits in plants have been studied with association studies, including kernel composition, size, color and taste, disease resistance, and starch quality. Genomewide association studies in domestic animals have identified chromosomal segments affecting body weight, body composition, reproductive traits, hormone levels, hair characteristics, and behaviors.

> **CONCEPTS**
>
> The availability of numerous genetic markers revealed by molecular methods makes it possible to map chromosome segments containing genes that contribute to polygenic characteristics. Genomewide association studies locate genes that affect quantitative traits by detecting associations between genetic markers and a trait within a population of individuals.

11.4 Genetically Variable Traits Change in Response to Selection

Evolution is genetic change among members of a population. Several different forces are potentially capable of producing evolution, and we will explore these forces and the process of evolution more fully in Chapter 25. Here, we consider how one of these forces—natural selection—may bring about genetic change in a quantitative characteristic.

Charles Darwin proposed the idea of natural selection in his book *On the Origin of Species* in 1859. **Natural selection** arises through the differential reproduction of individuals with different genotypes, allowing individuals with certain genotypes to produce more offspring than others. Natural selection is one of the most important of the forces that brings about evolutionary change. Through natural selection, organisms become genetically suited to their environments; as environments change, groups of organisms change in ways that make them better able to survive and reproduce.

For thousands of years, humans have practiced a form of selection by promoting the reproduction of organisms with traits perceived as desirable. This form of selection is **artificial selection**, and it has produced the domestic plants and animals that make modern agriculture possible. The power of artificial selection, the first application of genetic principles by humans, is illustrated by the tremendous diversity of shapes, colors, and behaviors of modern domesticated dogs (**Figure 11.20**).

Predicting the Response to Selection

When a quantitative characteristic is subjected to natural or artificial selection, it will frequently change with the passage of time, provided that there is genetic variation for that characteristic in the population. Suppose that a dairy farmer breeds only those cows in his herd that have the highest milk production. If there is genetic variation in milk production, the mean milk production in the offspring of the selected cows should be higher than the mean milk production of the original herd. This increased production is due to the fact that the selected cows possess more genes for high milk production than does the average cow, and these genes are passed on to the offspring. The offspring of the selected cows possess a higher proportion of genes for greater milk yield and therefore produce more milk than the average cow in the initial herd.

The extent to which a characteristic subjected to selection changes in one generation is termed the **response to selection**. Suppose that the average cow in a dairy herd produces 80 liters of milk per week. A farmer selects for increased milk production by breeding the highest milk producers, and the progeny of these selected cows produce 100 liters of milk per week on average. The response to selection is calculated by subtracting the mean phenotype of the original population (80 liters) from the mean phenotype of the offspring (100 liters), obtaining a response to selection of $100 - 80 = 20$ liters per week.

Factors influencing response to selection The response to selection is determined primarily by two factors. First, it is affected by narrow-sense heritability, which largely determines the degree of resemblance between parents and offspring. When the narrow-sense heritability is high, offspring will tend to resemble their parents; conversely, when the narrow-sense heritability is low, there will be little resemblance between parents and offspring.

The second factor that determines the response to selection is how much selection there is. If the farmer is very stringent in the choice of parents and breeds only the highest milk producers in the herd (say, the top 2 cows), then all the offspring will receive genes for high-quality milk production. If the farmer is less selective and breeds the top 20 milk producers in the herd, then the offspring will not carry as many superior genes for high milk production, and they will not, on average, produce as much milk as the offspring of the top 2 producers. The response to selection depends on the

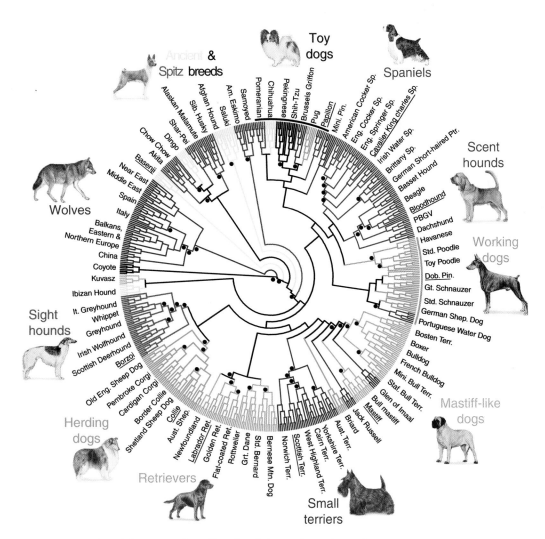

11.20 Artificial selection has produced the tremendous diversity of shape, size, color, and behavior seen today among breeds of domestic dogs. This diagram depicts the evolutionary relationships among wolves and different breeds of dogs from analyses of DNA sequences. [J. & C. Sohns/AgeFotostock.]

phenotypic difference of the individuals that are selected as parents; this phenotypic difference is measured by the **selection differential**, defined as the difference between the mean phenotype of the selected parents and the mean phenotype of the original population. If the average milk production of the original herd is 80 liters and the farmer breeds cows with an average milk production of 120 liters, then the selection differential is $120 - 80 = 40$ liters.

Calculation of response to selection The response to selection (R) depends on the narrow-sense heritability (h^2) and the selection differential (S):

$$R = h^2 \times S \qquad (11.21)$$

This equation can be used to predict the magnitude of change in a characteristic when a given selection differential is applied. G. A. Clayton and his colleagues estimated the response to selection that would take place in abdominal bristle number of *Drosophila melanogaster*. Using several different methods, including parent–offspring regression, they first estimated the narrow-sense heritability of abdominal bristle number in one population of fruit flies to be 0.52. The mean number of bristles in the original population was 35.3. They selected individual flies with a mean bristle number of 40.6 and intercrossed them to produce the next generation. The selection differential was $40.6 - 35.3 = 5.3$; so they predicted a response to selection to be

$$R = 0.52 \times 5.3 = 2.8$$

The response to selection of 2.8 is the expected increase in the characteristic of the offspring above the mean of the original population. They therefore expected the average number of abdominal bristles in the offspring of their selected flies to be $35.3 + 2.8 = 38.1$. Indeed, they found an average bristle number of 37.9 in these flies.

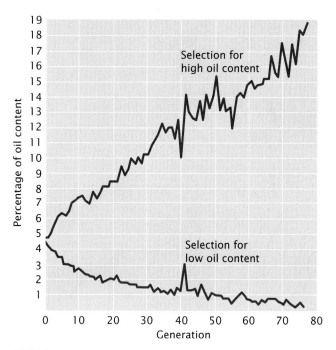

11.21 In a long-term response-to-selection experiment, selection for oil content in corn increased oil content in one line to about 20%, whereas it almost eliminated it in another line.

revealed that at least 20 loci take part in determining oil content, one of which we explored in the introduction to the chapter. TRY PROBLEM 38 →

CONCEPTS

The response to selection is influenced by narrow-sense heritability and the selection differential.

✔ CONCEPT CHECK 7

The narrow-sense heritability for a trait is 0.4 and the selection differential is 0.5. What is the predicted response to selection?

Limits to Selection Response

When a characteristic has been selected for many generations, the response eventually levels off, and the characteristic no longer responds to selection (**Figure 11.22**). A potential reason for this leveling off is that the genetic variation in the population may be exhausted; at some point, all individuals in the population have become homozygous for alleles that encode the selected trait. When there is no more additive genetic variation, heritability equals zero, and no further response to selection can take place.

The response to selection may level off even while some genetic variation remains in the population, however, because natural selection opposes further change in the characteristic. Response to selection for small body size in mice, for example, eventually levels off because the smallest animals are sterile and cannot pass on their genes for small body size. In this case, artificial selection for small size is

Estimating heritability from response to selection

Rearranging Equation 11.21 provides another way to calculate narrow-sense heritability:

$$h^2 = \frac{R}{S} \quad (11.22)$$

In this way, h^2 can be calculated by conducting a response-to-selection experiment. First, the selection differential is obtained by subtracting the population mean from the mean of selected parents. The selected parents are then interbred, and the mean phenotype of their offspring is measured. The difference between the mean of the offspring and that of the initial population is the response to selection, which can be used with the selection differential to estimate heritability. Heritability determined by a response-to-selection experiment is usually termed the **realized heritability**. If certain assumptions are met, the realized heritability is identical with the narrow-sense heritability.

One of the longest selection experiments is a study of oil and protein content in corn seeds (**Figure 11.21**). This experiment began at the University of Illinois on 163 ears of corn with an oil content ranging from 4% to 6%. Corn plants having high oil content and those having low oil content were selected and interbred. Response to selection for increased oil content (the upper line in Figure 11.21) reached about 20%, whereas response to selection for decreased oil content reached a lower limit near zero. Genetic analyses of the high- and low-oil-content strains

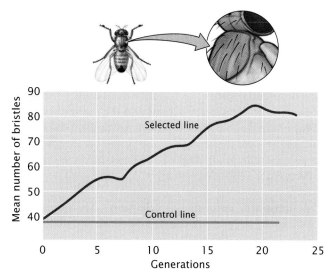

11.22 The response of a population to selection often levels off at some point in time. In a response-to-selection experiment for increased abdominal chaetae bristle number in female fruit flies, the number of bristles increased steadily for about 20 generations and then leveled off.

opposed by natural selection for fertility, and the population can no longer respond to the artificial selection.

Correlated Responses

Often when a specific trait is selected, other traits change at the same time. This type of associated response is due to the fact that the traits are encoded by the same genes.

Two or more characteristics are often correlated. Human height and weight exhibit a positive correlation: tall people, on the average, weigh more than short people. This correlation is a **phenotypic correlation** because the association is between two phenotypes of the same person. Phenotypic correlations may be due to environmental or genetic correlations. Environmental correlations refer to two or more characteristics that are influenced by the same environmental factor. Moisture availability, for example, may affect both the size of a plant and the number of seeds produced by the plant. Plants growing in environments with lots of water are large and produce many seeds, whereas plants growing in environments with limited water are small and have few seeds.

Alternatively, a phenotypic correlation may result from a **genetic correlation**, which means that the genes affecting two characteristics are associated. The primary genetic cause of phenotypic correlations is pleiotropy, which is due to the effect of one gene on two or more characteristics (see Chapter 5). In humans, for example, many body structures respond to growth hormone, and there are genes that affect the amount of growth hormone secreted by the pituitary gland. People with certain genes produce high levels of growth hormone, which increases both height and hand size. Others possess genes that produce lower levels of growth hormone, which leads to both short stature and small hands. Height and hand size are therefore phenotypically correlated in humans, and this correlation is due to a genetic correlation—the fact that both characteristics are affected by the same genes that control the amount of growth hormone. Genetically speaking, height and hand size are the same characteristic because they are the phenotypic manifestation of a single set of genes. When two characteristics are influenced by the same genes they are genetically correlated.

Genetic correlations are quite common (**Table 11.3**) and may be positive or negative. A positive genetic correlation between two characteristics means that genes that cause an increase in one characteristic also produce an increase in the other characteristic. Thorax length and wing length in *Drosophila* are positively correlated because the genes that increase thorax length also increase wing length. A negative genetic correlation means that genes that cause an increase in one characteristic produce a decrease in the other characteristic. Milk yield and percentage of butterfat are negatively correlated in cattle: genes that cause higher milk production result in milk with a lower percentage of butterfat.

Genetic correlations are important in animal and plant breeding because they produce a correlated response to selection, which means that, when one characteristic is selected, genetically correlated characteristics also change. Correlated responses to selection are due to the fact that both characteristics are influenced by the same genes; selection for one characteristic causes a change in the genes affecting that characteristic, and these genes also affect the second characteristic, causing it to change at the same time. Correlated responses may well be undesirable and may limit the ability to alter a characteristic by selection. From 1944 to 1964, domestic turkeys were subjected to intense selection for growth rate and body size. At the same time, fertility, egg production, and egg hatchability all declined. These correlated responses were due to negative genetic correlations between body size and fertility; eventually, these genetic correlations limited the extent to which the growth rate of turkeys could respond to selection. Genetic correlations may also limit the ability of natural populations to respond to selection in the wild and adapt to their environments.

TRY PROBLEM 42 →

Table 11.3 Genetic correlations in various organisms

Organism	Characteristics	Genetic Correlation
Cattle	Milk yield and percentage of butterfat	0.38
Pig	Weight gain and back-fat thickness	0.13
	Weight gain and efficiency	0.69
Chicken	Body weight and egg weight	0.42
	Body weight and egg production	−0.17
	Egg weight and egg production	−0.31
Mouse	Body weight and tail length	0.29
Fruit fly	Abdominal bristle number and sternopleural bristle number	0.41

Source: After D. S. Falconer, *Introduction to Quantitative Genetics* (London: Longman, 1981), p. 284.

CONCEPTS

Genetic correlations result from pleiotropy. When two characteristics are genetically correlated, selection for one characteristic will produce a correlated response in the other characteristic.

✔ CONCEPT CHECK 8

In a herd of dairy cattle, milk yield and the percentage of butterfat exhibit a genetic correlation of −0.38. If greater milk yield is selected in this herd, what will be the effect on the percentage of butterfat?

CONCEPTS SUMMARY

- Quantitative genetics focuses on the inheritance of complex characteristics whose phenotypes vary continuously. For many quantitative characteristics, the relation between genotype and phenotype is complex because many genes and environmental factors influence a characteristic.
- The individual genes that influence a polygenic characteristic follow the same Mendelian principles that govern discontinuous characteristics, but, because many genes participate, the expected ratios of phenotypes are obscured.
- A population is the group of interest, and a sample is a subset of the population used to describe it.
- A frequency distribution, in which the phenotypes are represented on one axis and the number of individuals possessing each phenotype is represented on the other, is a convenient means of summarizing phenotypes found in a group of individuals.
- The mean and variance provide key information about a distribution: the mean gives the central location of the distribution, and the variance provides information about how the phenotype varies within a group.
- The correlation coefficient measures the direction and strength of association between two variables. Regression can be used to predict the value of one variable on the basis of the value of a correlated variable.
- Phenotypic variance in a characteristic can be divided into components that are due to additive genetic variance, dominance genetic variance, genic interaction variance, environmental variance, and genetic–environmental interaction variance.
- Broad-sense heritability is the proportion of the phenotypic variance due to genetic variance; narrow-sense heritability is the proportion of the phenotypic variance due to additive genetic variance.
- Heritability provides information only about the degree to which variation in a characteristic results from genetic differences. Heritability is based on the variances present within a group of individuals, and an individual does not have heritability. The heritability of a characteristic varies among populations and among environments. Even if the heritability for a characteristic is high, the characteristic may still be altered by changes in the environment. Heritabilities provide no information about the nature of population differences in a characteristic.
- Quantitative trait loci are chromosome segments containing genes that control polygenic characteristics. QTLs can be mapped by examining the association between the inheritance of a quantitative characteristic and the inheritance of genetic markers. Genes influencing quantitative traits can also be located with the use of genomewide association studies.
- The amount that a quantitative characteristic changes in a single generation when subjected to selection (the response to selection) is directly related to the selection differential and narrow-sense heritability.
- A genetic correlation may be present when the same gene affects two or more characteristics (pleiotropy). Genetic correlations produce correlated responses to selection.

IMPORTANT TERMS

quantitative genetics (p. 303)
quantitative trait locus (QTL) (p. 303)
meristic characteristic (p. 306)
threshold characteristic (p. 306)
frequency distribution (p. 309)
normal distribution (p. 310)
population (p. 310)
sample (p. 310)
mean (p. 310)
variance (p. 311)
standard deviation (p. 311)

correlation (p. 312)
correlation coefficient (p. 312)
regression (p. 313)
regression coefficient (p. 314)
heritability (p. 316)
phenotypic variance (p. 316)
genetic variance (p. 317)
environmental variance (p. 317)
genetic–environmental interaction variance (p. 317)
additive genetic variance (p. 317)

dominance genetic variance (p. 317)
genic interaction variance (p. 318)
broad-sense heritability (p. 318)
narrow-sense heritability (p. 318)
natural selection (p. 324)
artificial selection (p. 324)
response to selection (p. 324)
selection differential (p. 325)
realized heritability (p. 326)
phenotypic correlation (p. 327)
genetic correlation (p. 327)

ANSWERS TO CONCEPT CHECKS

1. Cross two individuals that are each homozygous for different genes affecting the traits and then intercross the resulting F_1 progeny to produce the F_2. Determine what proportion of the F_2 progeny resembles one of the original homozygotes in the P generation. This proportion should be $(1/4)^n$, where n equals the number of loci with a segregating pair of alleles that affect the characteristic.

2. d

3. d
4. b
5. a
6. It indicates that about 40% of the differences in blood pressure among African Americans in Detroit are due to additive genetic differences. It neither provides information about the heritability of blood pressure in other groups of people nor indicates anything about the nature of differences in blood pressure between African Americans in Detroit and people in other groups.
7. 0.2
8. The percentage of butterfat will decrease.

WORKED PROBLEMS

1. Seed weight in a particular plant species is determined by pairs of alleles at two loci (a^+a^- and b^+b^-) that are additive and equal in their effects. Plants with genotype $a^-a^-\ b^-b^-$ have seeds that average 1 g in weight, whereas plants with genotype $a^+a^+\ b^+b^+$ have seeds that average 3.4 g in weight. A plant with genotype $a^-a^-\ b^-b^-$ is crossed with a plant of genotype $a^+a^+\ b^+b^+$.

a. What is the predicted weight of seeds from the F_1 progeny of this cross?

b. If the F_1 plants are intercrossed, what are the expected seed weights and proportions of the F_2 plants?

- **Solution**

The difference in average seed weight of the two parental genotypes is 3.4 g − 1 g = 2.4 g. These two genotypes differ in four genes; so, if the genes have equal and additive effects, each gene difference contributes an additional 2.4 g/4 = 0.6 g of weight to the 1-g weight of a plant ($a^-a^-\ b^-b^-$), which has none of the contributing genes.

The cross between the two homozygous genotypes produces the F_1 and F_2 progeny shown in the table below.

a. The F_1 are heterozygous at both loci ($a^+a^-\ b^+b^-$) and possess two genes that contribute an additional 0.6 g each to the 1-g weight of a plant that has no contributing genes. Therefore, the seeds of the F_1 should average 1 g + 2(0.6 g) = 2.2 g.

P $a^-a^-\ b^-b^-$ × $a^+a^+\ b^+b^+$
         1 g   3.4 g
                ↓
F_1 $a^+a^-\ b^+b^-$
                2.2 g
                ↓

	Genotype	Probability		Number of contributing genes	Average seed weight
	$a^-a^-\ b^-b^-$	$1/4 \times 1/4 = 1/16$		0	1 g + (0 × 0.6 g) = 1 g
	$a^+a^-\ b^-b^-$	$1/2 \times 1/4 = 1/8$	$\left.\begin{array}{l}\\\\\end{array}\right\}2/8 = 4/16$	1	1 g + (1 × 0.6 g) = 1.6 g
	$a^-a^-\ b^+b^-$	$1/4 \times 1/2 = 1/8$			
F_2	$a^+a^+\ b^-b^-$	$1/4 \times 1/4 = 1/16$			
	$a^-a^-\ b^+b^+$	$1/4 \times 1/4 = 1/16$	$\left.\begin{array}{l}\\\\\end{array}\right\}2/16 + 1/4 = 6/16$	2	1 g + (2 × 0.6 g) = 2.2 g
	$a^+a^-\ b^+b^-$	$1/2 \times 1/2 = 1/4$			
	$a^+a^+\ b^+b^-$	$1/4 \times 1/2 = 1/8$	$\left.\begin{array}{l}\\\\\end{array}\right\}2/8 = 4/16$	3	1 g + (3 × 0.6 g) = 2.8 g
	$a^+a^-\ b^+b^+$	$1/4 \times 1/2 = 1/8$			
	$a^+a^+\ b^+b^+$	$1/4 \times 1/4 = 1/16$		4	1 g + (4 × 0.6 g) = 3.4 g

b. The F_2 will have the following phenotypes and proportions: $^1/_{16}$ 1 g; $^4/_{16}$ 1.6 g; $^6/_{16}$ 2.2 g; $^4/_{16}$ 2.8 g; and $^1/_{16}$ 3.4 g.

2. Phenotypic variation is analyzed for milk production in a herd of dairy cattle and the following variance components are obtained:

Additive genetic variance (V_A)	$= 0.4$
Dominance genetic variance (V_D)	$= 0.1$
Genic interaction variance (V_I)	$= 0.2$
Environmental variance (V_E)	$= 0.5$
Genetic–environmental interaction variance (V_{GE})	$= 0.0$

a. What is the narrow-sense heritability of milk production?
b. What is the broad-sense heritability of milk production?

• **Solution**

To determine the heritabilities, we first need to calculate V_P and V_G:

$$V_P = V_A + V_D + V_I + V_E + V_{GE}$$
$$= 0.4 + 0.1 + 0.2 + 0.5 + 0$$
$$= 1.2$$
$$V_G = V_A + V_D + V_I$$
$$= 0.7$$

a. The narrow-sense heritability is:

$$h^2 = \frac{V_A}{V_P} = \frac{0.4}{1.2} = 0.33$$

b. The broad-sense heritability is:

$$H^2 = \frac{V_G}{V_P} = \frac{0.7}{1.2} = 0.58$$

3. The heights of parents and their offspring are measured for 10 families:

Mean height of parents (cm)	Mean height of offspring (cm)
150	152
157	163
188	193
165	163
160	152
142	157
170	183
183	175
152	163
173	180

From these data, determine

a. the mean, variance, and standard deviation of the height of parents and of offspring;

b. the correlation and regression coefficients for a regression of mean offspring height against mean parental height; and

c. the narrow-sense heritability of height in these families.

d. What conclusions can be drawn from the heritability value determined in part *c*?

• **Solution**

a. The best way to begin is by constructing a table, as shown on the facing page. To calculate the means, we need to sum the values of *x* and *y*, which are shown in the last rows of columns A and D of the table. For the mean of parental height,

$$\bar{x} = \frac{\sum x_i}{n} = \frac{1640}{10} = 164 \text{ cm}$$

For the mean of the offspring height,

$$\bar{y} = \frac{\sum y_i}{n} = \frac{1681}{10} = 168.1 \text{ cm}$$

For the variance, we subtract each *x* and *y* value from its mean (columns B and E) and square these differences (columns C and F). The sums of the these squared deviations are shown in the last row of columns C and F. For the regression, we need the covariance, which requires that we take the difference between each *x* value and its mean, multiply it by the difference between each *y* value and its mean [$(x_i - \bar{x})(y_i - \bar{y})$, column G], and then sum these products (last row of column G).

The variance is the sum of the squared deviations from the mean divided by $n - 1$, where *n* is the number of measurements:

$$s_x^2 = \frac{\sum (x_i - \bar{x})^2}{n-1} = \frac{1944}{9} = 216$$

$$s_y^2 = \frac{\sum (y_i - \bar{y})^2}{n-1} = \frac{1750.9}{9} = 194.54$$

The standard deviation is the square root of the variance:

$$s_x = \sqrt{s_x^2} = \sqrt{216} = 14.70$$

$$s_y = \sqrt{s_y^2} = \sqrt{194.54} = 13.95$$

b. To calculate the correlation coefficient and regression coefficient, we need the covariance:

$$\text{cov}_{xy} = \frac{\sum (x_i - \bar{x})(y_i - \bar{y})}{n-1} = \frac{1551}{9} = 172.33 \text{ cm}$$

The correlation coefficient is the covariance divided by the standard deviation of *x* and the standard deviation of *y*:

$$r = \frac{\text{cov}_{xy}}{s_x s_y} = \frac{172.33}{(14.70)(13.95)} = 0.84$$

The regression coefficient is the covariance divided by the variance of *x*:

$$r = \frac{\text{cov}_{xy}}{s_x^2} = \frac{172.33}{216} = 0.80$$

A Mean height parents (cm) x_i	B $x_i - \bar{x}$	C $(x_i - \bar{x})^2$	D Mean height offspring (cm) y_i	E $y_i - \bar{y}$	F $(y_i - \bar{y})^2$	G $(x_i - \bar{x})(y_i - \bar{y})$
150	−14	196	152	−16.1	259.21	225.4
157	−7	49	163	−5.1	26.01	35.7
188	24	576	193	24.9	620.01	597.6
165	1	1	163	−5.1	26.01	−5.1
160	−4	16	152	−16.1	259.21	64.4
142	−22	484	157	−11.1	123.21	244.2
170	6	36	183	14.9	222.01	89.4
183	19	361	175	6.9	47.61	131.1
152	−12	144	163	−5.1	26.01	61.2
173	9	81	180	11.9	141.61	107.1
$\Sigma x_i = 1640$		$\Sigma(x_i - \bar{x})^2 = 1944$	$\Sigma y_i = 1681$		$\Sigma(y_i - \bar{y})^2 = 1750.9$	$\Sigma(x_i - \bar{x})(y_i - \bar{y}) = 1551$

c. In a regression of the mean phenotype of the offspring against the mean phenotype of the parents, the regression coefficient equals the narrow-sense heritability, which is 0.80.

d. We conclude that 80% of the variance in height among the members of these families results from additive genetic variance.

4. A farmer is raising rabbits. The average body weight in his population of rabbits is 3 kg. The farmer selects the 10 largest rabbits in his population, whose average body weight is 4 kg, and interbreeds them. If the heritability of body weight in the rabbit population is 0.7, what is the expected body weight among offspring of the selected rabbits?

• **Solution**

The farmer has carried out a response-to-selection experiment, in which the response to selection will equal the selection differential times the narrow-sense heritability. The selection differential equals the difference in average weights of the selected rabbits and the entire population: 4 kg − 3 kg = 1 kg. The narrow-sense heritability is given as 0.7; so the expected response to selection is: $R = h^2 \times S = 0.7 \times 1$ kb = 0.7 kg. This value is the increase in weight that is expected in the offspring of the selected parents; so the average weight of the offspring is expected to be: 3 kg + 0.7 kg = 3.7 kg.

COMPREHENSION QUESTIONS

Section 11.1

*1. How does a quantitative characteristic differ from a discontinuous characteristic?

2. Briefly explain why the relation between genotype and phenotype is frequently complex for quantitative characteristics.

*3. Why do polygenic characteristics have many phenotypes?

Section 11.2

*4. Explain the relation between a population and a sample. What characteristics should a sample have to be representative of the population?

5. What information do the mean and variance provide about a distribution?

6. How is the standard deviation related to the variance?

*7. What information does the correlation coefficient provide about the association between two variables?

8. What is regression? How is it used?

Section 11.3

*9. List all the components that contribute to the phenotypic variance and define each component.

*10. How do broad-sense and narrow-sense heritabilities differ?

11. Briefly outline some of the ways in which heritability can be calculated.

12. Briefly describe common misunderstandings or misapplications of the concept of heritability.

13. Briefly explain how genes affecting a polygenic characteristic are located with the use of QTL mapping.

Section 11.4

*14. How is the response to selection related to the narrow-sense heritability and the selection differential? What information does the response to selection provide?

15. Why does the response to selection often level off after many generations of selection?

APPLICATION QUESTIONS AND PROBLEMS

Section 11.1

*16. For each of the following characteristics, indicate whether it would be considered a discontinuous characteristic or a quantitative characteristic. Briefly justify your answer.

   a. Kernel color in a strain of wheat, in which two codominant alleles segregating at a single locus determine the color. Thus, there are three phenotypes present in this strain: white, light red, and medium red.

   b. Body weight in a family of Labrador retrievers. An autosomal recessive allele that causes dwarfism is present in this family. Two phenotypes are recognized: dwarf (less than 13 kg) and normal (greater than 13 kg).

   c. Presence or absence of leprosy. Susceptibility to leprosy is determined by multiple genes and numerous environmental factors.

   d. Number of toes in guinea pigs, which is influenced by genes at many loci.

   e. Number of fingers in humans. Extra (more than five) fingers are caused by the presence of an autosomal dominant allele.

*17. Assume that plant weight is determined by a pair of alleles at each of two independently assorting loci (A and a, B and b) that are additive in their effects. Further assume that each allele represented by an uppercase letter contributes 4 g to weight and that each allele represented by a lowercase letter contributes 1 g to weight.

   a. If a plant with genotype $AA\ BB$ is crossed with a plant with genotype $aa\ bb$, what weights are expected in the F_1 progeny?

   b. What is the distribution of weight expected in the F_2 progeny?

*18. Assume that three loci, each with two alleles (A and a, B and b, C and c), determine the differences in height between two homozygous strains of a plant. These genes are additive and equal in their effects on plant height. One strain ($aa\ bb\ cc$) is 10 cm in height. The other strain ($AA\ BB\ CC$) is 22 cm in height. The two strains are crossed, and the resulting F_1 are interbred to produce F_2 progeny. Give the phenotypes and the expected proportions of the F_2 progeny.

*19. A farmer has two homozygous varieties of tomatoes. One variety, called Little Pete, has fruits that average only 2 cm in diameter. The other variety, Big Boy, has fruits that average a whopping 14 cm in diameter. The farmer crosses Little Pete and Big Boy; he then intercrosses the F_1 to produce F_2 progeny. He grows 2000 F_2 tomato plants and doesn't find any F_2 offspring that produce fruits as small as Little Pete or as large as Big Boy. If we assume that the differences in fruit size of these varieties are produced by genes with equal and additive effects, what conclusion can we make about the minimum number of loci with pairs of alleles determining the differences in fruit size of the two varieties?

20. Seed size in a plant is a polygenic characteristic. A grower crosses two pure-breeding varieties of the plant and measures seed size in the F_1 progeny. She then backcrosses the F_1 plants to one of the parental varieties and measures seed size in the backcross progeny. The grower finds that seed size in the backcross progeny has a higher variance than does seed size in the F_1 progeny. Explain why the backcross progeny are more variable.

Section 11.2

*21. The following data are the numbers of digits per foot in 25 guinea pigs. Construct a frequency distribution for these data.

4, 4, 4, 5, 3, 4, 3, 4, 4, 5, 4, 4, 3, 2, 4, 4, 5, 6, 4, 4, 3, 4, 4, 4, 5

22. Ten male Harvard students were weighed in 1916. Their weights are given here in kilograms. Calculate the mean, variance, and standard deviation for these weights.

51, 69, 69, 57, 61, 57, 75, 105, 69, 63

23. Among a population of tadpoles, the correlation coefficient for size at metamorphosis and time required for metamorphosis is -0.74. On the basis of this correlation, what conclusions can you make about the relative sizes of tadpoles that metamorphose quickly and those that metamorphose more slowly?

*24. A researcher studying alcohol consumption in North American cities finds a significant, positive correlation between the number of Baptist preachers and alcohol consumption. Is it reasonable for the researcher to conclude that the Baptist preachers are consuming most of the alcohol? Why or why not?

25. Body weight and length were measured on six mosquito fish; these measurements are given in the following table. Calculate the correlation coefficient for weight and length in these fish.

[A. Hartl/AgeFotostock.]

Wet weight (g)	Length (mm)
115	18
130	19
210	22
110	17
140	20
185	21

*26. The heights of mothers and daughters are given in the following table:

Height of mother (in)	Height of daughter (in)
64	66
65	66
66	68
64	65
63	65
63	62
59	62
62	64
61	63
60	62

a. Calculate the correlation coefficient for the heights of the mothers and daughters.

b. Using regression, predict the expected height of a daughter whose mother is 67 inches tall.

Section 11.3

*27. Phenotypic variation in tail length of mice has the following components:

Additive genetic variance (V_A)	= 0.5
Dominance genetic variance (V_D)	= 0.3
Genic interaction variance (V_I)	= 0.1
Environmental variance (V_E)	= 0.4
Genetic–environmental interaction variance (V_{GE})	= 0.0

a. What is the narrow-sense heritability of tail length?

b. What is the broad-sense heritability of tail length?

28. The narrow-sense heritability of ear length in Reno rabbits is 0.4. The phenotypic variance (V_P) is 0.8, and the environmental variance (V_E) is 0.2. What is the additive genetic variance (V_A) for ear length in these rabbits?

*29. Assume that human ear length is influenced by multiple genetic and environmental factors. Suppose you measured ear length on three groups of people, in which group A consists of five unrelated persons, group B consists of five siblings, and group C consists of five first cousins.

a. With the assumption that the environment for each group is similar, which group should have the highest phenotypic variance? Explain why.

b. Is it realistic to assume that the environmental variance for each group is similar? Explain your answer.

30. A characteristic has a narrow-sense heritability of 0.6.

a. If the dominance variance (V_D) increases and all other variance components remain the same, what will happen to the narrow-sense heritability? Will it increase, decrease, or remain the same? Explain.

b. What will happen to the broad-sense heritability? Explain.

c. If the environmental variance (V_E) increases and all other variance components remain the same, what will happen to the narrow-sense heritability? Explain.

d. What will happen to the broad-sense heritability? Explain.

31. Flower color in the varieties of pea plants studied by Mendel is controlled by alleles at a single locus. A group of peas homozygous for purple flowers is grown in a garden. Careful study of the plants reveals that all their flowers are purple, but there is some variability in the intensity of the purple color. If heritability were estimated for this variation in flower color, what would it be? Explain your answer.

*32. A graduate student is studying a population of bluebonnets along a roadside. The plants in this population are genetically variable. She counts the seeds produced by 100 plants and measures the mean and variance of seed number. The variance is 20. Selecting one plant, the graduate student takes cuttings from it and cultivates these cuttings in the greenhouse, eventually producing many genetically identical clones of the same plant. She then transplants these clones into the roadside population, allows them to grow for 1 year, and then counts the number of seeds produced by each of the cloned plants. The graduate student finds that the variance in seed number among these cloned plants is 5. From the phenotypic variance of the genetically variable and genetically identical plants, she calculates the broad-sense heritability.

[Amy Lenise/Stock.xchng.]

a. What is the broad-sense heritability of seed number for the roadside population of bluebonnets?

b. What might cause this estimate of heritability to be inaccurate?

33. Many researchers have estimated the heritability of human traits by comparing the correlation coefficients of monozygotic and dizygotic twins (see p. 320). One of the assumptions in using this method is that two monozygotic twins experience environments that are no more similar to each other than those experienced by two dizygotic twins. How might this assumption be violated? Give some specific examples of ways in which the environments of two monozygotic twins might be more similar than the environments of two dizyotic twins.

34. A genetics researcher determines that the broad-sense heritability of height among Southwestern University undergraduate students is 0.90. Which of the following conclusions would be resonable? Explain your answer.

a. Because Sally is a Southwestern University undergraduate student, 10% of her height is determined by nongenetic factors.

b. Ninety percent of variation in height among all undergraduate students in the United States is due to genetic differences.

c. Ninety percent of the height of Southwestern University undergraduate students is determined by genes.

d. Ten percent of the variation in height among Southwestern University undergraduate students is determined by variation in nongenetic factors.

e. Because the heritability of height among Southwestern Unversity students is so high, any change in the students' environment will have minimal effect on their height.

*35. The length of the middle joint of the right index finger was measured on 10 sets of parents and their adult offspring. The mean parental lengths and the mean offspring lengths for each family are listed in the following table. Calculate the regression coefficient for regression of mean offspring length against mean parental length and estimate the narrow-sense heritability for this characteristic.

Mean parental length (mm)	Mean offspring length (mm)
30	31
35	36
28	31
33	35
26	27
32	30
31	34
29	28
40	38
33	34

36. *Drosophila buzzatii* is a fruit fly that feeds on the rotting fruits of cacti in Australia. Timothy Prout and Stuart Barker calculated the heritabilities of body size, as measured by thorax length, for a natural population of *D. buzzatii* raised in the wild and for a population of *D. buzzatii* collected in the wild but raised in the laboratory (T. Prout and J. S. F. Barker. 1989. *Genetics* 123:803–813). They found the following heritabilities.

Population	Heritability of body size (± standard error)
Wild population	0.0595 ± 0.0123
Laboratory-reared population	0.3770 ± 0.0203

Why do you think the heritability measured in the laboratory-reared population is higher than that measured in the natural population raised in the wild?

*37. Mr. Jones is a pig farmer. For many years, he has fed his pigs the food left over from the local university cafeteria, which is known to be low in protein, deficient in vitamins, and downright untasty. However, the food is free, and his pigs don't complain. One day a salesman from a feed company visits Mr. Jones. The salesman claims that his company sells a new, high-protein, vitamin-enriched feed that enhances weight gain in pigs. Although the food is expensive, the salesman claims that the increased weight gain of the pigs will more than pay for the cost of the feed, increasing Mr. Jones's profit. Mr. Jones responds that he took a genetics class when he went to the university and that he has conducted some genetic experiments on his pigs; specifically, he has calculated the narrow-sense heritability of weight gain for his pigs and found it to be 0.98. Mr. Jones says that this heritability value indicates that 98% of the variance in weight gain among his pigs is determined by genetic differences, and therefore the new pig feed can have little effect on the growth of his pigs. He concludes that the feed would be a waste of his money. The salesman doesn't dispute Mr. Jones' heritability estimate, but he still claims that the new feed can significantly increase weight gain in Mr. Jones' pigs. Who is correct and why?

Section 11.4

38. Joe is breeding cockroaches in his dorm room. He finds that the average wing length in his population of cockroaches is 4 cm. He chooses six cockroaches that have the largest wings; the average wing length among these selected cockroaches is 10 cm. Joe interbreeds these selected cockroaches. From earlier studies, he knows that the narrow-sense heritability for wing length in his population of cockroaches is 0.6.

   a. Calculate the selection differential and expected response to selection for wing length in these cockroaches.

   b. What should be the average wing length of the progeny of the selected cockroaches?

39. Three characteristics in beef cattle—body weight, fat content, and tenderness—are measured, and the following variance components are estimated:

	Body weight	Fat content	Tenderness
V_A	22	45	12
V_D	10	25	5
V_I	3	8	2
V_E	42	64	8
V_{GE}	0	0	1

In this population, which characteristic would respond best to selection? Explain your reasoning.

*40. A rancher determines that the average amount of wool produced by a sheep in her flock is 22 kg per year. In an attempt to increase the wool production of her flock, the rancher picks five male and five female sheep with the greatest wool production; the average amount of wool produced per sheep by those selected is 30 kg. She interbreeds these selected sheep and finds that the average wool production among the progeny of the selected sheep is 28 kg. What is the narrow-sense heritability for wool production among the sheep in the rancher's flock?

41. A strawberry farmer determines that the average weight of individual strawberries produced by plants in his garden is 2 g. He selects the 10 plants that produce the largest strawberries; the average weight of strawberries among these selected plants is 6 g. He interbreeds these selected strawberry plants. The progeny of these selected plants produce strawberries that weigh 5 g. If the farmer were to select plants that produce an average strawberry weight of 4 g, what would be the predicted weight of strawberries produced by the progeny of these selected plants?

42. The narrow-sense heritability of wing length in a population of *Drosophila melanogaster* is 0.8. The narrow-sense heritability of head width in the same population is 0.9. The genetic correlation between wing length and head width is −0.86. If a geneticist selects for increased wing length in these flies, what will happen to head width?

43. Pigs have been domesticated from wild boars. Would you expect to find higher heritability for weight among domestic pigs or among wild boars? Explain your answer.

CHALLENGE QUESTIONS

Section 11.1

44. Bipolar illness is a psychiatric disorder that has a strong hereditary basis, but the exact mode of inheritance is not known. Research has shown that siblings of patients with bipolar illness are more likely to develop the disorder than are siblings of unaffected persons. Findings from one study demonstrated that the ratio of bipolar brothers to bipolar sisters is higher when the patient is male than when the patient is female. In other words, relatively more brothers of bipolar patients also have the disease when the patient is male than when the patient is female. What does this observation suggest about the inheritance of bipolar illness?

Section 11.3

45. We have explored some of the difficulties in separating the genetic and environmental components of human behavioral characteristics. Considering these difficulties and what you know about calculating heritability, propose an experimental design for accurately measuring the heritability of musical ability.

46. A student who has just learned about quantitative genetics says, "Heritability estimates are worthless! They don't tell you anything about the genes that affect a characteristic. They don't provide any information about the types of offspring to expect from a cross. Heritability estimates measured in one population can't be used for other populations; so they don't even give you any general information about how much of a characteristic is genetically determined. I can't see that heritabilities do anything other than make undergraduate students sweat during tests." How would you respond to this statement? Is the student correct? What good are heritabilities, and why do geneticists bother to calculate them?

Section 11.4

47. Eugene Eisen selected for increased 12-day litter weight (total weight of a litter of offspring 12 days after birth) in a population of mice (E. J. Eisen. 1972. *Genetics* 72:129–142). The 12-day litter weight of the population steadily increased but then leveled off after about 17 generations. At generation 17, Eisen took one family of mice from the selected population and reversed the selection procedure: in this group, he selected for *decreased* 12-day litter size. This group immediately responded to decreased selection; the 12-day litter weight dropped 4.8 g within one generation and dropped 7.3 g after 5 generations. On the basis of the results of the reverse selection, what is the most likely explanation for the leveling off of 12-day litter weight in the original population?

[J & C Sohns/AgeFotostock.]

12 Population Genetics

Rocky Mountain bighorn sheep (*Ovis canadensis*). A population of bighorn sheep at the National Bison Range suffered the loss of genetic variation due to genetic drift; the introduction of sheep from other populations dramatically increased genetic variation and the fitness of the sheep. [Tom J. Ulrich/Visuals Unlimited.]

GENETIC RESCUE OF BIGHORN SHEEP

Rocky Mountain bighorn sheep (*Ovis canadensis*) are among North America's most spectacular animals, characterized by the male's magnificent horns that curve gracefully back over the ears, spiraling down and back up beside the face. Two hundred years ago, bighorn sheep were numerous throughout western North America, ranging from Mexico to southern Alberta and from Colorado to California. Meriwether Lewis and William Clark reported numerous sightings of these beautiful animals in their expedition across the western United States from 1804 to 1806. Before 1900, the number of bighorn sheep in North America was about 2 million.

Unfortunately, settlement of the west by Europeans was not kind to the bighorns. Beginning in the late 1800s, hunting, loss of habitat, competition from livestock, and diseases carried by domestic sheep decimated the bighorns. Today, fewer than 70,000 bighorn sheep remain, scattered across North America in fragmented and isolated populations.

In 1922, wildlife biologists established a population of bighorn sheep at the National Bison Range, an isolated tract of 18,000 acres nestled between the mountains of northwestern Montana. In that year, 12 bighorn sheep—4 males (rams) and 8 females (ewes)—were trapped at Banff National Park in Canada and transported to the National Bison Range. No additional animals were introduced to this population for the next 60 years.

At first, the population of bighorns at the National Bison Range flourished, protected there from hunting and livestock. Within 8 years, the population had grown to 90 sheep but then began to slowly decrease in size. Population size waxed and waned through the years, but the number of sheep had dropped to about 50 animals by 1985, and the population was in trouble. The amount of genetic variation was low compared with other native populations of bighorn sheep. The reproductive rate of both male and female sheep had dropped, and the size and survival of the sheep were lower than in more healthy populations. The population at the National Bison Range was suffering from genetic drift, an evolutionary force operating in small populations that causes random changes in the gene pool and the loss of genetic variation.

To counteract the negative effects of genetic drift, biologists added 5 new rams from other herds in Montana and Wyoming in 1985, mimicking the effects of natural migration among herds. Another 10 sheep were introduced between 1990 and 1994. This influx of new genes had a dramatic effect on the genetic health of the population. Genetic variation among individual sheep increased significantly. Outbred rams (those containing the new genes) were more dominant, were more likely to copulate, and were more likely to produce offspring. Outbred ewes had more than twice the annual reproductive success of inbred females. Adult survival increased after the introduction of new genes, and, slowly, the population grew in size, reaching 69 sheep by 2003.

The bighorn sheep at the National Bison Range illustrate an important principle of genetics: small populations lose genetic variation with the passage of time through genetic drift, often with catastrophic consequences for survival and reproduction. The introduction of new genetic variation into an inbred population, called **genetic rescue**, often dramatically improves the health of the population and can better ensure its long-term survival. These effects have important implications for wildlife management, as well as for how organisms evolve in the natural world.

This chapter introduces *population genetics*, the branch of genetics that studies the genetic makeup of groups of individuals and how a *group's* genetic composition changes with time. Population geneticists usually focus their attention on a **Mendelian population**, which is a group of interbreeding, sexually reproducing individuals that have a common set of genes, the **gene pool**. A population evolves through changes in its gene pool; therefore, population genetics is also the study of evolution. Population geneticists study the variation in alleles within and between groups and the evolutionary forces responsible for shaping the patterns of genetic variation found in nature. In this chapter, we will learn how the gene pool of a population is measured and what factors are responsible for shaping it.

12.1 Genotypic and Allelic Frequencies Are Used to Describe the Gene Pool of a Population

An obvious and pervasive feature of life is variability. Consider a group of students in a typical college class, the members of which vary in eye color, hair color, skin pigmentation, height, weight, facial features, blood type, and susceptibility to numerous diseases and disorders. No two students in the class are likely to be even remotely similar in appearance.

Humans are not unique in their extensive variability (**Figure 12.1a**); almost all organisms exhibit variation in phenotype. For instance, lady beetles are highly variable in their patterns of spots (**Figure 12.1b**), mice vary in body size, snails have different numbers of stripes on their shells, and plants vary in their susceptibility to pests. Much of this phenotypic variation is hereditary. Recognition of the extent of phenotypic variation led Charles Darwin to the idea of evolution

12.1 All organisms exhibit genetic variation. (a) Extensive variation among humans. (b) Variation in the spotting patterns of Asian lady beetles. [Part a: Paul Warner/AP.]

through natural selection. Genetic variation is the basis of all evolution, and the extent of genetic variation within a population affects its potential to adapt to environmental change.

In fact, even more genetic variation exists in populations than is visible in the phenotype. Much variation exists at the molecular level owing, in part, to the redundancy of the genetic code, which allows different codons to specify the same amino acid. Thus, two members of a population can produce the same protein even if their DNA sequences are different. DNA sequences between the genes and introns within genes do not encode proteins; much of the variation in these sequences also has little effect on the phenotype.

An important, but frequently misunderstood, tool used in population genetics is the mathematical model. Let's take a moment to consider what a model is and how it can be used. A mathematical model usually describes a process as an equation. Factors that may influence the process are represented by variables in the equation; the equation defines the way in which the variables influence the process. Most models are simplified representations of a process, because the simultaneous consideration of all of the influencing factors is impossible; some factors must be ignored in order to examine the effects of others. At first, a model might consider only one factor or a few factors, but, after their effects are understood, the model can be improved by the addition of more details. Importantly, even a simple model can be a source of valuable insight into how a process is influenced by key variables.

Before we can explore the evolutionary processes that shape genetic variation, we must be able to describe the genetic structure of a population. The usual way of describing this structure is to enumerate the types and frequencies of genotypes and alleles in a population.

Calculating Genotypic Frequencies

A frequency is simply a proportion or a percentage, usually expressed as a decimal fraction. For example, if 20% of the alleles at a particular locus in a population are A, we would say that the frequency of the A allele in the population is 0.20. For large populations, for which a determination of the genes of all individual members is impractical, a sample of the population is usually taken and the genotypic and allelic frequencies are calculated for this sample (see Chapter 11 for a discussion of samples). The genotypic and allelic frequencies of the sample are then used to represent the gene pool of the population.

To calculate a **genotypic frequency**, we simply add up the number of individuals possessing the genotype and divide by the total number of individuals in the sample (N). For a locus with three genotypes AA, Aa, and aa, the frequency (f) of each genotype is

$$f(AA) = \frac{\text{number of } AA \text{ individuals}}{N}$$

$$f(Aa) = \frac{\text{number of } Aa \text{ individuals}}{N} \quad (12.1)$$

$$f(aa) = \frac{\text{number of } aa \text{ individuals}}{N}$$

The sum of all the genotypic frequencies always equals 1.

Calculating Allelic Frequencies

The gene pool of a population can also be described in terms of the allelic frequencies. There are always fewer alleles than genotypes; so the gene pool of a population can be described in fewer terms when the allelic frequencies are used. In a sexually reproducing population, the genotypes are only temporary assemblages of the alleles: the genotypes break down each generation when individual alleles are passed to the next generation through the gametes, and so the types and numbers of alleles, rather than genotypes, have real continuity from one generation to the next and make up the gene pool of a population.

Allelic frequencies can be calculated from (1) the numbers or (2) the frequencies of the genotypes. To calculate the **allelic frequency** from the numbers of genotypes, we count the number of copies of a particular allele present in a sample and divide by the total number of all alleles in the sample:

$$\text{frequency of an allele} = \frac{\text{number of copies of the allele}}{\text{number of copies of all alleles at the locus}} \quad (12.2)$$

For a locus with only two alleles (A and a), the frequencies of the alleles are usually represented by the symbols p and q, and can be calculated as follows:

$$p = f(A) = \frac{2n_{AA} + n_{Aa}}{2N}$$

$$q = f(a) = \frac{2n_{aa} + n_{Aa}}{2N} \quad (12.3)$$

where n_{AA}, n_{Aa}, and n_{aa} represent the numbers of AA, Aa, and aa individuals, and N represents the total number of individuals in the sample. We divide by $2N$ because each diploid individual has two alleles at a locus. The sum of the allelic frequencies always equals 1 ($p + q = 1$); so, after p has been obtained, q can be determined by subtraction: $q = 1 - p$.

Alternatively, allelic frequencies can be calculated from the genotypic frequencies. To calculate an allelic frequency from genotypic frequencies, we add the frequency of the homozygote for each allele to half the frequency of the heterozygote (because half of the heterozygote's alleles are of each type):

$$p = f(A) = f(AA) + \tfrac{1}{2} f(Aa)$$

$$q = f(a) = f(aa) + \tfrac{1}{2} f(Aa) \quad (12.4)$$

We obtain the same values of p and q whether we calculate the allelic frequencies from the numbers of genotypes (see Equation 12.3) or from the genotypic frequencies (see Equation 12.4). A sample calculation of allelic frequencies is provided in the next Worked Problem.

TRY PROBLEM 16

Loci with multiple alleles We can use the same principles to determine the frequencies of alleles for loci with more than two alleles. To calculate the allelic frequencies from the numbers of genotypes, we count up the number of copies of an allele by adding twice the number of homozygotes to the number of heterozygotes that possess the allele and divide this sum by twice the number of individuals in the sample. For a locus with three alleles (A^1, A^2, and A^3) and six genotypes (A^1A^1, A^1A^2, A^2A^2, A^1A^3, A^2A^3, and A^3A^3), the frequencies (p, q, and r) of the alleles are

$$p = f(A^1) = \frac{2n_{A^1A^1} + n_{A^1A^2} + n_{A^1A^3}}{2N}$$

$$q = f(A^2) = \frac{2n_{A^2A^2} + n_{A^1A^2} + n_{A^2A^3}}{2N} \quad (12.5)$$

$$r = f(A^3) = \frac{2n_{A^3A^3} + n_{A^1A^3} + n_{A^2A^3}}{2N}$$

Alternatively, we can calculate the frequencies of multiple alleles from the genotypic frequencies by extending Equation 12.4. Once again, we add the frequency of the homozygote to half the frequency of each heterozygous genotype that possesses the allele:

$$p = f(A^1A^1) + \tfrac{1}{2}f(A^1A^2) + \tfrac{1}{2}f(A^1A^3)$$

$$q = f(A^2A^2) + \tfrac{1}{2}f(A^1A^2) + \tfrac{1}{2}f(A^2A^3) \quad (12.6)$$

$$r = f(A^3A^3) + \tfrac{1}{2}f(A^1A^3) + \tfrac{1}{2}f(A^2A^3)$$

X-linked loci To calculate allelic frequencies for genes at X-linked loci, we apply these same principles. However, we must remember that a female possesses two X chromosomes and therefore has two X-linked alleles, whereas a male has only a single X chromosome and has one X-linked allele.

Suppose there are two alleles at an X-linked locus, X^A and X^a. Females may be either homozygous (X^AX^A or X^aX^a) or heterozygous (X^AX^a). All males are hemizygous (X^AY or X^aY). To determine the frequency of the X^A allele (p), we first count the number of copies of X^A: we multiply the number of X^AX^A females by two and add the number of X^AX^a females and the number of X^AY males. We then divide the sum by the total number of alleles at the locus, which is twice the total number of females plus the number of males:

$$p = f(X^A) = \frac{2n_{X^AX^A} + n_{X^AX^a} + n_{X^AY}}{2n_{females} + n_{males}} \quad (12.7a)$$

Similarly, the frequency of the X^a allele is

$$q = f(X^a) = \frac{2n_{X^aX^a} + n_{X^AX^a} + n_{X^aY}}{2n_{females} + n_{males}} \quad (12.7b)$$

The frequencies of X-linked alleles can also be calculated from genotypic frequencies by adding the frequency of the females that are homozygous for the allele, half the frequency of the females that are heterozygous for the allele, and the frequency of males hemizygous for the allele:

$$p = f(X^A) = f(X^AX^A) + \tfrac{1}{2}f(X^AX^a) + f(X^AY)$$

$$q = f(X^a) = f(X^aX^a) + \tfrac{1}{2}f(X^AX^a) + f(X^aY) \quad (12.8)$$

If you remember the logic behind all of these calculations, you can determine allelic frequencies for any set of genotypes, and it will not be necessary to memorize the formulas. **TRY PROBLEM 18** →

CONCEPTS

Population genetics concerns the genetic composition of a population and how it changes with time. The gene pool of a population can be described by the frequencies of genotypes and alleles in the population.

✔ **CONCEPT CHECK 1**

What are some advantages of using allelic frequencies to describe the gene pool of a population instead of using genotypic frequencies?

Worked Problem

The human MN blood-type antigens are determined by two codominant alleles, L^M and L^N (see p. 102 in Chapter 5). The MN blood types and corresponding genotypes of 398 Finns from Karjala are tabulated here.

Phenotype	Genotype	Number
MM	L^ML^M	182
MN	L^ML^N	172
NN	L^NL^N	44

Source: W. C. Boyd, *Genetics and the Races of Man* (Boston: Little, Brown, 1950.)

Calculate the genotypic and allelic frequencies at the MN locus for the Karjala population.

• Solution

The genotypic frequencies for the population are calculated with the following formula:

genotypic frequency

$$= \frac{\text{number of individuals with genotype}}{\text{total number of individual in sample } (N)}$$

$$f(L^ML^M) = \frac{\text{number of } L^ML^M \text{ individuals}}{N} = \frac{182}{398} = 0.457$$

$$f(L^ML^N) = \frac{\text{number of } L^ML^M \text{ individuals}}{N} = \frac{172}{398} = 0.432$$

$$f(L^NL^N) = \frac{\text{number of } L^NL^N \text{ individuals}}{N} = \frac{44}{398} = 0.111$$

The allelic frequencies can be calculated from either the numbers or the frequencies of the genotypes. To calculate allelic frequencies from numbers of genotypes, we add the number of copies of the allele and divide by the number of copies of all alleles at that locus.

$$\text{frequency of an allele} = \frac{\text{number of copies of the allele}}{\text{number of copies all alleles}}$$

$$p = f(L^M) = \frac{(2n_{L^M L^M}) + (n_{L^M L^N})}{2N}$$
$$= \frac{2(182) + 172}{2(398)} = \frac{536}{796} = 0.673$$

$$q = f(L^N) = \frac{(2n_{L^N L^N}) + (n_{L^M L^N})}{2N}$$
$$= \frac{2(44) + 172}{2(398)} = \frac{260}{796} = 0.327$$

To calculate the allelic frequencies from genotypic frequencies, we add the frequency of the homozygote for that genotype to half the frequency of each heterozygote possessing that allele:

$$p = f(L^M) = f(L^M L^M) + \tfrac{1}{2} f(L^M L^N)$$
$$= 0.457 + \tfrac{1}{2}(0.432) = 0.673$$

$$q = f(L^N) = f(L^N L^N) + \tfrac{1}{2} f(L^M L^N)$$
$$= 0.111 + \tfrac{1}{2}(0.432) = 0.327$$

> ➔ Now try your hand at calculating genotypic and allelic frequencies by working Problem 17 at the end of the chapter.

12.2 The Hardy–Weinberg Law Describes the Effect of Reproduction on Genotypic and Allelic Frequencies

The primary goal of population genetics is to understand the processes that shape a population's gene pool. First, we must ask what effects reproduction and Mendelian principles have on the genotypic and allelic frequencies: How do the segregation of alleles in gamete formation and the combining of alleles in fertilization influence the gene pool? The answer to this question lies in the **Hardy–Weinberg law**, among the most important principles of population genetics.

The Hardy–Weinberg law was formulated independently by both G. H. Hardy and Wilhelm Weinberg in 1908. (Similar conclusions were reached by several other geneticists at about the same time.) The law is actually a mathematical model that evaluates the effect of reproduction on the genotypic and allelic frequencies of a population. It makes several simplifying assumptions about the population and provides two key predictions if these assumptions are met. For an autosomal locus with two alleles, the Hardy–Weinberg law can be stated as follows:

Assumptions If a population is large, randomly mating, and not affected by mutation, migration, or natural selection, then:

Prediction 1 the allelic frequencies of a population do not change; and

Prediction 2 the genotypic frequencies stabilize (will not change) after one generation in the proportions p^2 (the frequency of AA), $2pq$ (the frequency of Aa), and q^2 (the frequency of aa), where p equals the frequency of allele A and q equals the frequency of allele a.

The Hardy–Weinberg law indicates that, when the assumptions are met, reproduction alone does not alter allelic or genotypic frequencies and the allelic frequencies determine the frequencies of genotypes.

The statement that genotypic frequencies stabilize after one generation means that they may change in the first generation after random mating, because one generation of random mating is required to produce Hardy–Weinberg proportions of the genotypes. Afterward, the genotypic frequencies, like allelic frequencies, do not change as long as the population continues to meet the assumptions of the Hardy–Weinberg law. When genotypes are in the expected proportions of p^2, $2pq$, and q^2, the population is said to be in **Hardy–Weinberg equilibrium**.

CONCEPTS

The Hardy–Weinberg law describes how reproduction and Mendelian principles affect the allelic and genotypic frequencies of a population.

✔ CONCEPT CHECK 2

Which statement is not an assumption of the Hardy–Weinberg law?

a. The allelic frequencies (p and q) are equal.
b. The population is randomly mating.
c. The population is large.
d. Natural selection has no effect.

Genotypic Frequencies at Hardy–Weinberg Equilibrium

How do the conditions of the Hardy–Weinberg law lead to genotypic proportions of p^2, $2pq$, and q^2? Mendel's principle of segregation says that each individual organism possesses two alleles at a locus and that each of the two alleles has an equal probability of passing into a gamete. Thus, the frequencies of alleles in gametes will be the same as the frequencies of alleles in the parents. Suppose we have a Mendelian population in which the frequencies of alleles A and a are p and q, respectively. These frequencies will also be those in the gametes. If mating is random (one of the assumptions

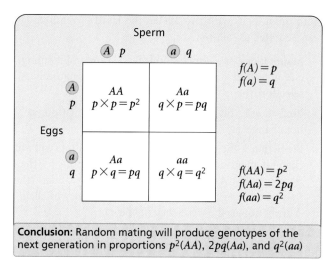

12.2 Random mating produces genotypes in the proportions p^2, $2pq$, and q^2.

of the Hardy–Weinberg law), the gametes will come together in random combinations, which can be represented by a Punnett square as shown in **Figure 12.2**.

The multiplication rule of probability can be used to determine the probability of various gametes pairing. For example, the probability of a sperm containing allele A is p and the probability of an egg containing allele A is p. Applying the multiplication rule, we find that the probability that these two gametes will combine to produce an AA homozygote is $p \times p = p^2$. Similarly, the probability of a sperm containing allele a combining with an egg containing allele a to produce an aa homozygote is $q \times q = q^2$. An Aa heterozygote can be produced in one of two ways: (1) a sperm containing allele A may combine with an egg containing allele a ($p \times q$) or (2) an egg containing allele A may combine with a sperm containing allele a ($p \times q$). Thus, the probability of alleles A and a combining to produce an Aa heterozygote is $2pq$. In summary, whenever the frequencies of alleles in a population are p and q, the frequencies of the genotypes in the next generation will be p^2, $2pq$, and q^2.

Closer Examination of the Assumptions of the Hardy–Weinberg Law

Before we consider the implications of the Hardy–Weinberg law, we need to take a closer look at the three assumptions that it makes about a population. First, it assumes that the population is large. How big is "large"? Theoretically, the Hardy–Weinberg law requires that a population be infinitely large in size, but this requirement is obviously unrealistic. In practice, many large populations are in the predicated Hardy–Weinberg proportions, and significant deviations arise only when population size is rather small. Later in the chapter, we will examine the effects of small population size on allelic frequencies.

The second assumption of the Hardy–Weinberg law is that members of the population mate randomly, which means that each genotype mates relative to its frequency. For example, suppose that three genotypes are present in a population in the following proportions: $f(AA) = 0.6$, $f(Aa) = 0.3$, and $f(aa) = 0.1$. With random mating, the frequency of mating between two AA homozygotes ($AA \times AA$) will be equal to the multiplication of their frequencies: $0.6 \times 0.6 = 0.36$, whereas the frequency of mating between two aa homozygotes ($aa \times aa$) will be only $0.1 \times 0.1 = 0.01$.

The third assumption of the Hardy–Weinberg law is that the allelic frequencies of the population are not affected by natural selection, migration, and mutation. Although mutation occurs in every population, its rate is so low that it has little short-term effect on the predictions of the Hardy–Weinberg law (although it may largely shape allelic frequencies over long periods of time when no other forces are acting). Although natural selection and migration are significant factors in real populations, we must remember that the purpose of the Hardy–Weinberg law is to examine only the effect of reproduction on the gene pool. When this effect is known, the effects of other factors (such as migration and natural selection) can be examined.

A final point is that the assumptions of the Hardy–Weinberg law apply to a single locus. No real population mates randomly for all traits; and a population is not completely free of natural selection for all traits. The Hardy–Weinberg law, however, does not require random mating and the absence of selection, migration, and mutation for all traits; it requires these conditions only for the locus under consideration. A population may be in Hardy–Weinberg equilibrium for one locus but not for others.

Implications of the Hardy–Weinberg Law

The Hardy–Weinberg law has several important implications for the genetic structure of a population. One implication is that a population cannot evolve if it meets the Hardy–Weinberg assumptions, because evolution consists of change in the allelic frequencies of a population. Therefore the Hardy–Weinberg law tells us that reproduction alone will not bring about evolution. Other processes such as natural selection, mutation, migration, or chance are required for populations to evolve.

A second important implication is that, when a population is in Hardy–Weinberg equilibrium, the genotypic frequencies are determined by the allelic frequencies. When a population is not in Hardy–Weinberg equilibrium, we have no basis for predicting the genotypic frequenices. Although we can always determine the allelic frequencies from the genotypic frequencies (see Equation 12.3), the reverse (determining the genotypic frequencies from the allelic frequencies) is possible only when the population is in Hardy–Weinberg equilibrium.

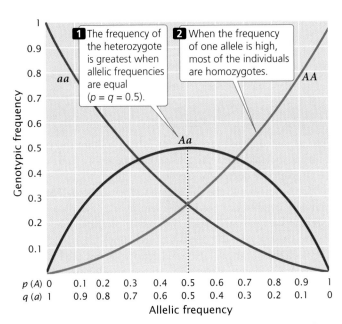

12.3 When a population is in Hardy–Weinberg equilibrium, the proportions of genotypes are determined by the frequencies of alleles.

www.whfreeman.com/pierce4e — Animation 25.1 examines the effect of changing allelic frequencies on genotypic frequencies when a population is in Hardy–Weinberg equilibrium.

For a locus with two alleles, the frequency of the heterozygote is greatest when allelic frequencies are between 0.33 and 0.66 and is at a maximum when allelic frequencies are each 0.5 (**Figure 12.3**). The heterozygote frequency also never exceeds 0.5 when the population is in Hardy–Weinberg equilibrium. Furthermore, when the frequency of one allele is low, homozygotes for that allele will be rare, and most of the copies of a rare allele will be present in heterozygotes. As you can see from Figure 12.3, when the frequency of allele a is 0.2, the frequency of the aa homozygote is only 0.04 (q^2), but the frequency of Aa heterozygotes is 0.32 ($2pq$); 80% of the a alleles are in heterozygotes.

A third implication of the Hardy–Weinberg law is that a single generation of random mating produces the equilibrium frequencies of p^2, $2pq$, and q^2. The fact that genotypes are in Hardy–Weinberg proportions does not prove that the population is free from natural selection, mutation, and migration. It means only that these forces have not acted since the last time random mating took place.

Extensions of the Hardy–Weinberg Law

The Hardy–Weinberg expected proportions can also be applied to multiple alleles and X-linked alleles (**Table 12.1**). With multiple alleles, the genotypic frequencies expected at equilibrium are the square of the allelic frequencies. For an autosomal locus with three alleles, the equilibrium genotypic frequencies will $(p + q + r)^2 = p^2 + 2pq + q^2 + 2pr + 2qr + r^2$. For an X-linked locus with two alleles, X^A and X^a, the equilibrium frequencies of the female genotypes are $(p + q)^2 = p^2 + 2pq + q^2$, where p^2 is the frequency of $X^A X^A$, $2pq$ is the frequency of $X^A X^a$, and q^2 is the frequency of $X^a X^a$. Males have only a single X-linked allele, and so the frequencies of the male genotypes are p (frequency of $X^A Y$) and q (frequency of $X^a Y$). These proportions are those of the genotypes among males and females rather than the proportions among the entire population. Thus, p^2 is the expected proportion of females with the genotype $X^A X^A$; if females make up 50% of the population, then the expected proportion of this genotype in the entire population is $0.5 \times p^2$.

The frequency of an X-linked recessive trait among males is q, whereas the frequency among females is q^2. When an X-linked allele is uncommon, the trait will therefore be much more frequent in males than in females. Consider hemophilia A, a clotting disorder caused by an X-linked recessive allele with a frequency (q) of approximately 1 in 10,000, or 0.0001. At Hardy–Weinberg equilibrium, this frequency will also be the frequency of the disease among males. The frequency of the disease among females, however, will be $q^2 = (0.0001)^2 = 0.00000001$, which is only 1 in 10 million. Hemophilia is 1000 times as frequent in males as in females.

Testing for Hardy–Weinberg Proportions

To determine if a population's genotypes are in Hardy–Weinberg equilibrium, the genotypic proportions expected under the Hardy–Weinberg law must be compared with the observed genotypic frequencies. To do so, we first calculate

Table 12.1 Extensions of the Hardy–Weinberg law

Situation	Allelic Frequencies	Genotypic Frequencies
Three alleles	$f(A^1) = p$	$f(A^1 A^1) = p^2$
	$f(A^2) = q$	$f(A^1 A^2) = 2pq$
	$f(A^3) = r$	$f(A^2 A^2) = q^2$
		$f(A^1 A^3) = 2pr$
		$f(A^2 A^3) = 2qr$
		$f(A^3 A^3) = r^2$
X-linked alleles	$f(X^1) = p$	$f(X^1 X^1 \text{ female}) = p^2$
	$f(X^2) = q$	$f(X^1 X^2 \text{ female}) = 2pq$
		$f(X^2 X^2 \text{ female}) = q^2$
		$f(X^1 Y \text{ male}) = p$
		$f(X^2 Y \text{ male}) = q$

Note: For X-linked female genotypes, the frequencies are the proportions among all females; for X-linked male genotypes, the frequencies are the proportions among all males.

the allelic frequencies, then find the expected genotypic frequencies by using the square of the allelic frequencies, and, finally, compare the observed and expected genotypic frequencies by using a chi-square test.

Worked Problem

Jeffrey Mitton and his colleagues found three genotypes (R^2R^2, R^2R^3, and R^3R^3) at a locus encoding the enzyme peroxidase in ponderosa pine trees growing at Glacier Lake, Colorado. The observed numbers of these genotypes were:

Genotypes	Number observed
R^2R^2	135
R^2R^3	44
R^3R^3	11

Are the ponderosa pine trees at Glacier Lake in Hardy–Weinberg equilibrium at the peroxidase locus?

• Solution

If the frequency of the R^2 allele equals p and the frequency of the R^3 allele equals q, the frequency of the R^2 allele is

$$p = f(R^2) = \frac{(2n_{R^2R^2}) + (n_{R^2R^3})}{2N} = \frac{2(135) + 44}{2(190)} = 0.826$$

The frequency of the R^3 allele is obtained by subtraction:

$$q = f(R^3) = 1 - p = 0.174$$

The frequencies of the genotypes expected under Hardy–Weinberg equilibrium are then calculated separately by using p^2, $2pq$, and q^2:

$$R^2R^2 = p^2 = (0.826)^2 = 0.683$$

$$R^2R^3 = 2pq = 2(0.826)(0.174) = 0.287$$

$$R^3R^3 = q^2 = (0.174)^2 = 0.03$$

Multiplying each of these expected genotypic frequencies by the total number of observed genotypes in the sample (190), we obtain the numbers expected for each genotype:

$$R^2R^2 = 0.683 \times 190 = 129.8$$

$$R^2R^3 = 0.287 \times 190 = 54.5$$

$$R^3R^3 = 0.03 \times 190 = 5.7$$

By comparing these expected numbers with the observed numbers of each genotype, we see that there are more R^2R^2 homozygotes and fewer R^2R^3 heterozygotes and R^3R^3 homozygotes in the population than we expect at equilibrium.

A goodness-of-fit chi-square test is used to determine whether the differences between the observed and the expected numbers of each genotype are due to chance:

$$X^2 = \sum \frac{(observerd - expected)^2}{expected}$$

$$= \frac{(135 - 129.8)^2}{129.8} + \frac{(44 - 54.5)^2}{54.5} + \frac{(11 - 5.7)^2}{5.7}$$

$$= 0.21 + 2.02 + 4.93 = 7.16$$

The calculated chi-square value is 7.16; to obtain the probability associated with this chi-square value, we determine the appropriate degrees of freedom.

Up to this point, the chi-square test for assessing Hardy–Weinberg equilibrium has been identical with the chi-square tests that we used in Chapter 3 to assess progeny ratios in a genetic cross, where the degrees of freedom were $n - 1$ and n equaled the number of expected genotypes. For the Hardy–Weinberg test, however, we must subtract an additional degree of freedom, because the expected numbers are based on the observed allelic frequencies; therefore, the observed numbers are not completely free to vary. In general, the degrees of freedom for a chi-square test of Hardy–Weinberg equilibrium equal the number of expected genotypic classes minus the number of associated alleles. For this particular Hardy–Weinberg test, the degree of freedom is $3 - 2 = 1$.

After we have calculated both the chi-square value and the degrees of freedom, the probability associated with this value can be sought in a chi-square table (see Table 3.5). With one degree of freedom, a chi-square value of 7.16 has a probability between 0.01 and 0.001. It is very unlikely that the peroxidase genotypes observed at Glacier Lake are in Hardy–Weinberg proportions.

→ For additional practice, determine whether the genotypic frequencies in Problem 20 at the end of the chapter are in Hardy–Weinberg equilibrium.

CONCEPTS

The observed number of genotypes in a population can be compared with the Hardy–Weinberg expected proportions by using a goodness-of-fit chi-square test.

✔ CONCEPT CHECK 3

What is the expected frequency of heterozygotes in a population with allelic frequencies x and y that is in Hardy–Weinberg equilibrium?

a. $x + y$ c. $2xy$
b. xy d. $(x - y)^2$

Estimating Allelic Frequencies with the Hardy–Weinberg Law

A practical use of the Hardy–Weinberg law is that it allows us to calculate allelic frequencies when dominance is present. For example, cystic fibrosis is an autosomal recessive disorder characterized by respiratory infections, incomplete digestion, and abnormal sweating (see p. 102 in Chapter 5). Among North American Caucasians, the incidence of the disease is approximately 1 person in 2000. The formula for calculating allelic frequency (see Equation 12.3) requires that we know the numbers of homozygotes and heterozygotes, but cystic fibrosis is a recessive disease and so we cannot easily distinguish between homozygous normal persons and heterozygous carriers. Although molecular tests are available for identifying heterozygous carriers of the cystic fibrosis gene, the low frequency of the disease makes widespread screening impractical. In such situations, the Hardy–Weinberg law can be used to estimate the allelic frequencies.

If we assume that a population is in Hardy–Weinberg equilibrium with regard to this locus, then the frequency of the recessive genotype (aa) will be q^2, and the allelic frequency is the square root of the genotypic frequency:

$$q = \sqrt{f(aa)} \qquad (12.9)$$

If the frequency of cystic fibrosis in North American Caucasians is approximately 1 in 2000, or 0.0005, then $q = \sqrt{0.0005} = 0.02$. Thus, about 2% of the alleles in the Caucasian population encode cystic fibrosis. We can calculate the frequency of the normal allele by subtracting: $p = 1 - q = 1 - 0.02 = 0.98$. After we have calculated p and q, we can use the Hardy–Weinberg law to determine the frequencies of homozygous normal people and heterozygous carriers of the gene:

$$f(AA) = p^2 = (0.98)^2 = 0.960$$
$$f(Aa) = 2pq = 2(0.02)(0.98) = 0.0392$$

Thus, about 4% (1 of 25) of Caucasians are heterozygous carriers of the allele that causes cystic fibrosis. **TRY PROBLEM 23**

CONCEPTS

Although allelic frequencies cannot be calculated directly for traits that exhibit dominance, the Hardy–Weinberg law can be used to estimate the allelic frequencies if the population is in Hardy–Weinberg equilibrium for that locus. The frequency of the recessive allele will be equal to the square root of the frequency of the recessive trait.

✔ CONCEPT CHECK 4

In cats, all-white color is dominant over not all-white. In a population of 100 cats, 19 are all-white cats. Assuming that the population is in Hardy–Weinberg equilibrium, what is the frequency of the all-white allele in this population?

12.3 Nonrandom Mating Affects the Genotypic Frequencies of a Population

An assumption of the Hardy–Weinberg law is that mating is random with respect to genotype. Nonrandom mating affects the way in which alleles combine to form genotypes and alters the genotypic frequencies of a population.

We can distinguish between two types of nonrandom mating. **Positive assortative mating** refers to a tendency for like individuals to mate. For example, humans exhibit positive assortative mating for height: tall people mate preferentially with other tall people; short people mate preferentially with other short people. **Negative assortative mating** refers to a tendency for unlike individuals to mate. If people engaged in negative assortative mating for height, tall and short people would preferentially mate. Assortative mating is usually for a particular trait and will affect only those genes that encode the trait (and genes closely linked to them).

One form of nonrandom mating is **inbreeding**, which is preferential mating between related individuals. Inbreeding is actually positive assortative mating for relatedness, but it differs from other types of assortative mating because it affects all genes, not just those that determine the trait for which the mating preference exists. Inbreeding causes a departure from the Hardy–Weinberg equilibrium frequencies of p^2, $2pq$, and q^2. More specifically, it leads to an increase in the proportion of homozygotes and a decrease in the proportion of heterozygotes in a population. **Outcrossing** is the avoidance of mating between related individuals.

In a diploid organism, a homozygous individual has two copies of the same allele. These two copies may be the same in *state*, which means that the two alleles are alike in structure and function but do not have a common origin. Alternatively, the two alleles in a homozygous individual may be the same because they are identical by *descent*; that is, the copies are descended from a single allele that was present in an ancestor (**Figure 12.4**). If we go back far enough in time, many alleles are likely to be identical by descent but, for calculating the effects of inbreeding, we consider identity by descent by going back only a few generations.

Inbreeding is usually measured by the **inbreeding coefficient**, designated F, which is a measure of the probability that two alleles are identical by descent. Inbreeding coefficients can range from 0 to 1. A value of 0 indicates that mating in a large population is random; a value of 1 indicates that all alleles are identical by descent. Inbreeding coefficients can be calculated from analyses of pedigrees or they can be determined from the reduction in the heterozygosity of a population. Although we will not go into the details of how F is calculated, an understanding of how inbreeding affects genotypic frequencies is important.

When inbreeding takes place, the proportion of heterozygotes decreases by $2Fpq$, and half of this value (Fpq) is *added* to the proportion of each homozygote each generation. The frequencies of the genotypes will then be

$$f(AA) = p^2 + Fpq$$
$$f(Aa) = 2pq - 2Fpq \qquad (12.10)$$
$$f(aa) = q^2 + Fpq$$

Consider a population that reproduces by self-fertilization (so $F = 1$). We will assume that this population begins

(a) Alleles identical by descent ### (b) Alleles identical by state

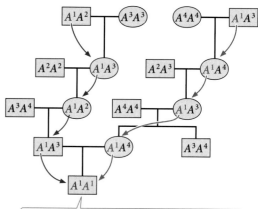

12.4 Individuals may be homozygous by descent or by state. Inbreeding is a measure of the probability that two alleles are identical by descent.

These two copies of the A^1 allele are descended from the same copy in a common ancestor; so they are identical by descent.

These two copies of the A^1 allele are the same in structure and function, but are descended from two different copies in ancestors; so they are identical in state.

with genotypic frequencies in Hardy–Weinberg proportions (p^2, $2pq$, and q^2). With selfing, each homozygote produces progeny only of the same homozygous genotype ($AA \times AA$ produces all AA; and $aa \times aa$ produces all aa), whereas only half the progeny of a heterozygote will be like the parent ($Aa \times Aa$ produces $^1/_4$ AA, $^1/_2$ Aa, and $^1/_4$ aa). Selfing therefore reduces the proportion of heterozygotes in the population by half with each generation, until all genotypes in the population are homozygous (**Table 12.2** and **Figure 12.5**).

TRY PROBLEM 26 →

For most outcrossing species, close inbreeding is harmful because it increases the proportion of homozygotes and thereby boosts the probability that deleterious and lethal recessive alleles will combine to produce homozygotes with a harmful trait. Assume that a recessive allele (a) that causes a genetic disease has a frequency (q) of 0.01. If the population mates randomly ($F = 0$), the frequency of individuals affected with the disease (aa) will be $q^2 = 0.01^2 = 0.0001$; so only 1 in 10,000 individuals will have the disease. However, if $F = 0.25$ (the equivalent of brother–sister mating), then the expected frequency of the homozygote genotype is $q^2 + Fpq = (0.01)^2 + (0.25)(0.99)(0.01) = 0.0026$; thus, the genetic

Table 12.2	Generational increase in frequency of homozygotes in a self-fertilizing population starting with $p = q = 0.5$		
	Genotypic Frequencies		
Generation	AA	Aa	aa
1	$^1/_4$	$^1/_2$	$^1/_4$
2	$^1/_4 + ^1/_8 = ^3/_8$	$^1/_4$	$^1/_4 + ^1/_8 = ^3/_8$
3	$^3/_8 + ^1/_{16} = ^7/_{16}$	$^1/_8$	$^3/_8 + ^1/_{16} = ^7/_{16}$
4	$^7/_{16} + ^1/_{32} = ^{15}/_{32}$	$^1/_{16}$	$^7/_{16} + ^1/_{32} = ^{15}/_{32}$
n	$\dfrac{1-(^1/_2)^n}{2}$	$(^1/_2)^n$	$\dfrac{1-(^1/_2)^n}{2}$
	$^1/_2$	0	$^1/_2$

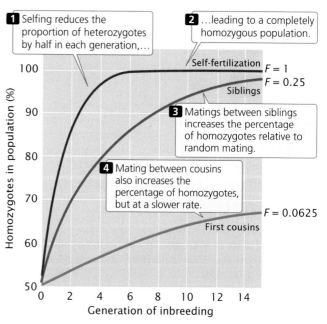

12.5 Inbreeding increases the percentage of homozygous individuals in a population.

www.whfreeman.com/pierce4e To see the effects of inbreeding on genotypic frequencies, view the Mini-Tutorial in Animation 25.1.

Table 12.3	Effects of inbreeding on Japanese children	
Genetic Relationship of Parents	F	Mortality of Children (through 12 years of age)
Unrelated	0	0.082
Second cousins	0.016 ($1/64$)	0.108
First cousins	0.0625 ($1/16$)	0.114

Source: After D. L. Hartl and A. G. Clark, *Principles of Population Genetics*, 2d ed. (Sunderland, Mass.: Sinauer, 1989), Table 2. Original data from W. J. Schull and J. V. Neel, *The Effects of Inbreeding on Japanese Children* (New York: Harper & Row, 1965).

12.7 Although inbreeding is generally harmful, a number of inbreeding organisms are successful. Shown here is the terrestrial slug *Arion circumscriptos*, an inbreeding species that causes damage in greenhouses and flower gardens. [William Leonard/DRK Photo.]

disease is 26 times as frequent at this level of inbreeding. This increased appearance of lethal and deleterious traits with inbreeding is termed **inbreeding depression**; the more intense the inbreeding, the more severe the inbreeding depression.

The harmful effects of inbreeding have been recognized by people for thousands of years and may be the basis of cultural taboos against mating between close relatives. William Schull and James Neel found that, for each 10% increase in F, the mean IQ of Japanese children dropped six points. Child mortality also increases with close inbreeding (**Table 12.3**); children of first cousins have a 40% increase in mortality over that seen among the children of unrelated people. Inbreeding also has deleterious effects on crops (**Figure 12.6**) and domestic animals.

Inbreeding depression is most often studied in humans, as well as in plants and animals reared in captivity, but the negative effects of inbreeding may be more severe in natural populations. Julie Jimenez and her colleagues collected wild mice from a natural population in Illinois and bred them in the laboratory for three to four generations. Laboratory matings were chosen so that some mice had no inbreeding, whereas others had an inbreeding coefficient of 0.25. When both types of mice were released back into the wild, the weekly survival of the inbred mice was only 56% of that of the noninbred mice. Inbred male mice also continously lost body weight after release into the wild, whereas noninbred male mice initially lost weight but then regained it within a few days after release.

In spite of the fact that inbreeding is generally harmful for outcrossing species, a number of plants and animals regularly inbreed and are successful (**Figure 12.7**). Inbreeding is commonly used to produce domesticated plants and animals having desirable traits. As stated earlier, inbreeding increases homozygosity, and eventually all individuals in the population become homozygous for the same allele. If a species undergoes inbreeding for a number of generations, many deleterious recessive alleles are weeded out by natural or artificial selection so that the population becomes homozygous for beneficial alleles. In this way, the harmful effects of inbreeding may eventually be eliminated, leaving a population that is homozygous for beneficial traits.

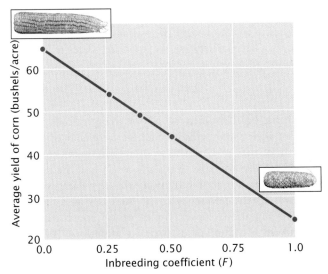

12.6 Inbreeding often has deleterious effects on crops. As inbreeding increases, the average yield of corn, for example, decreases.

CONCEPTS

Nonrandom mating alters the frequencies of the genotypes but not the frequencies of the alleles. Inbreeding is preferential mating between related individuals. With inbreeding, the frequency of homozygotes increases, whereas the frequency of heterozygotes decreases.

✔ **CONCEPT CHECK 5**

What is the effect of outcrossing on a population?

a. Allelic frequencies change.
b. There will be more heterozygotes than predicted by the Hardy–Weinberg law.
c. There will be fewer heterozygotes than predicted by the Hardy–Weinberg law.
d. Genotypic frequencies will equal those predicted by the Hardy–Weinberg law.

12.4 Several Evolutionary Forces Potentially Cause Changes in Allelic Frequencies

The Hardy–Weinberg law indicates that allelic frequencies do not change as a result of reproduction; thus, other processes must cause alleles to increase or decrease in frequency. Processes that bring about change in allelic frequency include mutation, migration, genetic drift (random effects due to small population size), and natural selection.

Mutation

Before evolution can take place, genetic variation must exist within a population; consequently, all evolution depends on processes that generate genetic variation. Although new *combinations* of existing genes may arise through recombination in meiosis, all genetic variants ultimately arise through mutation.

The effect of mutation on allelic frequencies Mutation can influence the rate at which one genetic variant increases at the expense of another. Consider a single locus in a population of 25 diploid individuals. Each individual possesses two alleles at the locus under consideration; so the gene pool of the population consists of 50 allelic copies. Let us assume that there are two different alleles, designated G^1 and G^2 with frequencies p and q, respectively. If there are 45 copies of G^1 and 5 copies of G^2 in the population, $p = 0.90$ and $q = 0.10$. Now suppose that a mutation changes a G^1 allele into a G^2 allele. After this mutation, there are 44 copies of G^1 and 6 copies of G^2, and the frequency of G^2 has increased from 0.10 to 0.12. Mutation has changed the allelic frequency.

If copies of G^1 continue to mutate to G^2, the frequency of G^2 will increase and the frequency of G^1 will decrease (**Figure 12.8**). The amount that G^2 will change (Δq) as a result of mutation depends on: (1) the rate of G^1-to-G^2 mutation (μ) and (2) p, the frequency of G^1 in the population. When p is large, there are many copies of G^1 available to mutate to G^2, and the amount of change will be relatively large. As more mutations occur and p decreases, there will be fewer copies of G^1 available to mutate to G^2. The change in G^2 as a result of mutation equals the mutation rate times the allelic frequency:

$$\Delta q = \mu p \qquad (12.11)$$

As the frequency of p decreases as a result of mutation, the change in frequency due to mutation will be less and less.

So far, we have considered only the effects of $G^1 \rightarrow G^2$ forward mutations. Reverse $G^2 \rightarrow G^1$ mutations also occur at rate ν, which will probably be different from the forward mutation rate, μ. Whenever a reverse mutation occurs, the frequency of G^2 decreases and the frequency of G^1 increases

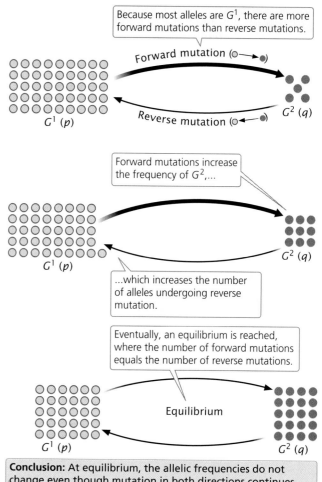

12.8 Recurrent mutation changes allelic frequencies. Forward and reserve mutations eventually lead to a stable equilibrium.

(see Figure 25.8). The rate of change due to reverse mutations equals the reverse mutation rate times the allelic frequency of G^2 ($\Delta q = \nu q$). The overall change in allelic frequency is a balance between the opposing forces of forward mutation and reverse mutation:

$$\Delta q = \mu p - \nu q \qquad (12.12)$$

Reaching equilibrium of allelic frequencies Consider an allele that begins with a high frequency of G^1 and a low frequency of G^2. In this population, many copies of G^1 are initially available to mutate to G^2, and the increase in G^2 due to forward mutation will be relatively large. However, as the frequency of G^2 increases as a result of forward mutations, fewer copies of G^1 are available to mutate; so the number of forward mutations decreases. On the other hand, few copies of G^2 are initially available to undergo a reverse mutation to G^1 but, as the frequency of G^2 increases, the number of copies of G^2 available to

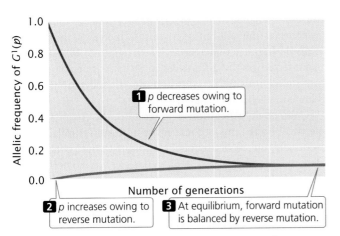

12.9 Change due to recurrent mutation slows as the frequency of p drops. Allelic frequencies are approaching mutational equilibrium at typical low mutation rates. The allelic frequency of G^1 decreases as a result of forward ($G^1 \rightarrow G^2$) mutation at rate m (0.0001) and increases as a result of reverse ($G^2 \rightarrow G^1$) mutation at rate n (0.00001). Owing to the low rate of mutations, eventual equilibrium takes many generations to be reached.

undergo reverse mutation to G^1 increases; so the number of genes undergoing reverse mutation will increase. Eventually, the number of genes undergoing forward mutation will be counterbalanced by the number of genes undergoing reverse mutation. At this point, the increase in q due to forward mutation will be equal to the decrease in q due to reverse mutation, and there will be no net change in allelic frequency ($\Delta q = 0$), in spite of the fact that forward and reverse mutations continue to occur. The point at which there is no change in the allelic frequency of a population is referred to as **equilibrium** (see Figure 12.8). At equilibrium, the frequency of G^2 ($\hat{q}$) will be

$$\hat{q} = \frac{\mu}{\mu + \nu} \qquad (12.13)$$

This final equation tells us that the allelic frequency at equilibrium is determined solely by the forward (μ) and reverse (ν) mutation rates. **TRY PROBLEM 28**

Summary of effects When the only evolutionary force acting on a population is mutation, allelic frequencies change with the passage of time because some alleles mutate into others. Eventually, these allelic frequencies reach equilibrium and are determined only by the forward and reverse mutation rates. When the allelic frequencies reach equilibrium, the Hardy–Weinberg law tells us that genotypic frequencies also will remain the same.

The mutation rates for most genes are low; so change in allelic frequency due to mutation in one generation is very small, and long periods of time are required for a population to reach mutational equilibrium. For example, if the forward and reverse mutation rates for alleles at a locus are 1×10^{-5} and 0.3×10^{-5} per generation, respectively (rates that have actually been measured at several loci in mice), and the allelic frequencies are $p = 0.9$ and $q = 0.1$, then the net change in allelic frequency per generation due to mutation is

$$\begin{aligned}\Delta q &= \mu p - \nu q \\ &= (1 \times 10^{-5})(0.9) - (0.3 \times 10^{-5})(0.1) \\ &= 8.7 \times 10^{-6} = 0.0000087\end{aligned}$$

Therefore, change due to mutation in a single generation is extremely small and, as the frequency of p drops as a result of mutation, the amount of change will become even smaller (**Figure 12.9**). The effect of typical mutation rates on Hardy–Weinberg equilibrium is negligible, and many generations are required for a population to reach mutational equilibrium. Nevertheless, if mutation is the only force acting on a population for long periods of time, mutation rates will determine allelic frequencies.

CONCEPTS

Recurrent mutation causes changes in the frequencies of alleles. At equilibrium, the allelic frequencies are determined by the forward and reverse mutation rates. Because mutation rates are low, the effect of mutation per generation is very small.

✔ CONCEPT CHECK 6

When a population is in equilibrium for forward and reverse mutation rates, which of the following is true:

a. The number of forward muations is greater than the number of reverse mutations.
b. No forward or reverse mutations occur.
c. The number of forward mutations is equal to the number of reverse mutations.
d. The population is in Hardy–Weinberg equilibrium.

Migration

Another process that may bring about change in allelic frequencies is the influx of genes from other populations, commonly called **migration** or **gene flow**. One of the assumptions of the Hardy–Weinberg law is that migration does not take place, but many natural populations do experience migration from other populations. The overall effect of migration is twofold: (1) it prevents populations from becoming genetically different from one another and (2) it increases genetic variation within populations.

The effect of migration on allelic frequencies Let us consider the effects of migration by looking at a simple, unidirectional model of migration between two populations that differ in the frequency of an allele a. Say that the frequency of this allele in population I is q_I and in population

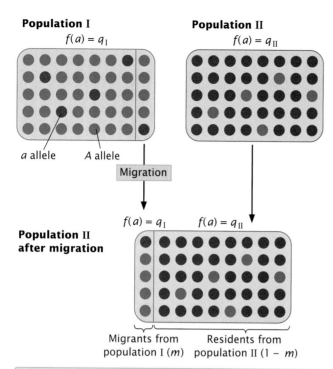

12.10 The amount of change in allelic frequency due to migration between populations depends on the difference in allelic frequency and the extent of migration. Shown here is a model of the effect of unidirectional migration on allelic frequencies. The frequency of allele a in the source population (population I) is q_I. The frequency of this allele in the recipient population (population II) is q_{II}.

II is q_{II} (**Figure 12.10**). In each generation, a representative sample of the individuals in population I migrates to population II and reproduces, adding its genes to population II's gene pool. Migration is only from population I to population II (is unidirectional), and all the conditions of the Hardy–Weinberg law apply, except the absence of migration.

After migration, population II consists of two types of individuals. Some are migrants; they make up proportion m of population II, and they carry genes from population I; so the frequency of allele a in the migrants is q_I. The other individuals in population II are the original residents. If the migrants make up proportion m of population II, then the residents make up $1 - m$; because the residents originated in population II, the frequency of allele a in this group is q_{II}. After migration, the frequency of allele a in the merged population II (q'_{II}) is

$$q'_{II} = q_I(m) + q_{II}(1 - m) \quad (12.14)$$

where $q_I(m)$ is the contribution to q made by the copies of allele a in the migrants and $q_{II}(1 - m)$ is the contribution to q made by copies of allele a in the residents. The change in the allelic frequency due to migration (Δq) will be

$$\Delta q = m(q_I - q_{II}) \quad (12.15)$$

Equation 12.15 summarizes the factors that determine the amount of change in allelic frequency due to migration. The amount of change in q is directly proportional to the migration (m); as the amount of migration increases, the change in allelic frequency increases. The magnitude of change is also affected by the differences in allelic frequencies of the two populations ($q_I - q_{II}$); when the difference is large, the change in allelic frequency will be large.

With each generation of migration, the frequencies of the two populations become more and more similar until, eventually, the allelic frequency of population II equals that of population I. When $q_I - q_{II} = 0$, there will be no further change in the allelic frequency of population II, in spite of the fact that migration continues. If migration between two populations takes place for a number of generations with no other evolutionary forces present, an equilibrium is reached at which the allelic frequency of the recipient population equals that of the source population.

The simple model of unidirectional migration between two populations just outlined can be expanded to accommodate multidirectional migration between several populations.

The overall effect of migration Migration has two major effects. First, it causes the gene pools of populations to become more similar. Later, we will see how genetic drift and natural selection lead to genetic differences between populations; migration counteracts this tendency and tends to keep populations homogeneous in their allelic frequencies. Second, migration adds genetic variation to populations. Different alleles may arise in different populations owing to rare mutational events, and these alleles can be spread to new populations by migration, increasing the genetic variation within the recipient population. **TRY PROBLEM 31** →

CONCEPTS

Migration causes changes in the allelic frequency of a population by introducing alleles from other populations. The magnitude of change due to migration depends on both the extent of migration and the difference in allelic frequencies between the source and the recipient populations. Migration decreases genetic differences between populations and increases genetic variation within populations.

✔ **CONCEPT CHECK 7**

Each generation, 10 random individuals migrate from population A to population B. What will happen to allelic frequency q as a result of migration when q is equal in populations A and B?

a. q in A will decrease.   c. q will not change in either A or B.
b. q in B will increase.   d. q in B will become q^2.

Genetic Drift

The Hardy–Weinberg law assumes random mating in an infinitely large population; only when population size is infinite will the gametes carry genes that perfectly represent the parental gene pool. But no real population is infinitely large, and when population size is limited, the gametes that unite to form individuals of the next generation carry a sample of alleles present in the parental gene pool. Just by chance, the composition of this sample will often deviate from that of the parental gene pool, and this deviation may cause allelic frequencies to change. The smaller the gametic sample, the greater the chance that its composition will deviate from that of the entire gene pool.

The role of chance in altering allelic frequencies is analogous to the flip of a coin. Each time we flip a coin, we have a 50% chance of getting a head and a 50% chance of getting a tail. If we flip a coin 1000 times, the observed ratio of heads to tails will be very close to the expected 50 : 50 ratio. If, however, we flip a coin only 10 times, there is a good chance that we will obtain not exactly 5 heads and 5 tails, but rather maybe 7 heads and 3 tails or 8 tails and 2 heads. This kind of deviation from an expected ratio due to limited sample size is referred to as **sampling error.**

Sampling error arises when gametes unite to produce progeny. Many organisms produce a large number of gametes but, when population size is small, a limited number of gametes unite to produce the individuals of the next generation. Chance influences which alleles are present in this limited sample and, in this way, sampling error may lead to **genetic drift**, or changes in allelic frequency. Because the deviations from the expected ratios are random, the direction of change is unpredictable. We can nevertheless predict the magnitude of the changes.

The magnitude of genetic drift The effect of genetic drift can be viewed in two ways. First, we can see how it influences the change in allelic frequencies of a single population with the passage of time. Second, we can see how it affects differences that accumulate among series of populations. Imagine that we have 10 small populations, all beginning with the exact same allelic frequencies of $p = 0.5$ and $q = 0.5$. When genetic drift occurs in a population, allelic frequencies within the population will change but, because drift is random, the way in which allelic frequencies change in each population will not be the same. In some populations, p may increase as a result of chance. In other populations, p may decrease as a result of chance. In time, the allelic frequencies in the 10 populations will become different: the populations will genetically diverge. As time passes, the change in allelic frequency within a population and the genetic divergence among populations are due to the same force—the random change in allelic frequencies. The magnitude of genetic drift can be assessed either by examining the change in allelic frequency within a single population or by examining the magnitude of genetic differences that accumulate among populations.

The amount of genetic drift can be estimated from the variance in allelic frequency. Variance, s^2, is a statistical measure that describes the degree of variability in a trait (see p. 311 in Chapter 11). Suppose that we observe a large number of separate populations, each with N individuals and allelic frequencies of p and q. After one generation of random mating, genetic drift expressed in terms of the variance in allelic frequency among the populations (s_p^2) will be

$$s_p^2 = \frac{pq}{2N} \qquad (12.16)$$

The amount of change resulting from genetic drift (the variance in allelic frequency) is determined by two parameters: the allelic frequencies (p and q) and the population size (N). Genetic drift will be maximal when p and q are equal (each 0.5). For example, assume that a population consists of 50 individuals. When the allelic frequencies are equal ($p = q = 0.5$), the variance in allelic frequency (s_p^2) will be $(0.5 \times 0.5)/(2 \times 50) = 0.0025$. In contrast, when $p = 0.9$ and $q = 0.1$, the variance in allelic frequency will be only 0.0009. Genetic drift will also be higher when the population size is small. If $p = q = 0.5$, but the population size is only 10 instead of 50, then the variance in allelic frequency becomes $(0.5 \times 0.5)/(2 \times 10) = 0.0125$, which is five times as great as when population size is 50.

This divergence of populations through genetic drift is strikingly illustrated in the results of an experiment carried out by Peter Buri on fruit flies (**Figure 12.11**). Buri examined the frequencies of two alleles (bw^{75} and bw) that affect eye color in the flies. He set up 107 populations, each consisting of 8 males and 8 females. He began each population with a frequency of bw^{75} equal to 0.5. He allowed the fruit flies within each replicate to mate randomly and, each generation, he randomly selected 8 male and 8 female flies to be the parents of the next generation. He followed the changes in the frequencies of the two alleles over 19 generations. In one population, the average frequency of bw^{75} (p) over the 19 generations was 0.5312. We can use Equation 12.16 to calculate the expected variance in allelic frequency due to genetic drift. The frequency of the bw allele (q) will be $1 - p = 1 - 0.53125 = 0.46875$. The population size ($N$) equals 16. The expected variance in allelic frequency will be

$$\frac{pq}{2N} = \frac{0.53125 \times 0.46875}{2 \times 16} = 0.0156$$

which was very close to the actual observed variance of 0.0151.

The effect of population size on genetic drift is illustrated by a study conducted by Luca Cavalli-Sforza and his colleagues. They studied variation in blood types among villagers in the Parma Valley of Italy, where the amount of migration between villages was limited. They found that variation in allelic frequency was greatest between small isolated villages in the upper valley but decreased between larger villages and towns farther down the valley. This result is exactly what we expect with genetic drift: there should be more genetic drift and thus more variation among villages when population size is small.

the equivalent number of breeding adults, the **effective population size** (N_e). Several factors determine the equivalent number of breeding adults, including the sex ratio, variation between individuals in reproductive success, fluctuations in population size, the age structure of the population, and whether mating is random.

CONCEPTS

Genetic drift is change in allelic frequency due to chance factors. The amount of change in allelic frequency due to genetic drift is inversely related to the effective population size (the equivalent number of breeding adults in a population).

✔ CONCEPT CHECK 8

Which of the following statements is an example of genetic drift?

a. Allele *g* for fat production increases in a small population because birds with more body fat have higher survivorship in a harsh winter.
b. Random mutation increases the frequency of allele *A* in one population but not in another.
c. Allele *R* reaches a frequency of 1.0 because individuals with genotype *rr* are sterile.
d. Allele *m* is lost when a virus kills all but a few individuals and just by chance none of the survivors possess allele *m*.

Causes of genetic drift All genetic drift arises from sampling error, but there are several different ways in which sampling error can arise. First, a population may be reduced in size for a number of generations because of limitations in space, food, or some other critical resource. Genetic drift in a small population for multiple generations can significantly affect the composition of a population's gene pool.

A second way that sampling error can arise is through the **founder effect**, which is due to the establishment of a population by a small number of individuals; the population of bighorn sheep at the National Bison Range, discussed in the introduction to this chapter, underwent a founder effect. Although a population may increase and become quite large, the genes carried by all its members are derived from the few genes originally present in the founders (assuming no migration or mutation). Chance events affecting which genes were present in the founders will have an important influence on the makeup of the entire population.

A third way in which genetic drift arises is through a **genetic bottleneck**, which develops when a population undergoes a drastic reduction in population size. A genetic bottleneck developed in northern elephant seals (**Figure 12.12**). Before 1800, thousands of elephant seals were found along the California coast, but the population was devastated by hunting between 1820 and 1880. By 1884, as few as 20 seals survived on a remote beach of Isla de Guadelupe west of Baja, California. Restrictions on hunting enacted by the United States and Mexico allowed the seals to recover, and there are now more

Experiment

Question: What effect does genetic drift have on the genetic composition of populations?

Methods Buri examined the frequencies of two alleles (bw^{75} and bw) that affect *Drosophila* eye color in 107 replicate small populations over 19 generations.

Results

Conclusion: As a result of genetic drift, allelic frequencies in the different populations diverged and often became fixed for one allele or the other.

12.11 Populations diverge in allelic frequency and become fixed for one allele as a result of genetic drift. In Buri's study of two eye-color alleles (bw^{75} and *bw*) in *Drosophila*, each population consisted of 8 males and 8 females and began with the frequency of bw^{75} equal to 0.5.

For ecological and demographic studies, population size is usually defined as the number of individuals in a group. The evolution of a gene pool depends, however, only on those individuals who contribute genes to the next generation. Population geneticists usually define population size as

12.12 Northern elephant seals underwent a severe genetic bottleneck between 1820 and 1880. Today, these seals have low levels of genetic variation. [PhotoDisc/Getty Images.]

A second effect of genetic drift is to reduce genetic variation within populations. Through random change, an allele may eventually reach a frequency of either 1 or 0, at which point all individuals in the population are homozygous for one allele. When an allele has reached a frequency of 1, we say that it has reached **fixation**. Other alleles are lost (reach a frequency of 0) and can be restored only by migration from another population or by mutation. Fixation, then, leads to a loss of genetic variation within a population. This loss can be seen in the northern elephant seals described earlier. Today, these seals have low levels of genetic variation; a study of 24 protein-encoding genes found no individual or population differences in these genes. A subsequent study of sequence variation in mitochondrial DNA also revealed low levels of genetic variation. In contrast, the southern elephant seals had much higher levels of mitochondrial DNA variation. The southern elephant seals also were hunted, but their population size never dropped below 1000; therefore, unlike the northern elephant seals, they did not experience a genetic bottleneck.

than 30,000 seals in the population. All seals in the population today are genetically similar, because they have genes that were carried by the few survivors of the population bottleneck.

The effects of genetic drift Genetic drift has several important effects on the genetic composition of a population. First, it produces change in allelic frequencies within a population. Because drift is random, allelic frequency is just as likely to increase as it is to decrease and will wander with the passage of time (hence the name genetic drift). **Figure 12.13** illustrates a computer simulation of genetic drift in five populations over 30 generations, starting with $q = 0.5$ and maintaining a constant population size of 10 males and 10 females. These allelic frequencies change randomly from generation to generation.

Given enough time, all small populations will become fixed for one allele or the other. Which allele becomes fixed is random and is determined by the initial frequency of the allele. If a population begins with two alleles, each with a frequency of 0.5, both alleles have an equal probability of fixation. However, if one allele is initially common, it is more likely to become fixed.

A third effect of genetic drift is that different populations diverge genetically with time. In Figure 12.13, all five populations begin with the same allelic frequency ($q = 0.5$) but, because drift is random, the frequencies in different populations do not change in the same way, and so populations gradually acquire genetic differences. Eventually, all the populations reach fixation; some will become fixed for one allele, and others will become fixed for the alternative allele.

The three results of genetic drift (allelic frequency change, loss of variation within populations, and genetic divergence between populations) take place simultaneously, and all result from sampling error. The first two results take place *within* populations, whereas the third takes place *between* populations. **TRY PROBLEM 32**

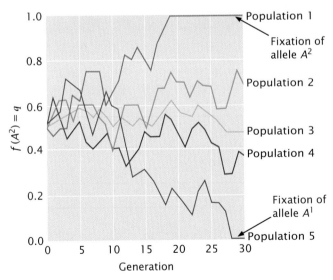

12.13 Genetic drift changes allelic frequencies within populations, leading to a reduction in genetic variation through fixation and genetic divergence among populations. Shown here is a computer simulation of changes in the frequency of allele A^2 (q) in five different populations due to random genetic drift. Each population consists of 10 males and 10 females and begins with $q = 0.5$.

> **CONCEPTS**
>
> Genetic drift results from continuous small population size, the founder effect (establishment of a population by a few founders), and the bottleneck effect (population reduction). Genetic drift causes a change in allelic frequencies within a population, a loss of genetic variation through the fixation of alleles, and genetic divergence between populations.

Natural Selection

A final process that brings about changes in allelic frequencies is natural selection, the differential reproduction of genotypes (see p. 324 in Chapter 11). Natural selection

12.14 Natural selection produces adaptations, such as those seen in the polar bears that inhabit the extreme Arctic environment. These bears blend into the snowy background, which helps them in hunting seals. The hairs of their fur stay erect even when wet, and thick layers of blubber provide insulation, which protects against subzero temperatures. Their digestive tracts are adapted to a seal-based carnivorous diet. [Digital Vision.]

takes place when individuals with adaptive traits produce a greater number of offspring than that produced by others in the population. If the adaptive traits have a genetic basis, they are inherited by the offspring and appear with greater frequency in the next generation. A trait that provides a reproductive advantage thereby increases with the passage of time, enabling populations to become better suited to their environments—to become better adapted. Natural selection is unique among evolutionary forces in that it promotes adaptation (**Figure 12.14**).

Fitness and the selection coefficient The effect of natural selection on the gene pool of a population depends on the fitness values of the genotypes in the population. **Fitness** is defined as the relative reproductive success of a genotype. Here, the term *relative* is critical: fitness is the reproductive success of one genotype compared with the reproductive successes of other genotypes in the population.

Fitness (W) ranges from 0 to 1. Suppose the average number of viable offspring produced by three genotypes is

Genotypes:	A^1A^1	A^1A^2	A^2A^2
Mean number of offspring produced:	10	5	2

To calculate fitness for each genotype, we take the mean number of offspring produced by a genotype and divide it by the mean number of offspring produced by the most prolific genotype:

Fitness (W):

$$W_{11} = \frac{10}{10} = 1.0 \quad W_{12} = \frac{5}{10} = 0.5 \quad W_{22} = \frac{2}{10} = 0.2 \quad (12.17)$$

$\qquad A^1A^1 \qquad\qquad A^1A^2 \qquad\qquad A^2A^2$

The fitness of genotype A^1A^1 is designated W_{11}, that of A^1A^2 is W_{12}, and that of A^2A^2 is W_{22}. A related variable is the **selection coefficient** (s), which is the relative intensity of selection against a genotype. We usually speak of selection for a particular genotype, but keep in mind that, when selection is *for* one genotype, selection is automatically *against* at least one other genotype. The selection coefficient is equal to $1 - W$; so the selection coefficients for the preceding three genotypes are

	A^1A^1	A^1A^2	A^2A^2
Selection coefficient ($1 - W$):	$s_{11} = 0$	$s_{12} = 0.5$	$s_{22} = 0.8$

CONCEPTS

Natural selection is the differential reproduction of genotypes. It is measured as fitness, which is the reproductive success of a genotype compared with other genotypes in a population.

✔ CONCEPT CHECK 9

The average numbers of offspring produced by three genotypes are: $GG = 6$; $Gg = 3$, $gg = 2$. What is the fitness of Gg?

a. 3   b. 0.5   c. 0.3   d. 0.27

Table 12.4 Method for determining changes in allelic frequency due to selection

	A^1A^1	A^1A^2	A^2A^2
Initial genotypic frequencies	p^2	$2pq$	q^2
Fitnesses	W_{11}	W_{12}	W_{22}
Proportionate contribution of genotypes to population	$p^2 W_{11}$	$2pq W_{12}$	$q^2 W_{22}$
Relative genotypic frequency after selection	$\dfrac{p^2 W_{11}}{\bar{w}}$	$\dfrac{2pq W_{12}}{\bar{w}}$	$\dfrac{q^2 W_{22}}{\bar{w}}$

Note: $\bar{w} = p^2 W_{11} + 2pq W_{12} + q^2 W_{22}$.

Allelic frequencies after selection: $p' = f(A^1) = f(A^1A^1) + \frac{1}{2}f(A^1A^2)$; $q' = 1 - p$.

Table 12.5 Formulas for calculating change in allelic frequencies with different types of selection

Type of Selection	Fitness Values			Change in q
	A^1A^1	A^1A^2	A^2A^2	
Selection against a recessive trait	1	1	$1-s$	$\dfrac{-spq^2}{1-sq^2}$
Selection against a dominant trait	1	$1-s$	$1-s$	$\dfrac{-spq^2}{1-s+sq^2}$
Selection against a trait with no dominance	1	$1-\frac{1}{2}s$	$1-s$	$\dfrac{-\frac{1}{2}spq}{1-sa}$
Selection against both homozygotes (overdominance)	$1-s_{11}$	1	$1-s_{22}$	$\dfrac{pq(s_{11}p - s_{22}q)}{1-s_{11}p^2 - s_{22}q^2}$

The general selection model With selection, differential fitness among genotypes leads to changes in the frequencies of the genotypes over time which, in turn, lead to changes in the frequencies of the alleles that make up the genotypes. We can predict the effect of natural selection on allelic frequencies by using a general selection model, which is outlined in **Table 12.4**. Use of this model requires knowledge of both the initial allelic frequencies and the fitness values of the genotypes. It assumes that mating is random and that the only force acting on a population is natural selection. The general selection model can be used to calculate the allelic frequencies after any type of selection. It is also possible to work out formulas for determining the change in allelic frequency when selection is against recessive, dominant, and codominant traits, as well as traits in which the heterozygote has highest fitness (**Table 12.5**).

CONCEPTS

The change in allelic frequency due to selection can be determined for any type of genetic trait by using the general selection model.

Worked Problem

Let's apply the general selection model in Table 12.4 to a real example. Alcohol is a common substance in rotting fruit, where fruit-fly larvae grow and develop; larvae use the enzyme alcohol dehydrogenase (ADH) to detoxify the effects of the alcohol. In some fruit-fly populations, two alleles are present at the locus that encodes ADH: Adh^F, which encodes a form of the enzyme that migrates rapidly (fast) on an electrophoretic gel; and Adh^S, which encodes a form of the enzyme that migrates slowly on an electrophoretic gel. Female fruit flies with different Adh genotypes produce the following numbers of offspring when alcohol is present:

Genotype	Mean number of offspring
Adh^F/Adh^F	120
Adh^F/Adh^S	60
Adh^S/Adh^S	30

a. Calculate the relative fitnesses of females having these genotypes.

b. If a population of fruit flies has an initial frequency of Adh^F equal to 0.2, what will the frequency be in the next generation when alcohol is present?

- **Solution**

a. First, we must calculate the fitnesses of the three genotypes. Fitness is the relative reproductive output of a genotype and is calculated by dividing the mean number of offspring produced by that genotype by the mean number of offspring produced by the most prolific genotype. The fitnesses of the three Adh genotypes therefore are:

Genotype	Mean number of offspring	Fitness
Adh^F/Adh^F	120	$W_{FF} = \dfrac{120}{120} = 1$
Adh^F/Adh^S	60	$W_{FS} = \dfrac{60}{120} = 0.5$
Adh^S/Adh^S	30	$W_{SS} = \dfrac{30}{120} = 0.25$

	$Adh^F Adh^F$	$Adh^F Adh^S$	$Adh^S Adh^S$
Initial genotypic frequencies:	$p^2 = (0.2)^2 = 0.04$	$2pq = 2(0.2)(0.8) = 0.32$	$q^2 = (0.8)^2 = 0.64$
Fitnesses:	$W_{FF} = 1$	$W_{FS} = 0.5$	$W_{SS} = 0.25$
Proportionate contribution of genotypes to population:	$p^2 W_{FF} = 0.04(1) = 0.04$	$2pq W_{FS} = (0.32)(0.5) = 0.16$	$q^2 W_{SS} = (0.64)(0.25) = 0.16$

b. To calculate the frequency of the Adh^F allele after selection, we can apply the table method. In the first row of the table above, we record the initial genotypic frequencies before selection has acted. If mating has been random (an assumption of the model), the genotypes will have the Hardy–Weinberg equilibrium frequencies of p^2, $2pq$, and q^2. In the second row of the table above, we put the fitness values of the corresponding genotypes. The proportion of the population represented by each genotype after selection is obtained by multiplying the initial genotypic frequency times its fitness (third row of Table 12.4). Now the genotypes are no longer in Hardy–Weinberg equilibrium.

The mean fitness ($\bar{w}$) of the population is the sum of the proportionate contributions of the three genotypes: $\bar{w} = p^2 W_{11} + 2pq W_{12} + q^2 W_{22} = 0.04 + 0.16 + 0.16 = 0.36$. The mean fitness $\bar{w}$ is the average fitness of all individuals in the population and allows the frequencies of the genotypes after selection to be obtained.

The frequency of a genotype after selection will be equal to its proportionate contribution divided by the mean fitness of the population ($p^2 W_{11}/\bar{w}$ for genotype $A^1 A^1$, $2pq W_{12}/\bar{w}$ for genotype $A^1 A^2$, and $q^2 W_{22}/\bar{w}$ for genotype $A^2 A^2$) as shown in the fourth line of Table 25.4. We can now add these values to our table as shown below:

	$Adh^F Adh^F$	$Adh^F Adh^S$	$Adh^S Adh^S$
Initial genotypic frequencies:	$p^2 = (0.2)^2 = 0.04$	$2pq = 2(0.2)(0.8) = 0.32$	$q^2 = (0.8)^2 = 0.64$
Fitnesses:	$W_{FF} = 1$	$W_{FS} = 0.5$	$W_{SS} = 0.25$
Proportionate contribution of genotypes to population:	$p^2 W_{FF} = 0.04(1) = 0.04$	$2pq W_{FS} = (0.32)(0.5) = 0.16$	$q^2 W_{SS} = (0.64)(0.25) = 0.16$
Relative genotypic frequency after selection:	$\dfrac{p^2 w_{FF}}{\bar{w}} = \dfrac{0.04}{0.36} = 0.11$	$\dfrac{2pq w_{FS}}{\bar{w}} = \dfrac{0.16}{0.36} = 0.44$	$\dfrac{q^2 w_{SS}}{\bar{w}} = \dfrac{0.16}{0.36} = 0.44$

After the new genotypic frequencies have been calculated, the new allelic frequency of Adh^F (p') can be determined by using the now-familiar formula of Equation 12.4:

$$p' = f(Adh^F) = f(Adh^F/Adh^F) + \tfrac{1}{2} f(Adh^F/Adh^S)$$
$$= 0.11 + \tfrac{1}{2}(0.44) = 0.33$$

and that of q' can be obtained by subtraction:

$$q' = 1 - p'$$
$$= 1 - 0.33 = 0.67$$

We predict that the frequency of Adh^F will increase from 0.2 to 0.33.

> ➡ For more practice with the selection model, try Problem 33 at the end of the chapter.

The results of selection The results of selection depend on the relative fitnesses of the genotypes. If we have three genotypes ($A^1 A^1$, $A^1 A^2$, and $A^2 A^2$) with fitnesses W_{11}, W_{12}, and W_{22}, we can identify six different types of natural selection (**Table 12.6**). In type 1 selection, a dominant allele A^1 confers a fitness advantage; in this case, the fitnesses of genotypes $A^1 A^1$ and $A^1 A^2$ are equal and higher than the fitness of $A^2 A^2$ ($W_{11} = W_{12} > W_{22}$). Because both the heterozygote and the $A^1 A^1$ homozygote have copies of the A^1 allele and produce more offspring than the $A^2 A^2$ homozygote does, the frequency of the A^1 allele will increase with time, and the frequency of the A^2 allele will decrease. This form of selection, in which one allele or trait is favored over another, is termed **directional selection**.

Type 2 selection (see Table 12.6) is directional selection against a dominant allele A^1 ($W_{11} = W_{12} < W_{22}$). In this case, the A^2 allele increases and the A^1 allele decreases. Type 3 and type 4 selection also are directional selection but, in these cases, there is incomplete dominance and the heterozygote has a fitness that is intermediate between the two homozygotes ($W_{11} > W_{12} > W_{22}$ for type 3; $W_{11} < W_{12} < W_{22}$ for type 4). When $A^1 A^1$ has the highest fitness (type 3), the A^1 allele increases and the A^2 allele decreases with the passage of time. When $A^2 A^2$ has the highest fitness (type 4), the A^2 allele increases and the A^1 allele decreases with time. Eventually, directional selection leads to fixation of the favored allele and elimination of the other allele, as long as no other evolutionary forces act on the population.

Table 12.6 Types of natural selection

Type	Fitness Relation	Form of Selection	Result
1	$W_{11} = W_{12} > W_{22}$	Directional selection against recessive allele A^2	A^1 increases, A^2 decreases
2	$W_{11} = W_{12} < W_{22}$	Directional selection against dominant allele A^1	A^2 increases, A^1 decreases
3	$W_{11} > W_{12} > W_{22}$	Directional selection against incompletely dominant allele A^2	A^1 increases, A^2 decreases
4	$W_{11} < W_{12} < W_{22}$	Directional selection against incompletely dominant allele A^1	A^2 increases, A^1 decreases
5	$W_{11} < W_{12} > W_{22}$	Overdominance	Stable equilibrium, both alleles maintained
6	$W_{11} > W_{12} < W_{22}$	Underdominance	Unstable equilibrium

Note: W_{11}, W_{12}, and W_{22} represent the fitnesses of genotypes A^1A^1, A^1A^2, and A^2A^2, respectively.

To see the effects of natural selection on allelic and genotypic frequencies, view the Mini-Tutorial in Animation 25.1.

Two types of selection (types 5 and 6) are special situations that lead to equilibrium, where there is no further change in allelic frequency. Type 5 selection is referred to as **overdominance** or heterozygote advantage. Here, the heterozygote has higher fitness than the fitnesses of the two homozygotes ($W_{11} < W_{12} > W_{22}$). With overdominance, both alleles are favored in the heterozygote, and neither allele is eliminated from the population. Initially, the allelic frequencies may change because one homozygote has higher fitness than the other; the direction of change will depend on the relative fitness values of the two homozygotes. The allelic frequencies change with overdominant selection until a stable equilibrium is reached, at which point there is no further change. The allelic frequency at equilibrium ($\hat{q}$) depends on the relative fitnesses (usually expressed as selection coefficients) of the two homozygotes:

$$\hat{q} = f(A^2) = \frac{s_{11}}{s_{11} + s_{22}} \qquad (12.18)$$

where s_{11} represents the selection coefficient of the A^1A^1 homozygote and s_{22} represents the selection coefficient of the A^2A^2 homozygote.

The last type of selection (type 6) is **underdominance**, in which the heterozygote has lower fitness than both homozygotes ($W_{11} > W_{12} < W_{22}$). Underdominance leads to an unstable equilibrium; here, allelic frequencies will not change as long as they are at equilibrium but, if they are disturbed from the equilibrium point by some other evolutionary force, they will move away from equilibrium until one allele eventually becomes fixed. **TRY PROBLEM 34**

CONCEPTS

Natural selection changes allelic frequencies; the direction and magnitude of change depend on the intensity of selection, the dominance relations of the alleles, and the allelic frequencies. Directional selection favors one allele over another and eventually leads to fixation of the favored allele. Overdominance leads to a stable equilibrium with maintenance of both alleles in the population. Underdominance produces an unstable equilibrium because the heterozygote has lower fitness than those of the two homozygotes.

✔ **CONCEPT CHECK 10**

How does overdominance differ from directional selection?

Change in the allelic frequency of a recessive allele due to natural selection The rate at which selection changes allelic frequencies depends on the allelic frequency itself. If an allele (A^2) is lethal and recessive, $W_{11} = W_{12} = 1$, whereas $W_{22} = 0$. The frequency of the A^2 allele will decrease with time (because the A^2A^2 homozygote produces no offspring), and the rate of decrease will be proportional to the frequency of the recessive allele. When the frequency of the allele is high, the change in each generation is relatively large but, as the frequency of the allele drops, a higher proportion of the alleles are in the heterozygous genotypes, where they are immune to the action of natural selection (the heterozygotes have the same phenotype as the favored homozygote). Thus, selection against a rare recessive allele is very inefficient and its removal from the population is slow.

The relation between the frequency of a recessive allele and its rate of change under natural selection has an important implication. Some people believe that the medical treatment of patients with rare recessive diseases will cause the disease gene to increase, eventually leading to degeneration of the human gene pool. This mistaken belief was the basis of eugenic laws that were passed in the early part of the twentieth century prohibiting the marriage of persons with certain genetic conditions and allowing the involuntary sterilization of others. However, most copies of rare recessive alleles are present in heterozygotes, and selection against the homozygotes will have little effect on the frequency of a recessive allele. Thus, whether the homozygotes for a recessive trait reproduce or not has little effect on the frequency of the disorder.

Mutation and natural selection Recurrent mutation and natural selection act as opposing forces on detrimental alleles; mutation increases their frequency and natural selection decreases their frequency. Eventually, these two forces reach an equilibrium, in which the number of alleles added by mutation is balanced by the number of alleles removed by selection.

The frequency of a recessive allele at equilibrium ($\hat{q}$) is equal to the square root of the mutation rate divided by the selection coefficient:

$$\hat{q} = \sqrt{\frac{\mu}{s}} \qquad (12.19)$$

For selection acting on a dominant allele, the frequency of the dominant allele at equilibrium can be shown to be

$$\hat{q} = \frac{\mu}{s} \qquad (12.20)$$

Achondroplasia is a common type of human dwarfism that results from a dominant gene. People with this condition are fertile, although they produce only about 74% as many children as are produced by people without achondroplasia. The fitness of people with achondroplasia therefore averages 0.74, and the selection coefficient (s) is $1 - W$, or 0.26. If we assume that the mutation rate for achondroplasia is about 3×10^{-5} (a typical mutation rate in humans), then we can predict that the equilibrium frequency for the achondroplasia allele will be

$$\hat{q} = \frac{0.00003}{0.26} = 0.0001153$$

This frequency is close to the actual frequency of the disease.

TRY PROBLEM 37

CONCEPTS

Mutation and natural selection act as opposing forces on detrimental alleles: mutation tends to increase their frequency and natural selection tends to decrease their frequency, eventually producing an equilibrium.

CONNECTING CONCEPTS

The General Effects of Forces That Change Allelic Frequencies

You now know that four processes bring about change in the allelic frequencies of a population: mutation, migration, genetic drift, and natural selection. Their short- and long-term effects on allelic frequencies are summarized in **Table 12.7**. In some cases, these changes continue until one allele is eliminated and the other becomes fixed in the population. Genetic drift and directional selection will eventually result in fixation, provided these forces are the only ones acting on a population. With the other evolutionary forces, allelic frequencies change until an equilibrium point is reached, and then there is no additional change in allelic frequency. Mutation, migration, and some forms of natural selection can lead to stable equilibria (see Table 12.7).

The different evolutionary forces affect both genetic variation within populations and genetic divergence between populations. Evolutionary forces that maintain or increase genetic variation within populations are listed in the upper-left quadrant of **Figure 12.15**. These forces include some types of natural selection, such as overdominance, in which both alleles are favored. Mutation and migration also increase genetic variation within populations because they introduce new alleles to the population. Evolutionary forces that decrease genetic variation within populations are listed in the lower-left quadrant of Figure 12.15. These forces include genetic drift, which decreases variation through the fixation of

Table 12.7 Effects of different evolutionary forces on allelic frequencies within populations

Force	Short-Term Effect	Long-Term Effect
Mutation	Change in allelic frequency	Equilibrium reached between forward and reverse mutations
Migration	Change in allelic frequency	Equilibrium reached when allelic frequencies of source and recipient population are equal
Genetic drift	Change in allelic frequency	Fixation of one allele
Natural selection	Change in allelic frequency	Directional selection: fixation of one allele Overdominant selection: equilibrium reached

	Within populations	Between populations
Increase genetic variation	Mutation Migration Some types of natural selection	Mutation Genetic drift Some types of natural selection
Decrease genetic variation	Genetic drift Some types of natural selection	Migration Some types of natural selection

12.15 Mutation, migration, genetic drift, and natural selection have different effects on genetic variation within populations and on genetic divergence between populations.

alleles, and some forms of natural selection such as directional selection.

The various evolutionary forces also affect the amount of genetic divergence between populations. Natural selection increases divergence between populations if different alleles are favored in the different populations, but it can also decrease divergence between populations by favoring the same allele in the different populations. Mutation almost always increases divergence between populations because different mutations arise in each population. Genetic drift also increases divergence between populations because changes in allelic frequencies due to drift are random and are likely to change in different directions in separate populations. Migration, on the other hand, decreases divergence between populations because it makes populations similar in their genetic composition.

Migration and genetic drift act in opposite directions: migration increases genetic variation within populations and decreases divergence between populations, whereas genetic drift decreases genetic variation within populations and increases divergence among populations. Mutation increases both variation within populations and divergence between populations. Natural selection can either increase or decrease variation within populations, and it can increase or decrease divergence between populations.

An important point to keep in mind is that real populations are simultaneously affected by many evolutionary forces. We have examined the effects of mutation, migration, genetic drift, and natural selection in isolation so that the influence of each process would be clear. However, in the real world, populations are commonly affected by several evolutionary forces at the same time, and evolution results from the complex interplay of numerous processes.

CONCEPTS SUMMARY

- Population genetics examines the genetic composition of groups of individuals and how this composition changes with time.
- A Mendelian population is a group of interbreeding, sexually reproducing individuals, whose set of genes constitutes the population's gene pool. Evolution takes place through changes in this gene pool.
- A population's genetic composition can be described by its genotypic and allelic frequencies.
- The Hardy–Weinberg law describes the effects of reproduction and Mendel's laws on the allelic and genotypic frequencies of a population. It assumes that a population is large, randomly mating, and free from the effects of mutation, migration, and natural selection. When these conditions are met, the allelic frequencies do not change and the genotypic frequencies stabilize after one generation in the Hardy–Weinberg equilibrium proportions p^2, $2pq$, and q^2, where p and q equal the frequencies of the alleles.
- Nonrandom mating affects the frequencies of genotypes but not those of alleles.
- Inbreeding, a type of positive assortative mating, increases the frequency of homozygotes while decreasing the frequency of heterozygotes. Inbreeding is frequently detrimental because it increases the appearance of lethal and deleterious recessive traits.
- Mutation, migration, genetic drift, and natural selection can change allelic frequencies.

- Recurrent mutation eventually leads to an equilibrium, with the allelic frequencies being determined by the relative rates of forward and reverse mutation. Change due to mutation in a single generation is usually very small because mutation rates are low.
- Migration, the movement of genes between populations, increases the amount of genetic variation within populations and decreases the number of differences between populations.
- Genetic drift is change in allelic frequencies due to chance factors. Genetic drift arises when a population consists of a small number of individuals, is established by a small number of founders, or undergoes a major reduction in size. Genetic drift changes allelic frequencies, reduces genetic variation within populations, and causes genetic divergence among populations.
- Natural selection is the differential reproduction of genotypes; it is measured by the relative reproductive successes (fitnesses) of genotypes. The effects of natural selection on allelic frequency can be determined by applying the general selection model. Directional selection leads to the fixation of one allele. The rate of change in allelic frequency due to selection depends on the intensity of selection, the dominance relations, and the initial frequencies of the alleles.
- Mutation and natural selection can produce an equilibrium, in which the number of new alleles introduced by mutation is balanced by the elimination of alleles through natural selection.

IMPORTANT TERMS

genetic rescue (p. 338)
Mendelian population (p. 338)
gene pool (p. 338)
genotypic frequency (p. 339)
allelic frequency (p. 339)
Hardy–Weinberg law (p. 341)
Hardy–Weinberg equilibrium (p. 341)
positive assortative mating (p. 345)
negative assortative mating (p. 345)

inbreeding (p. 345)
outcrossing (p. 345)
inbreeding coefficient (p. 345)
inbreeding depression (p. 347)
equilibrium (p. 349)
migration (gene flow) (p. 349)
sampling error (p. 351)
genetic drift (p. 351)
effective population size (p. 352)

founder effect (p. 352)
genetic bottleneck (p. 352)
fixation (p. 353)
fitness (p. 354)
selection coefficient (p. 354)
directional selection (p. 356)
overdominance (p. 357)
underdominance (p. 357)

ANSWERS TO CONCEPT CHECKS

1. There are fewer alleles than genotypes, and so the gene pool can be described by fewer parameters when allelic frequencies are used. Additionally, the genotypes are temporary assemblages of alleles that break down each generation; rather than the genotypes, the alleles are passed from generation to generation in sexually reproducing organisms.

2. a
3. c
4. 0.10
5. b
6. c
7. c
8. d
9. b
10. In overdominance, the heterozygote has the highest fitness. In directional selection, one allele or one trait is favored over another.

WORKED PROBLEM

1. A recessive allele for red hair (r) has a frequency of 0.2 in population I and a frequency of 0.01 in population II. A famine in population I causes a number of people in population I to migrate to population II, where they reproduce randomly with the members of population II. Geneticists estimate that, after migration, 15% of the people in population II consist of people who migrated from population I. What will be the frequency of red hair in population II after the migration?

• **Solution**

From Equation 12.14, the allelic frequency in a population after migration (q'_{II}) is

$$q'_{II} = q_{I}(m) + q_{II}(1 - m)$$

where q_I and q_{II} are the allelic frequencies in population I (migrants) and population II (residents), respectively, and m is the proportion of population II that consist of migrants. In this problem, the frequency of red hair is 0.2 in population I and 0.01 in population II. Because 15% of population II consists of migrants, $m = 0.15$. Substituting these values into Equation 12.14, we obtain

$$q'_{II} = 0.2(0.15) + (0.01)(1 - 0.15) = 0.03 + 0.0085 = 0.0385$$

which is the expected frequency of the allele for red hair in population II after migration. Red hair is a recessive trait; if mating is random for hair color, the frequency of red hair in population II after migration will be

$$f(rr) = q^2 = (0.0385)^2 = 0.0015$$

COMPREHENSION QUESTIONS

Section 12.1

1. What is a Mendelian population? How is the gene pool of a Mendelian population usually described?

Section 12.2

2. What are the predictions given by the Hardy–Weinberg law?

*3. What assumptions must be met for a population to be in Hardy–Weinberg equilibrium?

4. What is random mating?

*5. Give the Hardy–Weinberg expected genotypic frequencies for (a) an autosomal locus with three alleles, and (b) an X-linked locus with two alleles.

Section 12.3

6. Define inbreeding and briefly describe its effects on a population.

Section 12.4

7. What determines the allelic frequencies at mutational equilibrium?

*8. What factors affect the magnitude of change in allelic frequencies due to migration?

9. Define genetic drift and give three ways in which it can arise. What effect does genetic drift have on a population?

*10. What is effective population size? How does it affect the amount of genetic drift?

11. Define natural selection and fitness.

12. Briefly describe the differences between directional selection, overdominance, and underdominance. Describe the effect of each type of selection on the allelic frequencies of a population.

13. What factors affect the rate of change in allelic frequency due to natural selection?

*14. Compare and contrast the effects of mutation, migration, genetic drift, and natural selection on genetic variation within populations and on genetic divergence between populations.

APPLICATION QUESTIONS AND PROBLEMS

Section 12.1

15. How would you respond to someone who said that models are useless in studying population genetics because they represent oversimplifications of the real world?

*16. Voles (*Microtus ochrogaster*) were trapped in old fields in southern Indiana and were genotyped for a transferrin locus. The following numbers of genotypes were recorded, where T^E and T^F represent different alleles.

$T^E T^E$	$T^E T^F$	$T^F T^F$
407	170	17

[Tom McHugh/Photo Reseachers]

Calculate the genotypic and allelic frequencies of the transferrin locus for this population.

17. Jean Manning, Charles Kerfoot, and Edward Berger studied genotypic frequencies at the phosphoglucose isomerase (GPI) locus in the cladoceran *Bosmina longirostris*. At one location, they collected 176 animals from Union Bay in Seattle, Washington, and determined their GPI genotypes by using electrophoresis (J. Manning, W. C. Kerfoot, and E. M. Berger. 1978. *Evolution* 32:365–374).

Genotype	Number
$S^1 S^1$	4
$S^1 S^2$	38
$S^2 S^2$	134

Determine the genotypic and allelic frequencies for this population.

18. Orange coat color of cats is due to an X-linked allele (X^O) that is codominant with the allele for black (X^+). Genotypes of the orange locus of cats in Minneapolis and St. Paul, Minnesota, were determined, and the following data were obtained:

$X^O X^O$ females	11
$X^O X^+$ females	70
$X^+ X^+$ females	94
$X^O Y$ males	36
$X^+ Y$ males	112

Calculate the frequencies of the X^O and X^+ alleles for this population.

Section 12.2

19. A total of 6129 North American Caucasians were blood typed for the MN locus, which is determined by two codominant alleles, L^M and L^N. The following data were obtained:

Blood type	Number
M	1787
MN	3039
N	1303

Carry out a chi-square test to determine whether this population is in Hardy–Weinberg equilibrium at the MN locus.

20. Most black bears (*Ursus americanus*) are black or brown in color. However, occasional white bears of this species appear in some populations along the coast of British Columbia. Kermit Ritland and his colleagues determined that white coat color in these bears results from a recessive mutation (*G*)

[Marc Muench/Alamy.]

caused by a single nucleotide replacement in which guanine substitutes for adenine at the melanocortin-1 receptor locus (*mcr1*), the same locus responsible for red hair in humans (K. Ritland, C. Newton, and H. D. Marshall. 2001. *Current Biology* 11:1468–1472). The wild-type allele at this locus (*A*) encodes black or brown color. Ritland and his colleagues collected samples from bears on three islands and determined their genotypes at the *mcr1* locus.

Genotype	Number
AA	42
AG	24
GG	21

a. What are the frequencies of the *A* and *G* alleles in these bears?

b. Give the genotypic frequencies expected if the population is in Hardy–Weinberg equilibrium.

c. Use a chi-square test to compare the number of observed genotypes with the number expected under Hardy–Weinberg equilibrium. Is this population in Hardy–Weinberg equilibrium? Explain your reasoning.

21. Genotypes of leopard frogs from a population in central Kansas were determined for a locus (*M*) that encodes the enzyme malate dehydrogenase. The following numbers of genotypes were observed:

Genotype	Number
M^1M^1	20
M^1M^2	45
M^2M^2	42
M^1M^3	4
M^2M^3	8
M^3M^3	6
Total	125

a. Calculate the genotypic and allelic frequencies for this population.

b. What would the expected numbers of genotypes be if the population were in Hardy–Weinberg equilibrium?

22. Full color (*D*) in domestic cats is dominant over dilute color (*d*). Of 325 cats observed, 194 have full color and 131 have dilute color.

a. If these cats are in Hardy–Weinberg equilibrium for the dilution locus, what is the frequency of the dilute allele?

b. How many of the 194 cats with full color are likely to be heterozygous?

23. Tay–Sachs disease is an autosomal recessive disorder. Among Ashkenazi Jews, the frequency of Tay–Sachs disease is 1 in 3600. If the Ashkenazi population is mating randomly for the Tay–Sachs gene, what proportion of the population consists of heterozygous carriers of the Tay–Sachs allele?

24. In the plant *Lotus corniculatus*, cyanogenic glycoside protects the plant against insect pests and even grazing by cattle. This glycoside is due to a simple dominant allele. A population of *L. corniculatus* consists of 77 plants that possess cyanogenic glycoside and 56 that lack the compound. What is the frequency of the dominant allele responsible for the presence of cyanogenic glycoside in this population?

*25. Color blindness in humans is an X-linked recessive trait. Approximately 10% of the men in a particular population are color blind.

a. If mating is random for the color-blind locus, what is the frequency of the color-blind allele in this population?

b. What proportion of the women in this population are expected to be color blind?

c. What proportion of the women in the population are expected to be heterozygous carriers of the color-blind allele?

Section 12.3

*26. The human MN blood type is determined by two codominant alleles, L^M and L^N. The frequency of L^M in Eskimos on a small Arctic island is 0.80. If the inbreeding coefficient for this population is 0.05, what are the expected frequencies of the M, MN, and N blood types on the island?

27. Demonstrate mathematically that full-sib mating ($F = 1/4$) reduces the heterozygosity by $1/4$ with each generation.

Section 12.4

28. The forward mutation rate for piebald spotting in guinea pigs is 8×10^{-5}; the reverse mutation rate is 2×10^{-6}. If no other evolutionary forces are assumed to be present, what is the expected frequency of the allele for piebald spotting in a population that is in mutational equilibrium?

29. For a period of 3 years, Gunther Schlager and Margaret Dickie estimated the forward and reverse mutation rates for five loci in mice that encode various aspects of coat color by examining more than 5 million mice for spontaneous mutations (G. Schlager and M. M. Dickie. 1966. *Science* 151:205–206). The numbers of mutations detected at the dilute locus are as follows:

	Number of gametes examined	Number of mutations detected
Forward mutations	260,675	5
Reverse mutations	583,360	2

Calculate the forward and reverse mutation rates at this locus. If these mutations rates are representative of rates in natural populations of mice, what would the expected equilibrium frequency of dilute mutations be?

*30. In German cockroaches, curved wing (*cv*) is recessive to normal wing (cv^+). Bill, who is raising cockroaches in his dorm room, finds that the frequency of the gene for curved wings in his cockroach population is 0.6. In the apartment of his friend Joe, the frequency of the gene for curved wings is 0.2. One day Joe visits Bill in his dorm room, and several cockroaches jump out of Joe's hair and join the population in Bill's room. Bill estimates that, now, 10% of the cockroaches in his dorm room are individual roaches that jumped out of Joe's hair. What is the new frequency of curved wings among cockroaches in Bill's room?

31. A population of water snakes is found on an island in Lake Erie. Some of the snakes are banded and some are unbanded; banding is caused by an autosomal allele that is recessive to an allele for no bands. The frequency of banded snakes on the island is 0.4, whereas the frequency of banded snakes on the mainland is 0.81. One summer, a large number of snakes migrate from the mainland to the island. After this migration, 20% of the island population consists of snakes that came from the mainland.

   a. If both the mainland population and the island population are assumed to be in Hardy–Weinberg equilibrium for the alleles that affect banding, what is the frequency of the allele for bands on the island and on the mainland before migration?

   b. After migration has taken place, what is the frequency of the banded allele on the island?

32. Pikas are small mammals that live at high elevation in the talus slopes of mountains. Most populations located on mountain tops in Colorado and Montana in North America are isolated from one another, because the pikas don't occupy the low-elevation habitats that separate the mountain tops and don't venture far from the talus slopes. Thus, there is little gene flow between populations. Furthermore, each population is small in size and was founded by a small number of pikas.

   A group of population geneticists propose to study the amount of genetic variation in a series of pika populations and to compare the allelic frequencies in different populations. On the bases of the biology and the distribution of pikas, predict what the population geneticists will find concerning the within- and between-population genetic variation.

33. In a large, randomly mating population, the frequency of the allele (s) for sickle-cell hemoglobin is 0.028. The results of studies have shown that people with the following genotypes at the beta-chain locus produce the average numbers of offspring given:

Genotype	Average number of offspring produced
SS	5
Ss	6
ss	0

   a. What will the frequency of the sickle-cell allele (s) be in the next generation?

   b. What will the frequency of the sickle-cell allele be at equilibrium?

34. Two chromosomal inversions are commonly found in populations of *Drosophila pseudoobscura*: Standard (ST) and Arrowhead (AR). When treated with the insecticide DDT, the genotypes for these inversions exhibit overdominance, with the following fitnesses:

Genotype	Fitness
ST/ST	0.47
ST/AR	1
AR/AR	0.62

   What will the frequencies of ST and AR be after equilibrium has been reached?

*35. In a large, randomly mating population, the frequency of an autosomal recessive lethal allele is 0.20. What will the frequency of this allele be in the next generation if the lethality takes place before reproduction?

36. The fruit fly *Drosophila melanogaster* normally feeds on rotting fruit, which may ferment and contain high levels of alcohol. Douglas Cavener and Michael Clegg studied allelic frequencies at the locus for alcohol dehydrogenase (*Adh*) in experimental populations of *D. melanogaster* (D. R. Cavener and M. T. Clegg. 1981. *Evolution* 35:1–10). The experimental populations were established from wild-caught flies and were raised in cages in the laboratory. Two control populations (C1 and C2) were raised on a standard cornmeal–molasses–agar diet. Two ethanol populations (E1 and E2) were raised on a cornmeal–molasses–agar diet to which was added 10% ethanol. The four populations were periodically sampled to determine the allelic frequencies of two alleles at the alcohol dehydrogenase locus, Adh^S and Adh^F. The frequencies of these alleles in the experimental populations are shown in the graph.

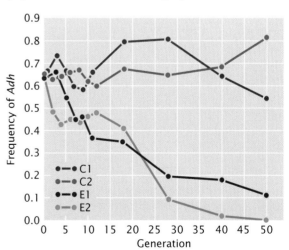

   a. On the basis of these data, what conclusion might you draw about the evolutionary forces that are affecting the *Adh* alleles in these populations?

b. Cavener and Clegg measured the viability of the different *Adh* genotypes in the alcohol environment and obtained the following values:

Genotype	Relative viability
Adh^F/Adh^F	0.932
Adh^F/Adh^S	1.288
Adh^S/Adh^S	0.596

Using these relative viabilities, calculate relative fitnesses for the three genotypes. If a population has an initial frequency of $p = f(Adh^F) = 0.5$, what will the expected frequency of Adh^F be in the next generation on the basis of these fitness values?

37. A certain form of congenital glaucoma is caused by an autosomal recessive allele. Assume that the mutation rate is 10^{-5} and that persons having this condition produce, on the average, only about 80% of the offspring produced by persons who do not have glaucoma.

a. At equilibrium between mutation and selection, what will the frequency of the gene for congenital glaucoma be?

b. What will the frequency of the disease be in a randomly mating population that is at equilibrium?

CHALLENGE QUESTION

Section 12.4

38. The Barton Springs salamander is an endangered species found only in a single spring in the city of Austin, Texas. There is growing concern that a chemical spill on a nearby freeway could pollute the spring and wipe out the species. To provide a source of salamanders to repopulate the spring in the event of such a catastrophe, a proposal has been made to establish a captive breeding population of the salamander in a local zoo. You are asked to provide a plan for the establishment of this captive breeding population, with the goal of maintaining as much of the genetic variation of the species as possible in the captive population. What factors might cause loss of genetic variation in the establishment of the captive population? How could loss of such variation be prevented? With the assumption that only a limited number of salamanders can be maintained in captivity, what procedures should be instituted to ensure the long-term maintenance of as much of the variation as possible?

[Danté B. Fenolio.]

13 Evolutionary Genetics

Some chimpanzees, like humans, have the ability to taste phenylthiocarbamide (PTC), whereas others cannot taste it. Recent research indicates that the PTC taste polymorphism evolved independently in humans and chimpanzees. [Jong & Petra Wegner/Animals Animals-Earth Scenes.]

TASTER GENES IN SPITTING APES

Almost every student of biology knows about the taster test. The teacher passes out small pieces of paper impregnated with a compound called phenylthiocarbamide (PTC), and the students, following the teacher's instructions, put the paper in their mouths. The reaction is always the same: a number of the students immediately spit the paper out, repelled by the bitter taste of PTC. A few students, however, can't taste the PTC and continue to suck on the paper, wondering what all the spitting is about. Variation among individuals in a trait such as the ability to taste PTC is termed a polymorphism.

The ability to taste PTC is inherited as an autosomal dominant trait in humans. The frequencies of taster and nontaster alleles have been estimated in hundreds of human populations worldwide. Almost all populations have both tasters and nontasters; the frequency of the two alleles varies widely.

PTC is not found in nature, but the ability to taste it is strongly correlated with the ability to taste other naturally occurring bitter compounds, some of which are toxic. The ability to taste PTC has also been linked to dietary preferences and may be associated with susceptibility to certain diseases, such as thyroid deficiency. These observations suggest that natural selection has played a role in the evolution of the taster trait. Some understanding of the mechanism of the evolution of the taster trait was gained when well-known population geneticist Ronald A. Fisher and his colleagues took a trip to the zoo.

Fisher wondered whether other primates also might have the ability to taste PTC. To answer this question, he prepared some drinks with different levels of PTC and set off for the zoo with his friends, fellow biologists Edmund (E. B.) Ford and Julian Huxley. At the zoo, the PTC-laced drinks were offered to eight chimpanzees and one orangutan. Fisher and his friends were initially concerned that they might not be able to tell whether the apes could taste the PTC. That concern disappeared, however, when the first one sampled the drink and immediately spat on Fisher. Of the eight chimpanzees tested, six were tasters and two were nontasters.

The observation that chimpanzees and humans both have the PTC taste polymorphism led Fisher and his friends to assume that the polymorphism arose in a common ancestor of humans and chimpanzees, which passed it on to both species. However, they had no way to test their hypothesis. Sixty-five years later, geneticists armed with the latest techniques of modern molecular biology were able to determine the origin of the PTC taste polymorphism and test the hypothesis of Fisher and his friends.

Molecular studies reveal that our ability to taste PTC is controlled by alleles at the *TAS2R38* locus, a 1000-bp gene found on chromosome 7. This locus encodes receptors for bitter compounds and is expressed in the cells of our taste buds. One common allele encodes a receptor that allows the ability to taste PTC; an alternative allele encodes a receptor that does not respond to PTC.

Recent research has demonstrated that PTC taste sensitivity in chimpanzees also is controlled by alleles at the *TAS2R38* locus. However, much to the surprise of the investigators, the taster alleles in humans and chimpanzees are not the same at the molecular level. In the human taster and nontaster alleles, nucleotide differences at three positions affect which amino acids are present in the taste-receptor protein. In chimpanzees, none of these nucleotide differences are present. Instead, a mutation in the initiation codon produces the nontaster allele. This substitution eliminates the normal initiation codon, and the ribosome initiates translation at an alternative downstream initiation codon, resulting in the production of a shortened protein receptor that fails to respond to PTC.

What these findings mean is that Fisher and his friends were correct in their observation that humans and chimpanzees both have PTC taste polymorphism but were incorrect in their hypothesis about its origin: humans and chimpanzees independently evolved the PTC taste polymorphism.

This chapter is about the genetic basis of evolution. As illustrated by the story of the PTC taster polymorphism, evolutionary genetics has a long history but has been transformed in recent years by the application of powerful techniques of molecular genetics. In Chapter 12, we considered the evolutionary forces that bring about change in the allelic frequencies of a population: mutation, migration, genetic drift, and selection. In this chapter, we examine some specific ways that these forces shape the genetic makeup of populations and bring about long-term evolution. We begin by looking at how the process of evolution depends on genetic variation and how much genetic variation is found in natural populations. We then turn to the evolutionary changes that bring about the appearance of new species and how evolutionary histories (phylogenies) are constructed. We end the chapter by taking a look at patterns of evolutionary change at the molecular level.

13.1 Organisms Evolve Through Genetic Change Taking Place Within Populations

Evolution is one of the foundational principles of all of biology. Theodosius Dobzhansky, an important early leader in the field of evolutionary genetics, once remarked "Nothing in biology makes sense except in the light of evolution." Indeed, evolution is an all-encompassing theory that helps to make sense of much of the natural world, from the sequences of DNA found in our cells to the types of plants and animals that surround us. The evidence for evolution is overwhelming. Evolution has been directly observed numerous times; for example, hundreds of different insects evolved resistance to common pesticides following their introduction after World War II. Evolution is supported by the fossil record, comparative anatomy, embryology, the distribution of plants and animals (biogeography), and molecular genetics.

Biological evolution In spite of its vast importance to all fields of biology, evolution is often misunderstood and misinterpreted. In our society, the term *evolution* frequently refers to any type of change. However, biological **evolution** refers only to a specific type of change—genetic change taking place in a group of organisms. Two aspects of this definition should be emphasized. First, evolution includes genetic change only. Many nongenetic changes take place in living organisms, such as the development of a complex intelligent person from a single-celled zygote. Although remarkable, this change isn't evolution, because it does not include changes in genes. The second aspect to emphasize is that evolution takes place in *groups* of organisms. An individual organism does not evolve; what evolves is the gene pool common to a group of organisms.

Evolution as a two-step process Evolution can be thought of as a two-step process. First, genetic variation arises. Genetic variation has its origin in the processes of mutation, which produces new alleles, and recombination, which shuffles alleles into new combinations. Both of these processes are random and produce genetic variation continually, regardless of evolution's need for it. The second step in the process of evolution is the increase and decrease in the frequencies of

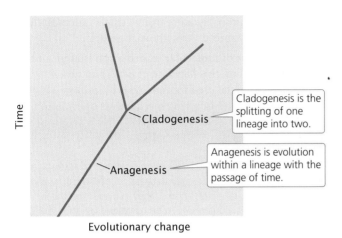

13.1 Anagenesis and cladogenesis are two different types of evolutionary change. Anagenesis is change within an evolutionary lineage; cladogenesis is the splitting of lineages (speciation).

genetic variants. Various evolutionary forces discussed in Chapter 12 cause some alleles in the gene pool to increase in frequency and other alleles to decrease in frequency. This shift in the composition of the gene pool common to a group of organisms constitutes evolutionary change.

Types of evolution We can differentiate between two types of evolution that take place within a group of organisms connected by reproduction. **Anagenesis** refers to evolution taking place in a single group (a lineage) with the passage of time (**Figure 13.1**). Another type of evolution is **cladogenesis**, the splitting of one lineage into two. When a lineage splits, the two branches no longer have a common gene pool and evolve independently of one another. New species arise through cladogenesis. TRY PROBLEM 22 →

CONCEPTS

Biological evolution is genetic change that takes place within a group of organisms. Anagenesis is evolution that takes place within a single lineage; cladogenesis is the splitting of one lineage into two.

✔ **CONCEPT CHECK 1**

Briefly describe how evolution takes place as a two-step process.

13.2 Many Natural Populations Contain High Levels of Genetic Variation

Because genetic variation must be present for evolution to take place, evolutionary biologists have long been interested in the amounts of genetic variation in natural populations and the forces that control the amount and nature of that variation. For many years, they could not examine genes directly and were limited to studying the phenotypes of organisms. Although genetic variation could not be quantified directly, studies of phenotypes suggested that many populations of organisms harbor considerable genetic variation. Populations of organisms in nature exhibit tremendous amounts of phenotypic variation: frogs vary in color pattern, birds differ in size, butterflies vary in spotting patterns, mice differ in coat color, humans vary in blood types, to mention just a few. Crosses revealed that a few of these traits were inherited as simple genetic traits (**Figure 13.2**) but, for most traits, the precise genetic basis was complex and unknown, preventing early evolutionary geneticists from quantifying the amount of genetic variation in natural populations.

As discussed in Chapter 11, the ability of a population to respond to selection (the response to selection) depends on narrow-sense heritability, which is a measure of the additive genetic variation of a trait within a population. Many organisms respond to artificial selection carried out by humans, suggesting that the populations of these organisms contain much additive genetic variation. For example, humans have used artificial selection to produce numerous breeds of dogs that vary tremendously in size, shape, color, and behavior

Normal homozygotes

Heterozygotes

Recessive bimacula phenotype

13.2 Early evolutionary geneticists were forced to rely on the phenotypic traits that had a simple genetic basis. Variation in the spotting patterns of the butterfly *Panaxia dominula* is an example.

(see Figure 11.20). Early studies of chromosome variations in *Drosophila* and plants also suggested that genetic variation in natural populations is plentiful and widespread.

Molecular Variation

Tremendous advances in molecular genetics have made it possible to investigate evolutionary change directly by analyzing protein and nucleic acid sequences. These molecular data offer a number of advantages for studying the process and pattern of evolution:

Molecular data are genetic. Evolution results from genetic change with time. Many anatomical, behavioral, and physiological traits have a genetic basis, but the relation between the underlying genes and the trait may be complex. Protein- and nucleic-acid-sequence variation has a clear genetic basis that is often easy to interpret.

Molecular methods can be used with all organisms. Early studies of population genetics relied on simple genetic traits such as human blood types, banding patterns in snails, or spotting patterns in butterflies (see Figure 13.2), which are restricted to a small group of organisms. However, all living organisms have proteins and nucleic acids; so molecular data can be collected from any organism.

Molecular methods can be applied to a huge amount of genetic variation. An enormous amount of data can be accessed by molecular methods. The human genome, for example, contains more than 3 billion base pairs of DNA, which constitutes a large pool of information about our evolution.

All organisms can be compared with the use of some molecular data. Trying to assess the evolutionary history of distantly related organisms is often difficult because they have few characteristics in common. The evolutionary relationships between angiosperms were traditionally assessed by comparing floral anatomy, whereas the evolutionary relationships of bacteria were determined by their nutritional and staining properties. Because plants and bacteria have so few structural characteristics in common, evaluating how they are related to one another was difficult in the past. All organisms have certain molecular traits in common, such as ribosomal RNA sequences and some fundamental proteins. These molecules offer a valid basis for comparisons among all organisms.

Molecular data are quantifiable. Protein and nucleic acid sequence data are precise, accurate, and easy to quantify, which facilitates the objective assessment of evolutionary relationships.

Molecular data often provide information about the process of evolution. Molecular data can reveal important clues about the process of evolution. For example, the results of a study of DNA sequences have revealed that one type of insecticide resistance in mosquitoes probably arose from a single mutation that subsequently spread throughout the world.

The database of molecular information is large and growing. Today, this database of DNA and protein sequences can be used for making evolutionary comparisons and inferring mechanisms of evolution.

> **CONCEPTS**
>
> Molecular techniques and data offer a number of advantages for evolutionary studies. Molecular data are genetic in nature and can be investigated in all organisms; they provide potentially large data sets, allow all organisms to be compared by using the same characteristics, are easily quantifiable, and provide information about the process of evolution.

Protein Variation

The initial breakthrough in quantifying genetic variation in natural populations came with the application of electrophoresis to population studies. This technique separates macromolecules, such as proteins or nucleic acids, on the basis of their size and charge. In 1966, Richard Lewontin and John Hubby extracted proteins from fruit flies, separated the proteins by electrophoresis, and stained them for specific enzymes. An examination of the pattern of bands on gels enabled them to assign genotypes to individual flies and to quantify the amount of genetic variation in natural populations. In the same year, Harry Harris quantified genetic variation in human populations by using the same technique. Protein variation has now been examined in hundreds of different species by using protein electrophoresis (**Figure 13.3**).

13.3 Molecular variation in proteins is revealed by electrophoresis. Tissue samples from *Drosophila pseudoobscura* were subjected to electrophoresis and stained for esterase. Esterases encoded by different alleles migrate different distances. Shown on the gel are homozygotes for three different alleles.

Table 13.1 Proportion of polymorphic loci and expected heterozygosity for different organisms, as determined by protein electrophoresis

Group	Number of Species	Proportion of Polymorphic Loci Mean	Proportion of Polymorphic Loci SD*	Expected Heterozygosity Mean	Expected Heterozygosity SD*
Plants	15	0.26	0.17	0.07	0.07
Invertebrates (excluding insects)	28	0.40	0.28	0.10	0.07
Insects (excluding *Drosophila*)	23	0.33	0.20	0.07	0.08
Drosophila	32	0.43	0.13	0.14	0.05
Fish	61	0.15	0.01	0.05	0.04
Amphibians	12	0.27	0.13	0.08	0.04
Reptiles	15	0.22	0.13	0.05	0.02
Birds	10	0.15	0.11	0.05	0.04
Mammals	46	0.15	0.10	0.04	0.02

*SD, standard deviation from the mean.

Source: After L. E. Mettler, T. G. Gregg, and H. E. Schaffer, *Population Genetics and Evolution*, 2d ed. (Englewood Cliffs, N.J.: Prentice Hall, 1988), Table 9.3. Original data from E. Nevo, Genetic variation in natural populations: Patterns and theory, *Theoretical Population Biology* 13:121–177, 1978

Measures of genetic variation The amount of genetic variation in populations is commonly measured by two parameters. The **proportion of polymorphic loci** is the proportion of examined loci in which more than one allele is present in a population. If we examined 30 different loci and found two or more alleles present at 15 of these loci, the percentage of polymorphic loci would be $^{15}/_{30} = 0.5$. The **expected heterozygosity** is the proportion of individuals that are expected to be heterozygous at a locus under the Hardy–Weinberg conditions, which is $2pq$ when there are two alleles present in the population. The expected heterozygosity is often preferred to the observed heterozygosity because expected heterozygosity is independent of the breeding system of an organism. For example, if a species self-fertilizes, it may have little or no heterozygosity but still have considerable genetic variation, which the expected heterozygosity will represent. Expected heterozygosity is typically calculated for a number of loci and is then averaged over all the loci examined.

The proportion of polymorphic loci and the expected heterozygosity (averaged over all loci measured) have been determined by protein electrophoresis for a number of species (**Table 13.1**). About one-third of all protein loci are polymorphic, and expected heterozygosity averages about 10%, although there is considerable diversity among species. These measures actually underestimate the true amount of genetic variation though, because protein electrophoresis does not detect some amino acid substitutions; nor does it detect genetic variation in DNA that does not alter the amino acids of a protein (synonymous codons and variation in noncoding regions of the DNA). **TRY PROBLEM 23**

Explanations for protein variation By the late 1970s, geneticists recognized that most populations possess large amounts of genetic variation, although the evolutionary significance of this fact was not at all clear. Two opposing hypotheses arose to account for the presence of the extensive molecular variation in proteins. The **neutral-mutation hypothesis** proposed that the molecular variation revealed by protein electrophoresis is adaptively neutral; that is, individuals with different molecular variants have equal fitness at realistic population sizes. This hypothesis does not propose that the proteins are functionless; rather, it suggests that most variants are functionally equivalent. Because these variants are functionally equivalent, natural selection does not differentiate between them, and their evolution is shaped largely by the random processes of genetic drift and mutation. The neutral-mutation hypothesis accepts that natural selection is an important force in evolution but views selection as a process that favors the "best" allele while eliminating others. It proposes that, when selection is important, there will be *little* genetic variation.

The **balance hypothesis** proposed, on the other hand, that the genetic variation in natural populations is maintained by selection that favors variation (balancing selection). Overdominance, in which the heterozygote has higher fitness than that of either homozygote, is one type of balancing selection. Under this hypothesis, the molecular variants are not physiologically equivalent and do not have the same fitness. Instead, genetic variation within natural populations is shaped largely by selection, and, when selection is important, there will be *much* variation.

Many attempts to prove one hypothesis or the other failed, because precisely how much variation was actually present was not clear (remember that protein electrophoresis detects only some genetic variation) and because both hypotheses are capable of explaining many different patterns of genetic variation. The results of recent studies that provide direct information about DNA sequence variation demonstrate that much variation at the level of DNA has little obvious effect on the phenotype and is therefore likely to be neutral.

CONCEPTS

The application of electrophoresis to the study of protein variation in natural populations revealed that most organisms possess large amounts of genetic variation. The neutral-mutation hypothesis proposes that most molecular variation is neutral with regard to natural selection and is shaped largely by mutation and genetic drift. The balance hypothesis proposes that genetic variation is maintained by balancing selection.

✔ CONCEPT CHECK 2

Which statement is true of the neutral-mutation hypothesis?

a. All proteins are functionless.
b. Natural selection plays no role in evolution.
c. Most molecular variants are functionally equivalent.
d. All of the above.

DNA Sequence Variation

The development of techniques for isolating, cutting, and sequencing DNA in the past 30 years has provided powerful tools for detecting, quantifying, and investigating genetic variation. The application of these techniques has provided a detailed view of genetic variation at the molecular level.

Restriction-site variation Among the first techniques for detecting and analyzing genetic variation in DNA sequences was the use of restriction enzymes. Each restriction enzyme recognizes and cuts a particular sequence of DNA nucleotides known as that enzyme's restriction site. Variation in the presence of a restriction site is called a restriction fragment length polymorphism (RFLP). Each restriction enzyme recognizes a limited number of nucleotide sites in a particular piece of DNA but, if a number of different restriction enzymes are used and the sites recognized by the enzymes are assumed to be random sequences, RFLPs can be used to estimate the amount of variation in the DNA and the proportion of nucleotides that differ between organisms. RFLPs can also be used to analyze the genetic structure of populations and to assess evolutionary relationships among organisms. RFLPs were widely used in evolutionary studies before the development of rapid and inexpensive methods for directly sequencing DNA, and restriction analysis is still employed today in studies of molecular evolution. However, the use of restriction enzymes to analyze DNA sequence variation gives an incomplete picture of the underlying variation, because it detects variation only at restriction sites.

In an evolutionary application of RFLPs, Nicholas Georgiadis and his colleagues studied genetic relationships among African elephants (*Loxodonta africana*; **Figure 13.4a**) from 10 protected areas of Africa. Mitochondrial DNA (mtDNA) was extracted from samples collected from 270 elephants and amplified with the polymerase chain reaction (PCR). The mtDNA was then cleaved with 10 different restriction enzymes, and RFLPs were detected with gel electrophoresis. The degree of genetic differences among elephants from different sites was measured from the sequence variation revealed by the RFLPs. The results of the study showed that the elephant populations are genetically differentiated across Africa, but there are no major differences in their genetic structure among regions (**Figure 13.4b**). From the patterns of variation, the researchers concluded that the elephants have a complex population history, with subdivided populations that exhibit intermittent gene flow.

(a)

(b)

Population Location	Haplotype									
	1	2	3	4	5	6	7	8	9	10
Kenya	■	■								
Kenya	■									
Tanzania	■		■							
Zimbabwe			■	■	■			■		
Zimbabwe			■							
Zimbabwe			■			■		■		
Botswana			■				■	■		
Botswana			■					■	■	
Botswana			■					■		■
South Africa			■					■		

13.4 Restriction fragment length polymorphisms have been used to study population structure and gene flow among populations of African elephants. (a) African elephant. (b) Geographic distribution of 10 mtDNA haplotypes determined by restriction fragment length analysis. Color indicates the presence of the restriction haplotype. [(a) Digital Vision; (b) data from N. Georgiadis et al., *Journal of Heredity* 85:100–104, 1994.]

Microsatellite variation Microsatellites are short DNA sequences that exist in multiple copies repeated in tandem. Variation in the number of copies of the repeats is common, with individual organisms often differing in the number of repeat copies. Microsatellites can be detected by using PCR. Pairs of primers are used that flank a region of repeated copies of the sequence. The DNA fragments that are synthesized in the PCR reaction vary in length, depending on the number of tandem repeats present. DNA from an individual organism with more repeats will produce a longer amplified segment. After PCR has been completed, the amplified fragments are separated with the use of gel electrophoresis and stained, producing a series of bands on a gel. The banding patterns that result represent different alleles (variants in the DNA sequence) and can be used to quantify genetic variation, assess genetic relationships among individual organisms, and quantify genetic differences in a population. An advantage of using microsatellites is that the PCR reaction can be used on very small amounts of DNA and is rapid. The amplified fragments can be fluorescently labeled and detected by a laser, allowing the process to be automated.

David Coltman and his colleagues used microsatellite variation to study paternity in bighorn sheep (**Figure 13.5**; also see the introduction to Chapter 12) and showed that sport hunting of trophy rams has reduced the weight and horn size of the animals. Samples of blood, hair, and ear tissue were collected from bighorn sheep at Ram Mountain in Alberta, Canada—a population that has been monitored since 1971. DNA was extracted from the tissue samples and amplified with PCR, revealing variation at 20 microsatellite loci. On the basis of the microsatellite variation, paternity was assigned to 241 rams, and the family relationships of the sheep were worked out. Using these family relationships and the quantitative genetic techniques described in Chapter 11, the geneticists were able to show that ram weight and horn size had high heritability and exhibited a strong positive genetic correlation (see p. 327 in Chapter 11). Trophy hunters selectively shoot rams with large horns, often before they are able to reproduce. This selective pressure has produced a response to selection: the rams are evolving smaller horns. Between 1971 and 2002, horn size in the population decreased by about one-quarter. Because of the positive genetic correlation between horn size and body size, the body size of rams also is decreasing. Unfortunately, the killing of trophy rams with large horns has led to a decrease in the very traits that are prized by the hunters. This research illustrates the use of microsatellites in evolutionary analysis that has practical application.

Variation detected by DNA sequencing The development of techniques for rapidly and inexpensively sequencing DNA have made this type of data an important tool in population and evolutionary studies. DNA sequence data often reveal processes that influence evolution and are invaluable for determining the evolutionary relationships of different organisms. The use of PCR for producing the DNA used in the sequencing reactions means that data can be obtained from a very small initial sample of DNA.

DNA sequencing was used to reassess the genetic relationships among African elephants. Recall that data from RFLPs found little evidence of regional subdivision among elephants in Africa. Alfred Roca and his colleagues obtained tissue samples from 195 elephants by shooting them with needlelike darts, which fell to the ground after hitting an elephant but retained a small plug of skin. From the skin samples, the scientists sequenced 1732 base pairs of DNA from four nuclear genes. Their analysis revealed large genetic differences between forest elephants and savannah elephants, suggesting that there is limited gene flow (migration) between these two groups of elephants. On the basis of these results, the scientists proposed that two different species of elephant exist in Africa.

Another example of the use of DNA sequence data to decipher evolutionary relationships is the unusual case of HIV infection in a dental practice in Florida. In July 1990, the U.S. Centers for Disease Control and Prevention (CDC) reported that a young woman in Florida (later identified as Kimberly Bergalis) had become HIV positive after undergoing an invasive dental procedure performed by a dentist who had AIDS. Bergalis had no known risk factors for HIV infection and no known contact with other HIV-positive persons. The CDC acknowledged that Bergalis might have acquired the infection from her dentist. Subsequently, the dentist wrote to all of his patients, suggesting that they be tested for HIV infection. By 1992, seven of the dentist's patients had tested positive for HIV, and this number eventually increased to ten.

13.5 Microsatellite variation has been used to study the response of bighorn sheep to selective pressure on horn size due to trophy hunting. [Eyewire.]

Table 13.2	HIV-positive persons included in study of HIV isolates from a Florida dental practice				
				Average Differences in DNA Sequences (%)	
Person	Sex	Known Risk Factors		From HIV from Dentist	From HIV from Controls
Dentist	M	Yes			11.0
Patient A	F	No		3.4	10.9
Patient B	F	No		4.4	11.2
Patient C	M	No		3.4	11.1
Patient E	F	No		3.4	10.8
Patient G	M	No		4.9	11.8
Patient D	M	Yes		13.6	13.1
Patient F	M	Yes		10.7	11.9

Source: After C. Y. Ou et al., *Science* 256:1165–1171, 1992, Table 1.

Originally diagnosed with HIV infection in 1986, the dentist began to develop symptoms of AIDS in 1987 but continued to practice dentistry for another 2 years. All of his HIV-positive patients had received invasive dental procedures, such as root canals and tooth extractions, in the period when the dentist was infected. Among the seven patients originally studied by the CDC (patients A–G, **Table 13.2**), two had known risk factors for HIV infection (intravenous drug use, homosexual behavior, or sexual relations with HIV-infected persons), and a third had possible but unconfirmed risk factors.

To determine whether the dentist had infected his patients, the CDC conducted a study of the molecular evolution of HIV isolates from the dentist and from the patients. HIV undergoes rapid evolution, making it possible to trace the path of its transmission. This rapid evolution also allows HIV to develop drug resistance quickly, making the development of a treatment for AIDS difficult.

Blood specimens were collected from the dentist, the patients, and a group of 35 local controls (other HIV-infected people who lived within 90 miles of the dental practice but who had no known contact with the dentist). DNA was extracted from white blood cells, and a 680-bp fragment of the *envelope* gene of the virus was amplified by PCR. The fragments from the dentist, the patients, and the local controls were then sequenced and compared.

The divergence between the viral sequences taken from the dentist, the seven patients, and the controls is shown in Table 13.2. Viral DNA taken from patients with no confirmed risk factors (patients A, B, C, E, and G) differed from the dentist's viral DNA by 3.4% to 4.9%, whereas the viral DNA from the controls differed from the dentist's by an average of 11%. The viral sequences collected from five patients (A, B, C, E, and G) were more closely related to the viral sequences collected from the dentist than to viral sequences from the general population, strongly suggesting that these patients acquired their HIV infection from the dentist. The viral isolates from patients D and F (patients with confirmed risk factors), however, differed from that of the dentist by 10.7% and 13.6%, suggesting that these two patients did not acquire their infection from the dentist.

An analysis of the evolutionary relationships of the viral sequences (**Figure 13.6**) confirmed that the virus taken from the dentist had a close evolutionary relationship to viruses taken from patients A, B, C, E, and G. The viruses from patients D and F, with known risk factors, were no more similar to the virus from the dentist than to viruses from local controls, indicating that the dentist most likely infected five of his patients, whereas the other two patients probably acquired their infections elsewhere. Of three additional HIV-positive patients that have been identified since 1992, only one has viral sequences that are closely related to those from the dentist.

The study of HIV isolates from the dentist and his patients provides an excellent example of the relevance of evolutionary studies to real-world problems. How the dentist infected his patients during their visits to his office remains a mystery, but this case is clearly unusual. A study of almost 16,000 patients treated by HIV-positive health-care workers failed to find a single case of confirmed transmission of HIV from the health-care worker to the patient.

CONCEPTS

Variation in DNA nucleotide sequence can be analyzed by using restriction fragment length polymorphisms, microsatellites, and data from direct DNA sequencing.

✔ CONCEPT CHECK 3

What are some of the advantages of using microsatellites for evolutionary studies?

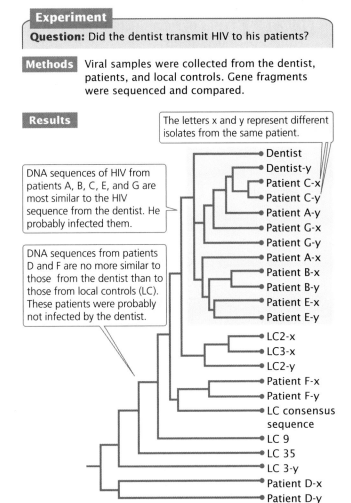

13.6 Evolutionary tree showing the relationships of HIV isolates from a dentist, seven of his patients (A through G), and other HIV-positive persons from the same region (local controls, LC). The phylogeny is based on DNA sequences taken from the *envelope* gene of the virus. [After C. Ou et al., Molecular epidemiology of HIV transmission in a dental practice, *Science* 256:1167, 1992.]

13.3 New Species Arise Through the Evolution of Reproductive Isolation

The term *species* literally means kind or appearance; **species** are different kinds or types of living organisms. In many cases species differences are easy to recognize: a horse is clearly a different species from a chicken. Sometimes, however, species differences are not so clear cut. Some species of *Plethodon* salamanders are so similar in appearance that they can be distinguished only by looking at their proteins or genes.

The concept of a species has two primary uses in biology. First, a species is a name given to a particular type of organism. For effective communication, biologists must use a standard set of names for the organisms that they study, and species names serve that purpose. When a geneticist talks about conducting crosses with *Drosophila melanogaster*, other biologists immediately understand which organism was used. The second use of the term species is in an evolutionary context: a species is considered an evolutionarily independent group of organisms.

The Biological Species Concept

What kinds of differences are required to consider two organisms different species? A widely used definition of species is the **biological species concept**, first fully developed by evolutionary biologist Ernst Mayr in 1942. Mayr was primarily interested in the biological characteristics that are responsible for separating organisms into independently evolving units. He defined a species as a group of organisms whose members are capable of interbreeding with one another but are reproductively isolated from the members of other species. In other words, members of the same species have the biological potential to exchange genes, and members of different species cannot exchange genes. Because different species do not exchange genes, each species evolves independently.

Not all biologists adhere to the biological species concept, and there are several problems associated with it. For example, reproductive isolation, on which the biological species concept is based, cannot be determined of fossils and, in practice, it is often difficult to determine even whether living species are biologically capable of exchanging genes. Furthermore, the biological species concept cannot be applied to asexually reproducing organisms, such as bacteria. In practice, most species are distinguished on the basis of phenotypic (usually anatomical) differences. Biologists often assume that phenotypic differences represent underlying genetic differences; if the phenotypes of two organisms are quite different, then they probably cannot and do not interbreed in nature.

Reproductive Isolating Mechanisms

The key to species differences under the biological species concept is reproductive isolation—biological characteristics that prevent genes from being exchanged between different species. Any biological factor or mechanism that prevents gene exchange is termed a **reproductive isolating mechanism**.

Prezygotic reproductive isolating mechanisms Prezygotic reproductive isolating mechanisms prevent gametes from two different species from fusing and forming a hybrid zygote. In **ecological isolation**, members of two species do not encounter one another and therefore do not reproduce with one another, because they have different ecological niches, living in different habitats and interacting with the environment in different ways. For example, some species of forest-dwelling birds feed and nest in the forest canopy, whereas other species confine their activities to the forest floor. Because they never come into contact, these birds are reproductively isolated from one another. Other species are separated by **behavioral isolation**, differences in behavior that prevent interbreeding. Many male frogs attract females of the same species by using a unique, species-specific call. Two closely related frogs may use the same pond but never interbreed, because females are attracted only to the call of their own species.

Another type of prezygotic reproductive isolation is **temporal isolation**, in which reproduction takes place at different times of the year. Some species of plants do not exchange genes, because they flower at different times of the year. **Mechanical isolation** results from anatomical differences that prevent successful copulation. This type of isolation is seen in many insects, in which closely related species differ in their male and female genitalia, and so copulation is physically impossible. Finally, some species are separated by **gametic isolation**, in which mating between individuals of different species takes place, but the gametes do not form zygotes. Male gametes may not survive in the female reproductive tract or may not be attracted to female gametes. In other cases, male and female gametes meet but are too incompatible to fuse to form a zygote. Gametic isolation is seen in many plants, where pollen from one species cannot fertilize the ovules of another species.

Postzygotic reproductive isolating mechanisms Other species are separated by **postzygotic reproductive isolating mechanisms**, in which gametes of two species fuse and form a zygote, but there is no gene flow between the two species, either because the resulting hybrids are inviable or sterile or because reproduction breaks down in subsequent generations.

If prezygotic reproductive isolating mechanisms fail or have not yet evolved, mating between two organisms of different species may take place, with the formation of a hybrid zygote containing genes from two different species. In many cases, such species are still separated by **hybrid inviability**, in which incompatibility between genomes of the two species prevents the hybrid zygote from developing. Hybrid inviability is seen in some groups of frogs, in which mating between different species and fertilization take place, but the resulting embryos never complete development.

Other species are separated by **hybrid sterility**, in which hybrid embryos complete development but are sterile, so that genes are not passed between species. Donkeys and horses frequently mate and produce a viable offspring—a mule—but most mules are sterile; thus, there is no gene flow between donkeys and horses (but see Problem 44 at the end of Chapter 9). Finally, some closely related species are capable of mating and producing viable and fertile F_1 progeny. However, genes do not flow between the two species because of **hybrid breakdown**, in which further crossing of the hybrids produces inviable or sterile offspring. The different types of reproductive isolating mechanisms are summarized in **Table 13.3**.

Table 13.3 Types of reproductive isolating mechanisms

Type	Characteristics
Prezygotic	**Before a zygote has formed**
Ecological	Differences in habitat; individuals do not meet
Temporal	Reproduction takes place at different times
Mechanical	Anatomical differences prevent copulation
Behavioral	Differences in mating behavior prevent mating
Gametic	Gametes incompatible or not attracted to each other
Postzygotic	**After a zygote has formed**
Hybrid inviability	Hybrid zygote does not survive to reproduction
Hybrid sterility	Hybrid is sterile
Hybrid breakdown	F_1 hybrids are viable and fertile, but F_2 are inviable or sterile

CONCEPTS

The biological species concept defines a species as a group of potentially interbreeding organisms that are reproductively isolated from the members of other species. Under this concept, species are separated by reproductive isolating mechanisms, which may intervene before a zygote is formed (prezygotic reproductive isolating mechanisms) or after a zygote is formed (postzygotic reproductive isolating mechanisms).

✔ CONCEPT CHECK 4

Which statement is an example of postzygotic reproductive isolation?

a. Sperm of species A dies in the oviduct of species B before fertilization can take place.
b. Hybrid zygotes between species A and B are spontaneously aborted early in development.
c. The mating seasons of species A and B do not overlap.
d. Males of species A are not attracted to the pheromones produced by the females of species B.

Modes of Speciation

Speciation is the process by which new species arise. In regard to the biological species concept, speciation comes about through the evolution of reproductive isolating mechanisms—mechanisms that prevent the exchange of genes between groups of organisms.

There are two principle ways in which new species arise. **Allopatric speciation** arises when a geographic barrier first splits a population into two groups and blocks the exchange of genes between them. The interruption of gene flow then leads to the evolution of genetic differences that result in reproductive isolation. **Sympatric speciation** arises in the absence of any external barrier to gene flow; reproductive isolating mechanisms evolve within a single population. We will take a more-detailed look at both of these mechanisms next.

Allopatric speciation Allopatric speciation is initiated when a geographic barrier splits a population into two or more groups and prevents gene flow between the isolated groups (**Figure 13.7a**). Geographic barriers can take a number of forms. Uplifting of a mountain range may split a population of lowland plants into separate groups on each side of the mountains. Oceans serve as effective barriers for many types of terrestrial organisms, separating individuals on different islands from one another and from those on the mainland. Rivers often separate populations of fish located in separate drainages. The erosion of mountains may leave populations of alpine plants isolated on separate mountain peaks.

After two populations have been separated by a geographic barrier that prevents gene flow between them, they evolve independently (**Figure 13.7b**). The genetic isolation allows each population to accumulate genetic differences that are not found in the other population; genetic differences arise through natural selection, unique mutations, and genetic drift (if the populations are small). Genetic differentiation eventually leads to prezygotic and postzygotic isolation. It is important to note that prezygotic isolation and postzygotic isolation arise simply as a consequence of genetic divergence.

If the geographic barrier that once separated the two populations disappears or individuals are able to disperse over it, the populations come into secondary contact (**Figure 13.7c**). At this point, several outcomes are possible. If limited genetic differentiation has taken place during the separation of the populations, reproductive isolating mechanisms may not have evolved or may be incomplete. Genes will flow between the two populations, eliminating any genetic differences that did arise, and the populations will remain a single species.

A second possible outcome is that genetic differentiation during separation leads to prezygotic reproductive isolating mechanisms; in this case, the two populations are different

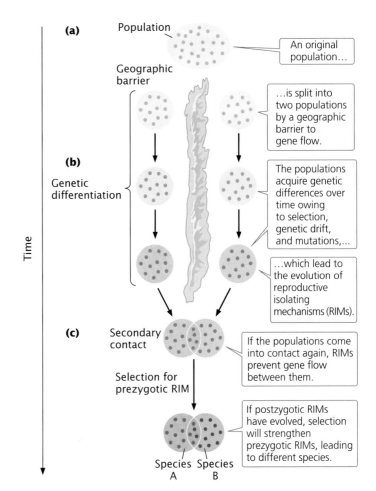

13.7 Allopatric speciation is initiated by a geographic barrier to gene flow between two populations.

species. A third possible outcome is that, during their time apart, some genetic differentiation took place between the populations, leading to incompatibility in their genomes and postzygotic isolation. If postzygotic isolating mechanisms have evolved, any mating between individuals from the different populations will produce hybrid offspring that are inviable or sterile. Individuals that mate only with members of the same population will have higher fitness than that of individuals that mate with members of the other population; so natural selection will increase the frequency of any trait that prevents interbreeding between members of the different populations. With the passage of time, prezygotic reproductive isolating mechanisms will evolve. In short, if some postzygotic reproductive isolation exists, natural selection will favor the evolution of prezygotic reproductive isolating mechanisms to prevent wasted reproduction by individuals mating with members of the other population. This process of postzygotic reproductive isolation leading to the evolution of prezygotic isolating mechanisms is termed *reinforcement*.

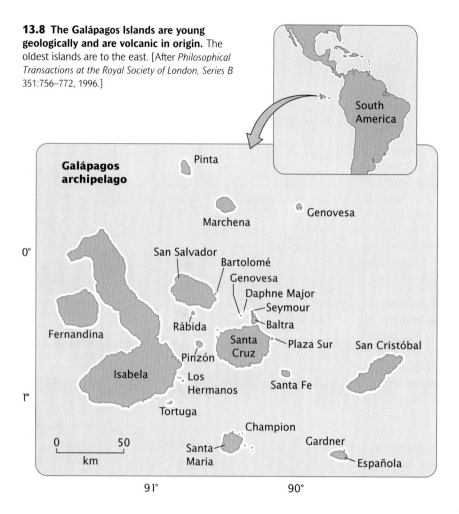

13.8 The Galápagos Islands are young geologically and are volcanic in origin. The oldest islands are to the east. [After *Philosophical Transactions at the Royal Society of London, Series B* 351:756–772, 1996.]

A number of variations in this general model of allopatric speciation are possible. Many new species probably arise when a small group of individuals becomes geographically isolated from the main population; for example, a few individuals of a mainland population might migrate to a geographically isolated island. In this situation, founder effect and genetic drift play a larger role in the evolution of genetic differences between the populations.

An excellent example of allopatric speciation can be found in Darwin's finches, a group of birds on the Galápagos Islands; these finches were discovered by Charles Darwin in his voyage aboard the *Beagle*. The Galápagos are an archipelago of islands located some 900 km off the coast of South America (**Figure 13.8**). Consisting of more than a dozen large islands and many smaller ones, the Galápagos formed from volcanoes that erupted over a geological hot spot that has remained stationary while the geological plate over it moved eastward in the past 3 million years. The movement of the geological plate pulled newly formed islands eastward, and so the islands to the east (San Cristóbal and Española) are older than those to the west (Isabela and Fernandina). With the passage of time, the number of islands in the archipelago increased as new volcanoes arose.

Darwin's finches consist of 14 species that are found on various islands in the Galápagos archipelago (**Figure 13.9**). The birds vary in the shape and sizes of their beaks, which are adapted for eating different types of food items. Recent studies of the development of finch embryos have helped to reveal some of the molecular details of how differences in beak shapes have evolved. Genetic studies have demonstrated that all the birds are closely related and evolved from a single ancestral species that migrated to the islands from the coast of South America some 2 million to 3 million years ago. The evolutionary relationships among the 14 species, based on studies of microsatellite data, are depicted in the evolutionary tree shown in **Figure 13.9**. Most of the species are separated by a behavioral isolating mechanism (song in particular), but some of the species can and occasionally do hybridize in nature.

The first finches to arrive in the Galápagos probably colonized one of the larger eastern islands. A breeding population became established and increased with time. At some point, a few birds dispersed to other islands, where they were effectively isolated from the original population, and established a new population. This population underwent genetic differentiation owing to genetic drift and adaptation to the local conditions of the island. It eventually became reproductively isolated from the original population. Individual birds from the new population then dispersed to other islands and gave rise to additional species. This process was repeated many times. Occasionally, newly evolved birds dispersed to an island where another species was already present, giving rise to secondary contact between the species. Today, many of the islands have more than one resident finch.

The age of the 14 species has been estimated with data from mitochondrial DNA. **Figure 13.10** shows that there is a strong correspondence between the number of bird species present at various times in the past and the number of islands in the archipelago. This correspondence is one of the most compelling pieces of evidence for the theory that the different species of finches arose through allopatric speciation.

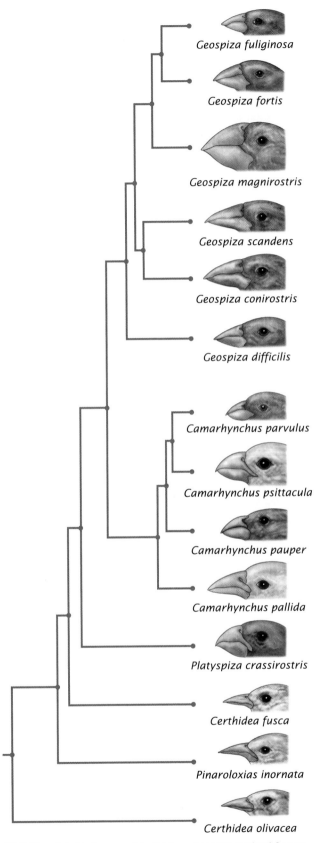

13.9 Darwin's finches consist of 14 species that evolved from a single ancestral species that migrated to the Galápagos Islands and underwent repeated allopatric speciation. [After B. R. Grant and P. R. Grant, *Bioscience* 53:965–975, 2003.]

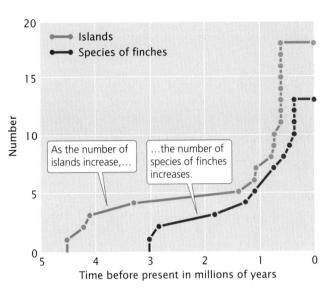

13.10 The number of species of Darwin's finches present at various times in the past corresponds with the number of islands in the Galápagos archipelago. [Data from P. R. Grant, B. R. Grant, and J. C. Deutsch, Speciation and hybridization in island birds, *Philosophical Transactions of the Royal Society of London, Series B* 351:765–772, 1996.]

CONCEPTS

Allopatric speciation is initiated by a geographic barrier to gene flow. A single population is split into two or more populations by a geographic barrier. With the passage of time, the populations evolve genetic differences, which bring about reproductive isolation. After postzygotic reproductive isolating mechanisms have evolved, selection favors the evolution of prezygotic reproductive isolating mechanisms.

✔ CONCEPT CHECK 5

What role does genetic drift play in allopatric speciation?

Sympatric speciation Sympatric speciation arises in the absence of any geographic barrier to gene flow; reproductive isolating mechanisms evolve within a single interbreeding population. Sympatric speciation has long been controversial within evolutionary biology. Ernst Mayr believed that sympatric speciation was impossible, and he demonstrated that many apparent cases of sympatric speciation could be explained by allopatric speciation. More recently, however, evidence has accumulated that sympatric speciation can and has arisen under special circumstances. The difficulty with sympatric speciation is that isolating mechanisms arise as a *consequence* of genetic differentiation, which takes place only if gene flow between groups is interrupted. But, without reproductive isolation (or some external barrier), how can gene flow be interrupted? How can genetic differentiation arise within a single group that is freely exchanging genes?

Most models of sympatric speciation assume that genetic differentiation is initiated by strong disruptive selection taking place within a single population. One homozygote (A^1A^1) is strongly favored on one resource (perhaps the plant species that is host to an insect) and the other homozygote (A^2A^2) is favored on a different resource (perhaps a different host plant). Heterozygotes (A^1A^2) have low fitness on both resources. In this situation, natural selection will favor genotypes at other loci that cause assortative mating (matings between like individuals, see Chapter 12), and so no matings take place between A^1A^1 and A^2A^2, which would produce A^1A^2 offspring with low fitness.

Now imagine that alleles at a second locus affect mating behavior, such that C^1C^1 individuals prefer mating only with other C^1C^1 individuals, and C^2C^2 individuals prefer mating with other C^2C^2 individuals. If alleles at the A locus are nonrandomly associated with alleles at the C locus so that only $A^1A^1 \; C^1C^1$ individuals and $A^2A^2 \; C^2C^2$ individuals exist, then gene flow will be restricted between individuals using the different resources, allowing the two groups to evolve further genetic differences that might lead to reproductive isolation and sympatric speciation.

The difficulty with this model is that recombination quickly breaks up the nonrandom associations between genotypes at the two loci, producing individuals such as $A^1A^1 \; C^2C^2$, which would prefer to mate with $A^2A^2 \; C^2C^2$ individuals. This mating would produce all A^1A^2 offspring, which do poorly on both resources. Thus, even limited recombination will prevent the evolution of the mating-preference genes.

Sympatric speciation is more probable if the genes that affect resource utilization also affect mating preferences. Genes that affect both resource utilization and mating preference are indeed present in some host races—populations of specialized insects that feed on different host plants. Guy Bush studied what appeared to be initial stages of speciation in host races of the apple maggot fly (*Rhagoletis pomonella*, **Figure 13.11**). The flies of this species feed on the fruits of a specific host tree. Mating takes place near the fruits, and the flies lay their eggs on the ripened fruits, where their larvae grow and develop. *R. pomnella* originally existed only on the fruits of hawthorn trees, which are native to North America; 150 years ago, *R. pomnella* was first observed on cultivated apples, which are related to hawthorns but a different species. Infestations of apples by this new apple host race of *R. pomnella* quickly spread, and, today, many apple trees throughout North America are infested with the flies.

The apple host race of *R. pomnella* probably originated when a few flies acquired a mutation that allowed them to feed on apples instead of the hawthorn fruits. Because mating takes place on and near the fruits, flies that use apples are more likely to mate with other flies using apples, leading to genetic isolation between flies using hawthorns and those using apples. Indeed, Bush found that some genetic differentiation has already taken place between the two host races. Flies lay their eggs on ripening fruit, and there has been strong selection for the flies to synchronize their reproduction with the period when their host species has ripening fruit. Apples ripen several weeks earlier than hawthorns. Correspondingly, the peak mating period of the apple host races is 3 weeks earlier than that of the hawthorn race. These differences in the timing of reproduction between apple and hawthorn races have further reduced gene flow—to about 2%—between the two host races and have led to significant genetic differentiation between them. These differences have evolved in the past 150 years and evolution appears to be ongoing. Although genetic differentiation has taken place between apple and hawthorn host races of *R. pomnella* and some degree of reproductive isolation has evolved between them, reproductive isolation is not yet complete and speciation has not fully taken place.

Speciation through polyploidy A special type of sympatric speciation takes place through polyploidy (see Chapter 9). Polyploid organisms have more than two genomes ($3n$, $4n$, $5n$, etc.). As discussed in Chapter 9, allopolyploidy often arises when two diploid species hybridize, producing $2n$ hybrid offspring. Nondisjunction in one of the hybrid offspring produces a $4n$ tetraploid. Because this tetraploid contains exactly two copies of each chromosome, it is usually fertile and will be reproductively isolated from the two parental species by differences in chromosome number (see Figure 9.28).

Numerous species of flowering plants are allopolyploids. Speciation through polyploidy was observed when it led to a new species of salt-marsh grass that arose along the coast of England about 1870. This polyploid contains genomes of the European salt grass *Spartina maritima* ($2n = 60$) and the American salt grass *S. alternaflora* ($2n = 62$; **Figure 13.12**). Seeds from the American salt grass were probably transported to England in the ballast of a ship. Regardless of how it got there, *S. alternaflora* grew in an English marsh and eventually crossed with *S. maritima*, producing a hybrid with $2n = 61$. Nondisjunction in the hybrid then led to chro-

13.11 Host races of the apple maggot fly, *Rhagoletis pomenella*, have evolved some reproductive isolation without any geographic barrier to gene flow. [Tom Murray.]

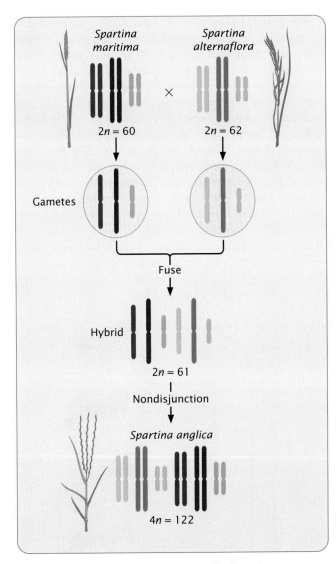

13.12 *Spartina anglica* arose sympatrically through allopolyploidy.

Table 13.4	Genetic similarity in groups of the *Drosophila willistoni* complex
Group	Mean Genetic Similarity
Geographic populations	0.970
Subspecies	0.795
Sibling species	0.517
Nonsibling species	0.352

mosome doubling, producing a new species *S. anglica* with $4n = 122$ (see Figure 13.12). This new species subsequently spread along the coast of England.

CONCEPTS

Sympatric speciation arises within a single interbreeding population without any geographic barrier to gene flow. Sympatric speciation may arise under special circumstances, such as when resource use is linked to mating preference (in host races) or when species hybridization leads to allopolyploidy.

Genetic Differentiation Associated with Speciation

As we have seen, genetic differentiation leads to the evolution of reproductive isolating mechanisms, which restrict gene flow between populations and lead to speciation. How much genetic differentiation is required for reproductive isolation to take place? This question has received considerable study by evolutionary geneticists, but, unfortunately, there is no universal answer. Some newly formed species differ in many genes, whereas others appear to have undergone divergence in just a few genes.

One group of organisms that have been extensively studied for genetic differences associated with speciation is the genus *Drosophila*. The *Drosophila willistoni* group consists of at least 12 species found in Central and South America in various stages in the process of speciation. Using protein electrophoresis, Francisco Ayala and his colleagues genotyped flies from different geographic populations (populations with limited genetic differences), subspecies (populations with considerable genetic differences), sibling species (newly arisen species), and nonsibling species (older species). For each group, they computed a measure of genetic similarity, which ranges from 1 to 0 and represents the overall level of genetic differentiation (**Table 13.4**). They found that there was a general decrease in genetic similarity as flies evolve from geographic populations to subspecies to sibling species to nonsibling species. These data suggest that considerable genetic differentiation at many loci is required for speciation to arise. A study of *D. simulans* and *D. melanogaster*, two species that produce inviable hybrids when crossed, suggested that at least 200 genes contribute to the inviability of hybrids between the two species.

However, other studies suggest that speciation may have arisen through changes in just a few genes. For example, *D. heteroneura* and *D. silvestris* are two species of Hawaiian fruit flies that exhibit behavioral reproductive isolation. The isolation is determined largely by differences in head shape; *D. heteroneura* has a hammer-shaped head with widely separated eyes that is recognized by females of the same species but rejected by *D. silvestris* females. Genetic studies indicate that only a few loci (about 10) determine the differences in head shape.

CONCEPTS

Some newly arising species have a considerable number of genetic differences; others have few genetic differences.

13.4 The Evolutionary History of a Group of Organisms Can Be Reconstructed by Studying Changes in Homologous Characteristics

The evolutionary relationships among a group of organisms are termed a **phylogeny**. Because most evolution takes place over long periods of time and is not amenable to direct observation, biologists must reconstruct phylogenies by inferring the evolutionary relationships among present-day organisms. The discovery of fossils of ancestral organisms can aid in the reconstruction of phylogenies, but the fossil record is often too poor to be of much help. Thus, biologists are often restricted to analyses of characteristics in present-day organisms to determine their evolutionary relationships. In the past, phylogenetic relationships were reconstructed on the basis of phenotypic characteristics—often, anatomical traits. Today, molecular data, including protein and DNA sequences, are frequently used to construct phylogenetic trees.

Phylogenies are reconstructed by inferring changes that have taken place in homologous characteristics. Such characteristics evolved from the same character in a common ancestor. For example, the front leg of a mouse and the wing of a bat are homologous structures, because both evolved from the forelimb of an early mammal that was an ancestor to both mouse and bat. Although these two anatomical features look different and have different functions, close examination of their structure and development reveals that they are indeed homologous. And, because mouse and bat have these homologous features and others in common, we know that they are both mammals. Similarly, DNA sequences are homologous if two present-day sequences evolved from a single sequence found in an ancestor. For example, all eukaryotic organisms have a gene for cytochrome c, an enzyme that helps carry out oxidative respiration. This gene is assumed to have arisen in a single organism in the distant past and was then passed down to descendants of that early ancestor. Today, all copies of the cytochrome c gene are homologous, because they all evolved from the same original copy in the distant ancestor of all organisms that possess this gene.

A graphical representation of a phylogeny is called a **phylogenetic tree**. As shown in **Figure 13.13**, a phylogenetic tree depicts the evolutionary relationships among different organisms, similarly to the way in which a pedigree represents the genealogical relationships among family members.

A phylogenetic tree consists of **nodes** that represent the different organisms being compared, which might be different individuals, populations, or species. Terminal nodes (those at the end of the outermost branches of the tree) represent organisms for which data have been obtained, usually present-day organisms. Internal nodes represent common ancestors that existed before divergence between organisms took place. In most cases, the internal nodes represent past ancestors that are inferred from the analysis. The nodes are connected by **branches**, which represent the evolutionary connections between organisms. In many phylogenetic trees, the lengths of the branches represent the amount of evolutionary divergence that has taken place between organisms. When one internal node represents a common ancestor to all other nodes on the tree, the tree is said to be **rooted**. Trees are often rooted by including in the analysis an organism that is distantly related to all the others; this distantly related organism is referred to as an outgroup.

13.13 A phylogenetic tree is a graphical representation of the evolutionary relationships among a group of organisms.

Phylogenetic trees are created to depict the evolutionary relationships among organisms; they are also created to depict the evolutionary relationships among DNA sequences. The latter type of phylogenetic tree is termed a **gene tree** (**Figure 13.14**). TRY PROBLEM 28 →

CONCEPTS

A phylogeny represents the evolutionary relationships among a group of organisms and is often depicted graphically by a phylogenetic tree, which consists of nodes representing the organisms and branches representing their evolutionary connections.

✔ CONCEPT CHECK 6

Which feature is found in a rooted tree but not in an unrooted tree?

a. Terminal nodes
b. Internal nodes
c. A common ancestor to all other nodes
d. Branch lengths that represent the amount of evolutionary divergence between nodes

The Alignment of Homologous Sequences

Today, phylogenetic trees are often constructed from DNA sequence da-ta. This construction requires that homologous sequences be compared. Thus, a first step in constructing phylogenetic trees from DNA sequence data is to identify homologous genes and properly align their nucleotide bases. Consider the following sequences that might be found in two different organisms:

Nucleotide position	1 2 3 4 5 6 7 8
Gene *X* from species A	5′–A T T G C G A A–3′
Gene *X* from species B	5′–A T G C C A A C–3′

These two sequences can be aligned in several possible ways. We might assume that there have been base substitutions at positions 3, 4, 6, and 8:

Nucleotide position	1 2 3 4 5 6 7 8
Gene *X* from species A	5′–A T T G C G A A–3′
Gene *X* from species B	5′–A T G C C A A C–3′

Alternatively, we might assume that a nucleotide at position 3 has been inserted or deleted, generating a gap in the sequence of species B, and that there has been a single nucleotide substitution at position 6:

Nucleotide position	1 2 3 4 5 6 7 8
Gene *X* from species A	5′–A T T G C G A A–3′
Gene *X* from species B	5′–A T – G C C A A C–3′

The second alignment requires fewer evolutionary steps (a deletion or insertion plus one base substitution) than does the first alignment (four base substitutions). Sequence alignments are usually made by computer programs that include

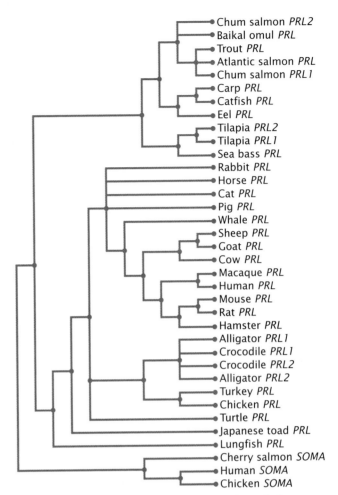

13.14 A gene tree can be used to represent the evolutionary relationships among a group of genes. This gene tree is a rooted tree, in which *PRL* represents a prolactin gene; *PRL1* and *PRL2* are two different prolactin genes found in the same organism; and *SOMA* represents a somatropin gene, which is related to prolactin genes. [After M. P. Simmons and J. V. Freudestein, Uninode coding vs. gene tree parsimony for phylogenetic reconstruction using duplicate genes, *Molecular Phylogenetics and Evolution* 23:488, 2002.]

assumptions about which types of change are more likely to take place. If two sequences have undergone much divergence, then generating alignments can be difficult. Most phylogenetic comparisons are based on alignments requiring the insertion of gaps into the sequence to maximize their similarity.

The Construction of Phylogenetic Trees

Consider a simple phylogeny that depicts the evolutionary relationships among three organisms—humans, chimpanzees, and gorillas. Charles Darwin originally proposed that chimpanzees and gorillas were closely related to humans and modern research supports a close relationship between these three species. There are three possible phylogenetic trees for humans, chimpanzees, and gorillas (**Figure 13.15**). The goal of the evolutionary biologist is to determine which of the trees is correct. Molecular data have been applied to this question;

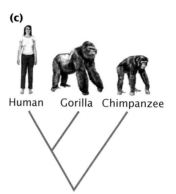

13.15 There are three potential phylogenetic trees for a group of three organisms.

they strongly suggest a close relationship between humans and chimpanzees. To understand the difficulty in constructing phylogenetic trees, let's consider for a moment the number of all possible trees that might exist for a group of organisms. The number of possible rooted trees for a group of organisms is

$$\text{number of rooted trees} = \frac{(2N-3)!}{2^{N-2}(N-2)!}$$

where N equals the number of organisms included in the phylogeny, and the ! symbol stands for factorial, the product of all the integers from N to 1. Substituting values of N into this equation, we find:

Number of organisms included in phylogeny (N)	Number of rooted trees
2	1
3	3
4	15
5	105
10	34,459,425
20	8.2×10^{21}

As the number of organisms in the phylogeny increases beyond just a few, the number of possible rooted trees becomes astronomically large. Clearly, choosing the best tree by directly comparing all the possibilities is impossible.

There are several different approaches to inferring evolutionary relationships and constructing phylogenetic trees. In one approach, termed the *distance approach*, evolutionary relationships are inferred on the basis of the overall degree of similarity between organisms. Typically, a number of different phenotypic characteristics or gene sequences are examined and the organisms are grouped on the basis of their overall similarity, taking into consideration all the examined characteristics and sequences. A second approach, called the *maximum parsimony approach*, infers phylogenetic relationships on the basis of the fewest number of evolutionary changes that must have taken place since the organisms last had an ancestor in common. A third approach, called *maximum likelihood* and *Baysian methods*, infers phylogenetic relationships on the basis of which phylogeny maximizes the probability of obtaining the set of characteristics exhibited by the organisms. In this approach, a phylogeny with a higher probability of producing the observed characters in the organisms studied is preferred over a phylogeny with a lower probability.

With all three approaches to constructing phylogenies, several different numerical methods are available for the construction of phylogenetic trees. These methods are beyond the scope of this book. All include certain assumptions that help limit the number of different trees that must be considered; most rely on computer programs that compare phenotypic characteristics or sequence data to sequentially group organisms in the construction of the tree.

> **CONCEPTS**
>
> Molecular data can be used to infer phylogenies (evolutionary histories) of groups of living organisms. The construction of phylogenies requires the proper alignment of homologous DNA sequences. Several different approaches are used to reconstruct phylogenies, including distance methods, maximum parsimony methods, and maximum likelihood and Baysian methods.

13.5 Patterns of Evolution Are Revealed by Changes at the Molecular Level

In recent years, the ability to analyze genetic variation at the molecular level has revealed a number of evolutionary processes and features that were formerly unsuspected. This section considers several aspects of evolution at the molecular level.

Rates of Molecular Evolution

Findings from molecular studies of numerous genes have demonstrated that different genes and different parts of the same gene often evolve at different rates.

Rates of nucleotide substituion Rates of evolutionary change in nucleotide sequences are usually measured as the rate of nucleotide substitution, which is the number of substitutions taking place per nucleotide site per year. To calculate the rate of nucleotide substitution, we begin by looking at homologous sequences from different organisms. We first align the homologous sequences and then compare the sequences and determine the number of nucleotides that differ between the two sequences. We might compare the growth-hormone sequences for mice and rats, which diverged from a common ancestor some 15 million years ago. From the number of different nucleotides in their growth-hormone genes, we compute the number of nucleotide substitutions that must have taken place since they diverged. Because the same site may have mutated more than once, the number of nucleotide substitutions is larger than the number of nucleotide differences in two sequences; so special mathematical methods have been developed for inferring the actual number of substitutions likely to have taken place.

When we have the number of nucleotide substitutions per nucleotide site, we divide by the amount of evolutionary time that separates the two organisms (usually obtained from the fossil record) to obtain an overall rate of nucleotide substitution. For the mouse and rat growth-hormone gene, the overall rate of nucleotide substitution is approximately 8×10^{-9} substitutions per site per year.

Nonsynonymous and synonymous rates of substitution Nucleotide changes in a gene that alter the amino acid sequence of a protein are referred to as nonsynonymous substitutions. Nucleotide changes, particularly those at the third position of a codon, that do not alter the amino acid sequence are called synonymous substitutions. The rate of nonsynonymous substitution varies widely among mammalian genes. The rate for the α-actin protein is only 0.01 $\times 10^{-9}$ substitutions per site per year, whereas the rate for interferon γ is 2.79×10^{-9}, about 1000 times as high. The rate of synonymous substitution also varies among genes, but not to the extent of variation in the nonsynonymous rate. For most protein-encoding genes, the synonymous rate of change is considerably higher than the nonsynonymous rate because synonymous mutations are tolerated by natural selection (**Table 13.5**). Nonsynonymous mutations, on the other hand, alter the amino acid sequence of the protein and, in many cases, are detrimental to the fitness of the organism; so most of these mutations are eliminated by natural selection.

Substituion rates for different parts of a gene Different parts of a gene also evolve at different rates, with the highest rates of substitutions in regions of the gene that have the least effect on function, such as the third position of a codon, flanking regions, and introns (**Figure 13.16**). The 5′ and 3′ flanking regions of genes are not transcribed into RNA; therefore, substitutions in these regions do not alter the amino acid sequence of the protein, although they may affect gene expression. Rates of substitution in introns are nearly as high. Although these nucleotides do not encode amino acids, introns must be spliced out of the pre-mRNA for a functional protein to be produced, and particular sequences are required at the 5′ splice site, 3′ splice site, and branch point for correct splicing.

Table 13.5 Rates of nonsynonymous and synonymous substitutions in mammalian genes based on human–rodent comparisons

Gene	Nonsynonymous Rate (per site per 10^9 years)	Synonymous Rate (per site per 10^9 years)
α-Actin	0.01	3.68
β-Actin	0.03	3.13
Albumin	0.91	6.63
Aldolase A	0.07	3.59
Apoprotein E	0.98	4.04
Creatine kinase	0.15	3.08
Erythropoietin	0.72	4.34
α-Globin	0.55	5.14
β-Globin	0.80	3.05
Growth hormone	1.23	4.95
Histone 3	0.00	6.38
Immunoglobulin heavy chain (variable region)	1.07	5.66
Insulin	0.13	4.02
Interferon α 1	1.41	3.53
Interferon γ	2.79	8.59
Luteinizing hormone	1.02	3.29
Somatostatin-28	0.00	3.97

Source: After W. Li and D. Graur, *Fundamentals of Molecular Evolution* (Sunderland, Mass.: Sinauer, 1991), p. 69, Table 1.

Substitution rates are somewhat lower in the 5′ and 3′ untranslated regions of a gene. These regions are transcribed into RNA but do not encode amino acids. The 5′ untranslated region contains the ribosome-binding site, which is essential for translation, and the 3′ untranslated region contains sequences that may function in regulating mRNA stability and translation; so substitutions in these regions may have deleterious effects on organismal fitness and may not be tolerated.

The lowest rates of substitution are seen in nonsynonymous changes in the coding region, because these substitutions always alter the amino acid sequence of the protein and are often deleterious. High rates of substitution occur in pseudogenes, which are duplicated nonfunctional copies of genes that have acquired mutations. Such genes no longer produce a functional product; so mutations in pseudogenes have no effect on the fitness of the organism. The highest rates of substitution are seen in noncoding regions of DNA found between genes.

In summary, there is a relation between the function of a sequence and its rate of evolution; higher rates are found

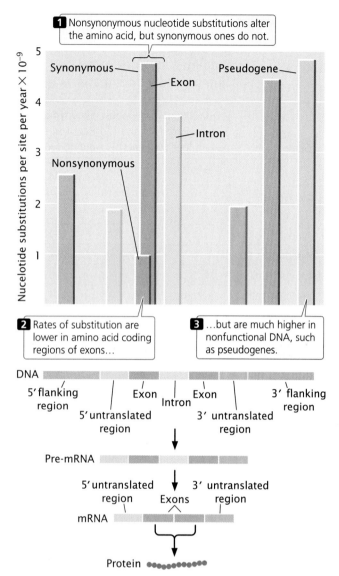

13.16 Different parts of genes evolve at different rates. The highest rates of nucleotide substitution are in sequences that have the least effect on protein function.

where they have the least effect on function. This observation fits with the neutral-mutation hypothesis, which predicts that molecular variation is not affected by natural selection.

The Molecular Clock

The neutral-mutation hypothesis proposes that evolutionary change at the molecular level takes place primarily through the fixation of neutral mutations by genetic drift. The rate at which one neutral mutation replaces another depends only on the mutation rate, which should be fairly constant for any particular gene. If the rate at which a protein evolves is roughly constant over time, the amount of molecular change that a protein has undergone can be used as a **molecular clock** to date evolutionary events.

For example, the enzyme cytochrome c could be examined in two organisms known from fossil evidence to have had a common ancestor 400 million years ago. By determining the number of differences in the cytochrome c amino acid sequences in each organism, we could calculate the number of substitutions that have occurred per amino acid site. The occurrence of 20 amino acid substitutions since the two organisms diverged indicates an average rate of 5 substitutions per 100 million years. Knowing how fast the molecular clock ticks allows us to use molecular changes in cytochrome c to date other evolutionary events: if we found that cytochrome c in two organisms differed by 15 amino acid substitutions, our molecular clock would suggest that they diverged some 300 million years ago. If we assumed some error in our estimate of the rate of amino acid substitution, statistical analysis would show that the true divergence time might range from 160 million to 440 million years. The molecular clock is analogous to geological dating based on the radioactive decay of elements.

The molecular clock was proposed by Emile Zuckerandl and Linus Pauling in 1965 as a possible means of dating evolutionary events on the basis of molecules in present-day organisms. A number of studies have examined the rate of evolutionary change in proteins (**Figure 13.17**), and the molecular clock has been widely used to date evolutionary events when the fossil record is absent or ambiguous. However, the results of several studies have shown that the molecular clock does not always tick at a constant rate, particularly over shorter time periods, and this method remains controversial.

> **CONCEPTS**
>
> Different genes and different parts of the same gene evolve at different rates. Those parts of genes that have the least effect on function tend to evolve at the highest rates. The idea of the molecular clock is that individual proteins and genes evolve at a constant rate and that the differences in the sequences of present-day organisms can be used to date past evolutionary events.
>
> ✔ **CONCEPT CHECK 7**
>
> In general, which types of sequences are expected to exhibit the slowest evolutionary change?
> a. Synonymous changes in amino acid coding regions of exons
> b. Nonsynonymous changes in amino acid coding regions of exons
> c. Introns
> d. Pseudogenes

Genome Evolution

The rapid growth of sequence data available in DNA databases has been a source of insight into evolutionary processes. Whole-genome sequences also are providing new information about how genomes evolve and the processes that shape the size, complexity, and organization of genomes.

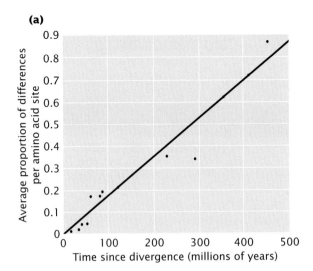

Exon shuffling Many proteins are composed of groups of amino acids, called domains, that specify discrete functions or contribute to the molecular structure of a protein.

Some genes elongated and evolved new functions when one or more exons duplicated and underwent divergence. For example, the human serum-albumin gene is made up of three copies of a sequence that encodes a protein domain consisting of 195 amino acids. Additionally, the genes that encode human immunoglobulins have undergone repeated tandem duplications, creating many similar *V*, *J*, *D*, and *C* segments that enable the immune system to respond to almost any foreign substance that enters the body.

A comparison of DNA sequences from different genes reveals that new genes have repeatedly evolved through a process called **exon shuffling**. In this process, exons of different genes are exchanged, creating genes that are mosaics of other genes. For example, tissue plasminogen activator (TPA) is an enzyme that contains four domains of three different types, called kringle, growth factor, and finger. Each domain is encoded by a different exon. The gene for TPA is believed to have acquired its exons from other genes that encode different proteins: the kringle exon came from the plasminogen gene, the growth-factor exon came from the epidermal growth-factor gene; and the finger exon came from the fibronectin gene. The mechanism by which exon shuffling takes place is poorly known, but new proteins with different combinations of functions encoded by other genes apparently have repeatedly evolved by this mechanism.

Gene duplication New genes have also evolved through the duplication of whole genes and their subsequent divergence. This process creates **multigene families**, sets of genes that are similar in sequence but encode different products. For example, humans possess 13 different genes found on chromosomes 11 and 16 that encode globinlike molecules, which take part in oxygen transport (**Figure 13.18**). All of these genes have a similar structure, with three exons separated by two introns, and are assumed to have evolved through repeated duplication and divergence from a single globin gene in a distant ancestor. This ancestral gene is thought to have been most similar to the present-day myoglobin gene and first duplicated to produce an α/β-globin precursor gene and the myoglobin gene. The α/β-globin gene then underwent another duplication to give rise to a primordial α-globin gene and a primordial β-globin gene. Subsequent duplications led to multiple α-globin and β-globin genes. Similarly, vertebrates contain four clusters of *Hox* genes, each cluster comprising from 9 to 11 genes. *Hox* genes play an important role in development.

Some gene families include genes that are arrayed in tandem on the same chromosome; others are dispersed

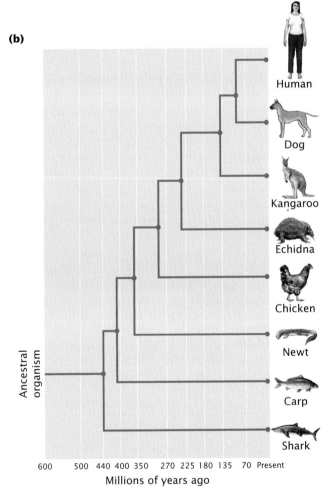

13.17 The molecular clock is based on the assumption of a constant rate of change in protein or DNA sequence. (a) Relation between the rate of amino acid substitution and time since divergence, based in part on amino acid sequences of alpha hemoglobin from the eight species shown in part *b*. The constant rate of evolution in protein and DNA sequences has been used as a molecular clock to date past evolutionary events. (b) Phylogeny of eight of the species that are plotted in part *a* and their approximate times of divergence based on the fossil record.

13.18 Human globin genes constitute a multigene family that has evolved through successive gene duplications.

among different chromosomes. Gene duplication is a common occurrence in eukaryotic genomes; for example, about 5% of the human genome consists of duplicated segments.

Gene duplication provides a mechanism for the addition of new genes with novel functions; after a gene duplicates, there are two copies of the sequence, one of which is free to change and potentially take on a new function. The extra copy of the gene may, for example, become active at a different time in development or be expressed in a different tissue or even diverge and encode a protein having different amino acids. However, the most common fate of gene duplication is that one copy acquires a mutation that renders it nonfunctional, giving rise to a pseudogene. Pseudogenes are common in the genomes of complex eukaryotes; the human genome is estimated to contain as many as 20,000 pseudogenes.

Whole-genome duplication In addition to the duplication of individual genes, whole genomes of some organisms have apparently duplicated in the past. For example, a comparison of the genome of the yeast *Saccharomyces cerevisiae* with the genomes of other fungi reveals that *S. cerevisiae* or one of its immediate ancestors underwent a whole-genome duplication, generating two copies of every gene. Many of the copies subsequently acquired new functions; others acquired mutations that destroyed the original function and then diverged into random DNA sequences. Whole-genome duplication can take place through polyploidy.

Horizontal gene transfer Most organisms acquire their genomes through vertical transmission—transfer through the reproduction of genetic information from parents to offspring. Most phylogenetic trees assume vertical transmission of genetic information. Findings from DNA sequence studies reveal that DNA sequences are sometimes exchanged by a horizontal gene transfer, in which DNA is transferred between individuals of different species (see Chapter 8). This process is especially common among bacteria, and there are a number of documented cases in which genes are transferred from bacteria to eukaryotes. The extent of horizontal gene transfer among eukaryotic organisms is controversial, with few well-documented cases. Horizontal gene transfer can obscure phylogenetic relationships and make the reconstruction of phylogenetic trees difficult.

> **CONCEPTS**
>
> New genes may evolve through the duplication of exons, shuffling of exons, duplication of genes, and duplication of whole genomes. Genes can be passed among distantly related organisms through horizontal gene transfer.

CONCEPTS SUMMARY

- Evolution is genetic change taking place within a group of organisms. It is a two-step process: (1) genetic variation arises and (2) genetic variants change in frequency.
- Anagenesis refers to change within a single lineage; cladogenesis is the splitting of one lineage into two.
- Molecular methods offer a number of advantages for the study of evolution.
- The use of protein electrophoresis to study genetic variation in natural populations showed that most natural populations have large amounts of genetic variation in their proteins. The neutral-mutation hypothesis proposes that molecular variation is selectively neutral and is shaped largely by mutation and genetic drift. The balance hypothesis proposes that molecular variation is maintained largely by balancing selection.
- Variation in DNA sequences can be assessed by analyzing restriction fragment length polymorphisms, microsatellites, and data from direct sequencing.

- A species can be defined as a group of organisms that are capable of interbreeding with one another and are reproductively isolated from the members of other species.
- Species are prevented from exchanging genes by reproductive isolating mechanisms, either before a zygote has formed (prezygotic reproductive isolation) or after a zygote has formed (postzygotic reproductive isolation).
- Allopatric speciation arises when a geographic barrier prevents gene flow between two populations. With the passage of time, the two populations acquire genetic differences that may lead to reproductive isolating mechanisms.
- Sympatric speciation arises when reproductive isolation exists in the absence of any geographic barrier. It may arise under special circumstances.
- Some species arise only after populations have undergone considerable genetic differences; others arise after changes have taken place in only a few genes.
- Evolutionary relationships (a phylogeny) can be represented by a phylogenetic tree, consisting of nodes that represent organisms and branches that represent their evolutionary connections.
- Approaches to constructing phylogenetic trees include the distance approach, the maximum parsimony approach, and the maximum likelihood (Baysian methods) approach.
- Different parts of the genome show different amounts of genetic variation. In general, those parts that have the least effect on function evolve at the highest rates.
- The molecular-clock hypothesis proposes a constant rate of nucleotide substitution, providing a means of dating evolutionary events by looking at nucleotide differences between organisms.
- Genome evolution takes place through the duplication and shuffling of exons, the duplication of genes to form gene families, whole-genome duplication, and the horizontal transfer of genes between organisms.

IMPORTANT TERMS

evolution (p. 366)
anagenesis (p. 367)
cladogenesis (p. 367)
proportion of polymorphic loci (p. 369)
expected heterozygosity (p. 369)
neutral-mutation hypothesis (p. 369)
balance hypothesis (p. 369)
species (p. 373)
biological species concept (p. 373)
reproductive isolating mechanism (p. 373)
prezygotic reproductive isolating mechanism (p. 373)

ecological isolation (p. 373)
behavioral isolation (p. 374)
temporal isolation (p. 374)
mechanical isolation (p. 374)
gametic isolation (p. 374)
postzygotic reproductive isolating mechanism (p. 374)
hybrid inviability (p. 374)
hybrid sterility (p. 374)
hybrid breakdown (p. 374)
speciation (p. 375)
allopatric speciation (p. 375)
sympatric speciation (p. 375)

phylogeny (p. 380)
phylogenetic tree (p. 380)
node (p. 380)
branch (p. 380)
rooted tree (p. 380)
gene tree (p. 381)
molecular clock (p. 384)
exon shuffling (p. 385)
multigene family (p. 385)

ANSWERS TO CONCEPT CHECKS

1. First genetic variation arises. Then various evolutionary forces cause changes in the frequency of genetic variants.
2. c
3. Microsatellites are often highly variable among individuals. They can be amplified with the use of PCR, and so they can be detected with a small amount of starting DNA. Finally, the detection and analysis of microsatellites can be automated.
4. b
5. Genetic drift can bring about changes in the allelic frequencies of populations and lead to genetic differences among populations. Genetic differentiation is the cause of postzygotic and prezygotic reproductive isolation between populations that leads to speciation.
6. c
7. b

COMPREHENSION QUESTIONS

Section 13.1

1. How is biological evolution defined?
*2. What are the two steps in the process of evolution?
3. How is anagenesis different from cladogenesis?

Section 13.2

*4. As a measure of genetic variation, why is the expected heterozygosity often preferred to the observed heterozygosity?

5. Why does protein variation, as revealed by electrophoresis, underestimate the amount of true genetic variation?
6. What are some of the advantages of using molecular data in evolutionary studies?
*7. What is the key difference between the neutral-mutation hypothesis and the balance hypothesis?
8. Describe some of the methods that have been used to study variation in DNA.

Section 13.3

*9. What is the biological species concept?
10. What is the difference between prezygotic and postzygotic reproductive isolating mechanisms. List the different types of each.
11. What is the basic difference between allopatric and sympatric modes of speciation?
*12. Briefly outline the process of allopatric speciation.
13. What are some of the difficulties with sympatric speciation?
*14. Briefly explain how switching from hawthorn fruits to apples has led to genetic differentiation and partial reproductive isolation in *Rhagoletis pomonella*.

Section 13.4

15. Draw a simple phylogenetic tree and identify a node, a branch, and an outgroup.
*16. Briefly describe differences among the distance approach, the maximum parsimony approach, and the maximum likelihood approach to the reconstruction of phylogenetic trees.

Section 13.5

17. Outline the different rates of evolution that are typically seen in different parts of a protein-encoding gene. What might account for these differences?
*18. What is the molecular clock?
19. What is exon shuffling? How can it lead to the evolution of new genes?
20. What is a multigene family? What processes produce multigene families?
*21. Define horizontal gene transfer. What problems does it cause for evolutionary biologists?

APPLICATION QUESTIONS AND PROBLEMS

Section 13.1

22. The following illustrations represent two different patterns of evolution. Briefly discuss the differences in these two patterns, particularly in regard to the role of cladogenesis in evolutionary change.

Section 13.2

*23. Donald Levin used protein electrophoresis to study genetic variation in two species of wildflowers in Texas, *Phlox drummondi* and *P. cuspidate* (D. Levin. 1978. *Evolution* 32:245–263). *P. drummondi* reproduces only by cross fertilization, whereas *P. cuspidate* is partly self-fertilizing. The table below gives allelic frequencies for several loci and populations studied by Levin.

a. Calculate the percentage of polymorphic loci and the expected heterozygosity for each species. For the expected heterozygosity, calculate the mean for all loci in a population, and then calculate the mean for all populations.

b. What tentative conclusions can you draw about the effect of self-fertilization on genetic variation in these flowers?

		Allelic frequencies							
		Adh		Got-2		Pgi-2		Pgm-2	
Species	Population	a	b	a	b	a	b	a	b
P. drummondii	1	0.10	0.90	0.02	0.98	0.04	0.96	1.0	0.0
P. drummondii	2	0.11	0.89	0.0	1.0	0.31	0.69	0.83	0.17
P. drummondii	3	0.26	0.74	1.0	0.0	0.14	0.86	1.0	0.0
P. cuspidate	1	0.0	1.0	0.0	1.0	0.0	1.0	0.0	1.0
P. cuspidate	2	0.0	1.0	0.0	1.0	0.0	1.0	0.0	1.0
P. cuspidate	3	0.91	0.09	0.0	1.0	0.0	1.0	0.0	1.0

Note: *a* and *b* represent different alleles at each locus.

Section 13.3

24. Which of the isolating mechanisms listed in Table 13.3 have partly evolved between apple and hawthorn host races of *Rhagoletis pomenella*, the apple maggot fly?

25. In Section 13.3, we considered the sympatric evolution of reproductive isolating mechanisms in host races of *Rhagoletis pomenella*, the apple maggot fly. The wasp *Diachasma alloeum* parasitizes apple maggot flies, laying its eggs on the larvae of the flies. Immature wasps hatch from the eggs and feed on the fly larvae. Research by Andrew Forbes and his colleagues (Forbes et al. 2009. *Science* 323:776–779) demonstrated that wasps that parasitize apple races of *R. pomenella* are genetically differentiated from those that parasitize hawthorn races of *R. pomenella*. They also found that wasps that prey on the apple race of the flies are attracted to odors from apples, whereas wasps that prey on the hawthorn race are attracted to odors from hawthorn fruit. (a) Propose an explanation for how genetic differences might have evolved between the wasps that parasitize the two races of *R. pomenella*. (b) How might these differences lead to speciation in the wasps?

Section 13.4

*26. How many rooted trees are theoretically possible for a group of 7 organisms? How many for 12 organisms?

27. Align the following sequence so as to maximize their similarity. What is the minimum number of evolutionary steps that that separate these two sequences?

   TTGCAAAC

   TGAAACTG

28. Michael Bunce and his colleagues in England, Canada, and the United States extracted and sequenced mitochondrial DNA from fossils of Haast's eagle, a gigantic eagle that was driven to extinction 700 years ago when humans first arrived in New Zealand (M. Bunce et al. 2005. *Plos Biology* 3:44–46). Using mitochondrial DNA sequences from living eagles and those from Haast eagle fossils, they created the phylogenetic tree at the right.
   On this phylogenetic tree, identify (a) all terminal nodes; (b) all internal nodes; (c) one example of a branch; and (d) the outgroup.

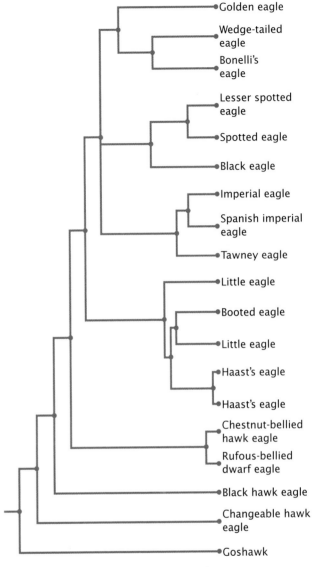

[After M. Bunce et al., *Plos Biology* 3:44–46, 2005.]

29. On the basis of the phylogeny of Darwin's finches shown in Figure 13.9, predict which two species in each of the following groups will be the most similar genetically.

   a. *Camarhynchus parvulus, Camarhynchus psittacula, Camarhynchus pallida*
   b. *Camarhynchus parvulus, Camarhynchus pallida, P. crassirostris*
   c. *Geospiza difficilis, Geospiza conirostris, Geospiza scandens*
   d. *Camarhynchus parvulus, Certhidea fusca, Pinaroloxias inornata*

CHALLENGE QUESTIONS

Section 13.3

30. Explain why natural selection may cause prezygotic reproductive isolating mechanisms to evolve if postzygotic reproductive isolating mechanisms are already present but natural selection can never cause the evolution of postzygotic reproductive isolating mechanisms.

31. Polyploidy is very common in flowering plants: approximately 40% of all flowering plant species are polyploids. Although polyploidy exists in many different animal groups, it is much less common. Why is polyploidy more common in plants than in animals? Give one or more possible reasons.

[Wally Eberhart/Visuals Unlimited.]

[Sinclair Stammers/Photo Researchers.]

[Eye of Science/Photo Researchers.]

[Steve Gschmeissner/Science Photo Library/Alamy.]

[Biophoto Associates/Photo Researchers.]

[Namboori B. Raju.]

Model genetic organisms possess characteristics that make them useful for genetic studies. Shown are several organisms commonly used in genetic studies.

Reference Guide to Model Genetic Organisms

What do Lou Gehrig, the finest first baseman in major league history, and Stephen Hawking, the world's most famous theoretical physicist, have in common? They both suffered or suffer from amyotrophic lateral sclerosis (ALS, also known as Lou Gehrig disease), a degenerative neurological disease that leads to progressive weakness and wasting of skeletal muscles. Interestingly, recent research findings reveal that some athletes with ALS symptoms in fact suffer from brain trauma due to sports injuries; so Lou Gehrig might not have had the disease named for him.

Most cases of ALS are sporadic, appearing in people who have no family history of the disease. However, about 10% of the cases run in families and are inherited. In 2004, geneticists discovered a large Brazilian family with multiple cases of ALS. Genetic analysis revealed that ALS in this family is due to a mutation in a gene called *VABP*, which encodes a vesicle-associated membrane protein.

To better understand how mutations in *VABP* lead to the symptoms of ALS, geneticists turned to an unlikely subject—the fruit fly *Drosophila melanogaster*. Fruit flies don't have ALS, but they do possess a gene very similar to *VABP*. Using a wide array of techniques that have been developed for genetically manipulating fruit flies, geneticists created transgenic flies with the mutant sequence of the *VABP* gene that causes ALS in humans. These flies are a disease model for ALS and are being used to better understand what the gene *VABP* does normally and how its disruption can lead to ALS.

The field of genetics has been greatly influenced and shaped by a few key organisms—called model genetic organisms—whose characteristics make them particularly amenable to genetic studies. The use of *Drosophila* for studying ALS in humans illustrates the power of this approach. Because features of genetic systems are common to many organisms, research conducted on one species can often be a source of insight into the genetic systems of other species. This commonality of genetic function means that geneticists can focus their efforts on model organisms that are easy to work with and likely to yield results.

Model genetic organisms possess life cycles and genomic features that make them well suited to genetic study and analysis. Some key features possessed by many model genetic organisms include:

- a short generation time, and so several generations of genetic crosses can be examined in reasonable time;
- the production of numerous progeny, and so genetic ratios can be easily observed;
- the ability to carry out and control genetic crosses in the organism;
- the ability to be reared in a laboratory environment, requiring little space and few resources to maintain;
- the availability of numerous genetic variants; and
- an accumulated body of knowledge about their genetic systems.

In recent years, the genomes of many model genetic organisms have been completely sequenced, greatly facilitating their use in genetic research.

Not all model organisms possess all of these characteristics. However, each model genetic organism has one or more features that make it useful for genetic analysis. For example, corn cannot be easily grown in the laboratory (and usually isn't) and it has a relatively long generation time, but it produces numerous progeny and there are many genetic variants of corn available for study.

This reference guide highlights six model genetic organisms with important roles in the development of genetics: the fruit fly *(Drosophila melanogaster)*, bacterium *(Escherichia coli)*, roundworm *(Caenorhabditis elegans)*, thale cress plant *(Arabidopsis thaliana)*, house mouse *(Mus musculus)*, and yeast *(Saccharomyces cerevisiae)*. These six organisms have been widely used in genetic research and instruction. A number of other organisms also are used as model systems in genetics, including corn *(Zea mays)*, zebrafish *(Danio rerio)*, clawed frog *(Xenopus lavis)*, bread mold *(Neurospora crassa)*, rat *(Rattus norvegicus)*, and Rhesus macaque *(Macaca mulatta)*, just to mention a few.

The Fruit Fly *Drosophila melanogaster*

Drosophila melanogaster, a fruit fly, was among the first organisms used for genetic analysis and, today, it is one of the most widely used and best known genetically of all organisms. It has played an important role in studies of linkage, epistasis, chromosome genetics, development, behavior, and evolution. Because all organisms use a common genetic system, understanding a process such as replication or transcription in fruit flies helps us to understand these same processes in humans and other eukaryotes.

Drosophila is a genus of more than 1000 described species of small flies (about 1 to 2 mm in length) that frequently feed and reproduce on fruit, although they rarely cause damage and are not considered economic pests. The best known and most widely studied of the fruit flies is *D. melanogaster*, but genetic studies have been extended to many other species of the genus as well.

D. melanogaster first began to appear in biological laboratories about 1900. After first taking up breeding experiments with mice and rats, Thomas Hunt Morgan began using fruit flies in experimental studies of heredity at Columbia University. Morgan's laboratory, located on the top floor of Schermerhorn Hall, became known as the Fly Room (see Figure 4.11b). To say that the Fly Room was unimpressive is an understatement. The cramped room, only about 16 by 23 feet, was filled with eight desks, each occupied by a student and his experiments. The primitive laboratory equipment consisted of little more than milk bottles for rearing the flies and hand-held lenses for observing their traits. Later, microscopes replaced the hand-held lenses, and crude incubators were added to maintain the fly cultures, but even these additions did little to increase the physical sophistication of the laboratory. Morgan and his students were not tidy: cockroaches were abundant (living off spilled *Drosophila* food), dirty milk bottles filled the sink, ripe bananas—food for the flies—hung from the ceiling, and escaped fruit flies hovered everywhere. In spite of its physical limitations, the Fly Room was the source of some of the most important research in the history of biology. There was daily excitement among the students, some of whom initially came to the laboratory as undergraduates. The close quarters facilitated informality and the free flow of ideas. Morgan and the Fly Room illustrate the tremendous importance of "atmosphere" in producing good science. Morgan and his students eventually used *Drosophila* to elucidate many basic principles of heredity, including sex-linked inheritance, epistasis, multiple alleles, and gene mapping.

Advantages of *D. melanogaster* as a model genetic organism
Drosophila's widespread use in genetic studies is no accident. The fruit fly has a number of characteristics that make it an ideal subject for genetic investigations. Compared with other organisms, it has a relatively short generation time; fruit flies will complete an entire generation in about 10 days at room temperature, and so several generations can be studied within a few weeks. Although *D. melanogaster* has a short generation time, it possesses a complex life cycle, passing through several different developmental stages, including egg, larva, pupa, and adult. A female fruit fly is capable of mating within 8 hours of emergence and typically begins to lay eggs after about 2 days. Fruit flies also produce a large number of offspring, laying as many as 400 to 500 eggs in a 10-day period. Thus, large numbers of progeny can be obtained from a single genetic cross.

Another advantage is that fruit flies are easy to culture in the laboratory. They are usually raised in small glass vials or bottles and are fed easily prepared, pastelike food consisting of bananas or corn meal and molasses. Males and females are readily distinguished and virgin females are easily isolated, facilitating genetic crosses. The flies are small, requiring little space—several hundred can be raised in a half-pint bottle—but they are large enough for many mutations to be easily observed with a hand lens or a dissecting microscope.

Finally, *D. melanogaster* is the organism of choice for many geneticists because it has a relatively small genome consisting of 175 million base pairs of DNA, which is only about 5% of the size of the human genome. It has four pairs of chromosomes: three pairs of autosomes and one pair of sex chromosomes. The X chromosome (designated chromosome 1) is large and acrocentric, whereas the Y chromosome is large and submetacentric, although it contains very little genetic information. Chromosomes 2 and 3 are large and metacentric; chromosome 4 is a very small acrocentric chromosome. In the salivary glands, the chromosomes are very large, making *Drosophila* an excellent subject for chromosome studies. In 2000, the complete genome of *D. melanogaster* was sequenced, followed by the sequencing of the genome of *D. pseudoobscura* in 2005 and the genomes of 10 additional *Drosophila* genomes in 2007. *Drosophila* continues today to be one of the most versatile and powerful of all genetic model organisms. ∎

The Fruit Fly
Drosophila melanogaster

ADVANTAGES

- Small size
- Short generation time of 10 days at room temperature
- Each female lays 400–500 eggs
- Easy to culture in laboratory
- Small genome
- Large chromosomes
- Many mutations available

STATS

Taxonomy:	Insect
Size:	2–3 mm in length
Anatomy:	3 body segments, 6 legs, 1 pair of wings
Habitat:	Feeds and reproduces on fruit

Life Cycle

Autosomes

Sex chromosomes

GENOME

Chromosomes:	3 pairs of autosomes and X and Y ($2n = 8$)
Amount of DNA:	175 million base pairs
Number of genes:	14,000
Percentage of genes in common with humans:	~50%
Average gene size:	3000 base pairs
Genome sequenced in year:	2000

CONTRIBUTIONS TO GENETICS

- Basic principles of heredity including sex-linked inheritance, multiple alleles, epistatsis, gene mapping, etc.
- Mutation research
- Chromosome variation and behavior
- Population genetics
- Genetic control of pattern formation
- Behavioral genetics

The Bacterium *Escherichia coli*

The most widely studied prokaryotic organism and one of the best genetically characterized of all species is the bacterium *Escherichia coli*. Although some strains of *E. coli* are toxic and cause disease, most are benign and reside naturally in the intestinal tracts of humans and other warm-blooded animals. *E. coli* was first described by Theodore Escherich in 1885 but, for many years, the assumption was that all bacteria reproduced only asexually and that genetic crosses were impossible. In 1946, Joshua Lederberg and Edward Tatum demonstrated that *E. coli* undergoes a type of sexual reproduction; their finding initiated the use of *E. coli* as a model genetic organism. A year later, Lederberg published the first genetic map of *E. coli* based on recombination frequencies and, in 1952, William Hays showed that mating between bacteria is asymmetrical, with one bacterium serving as genetic donor and the other as genetic recipient.

Advantages of *E. coli* as a model genetic organism *Escherichia coli* is one of the true workhorses of genetics; its twofold advantage is rapid reproduction and small size. Under optimal conditions, this organism can reproduce every 20 minutes and, in a mere 7 hours, a single bacterial cell can give rise to more than 2 million descendants. One of the values of rapid reproduction is that enormous numbers of cells can be grown quickly, and so even very rare mutations will appear in a short period. Consequently, numerous mutations in *E. coli*, affecting everything from colony appearance to drug resistance, have been isolated and characterized.

Escherichia coli is easy to culture in the laboratory in liquid medium (see Figure 8.1a) or on solid medium within petri plates (see Figure 8.1b). In liquid culture, *E. coli* cells will grow to a concentration of a billion cells per milliliter, and trillions of bacterial cells can be easily grown in a single test tube. When *E. coli* cells are diluted and spread onto the solid medium of a petri dish, individual bacteria reproduce asexually, giving rise to a concentrated clump of 10 million to 100 million genetically identical cells, called a colony. This colony formation makes it easy to isolate genetically pure strains of the bacteria.

The *E. coli* genome The *E. coli* genome is on a single chromosome and—compared with those of humans, mice, plants, and other multicellular organisms—is relatively small, consisting of 4,638,858 base pairs. If stretched out straight, the DNA molecule in the single *E. coli* chromosome would be 1.6 mm long, almost a thousand times as long as the *E. coli* cell within which it resides. To accommodate this huge amount of DNA within the confines of a single cell, the *E. coli* chromosome is highly coiled and condensed. The information within the *E. coli* chromosome also is compact, having little noncoding DNA between and within the genes and having few sequences for which there is more than one copy. The *E. coli* genome contains an estimated 4300 genes, more than half of which have no known function. These "orphan genes" may play important roles in adapting to unusual environments, coordinating metabolic pathways, organizing the chromosome, or communicating with other bacterial cells. The haploid genome of *E. coli* makes it easy to isolate mutations because there are no dominant genes at the same loci to suppress and mask recessive mutations.

Life cycle of *E. coli* Wild-type *E. coli* is prototrophic and can grow on minimal medium that contains only glucose and some inorganic salts. Under most conditions, *E. coli* divides about once an hour, although, in a richer medium containing sugars and amino acids, it will divide every 20 minutes. It normally reproduces through simple binary fission, in which the single chromosome of a bacterium replicates and migrates to opposite sides of the cell, followed by cell division, giving rise to two identical daughter cells (see Figure 2.5). Mating between bacteria, called conjugation, is controlled by fertility genes normally located on the F plasmid (see p. 211). In conjugation, one bacterium donates genetic material to another bacterium, followed by genetic recombination that integrates new alleles into the bacterial chromosome. Genetic material can also be exchanged between strains of *E. coli* through transformation and transduction (see Figure 8.7).

Genetic techniques with *E. coli* *Escherichia coli* is used in a number of experimental systems in which fundamental genetic processes are studied in detail. For example, in vitro translation systems contain within a test tube all the components necessary to translate the genetic information of a messenger RNA molecule into a polypeptide chain. Similarly, in vitro systems containing components from *E. coli* cells allow transcription, replication, gene expression, and many other important genetic functions to be studied and analyzed under controlled laboratory conditions.

Escherichia coli is also widely used in genetic engineering (recombinant DNA). Plasmids have been isolated from *E. coli* and genetically modified to create effective vectors for transferring genes into bacteria and eukaryotic cells. Often, new genetic constructs (DNA sequences created in the laboratory) are assembled and cloned in *E. coli* before transfer to other organisms. Methods have been developed to introduce specific mutations within *E. coli* genes, and so genetic analysis no longer depends on the isolation of randomly occurring mutations. New DNA sequences produced by recombinant DNA can be introduced by transformation into special strains of *E. coli* that are particularly efficient (competent) at taking up DNA.

Because of its powerful advantages as a model genetic organism, *E. coli* has played a leading role in many fundamental discoveries in genetics, including elucidation of the genetic code, probing the nature of replication, and working out the basic mechanisms of gene regulation. ∎

Bacterium
Escherichia coli

ADVANTAGES
- Small size
- Rapid reproduction, dividing every 20 minutes under optimal conditions
- Easy to culture in liquid medium or on petri plates
- Small genome
- Many mutants available
- Numerous methods available for genetic engineering

STATS
Taxonomy:	Eubacteria
Size:	1–2 μm in length
Anatomy:	Single cell surrounded by cell wall with nucleoid region
Habitat:	Intestinal tract of warm-blooded animals

Life Cycle

Chromosome

GENOME
Chromosomes:	1 circular chromosome
Amount of DNA:	4.64 million base pairs
Number of genes:	4300
Percentage of genes in common with humans:	8%
Average gene size:	1000 base pairs
Genome sequenced in:	1997

CONTRIBUTIONS TO GENETICS
- Gene regulation
- Molecular biology and biochemistry of genetic processes, such as replication, transcription, translation, recombination
- Gene structure and organization in bacteria
- Workhorse of recombinant DNA
- Gene mutations

The Nematode Worm *Caenorhabditis elegans*

You may be asking, What is a nematode, and why is it a model genetic organism? Although rarely seen, nematodes are among the most abundant organisms on Earth, inhabiting soils throughout the world. Most are free living and cause no harm, but a few are important parasites of plants and animals, including humans. Although *Caenorhabditis elegans* has no economic importance, it has become widely used in genetic studies because of its simple body plan, ease of culture, and high reproductive capacity. First introduced to the study of genetics by Sydney Brenner, who formulated plans in 1962 to use *C. elegans* for the genetic dissection of behavior, this species has made important contributions to the study of development, cell death, aging, and behavior.

Advantages of *C. elegans* as a model genetic organism An ideal genetic organism, *C. elegans* is small, easy to culture, and produces large numbers of offspring. The adult *C. elegans* is about 1 mm in length. Most investigators grow *C. elegans* on agar-filled petri plates that are covered with a lawn of bacteria, which the nematodes devour. Thousands of worms can be easily cultured in a single laboratory. Compared with most multicellular animals, they have a very short generation time, about 3 days at room temperature. And they are prolific reproducers, with a single female producing from 250 to 1000 fertilized eggs in 3 to 4 days.

Another advantage of *C. elegans*, particularly for developmental studies, is that the worm is transparent, allowing easy observation of internal development at all stages. It has a simple body structure, with a small, invariant number of somatic cells: 959 cells in a mature hermaphroditic female and 1031 cells in a mature male.

Life cycle of *C. elegans* Most mature adults are hermaphrodites, with the ability to produce both eggs and sperm and undergo self-fertilization. A few are male, which produce only sperm and mate with hermaphrodites. The hermaphrodites have two sex chromosomes (XX); the males possess a single sex chromosome (XO). Thus, hermaphrodites that self-fertilize produce only hermaphrodites (with the exception of a few males that result from nondisjunction of the X chromosomes). When hermaphrodites mate with males, half of the progeny are XX hermaphrodites and half are XO males.

Eggs are fertilized internally, either from sperm produced by the hermaphrodite or from sperm contributed by a male. The eggs are then laid, and development is completed externally. Approximately 14 hours after fertilization, a larva hatches from the egg and goes through four larval stages—termed L1, L2, L3, and L4—that are separated by molts. The L4 larva undergoes a final molt to produce the adult worm. Under normal laboratory conditions, worms will live for 2 to 3 weeks.

The *C. elegans* genome Geneticists began developing plans in 1989 to sequence the genome of *C. elegans*, and the complete genome sequence was obtained in 1998. Compared with the genomes of most multicellular animals, that of *C. elegans*, at 103 million base pairs of DNA, is small, which facilitates genomic analysis. The availability of the complete genome sequence provides a great deal of information about gene structure, function, and organization in this species. For example, the process of programmed cell death (apoptosis) plays an important role in development and in the suppression of cancer. Apoptosis in *C. elegans* is remarkably similar to that in humans. Having the complete genome sequence of *C. elegans*, and given its ease of genetic manipulation, geneticists have identified genes that participate in apoptosis, which has increased our understanding of apoptosis in humans and its role in cancer.

Genetic techniques with *C. elegans* Chemical mutagens are routinely used to generate mutations in *C. elegans*—mutations that are easy to identify and isolate. The ability of hermaphrodites to self-fertilize means that progeny homozygous for recessive mutations can be obtained in a single generation; the existence of males means that genetic crosses can be carried out.

Developmental studies are facilitated by the transparent body of the worms. As stated earlier, *C. elegans* has a small and exact number of somatic cells. Researchers studying the development of *C. elegans* have meticulously mapped the entire cell lineage of the species, and so the developmental fate of every cell in the adult body can be traced to the original single-celled fertilized egg. Developmental biologists often use lasers to destroy (ablate) specific cells in a developing worm and then study the effects on physiology, development, and behavior.

RNA interference has proved to be an effective tool for turning off genes in *C. elegans*. Geneticists inject double-stranded

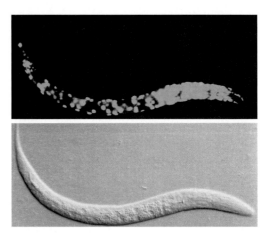

Figure 1 A sequence for the green fluorescent protein (GFP) has been used to visually determine the expression of genes inserted into *C. elegans* (lower photograph). The gene for GFP is injected into the ovary of a worm and becomes incorporated into the worm genome. The expression of this transgene produces GFP, which fluoresces green (upper photograph). [From Huaqi Jiang et al., *PNAS* 98:7916–7921, 2001. Copyright 2001 National Academy of Sciences, U.S.A.]

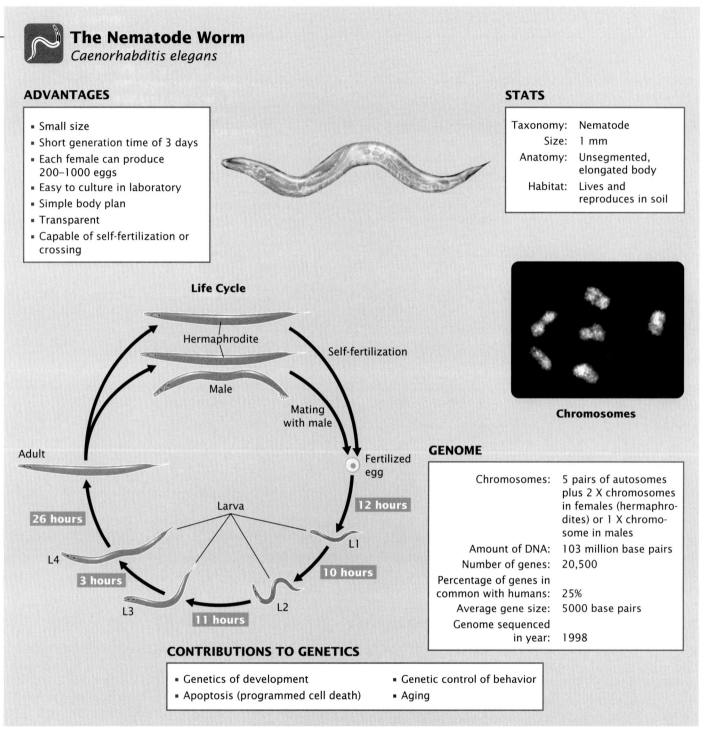

[Photography courtesy of William Goodyer and Monique Zetka.]

copies of RNA that is complementary to specific genes; the double-stranded RNA then silences the expression of these genes through the RNAi process. The worms can even be fed bacteria that have been genetically engineered to express the double-stranded RNA, thus avoiding the difficulties of microinjection.

Transgenic worms can be produced by injecting DNA into the ovary, where the DNA becomes incorporated into the oocytes. Geneticists have created a special reporter gene that produces the jellyfish green fluorescent protein (GFP). When this reporter gene is injected into the ovary and becomes inserted into the worm genome, its expression produces GFP, which fluoresces green, allowing the expression of the gene to be easily observed (**Figure 1**). ■

The Plant *Arabidopsis thaliana*

Much of the early work in genetics was carried out on plants, including Mendel's seminal discoveries in pea plants as well as important aspects of heredity, gene mapping, chromosome genetics, and quantitative inheritance in corn, wheat, beans, and other plants. However, by the mid-twentieth century, many geneticists had turned to bacteria, viruses, yeast, *Drosophila*, and mouse genetic models. Because a good genetic plant model did not exist, plants were relatively neglected, particularly for the study of molecular genetic processes.

This changed in the last part of the twentieth century with the widespread introduction of a new genetic model organism, the plant *Arabidopsis thaliana*. *A. thaliana* was identified in the sixteenth century, and the first mutant was reported in 1873; but this species was not commonly studied until the first detailed genetic maps appeared in the early 1980s. Today, *Arabidopsis* figures prominently in the study of genome structure, gene regulation, development, and evolution in plants, and it provides important basic information about plant genetics that is applied to economically important plant species.

Advantages of *Arabidopsis* as a model genetic organism
The thale cress *Arabidopsis thaliana* is a member of the Brassicaceae family and grows as a weed in many parts of the world. Except in its role as a model genetic organism, *Arabidopsis* has no economic importance, but it has a number of characteristics that make it well suited to the study of genetics. As an angiosperm, it has features in common with other flowering plants, some of which play critical roles in the ecosystem or are important sources of food, fiber, building materials, and pharmaceutical agents. *Arabidopsis*'s chief advantages are its small size (maximum height of 10–20 cm), prolific reproduction, and small genome.

Arabidopsis thaliana completes development—from seed germination to seed production—in about 6 weeks. Its small size and ability to grow under low illumination make it ideal for laboratory culture. Each plant is capable of producing from 10,000 to 40,000 seeds, and the seeds typically have a high rate of germination; so large numbers of progeny can be obtained from single genetic crosses.

The *Arabidopsis* genome A key advantage for molecular studies is *Arabidopsis*'s small genome, which consists of only 125 million base pairs of DNA on five pairs of chromosomes, compared with 2.5 billion base pairs of DNA in the maize genome and 16 billion base pairs in the wheat genome. The genome of *A. thaliana* was completely sequenced in 2000, providing detailed information about gene structure and organization in this species. A number of variants of *A. thaliana*—called ecotypes—that vary in shape, size, physiological characteristics, and DNA sequence are available for study.

Life cycle of *Arabidopsis* The *Arabidopsis* life cycle is fairly typical of most flowering plants (see Figure 2.22). The main, vegetative part of the plant is diploid; haploid gametes are produced in the pollen and ovaries. When a pollen grain lands on the stigma of a flower, a pollen tube grows into the pistil and ovary. Two haploid sperm nuclei contained in each pollen grain travel down the pollen tube and enter the embryo sac. There, one of the haploid sperm cells fertilizes the haploid egg cell to produce a diploid zygote. The other haploid sperm cell fuses with two haploid nuclei to form the $3n$ endosperm, which provides tissue that will nourish the growing embryonic plant. The zygotes develop within the seeds, which are produced in a long pod.

Under appropriate conditions, the embryo germinates and begins to grow into a plant. The shoot grows upward and the roots downward, a compact rosette of leaves is produced and, under the right conditions, the shoot enlarges and differentiates into flower structures. At maturity, *A. thaliana* is a low-growing plant with roots, a main shoot with branches that bear mature leaves, and small white flowers at the tips of the branches.

Genetic techniques with *Arabidopsis* A number of traditional and modern molecular techniques are commonly used with *Arabidopsis* and provide it with special advantages for genetic studies. *Arabidopsis* can self-fertilize, which means that any recessive mutation appearing in the germ line can be recovered in the immediate progeny. Cross-fertilization also is possible by removing the anther from one plant and dusting pollen on the stigma of another plant—essentially the same technique used by Gregor Mendel with pea plants (see Figure 3.4).

As already mentioned, many naturally occurring variants of *Arabidopsis* are available for study, and new mutations can be produced by exposing its seeds to chemical mutagens, radiation, or transposable elements that randomly insert into genes. The large number of offspring produced by *Arabidopsis* facilitates screening for rare mutations.

Genes from other organisms can be transferred to *Arabidopsis* by the Ti plasmid from the bacterium *Agrobacterium tumefaciens*, which naturally infects plants and transfers the Ti plasmid to plant cells. After transfer to a plant cell, the Ti plasmid randomly

[Photography courtesy of Anand P. Tyazi and Luca Comai, Dept. of Biology, University of Washington, Seattle.]

inserts into the DNA of the plant that it infects, thereby generating mutations in the plant DNA in a process called *insertional mutagenesis*. Geneticists have modified the Ti plasmid to carry a *GUS* gene, which has no promoter of its own. The *GUS* gene encodes an enzyme that converts a colorless compound (X-Glu) into a blue dye. Because the *GUS* gene has no promoter, it is expressed only when inserted into the coding sequence of a plant gene. When that happens, the enzyme encoded by *GUS* is synthesized and converts X-Glu into a blue dye that stains the cell. This dye provides a means to visually determine the expression pattern of a gene that has been interrupted by Ti DNA, producing information about the expression of genes that are mutated by insertional mutagenesis. ∎

The Mouse *Mus musculus*

The common house mouse, *Mus musculus,* is among the oldest and most valuable subjects for genetic study. It's an excellent genetic organism—small, prolific, and easy to keep, with a short generation time.

Advantages of the mouse as a model genetic organism Foremost among many advantages that *Mus musculus* has as a model genetic organism is its close evolutionary relationship to humans. Being a mammal, the mouse is genetically, behaviorally, and physiologically more similar to humans than are other organisms used in genetics studies, making the mouse the model of choice for many studies of human and medical genetics. Other advantages include a short generation time compared with that of most other mammals. *Mus musculus* is well adapted to life in the laboratory and can be easily raised and bred in cages that require little space; thus several thousand mice can be raised within the confines of a small laboratory room. Mice have large litters (8–10 pups), and are docile and easy to handle. Finally, a large number of mutations have been isolated and studied in captive-bred mice, providing an important source of variation for genetic analysis.

Life cycle of the mouse The production of gametes and reproduction in the mouse are very similar to those in humans. Diploid germ cells in the gonads undergo meiosis to produce sperm and oocytes, as outlined in Chapter 2. Male mice begin producing sperm at puberty and continue sperm production throughout the remainder of their lives. Starting at puberty, female mice go through an estrus cycle about every 4 days. If mating takes place during estrus, sperm are deposited into the vagina and swim into the oviduct, where one sperm penetrates the outer layer of the ovum, and the nuclei of sperm and ovum fuse. After fertilization, the diploid embryo implants into the uterus. Gestation typically takes about 21 days. Mice reach puberty in about 5 to 6 weeks and will live for about 2 years. A complete generation can be completed in about 8 weeks.

The mouse genome The mouse genome contains about 2.6 billion base pairs of DNA, which is similar in size to the human genome. Mice and humans also have similar numbers of genes. For most human genes, there are homologous genes in the mouse. An important tool for determining the function of an unknown gene in humans is to search for a homologous gene whose function has already been determined in the mouse. Furthermore, the linkage relations of many mouse genes are similar to those in humans, and the linkage relations of genes in mice often provide important clues to linkage relations among genes in humans. The mouse genome is distributed across 19 pairs of autosomes and one pair of sex chromosomes.

Genetic techniques with the mouse A number of powerful techniques have been developed for use in the mouse. They include the creation of transgenic mice by the injection of DNA into a mouse embryo, the ability to disrupt specific genes by the creation of knockout mice, and the ability to insert specific sequences into specific loci. These techniques are made possible by the ability to manipulate the mouse reproductive cycle, including the ability to hormonally induce ovulation, isolate unfertilized oocytes from the ovary, and implant fertilized embryos back into the uterus of a surrogate mother. The ability to create transgenic, knockout, and knock-in mice has greatly facilitated the study of human genetics, and these techniques illustrate the power of the mouse as a model genetic organism.

Mouse and human cells can be fused, allowing somatic-cell hybridization techniques (see Chapter 7) that have been widely used to assign human genes to specific chromosomes. Mice also tolerate inbreeding well, and inbred strains of mice are easily created by brother–sister mating. Members of an inbred strain are genetically very similar or identical, allowing researchers to examine the effects of environmental factors on a trait.

The use of mice as a model genetic organism has led to many important genetic discoveries. In the early twentieth century, mice were used to study the genetic basis of coat-color variation in mammals. More recently, they have figured prominently in research on the genetic basis of cancer, and potential carcinogens are often tested in mice. Mice have been used to study genes that influence mammalian development, including mutations that produce birth defects in humans. A large number of mouse models of specific human diseases have been created—in some cases, by isolating and inbreeding mice with naturally occurring mutations and, in other cases, by using knockout and knock-in techniques to disable and modify specific genes. ■

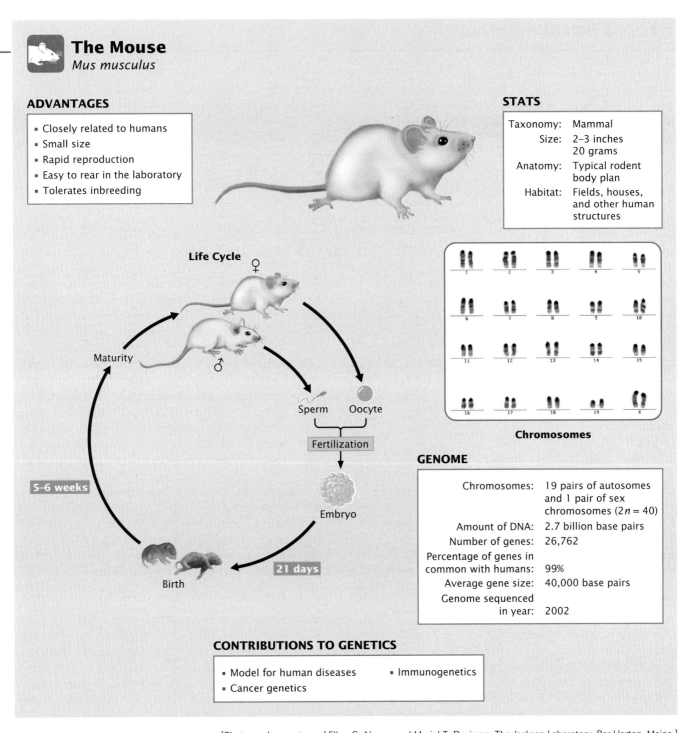

The Yeast *Saccharomyces cerevisiae*

Common baker's yeast (*Saccharomyces cerevisiae*) has been widely adopted as a simple model system for the study of eukaryotic genetics. Long used for baking bread and making beer, yeast has more recently been utilized for the production of biofuels. Louis Pasteur identified *S. cerevisiae* as the microorganism responsible for fermentation in 1857, and its use in genetic analysis began about 1935. Having been the subject of extensive studies in classical genetics for many years, yeast genes are well known and characterized. At the same time, yeast's unicellular nature makes it amenable to molecular techniques developed for bacteria. Thus, yeast combines both classical genetics and molecular biology to provide a powerful model for the study of eukaryotic genetic systems.

Advantages of yeast as a model genetic organism The great advantage of yeast is that it not only is a eukaryotic organism, with genetic and cellular systems similar to those of other, more-complex eukaryotes such as humans, but also is unicellular, with many of the advantages of manipulation found with bacterial systems. Like bacteria, yeast cells require little space, and large numbers of cells can be grown easily and inexpensively in the laboratory.

Yeast exists in both diploid and haploid forms. When haploid, the cells possess only a single allele at each locus, which means that the allele will be expressed in the phenotype; unlike the situation in diploids, there is no dominance through which some alleles mask the expression of others. Therefore, recessive alleles can be easily identified in haploid cells, and then the interactions between alleles can be examined in the diploid cells.

Another feature that makes yeast a powerful genetic model system is that, subsequent to meiosis, all of the products of a meiotic division are present in a single structure called an ascus (see the next subsection) and remain separate from the products of other meiotic divisions. The four cells produced by a single meiotic division are termed a **tetrad**. In most organisms, the products of different meiotic divisions mix, and so identification of the results of a single meiotic division is impossible. For example, if we were to isolate four sperm cells from the testes of a mouse, it is extremely unlikely that all four would have been produced by the same meiotic division. Having tetrads separate in yeast allows us to directly observe the effects of individual meiotic divisions on the types of gametes produced and to more easily identify crossover events. The genetic analysis of a tetrad is termed **tetrad analysis**.

Yeast has been subjected to extensive genetic analysis, and thousands of mutants have been identified. In addition, many powerful molecular techniques developed for manipulating genetic sequences in bacteria have been adapted for use in yeast. Yet, in spite of a unicellular structure and ease of manipulation, yeast cells possess many of the same genes found in humans and other complex multicellular eukaryotes, and many of these genes have identical or similar functions in these eukaryotes. Thus, the genetic study of yeast cells often contributes to our understanding of other, more-complex eukaryotic organisms, including humans.

Life cycle of yeast As stated earlier, *Saccharomyces cerevisiae* can exist as either haploid or diploid cells. Haploid cells usually exist when yeast is starved for nutrients and reproduce mitotically, producing identical, haploid daughter cells through budding. Yeast cells can also undergo sexual reproduction. There are two mating types, **a** and α; haploid cells of different mating types fuse and then undergo nuclear fusion to create a diploid cell. The diploid cell is capable of budding mitotically to produce genetically identical diploid cells. Starvation induces the diploid cells to undergo meiosis, resulting in four haploid nuclei, which become separated into different cells, producing haploid spores. Because the four products of meiosis (a tetrad) are enclosed in a common structure, the ascus, all the products of a single meiosis can be isolated (tetrad analysis).

The yeast genome *Saccharomyces cerevisiae* has 16 pairs of typical eukaryotic chromosomes. The rate of recombination is high, giving yeast a relatively long genetic map compared with those of other organisms. The genome of *S. cerevisiae* contains 12 million base pairs, plus the 2 million to 3 million base pairs of rRNA genes. In 1996, *S. cerevisiae* was the first eukaryotic organism whose genome was completely sequenced.

Genetic techniques with yeast One advantage of yeast to researchers is the use of plasmids to transfer genes or DNA sequences of interest into cells. Yeast cells naturally possesses a circular plasmid, named 2μ, that is 6300 bp long and is transmitted to daughter cells in mitosis and meiosis. This plasmid has an origin of replication recognized by the yeast replication system, and so it replicates autonomously in the cell. The 2μ plasmid has been engineered to provide an efficient vector for transferring genes into yeast. In other cases, bacterial plasmids have been adapted for use in yeast. Some of them undergo homologous recombination with the yeast chromosome, transferring their sequences to the yeast chromosome. Shuttle vectors, which can be propagated in both bacteria and yeast, are particularly effective. Such vectors make it possible to construct and manipulate gene sequences in bacteria, where often more-powerful tech-

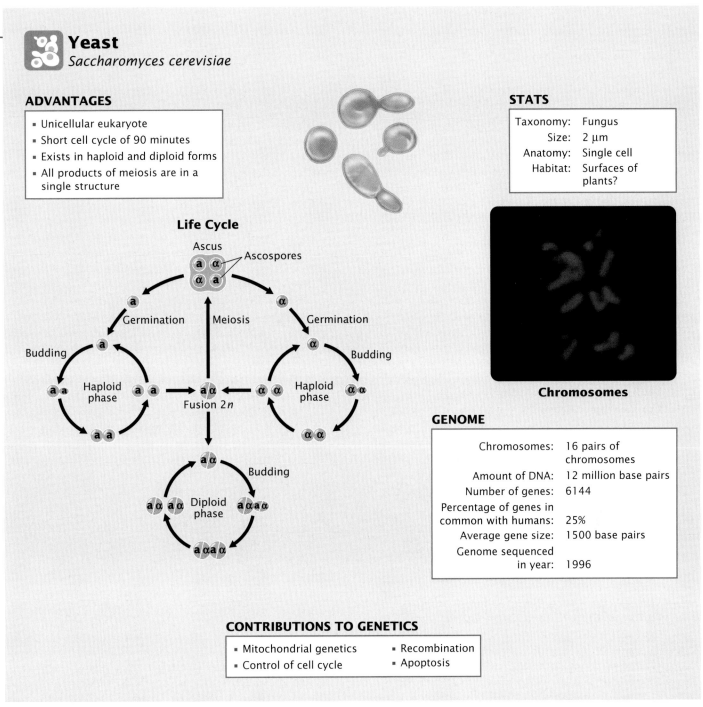

[Photography courtesy of Mara Stewart and Dean Dawson, Dept. of Microbiology and Molecular Biology, Sackler School of Biomedical Sciences, Tufts Univeristy.]

niques are available for genetic manipulation and selection, and then transfer the gene sequences into yeast cells, where their function can be tested.

Plasmids are limited in the size of DNA fragments that they can carry. Yeast artificial chromosomes (YACs) overcome this limitation; YACs can hold DNA fragments as large as several hundred thousand base pairs. Yeast artificial chromosomes are engineered DNA fragments that contain centromeric and telomeric sequences and segregate like chromosomes in meiosis and mitosis. ∎

Glossary

acceptor arm The arm in tRNA to which an amino acid attaches.

acentric chromatid Lacks a centromere; produced when crossing over takes place within a paracentric inversion. The acentric chromatid does not attach to a spindle fiber and does not segregate in meiosis or mitosis; so it is usually lost after one or more rounds of cell division.

acidic activation domain Commonly found in some transcriptional activator proteins, a domain that contains multiple amino acids with negative charges and stimulates the transcription of certain genes.

acrocentric chromosome Chromosome in which the centromere is near one end, producing a long arm at one end and a knob, or satellite, at the other end.

activator *See* **transcriptional activator protein**.

adaptive mutation Process by which a specific environment induces mutations that enable organisms to adapt to the environment.

addition rule States that the probability of any of two or more mutually exclusive events occurring is calculated by adding the probabilities of the individual events.

additive genetic variance Component of the genetic variance that can be attributed to the additive effect of different genotypes.

adenine (A) Purine base in DNA and RNA.

adenosine-3′,5′-cyclic monophosphate (cAMP) Modified nucleotide that functions in catabolite repression. Low levels of glucose stimulate high levels of cAMP; cAMP then attaches to CAP, which binds to the promoter of certain operons and stimulates transcription.

adjacent-1 segregation Type of segregation that takes place in a heterozygote for a translocation. If the original, nontranslocated chromosomes are N_1 and N_2 and the chromosomes containing the translocated segments are T_1 and T_2, then adjacent-1 segregation takes place when N_1 and T_2 move toward one pole and T_1 and N_2 move toward the opposite pole.

adjacent-2 segregation Type of segregation that takes place in a heterozygote for a translocation. If the original, nontranslocated chromosomes are N_1 and N_2 and the chromosomes containing the translocated segments are T_1 and T_2, then adjacent-2 segregation takes place when N_1 and T_1 move toward one pole and T_2 and N_2 move toward the opposite pole.

A-DNA Right-handed helical structure of DNA that exists when little water is present.

affinity capture Use of an antibody to capture one protein from a complex mixture of proteins. The captured protein will "pull down" with it any proteins with which it interacts, which can then be analyzed by mass spectrometry to identify these proteins.

allele One of two or more alternate forms of a gene.

allelic frequency Proportion of a particular allele in a population.

allopatric speciation Speciation that arises when a geographic barrier first splits a population into two groups and blocks the exchange of genes between them.

allopolyploidy Condition in which the sets of chromosomes of a polyploid individual possessing more than two haploid sets are derived from two or more species.

allosteric protein Protein that changes its conformation on binding with another molecule.

alternate segregation Type of segregation that takes place in a heterozygote for a translocation. If the original, nontranslocated chromosomes are N_1 and N_2 and the chromosomes containing the translocated segments are T_1 and T_2, then alternate segregation takes place when N_1 and N_2 move toward one pole and T_1 and T_2 move toward the opposite pole.

alternation of generations Complex life cycle in plants that alternates between the diploid sporophyte stage and the haploid gametophyte stage.

alternative processing pathway One of several pathways by which a single pre-mRNA can be processed in different ways to produce alternative types of mRNA.

alternative splicing Process by which a single pre-mRNA can be spliced in more than one way to produce different types of mRNA.

Ames test Test in which special strains of bacteria are used to evaluate the potential of chemicals to cause cancer.

amino acid Repeating unit of proteins; consists of an amino group, a carboxyl group, a hydrogen atom, and a variable R group.

aminoacyl (A) site One of three sites in a ribosome occupied by a tRNA in translation. All charged tRNAs (with the exception of the initiator tRNA) first enter the A site in translation.

aminoacyl-tRNA synthetase Enzyme that attaches an amino acid to a tRNA. Each aminoacyl-tRNA synthetase is specific for a particular amino acid.

amniocentesis Procedure used for prenatal genetic testing to obtain a sample of amniotic fluid from a pregnant woman. A long sterile needle is inserted through the abdominal wall into the amniotic sac to obtain the fluid.

amphidiploidy Type of allopolyploidy in which two different diploid genomes are combined such that every chromosome has one and only one homologous partner and the genome is functionally diploid.

anagenesis Evolutionary change within a single lineage.

anaphase Stage of mitosis in which chromatids separate and move toward the spindle poles.

anaphase I Stage of meiosis I. In anaphase I, homologous chromosomes separate and move toward the spindle poles.

anaphase II Stage of meiosis II. In anaphase II, chromatids separate and move toward the spindle poles.

aneuploidy Change from the wild type in the number of chromosomes; most often an increase or decrease of one or two chromosomes.

annotation Linking the sequence information of a gene that has been identified to other information about the gene's function and expression, about the protein encoded by the gene, and about similar genes in other species.

***Antennapedia* complex** Cluster of five homeotic genes in fruit flies that affects the development of the adult fly's head and anterior thoracic segments.

antibody Produced by a B cell, a protein that circulates in the blood and other body fluids. An antibody binds to a specific antigen and marks the antigen for destruction by a phagocytic cell.

anticipation Increasing severity or earlier age of onset of a genetic trait in succeeding generations. For example, symptoms of a genetic disease may become more severe as the trait is passed from generation to generation.

anticodon Sequence of three nucleotides in transfer RNA that pairs with the corresponding codon in messenger RNA in translation.

antigen Substance that is recognized by the immune system and elicits an immune response.

antigenic drift Used in reference to a rapidly evolving virus, with new strains appearing frequently because of mutations.

antigenic shift Used in reference to a virus that has undergone major changes in its genome through the reassortment of genetic material from two different strains of the virus.

antiparallel Refers to a characteristic of the DNA double helix in which the two polynucleotide strands run in opposite directions.

antisense RNA Small RNA molecule that base pairs with a complementary DNA or RNA sequence and affects its functioning.

antiterminator Protein or DNA sequence that inhibits the termination of transcription.

apoptosis Programmed cell death, in which a cell degrades its own DNA, the nucleus and cytoplasm shrink, and the cell undergoes phagocytosis by other cells without leakage of its contents.

archaea One of the three primary divisions of life. Archaea consist of unicellular organisms with prokaryotic cells.

artificial selection Selection practiced by humans.

attenuation Type of gene regulation in some bacterial operons, in which transcription is initiated but terminates prematurely before the transcription of the structural genes.

attenuator Secondary structure that forms in the 5′ untranslated region of some operons and causes the premature termination of transcription.

autoimmune disease Characterized by an abnormal immune response to a person's own (self) antigen.

autonomous element Transposable element that is fully functional and able to transpose on its own.

autonomously replicating sequence (ARS) DNA sequence that confers the ability to replicate; contains an origin of replication.

autopolyploidy Condition in which all the sets of chromsomes of a polyploid individual possessing more than two haploid sets are derived from a single species.

autoradiography Method for visualizing DNA or RNA molecules labeled with radioactive substances. A piece of X-ray film is placed on top of a slide, gel, or other substance that contains DNA labeled with radioactive chemicals. Radiation from the labeled DNA exposes the film, providing a picture of the labeled molecules.

autosome Chromosome that is the same in males and females; nonsex chromosome.

auxotrophic bacterium A bacterium or fungus that possesses a nutritional mutation that disrupts its ability to synthesize an essential biological molecule; cannot grow on minimal medium but can grow on minimal medium to which has been added the biological molecule that it cannot synthesize.

backcross Cross between an F_1 individual and one of the parental (P) genotypes.

bacterial artificial chromosome (BAC) Cloning vector used in bacteria that is capable of carrying DNA fragments as large as 500 kb.

bacterial colony Clump of genetically identical bacteria derived from a single bacterial cell that undergoes repeated rounds of division.

bacteriophage Virus that infects bacterial cells.

balance hypothesis Proposes that much of the molecular variation seen in natural populations is maintained by balancing selection that favors genetic variation.

Barr body Condensed, darkly staining structure that is found in most cells of female placental mammals and is an inactivated X chromosome.

basal transcription apparatus Complex of transcription factors, RNA polymerase, and other proteins that assemble on the promoter and are capable of initiating minimal levels of transcription.

base *See* **nitrogenous base.**

base analog Chemical substance that has a structure similar to that of one of the four standard bases of DNA and may be incorporated into newly synthesized DNA molecules in replication.

base-excission repair DNA repair that first excises modified bases and then replaces the entire nucleotide.

base substitution Mutation in which a single pair of bases in DNA is altered.

B cell Particular type of lymphocyte that produces humoral immunity; matures in the bone marrow and produces antibodies.

B-DNA Right-handed helical structure of DNA that exists when water is abundant; the secondary structure described by Watson and Crick and probably the most common DNA structure in cells.

behavioral isolation Reproductive isolation due to differences in behavior that prevent interbreeding.

bidirectional replication Replication at both ends of a replication bubble.

bioinformatics Synthesis of molecular biology and computer science that develops databases and computational tools to store, retrieve, and analyze nucleic-acid- and protein-sequence data.

biological species concept Defines a species as a group of organisms whose members are capable of interbreeding with one another but are reproductively isolated from the members of other species. Because different species do not exchange genes, each species evolves independently. Not all biologists adhere to this concept.

biotechnology Use of biological processes, particularly molecular genetics and recombinant DNA technology, to produce products of commercial value.

***bithorax* complex** Cluster of three homeotic genes in fruit flies that influences the adult fly's posterior thoracic and abdominal segments.

bivalent Refers to a synapsed pair of homologous chromosomes.

blending inheritance Early concept of heredity proposing that offspring possess a mixture of the traits from both parents.

branch Evolutionary connections between organisms in a phylogenetic tree.

branch migration Movement of a cross bridge along two DNA molecules.

branch point Adenine nucleotide in nuclear pre-mRNA introns that lies from 18 to 40 nucleotides upstream of the 3′ splice site.

broad-sense heritability Proportion of the phenotypic variance that can be attributed to genetic variance.

caspase Enzyme that cleaves other proteins and regulates apoptosis. Each caspase is synthesized as a large, inactive precursor (a procaspase) that is activated by cleavage, often by another caspase.

catabolite activator protein (CAP) Protein that functions in catabolite repression. When bound with cAMP, CAP binds to the promoter of certain operons and stimulates transcription.

catabolite repression System of gene control in some bacterial operons in which glucose is used preferentially and the metabolism of other sugars is repressed in the presence of glucose.

cDNA (complementary DNA) library Collection of bacterial colonies or phage colonies containing DNA fragments that have been produced by reverse transcription of cellular mRNA.

cell cycle Stages through which a cell passes from one cell division to the next.

cell line Genetically identical cells that divide indefinitely and can be cultured in the laboratory.

cell theory States that all life is composed of cells, that cells arise only from other cells, and that the cell is the fundamental unit of structure and function in living organisms.

cellular immunity Type of immunity resulting from T cells, which recognize antigens found on the surfaces of self cells.

centiMorgan Another name for map unit.

central dogma Concept that genetic information passes from DNA to RNA to protein in a one-way information pathway.

centriole Cytoplasmic organelle consisting of microtubules; present at each pole of the spindle apparatus in animal cells.

centromere Constricted region on a chromosome that stains less strongly than the rest of the chromosome; region where spindle microtubules attach to a chromosome.

centromeric sequence DNA sequence found in functional centromeres.

centrosome Structure from which the spindle apparatus develops; contains the centriole.

Chargaff's rules Rules developed by Erwin Chargaff and his colleagues concerning the ratios of bases in DNA.

checkpoint A key transition point at which progression to the next stage in the cell cycle is regulated.

chiasma (pl., **chiasmata**) Point of attachment between homologous chromosomes at which crossing over took place.

chloroplast DNA (cpDNA) DNA in chloroplasts; has many characteristics in common with eubacterial DNA and typically consists of a circular molecule that lacks histone proteins and encodes some of the rRNAs, tRNAs, and proteins found in chloroplasts.

chorionic villus sampling (CVS) Procedure used for prenatal genetic testing in which a small piece of the chorion (the outer layer of the placenta) is removed from a pregnant woman. A catheter is inserted through the vagina and cervix into the uterus. Suction is then applied to remove the sample.

chromatin Material found in the eukaryotic nucleus; consists of DNA and proteins.

chromatin-remodeling complex Complex of proteins that alters chromatin structure without acetylating histone proteins.

chromatin-remodeling protein Binds to a DNA sequence and disrupts chromatin structure, causing the DNA to become more accessible to RNA polymerase and other proteins.

chromatosome Consists of a nucleosome and an H1 histone protein.

chromosomal puff Localized swelling of a polytene chromosome; a region of chromatin in which DNA has unwound and is undergoing transcription.

chromosome Structure consisting of DNA and associated proteins that carries and transmits genetic information.

chromosome deletion Loss of a chromosome segment.

chromosome duplication Mutation that doubles a segment of a chromosome.

chromosome inversion Rearrangement in which a segment of a chromosome has been inverted 180 degrees.

chromosome jumping Method of moving from a gene on a cloned fragment to sequences on distantly linked fragments.

chromosome mutation Difference from the wild type in the number or structure of one or more chromosomes; often affects many genes and has large phenotypic effects.

chromosome rearrangement Change from the wild type in the structure of one or more chromosomes.

chromosome theory of heredity States that genes are located on chromosomes.

chromosome walking Method of locating a gene by using partly overlapping genomic clones to move in steps from a previously cloned, linked gene to the gene of interest.

cis configuration Arrangement in which two or more wild-type genes are on one chromosome and their mutant alleles are on the homologous chromosome; also called coupling configuration.

cladogenesis Evolution in which one lineage is split into two.

clonal evolution Process by which mutations that enhance the ability of cells to proliferate predominate in a clone of cells, allowing the clone to become increasingly rapid in growth and increasingly aggressive in proliferation properties.

cloning vector Stable, replicating DNA molecule to which a foreign DNA fragment can be attached and transferred to a host cell.

cloverleaf structure Secondary structure common to all tRNAs.

coactivator Protein that cooperates with an activator of transcription. In eukaryotic transcriptional control, coactivators often physically interact with transcriptional activators and the basal transcription apparatus.

codominance Type of allelic interaction in which the heterozygote simultaneously expresses traits of both homozygotes.

codon Sequence of three nucleotides that encodes one amino acid in a protein.

coefficient of coincidence Ratio of observed double crossovers to expected double crossovers.

cohesin Molecule that holds the two sister chromatids of a chromosome together. The breakdown of cohesin at the centromeres enables the chromatids to separate in anaphase of mitosis and anaphase II of meiosis.

cohesive end Short, single-stranded overhanging end on a DNA molecule produced when the DNA is cut by certain restriction enzymes. Cohesive ends are complementary and can spontaneously pair to rejoin DNA fragments that have been cut with the same restriction enzyme.

cointegrate structure Produced in replicative transposition, an intermediate structure in which two DNA molecules with two copies of the transposable element are fused.

colinearity Concept that there is a direct correspondence between the nucleotide sequence of a gene and the continuous sequence of amino acids in a protein.

colony *See* bacterial colony.

comparative genomics Comparative studies of the genomes of different organisms.

competent cell Capable of taking up DNA from its environment (capable of being transformed).

complementary DNA strands The relation between the two nucleotide strands of DNA in which each purine on one strand pairs with a specific pyrimidine on the opposite strand (A pairs with T, and G pairs with C).

complementation Two different mutations in the heterozygous condition are exhibited as the wild-type phenotype; indicates that the mutations are at different loci.

complementation test Test designed to determine whether two different mutations are at the same locus (are allelic) or at different loci (are nonallelic). Two individuals that are homozygous for two independently derived mutations are crossed, producing F_1 progeny that are heterozygous for the mutations. If the mutations are at the same locus, the F_1 will have a mutant phenotype. If the mutations are at different loci, the F_1 will have a wild-type phenotype.

complete dominance Refers to an allele or a phenotype that is expressed in homozygotes (*AA*) and in heterozygotes (*Aa*); only the dominant allele is expressed in a heterozygote phenotype.

complete linkage Linkage between genes that are located close together on the same chromosome with no crossing over between them.

complete medium Used to culture bacteria or some other microorganism; contains all the nutrients required for growth and synthesis, including those normally synthesized by the organism. Nutritional mutants can grow on complete medium.

composite transposon Type of transposable element in bacteria that consists of two insertion sequences flanking a segment of DNA.

concept of dominance Principle of heredity discovered by Mendel stating that, when two different alleles are present in a genotype, only one allele may be expressed in the phenotype. The dominant allele is the allele that is expressed, and the recessive allele is the allele that is not expressed.

concordance Percentage of twin pairs in which both twins have a particular trait.

concordant Refers to a pair of twins both of whom have the trait under consideration.

conditional mutation Expressed only under certain conditions.

conjugation Mechanism by which genetic material can be exchanged between bacterial cells. In conjugation, two bacteria lie close together and a cytoplasmic connection forms between them. A plasmid or sometimes a part of the bacterial chromosome passes through this connection from one cell to the other.

consanguinity Mating between related individuals.

consensus sequence Comprises the most commonly encountered nucleotides found at a specific location in DNA or RNA.

−10 consensus sequence (Pribnow box) Consensus sequence (TATAAT) found in most bacterial promoters approximately 10 bp upstream of the transcription start site.

−35 consensus sequence Consensus sequence (TTGACA) found in many bacterial promoters approximately 35 bp upstream of the transcription start site.

constitutive gene A gene that is not regulated and is expressed continually.

constitutive mutation Causes the continuous transcription of one or more structural genes.

contig Set of overlapping DNA fragments that have been assembled in the correct order to form a continuous stretch of DNA sequence.

continuous characteristic Displays a large number of possible phenotypes that are not easily distinguished, such as human height.

continuous replication Replication of the leading strand in the same direction as that of unwinding, allowing new nucleotides to be added continuously to the 3′ end of the new strand as the template is exposed.

coordinate induction Simultaneous synthesis of several enzymes that is stimulated by a single environmental factor.

copy-number variation (CNV) Difference among individual organisms in the number of copies of any large DNA sequence (larger than 1000 bp).

core enzyme Part of bacterial RNA polymerase that, during transcription, catalyzes the elongation of the RNA molecule by the addition of RNA nucleotides; consists of four subunits: two copies of alpha (α), a single copy of beta (β), and a single copy of beta prime (β′).

corepressor Substance that inhibits transcription in a repressible system of gene regulation; usually a small molecule that binds to a repressor protein and alters it so that the repressor is able to bind to DNA and inhibit transcription.

core promoter Located immediately upstream of the eukaryotic promoter, DNA sequences to which the basal transcription apparatus binds.

correlation Degree of association between two or more variables.

correlation coefficient Statistic that measures the degree of association between two or more variables. A correlation coefficient can range from −1 to +1. A positive value indicates a direct relation between the variables; a negative correlation indicates an inverse relation. The absolute value of the correlation coefficient provides information about the strength of association between the variables.

cosmid Cloning vector that combines the properties of plasmids and phage vectors and is used to clone large pieces of DNA in bacteria. A cosmid is a small plasmid that carries a λ *cos* site, allowing the plasmid to be packaged into a viral coat.

cotransduction Process in which two or more genes are transferred together from one bacterial cell to another. Only genes located close together on a bacterial chromosome will be cotransduced.

cotransformation Process in which two or more genes are transferred together during cell transformation.

coupling configuration *See* cis configuration.

CpG island DNA region that contains many copies of a cytosine base followed by a guanine base; often found near transcription start sites in eukaryotic DNA. The cytosine bases in CpG islands are commonly methylated when genes are inactive but are demethylated before the initiation of transcription.

cross bridge In a heteroduplex DNA molecule, the point at which each nucleotide strand passes from one DNA molecule to the other.

crossing over Exchange of genetic material between homologous but nonsister chromatids.

cruciform Structure formed by the pairing of inverted repeats on both strands of double-stranded DNA.

C value Haploid amount of DNA found in a cell of an organism.

cyclin A key protein in the control of the cell cycle; combines with a cyclin-dependent kinase (CDK). The levels of cyclin rise and fall in the course of the cell cycle.

cyclin-dependent kinase (CDK) A key protein in the control of the cell cycle; combines with cyclin.

cytokinesis Process by which the cytoplasm of a cell divides.

cytoplasmic inheritance Inheritance of characteristics encoded by genes located in the cytoplasm. Because the cytoplasm is usually contributed entirely by only one parent, most cytoplasmically inherited characteristics are inherited from a single parent.

cytosine (C) Pyrimidine base in DNA and RNA.

deamination Loss of an amino group (NH_2) from a base.

degenerate genetic code Refers to the fact that the genetic code contains more information than is needed to specify all 20 common amino acids.

deletion Mutation in which one or more nucleotides are deleted from a DNA sequence.

deletion mapping Technique for determining the chromsomal location of a gene by studying the association of its phenotype or product with particular chromosome deletions.

delta sequence Long terminal repeat in *Ty* elements of yeast.

denaturation (melting) Process that separates the strands of double-stranded DNA when DNA is heated.

deoxyribonucleotide Basic building block of DNA, consisting of deoxyribose, a phosphate, and a nitrogenous base.

deoxyribose Five-carbon sugar in DNA; lacks a hydroxyl group on the 2′-carbon atom.

depurination Break in the covalent bond connecting a purine base to the 1′-carbon atom of deoxyribose, resulting in the loss of the purine base.

determination Process by which a cell becomes committed to developing into a particular cell type.

diakinesis Fifth substage of prophase I in meiosis. In diakinesis, chromosomes contract, the nuclear membrane breaks down, and the spindle forms.

dicentric bridge Structure produced when the two centromeres of a dicentric chromatid are pulled toward opposite poles, stretching the dicentric chromosome across the center of the nucleus. Eventually, the dicentric bridge breaks as the two centromeres are pulled apart.

dicentric chromatid Chromatid that has two centromeres; produced when crossing over takes place within a paracentric inversion. The two centromeres of the dicentric chromatid are frequently pulled toward opposite poles in mitosis or meiosis, breaking the chromosome.

dideoxyribonucleoside triphosphate (ddNTP) Special substrate for DNA synthesis used in the Sanger dideoxy sequencing method; identical with dNTP (the usual substrate for DNA synthesis) except that it lacks a 3′-OH group. The incorporation of a ddNTP into DNA terminates DNA synthesis.

dihybrid cross A cross between two individuals that differ in two characteristics—more specifically, a cross between individuals that are homozygous for different alleles at the two loci ($AA\ BB \times aa\ bb$); also refers to a cross between two individuals that are both heterozygous at two loci ($Aa\ Bb \times Aa\ Bb$).

dioecious organism Belongs to a species whose members have either male or female reproductive structures.

diploid Possessing two sets of chromosomes (two genomes).

diplotene Fourth substage of prophase I in meiosis. In diplotene, centromeres of homologous chromosomes move apart, but the homologs remain attached at chiasmata.

directional selection Selection in which one trait or allele is favored over another.

direct repair DNA repair in which modified bases are changed back into their original structures.

direct-to-consumer genetic test Test for a genetic condition; the test can be purchased directly by a consumer, without the involvement of a physician or other health care provider.

discontinuous characteristic Exhibits only a few, easily distinguished phenotypes. An example is seed shape in which seeds are either round or wrinkled.

discontinuous replication Replication of the lagging strand in the direction opposite that of unwinding, which means that DNA must be synthesized in short stretches (Okazaki fragments).

discordant Refers to a pair of twins of whom one twin has the trait under consideration and the other does not.

displaced duplication Chromosome rearrangement in which the duplicated segment is some distance from the original segment, either on the same chromosome or on a different one.

dizygotic twins Nonidentical twins that arise when two different eggs are fertilized by two different sperm; also called fraternal twins.

D loop Region of mitochondrial DNA that contains an origin of replication and promoters; it is displaced during the initiation of replication, leading to the name displacement, or D, loop.

DNA fingerprinting Technique used to identify individuals by examining their DNA sequences.

DNA gyrase *E. coli* topoisomerase enzyme that relieves the torsional strain that builds up ahead of the replication fork.

DNA helicase Enzyme that unwinds double-stranded DNA by breaking hydrogen bonds.

DNA library Collection of bacterial colonies containing all the DNA fragments from one source.

DNA ligase Enzyme that catalyzes the formation of a phosphodiester bond between adjacent 3′-OH and 5′-phosphate groups in a DNA molecule.

DNA methylation Modification of DNA by the addition of methyl groups to certain positions on the bases.

DNA polymerase Enzyme that synthesizes DNA.

DNA polymerase I Bacterial DNA polymerase that removes and replaces RNA primers with DNA nucleotides.

DNA polymerase II Bacterial DNA polymerase that takes part in DNA repair; restarts replication after synthesis has halted because of DNA damage.

DNA polymerase III Bacterial DNA polymerase that synthesizes new nucleotide strands by using the 3′-OH group provided by the primer.

DNA polymerase IV Bacterial DNA polymerase; probably takes part in DNA repair.

DNA polymerase V Bacterial DNA polymerase; probably takes part in DNA repair.

DNA polymerase α Eukaryotic DNA polymerase that initiates replication.

DNA polymerase β Eukaryotic DNA polymerase that participates in DNA repair.

DNA polymerase δ Eukaryotic DNA polymerase that replicates the lagging strand during DNA synthesis; also carries out DNA repair and translesion in DNA synthesis.

DNA polymerase ε Eukaryotic DNA polymerase that replicates the leading strand during DNA synthesis.

DNA polymerase γ Eukaryotic DNA polymerase that replicates mitochondrial DNA. A γ-like DNA polymerase replicates chloroplast DNA.

DNase I hypersensitive site Chromatin region that becomes sensitive to digestion by the enzyme DNase I.

DNA sequencing Process of determining the sequence of bases along a DNA molecule.

DNA transposon *See* **transposable element**.

domain Functional part of a protein.

dominance genetic variance Component of the genetic variance that can be attributed to dominance (interaction between genes at the same locus).

dominant Refers to an allele or a phenotype that is expressed in homozygotes (*AA*) and in heterozygotes (*Aa*); only the dominant allele is expressed in a heterozygote phenotype.

dosage compensation Equalization in males and females of the amount of protein produced by X-linked genes. In placental mammals, dosage compensation is accomplished by the random inactivation of one X chromosome in the cells of females.

double fertilization Fertilization in plants; includes the fusion of a sperm cell with an egg cell to form a zygote and the fusion of a second sperm cell with the polar nuclei to form an endosperm.

double-strand-break model Model of homologous recombination in which a DNA molecule undergoes double-strand breaks.

down mutation Decreases the rate of transcription.

downstream core promoter element Consensus sequence [RG(A or T)CGTG] found in some eukaryotic RNA polymerase II core promoters; usually located approximately 30 bp downstream of the transcription start site.

Down syndrome (trisomy 21) Characterized by variable degrees of mental retardation, characteristic facial features, some retardation of growth and development, and an increased incidence of heart defects, leukemia, and other abnormalities; caused by the duplication of all or part of chromosome 21.

ecological isolation Reproductive isolation in which different species live in different habitats and interact with the environment in different ways. Thus, their members do not encounter one another and do not reproduce with one another.

Edward syndrome (trisomy 18) Characterized by severe retardation, low-set ears, a short neck, deformed feet, clenched fingers, heart problems, and other disabilities; results from the presence of three copies of chromosome 18.

effective population size Effective number of breeding adults in a population; influenced by the number of individuals contributing genes to the next generation, their sex ratio, variation between individuals in reproductive success, fluctuations in population size, the age structure of the population, and whether mating is random.

egg Female gamete.

egg-polarity gene Determines the major axes of development in an early fruit-fly embryo. One set of egg-polarity genes determines the anterior–posterior axis and another determines the dorsal–ventral axis.

elongation factor G (EF-G) Protein that combines with GTP and is required for movement of the ribosome along the mRNA during translation.

elongation factor Ts (EF-Ts) Protein that regenerates elongation factor Tu in the elongation stage of protein synthesis.

elongation factor Tu (EF-Tu) Protein taking part in the elongation stage of protein synthesis; forms a complex with GTP and a charged amino acid and then delivers the charged tRNA to the ribosome.

endosymbiotic theory States that some membrane-bounded organelles, such as mitochondria and chloroplasts, in eukaryotic cells originated as free-living eubacterial cells that entered into an endosymbiotic relation with a eukaryotic host cell and evolved into the present-day organelles; supported by a number of similarities in structure and sequence between organelle and eubacterial DNAs.

enhancer Sequence that stimulates maximal transcription of distant genes; affects only genes on the same DNA molecule (is cis acting), contains short consensus sequences, is not fixed in relation to the transcription start site, can stimulate almost any promoter in its vicinity, and may be upstream or downstream of the gene. The function of an enhancer is independent of sequence orientation.

environmental variance Component of the phenotypic variance that is due to environmental differences among individual members of a population.

epigenetic change Stable alteration of chromatin structure that may be passed on to other cells or to an individual organism. *See also* **epigenetics.**

epigenetics Phenonmena due to alterations to DNA that do not include changes in the base sequence; often affect the way in which the DNA sequences are expressed. Such alterations are often stable and heritable in the sense that they are passed from one cell to another.

epigenome All epigenetic modifications within the genome of an individual organism.

episome Plasmid capable of integrating into a bacterial chromosome.

epistasis Type of gene interaction in which a gene at one locus masks or suppresses the effects of a gene at a different locus.

epistatic gene Masks or suppresses the effect of a gene at a different locus.

equilibrium Situation in which no further change takes place; in population genetics, refers to a population in which allelic frequencies do not change.

equilibrium density gradient centrifugation Method used to separate molecules or organelles of different density by centrifugation.

eubacteria One of the three primary divisions of life. Eubacteria consist of unicellular organisms with prokaryotic cells and include most of the common bacteria.

euchromatin Chromatin that undergoes condensation and decondensation in the course of the cell cycle.

eukaryote Organism with a complex cell structure including a nuclear envelope and membrane-bounded organelles. One of the three primary divisions of life, eukaryotes include unicellular and multicellular forms.

evolution Genetic change taking place in a group of organisms.

exit (E) site One of three sites in a ribosome occupied by a tRNA. In the elongation stage of translation, the tRNA moves from the peptidyl (P) site to the E site from which it then exits the ribosome.

exon Coding region of a split gene (a gene that is interrupted by introns). After processing, the exons remain in messenger RNA.

exon shuffling Process, important in the evolution of eukaryotic genes, by which exons of different genes are exchanged and mixed into new combinations, creating new genes that are mosaics of other preexisting genes.

expanding trinucleotide repeat Mutation in which the number of copies of a set of nucleotides (most often three nucleotides) increases in succeeding generations.

expected heterozygosity Proportion of individuals that are expected to be heterozygous at a locus when the Hardy–Weinberg assumptions are met.

expressed-sequence tag (EST) Unique fragment of DNA from the coding region of a gene, produced by the reverse transcription of cellular RNA. Parts of the fragments are sequenced so that they can be identified.

expression vector Cloning vector containing DNA sequences such as a promoter, a ribosome-binding site, and transcription initiation and termination sites that allow DNA fragments inserted into the vector to be transcribed and translated.

expressivity Degree to which a trait is expressed.

familial Down syndrome Caused by a Robertsonian translocation in which the long arm of chromosome 21 is translocated to another chromosome; tends to run in families.

fertilization Fusion of gametes (sex cells) to form a zygote.

fetal cell sorting Separation of fetal cells from maternal blood. Genetic testing on the fetal cells can provide information about genetic diseases and disorders in the fetus.

F factor Episome of *E. coli* that controls conjugation and gene exchange between *E. coli* cells. The F factor contains an origin of replication and genes that enable the bacterium to undergo conjugation.

F_1 (first filial) generation Offspring of the initial parents (P) in a genetic cross.

F_2 (second filial) generation Offspring of the F_1 generation in a genetic cross; the third generation of a genetic cross.

first polar body One of the products of meiosis I in oogenesis; contains half the chromosomes but little of the cytoplasm.

fitness Reproductive success of a genotype compared with that of other genotypes in a population.

5′ cap Modified 5′ end of eukaryotic mRNA, consisting of an extra nucleotide (methylated) and methylation of the 2′ position of the ribose sugar in one or more subsequent nucleotides; plays a role in the binding of the ribosome to mRNA and affects mRNA stability and the removal of introns.

5′ end End of the polynucleotide chain where a phosphate is attached to the 5′-carbon atom of the nucleotide.

5′ splice site The 5′ end of an intron where cleavage takes place in RNA splicing.

5′ untranslated (5′ UTR) region Sequence of nucleotides at the 5′ end of mRNA; does not encode the amino acids of a protein.

fixation Point at which one allele reaches a frequency of 1. At this point, all members of the population are homozygous for the same allele.

flanking direct repeat Short, directly repeated sequence produced on either side of a transposable element when the element inserts into DNA.

forward genetics Traditional approach to the study of gene function that begins with a phenotype (a mutant organism) and proceeds to a gene that encodes the phenotype.

forward mutation Alters a wild-type phenotype.

founder effect Sampling error that arises when a population is established by a small number of individuals; leads to genetic drift.

fragile site Constriction or gap that appears at a particular location on a chromosome when cells are cultured under special conditions. One fragile site on the human X chromosome is associated with mental retardation (fragile-X syndrome) and results from an expanding trinucleotide repeat.

fragile-X syndrome A form of X-linked mental retardation that appears primarily in males; results from an increase in the number of repeats of a CGG trinucleotide.

frameshift mutation Alters the reading frame of a gene.

fraternal twins Nonidentical twins that arise when two different eggs are fertilized by two different sperm; also called dizygotic twins.

frequency distribution Graphical way of representing values. In genetics, usually the phenotypes found in a group of individuals are displayed as a frequency distribution. Typically, the phenotypes are plotted on the horizontal (x) axis and the numbers (or proportions) of individuals with each phenotype are plotted on the vertical (y) axis.

functional genomics Area of genomics that studies the functions of genetic information contained within genomes.

G_0 (gap 0) Nondividing stage of the cell cycle.

G_1 (gap 1) Stage in interphase of the cell cycle in which the cell grows and develops.

G_2 (gap 2) Stage of interphase in the cell cycle that follows DNA replication. In G_2, the cell prepares for division.

gain-of-function mutation Produces a new trait or causes a trait to appear in inappropriate tissues or at inappropriate times in development.

gametic isolation Reproductive isolation due to the incompatibility of gametes. Mating between members of different species takes place, but the gametes do not form zygotes. Seen in many plants, where pollen from one species cannot fertilize the ovules of another species.

gametophyte Haploid phase of the life cycle in plants.

gap genes In fruit flies, set of segmentation genes that define large sections of the embryo. Mutations in these genes usually eliminate whole groups of adjacent segments.

gel electrophoresis Technique for separating charged molecules (such as proteins or nucleic acids) on the basis of molecular size or charge or both.

gene Genetic factor that helps determine a trait; often defined at the molecular level as a DNA sequence that is transcribed into an RNA molecule.

gene cloning Insertion of DNA fragments into bacteria in such a way that the fragments will be stable and copied by the bacteria.

gene conversion Process of nonreciprocal genetic exchange that can produce abnormal ratios of gametes following meiosis.

gene desert In reference to the density of genes in the genome, a region that is gene poor—that is, a long stretch of DNA possibly consisting of hundreds of thousands to millions of base pairs completely devoid of any known genes or other functional sequences.

gene family *See* **multigene family**.

gene flow Movement of genes from one population to another; also called migration.

gene interaction Interaction between genes at different loci that affect the same characteristic.

gene mutation Affects a single gene or locus.

gene pool Total of all genes in a population.

generalized transduction Transduction in which any gene can be transferred from one bacterial cell to another by a virus.

general transcription factor Protein that binds to eukaryotic promoters near the start site and is a part of the basal transcription apparatus that initiates transcription.

gene regulation Mechanisms and processes that control the phenotypic expression of genes.

gene therapy Use of recombinant DNA to treat a disease or disorder by altering the genetic makeup of the patient's cells.

genetic bottleneck Sampling error that arises when a population undergoes a drastic reduction in population size; leads to genetic drift.

genetic conflict hypothesis Suggests that genomic imprinting evolved because different and conflicting pressures act on maternal and paternal alleles for genes that affect fetal growth. For example, paternally derived alleles often favor maximum fetal growth, whereas maternally derived alleles favor less than maximum fetal growth because of the high cost of fetal growth to the mother.

genetic correlation Phenotypic correlation due to the same genes affecting two or more characteristics.

genetic counseling Educational process that attempts to help patients and family members deal with all aspects of a genetic condition.

genetic drift Change in allelic frequency due to sampling error.

genetic engineering Common term for recombinant DNA technology.

genetic–environmental interaction variance Component of the phenotypic variance that results from an interaction between genotype and environment. Genotypes are expressed differently in different environments.

Genetic Information Nondiscrimination Act (GINA) U.S. law prohibiting health insurers from using genetic information to make decisions about health-insurance coverage and rates; prevents employers from using genetic information in employment decisions; also prevents health insurers and employers from asking or requiring a person to take a genetic test.

genetic (linkage) map Map of the relative distances between genetic loci, markers, or other chromosome regions determined by rates of recombination; measured in percent recombination or map units.

genetic marker Any gene or DNA sequence used to identify a location on a genetic or physical map.

genetic maternal effect Determines the phenotype of an offspring. With genetic maternal effect, an offspring inherits genes for the characteristics from both parents, but the offspring's phenotype is determined not by its own genotype but by the nuclear genotype of its mother.

genetic rescue Introduction of new genetic variation into an inbred population that often dramatically improves the health of the population in an effort to increase its chances of long-term survival.

genetic variance Component of the phenotypic variance that is due to genetic differences among individual members of a population.

gene tree Phylogenetic tree representing the evolutionary relationships among a set of genes.

genic balance system Sex-determining system in which sexual phenotype is controlled by a balance between genes on the X chromosome and genes on the autosomes.

genic interaction variance Component of the genetic variance that can be attributed to genic interaction (interaction between genes at different loci).

genic sex determination Sex determination in which the sexual phenotype is specified by genes at one or more loci, but there are no obvious differences in the chromosomes of males and females.

genome Complete set of genetic instructions for an organism.

genomewide association studies Studies that look for nonrandom associations between the presence of a trait and alleles at many different loci scattered across a genome—that is, for associations between traits and particular suites of alleles in a population.

genomic imprinting Differential expression of a gene that depends on the sex of the parent that transmitted the gene. If the gene is inherited from the father, its expression differs from that if the gene is inherited from the mother.

genomic library Collection of bacterial or phage colonies containing DNA fragments that consist of the entire genome of an organism.

genomics Study of the content, organization, and function of genetic information in whole genomes.

genotype The set of genes possessed by an individual organism.

genotypic frequency Proportion of a particular genotype within a population.

germ-line mutation Mutation in a germ-line cell (one that gives rise to gametes).

germ-plasm theory States that cells in the reproductive organs carry a complete set of genetic information.

G_2/M checkpoint Important point in the cell cycle near the end of G_2. After this checkpoint has been passed, the cell undergoes mitosis.

goodness-of-fit chi-square test Statistical test used to evaluate how well a set of observed values fit the expected values. The probability associated with a calculated chi-square value is the probability that the differences between the observed and the expected values may be due to chance.

G overhang A guanine-rich sequence of nucleotides that protrudes beyond the complementary C-rich strand at the end of a chromosome.

group I intron Belongs to a class of introns in some ribosomal RNA genes that are capable of self-splicing.

group II intron Belongs to a class of introns in some protein-encoding genes that are capable of self-splicing and are found in mitochondria, chloroplasts, and a few eubacteria.

G_1/S checkpoint Important point in the cell cycle. After the G_1/S checkpoint has been passed, DNA replicates and the cell is committed to dividing.

guanine (G) Purine base in DNA and RNA.

guide RNA (gRNA) RNA molecule that serves as a template for an alteration made in mRNA during RNA editing.

gynandromorph Individual organism that is a mosaic for the sex chromosomes, possessing tissues with different sex-chromosome constitutions.

gyrase *See* **DNA gyrase.**

hairpin Secondary structure formed when sequences of nucleotides on the same strand are complementary and pair with each other.

haploid Possessing a single set of chromosomes (one genome).

haploinsufficiency The appearance of a mutant phenotype in an individual cell or organism that is heterozygous for a normally recessive trait.

haploinsufficient gene Must be present in two copies for normal function. If one copy of the gene is missing, a mutant phenotype is produced.

haplotype A specific set of linked genetic variants or alleles on a single chromosome or on part of a chromosome.

Hardy–Weinberg equilibrium Frequencies of genotypes when the conditions of the Hardy–Weinberg law are met.

Hardy–Weinberg law Important principle of population genetics stating that, in a large, randomly mating population not affected by mutation, migration, or natural selection, allelic frequencies will not change and genotypic frequencies stabilize after one generation in the proportions p^2 (the frequency of *AA*), $2pq$ (the frequency of *Aa*), and q^2 (the frequency of *aa*), where *p* equals the frequency of allele *A* and *q* equals the frequency of allele *a*.

heat-shock protein Produced by many cells in response to extreme heat and other stresses; helps cells prevent damage from such stressing agents.

helicase *See* **DNA helicase.**

hemizygosity Possession of a single allele at a locus. Males of organisms with XX-XY sex determination are hemizygous for X-linked loci because their cells possess a single X chromosome.

heritability Proportion of phenotypic variation due to genetic differences. *See also* **broad-sense heritability** and **narrow-sense heritability.**

hermaphroditism Condition in which an individual organism possesses both male and female reproductive structures. True hermaphrodites produce both male and female gametes.

heterochromatin Chromatin that remains in a highly condensed state throughout the cell cycle; found at the centromeres and telomeres of most chromosomes.

heteroduplex DNA DNA consisting of two strands, each of which is from a different chromosome.

heterogametic sex The sex (male or female) that produces two types of gametes with respect to sex chromosomes. For example, in the XX-XY sex-determining system, the male produces both X-bearing and Y-bearing gametes.

heterokaryon Cell possessing two nuclei derived from different cells through cell fusion.

heteroplasmy Presence of two or more distinct variants of DNA within the cytoplasm of a single cell.

heterozygote screening Tests members of a population to identify heterozygous carriers of a disease-causing allele who are healthy but have the potential to produce children who have the disease.

heterozygous Refers to an individual organism that possesses two different alleles at a locus.

highly repetitive DNA DNA that consists of short sequences that are present in hundreds of thousands to millions of copies; clustered in certain regions of chromosomes.

histone Low-molecular-weight protein found in eukaryotes that complexes with DNA to form chromosomes.

histone code Modification of histone proteins, such as the addition or removal of phosphate groups, methyl groups, or acetyl groups, that encode information affecting how genes are expressed.

Holliday intermediate Structure that forms in homologous recombination; consists of two duplex molecules connected by a cross bridge.

Holliday junction Model of homologous recombination that is initiated by single-strand breaks in a DNA molecule.

holoenzyme Complex of an enzyme and other protein factors necessary for complete function.

homeobox Conserved subset of nucleotides in homeotic genes. In *Drosophila*, it consists of 180 nucleotides that encode 60 amino acids of a DNA-binding domain related to the helix-turn-helix motif.

homeotic complex (HOM-C) Major cluster of homeotic genes in fruit flies; consists of the *Antennapedia* complex, which affects the development of the adult fly's head and anterior segments, and the *bithorax* complex, which affects the adult fly's posterior thoracic and abdominal segments.

homeotic gene Determines the identity of individual segments or parts in an early embryo. Mutations in such genes cause body parts to appear in the wrong places.

homogametic sex The sex (male or female) that produces gametes that are all alike with regard to sex chromosomes. For example, in the XX-XY sex-determining system, the female produces only X-bearing gametes.

homologous genes Evolutionarily related genes, having descended from a gene in a common ancestor.

homologous pair of chromosomes Two chromosomes that are alike in structure and size and that carry genetic information for the same set of hereditary characteristics. One chromosome of a homologous pair is inherited from the male parent and the other is inherited from the female parent.

homologous recombination Exchange of genetic information between homologous DNA molecules.

homoplasmy Presence of only one version of DNA within the cytoplasm of a single cell.

homozygous Refers to an individual organism that possesses two identical alleles at a locus.

horizontal gene transfer Transfer of genes from one organism to another by a mechanism other than reproduction.

Hox **gene** Gene that contains a homeobox.

human papilloma virus (HPV) Virus associated with cervical cancer.

Human Proteome Project Project with the goal of identifying and characterizing all proteins in the human body.

humoral immunity Type of immunity resulting from antibodies produced by B cells.

hybrid breakdown Reproductive isolating mechanism in which closely related species are capable of mating and producing viable and fertile F_1 progeny, but genes do not flow between the two species, because further crossing of the hybrids produces inviable or sterile offspring.

hybrid dysgenesis Sudden appearance of numerous mutations, chromosome aberrations, and sterility in the offspring of a cross between a male fly that possesses *P* elements and a female fly that lacks them.

hybrid inviability Reproductive isolating mechanism in which mating between two organisms of different species take place and hybrid offspring are produced but are not viable.

hybridization Pairing of two partly or fully complementary single-stranded nucleotide chains.

hybrid sterility Hybrid embryos complete development but are sterile; exemplified by mating between donkeys and horses to produce a mule, a viable but usually sterile offspring.

hypostatic gene Gene that is masked or suppressed by the action of a gene at a different locus.

identical twins Twins that arise when a single egg fertilized by a single sperm splits into two separate embryos; also called monozygotic twins.

inbreeding Mating between related individuals that takes place more frequently than expected on the basis of chance.

inbreeding coefficient Measure of inbreeding; the probability (ranging from 0 to 1) that two alleles are identical by descent.

inbreeding depression Decreased fitness arising from inbreeding; often due to the increased expression of lethal or deleterious recessive traits.

incomplete dominance Refers to the phenotype of a heterozygote that is intermediate between the phenotypes of the two homozygotes.

incomplete linkage Linkage between genes that exhibit some crossing over; intermediate in its effects between independent assortment and complete linkage.

incomplete penetrance Refers to a genotype that does not always express the expected phenotype. Some individuals possess the genotype for a trait but do not express the phenotype.

incorporated error Incorporation of a damaged nucleotide or mismatched base pair into a DNA molecule.

independent assortment Independent separation of chromosome pairs in anaphase I of meiosis; contributes to genetic variation.

induced mutation Results from environmental agents, such as chemicals or radiation.

inducer Substance that stimulates transcription in an inducible system of gene regulation; usually a small molecule that binds to a repressor protein and alters that repressor so that it can no longer bind to DNA and inhibit transcription.

inducible operon Operon or other system of gene regulation in which transcription is normally off. Something must take place for transcription to be induced, or turned on.

induction Stimulation of the synthesis of an enzyme by an environmental factor, often the presence of a particular substrate.

in-frame deletion Deletion of some multiple of three nucleotides, which does not alter the reading frame of the gene.

in-frame insertion Insertion of some multiple of three nucleotides, which does not alter the reading frame of the gene.

inheritance of acquired characteristics Early notion of inheritance proposing that acquired traits are passed to descendants.

initiation codon The codon in mRNA that specifies the first amino acid (fMet in bacterial cells; Met in eukaryotic cells) of a protein; most commonly AUG.

initiation factor 1 (IF-1) Protein required for the initiation of translation in bacterial cells; enhances the dissociation of the large and small subunits of the ribosome.

initiation factor 2 (IF-2) Protein required for the initiation of translation in bacterial cells; forms a complex with GTP and the charged initiator tRNA and then delivers the charged tRNA to the initiation complex.

initiation factor 3 (IF-3) Protein required for the initiation of translation in bacterial cells; binds to the small subunit of the ribosome and prevents the large subunit from binding during initiation.

initiator protein Binds to an origin of replication and unwinds a short stretch of DNA, allowing helicase and other single-strand-binding proteins to bind and initiate replication.

insertion Mutation in which nucleotides are added to a DNA sequence.

insertion sequence Simple type of transposable element found in bacteria and their plasmids that contains only the information necessary for its own movement.

in situ hybridization Method used to determine the chromosomal location of a gene or other specific DNA fragment or the tissue distribution of an mRNA by using a labeled probe that is complementary to the sequence of interest.

insulator DNA sequence that blocks or insulates the effect of an enhancer; must be located between the enhancer and the promoter to have blocking activity; also may limit the spread of changes in chromatin structure.

integrase Enzyme that inserts prophage, or proviral, DNA into a chromosome.

interactome Complete set of protein interactions in a cell.

intercalating agent Chemical substance that is about the same size as a nucleotide and may become sandwiched between adjacent bases in DNA, distorting the three-dimensional structure of the helix and causing single-nucleotide insertions and deletions in replication.

interchromosomal recombination Recombination among genes on different chromosomes.

interference Degree to which one crossover interferes with additional crossovers.

intergenic suppressor mutation Suppressor mutation that occurs in a gene (locus) that is different from the gene containing the original mutation.

interkinesis Period between meiosis I and meiosis II.

internal promoter Promoter located within the sequences of DNA that are transcribed into RNA.

interphase Period in the cell cycle between the cell divisions. In interphase, the cell grows, develops, and prepares for cell division.

interspersed repeat sequence Repeated sequence at multiple locations throughout the genome.

intrachromosomal recombination Recombination among genes located on the same chromosome.

intragenic mapping Maps the locations of mutations within a single locus.

intragenic suppressor mutation Suppressor mutation that occurs in the same gene (locus) as the mutation that it suppresses.

intron Intervening sequence in a split gene; removed from the RNA after transcription.

inverted repeats Sequences on the same strand that are inverted and complementary.

isoaccepting tRNAs Different tRNAs with different anticodons that specify the same amino acid.

isotopes Different forms of an element that have the same number of protons and electrons but differ in the number of neutrons in the nucleus.

junctional diversity Addition or deletion of nucleotides at the junctions of gene segments brought together in the somatic recombination of genes that encode antibodies and T-cell receptors.

karyotype Picture of an individual organism's complete set of metaphase chromosomes.

kinetochore Set of proteins that assemble on the centromere, providing the point of attachment for spindle microtubules.

Klinefelter syndrome Human condition in which cells contain one or more Y chromosomes along with multiple X chromosomes (most commonly XXY but may also be XXXY, XXXXY, or XXYY). Persons with Klinefelter syndrome are male in appearance but frequently possess small testes, some breast enlargement, and reduced facial and pubic hair; often taller than normal and sterile, most have normal intelligence.

knock-in mouse Mouse that carries a foreign sequence inserted at a specific chromosome location.

knockout mouse Mouse in which a normal gene has been disabled ("knocked out").

labeling Method for adding a radioactive or chemical label to a molecule such as DNA or RNA.

lagging strand DNA strand that is replicated discontinuously.

large ribosomal subunit The larger of the two subunits of a functional ribosome.

lariat Looplike structure created in the splicing of nuclear pre-mRNA in which the 5′ end of an intron is attached to a branch point in pre-mRNA.

leading strand DNA strand that is replicated continuously.

leptotene First substage of prophase I in meiosis. In leptotene, chromosomes contract and become visible.

lethal allele Causes the death of an individual organism, often early in development, and so the organism does not appear in the progeny of a genetic cross. Recessive lethal alleles kill individual organisms that are homozygous for the allele; dominant lethals kill both heterozygotes and homozygotes.

lethal mutation Causes premature death.

LINE *See* **long interspersed element.**

linkage analysis Gene mapping based on the detection of physical linkage between genes, as measured by the rate of recombination, in progeny from a cross.

linkage disequilibrium Nonrandom association between genetic variants within a haplotype.

linkage group Genes located together on the same chromosome.

linked genes Genes located on the same chromosome.

linker Small, synthetic DNA fragment that contains one or more restriction sites. Can be attached to the ends of any piece of DNA and used to insert a gene into a plasmid vector when restrictions sites are not available.

linker DNA Stretch of DNA separating two nucleosomes.

locus Position on a chromosome where a specific gene is located.

lod (logarithm of odds) score Logarithm of the ratio of the probability of obtaining a set of observations, assuming a specified degree of linkage, to the probability of obtaining the same set of observations with independent assortment; used to assess the likelihood of linkage between genes from pedigree data.

long interspersed element (LINE) Long DNA sequence repeated many times and interspersed throughout the genome.

loss-of-function mutation Causes the complete or partial absence of normal function.

loss of heterozygosity At a locus having a normal allele and a mutant allele, inactivation or loss of the normal allele.

Lyon hypothesis Proposed by Mary Lyon in 1961, this hypothesis proposes that one X chromosome in each female cell becomes inactivated (a Barr body) and suggests that which of the X chromosomes becomes inactivated is random and varies from cell to cell.

lysogenic cycle Life cycle of a bacteriophage in which phage genes first integrate into the bacterial chromosome and are not immediately transcribed and translated.

lytic cycle Life cycle of a bacteriophage in which phage genes are transcribed and translated, new phage particles are produced, and the host cell is lysed.

major histocompatibility complex (MHC) antigen Belongs to a large and diverse group of antigens found on the surfaces of cells that mark those cells as self; encoded by a large cluster of genes known as the major histocompatibility complex. T cells simultaneously bind to foreign and MHC antigens.

malignant tumor Consists of cells that are capable of invading other tissues.

map-based sequencing Method of sequencing a genome in which sequenced fragments are ordered into contigs with the use of genetic or physical maps.

mapping function Relates recombination frequencies to actual physical distances between genes.

map unit (m.u.) Unit of measure for distances on a genetic map; 1 map unit equals 1% recombination.

mass spectrometry Method for precisely determining the molecular mass of a molecule by using the migration rate of an ionized molecule in an electrical field.

maternal blood screening test Tests for genetic conditions in a fetus by analyzing the blood of the mother. For example, the level of α-fetoprotein in maternal blood provides information about the probability that a fetus has a neural-tube defect.

mean Statistic that describes the center of a distribution of measurements; calculated by dividing the sum of all measurements by the number of measurements; also called the average.

mechanical isolation Reproductive isolation resulting from anatomical differences that prevent successful copulation.

mediator Complex of proteins that is one of the components of the basal transcription apparatus.

megaspore One of the four products of meiosis in plants.

megasporocyte In the ovary of a plant, a diploid reproductive cell that undergoes meiosis to produce haploid macrospores.

meiosis Process in which chromosomes of a eukaryotic cell divide to give rise to haploid reproductive cells. Consists of two divisions: meiosis I and meiosis II.

meiosis I First phase of meiosis. In meiosis I, chromosome number is reduced by half.

meiosis II Second phase of meiosis. Events in meiosis II are essentially the same as those in mitosis.

melting *See* **denaturation.**

melting temperature Midpoint of the melting range of DNA.

memory cell Long-lived lymphocyte among the clone of cells generated when a foreign antigen is encountered. If the same antigen is encountered again, the memory cells quickly divide and give rise to another clone of cells specific for that particular antigen.

Mendelian population Group of interbreeding, sexually reproducing individuals.

meristic characteristic Characteristic whose phenotype varies in whole numbers, such as number of vertebrae.

merozygote Bacterial cell that has two copies of some genes—one copy on the bacterial chromosome and a second copy on an introduced F plasmid; also called partial diploid.

messenger RNA (mRNA) RNA molecule that carries genetic information for the amino acid sequence of a protein.

metacentric chromosome Chromosome in which the two chromosome arms are approximately the same length.

metagenomics An emerging field of sequencing technology in which the genome sequences of an entire group of organisms inhabiting a common environment are sampled and determined.

metaphase Stage of mitosis. In metaphase, chromosomes align in the center of the cell.

metaphase I Stage of meiosis I. In metaphase I, homologous pairs of chromosomes align in the center of the cell.

metaphase II Stage of meiosis II. In metaphase II, individual chromosomes align on the metaphase plate.

metaphase plate Plane in a cell between two spindle poles. In metaphase, chromosomes align on the metaphase plate.

metastasis Refers to cells that separate from malignant tumors and travel to other sites, where they establish secondary tumors.

5′-methylcytosine Modified nucleotide, consisting of cytosine to which a methyl group has been added; predominate form of methylation in eukaryotic DNA.

microarray Ordered array of DNA fragments fixed to a solid support, which serve as probes to detect the presence of complementary sequences; often used to assess the expression of genes in various tissues and under different conditions.

microRNA (miRNA) Small RNA molecule, typically 21 or 22 bp in length, produced by cleavage of double-stranded RNA arising from small hairpins within RNA that is mostly single stranded. The miRNAs combine with proteins to form a complex that binds (imperfectly) to mRNA molecules and inhibits their translation.

microsatellite *See* **variable number of tandem repeats.**

microspore Haploid product of meiosis in plants.

microsporocyte Diploid reproductive cell in the stamen of a plant; undergoes meiosis to produce four haploid microspores.

microtubule Long fiber composed of the protein tubulin; plays an important role in the movement of chromosomes in mitosis and meiosis.

migration Movement of genes from one population to another; also called gene flow.

minimal medium Used to culture bacteria or some other microorganism; contains only the nutrients required by prototrophic (wild-type) cells—typically, a carbon source, essential elements such as nitrogen and phosphorus, certain vitamins, and other required ions and nutrients.

mismatch repair Process that corrects mismatched nucleotides in DNA after replication has been completed. Enzymes excise incorrectly paired nucleotides from the newly synthesized strand and use the original nucleotide strand as a template when replacing them.

missense mutation Alters a codon in mRNA, resulting in a different amino acid in the protein encoded.

mitochondrial DNA (mtDNA) DNA in mitochondria; has some characteristics in common with eubacterial DNA and typically consists of a circular molecule that lacks histone proteins and encodes some of the rRNAs, tRNAs, and proteins found in mitochondria.

mitosis Process by which the nucleus of a eukaryotic cell divides.

mitosis-promoting factor (MPF) Protein functioning in the control of the cell cycle; consists of a cyclin combined with cyclin-dependent kinase (CDK). Active MPF stimulates mitosis.

mitotic spindle Array of microtubules that radiate from two poles; moves chromosomes in mitosis and meiosis.

model genetic organism An organism that is widely used in genetic studies because it has characteristics, such as short generation time and large numbers of progeny, that make it particularly useful for genetic analysis.

moderately repetitive DNA DNA consisting of sequences that are from 150 to 300 bp in length and are repeated thousands of times.

modified base Rare base found in some RNA molecules. Such bases are modified forms of the standard bases (adenine, guanine, cytosine, and uracil).

molecular chaperone Molecule that assists in the proper folding of another molecule.

molecular clock Refers to the use of molecular differences to estimate the time of divergence between organisms; assumes a roughly constant rate at which one neutral mutation replaces another.

molecular genetics Study of the chemical nature of genetic information and how it is encoded, replicated, and expressed.

molecular motor Specialized protein that moves cellular components.

monoecious organism Individual organism that has both male and female reproductive structures.

monohybrid cross A cross between two individuals that differ in a single characteristic—more specifically, a cross between individuals that are homozygous for different alleles at the same locus ($AA \times aa$); also refers to a cross between two individuals that are heterozygous for two alleles at a single locus ($Aa \times Aa$).

monosomy Absence of one of the chromosomes of a homologous pair.

monozygotic twins Identical twins that arise when a single egg fertilized by a single sperm splits into two separate embryos.

Morgan 100 map units.

morphogen Molecule whose concentration gradient affects the developmental fate of surrounding cells.

mosaicism Condition in which regions of tissue within a single individual have different chromosome constitutions.

M (mitotic) phase Period of active cell division; includes mitosis (nuclear division) and cytokinesis (cytoplasmic division).

mRNA surveillance Mechanisms for the detection and elimination of mRNAs that contain errors that may create problems in the course of translation.

multifactorial characteristic Determined by multiple genes and environmental factors.

multigene family Set of genes similar in sequence that arose through repeated duplication events; often encode different proteins.

multiple alleles Presence in a group of individuals of more than two alleles at a locus. Although, for the group, the locus has more than two alleles, each member of the group has only two of the possible alleles.

multiple 3′ cleavage sites Refers to the presence of more than one 3′ cleavage site on a single pre-mRNA, which allows cleavage and polyadenylation to take place at different sites, producing mRNAs of different lengths.

multiplication rule States that the probability of two or more independent events occurring together is calculated by multiplying the probabilities of each of the individual events.

mutagen Any environmental agent that significantly increases the rate of mutation above the spontaneous rate.

mutagenesis screen Method for identifying genes that influence a specific phenotype. Random mutations are induced in a population of organisms, and individual organisms with mutant phenotypes are identified. These individual organisms are crossed to determine the genetic basis of the phenotype and to map the location of mutations that cause the phenotype.

mutation Heritable change in genetic information.

mutation rate Frequency with which a gene changes from the wild type to a specific mutant; generally expressed as the number of mutations per biological unit (that is, mutations per cell division, per gamete, or per round of replication).

narrow-sense heritability Proportion of the phenotypic variance that can be attributed to additive genetic variance.

natural selection Differential reproduction of genotypes.

negative assortative mating Mating between unlike individuals that is more frequent than would be expected on the basis of chance.

negative control Gene regulation in which the binding of a regulatory protein to DNA inhibits transcription (the regulatory protein is a repressor).

negative supercoiling *See* **supercoiling**.

neutral mutation Changes the amino acid sequence of a protein but does not alter the function of the protein.

neutral-mutation hypothesis Proposes that much of the molecular variation seen in natural populations is adaptively neutral and unaffected by natural selection. Under this hypothesis, individuals with different molecular variants have equal fitnesses.

newborn screening Tests newborn infants for certain genetic disorders.

next-generation sequencing technologies Sequencing methods, such as pyrosequencing, that are capable of simultaneously determining the sequences of many DNA fragments; the technologies are much faster and less expensive than the Sanger method.

nitrogenous base Nitrogen-containing base that is one of the three parts of a nucleotide.

node Point in a phylogenetic tree that represents an organism. Terminal nodes are those that are at the outermost branches of the tree and represent organisms for which data have been obtained. Internal nodes represent ancestors common to organisms on different branches of the tree.

nonautonomous element Transposable element that cannot transpose on its own but can transpose in the presence of an autonomous element of the same family.

nondisjunction Failure of homologous chromosomes or sister chromatids to separate in meiosis or mitosis.

nonhistone chromosomal protein One of a heterogeneous assortment of nonhistone proteins in chromatin.

nonidentical twins Twins that arise when two different eggs are fertilized by two different sperm; also called dizygotic twins or fraternal twins.

nonoverlapping genetic code Refers to the fact that, generally, each nucleotide is a part of only one codon and encodes only one amino acid in a protein.

nonreciprocal translocation Movement of a chromosome segment to a nonhomologous chromosome or region without any (or with unequal) reciprocal exchange of segments.

nonrecombinant (parental) gamete Contains only the original combinations of genes present in the parents.

nonrecombinant (parental) progeny Possesses the original combinations of traits possessed by the parents.

nonreplicative transposition Type of transposition in which a transposable element excises from an old site and moves to a new site, resulting in no net increase in the number of copies of the transposable element.

nonsense codon Codon in mRNA that signals the end of translation; also called a stop codon or termination codon. The three common nonsense codons are UAA, UAG, and UGA.

nonsense-mediated mRNA decay (NMD) Process that brings about the rapid elimination of mRNA that has a premature stop codon.

nonsense mutation Changes a sense codon (one that specifies an amino acid) into a stop codon.

nonstop RNA decay Mechanism in eukaryotic cells for dealing with ribosomes stalled at the 3′ end of an mRNA that lacks a termination codon. A protein binds to the A site of the stalled ribosome and recruits other proteins that degrade the mRNA from the 3′ end.

nontemplate strand The DNA strand that is complementary to the template strand; not ordinarily used as a template during transcription.

normal distribution Common type of frequency distribution that exhibits a symmetrical, bell-shaped curve; usually arises when a large number of independent factors contribute to the measurement.

Northern blotting Process by which RNA is transferred from a gel to a solid support such as a nitrocellulose or nylon filter.

nuclear envelope Membrane that surrounds the genetic material in eukaryotic cells to form a nucleus; segregates the DNA from other cellular contents.

nuclear matrix Network of protein fibers in the nucleus; holds the nuclear contents in place.

nuclear pre-mRNA intron Belongs to a class of introns in protein-encoding genes that reside in the nuclei of eukaryotic cells; removed by spliceosomal-mediated splicing.

nucleoid Bacterial DNA confined to a definite region of the cytoplasm.

nucleoside Ribose or deoxyribose bonded to a base.

nucleosome Basic repeating unit of chromatin, consisting of a core of eight histone proteins (two each of H2A, H2B, H3, and H4) and about 146 bp of DNA that wraps around the core about two times.

nucleotide Repeating unit of DNA or RNA made up of a sugar, a phosphate, and a base.

nucleotide-excision repair DNA repair that removes bulky DNA lesions and other types of DNA damage.

nucleus Space in eukaryotic cells that is enclosed by the nuclear envelope and contains the chromosomes.

nullisomy Absence of both chromosomes of a homologous pair $(2n - 2)$.

Okazaki fragment Short stretch of newly synthesized DNA. Produced by discontinuous replication on the lagging strand, these fragments are eventually joined together.

oligonucleotide-directed mutagenesis Method of site-directed mutagenesis that uses an oligonucleotide to introduce a mutant sequence into a DNA molecule.

oncogene Dominant-acting gene that stimulates cell division, leading to the formation of tumors and contributing to cancer; arises from mutated copies of a normal cellular gene (proto-oncogene).

one gene, one enzyme hypothesis Idea proposed by Beadle and Tatum that each gene encodes a separate enzyme.

one gene, one polypeptide hypothesis Modification of the one gene, one enzyme hypothesis; proposes that each gene encodes a separate polypeptide chain.

oogenesis Egg production in animals.

oogonium Diploid cell in the ovary; capable of undergoing meiosis to produce an egg cell.

open reading frame (ORF) Continuous sequence of DNA nucleotides that contains a start codon and a stop codon in the same reading frame; is assumed to be a gene that encodes a protein but, in many cases, the protein has not yet been identified.

operator DNA sequence in the operon of a bacterial cell. A regulator protein binds to the operator and affects the rate of transcription of structural genes.

operon Set of structural genes in a bacterial cell along with a common promoter and other sequences (such as an operator) that control the transcription of the structural genes.

origin of replication Site where DNA synthesis is initiated.

orthologous genes Homologous genes found in different species because the two species have a common ancestor that also possessed the gene.

outcrossing Mating between unrelated individuals that is more frequent than would be expected on the basis of chance.

overdominance Selection in which the heterozygote has higher fitness than that of either homozygote; also called heterozygote advantage.

ovum Final product of oogenesis.

pachytene Third substage of prophase I in meiosis. The synaptonemal complex forms during pachytene.

pair-rule genes Set of segmentation genes in fruit flies that define regional sections of the embryo and affect alternate segments. Mutations in these genes often cause the deletion of every other segment.

palindrome Sequence of nucleotides that reads the same on complementary strands; inverted repeats.

pangenesis Early concept of heredity proposing that particles carry genetic information from different parts of the body to the reproductive organs.

paracentric inversion Chromosome inversion that does not include the centromere in the inverted region.

paralogous genes Homologous genes in the same species that arose through the duplication of a single ancestral gene.

paramutation Epigenetic change in which one allele of a genotype alters the expression of another allele; the altered expression persists for several generations, even after the altering allele is no longer present.

parental gamete *See* **nonrecombinant (parental) gamete.**

parental progeny *See* **nonrecombinant (parental) progeny.**

partial diploid Bacterial cell that possesses two copies of genes, including one copy on the bacterial chromosome and the other on an extra piece of DNA (usually a plasmid); also called merozygote.

Patau syndrome (trisomy 13) Characterized by severe mental retardation, a small head, sloping forehead, small eyes, cleft lip and palate, extra fingers and toes, and other disabilities; results from the presence of three copies of chromosome 13.

pedigree Pictorial representation of a family history outlining the inheritance of one or more traits or diseases.

penetrance Percentage of individuals with a particular genotype that express the phenotype expected of that genotype.

pentaploidy Possession of five haploid sets of chromosomes ($5n$).

peptide bond Chemical bond that connects amino acids in a protein.

peptidyl (P) site One of three sites in a ribosome occupied by a tRNA in translation. In the elongation stage of protein synthesis, tRNAs move from the aminoacyl (A) site into the P site.

peptidyl transferase Activity in the ribosome that creates a peptide bond between two amino acids. Evidence suggests that this activity is carried out by one of the RNA components of the ribosome.

pericentric inversion Chromosome inversion that includes the centromere in the inverted region.

P (parental) generation First set of parents in a genetic cross.

phage *See* **bacteriophage.**

phenocopy Phenotype that is produced by environmental effects and is the same as the phenotype produced by a genotype.

phenotype Appearance or manifestation of a characteristic.

phenotypic correlation Correlation between two or more phenotypes in the same individual.

phenotypic variance Measures the degree of phenotypic differences among a group of individuals; composed of genetic, environmental, and genetic–environmental interaction variances.

phenylketonuria (PKU) Genetic disease characterized by mental retardation, light skin, and eczema; caused by mutations in the gene that encodes phenylalanine hydroxylase (PAH), a liver enzyme that normally metabolizes the amino acid phenylalanine. When the enzyme is defective, phenylalanine is not metabolized and builds up to high levels in the body, eventually causing mental retardation and other characteristics of the disease. The disease is inherited as an autosomal recessive disorder and can be effectively treated by limiting phenylalanine in the diet.

phosphate group A phosphorus atom attached to four oxygen atoms; one of the three components of a nucleotide.

phosphodiester Molecule containing R–O–P–O–R, where R is a carbon-containing group, O is oxygen, and P is phosphorus.

phosphodiester linkage Phosphodiester bond connecting two nucleotides in a polynucleotide strand.

phylogenetic tree Graphical represenation of the evolutionary connections between organisms or genes.

phylogeny Evolutionary relationships among a group of organisms or genes, usually depicted as a family tree or branching diagram.

physical map Map of physical distances between loci, genetic markers, or other chromosome segments; measured in base pairs.

pilus (pl., pili) Extension of the surface of some bacteria that allows conjugation to take place. When a pilus on one cell makes contact with a receptor on another cell, the pilus contracts and pulls the two cells together.

Piwi-interacting RNA (piRNA) Small RNA molecule belonging to a class named after Piwi proteins with which these molecules interact; similar to microRNA and small interfering RNA and thought to have a role in the regulation of sperm development.

plaque Clear patch of lysed cells on a continuous layer of bacteria on the agar surface of a petri plate. Each plaque represents a single original phage that multiplied and lysed many cells.

plasmid Small, circular DNA molecule found in bacterial cells that is capable of replicating independently from the bacterial chromosome.

pleiotropy A single genotype influences multiple phenotypes.

pluripotency In embryonic stem cells, the property of being undifferentiated, with the capacity to form every type of cell in an organism.

poly(A)-binding protein (PABP) Binds to the poly(A) tail of eukaryotic mRNA and makes the mRNA more stable. There are several types of PABPs, one of which is PABII.

poly(A) tail String of adenine nucleotides added to the 3′ end of a eukaryotic mRNA after transcription.

polycistronic mRNA Single bacterial RNA molecule that encodes more than one polypeptide chain; uncommon in eukaryotes.

polygenic characteristic Encoded by genes at many loci.

polymerase chain reaction (PCR) Method of enzymatically amplifying DNA fragments.

polynucleotide strand Series of nucleotides linked together by phosphodiester bonds.

polypeptide Chain of amino acids linked by peptide bonds; also called a protein.

polyploidy Possession of more than two haploid sets of chromosomes.

polyribosome Messenger RNA molecule with several ribosomes attached to it.

polytene chromosome Giant chromosome in the salivary glands of *Drosophila melanogaster*. Each polytene chromosome consists of a number of DNA molecules lying side by side.

population The group of interest; often represented by a subset called a sample. Also, a group of members of the same species.

population genetics Study of the genetic composition of populations (groups of members of the same species) and how a population's collective group of genes changes with the passage of time.

positional cloning Method that allows for the isolation and identification of a gene by examining the cosegregation of a phenotype with previously mapped genetic markers.

position effect Dependence of the expression of a gene on the gene's location in the genome.

positive assortative mating Mating between like individuals that is more frequent than would be expected on the basis of chance.

positive control Gene regulation in which the binding of a regulatory protein to DNA stimulates transcription (the regulatory protein is an activator).

positive supercoiling *See* **supercoiling.**

posttranslational modification Alteration of a protein after translation; may include cleavage from a larger precursor protein, the removal of amino acids, and the attachment of other molecules to the protein.

postzygotic reproductive isolating mechanism Reproductive isolation that arises after a zygote is formed, either because the resulting hybrids are inviable or sterile or because reproduction breaks down in subsequent generations.

preformationism Early concept of inheritance proposing that a miniature adult (homunculus) resides in either the egg or the sperm and increases in size in development, with all traits being inherited from the parent that contributes the homunculus.

preimplantation genetic diagnosis (PGD) Genetic testing on an embryo produced by in vitro fertilization before implantation of the embryo in the uterus.

pre-messenger RNA (pre-mRNA) Eukaryotic RNA molecule that is modified after transcription to become mRNA.

presymptomatic genetic testing Tests people to determine whether they have inherited a disease-causing gene before the symptoms of the disease have appeared.

prezygotic reproductive isolating mechanism Reproductive isolation in which gametes from two different species are prevented from fusing and forming a hybrid zygote.

primary Down syndrome Caused by the presence of three copies of chromosome 21.

primary immune response Initial clone of cells specific for a particular antigen and generated when the antigen is first encountered by the immune system.

primary oocyte Oogonium that has entered prophase I.

primary spermatocyte Spermatogonium that has entered prophase I.

primary structure of a protein The amino acid sequence of a protein.

primase Enzyme that synthesizes a short stretch of RNA on a DNA template; functions in replication to provide a 3′-OH group for the attachment of a DNA nucleotide.

primer Short stretch of RNA on a DNA template; provides a 3′-OH group for the attachment of a DNA nucleotide at the initiation of replication.

principle of independent assortment (Mendel's second law) Important principle of heredity discovered by Mendel that states that genes encoding different characteristics (genes at different loci) separate independently; applies only to genes located on different chromosomes or to genes far apart on the same chromosome.

principle of segregation (Mendel's first law) Important principle of heredity discovered by Mendel that states that each diploid individual possesses two alleles at a locus and that these two alleles separate when gametes are formed, one allele going into each gamete.

probability Likelihood of the occurrence of a particular event; more formally, the number of times that a particular event occurs divided by the number of all possible outcomes. Probability values range from 0 to 1.

proband A person having a trait or disease for whom a pedigree is constructed.

probe Known sequence of DNA or RNA that is complementary to a sequence of interest and will pair with it; used to find specific DNA sequences.

prokaryote Unicellular organism with a simple cell structure. Prokaryotes include eubacteria and archaea.

prometaphase Stage of mitosis. In prometaphase, the nuclear membrane breaks down and the spindle microtubules attach to the chromosomes.

promoter DNA sequence to which the transcription apparatus binds so as to initiate transcription; indicates the direction of transcription, which of the two DNA strands is to be read as the template, and the starting point of transcription.

proofreading Ability of DNA polymerases to remove and replace incorrectly paired nucleotides in the course of replication.

prophage Phage genome that is integrated into a bacterial chromosome.

prophase Stage of mitosis. In prophase, the chromosomes contract and become visible, the cytoskeleton breaks down, and the mitotic spindle begins to form.

prophase I Stage of meiosis I. In prophase I, chromosomes condense and pair, crossing over takes place, the nuclear membrane breaks down, and the spindle forms.

prophase II Stage of meiosis after interkinesis. In prophase II, chromosomes condense, the nuclear membrane breaks down, and the spindle forms. Some cells skip this stage.

proportion of polymorphic loci Percentage of loci in which more than one allele is present in a population.

protein-coding region The part of mRNA consisting of the nucleotides that specify the amino acid sequence of a protein.

protein domain Region of a protein that has a specific shape or function.

protein kinase Enzyme that adds phosphate groups to other proteins.

protein microarray Large number of different proteins applied to a glass slide as a series of spots, each spot containing a different protein; used to analyze protein–protein interactions.

proteome Set of all proteins encoded by a genome.

proteomics Study of the proteome, the complete set of proteins found in a given cell.

proto-oncogene Normal cellular gene that controls cell division. When mutated, it may become an oncogene and contribute to cancer progression.

prototrophic bacterium A wild-type bacterium that can use a carbon source, essential elements such as nitrogen and phosphorus, certain vitamins, and other required ions and nutrients to synthesize all the compounds that they need for growth and reproduction.

provirus DNA copy of viral DNA or viral RNA; integrated into the host chromosome and replicated along with the host chromosome.

pseudoautosomal region Small region of the X and Y chromosomes that contains homologous gene sequences.

pseudodominance Expression of a normally recessive allele owing to a deletion on the homologous chromosome.

Punnett square Shorthand method of determining the outcome of a genetic cross. On a grid, the gametes of one parent are written along the upper edge and the gametes of the other parent are written along the left-hand edge. Within the cells of the grid, the alleles in the gametes are combined to form the genotypes of the offspring.

purine Type of nitrogenous base in DNA and RNA. Adenine and guanine are purines.

pyrimidine Type of nitrogenous base in DNA and RNA. Cytosine, thymine, and uracil are pyrimidines.

pyrimidine dimer Structure in which a bond forms between two adjacent pyrimidine molecules on the same strand of DNA; disrupts normal hydrogen bonding between complementary bases and distorts the normal configuration of the DNA molecule.

quantitative characteristic Continuous characteristic; displays a large number of possible phenotypes, which must be described by a quantitative measurement.

quantitative genetics Genetic analysis of complex characteristics or characteristics influenced by multiple genetic factors.

quantitative trait locus (QTL) A gene or chromosomal region that contributes to the expression of quantitative characteristics.

quaternary structure of a protein Interaction of two or more polypeptides to form a functional protein.

reading frame Particular way in which a nucleotide sequence is read in groups of three nucleotides (codons) in translation. Each reading frame begins with a start codon and ends with a stop codon.

realized heritability Narrow-sense heritability measured from a response-to-selection experiment.

real-time PCR Modification of the polymerase chain reaction that quantitatively determines the amount of starting nucleic acid; the amount of DNA amplified is measured as the reaction proceeds.

reannealing *See* **renaturation**.

recessive Refers to an allele or phenotype that is expressed only when the recessive allele is present in two copies (homozygous). The recessive allele is not expressed in the heterozygote phenotype.

reciprocal crosses Crosses in which the phenotypes of the male and female parents are reversed. For example, in one cross, a tall male is crossed with a short female and, in the other cross, a short male is crossed with a tall female.

reciprocal translocation Reciprocal exchange of segments between two nonhomologous chromosomes.

recombinant DNA technology Set of molecular techniques for locating, isolating, altering, combining, and studying DNA segments.

recombinant gamete Possesses new combinations of genes.

recombinant progeny Possesses new combinations of traits formed from recombinant gametes.

recombination Process that produces new combinations of alleles.

recombination frequency Proportion of recombinant progeny produced in a cross.

regression Analysis of how one variable changes in response to another variable.

regression coefficient Statistic that measures how much one variable changes, on average, with a unit change in another variable.

regulator gene Gene associated with an operon in bacterial cells that encodes a protein or RNA molecule that functions in controlling the transcription of one or more structural genes.

regulator protein Produced by a regulator gene, a protein that binds to another DNA sequence and controls the transcription of one or more structural genes.

regulatory element DNA sequence that affects the transcription of other DNA sequences to which it is physically linked.

regulatory gene DNA sequence that encodes a protein or RNA molecule that interacts with DNA sequences and affects their transcription or translation or both.

regulatory promoter DNA sequence located immediately upstream of the core promoter that affects transcription; contains consensus sequences to which transcriptional regulator proteins bind.

relaxed state of DNA Energy state of a DNA molecule when there is no structural strain on the molecule.

release factor Protein required for the termination of translation; binds to a ribosome when a stop codon is reached and stimulates the release of the polypeptide chain, the tRNA, and the mRNA from the ribosome. Eukaryotic cells require two release factors (eRF_1 and eRF_2), whereas *E. coli* requires three (RF_1, RF_2, and RF_3).

renaturation The process by which two complementary single-stranded DNA molecules pair; also called reannealing.

repetitive DNA Sequences that exist in multiple copies in a genome.

replication Process by which DNA is synthesized from a single-stranded nucleotide template.

replication bubble Segment of a DNA molecule that is unwinding and undergoing replication.

replication error Replication of an incorporated error in which a change in the DNA sequence has been replicated and all base pairings in the new DNA molecule are correct.

replication fork Point at which a double-stranded DNA molecule separates into two single strands that serve as templates for replication.

replication licensing factor Protein that ensures that replication takes place only once at each origin; required at the origin before replication can be initiated and removed after the DNA has been replicated.

replication origin Sequence of nucleotides where replication is initiated.

replication terminus Point at which replication stops.

replicative segregation Random segregation of organelles into progeny cells in cell division. If two or more versions of an organelle are present in the original cell, chance determines the proportion of each type that will segregate into each progeny cell.

replicative transposition Type of transposition in which a copy of the transposable element moves to a new site while the original copy remains at the old site; increases the number of copies of the transposable element.

replicon Unit of replication, consisting of DNA from the origin of replication to the point at which replication on either side of the origin ends.

repressible operon Operon or other system of gene regulation in which transcription is normally on. Something must take place for transcription to be repressed, or turned off.

repressor Regulatory protein that binds to a DNA sequence and inhibits transcription.

reproductive isolating mechanism Any biological factor or mechanism that prevents gene exchange.

repulsion *See* **trans configuration.**

resolvase Enzyme required for some types of transposition; brings about resolution—that is, crossing over between sites located within the transposable element. Resolvase may be encoded by the transposable element or by a cellular enzyme that normally functions in homologous recombination.

response element Common DNA sequence found upstream of some groups of eukaryotic genes. A regulatory protein binds to a response element and stimulates the transcription of a gene. The presence of the same response element in several promoters or enhancers allows a single factor to simultaneously stimulate the transcription of several genes.

response to selection The amount that a characteristic changes in one generation owing to selection; equals the selection differential times the narrow-sense heritability.

restriction endonuclease Technical term for a restriction enzyme, which recognizes particular base sequences in DNA and makes double-stranded cuts nearby.

restriction enzyme Recognizes particular base sequences in DNA and makes double-stranded cuts nearby; also called restriction endonuclease.

restriction fragment length polymorphism (RFLP) Variation in the pattern of fragments produced when DNA molecules are cut with the same restriction enzyme; represents a heritable difference in DNA sequences and can be used in genetic analyses.

restriction mapping Determines the locations of sites cut by restriction enzymes in a piece of DNA.

retrotransposon Type of transposable element in eukaryotic cells that possesses some characteristics of retroviruses and transposes through an RNA intermediate.

retrovirus RNA virus capable of integrating its genetic material into the genome of its host. The virus injects its RNA genome into the host cell, where reverse transcription produces a complementary, double-stranded DNA molecule from the RNA template. The DNA copy then integrates into the host chromosome to form a provirus.

reverse duplication Duplication of a chromosome segment in which the sequence of the duplicated segment is inverted relative to the sequence of the original segment.

reverse genetics A molecular approach that begins with a genotype (a DNA sequence) and proceeds to the phenotype by altering the sequence or by inhibiting its expression.

reverse mutation (reversion) Mutation that changes a mutant phenotype back into the wild type.

reverse transcriptase Enzyme capable of synthesizing complementary DNA from an RNA template.

reverse transcription Synthesis of DNA from an RNA template.

reverse-transcription PCR Amplifies sequences corresponding to RNA. Reverse transcriptase is used to convert RNA into complementary DNA, which can then be amplified by the usual polymerase chain reaction.

rho-dependent terminator Sequence in bacterial DNA that requires the presence of the rho subunit of RNA polymerase to terminate transcription.

rho factor Subunit of bacterial RNA polymerase that facilitates the termination of transcription of some genes.

rho-independent terminator Sequence in bacterial DNA that does not require the presence of the rho subunit of RNA polymerase to terminate transcription.

ribonucleoside triphosphate (rNTP) Substrate of RNA synthesis; consists of ribose, a nitrogenous base, and three phosphates linked to the $5'$-carbon atom of the ribose. In transcription, two of the phosphates are cleaved, producing an RNA nucleotide.

ribonucleotide Nucleotide containing ribose; present in RNA.

ribose Five-carbon sugar in RNA.

ribosomal RNA (rRNA) RNA molecule that is a structural component of the ribosome.

riboswitch Regulatory sequences in an RNA molecule. When an inducer molecule binds to the riboswitch, the binding changes the configuration of the RNA molecule and alters the expression of the RNA, usually by affecting the termination of transcription or affecting translation.

ribozyme RNA molecule that can act as a biological catalyst.

RNA-coding region Sequence of DNA nucleotides that encodes an RNA molecule.

RNA editing Process in which the protein-coding sequence of an mRNA is altered after transcription. The amino acids specified by the altered mRNA are different from those predicted from the nucleotide sequence of the gene encoding the protein.

RNA-induced silencing complex (RISC) Combination of a small interfering RNA (siRNA) molecule or a microRNA (miRNA) molecule and proteins that can cleave mRNA, leading to the degradation of the mRNA, or affect transcription or repress translation of the mRNA.

RNA interference (RNAi) Process in which cleavage of double-stranded RNA produces small interfering RNAs (siRNAs) that bind to mRNAs containing complementary sequences and bring about their cleavage and degradation.

RNA polymerase Enzyme that synthesizes RNA from a DNA template during transcription.

RNA polymerase I Eukaryotic RNA polymerase that transcribes large ribosomal RNA molecules (18 S rRNA and 28 S rRNA).

RNA polymerase II Eukaryotic RNA polymerase that transcribes pre-messenger RNA, some small nuclear RNAs, and some microRNAs.

RNA polymerase III Eukaryotic RNA polymerase that transcribes transfer RNA, small ribosomal RNAs (5 S rRNA), some small nuclear RNAs, and some microRNAs.

RNA polymerase IV Transcribes small interfering RNAs in plants.

RNA polymerase V Transcribes RNA that has a role in heterochromatin formation in plants.

RNA replication Process in some viruses by which RNA is synthesized from an RNA template.

RNA silencing Mechanism by which double-stranded RNA is cleaved and processed to yield small single-stranded interfering RNAs (siRNAs), which bind to complementary sequences in mRNA and bring about the cleavage and degradation of mRNA; also known as RNA interference and posttranscriptional RNA gene silencing. Some siRNAs also bind to complementary sequences in DNA and guide enzymes to methylate the DNA.

RNA splicing Process by which introns are removed and exons are joined together.

Robertsonian translocation Translocation in which the long arms of two acrocentric chromosomes become joined to a common centromere, resulting in a chromosome with two long arms and usually another chromosome with two short arms.

rolling-circle replication Replication of circular DNA that is initiated by a break in one of the nucleotide strands, producing a double-stranded circular DNA molecule and a single-stranded linear DNA molecule, the latter of which may circularize and serve as a template for the synthesis of a complementary strand.

rooted tree Phylogenetic tree in which one internal node represents the common ancestor of all other organisms (nodes) on the tree. In a rooted tree, all the organisms depicted have a common ancestor.

R plasmid (R factor) Plasmid having genes that confer antibiotic resistance to any cell that contains the plasmid.

sample Subset used to describe a population.

sampling error Deviations from expected ratios due to chance occurrences when the number of events is small.

secondary immune response Clone of cells generated when a memory cell encounters an antigen; provides long-lasting immunity.

secondary oocyte One of the products of meiosis I in female animals; receives most of the cytoplasm.

secondary spermatocyte Product of meiosis I in male animals.

secondary structure of a protein Regular folding arrangement of amino acids in a protein. Common secondary structures found in proteins include the alpha helix and the beta pleated sheet.

second polar body One of the products of meiosis II in oogenesis; contains a set of chromosomes but little of the cytoplasm.

segmental duplications Regions larger than 1000 bp that are almost identical in sequence in eukaryotic genomes.

segmentation genes Set of about 25 genes in fruit flies that control the differentiation of the embryo into individual segments, affecting the number and organization of the segments. Mutations in these genes usually disrupt whole sets of segments.

segment-polarity genes Set of segmentation genes in fruit flies that affect the organization of segments. Mutations in these genes cause part of each segment to be deleted and replaced by a mirror image of part or all of an adjacent segment.

selection coefficient Measure of the relative intensity of selection against a genotype; equals 1 minus fitness.

selection differential Difference in phenotype between the selected individuals and the average of the entire population.

semiconservative replication Replication in which the two nucleotide strands of DNA separate, each serving as a template for the synthesis of a new strand. All DNA replication is semiconservative.

sense codon Codon that specifies an amino acid in a protein.

separase Molecule that cleaves cohesin molecules, which hold the sister chromatids together.

sequential hermaphroditism Phenomenon in which the sex of an individual organism changes in the course of its lifetime; the organism is male at one age or developmental stage and female at a different age or stage.

70S initiation complex Final complex formed in the initiation of translation in bacterial cells; consists of the small and large subunits of the ribosome, mRNA, and initiator tRNA charged with fMet.

sex Male or female.

sex chromosomes Chromosomes that differ morphologically or in number in males and females.

sex determination Specification of sex (male or female). Sex-determining mechanisms include chromosomal, genic, and environmental sex-determining systems.

sex-determining region Y (*SRY*) gene On the Y chromosome, a gene that triggers male development.

sex-influenced characteristic Characteristic encoded by autosomal genes that are more readily expressed in one sex. For example, an autosomal dominant gene may have higher penetrance in males than in females or an autosomal gene may be dominant in males but recessive in females.

sex-limited characteristic Characteristic encoded by autosomal genes and expressed in only one sex. Both males and females carry genes for sex-limited characteristics, but the characteristics appear in only one of the sexes.

sex-linked characteristic Characteristic determined by a gene or genes on sex chromosomes.

shelterin A multiprotein complex that binds to mammalian telomeres and protects the ends of the DNA from being inadvertently repaired as a double-stranded break in the DNA.

Shine–Dalgarno sequence Consensus sequence found in the bacterial 5′ untranslated region of mRNA; contains the ribosome-binding site.

short interspersed element (SINE) Short DNA sequence repeated many times and interspersed throughout the genome.

short tandem repeat (STR) Very short DNA sequence repeated in tandem and found widely in the human genome.

shuttle vector Cloning vector that allows DNA to be transferred to more than one type of host cell.

sigma factor Subunit of bacterial RNA polymerase that allows the RNA polymerase to recognize a promoter and initiate transcription.

signal sequence From 15 to 30 amino acids that are found at the amino end of some eukaryotic proteins and direct the protein to specific locations in the cell; usually cleaved from the protein.

signal-transduction pathway System in which an external signal (initiated by a hormone or growth factor) triggers a cascade of intracellular reactions that ultimately produce a specific response.

silencer Sequence that has many of the properties possessed by an enhancer but represses transcription.

silent mutation Change in the nucleotide sequence of DNA that does not alter the amino acid sequence of a protein.

single-nucleotide polymorphism (SNP) Single-base-pair differences in DNA sequence between individual members of a species.

single-strand-binding (SSB) protein Binds to single-stranded DNA in replication and prevents it from annealing with a complementary strand and forming secondary structures.

sister chromatids Two copies of a chromosome that are held together at the centromere. Each chromatid consists of a single DNA molecule.

site-directed mutagenesis Produces specific nucleotide changes at selected sites in a DNA molecule.

small cytoplasmic RNA (scRNA) Small RNA molecule found in the cytoplasm of eukaryotic cells.

small interfering RNA (siRNA) Single-stranded RNA molecule (usually from 21 to 25 nucleotides in length) produced by the cleavage and processing of double-stranded RNA; binds to complementary sequences in mRNA and brings about the cleavage and degradation of the mRNA. Some siRNAs bind to complementary sequences in DNA and bring about their methylation.

small nuclear ribonucleoprotein (snRNP) Structure found in the nuclei of eukaryotic cells that consists of small nuclear RNA (snRNA) and protein; functions in the processing of pre-mRNA.

small nuclear RNA (snRNA) Small RNA molecule found in the nuclei of eukaryotic cells; functions in the processing of pre-mRNA.

small nucleolar RNA (snoRNA) Small RNA molecule found in the nuclei of eukaryotic cells; functions in the processing of rRNA and in the assembly of ribosomes.

small ribosomal subunit The smaller of the two subunits of a functional ribosome.

somatic-cell hybridization Fusion of different cell types.

somatic hypermutation High rate of somatic mutation such as that in genes encoding antibodies.

somatic mutation Mutation in a cell that does not give rise to a gamete.

somatic recombination Recombination in somatic cells, such as maturing lymphocytes, among segments of genes that encode antibodies and T-cell receptors.

SOS system System of proteins and enzymes that allow a cell to replicate its DNA in the presence of a distortion in DNA structure; makes numerous mistakes in replication and increases the rate of mutation.

Southern blotting Process by which DNA is transferred from a gel to a solid support such as a nitrocellulose or nylon filter.

specialized transduction Transduction in which genes near special sites on the bacterial chromosome are transferred from one bacterium to another; requires lysogenic bacteriophages.

speciation Process by which new species arise. *See also* **biological species concept, allopatric speciation,** and **sympatric speciation.**

species Term applied to different kinds or types of living organisms.

spermatid Immediate product of meiosis II in spermatogenesis; matures to sperm.

spermatogenesis Sperm production in animals.

spermatogonium Diploid cell in the testis; capable of undergoing meiosis to produce a sperm.

S (synthesis) phase Stage of interphase in the cell cycle. In S phase, DNA replicates.

spindle microtubule Microtubule that moves chromosomes in mitosis and meiosis.

spindle pole Point from which spindle microtubules radiate.

spliceosome Large complex consisting of several RNAs and many proteins that splices protein-encoding pre-mRNA; contains five small ribonucleoprotein particles (U1, U2, U4, U5, and U6).

spontaneous mutation Arises from natural changes in DNA structure or from errors in replication.

sporophyte Diploid phase of the life cycle in plants.

SR proteins Group of serine- and arginine-rich proteins that regulate alternative splicing of pre-mRNA.

standard deviation Statistic that describes the variability of a group of measurements; the square root of the variance.

stop (termination or nonsense) codon Codon in mRNA that signals the end of translation. The three common stop codons are UAA, UAG, and UGA.

strand slippage Slipping of the template and newly synthesized strands in replication in which one of the strands loops out from the other and nucleotides are inserted or deleted on the newly synthesized strand.

structural gene DNA sequence that encodes a protein that functions in metabolism or biosynthesis or that has a structural role in the cell.

structural genomics Area of genomics that studies the organization and sequence of information contained within genomes; sometimes used by protein chemists to refer to the determination of the three-dimensional structure of proteins.

submetacentric chromsome Chromosome in which the centromere is displaced toward one end, producing a short arm and a long arm.

supercoiling Coiled tertiary structure that forms when strain is placed on a DNA helix by overwinding or underwinding of the helix. An overwound DNA exhibits positive supercoiling; an underwound DNA exhibits negative supercoiling.

suppressor mutation Mutation that hides or suppresses the effect of another mutation at a site that is distinct from the site of the original mutation.

sympatric speciation Speciation arising in the absence of any geographic barrier to gene flow; reproductive isolating mechanisms evolve with a single interbreeding population.

synapsis Close pairing of homologous chromosomes.

synaptonemal complex Three-part structure that develops between synapsed homologous chromosomes.

synonymous codons Different codons that specify the same amino acid.

tag single-nucleotide polymorphism (tag-SNP) Used to identify a haplotype.

tandem duplication Chromosome rearrangement in which a duplicated chromosome segment is adjacent to the original segment.

tandem repeat sequences DNA sequences repeated one after another; tend to be clustered at specific locations on a chromosome.

***Taq* polymerase** DNA polymerase commonly used in PCR reactions. Isolated from the bacterium *Thermus aquaticus*, the enzyme is stable at high temperatures, and so it is not denatured during the strand-separation step of the cycle.

TATA-binding protein (TBP) Polypeptide chain found in several different transcription factors that recognizes and binds to sequences in eukaryotic promoters.

TATA box Consensus sequence (TATAAAA) commonly found in eukaryotic RNA polymerase II promoters; usually located from 25 to 30 bp upstream of the transcription start site. The TATA box determines the start point for transcription.

T cell Particular type of lymphocyte that produces cellular immunity; originates in the bone marrow and matures in the thymus.

T-cell receptor Found on the surface of a T cell, a receptor that simultaneously binds a foreign and a self-antigen on the surface of a cell.

telocentric chromosome Chromosome in which the centromere is at or very near one end.

telomerase Enzyme that is made up of both protein and RNA and replicates the ends (telomeres) of eukaryotic chromosomes. The RNA part of the enzyme has a template that is complementary to repeated sequences in the telomere and pairs with them, providing a template for the synthesis of additional copies of the repeats.

telomere Stable end of a chromosome.

telomere-associated sequence Sequence found at the ends of a chromosome next to the telomeric sequence; consists of long, complex repeated sequences.

telomeric sequence Sequence found at the ends of a chromosome; consists of many copies of short, simple sequences repeated one after the other.

telophase Stage of mitosis. In telophase, the chromosomes arrive at the spindle poles, the nuclear membrane re-forms, and the chromosomes relax and lengthen.

telophase I Stage of meiosis I. In telophase I, chromosomes arrive at the spindle poles.

telophase II Stage of meiosis II. In telophase II, chromosomes arrive at the spindle poles.

temperate phage Bacteriophage that utilizes the lysogenic cycle, in which the phage DNA integrates into the bacterial chromosome and remains in an inactive state.

temperature-sensitive allele Expressed only at certain temperatures.

template strand The strand of DNA that is used as a template during transcription. The RNA synthesized during transcription is complementary and antiparallel to the template strand.

temporal isolation Reproductive isolation in which the reproduction of different groups takes place at different times of the year, and so there is no gene flow between groups; exemplified by species of plants that flower at different times of the year and thus do not exchange genes.

terminal inverted repeats Sequences found at both ends of a transposable element that are inverted complements of one another.

termination codon Codon in mRNA that signals the end of translation; also called nonsense codon or stop codon. The three common termination codons are UAA, UAG, and UGA.

terminator Sequence of DNA nucleotides that causes the termination of transcription.

tertiary structure of a protein Higher-order folding of amino acids in a protein to form the overall three-dimensional shape of the molecule.

testcross A cross between an individual with an unknown genotype and an individual with the homozygous recessive genotype.

tetrad The four products of meiosis; all four chromatids of a homologous pair of chromosomes.

tetrad analysis Genetic analysis of a tetrad, the products of a single meiosis.

tetraploidy Possession of four haploid sets of chromosomes ($4n$).

tetrasomy Presence of two extra copies of a chromosome ($2n + 2$).

TFIIB recognition element (BRE) Consensus sequence [(G or C)(G or C)(G or C)CGCC] found in some RNA polymerase II core promoters; usually located from 32 to 38 bp upstream of the transcription start site.

theory of clonal selection Explains the generation of primary and secondary immune responses. The binding of a B cell to an antigen stimulates the cell to divide, giving rise to a clone of genetically identical cells, all of which are specific for the antigen.

theta replication Replication of circular DNA that is initiated by the unwinding of the two nucleotide strands, producing a replication bubble. Unwinding continues at one or both ends of the bubble, making it progressively larger. DNA replication on both of the template strands is simultaneous with unwinding until the two replication forks meet.

30S initiation complex Initial complex formed in the initiation of translation in bacterial cells; consists of the small subunit of the ribosome, mRNA, initiator tRNA charged with fMet, GTP, and initiation factors 1, 2, and 3.

three-point testcross Cross between an individual heterozygous at three loci and an individual homozygous for recessive alleles at those loci.

3′ end End of a polynucleotide chain where an OH group is attached to the 3′-carbon atom of the nucleotide.

3′ splice site The 3′ end of an intron where cleavage takes place in RNA splicing.

3′ untranslated (3′ UTR) region Sequence of nucleotides at the 3′ end of mRNA; does not encode the amino acids of a protein but affects both the stability of the mRNA and its translation.

threshold characteristic Discontinuous characteristic whose expression depends on an underlying susceptibility that varies continuously.

thymine (T) Pyrimidine base in DNA but not in RNA.

Ti plasmid Large plasmid isolated from the bacterium *Agrobacterium tumefaciens* and used to transfer genes to plant cells.

topoisomerase Enzyme that adds or removes rotations in a DNA helix by temporarily breaking nucleotide strands; controls the degree of DNA supercoiling.

totipotency The potential of a cell to develop into any other cell type.

trans configuration Arrangement in which each chromosome contains one wild-type (dominant) gene and one mutant (recessive) gene; also called repulsion.

transcription Process by which RNA is synthesized from a DNA template.

transcriptional activator protein Protein in eukaryotic cells that binds to consensus sequences in regulatory promoters or enhancers and affects transcription initiation by stimulating the assembly of the basal transcription apparatus.

transcription bubble Region of a DNA molecule that has unwound to expose a single-stranded template, which is being transcribed into RNA.

transcription factor Protein that binds to DNA sequences in eukaryotic cells and affects transcription.

transcription start site The first DNA nucleotide that is transcribed into an RNA molecule.

transcription unit Sequence of nucleotides in DNA that encodes a single RNA molecule, along with the sequences necessary for its transcription; normally contains a promoter, an RNA-coding sequence, and a terminator.

transcriptome Set of all RNA molecules transcribed from a genome.

transducing phage Contains a piece of the bacterial chromosome inside the phage coat. *See also* **generalized transduction.**

transductant Bacterial cell that has received genes from another bacterium through transduction.

transduction Type of gene exchange that takes place when a virus carries genes from one bacterium to another. After it is inside the cell, the newly introduced DNA may undergo recombination with the bacterial chromosome.

transesterification Chemical reaction in some RNA-splicing reactions.

transfer–messenger RNA (tmRNA) An RNA molecule that has properties of both mRNA and tRNA; functions in rescuing ribosomes that are stalled at the end of mRNA.

transfer RNA (tRNA) RNA molecule that carries an amino acid to the ribosome and transfers it to a growing polypeptide chain in translation.

transfer RNA intron Belongs to a class of introns in tRNA genes. The splicing of these genes relies on enzymes.

transformant Cell that has received genetic material through transformation.

transformation Mechanism by which DNA found in the medium is taken up by the cell. After transformation, recombination may take place between the introduced genes and the cellular chromosome.

transforming principle Substance responsible for transformation. DNA is the transforming principle.

transgene Foreign gene or other DNA fragment carried in germ-line DNA.

transgenic mouse Mouse whose genome contains a foreign gene or genes added by employing recombinant DNA methods.

transition Base substitution in which a purine is replaced by a different purine or a pyrimidine is replaced by a different pyrimidine.

translation Process by which a protein is assembled from information contained in messenger RNA.

translocation Movement of a chromosome segment to a nonhomologous chromosome or to a region within the same chromosome. Also, movement of a ribosome along mRNA in the course of translation.

translocation carrier Individual organism heterozygous for a chromosome translocation.

transmission genetics Field of genetics that encompasses the basic principles of genetics and how traits are inherited.

transposable element DNA sequence capable of moving from one site to another within the genome through a mechanism that differs from that of homologous recombination.

transposase Enzyme encoded by many types of transposable elements that is required for their transposition. The enzyme makes single-strand breaks at each end of the transposable element and on either side of the target sequence where the element inserts.

transposition Movement of a transposable genetic element from one site to another. Replicative transposition increases the number of copies of the transposable element; nonreplicative transposition does not increase the number of copies.

trans-splicing The process of splicing together exons from two or more pre-mRNAs.

transversion Base substitution in which a purine is replaced by a pyrimidine or a pyrimidine is replaced by a purine.

trihybrid cross A cross between two individuals that differ in three characteristics ($AA\ BB\ CC \times aa\ bb\ cc$); also refers to a cross between two individuals that are both heterozygous at three loci ($Aa\ Bb\ Cc \times Aa\ Bb\ Cc$).

triplet code Refers to the fact that three nucleotides encode each amino acid in a protein.

triploidy Possession of three haploid sets of chromosomes ($3n$).

triplo-X syndrome Human condition in which cells contain three X chromosomes. A person with triplo-X syndrome has a female phenotype without distinctive features other than a tendency to be tall and thin; a few such women are sterile, but many menstruate regularly and are fertile.

trisomy Presence of an additional copy of a chromosome ($2n + 1$).

trisomy 8 Presence of three copies of chromosome 8; in humans, results in mental retardation, contracted fingers and toes, low-set malformed ears, and a prominent forehead.

trisomy 13 Presence of three copies of chromosome 13; in humans, results in Patau syndrome.

trisomy 18 Presence of three copies of chromosome 18; in humans, results in Edward syndrome.

trisomy 21 Presence of three copies of chromosome 21; in humans, results in Down syndrome.

tRNA charging Chemical reaction in which an aminoacyl-tRNA synthetase attaches an amino acid to its corresponding tRNA.

tRNA-modifying enzyme Creates a modified base in RNA by catalyzing a chemical change in the standard base.

tubulin Protein found in microtubules.

tumor-suppressor gene Gene that normally inhibits cell division. Recessive mutations in such genes often contribute to cancer.

Turner syndrome Human condition in which cells contain a single X chromosome and no Y chromosome (XO). Persons with Turner syndrome are female in appearance but do not undergo puberty and have poorly developed female secondary sex characteristics; most are sterile but have normal intelligence.

two-dimensional polyacrylamide gel electrophorsis (2D-PAGE) Method for separating proteins into spots in which the proteins are separated in one dimension by charge, separated in a second dimension by mass, and then stained. Each spot is proportional to the amount of protein present.

two-point testcross Cross between an individual heterozygous at two loci and an individual homozygous for recessive alleles at those loci.

ultrasonography Procedure for visualizing the fetus. High-frequency sound is beamed into the uterus. Sound waves that encounter dense tissue bounce back and are transformed into a picture of the fetus.

unbalanced gamete Gamete that has a variable number of chromosomes; some chromosomes may be missing and others may be present in more than one copy.

underdominance Selection in which the heterozygote has lower fitness than that of either homozygote.

unequal crossing over Misalignment of the two DNA molecules during crossing over, resulting in one DNA molecule with an insertion and the other with a deletion.

uniparental disomy Inheritance of both chromosomes of a homologous pair from a single parent.

unique-sequence DNA Sequence present only once or a few times in a genome.

universal genetic code Refers to the fact that particular codons specify the same amino acids in almost all organisms.

up mutation Mutation that increases the rate of transcription.

upstream element Consensus sequence found in some bacterial promoters that contains a number of A–T pairs and is found about 40 to 60 bp upstream of the transcription start site.

uracil (U) Pyrimidine base in RNA but not normally in DNA.

variable number of tandem repeats (VNTRs) Short sequences repeated in tandem that vary greatly in number among individuals; also called microsatellites. Because they are quite variable, VNTRs are commonly used in DNA fingerprinting.

variance Statistic that describes the variability of a group of measurements.

virulent phage Bacteriophage that reproduces only through the lytic cycle and kills its host cell.

virus Noncellular replicating agent consisting of nucleic acid surrounded by a protein coat; can replicate only within its host cell.

Western blotting Process by which protein is transferred from a gel to a solid support such as a nitrocellulose or nylon filter.

whole-genome shotgun sequencing Method of sequencing a genome in which sequenced fragments are assembled into the correct sequence in contigs by using only the overlaps in sequence.

wild type The trait or allele that is most commonly found in natural (wild) populations.

wobble Base pairing between codon and anticodon in which there is nonstandard pairing, usually at the third (3′) position of the codon; allows more than one codon to pair with the same anticodon.

X : A : ratio Ratio of the number of X chromosomes to the number of haploid autosomal sets of chromosomes; determines sex in fruit flies.

X-linked characteristic Characteristic determined by a gene or genes on the X chromosome.

X-ray diffraction Method for analyzing the three-dimensional shape and structure of chemical substances. Crystals of a substance are bombarded with X-rays, which hit the crystals, bounce off, and produce a diffraction pattern on a detector. The pattern of the spots produced on the detector provides information about the molecular structure.

yeast artificial chromosome (YAC) Cloning vector consisting of a DNA molecule with a yeast origin of replication, a pair of telomeres, and a centromere. YACs can carry very large pieces of DNA (as large as several hundred thousand base pairs) and replicate and segregate as yeast chromosomes do.

Y-linked characteristic Characteristic determined by a gene or genes on the Y chromosome.

Z-DNA Secondary structure of DNA characterized by 12 bases per turn, a left-handed helix, and a sugar–phosphate backbone that zigzags back and forth.

zygotene Second substage of prophase I in meiosis. In zygotene, chromosomes enter into synapsis.

Answers to Selected Questions and Problems

Chapter 1

1. In the Hopi culture, albinos were considered special and given special status. Because extensive exposure to sunlight could be damaging or deadly, Hopi male albinos did no agricultural work. Albinism was a considered a positive trait rather than a negative physical condition, allowing albinos to have more children and thus increasing the frequency of the allele. Finally, the small population size of the Hopi tribe may have helped increase the allele frequency of the albino gene owing to chance.

3. Genetics plays important roles in the diagnosis and treatment of hereditary diseases, in breeding plants and animals for improved production and disease resistance, and in producing pharmaceuticals and novel crops through genetic engineering.

5. Transmission genetics: The inheritance of genes from one generation to the next, gene mapping, characterization of the phenotypes produced by mutations.

   Molecular genetics: The structure, organization, and function of genes at the molecular level.

   Population genetics: Genes and changes in genes in populations.

8. Pangenesis proposes that information for creating each part of an offspring's body originates in each part of the parent's body and is passed through the reproductive organs to the embryo at conception. Pangenesis suggests that changes in parts of the parent's body may be passed to the offspring's body. The germ-plasm theory, in contrast, states that the reproductive cells possess all of the information required to make the complete body; the rest of the body contributes no information to the next generation.

10. Preformationism is the idea that an offspring results from a miniature adult form that is already preformed in the sperm or the egg. All traits are thus inherited from only one parent, either the father or the mother, depending on whether the homunculus (the preformed miniature adult) resides in the sperm or the egg.

13. Gregor Mendel

16. Genes are composed of DNA nucleotide sequences and are located at specific positions in chromosomes.

18. (a) Transmission genetics; (b) population genetics; (c) population genetics; (d) molecular genetics; (e) molecular genetics; (f) transmission genetics.

21. Genetics is old in the sense that humans have been aware of hereditary principles for thousands of years and have applied them since the beginning of agriculture and the domestication of plants and animals. It is very young in the sense that the fundamental principles were not uncovered until Mendel's time and the structure of DNA and the principles of recombinant DNA were discovered within the past 60 years.

23. (a) Pangenesis postulates that specific particles (called gemmules) carry genetic information from all parts of the body to the reproductive organs, and then the genetic information is conveyed to the embryo where each unit directs the formation of its own specific part of the body. According to the germ-plasm theory, gamete-producing cells found within the reproductive organs contain a complete set of genetic information that is passed to the gametes. The two concepts are similar in that both propose that genetic information is contained in discrete units that are passed on to offspring. They differ in where that genetic information resides. In pangenesis, it resides in different parts of the body and must travel to the reproductive organs. In the germ-plasm theory, all the genetic information is already in the reproductive cells.
    (b) Preformationism holds that the sperm or egg contains a miniature preformed adult called a homunculus. In development, the homunculus grows to produce an offspring. Only one parent contributes genetic traits to the offspring. Blending inheritance requires contributions of genetic material from both parents. The genetic contributions from the parents blend to produce the genetic material of the offspring. Once blended, the genetic material cannot be separated for future generations.
    (c) The inheritance of acquired characteristics postulates that traits acquired in a person's lifetime alter the genetic material and can be transmitted to offspring. Our modern theory of heredity indicates that offspring inherit genes located on chromosomes passed from their parents. These chromosomes segregate in meiosis in the parent's germ cells and are passed into the gametes.

24. (a) Both cell types have lipid bilayer membranes, DNA genomes, and machinery for DNA replication, transcription, translation, energy metabolism, response to stimuli, growth, and reproduction. Eukaryotic cells have a nucleus containing chromosomal DNA and possess internal membrane-bounded organelles.
    (b) A gene is a basic unit of hereditary information, usually encoding a functional RNA or polypeptide. Alleles are variant forms of a gene, arising through mutation.
    (c) The genotype is the set of genes or alleles inherited by an organism from its parent(s). The expression of the genes of a particular genotype, through interaction with environmental factors, produces the phenotype, the observable trait.
    (d) Both are nucleic acid polymers. RNA contains a ribose sugar, whereas DNA contains a deoxyribose sugar. RNA also contains uracil as one of the four bases, whereas DNA contains thymine. The other three bases are common to both DNA and RNA. Finally, DNA is usually double stranded, consisting of two complementary strands, whereas RNA is single stranded.
    (e) Chromosomes are structures consisting of DNA and associated proteins. The DNA contains the genetic information.

27. All genomes must have the ability to store complex information and to vary. The blueprint for an entire organism must be contained within the genome of each reproductive cell. The information has to be in the form of a code that can be used as a set of instructions for assembling the components of the cells. The genetic material of any organism must be stable, be replicated precisely, and be transmitted faithfully to the progeny but must be capable of mutating.

29. Legally, the woman is not required to inform her children or other relatives, but people may have different opinions about her moral and parental responsibilities. On the one hand, she has the legal right to keep private the results of any medical information,

including the results of genetic testing. On the other hand, her children may be at increased risk of developing these disorders and might benefit from that information. For example, the risk of colon cancer can be reduced by regular exams, in which any tumors can be detected and removed before they become cancerous. Some people might argue that her parental responsibilities include providing her children with information about possible medical problems. Another issue to consider is the possibility that her children or other relatives might not want to know their genetic risk, particularly for a disorder such as Alzheimer disease, for which there is no cure.

30. **(a)** Having the genetic test removes doubt about the potential for the disorder: you are either susceptible or not. Knowing about the potential of a genetic disorder enables you to make life-style changes that might lessen the effect of the disease or lessen the risk. The types and nature of future medical tests could be positively affected by the genetic testing, thus allowing for early warning and screening for the disease. The knowledge could also enable you to make informed decisions regarding offspring and the potential of passing the trait to your offspring. Additionally, by knowing what to expect, you could plan your life accordingly.

Reasons for not having the test typically concern the potential for testing positive for the susceptibility to the genetic disease. If the susceptibility is detected, there is potential for discrimination. For example, your employer (or possibly future employer) might consider you a long-term liability, thus affecting employment options. Insurance companies may not want to insure you for the condition or its symptoms, and social stigmatization regarding the disease could be a factor. Knowledge of the potential future condition could lead to psychological difficulties in coping with the anxiety of waiting for the disease to manifest.

(b) There is no "correct" answer, but some of the reasons for wanting to be tested are: the test would remove doubt about the susceptibility, particularly if family members have had the genetic disease; either a positive or a negative result would allow for informed planning of life style, medical testing, and family choices in the future.

Chapter 2

1.
Prokaryotic cell	Eukaryotic cell
No nucleus	Nucleus present
No paired chromosomes	Paired chromosomes common
Typically a single circular chromosome containing a single origin of replication	Typically multiple linear chromosomes containing centromeres, telomeres, and multiple origins of replication
Single chromosome is replicated, with each copy moving to opposite sides of the cell	Chromosomes require mitosis or meiosis to segregate to the proper location
No histone proteins complexed to DNA	Histone proteins are complexed to DNA

3. The events are (i) a cell's genetic information must be copied; (ii) the copies of the genetic information must be separated from each other; and (iii) the cell must divide.

6.

8. Prophase: The chromosomes condense and become visible, the centrosomes move apart, and microtubule fibers form from the centrosomes.

Prometaphase: The nucleoli disappear and the nuclear envelope begins to disintegrate, allowing the cytoplasm and nucleoplasm to join. The sister chromatids of each chromosome are attached to microtubules from the opposite centrosomes.

Metaphase: The spindle microtubules are clearly visible, and the chromosomes arrange themselves on the metaphase plate of the cell.

Anaphase: The sister chromatids separate at the centromeres after the breakdown of cohesin protein, and the newly formed daughter chromosomes move to the opposite poles of the cell.

Telophase: The nuclear envelope re-forms around each set of daughter chromosomes. Nucleoli reappear. Spindle microtubules disintegrate.

10. In the mitotic cell cycle, the genetic material is precisely copied, and so the two resulting cells contain the same genetic information. In other words, the daughter cells have genomes identical with each other and with that of the mother cell.

14. Meiosis includes two cell divisions, thus producing four new cells (in many species). The chromosome number of a haploid cell produced by meiosis I is half the chromosome number of the original diploid cell. Finally, the cells produced by meiosis are genetically different from the original cell and genetically different from each other.

16.
Mitosis	Meiosis
A single cell division produces two genetically identical progeny cells.	Two cell divisions usually result in four progeny cells that are not genetically identical.
Chromosome number in progeny cells and the original cell remains the same.	Daughter cells are haploid and have half the chromosomal complement of the original diploid cell as a result of the separation of homologous pairs in anaphase I.
Daughter cells and the original cell are genetically identical. No crossing over or separation of homologous chromosomes takes place.	Crossing over in prophase I and separation of homologous pairs in anaphase I produce daughter cells that are genetically different from each other and from the original cell.
Homologous chromosomes do not synapse.	Homologous chromosomes synapse in prophase I.
In metaphase, individual chromosomes line up on the metaphase plate.	In metaphase I, homologous pairs of chromosomes line up on the metaphase plate. Individual chromosomes line up in metaphase II.
In anaphase, sister chromatids separate.	In anaphase I, homologous chromosomes separate. Sister chromatids separate in anaphase II.

The most important difference is that mitosis produces cells genetically identical with each other and with the original cell, resulting in the orderly passage of information from one cell to its progeny. In contrast, by producing progeny that do not contain pairs of homologous chromosomes, meiosis results in the reduction of chromosome number from that of the original cell. Meiosis also allows for genetic variation through crossing over and the random assortment of homologous chromosomes.

20. **(a)** The two chromatids of a chromosome
(b) The two chromosomes of a homologous pair
(c) Cohesin
(d) The enzyme separase
(e) The hands of the two blind men
(f) If one man failed to grasp his sock, use of the knife to cut the string holding them together would be difficult. The two socks of a

pair would not be separated, and both would end up in one man's bag. Similarly, if each chromatid is not attached to spindle fibers and pulled in opposite directions, the two chromatids will not separate and both will migrate to the same cell. This cell will have two copies of one chromosome.

23. (a) 12 chromosomes and 24 DNA molecules; (b) 12 chromosomes and 24 DNA molecules; (c) 12 chromosomes and 24 DNA molecules; (d) 12 chromosomes and 24 DNA molecules; (e) 12 chromosomes and 12 DNA molecules; (f) 6 chromosomes and 12 DNA molecules; (g) 12 chromosomes and 12 DNA molecules; (h) 6 chromosomes and 6 DNA molecules.

25. The diploid number of chromosomes is six. The left-hand cell is in anaphase I of meiosis; the middle cell is in anaphase of mitosis; the right-hand cell is in anaphase II of meiosis.

27. (a) Cohesin is needed to hold the sister chromatids together until anaphase of mitosis. If cohesin fails to form early in mitosis, the sister chromatids could separate before anaphase. As a result, chromosomes would not segregate properly to daughter cells.
(b) Shugoshin protects cohesin proteins from degradation at the centromere during meiosis I. Cohesin at the arms of the homologous chromosomes is not protected by shugoshin and is broken in anaphase I, enabling the two homologs to separate. If shugoshin is absent during meiosis, then the cohesin at the centromere may be broken, allowing for the separation of sister chromatids along with the homologs in anaphase I and leading to the improper segregation of chromosomes to daughter cells.
(c) If shugoshin does not break down, the cohesin proteins at the centromere remain protected from degradation. The intact cohesin prevents the sister chromatids from separating in anaphase II of meiosis; as a result, sister chromatids would not separate, leading to daughter cells that have too many chromosomes or too few.
(d) If separase is defective, homologous chromosomes and sister chromatids would not separate in meiosis and mitosis, resulting in some cells that have too few chromosomes and some cells that have too many chromosomes.

30. The progeny of an organism with the larger number of homologous pairs of chromosomes should be expected to exhibit more variation. The number of different combinations of chromosomes that are possible in the gametes is $2n$, where n is equal to the number of homologous pairs of chromosomes. For the fruit fly, which has four pairs of chromosomes, the number of possible combinations is $2^4 = 16$. For the house fly, which has six pairs of chromosomes, the number of possible combinations is $2^6 = 64$.

31. (a) Metaphase I

(b) Gametes

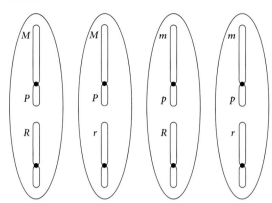

34. (a) No. The first polar body and the secondary oocyte are the result of meiosis I. In meiosis I, homologous chromosomes segregate; thus, both the first polar body and the secondary oocyte will contain only one homolog of each original chromosome pair, and the alleles of some genes in these homologs will differ. Additionally, crossing over in prophase I will have generated new and different arrangements of genetic material for each homolog of the pair.
(b) No. The second polar body and the ovum will contain the same members of the homologous pairs of chromosomes that were separated in meiosis I and produced by the separation of sister chromatids in anaphase II. However, the sister chromatids of each pair are no longer identical, because they underwent crossing over in prophase I and thus contain genetic information that is not identical.

37. Most male animals produce sperm by meiosis. Because meiosis takes place only in diploid cells, haploid male bees do not undergo meiosis. Male bees can produce sperm but only through mitosis. Haploid cells that divide mitotically produce haploid cells.

Chapter 3

1. Mendel was successful for several reasons. He chose to work with a plant, *Pisum sativum*, that was easy to cultivate, grew relatively rapidly compared with other plants, and produced many offspring whose phenotypes were easy to determine, which allowed Mendel to detect mathematical ratios of progeny phenotypes. The seven characteristics that he chose to study exhibited only a few distinct phenotypes and did not show a range of variation. Finally, by looking at each trait separately and counting the numbers of the different phenotypes, Mendel adopted a reductionist experimental approach and applied the scientific method. From his observations, he proposed hypotheses that he was then able to test empirically.

3. The principle of segregation states that an organism possesses two alleles for any particular characteristic. These alleles separate in the formation of gametes. In other words, one allele goes into each gamete. The principle of segregation is important because it explains how the genotypic ratios in the haploid gametes are produced.

8. Walter Sutton's chromosome theory of heredity states that genes are located on chromosomes. The independent segregation of pairs of homologous chromosomes in meiosis provides the biological basis for Mendel's two principles of heredity.

9. The principle of independent assortment states that alleles at different loci segregate independently of one another. The principle of independent assortment is an extension of the principle of segregation: the principle of segregation states that the two alleles at a locus separate; according to the principle of independent assortment, when these two alleles separate, their separation is independent of the separation of alleles at other loci.

14. (a) The parents are *RR* (orange fruit) and *rr* (cream fruit). All the F$_1$ are *Rr* (orange). The F$_2$ are 1 *RR* : 2 *Rr* : 1 *rr* and have an orange-to-cream phenotypic ratio of 3 : 1.
    (b) Half of the progeny are homozygous for orange fruit (*RR*) and half are heterozygous for orange fruit (*Rr*).
    (c) Half of the progeny are heterozygous for orange fruit (*Rr*) and half are homozygous for cream fruit (*rr*).

16. (a) Although the white female guinea pig gave birth to the offspring, her eggs were produced by the ovary from the black female guinea pig. The transplanted ovary produced only eggs containing the allele for black coat color. Like most mammals, guinea pig females produce primary oocytes early in development, and thus the transplanted ovary already contained primary oocytes produced by the black female guinea pig.
    (b) The white male guinea pig contributed a *w* allele, and the white female guinea pig contributed the *W* allele from the transplanted ovary. The offspring are thus *Ww*.
    (c) The transplant experiment supports the germ-plasm theory. According to the germ-plasm theory, only the genetic information in the germ-line tissue in the reproductive organs is passed to the offspring. The production of black guinea pig offspring suggests that the allele for black coat color was passed to the offspring from the transplanted ovary in agreement with the germ-plasm theory. According to the pangenesis theory, the genetic information passed to the offspring originates at various parts of the body and travels to the reproductive organs for transfer to the gametes. If pangenesis were correct, then the guinea pig offspring should have been white. The white-coat alleles would have traveled to the transplanted ovary and then into the white female's gametes. The absence of any white offspring indicates that the pangenesis hypothesis is invalid.

17. (a) Female parent is $i^B i^B$; male parent is $I^A i^B$.
    (b) Both parents are $i^B i^B$.
    (c) Male parent is $i^B i^B$; female parent is $I^A I^A$ or, possibly, $I^A i^B$, but a heterozygous female in this mating is unlikely to have produced eight blood-type-A kittens owing to chance alone.
    (d) Both parents are $I^A i^B$.
    (e) Either both parents are $I^A I^A$ or one parent is $I^A I^A$ and the other parent is $I^A i^B$. The blood type of the offspring does not allow a determination of the precise genotype of either parent.
    (f) Female parent is $i^B i^B$; male parent is $I^A i^B$.

20. (a) Sally (*Aa*), Sally's mother (*Aa*), Sally's father (*aa*), and Sally's brother (*aa*); (b) $1/2$; (c) $1/2$.

22. Use *h* for the hairless allele and *H* for the dominant allele for the presence of hair. Because *H* is dominant over *h*, a rat terrier with hair could be either homozygous (*HH*) or heterozygous (*Hh*). To determine which genotype is present in the rat terrier with hair, cross this dog with a hairless rat terrier (*hh*). If the terrier with hair is homozygous (*HH*), then no hairless offspring will be produced by the testcross. However, if the terrier is heterozygous (*Hh*), then $1/2$ of the offspring will be hairless.

24. (a) $1/18$; (b) $1/36$; (c) $11/36$; (d) $1/6$; (e) $1/4$; (f) $3/4$.

25. (a) $1/128$; (b) $1/64$; (c) $7/128$; (d) $35/128$; (e) $35/128$.

27. Parents:

F$_1$ generation:

F$_2$ generation:

28. (a) In the F$_1$ black guinea pigs (*Bb*), only one chromosome possesses the black allele, and so the number of copies present at each stage are: G$_1$, one black allele; G$_2$, two black alleles; metaphase of mitosis, two black alleles; metaphase I of meiosis, two black alleles; after cytokinesis of meiosis, one black allele but only in half of the cells produced by meiosis. (The remaining half will not contain the black allele.)
    (b) In the F$_1$ brown guinea pigs (*bb*), both homologs possess the brown allele, and so the number of copies present at each stage are: G$_1$, two brown alleles; G$_2$, four brown alleles; metaphase of mitosis, four brown alleles; metaphase I of meiosis, four brown alleles; metaphase II, two brown alleles; after cytokinesis of meiosis, one brown allele.

30. (a) $9/16$ black and curled, $3/16$ black and normal, $3/16$ gray and curled, and $1/16$ gray and normal.
    (b) $1/4$ black and curled, $1/4$ black and normal, $1/4$ gray and curled, $1/4$ gray and normal.

31. (a) $1/2$ (*Aa*) × $1/2$ (*Bb*) × $1/2$ (*Cc*) × $1/2$ (*Dd*) × $1/2$ (*Ee*) = $1/32$
    (b) $1/2$ (*Aa*) × $1/2$ (*bb*) × $1/2$ (*Cc*) × $1/2$ (*dd*) × $1/4$ (*ee*) = $1/64$
    (c) $1/4$ (*aa*) × $1/2$ (*bb*) × $1/4$ (*cc*) × $1/2$ (*dd*) × $1/4$ (*ee*) = $1/256$
    (d) No offspring having this genotype. The *Aa Bb Cc dd Ee* parent cannot contribute a *D* allele, the *Aa bb Cc Dd Ee* parent cannot contribute a *B* allele. Therefore, their offspring cannot be homozygous for the *BB* and *DD* gene loci.

34. (a) Gametes from *Aa Bb* individual:

Gametes from *aa bb* individual:

(b) Progeny at G$_1$:

Progeny at G$_2$:

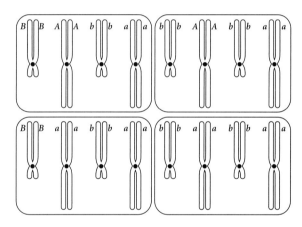

Progeny at prophase of mitosis:

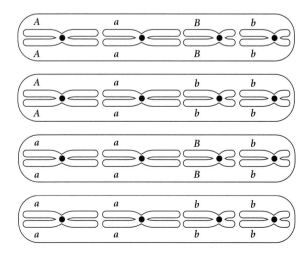

The order of chromosomes on the metaphase plate can vary.

35. (a) The burnsi × burnsi cross produced both burnsi and pipiens offspring, suggesting that the parents were heterozygous with each possessing a burnsi allele and a pipiens allele. The cross also suggests that the burnsi allele is dominant over the pipiens allele. The progeny of the burnsi × pipiens crosses suggest that each of the crosses was between a homozygous recessive frog (pipiens) and a heterozygous dominant frog (burnsi). The results of both crosses are consistent with the burnsi phenotype being recessive to the pipiens phenotype.
(b) Let B represent the burnsi allele and B^+ represent the pipiens allele.
$$\text{burnsi } (BB^+) \times \text{burnsi } (BB^+)$$
$$\text{burnsi } (BB^+) \times \text{pipiens } (B^+B^+)$$
$$\text{burnsi } (BB^+) \times \text{pipiens } (B^+B^+)$$
c. For the burnsi × burnsi ($BB^+ \times BB^+$) cross, we would expect a phenotypic ratio of 3 : 1 in the offspring. A chi-square test to evaluate the fit of the observed numbers of progeny with an expected 3 : 1 ratio gives a chi-square value of 2.706 with 1 degree of freedom. The probability associated with this chi-square value is between 0.1 and 0.05, indicating that the differences between what we expected and what we observed could have been due to chance.

For the first burnsi × pipiens ($BB^+ \times B^+B^+$) cross, we would expect a phenotypic ratio of 1 : 1. A chi-square test comparing observed and expected values yields $\chi^2 = 1.78$, df = 1, $P > 0.05$. For the second burnsi × pipiens ($BB^+ \times B^+B^+$) cross, we would expect a phenotypic ratio of 1 : 1. A chi-square test of the fit of the observed numbers with those expected with a 1 : 1 ratio yields $\chi^2 = 0.46$, df = 1, $P > 0.05$. Thus, all three crosses are consistent with the predication that the burnsi allele is dominant over the pipiens allele.

38. (a) For the cross between a heterozygous F$_1$ plant (*Cc Ff*) and a homozygous recessive plant (*cc ff*), we would expect a phenotypic ratio of 1 : 1 : 1 : 1 for the different phenotypic classes. A chi-square test comparing the fit of the observed data with the expected 1 : 1 : 1 : 1 ratio yields a chi-square value of 35 with df = 3 and $P < 0.005$.
(b) From the chi-square value, it is unlikely that chance produced the differences between the observed and the expected ratios, indicating that the progeny are not in a 1 : 1 : 1 : 1 ratio.
(c) The number of plants with the *cc ff* genotype is fewer than expected. The *cc ff* genotype is possibly sublethal. In other words, California poppies with the homozygous recessive genotypes are possibly less viable than the other possible genotypes.

40. The first geneticist has identified an allele for obesity that he believes to be recessive. Let's call his allele for obesity o_1 and the normal allele O_1. On the basis of the crosses that the geneticist performed, the allele for obesity appears to be recessive.

Cross 1 with possible genotype:
$$\text{Obese } (o_1o_1) \times \text{Normal } (O_1O_1)$$
$$\downarrow$$
F$_1$ All normal (O_1o_1)

Cross 2 with possible genotypes:
F$_1$ $\quad\quad$ Normal (O_1o_1) × Normal (O_1o_1)
$$\downarrow$$
F$_2$ $\quad\quad$ 8 normal (O_1O_1 and O_1o_1)
$\quad\quad\quad\quad$ 2 obese (o_1o_1)

Cross 3 with possible genotypes:
$$\text{Obese } (o_1o_1) \times \text{Obese } (o_1o_1)$$
$$\downarrow$$
F$_1$ $\quad\quad$ All obese (o_1o_1)

Let's call the second geneticist's allele for obesity o_2 and the normal allele O_2. The cross between obese mice from the two laboratories produced only normal mice. The alleles for obesity from both laboratories are recessive. However, they are located at different gene loci. Essentially, the obese mice from the different laboratories have separate obesity genes that are independent of one another.

The likely genotypes of the obese mice are as follows:

$$\text{Obese mouse 1 } (o_1 o_1\ O_2 O_2) \times \text{Obese mouse 2 } (O_1 O_1\ o_2 o_2)$$
$$\downarrow$$
$$F_1 \quad \text{All normal } (O_1 o_1\ O_2 o_2)$$

Chapter 4

1. Females produce larger gametes than those produced by males.

5. The pseudoautosomal region is a region of similarity between the X and the Y chromosomes that is responsible for pairing the X and Y chromosomes in meiotic prophase I. Genes in this region are present in two copies in males and females and are thus inherited in the same way as autosomal genes are inherited, whereas other Y-linked genes are passed only from father to son.

9. Males show the phenotypes of all X-linked traits, regardless of whether the X-linked allele is normally recessive or dominant. Males inherit X-linked traits from their mothers, pass X-linked traits to all of their daughters, and, through their daughters, to their daughters' descendants but not to their sons or their sons' descendants.

11. Y-linked traits appear only in males and are always transmitted from fathers to sons, thus following a strict paternal lineage. Autosomal male-limited traits also appear only in males, but they can be transmitted to sons through their mothers.

15. (a) Female; (b) male; (c) male, sterile; (d) female; (e) male; (f) female; (g) metafemale; (h) male; (i) intersex; (j) female; (k) metamale, sterile; (l) metamale; (m) intersex.

20. Her father's mother and her father's father, but not her mother's mother or her mother's father.

22. (a) Yes; (b) yes; (c) no; (d) no.

23. (a) F_1: $1/2\ X^+Y$ (gray males), $1/2\ X^+X^y$ (gray females); F_2: $1/4\ X^+Y$ (gray males), $1/4\ X^yY$ (yellow males), $1/4\ X^+X^y$ (gray females), $1/4\ X^+X^+$ (gray females).
(b) F_1: $1/2\ X^yY$ (yellow males), $1/2\ X^+X^y$ (gray females); F_2: $1/4\ X^+Y$ (gray males), $1/4\ X^yY$ (yellow males), $1/4\ X^+X^y$ (gray females), $1/4\ X^yX^y$ (yellow females).
(c) F_2: $1/4\ X^+Y$ (gray males), $1/4\ X^yY$ (yellow males), $1/4\ X^+X^+$ (gray females), $1/4\ X^+X^y$ (gray females).
(d) F_3: $1/8$ gray males, $3/8$ yellow males, $5/16$ gray females, and $3/16$ yellow females.

25. Because color blindness is a recessive trait, the color-blind daughter must be homozygous recessive. Because the color blindness is X-linked, John has grounds for suspicion. Normally, their daughter would have inherited John's X chromosome. Because John is not color blind, he could not have transmitted an X chromosome with a color-blind allele to his daughter.

    A remote alternative possibility is that the daughter is XO, having inherited a recessive color-blind allele from her mother and no sex chromosome from her father. In that case, the daughter would have Turner syndrome. A new X-linked color-blind mutation also is possible, albeit even less likely.

    If Cathy had given birth to a color-blind son, then John would have no grounds for suspicion. The son would have inherited John's Y chromosome and the color-blind X chromosome from Cathy.

27. Because Bob must have inherited the Y chromosome from his father and his father has normal color vision, a nondisjunction event in the paternal lineage cannot account for Bob's genotype. Bob's mother must be heterozygous X^+X^c because she has normal color vision, and she must have inherited a color-blind X chromosome from her color-blind father. For Bob to inherit two color-blind X chromosomes from his mother, the egg must have arisen from a nondisjunction in meiosis II. In meiosis I, the homologous X chromosomes separate, and so one cell has the X^+ chromosome and the other has X^c. The failure of sister chromatids to separate in meiosis II would then result in an egg with two copies of X^c.

31. (a) Male parent is X^+Y; female parent is X^+X^m.
(b) Male parent is X^mY; female parent is X^+X^+.
(c) Male parent is X^mY; female parent is X^+X^m.
(d) Male parent is X^+Y; female parent is X^mX^m.
(e) Male parent is X^+Y; female parent is X^+X^+.

32. F_1: $1/2\ Z^bZ^+$ (normal males); $1/2\ Z^bW$ (bald females).
F_2: $1/4\ Z^+Z^b$ (normal males), $1/4\ Z^+W$ (normal females), $1/4\ Z^bZ^b$ (bald males), $1/4\ Z^bW$ (bald females).

36. (a) F_1: all males are have miniature wings and red eyes ($X^mY\ s^+s$), and all females have long wings and red eyes ($X^+X^m\ s^+s$).
F_2: $3/16$ male, normal, red; $1/16$ male, normal, sepia; $3/16$ male, miniature, red; $1/16$ male, miniature, sepia; $3/16$ female, normal, red; $1/16$ female, normal, sepia; $3/16$ female, miniature, red; $1/16$ female, miniature, sepia.
(b) F_1: all females have long wings and red eyes ($X^{m+}X^m\ s^+s$), and all males have long wings and red eyes ($X^{m+}Y\ s^+s$). F_2: $3/16$ males, long wings, red eyes; $1/16$ males, long wings, sepia eyes; $3/16$ males, mini wings, red eyes; $1/16$ males, mini wings, sepia eyes; $6/16$ females, long wings, red eyes; $2/16$ females, long wings, sepia eyes.

38. The trivial explanation for these observations is that this form of color blindness is an autosomal recessive trait. In that case, the father would be a heterozygote, and we would expect equal proportions of color-blind and normal children of either sex.

    If, on the other hand, we assume that this form of color blindness is an X-linked trait, then the mother is X^cX^c and the father must be X^+Y. Normally, all the sons would be color blind, and all the daughters should have normal vision. The most likely way to have a daughter who is color blind would be for her not to have inherited an X^+ from her father. The observation that the color-blind daughter is short in stature and has failed to undergo puberty is consistent with Turner syndrome (XO). The color-blind daughter would then be X^cO.

39. (a) 1; (b) 0; (c) 0; (d) 1; (e) 1; (f) 2; (g) 0; (h) 2; (i) 3.

Chapter 5

1. In incomplete dominance, the phenotype of the heterozygote is intermediate between the phenotypes of the two homozygotes. In codominance, both alleles are expressed and both phenotypes are manifested simultaneously.

2. In incomplete penetrance, the expected phenotype of a particular genotype is not expressed. Environmental factors, as well as the effects of other genes, may alter the phenotypic expression of a particular genotype.

5. A complementation test is used to determine whether two different recessive mutations are at the same locus (are allelic) or at different loci. The two mutations are introduced into the same individual organism by crossing homozygotes for each of the mutants. If the progeny show a mutant phenotype, then the mutations are allelic (at the same locus). If the progeny show a wild-type (dominant) phenotype, then the mutations are at different loci and are said to complement each other because each of the mutant parents can supply a functional copy (or dominant allele) of the gene mutated in the other parent.

6. Cytoplasmically inherited traits are encoded by genes in the cytoplasm. Because the cytoplasm is usually inherited from a single (most often the female) parent, reciprocal crosses do not show the same results. Cytoplasmically inherited traits often show great variability because different egg cells (female gametes) may have differing proportions of cytoplasmic alleles owing to random sorting of mitochondria (or plastids in plants).

11. Continuous characteristics, also called quantitative characteristics, exhibit many phenotypes with a continuous distribution. They result from the interaction of multiple genes (polygenic traits), the influence of environmental factors on the phenotype, or both.

14. (a) The results of the crosses indicate that cremello and chestnut are pure-breeding traits (homozygous). Palomino is a heterozygous trait that produces a 1 : 2 : 1 ratio when palominos are crossed with each other. The simplest hypothesis consistent with these results is incomplete dominance, with palomino as the phenotype of the heterozygotes resulting from chestnuts crossed with cremellos.
(b) Let C^B = chestnut, C^W = cremello. The parents and offspring of these crosses have the following genotypes: chestnut = $C^B C^B$; cremello = $C^W C^W$; palomino = $C^B C^W$.

15. (a) $1/2\ L^M L^M$ (type M), $1/2\ L^M L^N$ (type MN); (b) all $L^N L^N$ (type N); (c) $1/2\ L^M L^N$ (type MN), $1/4\ L^M L^M$ (type M), $1/4\ L^N L^N$ (type N); (d) $1/2\ L^M L^N$ (type MN), $1/2\ L^N L^N$ (type N); (e) all $L^M L^N$ (type MN).

18. (a) The 2 : 1 ratio in the progeny of two spotted hamsters suggests lethality, and the 1 : 1 ratio in the progeny of a spotted hamster and a hamster without spots indicates that spotted is a heterozygous phenotype. If S and s represent alleles at the locus for white spotting, spotted hamsters are Ss and solid-colored hamsters are ss. One-quarter of the zygotes expected from a mating of two spotted hamsters are SS—embryonic lethal—and missing from the progeny, resulting in the 2 : 1 ratio of spotted to solid progeny.
(b) Because spotting is a heterozygous phenotype, obtaining Chinese hamsters that breed true for spotting is impossible.

24. The child's genotype has an allele for blood-type B and an allele for blood-type N that could not have come from the mother and must have come from the father. Therefore, the child's father must have an allele for type B and an allele for type N. George, Claude, and Henry are eliminated as possible fathers because they lack an allele for either type B or type N.

25. (a) All walnut ($Rr\ Pp$); (b) $1/4$ walnut ($Rr\ Pp$), $1/4$ rose ($Rr\ pp$), $1/4$ pea ($rr\ Pp$), $1/4$ single ($rr\ pp$); (c) $9/16$ walnut ($R\_\ P\_$), $3/16$ rose ($R\_\ pp$), $3/16$ pea ($rr\ P\_$), $1/16$ single ($rr\ pp$); (d) $3/4$ rose ($R\_\ pp$), $1/4$ single ($rr\ pp$); (e) $1/4$ walnut ($Rr\ Pp$), $1/4$ rose ($Rr\ pp$), $1/4$ pea ($rr\ Pp$), $1/4$ single ($rr\ pp$); (f) $1/2$ rose ($Rr\ pp$), $1/2$ single ($rr\ pp$).

26. (a) Each of the backcross progeny received recessive alleles b and r from the homozygous white parent. Their phenotype is therefore determined by the alleles received from the F_1 parent: brown fish are $Bb\ Rr$; blue fish are $Bb\ rr$; red fish are $bb\ Rr$; and white fish are $bb\ rr$.
(b) We expect a 1 : 1: 1: 1 ratio of the four phenotypes

	Observed	Expected	O − E	(O − E)²/E
Brown	228	229.25	−1.25	0.007
Blue	230	229.25	0.75	0.002
Red	237	229.25	7.75	0.262
White	222	229.25	−7.25	0.229
Total	917	917		0.5 = χ^2

df = 4 − 1 = 3; 0.9 < P < 0.975
we cannot reject the hypothesis.

(c) The homozygous red fish would be $bb\ RR$, crossed with $bb\ rr$. All progeny would be $bb\ Rr$, or red fish.
(d) Homozygous red fish $bb\ RR$ × homozygous blue fish $BB\ rr$

F_1 will be all brown: $Bb\ Rr$ backcrossed to $bb\ RR$ (homozygous red parent).

Backcross progeny will be in equal proportions $Bb\ RR$ (brown); $Bb\ Rr$ (brown); $bb\ RR$ (red); and $bb\ Rr$ (red). Overall, $1/2$ brown and $1/2$ red.

29. (a) Labrador retrievers vary in two loci, B and E. Black dogs have dominant alleles at both loci ($B\_\ E\_$), brown dogs have $bb\ E\_$, and yellow dogs have $B\_\ ee$ or $bb\ ee$. Because all the puppies were black, all of them must have inherited a dominant B allele from the yellow parent and a dominant E allele from the brown parent. The brown female parent must have been $bb\ EE$, and the yellow male must have been $BB\ ee$. All the black puppies were $Bb\ Ee$.
(b) Mating two yellow Labradors will produce yellow puppies only. Mating two brown Labradors will produce either brown puppies only, if at least one of the parents is homozygous EE, or $3/4$ brown and $1/4$ yellow if both parents are heterozygous Ee.

31. Let A and B represent the two loci. The F_1 heterozygotes are $Aa\ Bb$. The F_2 are: $A\_\ B\_$ disc-shaped, $A\_\ bb$ spherical, $aa\ B\_$ spherical, $aa\ bb$ long.

Chapter 6

1. The three factors are: (i) controlled mating experiments are impossible; (ii) humans have a long generation time, and so tracking the inheritance of traits for more than one generation takes a long time; and (iii) the number of progeny per mating is limited, and so phenotypic ratios are uncertain.

3. Autosomal recessive trait: affected males and females arise with equal frequency from unaffected parents; often appears to skip generations; unaffected child of an affected parent is a carrier.

   Autosomal dominant trait: affected males and females arise with equal frequency from a single affected parent; does not usually skip generations.

   X-linked recessive trait: affects males predominantly and is passed from an affected male through his unaffected daughter to his grandson; not passed from father to son.

   X-linked dominant trait: affects males and females; is passed from an affected male to all his daughters but not to his sons; is passed from an affected woman (usually heterozygous for a rare dominant trait) equally to half her daughters and half her sons.

   Y-linked trait: affects males exclusively; is passed from father to all sons.

6. Monozygotic and dizygotic. Monozygotic twins arise when a single fertilized egg splits into two embryos in early embryonic development divisions. They are genetically identical. Dizygotic twins arise from two different eggs fertilized by two different sperm. They have, on the average, 50% of their genes in common.

9. Genetic counseling provides assistance to clients by interpreting the results of genetic testing and diagnosis; provides information about relevant disease symptoms, treatment, and progression; assesses and calculates the various genetic risks facing the person or couple; and helps clients and family members cope with the stress of decision-making and facing up to the drastic changes in their lives that may be precipitated by a genetic condition.

21. (a)

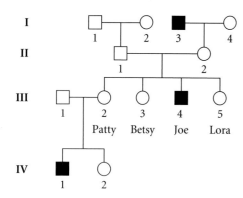

(b) X-linked recessive; (c) zero; (d) $1/4$; (e) $1/4$.

25. (a) Autosomal dominant. The trait must be autosomal because affected males pass the trait to both sons and daughters. It is dominant because it does not skip generations, all affected individuals have affected parents, and it is extremely unlikely that multiple unrelated individuals mating into the pedigree are carriers of a rare trait.
(b) X-linked dominant. Superficially, this pedigree appears to be similar to the pedigree in part a in that both males and females are affected, and it appears to be a dominant trait. However, closer inspection reveals that, whereas affected females can pass the trait to either sons or daughters, affected males pass the trait only to all daughters.
(c) Y-linked. The trait affects only males and is passed from father to son. All sons of an affected male are affected.
(d) X-linked recessive or sex-limited autosomal dominant. Because only males show the trait, the trait could be X-linked recessive, Y-linked, or sex-limited. We can eliminate Y-linkage because affected males do not pass the trait to their sons. X-linked recessive inheritance is consistent with the pattern of unaffected female carriers producing both affected and unaffected sons and affected males producing unaffected female carriers, but no affected sons. Sex-limited autosomal dominant inheritance is also consistent with unaffected heterozygous females producing affected heterozygous sons, unaffected homozygous recessive sons, and unaffected heterozygous or homozygous recessive daughters. The two remaining possibilities of X-linked recessive versus sex-limited autosomal dominant could be distinguished if we had enough data to determine whether affected males have both affected and unaffected sons, as expected from autosomal dominant inheritance, or whether affected males have only unaffected sons, as expected from X-linked recessive inheritance. Unfortunately, this pedigree shows only two sons from affected males. In both cases, the sons are unaffected, consistent with X-linked recessive inheritance, but two male progeny are not enough to conclude that affected males cannot produce affected sons.
(e) Autosomal recessive. Unaffected parents produce affected progeny, and so the trait is recessive. The affected daughter must have inherited recessive alleles from both unaffected parents, and so the trait must be autosomal. If it were X-linked, her father would show the trait.

27. (a) X-linked recessive; (b) $1/4$; (c) $1/2$.

31. Migraine headaches: genetic and environmental. Markedly greater concordance in monozygotic twins, who are 100% genetically identical, than in dizygotic twins, who are 50% genetically identical, is indicative of a genetic influence. However, only 60% concordance for monozygotic twins indicates that environmental factors also play a role.

Eye color: genetic. The concordance is greater in monozygotic twins than in dizygotic twins. Moreover, the monozygotic twins have 100% concordance for this trait, indicating that environment has no detectable influence.
Measles: no detectable genetic influence. There is no difference in concordance between monozygotic and dizygotic twins. Some environmental influence can be detected because monozygotic twins show less than 100% concordance.
Clubfoot: genetic and environmental. The reasoning for migraine headaches applies here. A strong environmental influence is indicated by the high discordance in monozygotic twins.
High blood pressure: genetic and environmental. Reasoning is similar to that for clubfoot.
Handedness: no genetic influence. The concordance is the same in monozygotic and dizygotic twins. Environmental influence is indicated by the less than 100% concordance in monozygotic twins.
Tuberculosis: no genetic influence. Concordance is the same in monozygotic and dizygotic twins. The importance of environmental influence is indicated by the very low concordance in monozygotic twins.

Chapter 7

1. Recombination means that meiosis generates gametes with allelic combinations that differ from the original gametes inherited by an organism. Recombination may be caused by the independent assortment of loci on different chromosomes or by a physical crossing over between two loci on the same chromosome.

2. (a) Nonrecombinant gametes; (b) 50% recombinant gametes and 50% nonrecombinant; (c) more than 50% of the gametes nonrecombinant, and less than 50% recombinant.

5. For genes in coupling configuration, two wild-type alleles are on the same chromosome and the two mutant alleles are on the homologous chromosome. For genes in repulsion, the wild-type allele of one gene and the mutant allele of the other gene are on the same chromosome, and vice versa on the homologous chromosome. The two arrangements have opposite effects on the results of a cross. For genes in coupling configuration, most of the progeny will be either wild type for both genes or mutant for both genes, with relatively few that are wild type for one gene and mutant for the other. For genes in repulsion, most of the progeny will be mutant for only one gene and wild type for the other, with relatively few recombinants that are wild type for both or mutant for both.

8. The farther apart two loci are, the more likely there will be double crossovers between them. The calculated recombination frequency will underestimate the true crossover frequency because the double-crossover progeny are not counted as recombinants.

10. Positive interference indicates that a crossover inhibits or interferes with the occurrence of a second crossover nearby. Negative interference suggests that a crossover event can stimulate additional crossover events in the same region of the chromosome.

16. The genes are linked and have not assorted independently.

18. (a) Both plants are $Dd\ Pp$.
(b) Yes. Recombination frequency = 3.8%.
(c) The two plants have different coupling configurations. In plant A, the dominant alleles D and P are coupled; one chromosome is $\underline{\ D\quad P\ }$ and the other is $\underline{\ d\quad p\ }$. In plant B, they are in repulsion; its chromosomes are $\underline{\ D\quad p\ }$ and $\underline{\ d\quad P\ }$.

21. The distances between the genes are indicated by the recombination rates. Because loci R and L_2 have the greatest recombination rate, they must be the farthest apart, and W_2 is in the middle. The order of the genes is: R, W_2, L_2.

24.

	Genotype	Body color	Eyes	Bristles	Proportion
(a)	$e^+\ ro^+\ f^+$	normal	normal	normal	20%
	$e^+\ ro^+\ f$	normal	normal	forked	20%
	$e\ ro\ f^+$	ebony	rough	normal	20%
	$e\ ro\ f$	ebony	rough	forked	20%
	$e^+\ ro\ f^+$	normal	rough	normal	5%
	$e^+\ ro\ f$	normal	rough	forked	5%
	$e\ ro^+\ f^+$	ebony	normal	normal	5%
	$e\ ro^+\ f$	ebony	normal	forked	5%
(b)	$e^+\ ro^+\ f^+$	normal	normal	normal	5%
	$e^+\ ro^+\ f$	normal	normal	forked	5%
	$e\ ro\ f^+$	ebony	rough	normal	5%
	$e\ ro\ f$	ebony	rough	forked	5%
	$e^+\ ro\ f^+$	normal	rough	normal	20%
	$e^+\ ro\ f$	normal	rough	forked	20%
	$e\ ro^+\ f^+$	ebody	normal	normal	20%
	$e\ ro^+\ f$	ebony	normal	forked	20%

26.

```
    a         g              d
    |---------|--------------|
       4 m.u.      8 m.u.


    c              b                    e
    |--------------|--------------------|
         10 m.u.           18 m.u.
```

Gene f is unlinked to either of these groups; it is on a third linkage group.

29. (a) V is the middle gene.
(b) The Wx–V distance = 7 m.u. and the Sh–V distance = 30 m.u. The Wx–Sh distance is the sum of these two distances, or 37 m.u.
(c) Coefficient of coincidence = 0.80; interference = 0.20

32.

Genotype	Body	Eyes	Wings	Proportion
$b^+\ pr^+\ vg^+$	normal	normal	normal	40.7%
$b\ pr\ vg$	black	purple	vestigial	40.7%
$b^+\ pr^+\ vg$	normal	normal	vestigial	6.3%
$b\ pr\ vg^+$	black	purple	normal	6.3%
$b^+\ pr\ vg$	normal	purple	vestigial	2.8%
$b\ pr^+\ vg^+$	black	normal	normal	2.8%
$b^+\ pr\ vg^+$	normal	purple	normal	0.2%
$b\ pr^+\ vg$	black	normal	vestigial	0.2%

34.

```
                      Chromosome
         Deletion 1 ─────
              Deletion 2 ────
                  Deletion 3 ────
                  Deletion 4 ──
                      Deletion 5 ────
                          Deletion 6 ──

         f?    a   f?   c    e   d    b
```

The location of f is ambiguous; it could be in either location shown in the deletion map.

36. Enzyme 1 is on chromosome 9; enzyme 2 is on chromosome 4; enzyme 3 is on the X chromosome.

Chapter 8

3.
Types of matings	Outcomes
$F^+ \times F^-$	Two F^+ cells
$Hfr \times F^-$	One F^+ cell and one F^- cell
$F' \times F^-$	Two F' cells

The F factor contains a number of genes that take part in the conjugation process, including genes necessary for the synthesis of the sex pilus. The F factor has an origin of replication that enables the factor to be replicated in the conjugation process and has genes for opening the plasmid and initiating the chromosome transfer.

4. To map genes by conjugation, an Hfr strain is mixed with an F^- strain. The two strains must have different genotypes and must remain in physical contact for the transfer to take place. The conjugation process is interrupted at regular intervals. The chromosomal transfer from the Hfr strain always begins with a part of the integrated F factor and proceeds in a linear fashion. Transfer of the entire chromosome takes approximately 100 minutes. The time required for individual genes to be transferred is relative to their positions on the chromosome and the direction of transfer initiated by the F factor. Gene distances are typically mapped in minutes. The genes that are transferred by conjugation to the recipient must be incorporated into the recipient's chromosome by recombination to be expressed.

In transformation, the relative frequency at which pairs of genes are transferred, or cotransformed, indicates the distance between the two genes. Closer gene pairs are cotransformed more frequently. As in interrupted conjugation, the donor DNA must be incorporated into the recipient cell's chromosome by recombination. Physical contact of the donor and recipient cells is not needed. The recipient cell takes up the DNA directly from the environment, and so the DNA from the donor strain must be isolated and broken up before transformation can take place.

The transfer of DNA by transduction requires a viral vector. DNA from the donor cell is packaged into a viral protein coat. The viral particle containing the bacterial donor DNA then infects a recipient bacterial cell. The donor bacterial DNA is incorporated into the recipient cell's chromosome by recombination. Only genes that are close together on the bacterial chromosome can be cotransduced. Therefore, the rate of cotransduction, like the rate of cotransformation, is an indication of the physical distances between genes on the chromosome.

In all three processes, the recipient cell takes up a piece of the donor chromosome and incorporates some of that piece into its own chromosome by recombination. In all three processes, mapping distance is calculated by measuring the frequency with which recipient cells are transformed. The processes use different methods to incorporate donor DNA in the recipient cell.

6. Reproduction is rapid, asexual, and produces lots of progeny. Their genomes are small. They are easy to grow in the laboratory. Techniques are available for isolating and manipulating their genes. Mutant phenotypes, especially auxotrophic phenotypes, are easy to measure.

10. In generalized transduction, randomly selected bacterial genes are transferred from one bacterial cell to another by a virus. In specialized transduction, only genes from a particular locus on the bacterial chromosome are transferred to another bacterium. The process

of specialized transduction requires lysogenic phages that integrate into specific locations on the host cell's chromosome. When the phage DNA excises from the host chromosome and the excision process is imprecise, the phage DNA will carry a small part of the bacterial DNA. The hybrid DNA must be injected by the phage into another bacterial cell in another round of infection.

The transfer of DNA by generalized transduction requires that the host DNA be broken down into smaller pieces and that a piece of the host DNA, instead of phage DNA, be packaged into a phage coat. The defective phage cannot produce new phage particles in a subsequent infection, but it can inject the bacterial DNA into another bacterium or recipient. Through a double-crossover event, the donor DNA can become incorporated into the bacterial recipient's chromosome.

11. To conduct the complementation test, Benzer infected cells of *E. coli* K with large numbers of the two mutant phage types. For successful infection, each mutant phage must supply the gene product or protein missing in the other mutant phage. Complementation will result only if the mutations are at separate loci. If the two mutations are at the same locus, there will be no complementation of gene products, and no plaques will be produced on the *E. coli* K lawns.

12. A retrovirus is able to integrate its RNA genome into its host cell's DNA genome through the action of the enzyme reverse transcriptase, which can synthesize complementary DNA from either an RNA or a DNA template. Reverse transcriptase uses the retroviral single-stranded RNA as a template to synthesize double-stranded DNA. The newly synthesized DNA molecule can then integrate into the host chromosome to form a provirus.

17. By using low doses of antibiotics for 5 years, Farmer Smith selected for bacteria that are resistant to the antibiotics. The doses used killed sensitive bacteria but not moderately sensitive or slightly resistant bacteria. As time passed, only resistant bacteria remained in his pigs because any sensitive bacteria were eliminated by the low doses of antibiotics.

In the future, Farmer Smith should continue to use the vitamins but should use the antibiotics only when a sick pig requires them. In this manner, he will not be selecting for antibiotic-resistant bacteria, and the chances of successful treatment of his sick pigs will be greater.

20. In each of the Hfr strains, the F factor has been inserted into a different location in the chromosome. The orientation of the F factor in the strains varies as well.

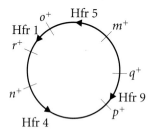

21. Distances between genes are in minutes.

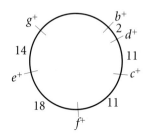

30. Number of plaques produced by $c^+ m^+ = 460$; by $c^- m^- = 460$; by $c^+ m^- = 40$ (recombinant); by $c^- m^+ = 40$ (recombinant).

32. (a) The recombination frequency between r_2 and h is $^{13}/_{180} = 0.072$, or 7.2%. The RF between r_{13} and h is $^{3}/_{216} = 0.014$, or 1.4%.

(b)

33. (a)

(b)

h^+ t^+ c^+

7.1 m.u.   24.1 m.u.

(c) Coefficient of coincidence = $(6 + 5)/(0.071 \times 0.241 \times 942) = 0.68$. Interference = $1 - 0.68 = 0.32$.

Chapter 9

1. Chromosome rearrangements:

   Deletion: Loss of a part of a chromosome.

   Duplication: Addition of an extra copy of a part of a chromosome.

   Inversion: A part of a chromosome is reversed in orientation.

   Translocation: A part of one chromosome becomes incorporated into a different (nonhomologous) chromosome or into the same chromosome.

   Aneuploidy: Loss or gain of one or more chromosomes, causing the chromosome number to deviate from $2n$ or the normal euploid complement.

   Polyploidy: Gain of entire sets of chromosomes, causing the chromosome number to change from $2n$ to $3n$ (triploid), $4n$ (tetraploid), and so on.

2. The expression of some genes is balanced with the expression of other genes; the ratios of their gene products, usually proteins, must be maintained within a narrow range for proper cell function. Extra copies of one of these genes cause that gene to be expressed at proportionately higher levels, thereby upsetting the balance of gene products.

5. A paracentric inversion does not include the centromere; a pericentric inversion includes the centromere.

9. Translocations can produce phenotypic effects if the translocation breakpoint disrupts a gene or if a gene near the breakpoint is altered in its expression because of relocation to a different chromosomal environment (a position effect).

13. Primary Down syndrome is caused by the spontaneous, random nondisjunction of chromosome 21, leading to trisomy 21. Familial Down syndrome most frequently arises as a result of a Robertsonian translocation of chromosome 21 with another chromosome, usually chromosome 14.

14. Uniparental disomy is the inheritance of both copies of a chromosome from the same parent. It may arise from a trisomy condition

in which the early embryo loses one of the three chromosomes, and the two remaining chromosomes are from the same parent.

16. In autopolyploidy, all sets of chromosomes are derived from a single species. In allopolyploidy, the sets of chromosomes are derived from two or more different species. Autopolyploidy may arise through nondisjunction in an early 2n embryo or through meiotic nondisjunction that produces a gamete with extra sets of chromosomes. Allopolyploidy is usually preceded by hybridization between two different species, followed by chromosome doubling.

18. (a) Duplications; (b) polyploidy; (c) deletions; (d) inversions; (e) translocations.

19. (a) Tandem duplication of AB; (b) displaced duplication of AB; (c) paracentric inversion of DEF; (d) deletion of B; (e) deletion of FG; (f) paracentric inversion of CDE; (g) pericentric inversion of ABC; (h) duplication and inversion of DEF; (i) duplication of CDEF, inversion of EF.

22. (a) $1/3$ Notch white-eyed females, $1/3$ wild-type females, and $1/3$ wild-type males; (b) $1/3$ Notch red-eyed females, $1/3$ wild-type females, and $1/3$ white-eyed males; (c) $1/3$ Notch white-eyed females, $1/3$ white-eyed females, and $1/3$ white-eyed males.

24. (a)

(b)

(c)

(d)

27. (a)

(b) Alternate:

Adjacent-1:

Adjacent-2:

(c) Alternate: Gametes contain either both normal or both translocation chromosomes, and are all viable.

$$\underline{A B \cdot C D E F G} + \underline{R S \cdot T U V W X}$$

and

$$\underline{A B \cdot C D V W X} + \underline{R S \cdot T U E F G}$$

Adjacent-1: Gametes contain one normal and one translocation chromosome, resulting in the duplication of some genes and a deficiency of others.

$$\underline{A B \cdot C D E F G} + \underline{R S \cdot T U E F G}$$

and

$$\underline{A B \cdot C D V W X} + \underline{R S \cdot T U V W X}$$

Adjacent-2 (rare): Gametes contain one normal and one translocation chromosome, with the duplication of some genes and a deficiency of others.

and
$$A\ B\cdot C\ D\ E\ F\ G + A\ B\cdot C\ D\ V\ W\ X$$
$$R\ S\cdot T\ U\ V\ W\ X + R\ S\cdot T\ U\ E\ F\ G$$

31. The high incidence of Down syndrome in Bill's family and among Bill's relatives is consistent with familial Down syndrome, caused by a Robertsonian translocation of chromosome 21. Bill and his sister, who are unaffected, are phenotypically normal carriers of the translocation and have 45 chromosomes. Their children and Bill's brother, who have Down syndrome, have 46 chromosomes. From the information given, there is no reason to suspect that Bill's wife Betty has any chromosomal abnormalities. Therefore, the statement in part *d* is most likely correct. All other statements are incorrect.

32. In mammals, the higher frequency of sex-chromosome aneuploids compared with that of autosomal aneuploids is due to X-chromosome inactivation and the lack of essential genes on the Y chromosome. If fish do not have X-chromosome inactivation and if both of their sex chromosomes have numerous essential genes, then the frequency of aneuploids should be similar for both sex chromosomes and autosomes.

33. (a) Possible gamete types: (i) normal chromosome 13 and normal chromosome 22; (ii) translocated chromosome 13 + 22; (iii) translocated chromosome 13 + 22 and normal chromosome 22; (iv) normal chromosome 13; (v) normal chromosome 13 and translocated chromosome 13 + 22; (vi) normal chromosome 22.
(b) Zygotes types: (i) 13, 13, 22, 22; normal; (ii) 13, 13 + 22, 22; translocation carrier; (iii) 3, 13 + 22, 22, 22; trisomy 22; (iv) 13, 13, 22; monosomy 22; (v) 13, 13, 13 + 22, 22; trisomy 13; (vi) 13, 22, 22; monosomy 13.
(c) 50%

35. (a) 15; (b) 24; (c) 32; (d) 17; (e) 14; (f) 14; (g) 40; (h) 18.

Chapter 10

1. Germ-line mutations are changes in the DNA of reproductive (germ) cells and can be passed to offspring. Somatic mutations are changes in the DNA of an organism's somatic-tissue cells and cannot be passed to offspring.

2. Transition mutations result from purine-to-purine or pyrimidine-to-pyrimidine base substitutions. Transversions result from purine-to-pyrimidine or pyrimidine-to-purine base substitutions. Transition mutations are more common because spontaneous mutations typically result in transition mutations rather than transversions.

3. Expanding nucleotide repeats result when a DNA insertion mutation increases the number of copies of a nucleotide repeat sequence. The increase may be due to errors in replication or to unequal recombination. Within a given family, a particular type of nucleotide repeat may increase in number from generation to subsequent generation, increasing the severity of the mutation in a process called anticipation.

6. Intergenic suppressor mutations restore the wild-type phenotype. However, they do not revert the original mutation. The suppression is the result of a mutation in a gene other than the gene containing the original mutation and reverses the effect of the original mutation. For example, because many proteins interact with other proteins, the original mutation may have altered one protein in a way that disrupts a protein–protein interaction, and the second mutation alters another protein to restore that interaction. Another type of intergenic suppression is due to a mutation within the anticodon region of a tRNA molecule that allows the tRNA to pair with the codon containing the original mutation and results in the substitution of a functional amino acid in the protein.

7. Two types of events have been proposed that could lead to DNA replication errors: (i) mispairing due to tautomeric shifts in nucleotides and (ii) mispairing through wobble or flexibility of the DNA molecule. Mispairing through wobble caused by flexibility in the DNA helix is now thought to be the most likely cause.

9. Base analogs have structures similar to those of the nucleotides and may be incorporated into the DNA in the course of replication. Many analogs tend to mispair, which can lead to mutations. DNA replication is required for the base-analog-induced mutations to be incorporated into the DNA.

12. The SOS system is an error-prone DNA-repair system consisting of at least 25 genes. Induction of the SOS system causes damaged DNA regions to be bypassed, which allows for DNA replication across the damaged regions. However, the insertion of bases into the new DNA strand in the absence of bases on the template strand results in a less-accurate replication process; thus, more mutations will occur.

14. Mismatch repair: Repairs replication errors that result from base-pair mismatches. Mismatch-repair enzymes recognize distortions in the DNA structure due to mispairing and detect the newly synthesized strand by its lack of methylation. The distorted segment is excised, and DNA polymerase and DNA ligase fill in the gap.

Direct repair: Repairs DNA damage by directly changing the damaged nucleotide back into its original structure.

Base-excision repair: Excises the damaged base and then replaces the entire nucleotide.

Nucleotide-excision repair: Relies on repair enzymes to recognize distortions of the DNA double helix. These enzymes excise a damaged region by cleaving phosphodiester bonds on either side of the damaged region. The gap created by the excision is filled in by DNA polymerase.

17. Transversion at the first position: GGA→UGA

18. (a) Leucine, serine, or phenylalanine; (b) isoleucine, tyrosine, leucine, valine, or cysteine; (c) phenylalanine, proline, serine, or leucine; (d) methionine, phenylalanine, valine, arginine, tryptophan, leucine, isoleucine, tyrosine, histidine, or glutamine, or a stop codon could result as well.

20. (a) amino–Met-Thr-Gly-**Ser**-Gln-Leu-Tyr–carboxyl
(b) amino–Met-Thr-Gly-Asn–carboxyl
(c) amino–Met-Thr-**Ala-Ile-Asn-Tyr-Ile**-carboxyl
(d) amino–Met-Thr-Gly-Asn-**His**-Leu-Tyr–carboxyl
(e) amino–Met-Thr-**Thr**-Gly-Asn-Gln-Leu-Tyr–carboxyl
(f) amino–Met-Thr-**Gly**-Asn-Gln-Leu-Tyr–carboxyl

22. Four of the six Arg codons could be mutated by a single-base substitution to produce a Ser codon. However, only two of the Arg codons mutated to form Ser codons could subsequently be mutated at a second position by a single-base substitution to regenerate the Arg codon. In both events, the mutations are transversions.

Original Arg codon	**Ser codon**	**Restored Arg codon**
CGU	AGU	AGG or AGA
CGC	AGC	AGG or AGA

25. No, hydroxylamine cannot reverse nonsense mutations. Hydroxylamine modifies cytosine-containing nucleotides and can result only in GC-to-AT transition mutations. In a stop codon, the GC-to-AT transition will result only in a different stop codon.

26. | **Original sequence** | **Mutated sequence** |
|---|---|
| 5′–ATGT–3′ | 5′–ATAT–3′ |
| 3′–TACA–5′ | 3′–TATA–5′ |

29. PFI1 causes transitions, PFI2 causes transversions or large deletions, PFI3 causes transitions, and PFI4 causes single-base insertions or deletions.

32. By looking for plants that have increased levels of mutations either in their germ-line or somatic tissues. Potentially mutant plants may have been exposed to standard mutagens that damage DNA. If they are defective in DNA repair, they should have higher rates of mutation. For example, tomato plants with defective DNA-repair systems should have an increased mutation rate when exposed to high levels of ultraviolet light. Therefore, they need to be grown in an environment that has lower levels of sunlight.

Chapter 11

1. Discontinuous characteristics have only a few distinct phenotypes. In contrast, a quantitative characteristic shows a continuous variation in phenotype.

3. Many genotypes are possible with multiple genes. Even for the simplest two-allele loci, the number of possible genotypes is equal to 3^n, where n is the number of loci, or genes. Thus, for three genes, there are 27 genotypes; four genes yield 81 genotypes; and so forth. If each genotype corresponds to a unique phenotype, the the number of phenotypes is the same: 27 for three genes and 81 for four genes. Moreover, the phenotype for a given genotype may be influenced by environmental factors, leading to an even greater array of phenotypes.

4. A sample is a subset of a population. To be representative of the population, the sample should be randomly selected and sufficiently large to minimize random differences between members of the sample and the population.

7. The magnitude or absolute value of the correlation coefficient reports the strength of the association between the two variables. A value close to $+1$ or -1 indicates a strong association; values close to zero indicate weak association.

9. V_G, component of variance due to variation in genotype; V_A, component of variance due to additive genetic variance; V_D, component of variance due to dominance genetic variance; V_I, component of variance due to genic interaction variance; V_E, component of variance due to environmental differences; V_{GE}, component of variance due to interaction between genes and environment.

10. Broad-sense heritability is the part of phenotypic variance that is due to all types of genetic variance, including additive, dominance, and genic-interaction variances. Narrow-sense heritability is just that part of the phenotypic variance due to additive genetic variance.

14. The response to selection (R) = narrow-sense heritability (h^2) × selection differential (S). The value of R predicts how much the mean quantitative phenotype will change with different selection in a single generation.

16. (a) Discontinuous characteristic because only a few distinct phenotypes are present and it is determined by alleles at a single locus.
(b) Discontinuous characteristic because there are only two phenotypes (dwarf and normal) and a single locus determines the characteristic.
(c) Quantitative characteristic because susceptibility is a continuous trait determined by multiple genes and environmental factors (an example of a quantitative phenotype with a threshold effect).
(d) Quantitative characteristic because it is determined by many loci (an example of a meristic characteristic).
(e) Discontinuous characteristic because only a few distinct phenotypes are determined by alleles at a single locus.

17. (a) All weigh 10 grams; (b) $1/16$ weighing 16 grams, $4/16$ weighing 13 grams, $6/16$ weighing 10 grams, $4/16$ weighing 7 grams, and $1/16$ weighing 4 grams.

18. $1/64$ 22 cm tall; $6/64$ 20 cm tall; $15/64$ 18 cm tall; $20/64$ 16 cm tall; $15/64$ 14 cm tall; $6/64$ 12 cm tall; $1/64$ 10 cm tall.

19. That six or more loci take part.

21.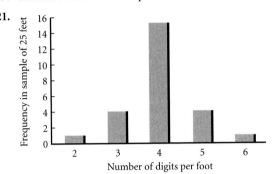

24. No. Correlation does not mean causation. A number of other factors may be responsible for the observed correlation. For example, Baptist preachers may find that their services are needed in areas of high alcohol consumption.

26. (a) $r = 0.80$; (b) 67.8 inches.

27. (a) 0.38; (b) 0.69.

29. (a) Group A, because unrelated people have the greatest genetic variance; (b) no, because siblings raised in the same house should have smaller environmental variance than that of unrelated people.

32. (a) 0.75; (b) its inaccuracy might be due to a difference between the environmental variance of the genetically identical population and that of the genetically diverse population.

35. 0.8

37. The salesman is correct because Mr. Jones's determination of heritability was conducted for a population of pigs under one environmental condition: low nutrition. His findings do not apply to any other population or even to the same population under different environmental conditions. High heritability for a trait does not mean that environmental changes will have little effect.

40. 0.75

Chapter 12

3. Large population, random mating, and not affected by migration, selection, or mutation.

5. (a) If the frequencies of alleles A^1, A^2, and A^3 are defined as p, q, and r, respectively: $f(A^1A^1) = p^2$; $f(A^1A^2) = 2pq$; $f(A^2A^2) = q^2$; $f(A^1A^3) = 2pr$; $f(A^2A^3) = 2qr$; $f(A^3A^3) = r^2$.

(b) For an X-linked locus with two alleles:

$f(X^1X^1) = p^2$ among females; $p^2/2$ for the whole population

$f(X^1X^2) = 2pq$ among females; pq for the whole population

$f(X^2X^2) = q^2$ among females; $q^2/2$ for the whole population

$f(X^1Y) = p$ among males; $p/2$ for the whole population

$f(X^2Y) = q$ among males; $q/2$ for the whole population

8. The proportion of the population due to migrants and the difference in allelic frequencies between the migrant population and the original resident population.

10. The effective population size N_E differs from the actual population size if the sex ratio is disproportionate or if a few dominant individuals of either sex contribute disproportionately to the gene pool of the next generation.

14. Mutation increases genetic variation within populations and increases divergence between populations because different mutations arise in each population.

    Migration increases genetic variation within a population by introducing new alleles, but it decreases divergence between populations.

    Genetic drift decreases genetic variation within populations because it causes alleles to eventually become fixed, but it increases divergence between populations because the effects of drift differ in each population.

    Natural selection may either increase or decrease genetic variation, depending on whether the selection is directional or balanced. It may increase or decrease divergence between populations, depending on whether different populations have similar or different selection pressures.

16. $f(T^E T^E) = 0.685; f(T^E T^F) = 0.286; f(T^F T^F) = 0.029;$ $f(T^E) = 0.828; f(T^F) = 0.172.$
25. (a) 0.1; (b) 1%; (c) 18%.
26. $f(L^M L^M) = 0.648; f(L^M L^N) = 0.304; f(L^N L^N) = 0.048.$
30. 0.31
35. 0.17

Chapter 13

2. Mutation and recombination.
4. Because the expected heterozygosity is independent of the breeding system.
7. Neutral-mutation hypothesis: Most molecular variation is adaptively neutral.

   Balance hypothesis: Most genetic variation is maintained by balanced selection, favoring heterozygosity at most loci.
9. The biological species concept defines a species as a group of organisms whose members can potentially interbreed with one another other but are reproductively isolated from members of other species. One problem with this concept is that it does not apply to asexually reproducing organisms. Another is that, in practice, most species are defined by phenotypic or anatomical differences, and determining their potential for interbreeding may be difficult or impossible.
12. First, a geographic barrier splits a population into, say, two groups and prevents gene flow between the groups on either side of the barrier. The two groups then evolve independently. Through natural selection, genetic drift, and mutation the two groups become genetically different. These genetic differences lead to reproductive isolation, and the two groups will have undergone allopatric speciation.
14. The flies that feed on apples mate on or near apples, whereas flies that feed on hawthorn fruit mate on or near the hawthorn fruit. Because apples ripen several weeks earlier than do hawthorn fruit, flies that feed on apples mate earlier than flies that feed on hawthorn fruit. These differences in hosts and time of mating lead to reproductive isolation and an accumulation of genetic differences between the two groups of flies.
16. The distance approach relies on the degree of overall similarity between phenotypic characteristics or gene sequences; the most-similar species are grouped together. The parsimony approach tries to reconstruct an evolutionary pathway on the basis of the minimum number of evolutionary changes that must have taken place since the species last had an ancestor in common. Maximum likelihood and Bayesian methods infer phylogenetic relationship on the basis of which phylogeny maximize the probability of obtaining the set of characteristics exhibited by the organisms.
18. The molecular clock estimates the rate at which nucleotide changes take place in a DNA sequence.
21. Horizontal gene transfer, also called lateral gene transfer, is the transmission of genetic information across species boundaries. Horizontal gene transfer takes place frequently in bacteria, through transformation and phage-mediated transduction. In eukaryotes, horizontal gene transfer may take place through endosymbiosis (e.g., mitochondrial and chloroplast genes), viral infections, parasitic infections, and human intervention (genetic engineering). Horizontal gene transfers create problems in the construction of evolutionary relationships with the use of molecular sequence data because the evolutionary history of horizontally transferred genes differs from the evolutionary history of other genes.
23. (a) For *P. drummondii*, the percentage of polymorphic loci ranges from 50% in population 3 to 75% in populations 1 and 2. The mean expected heterozygosities in each population are 0.074 for population 1, 0.227 for population 2, and 0.156 for population 3. The mean expected heterozygosity for all populations is 0.15.

    For *P. cuspidate*, the percentage of polymorphic loci ranges from 0% in populations 1 and 2 to 25% in population 3. The mean expected heterozygosities are 0.0 for populations 1 and 2 and 0.16 for population 3. The mean expected heterozygosity for all populations is 0.055.

    (b) Self-fertilization reduces genetic variation.
26. For 7 organisms, 10,395 rooted trees; for 12 organisms, 13,749,310,575 rooted trees.

Index

Note: Page numbers followed by f indicate figures; those followed by t indicate tables. Page numbers preceded by A refer to the *Reference Guide to Model Genetic Organisms*.

ABO blood group antigens, 105, 105f, 108–109
   Bombay phenotype and, 108–109
   nail-patella syndrome and, 185–186, 185f
Acentric chromatids, 247
Achondroplasia, 148t, 358
Acquired immunodeficiency syndrome, 227–229, 371–372
Acquired traits, inheritance of, 47
Acridine orange, as mutagen, 286–287, 287f
Acrocentric chromosomes, 20, 20f, 240
Adaptive mutations, 280–281
Addition rule, 52f, 53
Additive genetic variance, 317
Adenosine triphosphate. *See* ATP
Adjacent-1 segregation, 249, 250f
Adjacent-2 segregation, 249, 250f
Adoption studies, 146
   of obesity, 146, 146f
Agar plates, 205, 205f
Age, maternal, aneuploidy and, 256, 257f
Aging
   premature
      in Hutchinson–Gilford progeria syndrome, 135–136, 135f, 136f
Agriculture. *See also* Breeding; Crop plants
   genetics in, 3, 3f, 7–8, 8f, 9
   quantitative genetics in, 303–304
AIDS, 227–229, 371–372
Albinism. *See also* Color/pigmentation
   gene interaction and, 109–110
   in Hopi, 1–2, 2f
   inheritance of, 53–55
   in snails, 110, 111f
Alkylating agents, as mutagens, 286
Alleles, 11, 19. *See also* Gene(s)
   in crossing-over, 27, 28f
   definition of, 46, 46t
   dominant, 48f, 49, 100–103
   lethal, 100f, 103–104
   letter notation for, 48, 55
   multiple, 104, 104f, 105f
   recessive, 49
   segregation of, 9, 50f, 57f, 59–60, 162–163
   temperature-sensitive, 123
   wild-type, 55
Allelic fixation, 353
Allelic frequency
   calculation of, 339–340
   at equilibrium, 341–342, 348–349, 348f
   estimation of, 344–345
   fixation and, 353
   genetic drift and, 350–353, 353f, 358–359, 358t
   Hardy–Weinberg law and, 344–345
   migration and, 349–350, 350f, 358–359, 358t
   for multiple alleles, 340
   mutations and, 348–349, 348f, 358–359, 358t
   natural selection and, 353–358, 354t, 355t, 357t–359t
   nonrandom mating and, 345–347
   rate of change for, 357–358, 359f
   for X-linked alleles, 340
Allelic series, 104
Allopatric speciation, 375, 375f
Allopolyploidy, 260, 260f, 262t
   speciation and, 378–379
α-Fetoprotein, 150
Alternate segregation, 249, 250f
Alternation of generations, 35, 35f
Ames, Bruce, 288
Ames test, 288, 289f
2-Aminopurine, as mutagen, 285–286
Amniocentesis, 148–149, 149f
Amphidiploidy, 260
Anagenesis, 367
Anaphase
   in meiosis, 27, 27t, 28f
   in mitosis, 22–24, 23f, 24t, 27
Androgen insensitivity syndrome, 80–81
Anemia, Fanconi, 296t
Aneuploidy, 241, 241f, 252–257, 262t
   autosomal, 254–256, 255f, 257f
   cancer, 256
   causes of, 252, 253f
   definition of, 252
   in Down syndrome, 254–255, 255f, 257f
   in Edward syndrome, 256
   effects of, 252–254
   in evolution, 262–263
   in humans, 254–256
   maternal age and, 256, 257f
   mosaicism and, 257, 258f
   in Patau syndrome, 256
   rate of, 253
   of sex chromosomes, 254
      in Klinefelter syndrome, 79, 79f, 90
      in Turner syndrome, 79, 79f, 90
   in trisomy 8, 256
   types of, 252
   uniparental disomy and, 257
Angelman syndrome, 121
Angiosperms. *See also Arabidopsis thaliana* (mustard plant); Plants
   flower color in
      inheritance of, 100–101, 101f, 102
      lethal alleles and, 103–104
   flower length in, inheritance of, 315–316, 316f
Animals
   breeding of, 7–8
      artificial selection and, 324, 325f
      genetic correlations in, 327, 327t
      inbreeding in, 347
   coat color in. *See* Coat color
   sexual reproduction in, 33–35, 34f
Antibiotics
   resistance to, gene transfer and, 215–216
Anticipation, 122–123, 275
Antigenic drift, 230
Antigenic shift, 230
   blood group, 105–106, 105f
      ABO, 105–106, 105f, 108–109, 185
      Bombay phenotype and, 108–109
      MN, 102
      nail–patella syndrome and, 185–186, 185f
Arabidopsis thaliana (mustard plant)
   as model genetic organism, 5, 7f, A8–A9
Archaea, 17, 204. *See also* Bacteria; Prokaryotes
Artificial selection, 324, 325f. *See also* Breeding
Asexual reproduction, polyploidy and, 261
Assortative mating, 345–347
   in sympatric speciation, 378
Asthma, twin studies of, 145, 145t
Ataxia-telangiectasia, 296t
Atomic bomb, radiation exposure from
   in Japan, 289, 290, 290f
   in Russia, 289–290
Auerbach, Charlotte, 285
Autopolyploidy, 258–259, 258f, 259f, 262t
Autosomal aneuploidy, 254–256, 255f, 257f
Autosomal traits
   dominant, 137, 139, 143t
   recessive, 138–139, 138f, 143t
   X-linked, 81–90
Autosomes, 76. *See also* Sex chromosomes
Auxotrophs, 205, 205f
Avian influenza, 230, 230f
Ayala, Francisco, 379

Backcross, 51
Bacteria, 203–219. *See also Escherichia coli* (bacterium)
   antibiotic resistance in, 215–216
   archaea, 17, 204. *See also* Archaea
   auxotrophic, 205, 205f
   cell reproduction in, 18, 19f. *See also* Replication
   cell structure in, 16f, 17
   conjugation in, 208–215, 209f–214f, 213t. *See also* Conjugation
   culture of, 205–206, 205f
   diversity of, 203–205
   DNA in, 17, 17f,
   eubacteria, 17, 204. *See also* Eubacteria
   evolution of, 203–204
   gene mapping in, 213–215, 214f, 217–218, 217f, 220–227, 222f
   gene transfer in, 208–219
      by conjugation, 208–215, 209f–213f, 214f
      horizontal, 218–219, 386
      by transduction, 208, 209f, 220–223, 222f
      by transformation, 208, 209f, 217–218, 217f
   genes of, 17, 17f, 206, 207f, 218
   in genetic studies, 204t
   mutant strains of, 205, 205f
   overview of, 203–204

D1

Bacteria *(cont.)*
  phages and, 219–227. *See also* Bacteriophage(s)
  population of, 203
  prokaryotic, 11, 17–18. *See also* Prokaryotes
  prototrophic, 205
  transduction in, 208, 209f, 220–223, 222f. *See also* Transduction
  transformation in, 208, 209f, 217–218, 217f
  vs. eukaryotes, 204–205
  wild-type, 205
Bacterial chromosomes, 17, 206, 207f, 215, 217f,
Bacterial colonies, 205, 205f
Bacterial cultures, 205–206, 205f
Bacterial genome, 16f, 17f, 18, 206, 207f, 218
Bacterial plasmids, 206–207, 207f. *See also* Plasmid(s)
Bacteriophage(s), 219–227
  in bacterial gene mapping, 220–227, 222f
  complementation tests in, 226–227, 226f
  culture of, 219, 220f
  experimental advantages of, 219
  experimental techniques with, 219
  fine-structure analysis of genes in, 224–227
  gene mapping in, 223–224, 224f
  life cycle of, 219, 220f
  prophage, 219, 220f
  temperate, 219, 220f
  transducing, 220–223
  virulent, 219, 220f
Balance hypothesis, 369
Baldness, male-pattern baldness, 161–162
Banding, chromosome, 241
Bar mutations, 243, 243f
Barr bodies, 89, 89f, 89t
Barry, Joan, 105–106
Base analogs, as mutagens, 285–286, 285f
Base pairing
  deamination and, 284, 284f
  depurination and, 283–284, 284f
  mutations and, 281–283, 292f
    Ames test and, 288, 289f
  nonstandard, 281–282, 282f
  tautomeric shifts in, 281, 282f
  wobble in, 281–282, 283f
Base substitution, 273, 273f, 275–276, 276f, 278f
Base-excision repair, 290t, 292–293, 293f
Bateson, William, 163
Baur, Erwin, 103–104
Baysian method, 382
Bees
  epigenetics and, 122
Behavioral isolation, 374, 375t
Belling, John, 253
Benzer, Seymour, 224
Benzer's mapping technique, 224–227, 225f
Bergalis, Kimberly, 371
Berns, Sam, 135
Berns, Scott, 135
Bighorn sheep, 337–338, 371, 371f
Binomial expansion, 53–54
Biological species concept, 373
Biology, genetics in, 4
Biotechnology, 3–4. *See also* Recombinant DNA technology
Bird flu, 230, 230f

Bivalent, 27
Blakeslee, A. Francis, 253
Blending inheritance, 9, 10t
Blind men's riddle, 15–16
Blood group antigens
  ABO, 105–106, 105f, 108–109
    Bombay phenotype and, 108–109
    nail–patella syndrome and, 185–186, 185f
  MN, 102
Blood tests, maternal, 149–150
Blood transfusions, ABO antigens and, 105, 105f
Bodmer, Rolf, 271
Body weight, adoption studies of, 146, 146f
Body-mass index (BMI), 146
Bombay phenotype, 108–109
Boveri, Theodor, 50
Brain size, evolution of, 154
Branch diagrams, 57–58, 58f, 59f
Branches, of phylogenetic trees, 380, 381f
Bread mold *(Neurospora crassa)*, 5
Breeding
  of animals, 7–8
  artificial selection and, 324, 325f
  genetic correlations in, 327, 327t
  inbreeding and, 345–347, 346f, 347f
  of plants, 3, 3f, 7–8, 8f, 9, 260, 260f
  quantitative genetics in, 303–304
Bridges, Calvin, 82–83, 169, 253
Brittle hair syndrome, 295, 296t
Broad-sense heritability, 318
5-Bromouracil, as mutagen, 285, 285t
Brown, Robert, 9
Bulbar muscular atrophy, 274t
Buri, Peter, 351

C banding, 241, 241f
Caenorhabditis elegans (nematode)
  dosage compensation in, 89
  as model genetic organism, 5, 6f, A6–A7
Cancer
  aneuploidy and, 256
  faulty DNA repair in, 295, 296t
  retroviruses and, 227
Cancer-causing agents, Ames test for, 289, 289f
Capsicum annuum (pepper), fruit color in, 106–107, 107f
  Ames test for, 288, 289f
Carriers, translocation, 255, 257f
Castle, William Ernest, 99–100
Cats, coat color in, 89–90, 90f
Cell(s)
  competent, 217
  diploid, 19, 20f
  eukaryotic, 11, 16f, 17, 17f. *See also* Eukaryotes
  genetic material in, 16f, 17, 17f
  haploid, 19
  prokaryotic, 11, 16f, 17. *See also* Prokaryotes
  reproduction of, 18–21
  structure of, 16f, 17
  transformant, 218
Cell cycle. *See also* Meiosis; Mitosis
  checkpoints in, 21, 21f
  chromosome movement in, 22–24, 234f
  chromosome number in, 25, 25f
  definition of, 21

DNA molecule number in, 25, 25f
DNA synthesis in, 22, 24, 25
G_0 phase of, 21f, 22, 24t
G_1 phase of, 21f, 22, 24t
G_2 phase of, 21f, 22, 24t
genetic consequences of, 24
interphase in, 21–22, 21f, 23f, 24t
M phase of, 21, 21f, 22–24, 23f, 24, 24t
molecular motors in, 24
overview of, 21–24, 21f, 24t
replication in, 21f, 22, 24t. *See also* Replication
S phase of, 21f, 22, 24, 24t
stages of, 21–24, 21f, 23f, 24t
Cell division
  in cytokinesis, 21, 21f
  cytoplasmic, 21, 21f
  in eukaryotes, 19–24. *See also* Cell cycle
  in meiosis, 25–37
  in mitosis, 21f, 22–24, 24t, 32, 32f
  nuclear, 21, 21f
  in prokaryotes, 17f, 18
Cell lines, 188–190
Cell theory, 9, 10t
CentiMorgan, 174
Centriole, 22
Centromeres, 20, 20f
Centrosomes, 22, 23f
Cerebellar ataxia, 274t
Chaplin, Charlie, 105–106
Characteristics. *See also* Traits
  definition of, 46t
  vs. traits, 46
Checkpoints, in cell cycle, 21, 21f
Cheng, Keith, 6
Chiasma, 27
Chickens, feathering patterns in, 116, 117f
Chimpanzees
  chromosomes in, 249–251, 251f
  genome in, vs. human genome, 153–154
Chi-square test
  for crosses/linkage, 61–63
  for crosses/linkages, 62t, 63f, 172–174, 173f
  goodness-of-fit, 61–63, 63f, 172–173
  for Hardy–Weinberg proportions, 344
  of independence, 173–174, 173f
  significant, 172–173
Chloroplast DNA (cpDNA), 118
  mutations in, 118–119
Cholesterol
  in familial hypercholesterolemia, 139, 140f
Chorionic villus sampling, 149, 150f
Chromatids
  acentric, 247
  dicentric, 247
  nonsister, in crossing over, 28f, 29–31
  sister, 20f, 21
    separation of, 22–24, 23f, 24t, 30, 32
Chromatin, 17, 17f
  structure of, 17, 17f
Chromosome(s)
  acrocentric, 20, 20f, 240
  bacterial, 17, 17f, 206, 207f, 215, 215f, 217f,
  in chimpanzees vs. humans, 249–251, 251f
  chromatin in. *See* Chromatin
  condensation of, 26–27
  counting of, 25, 25f

coupling configurations for, 168, 169f
crossing over of. *See* Crossing over
daughter, formation of, 22, 23f, 24
in diploid organisms, 19, 20f
DNA packaging in, 19–20
ends of, 20–21, 20f, 24, 24f
eukaryotic, 17, 17f, 18–21
fragile sites on, 251, 251f
gene location on, 12, 12f, 170–171. *See also*
Gene loci
in haploid organisms, 19
homologous pairs of, 19, 20f, 30, 32
random separation of, 30
karyotypes and. *See* Karyotypes
mapping of. *See* Gene mapping
in meiosis, 25–37, 48f, 49
metacentric, 20, 20f, 240
in mitosis, 22–24, 23f, 24t
morphology of, 240–241, 240f, 241f
movement of, 22–24, 23f
nondisjunction of. *See* Nondisjunction
nonsex, 76
number of, 19, 20f
abnormal, 252–262. *See also* Aneuploidy;
Polyploidy
origins of replication of, 21
prokaryotic, 17, 17f, 18
random distribution of, 30, 31f
replication of, 20f, 21, 23f. *See also* Cell cycle
in repulsion, 168, 169f
segregation of. *See* Segregation
sets of, 19, 20f
sex. *See* Sex chromosomes
staining of, 240, 241f
structure of, 19–21, 20f, 240–241, 240f, 241f
submetacentric, 20, 20f, 240
synapsis of, 26
telocentric, 20, 20f, 240
telomeric sequences in, 20–21, 20f
Chromosome banding, 241, 241f
Chromosome mutations, 242–252, 262t, 273.
See also Mutations
aneuploid, 241, 241f, 252. *See also* Aneuploidy
definition of, 240
fragile-site, 251, 251f
lethality of, 253, 254
maternal age and, 256, 257f
polyploid, 241, 241f, 252, 258–262. *See also*
Polyploidy
rate of, 253
rearrangement, 242–252. *See also*
Chromosome rearrangements
types of, 241, 241f, 262t
unequal crossing over and, 243, 243f, 283, 283f
Chromosome rearrangements, 242–252, 245t, 262t. *See also* Mutations
copy-number variations, 252
deletions, 244–246, 244f–246f, 245t, 262t, 273f, 274, 279t, 282–283, 283f
duplications, 241f, 242–244, 245t, 262, 262t, 263. *See also* Duplications
in evolution, 262–263
evolutionary significance of, 249–251, 251f
inversions, 246–248, 246f–248f, 262–263, 262t

phenotypic consequences of, 242–244, 244f, 245t
translocations, 248–251, 251f, 262t. *See also*
Translocation
Chromosome theory of heredity, 50, 75, 163, 170–171
nondisjunction and, 82–83, 83f
Circular chromosome, bacterial, 206, 215
Circular DNA
bacterial, 206, 207f
Cis configuration, 168, 169f
Cis–trans test, 226–227, 226f
Cistron, 227
Cladogenesis, 367
Coat color
in cats, 90f
in dogs, 107–108, 113–115, 114f
gene interaction and, 107–108, 113–115, 114f
lethal alleles and, 99–100, 103–104
in mice, 99–100, 99f, 100f, 103–104
in rabbits, 123, 124f
Cock feathering, 116, 117f
Cockayne syndrome, 295, 296t
Codominance, 101t, 102
Coefficient of coincidence, 182
Cohesin, 31, 33f
Colon cancer
faulty DNA repair in, 295, 296t
Colonies, bacterial, 205, 205f
Color blindness, 84–85, 84f
Color/pigmentation
coat. *See* Coat color
eye
in *D. melanogaster,* 81–83, 82f, 83f, 113, 176–182, 178f–181f, 351, 352f
feather, 85–86, 86f, 104, 104f
flower
inheritance of, 100–101, 101f, 102
lethal alleles and, 103–104
fruit
epistasis and, 109, 110f
gene interaction for, 106–107, 109, 110f
incomplete dominance and, 101, 101f
gene interaction and, 106–110, 110f
of hair, 43–44
inheritance of, 43–44, 53–55, 307–308, 308f
leaf, cytoplasmic inheritance and, 118–119, 119f
in snails, 110, 111f
temperature-dependent, 123, 124f
wheat kernel, 307–308, 308f
Coltman, David, 371
Common slipper limpet, sex determination in, 77, 77f
Competence
cellular, 217
Complementation, 113
Complementation tests, 113
in bacteriophages, 226–227, 226f
Complete dominance, 100, 101f, 101t
Complete linkage, 165f, 167f, 169
Complete medium, 205
Concordance, in twin studies, 144–145, 144t, 145t, 146t
Condensation, chromosomal, 25
Conditional mutations, 276

Congenital heart disease, 271
Conjugation, 208–215, 209f–213f, 213t, 214f
in Davis' U-tube experiment, 210, 210f
definition of, 208
F′ cells in, 211f–213f, 212–213, 213t
F$^+$ cells in, 210f, 211–212, 211f1, 211f–213f, 213t
F$^-$ cells in, 210f, 211–212, 211f–213f, 213t
F factor in, 207, 208f, 211–213, 211f–213f, 213t, 214f
Hfr cells in, 211–212, 212f, 213t
interrupted, in gene mapping, 213–215
in Lederberg–Tatum experiment, 208–210
Consanguinity, 138, 138f
Continuous characteristics. *See* Quantitative trait(s)
Copy-number variations, 252. *See also* Deletion(s); Duplications
Corn. *See* Maize
Correlation, 312–313, 312f
Correlation coefficient, 312–313, 312f
Correns, Carl, 118
Cotransduction, 221
Cotransformation, 217f, 218
Coupling (cis) configuration, 168, 169f
cpDNA. *See* Chloroplast DNA (cpDNA)
Creighton, Harriet, 170–171, 170f
Crepidula fornicata (limpet), sex determination in, 77, 77f
Cri-du-chat syndrome, 245t, 246
Crop plants. *See also* Plants
breeding of, 3, 3f, 7–8, 8f, 260, 260f
genetically modified, 3, 3f
inbreeding in, 347f
Crosses, 27t, 29–30, 29f–32f, 32f
addition rule for, 52f, 53
backcross, 51
bacteriophage, 223–224
branch diagrams for, 57–58, 58f, 59f
chi-square test for, 61–63, 62t, 63f
cis configuration in, 168, 169f
dihybrid, 56–57, 57–58, 57f–59f
genotypic ratios in, 55–56, 56t
with linked genes, 163–176. *See also* Linkage
coupling in, 168, 169f
notation for, 164
predicting outcome of, 171–172
recombination frequency for, 167
in repulsion, 168, 169f
testcrosses for, 165–166, 165f
meiosis and, 49–51
monohybrid, 47–56, 55
multiple-loci, 56–61
multiplication rule for, 52–53, 52f
notation for, 49–50, 55, 164
outcome prediction for, 51–55, 171–172
phenotypic ratios for, 55–56, 55t
observed vs. expected, 61–63, 63f
probability rules for, 52–55, 52f, 57–58, 58f, 59f
Punnett square for, 51, 51f
reciprocal, 48
recombination frequencies for, 167, 174–175, 180–181
testcrosses, 55, 59, 59f, 165–166, 165f, 172–183, 173f. *See also* Testcrosses

Crosses *(cont.)*
    three-point, 176–183
    two-point, 175–176
        gene mapping with, 175–176
    with unlinked genes, 61–63, 62t, 63f
Crossing over, 27t, 28f–32f, 29–30. *See also* Crosses; Crossovers
    among three genes, 176–183
    coupling configurations in, 168, 169f
    definition of, 27
    dominance and, 48f, 49
    genetic diversity and, 29–31
    with incompletely linked genes, 166–167, 169–170
    independent assortment and, 56–57, 57f, 162–163, 164f, 165f, 166–167, 169, 172–174, 173f
    within inversions, 247–248, 247f, 248f
    with linked genes, 163–176, 165f, 169. *See also* Linked genes
    nonindependent assortment and, 163, 163f, 164f
    recombination frequencies and, 167, 174–175, 180–181
    recombination in, 164f, 167, 169–170. *See also* Recombination
    repulsion and, 168, 169f
    segregation in, 49, 51f, 162–163. *See also* Segregation
    trans configuration in, 168, 169f
    transformation and, 218
    unequal, chromosome mutations and, 243, 243f, 283, 283f
Crossovers
    double, 175, 175f, 182
        coefficient of coincidence for, 182
        four-strand, 184, 184f
        within inversions, 248, 248f
        three-strand, 184, 184f
        two-strand, 184–185, 184f
    hot spots for, 191
    interference between, 182
    location of, mapping of, 178f, 179–180, 180f
    multiple, 184–185, 184f
    predicted number of, 181–182
    three-gene, 176–183, 177f
    two-strand, 184–185, 184f
Cuénot, Lucien, 99–100, 103
Cultures
    bacterial, 205f
    bacteriophage, 219, 220f
Cystic fibrosis, prenatal diagnosis of, 148t, 344–345
    inheritance of, 102, 257
Cytochrome *c*, in evolutionary studies, 384
Cytokinesis, 21, 24t
Cytology, history of, 9–10
Cytoplasmic inheritance, 117–119, 118f, 119f, 122t

Danio rerio (zebra fish), 5–6, 6f
Darwin, Charles, 9, 324, 338, 376, 381
    allopatric speciation in, 376, 376f, 377f
Datura stramonium (Jimson weed), aneuploidy in, 253, 254f
Daughter chromosomes, formation of, 22, 23f, 24

Davenport, Charles, 43
Davenport, Gertrude, 43
Davis, Bernard, 210
Dawkins, Richard, 4
De Vries, Hugo, 45
Deamination
    induced, 286, 286f
    spontaneous, 284, 284f
Deletion(s), 242–246, 244f, 245t, 246f, 262t, 273f, 284, 279t, 282–283, 283f
    in-frame, 274, 279t
Deletion mapping, 188, 188f
Dentatorubral-pallidoluysian atrophy, 274t
Depurination, mutations and, 283–284, 284f
Development
    gene regulation in, 4, 4f
    sexual
        precocious puberty in, 117, 117f
        sequential hermaphroditism in, 77, 77f
Diagnosis, of genetic diseases, 148–153. *See also* Genetic testing
Diakinesis, 27, 29f
Diastrophic dysplasia, 3f
Dicentric bridge, 247
Dicentric chromatids, 247
Diet, epigenetics and, 122
Dihybrid crosses, 56–61, 57f–59f
    phenotypic ratios from, 111–112, 111t
Dillon, Robert T., 110
Dioecious organisms, 75
Diploid cells, 19, 20f
Diplotene, 27, 29f
Direct repair, 290t, 292
Directional gene transfer, 215, 215f, 217f
Directional selection, 356
Direct-to-consumer genetic tests, 153
Discontinuous characteristics, 124, 306
Discontinuous traits, 304f
Discordance, 144
    genetic. *See* Genetic diseases
Disomy, uniparental, 257
Displaced duplications, 242
Distance approach, for evolutionary relationships, 382
Distributions, in statistical analysis, 309–310, 310f
Dizygotic twins, 143, 144f
DNA, 12
    bacterial, 17, 17f
    during cell cycle, 21f, 22, 24t, 25, 25f
    chloroplast, 118, 119. *See also* Chloroplast DNA (cpDNA)
    circular
        bacterial, 206, 207f
        coiling of, 20
    damage to
        repair of. *See* DNA repair
    eukaryotic, 17, 17f
    fetal, in prenatal diagnosis, 150
    measurement of, 25, 25f
    microsatellite
        variation in, 371
    mitochondrial, 118. *See also* Mitochondrial DNA (mtDNA)
    packaging of, 19–20
    palindromes in, 87

    polynucleotide strands of
        slippage of, 283, 283f
    prokaryotic, 17, 17f
    replication of. *See* Replication
    synthesis of. *See* Replication
    in transformation, 217
DNA methylation
    in genomic imprinting, 121, 122
DNA polymerase(s)
    in base-excision repair, 292–293, 292f
    translesion, 294–295
DNA probes
    in in situ hybridization, 190, 190f
DNA proofreading, 291
DNA repair, 290–295
    base-excision, 290t, 292–293, 292f
    direct, 290t, 292
    of double-strand breaks, 294
    in genetic diseases, 295, 296t
    mismatch, 290, 291, 291f
    nonhomologous end joining in, 294
    nucleotide-excision, 290t, 293–294, 293f
        in genetic disease, 296t
    SOS system in, 287–288
    steps in, 294
DNA sequence alignments, in phylogenetic trees, 381–382
DNA sequence analysis
    in gene mapping, 190
DNA sequence variation, 371
    restriction-site, 370
    in gene mapping, 190
Dobzhansky, Theodosius, 366
Dogs, coat color in, 107–108, 113–115, 114f, 115t
Dominance, 48f, 49, 100–103, 101f, 101t
    characteristics of, 102–103
    codominance and, 101f, 101t, 102
    complete, 100, 101f, 101t
    incomplete, 100–101, 101f, 101t
Dominance genetic variance, 317
Dominant epistasis, 109, 110f
Dominant traits
    autosomal, 139, 139f, 143t
    inheritance of, 49
    X-linked, 141f, 143t
Dosage compensation, 88–90, 254
double Bar mutations, 243, 243f
Double crossovers, 175, 175f, 182
    coefficient of coincidence for, 182
    within inversions, 248
Double fertilization, 37
Down, John Langdon, 239
Down syndrome, 239–240, 254–256, 255f, 256f
    familial, 255, 255f, 257f
    maternal age and, 256, 257f
    prenatal diagnosis of, 150
    primary, 254, 255f
Drosophila heteroneura, 379
Drosophila melanogaster (fruit fly)
    Bar mutations in, 243, 243f
    eye color in, 81–83, 82f, 83f, 113
        gene mapping for, 176–182, 178f, 180f, 181f
        genetic drift and, 351, 352f
    eye size in, 243, 243f
    genetic drift in, 351, 352f

genetic map for, 176–182, 178f, 180f, 181f
gynandromorphic, 257
as model genetic organism, 5, 6f, A2–A3
mosaicism in, 257, 258f
Notch mutation in, 245–246, 246f
sex determination in, 78, 79t
speciation and, 379
temperature-dependent gene expression in, 123, 123f
X-linked characteristics in, 81–83, 82f, 83f
Drosophila silvestris, 379
Drosophila simulans, 379
Drosophila spp., speciation and, 379
Drosophila willistoni, 379, 379t
Drug development
genetics in, 3
Drug resistance, gene transfer and, 215–216
Ducks, feather color in, 104, 104f
Duplicate recessive epistasis, 109–110
Duplications, 241f, 242
chromosome, 242–244, 245t, 262t
displaced, 242
in evolution, 262, 263
gene, 385–386
reverse, 242
segmental, 244
tandem, 242
whole-genome, 386

East, Edward, 307, 315
Ecological isolation, 373–374, 374t
Edward syndrome, 256
Effective population size, 352
Eggplant, fruit color in, 101, 101f
Elephants, genetic variation in, 371
env gene, 227
Environmental factors
in adoption studies, 146
in gene expression, 123–125, 123f, 124f
genotype–phenotype relationship and, 305–306. See also Genotype–phenotype relationship
in heritability, 321
in twin studies, 145
Environmental sex determination, 77–78, 77f
Environmental variance, 317
Enzyme(s). See also *specific enzymes*
deficiencies of, 123–124. See also Genetic diseases
Epigenetics, 11, 122
Epilepsy, in twins, 144, 144t
Episomes, 207
Epistasis, 107–113
definition of, 107
dominant, 109, 110f
recessive, 107–109
duplicate, 109–110
Epistatic gene, 107
Equilibrium
for allelic frequencies, 348–349, 348f
Hardy–Weinberg, 341
unstable, 357
Escherichia coli (bacterium). See also Bacteria
conjugation in, 208–215, 208f–213f, 213t. See also Conjugation
DNA mismatch repair in, 292

as model genetic organism, 5, 6f, A4–A5
transformation in, 208, 209f
Ethical issues, in genetic testing, 152, 153
Ethylmethanesulfonate (EMS), as mutagen, 286, 286f
Eubacteria, 17, 204. See also Bacteria; Prokaryotes
Eukaryotes, 11, 16f, 17–21, 17f
cell reproduction in, 18–21. See also Cell cycle
cell structure in, 16f, 17
definition of, 17
DNA in, 17, 17f
genes of, 17
genome of, 17f
sexual reproduction in, 74, 75–77
vs. bacteria, 204–205
Eukaryotic bacteria, 17
Evolution
anagentic, 367
of brain size, 154
chromosome rearrangements and, 249–251, 251f
cladogenetic, 367
Darwinian theory and, 9
definition of, 366
DNA sequence variation and, 371
as genetic change, 12
genetic drift in, 337–338
genetic variation and, 4, 338–339, 358–359. See also Population genetics
genome, 384–386, 386f
of language, 154
molecular clock and, 384, 385f
molecular data for, 368
mutations in, 262–263
natural selection in, 324, 353–358. See also Natural selection; Selection
phylogenetics and, 380–382
protein variation and, 368–370, 369t
rate of, 382–384
reproductive isolation in, 373–379
selection response in, 324–327
speciation in, 375–379. See also Speciation
as two-step process, 366–367
of viruses, 228, 229f
Evolutionary relationships, phylogenetic trees for, 380–382, 380f–382f
Exon shuffling, 385
Expanding trinucleotide repeats, 274–275, 274t, 275f, 279t
Expected heterozygosity, 369
Expressivity, 103, 103f
Eye color
in *D. melanogaster*, 81–83, 82f, 83f, 113
gene mapping for, 176–182, 178f, 180f, 181f
genetic drift and, 351, 352f
as X-linked characteristic, 81–83, 82f, 83f
Eye location, in *D. melanogaster*, 243, 243f
Eye size, in *D. melanogaster*, 243, 243f

F′ cells, 212–213, 213f, 213t
F⁺ cells, 210f, 211–212, 211f, 213t
F⁻ cells, 210f, 211–212, 211f, 213t
F factor, 207, 208f, 211–213, 211f–213f, 213t, 214f

F (filial) generations, in monohybrid crosses, 47f, 48–49, 48f
Familial Down syndrome, 255, 255f, 256f. See also Down syndrome
Familial hypercholesterolemia, 139, 140f
Familial vitamin D-resistant rickets, 142
Fanconi anemia, 296t
Feather color
in ducks, 104, 104f
in Indian blue peafowl, 85–86, 86f
Feathering, in cock vs. hen, 116, 117f
Fertility (F) factor, 207, 208f, 211–213, 211f–213f, 213t, 214f
Fertilization, 75f
in animals, 34, 34f
definition of, 26
double, 36f, 37
in plants, 37
Fetal cell sorting, 150
Fetal DNA, in prenatal diagnosis, 150
Fetal ultrasonography, 148, 149f
Finches, Darwin's
allopatric speciation in, 376, 376f, 377f
First polar body, 34, 34f
FISH (fluorescence in situ hybridization) in gene mapping, 190, 190f
Fisher, Ronald A., 10, 307, 365
Fitness, 354
Fixation, allelic, 353
Flemming, Walther, 9
Flower color
inheritance of, 100–102
lethal alleles and, 103–104
Flower length, inheritance of, 315–316, 316f
Flowering plants. See Angiosperms; Plants
Fluorescence in situ hybridization (FISH), in gene mapping, 190, 190f
Fly Room, 81, 81f
Ford, Charles, 79
Ford, Edmund (E. B.), 365
Founder effect, 352
FOXP2, 154
Fragile sites, 251, 251f
Fragile-X syndrome, 251, 274, 274f, 274t
Frameshift mutations, 284, 279t
in Ames test, 288
Franklin, Rosalind, 10
Free radicals
as mutagens, 286, 287f
Frequency
allelic, 339–340
definition of, 339
genotypic, 339
Frequency distribution, 309–310, 309f
Friedreich's ataxia, 274t
Fruit color
epistasis and, 109, 110f
gene interaction for, 106–107, 109, 110f
incomplete dominance and, 101, 101f
Fruit fly. See *Drosophila melanogaster* (fruit fly)
Functional genes, complementation testing for, 226–227, 226f

G banding, 241, 241f
G_0 phase, 21f, 22, 24t

G_1 phase, 21f, 22, 24t
G_1/S checkpoint, 21f, 22
G_2 phase, 21f, 22, 24t
G_2/M checkpoint, 21f, 22
gag gene, 227
Gain-of-function mutations, 276, 279t
Galápagos Islands, Darwin's finches of, 376
Gallo, Robert, 227–228
Galton, Francis, 306
Gametes
    nonrecombinant (parental), 165f, 166–167
    recombinant, 165f, 166–167
    size of, 74, 76f
    unbalanced, 259
Gametic isolation, 374, 374t
Gametophyte, 35, 35f
Garrod, Archibald, 1
Gene(s). *See also* Genome(s); Protein(s) *and specific genes*
    allelic, 12, 46. *See also* Alleles
    bacterial, 17, 17f, 206, 207f, 218
    bacteriophage, fine-structural analysis of, 224–227
    definition of, 46, 46t
    distance between, recombination frequencies and, 174–175, 180–181
    dosage of, 243–244
    duplication of, 385–386
    epistatic, 107
    eukaryotic, 17
    evolution of, 384–386, 386f
    functional, complementation testing for, 226–227
    as fundamental unit of heredity, 11
    haploinsufficient, 245–246, 246f
    hypostatic, 107
    linked, 163–174. *See also* Linkage; Linked genes
    location of. *See* Gene loci
    nucleotide substitutions in, rate of, 383–384, 383t
    number of
        for polygenic trait, 308–309
    oncogenes, 227
    prokaryotic
    taster, 365–366
    unique human, 153–154
    viral, 18, 18f
    vs. traits, 11–12. *See also* Genotype-phenotype relationship
Gene dosage, unbalanced, 243–244
Gene expression, 4, 4f
    environmental effects on, 123–125
    epistatic, 107–113
    expressivity in, 103, 103f
    genomic imprinting and, 120–122, 122t
    penetrance in, 102f, 103
    phenotype and. *See* Genotype–phenotype relationship
    regulation of. *See* Gene regulation
Gene flow, 349–350, 350f
Gene interaction, 106–115
    albinism and, 110, 111f
    definition of, 106
    dog coat color and, 113–115, 114f, 115t
    epistasis and, 107–113

    novel phenotypes from, 106–107, 107f
    phenotypic ratios from, 111–112, 111t
Gene loci, 12, 46, 46t, 170–171
    definition of, 46t
    mapping of. *See* Gene mapping
    polymorphic, proportion of, 369, 369t
    quantitative trait, 303–304, 307, 322–324, 323f, 323t
        definition of, 303
        mapping of, 322–324, 323t
Gene mapping, 161–191
    in bacteria, 213–215, 214f, 217–218, 220–223, 222f
    Benzer's technique for, 224–225, 225f
    coefficient of coincidence in, 182
    with cotransformation, 217f, 218
    crossover locations in, 178f, 179–180, 180f
    in *D. melanogaster*, 176–182, 178f, 180f, 181f
    deletion, 188, 188f
    DNA sequencing in, 190
    double crossovers and, 175, 175f, 182
        four-strand, 184, 184f
        three-strand, 184, 184f
        two-strand, 184–185, 184f
    in eukaryotes, 161–191
    gene order in, 178–179, 215, 216f
    genetic maps in, 174–186
    genetic markers in, 186
    in humans, 185–187
    interference in, 182
    with interrupted conjugation, 213–215, 223
    intragenic, 224–225, 225f
    lod scores and, 186
    map units for, 174, 558
    mutations in, 224–225, 225f, 226f
    in phages, 223–224, 224f
    physical maps in, 174, 188–190
    quantitative trait loci in, 322–324, 323t
    recombination frequencies in, 174–175, 180–181
    in situ hybridization in, 190, 190f
    somatic-cell hybridization in, 188–190, 189f
    with three-point testcrosses, 176–183
    with transduction, 220–223, 222f
    with transformation, 217–218, 217f, 223
    with two-point testcrosses, 175–176
    in viruses, 223–224, 224f
Gene mutations, 273. *See also* Mutations
Gene pool, 338
Gene therapy, 4, 10
Gene transfer
    antibiotic resistance and, 215–216
    bacterial, 208–219, 386
    in biotechnology. *See* Recombinant DNA technology
    by conjugation, 208–215, 209f–213f, 213t, 214f, 223
    directional, 215, 216f
    gene order in, 215, 216f
    horizontal, 218–219, 742
    natural, 215–216
    by transduction, 208, 209f, 220–223, 222f
    by transformation, 208, 209f, 217–218, 217f, 223
Gene trees, 381, 381f

Gene–environment interactions, 123–125, 123f, 124f
General selection model, 355
Generalized transduction, 220–221, 222f
Genetic bottleneck, 352–353
Genetic correlation, 327, 327t
Genetic counseling, 146–148, 147t
Genetic crosses. *See* Crosses
Genetic diagnosis. *See* Genetic testing
Genetic differentiation, speciation and, 373–379. *See also* Speciation
Genetic diseases
    achondroplasia, 148t, 358
    albinism, 1–2
    cystic fibrosis, 102, 148t, 344–345
    cytoplasmic inheritance of, 119
    diagnosis of, 148–153, 148t
        postnatal, 148t, 151–152
        prenatal, 148–151, 148t
    environmental factors in, 123–124
    expanding trinucleotide repeats in, 274–275, 274t
    familial hypercholesterolemia, 139, 140f
    faulty DNA repair in, 295, 296t
    gene therapy for, 4, 10
    genetic counseling for, 146–148, 147t
    genetic testing for. *See* Genetic testing
    genomic imprinting and, 120–122, 122t
    hemophilia, 8, 140, 141f, 148t
    Hutchinson–Gilford progeria syndrome, 135–136, 135f, 136f
    hypophosphatemia, 142
    Leber hereditary optic neuropathy, 119
    Lesch–Nyhan syndrome, 148t
    mitochondrial, 119
    nail–patella syndrome, 185–186, 185f
    neural tube defects, 148t
    neurofibromatosis, 249
    osteogenesis imperfecta, 148t
    pedigree analysis of, 137–143
    phenylketonuria (PKU), 123–124, 148t
    screening for. *See* Genetic testing
    sickle-cell anemia, 148t
    Tay–Sachs disease, 139, 148t
    Waardenburg syndrome, 137, 138f
    xeroderma pigmentosum, 295, 295f, 295t
Genetic dissection, 272
Genetic diversity. *See* Genetic variation
Genetic drift, 337–338, 350–353, 352f, 353f, 358–359, 358t
    allelic frequencies and, 353, 353f, 358t
    causes of, 352–353
    definition of, 351
    effects of, 353, 358t
    magnitude of, 351
Genetic Information Nondiscrimination Act, 153
Genetic maps, 174–186. *See also* Gene mapping
Genetic markers
    in gene mapping, 186
    Y-linked, 87–88
Genetic maternal effect, 119–120, 120f, 122t
Genetic mutations. *See* Mutations
Genetic recombination. *See* Recombination
Genetic rescue, 338
Genetic studies, 135–154

bacteria in, 204t. *See also* Bacteria
human
    adoption studies, 146, 146f
    difficulties in, 136
    pedigree analysis, 137–143
    twin studies, 144–145
model organisms for, 5–6, 6f, 7f. *See also* Model genetic organisms
viruses in, 204t. *See also* Viruses
Genetic testing, 148–153, 148t
    direct-to-consumer, 153
    discrimination and, 153
    ethical issues in, 152, 153
    legal issues in, 152, 153
    postnatal, 151–152
    prenatal, 148–151, 148t, 149f, 150f
    presymptomatic, 152
    privacy and, 153
    test interpretation in, 152–153
Genetic variance, 317
    additive, 317
    dominance, 317
Genetic variation, 4, 29–30, 338–339
    allelic fixation and, 353
    balance hypothesis for, 369
    chromosome distribution and, 30, 31f
    crossing over and, 29–30
    evolution and, 4, 358–359, 366–373, 369t
    expected heterozygosity and, 369
    genetic drift and, 350–353
    loss of, 337–338
    measures of, 369
    migration and, 349–350, 350f
    molecular evolution and, 369t
    mutations and, 280, 348–349. *See also* Mutations
    neutral-mutation hypothesis for, 369
    proportion of polymorphic loci and, 369, 369t
    random separation of homologous chromosomes and, 30
    sexual reproduction and, 29–30
    theories of, 369–370
    universality of, 338
Genetically modified plants, 3, 3f
Genetic-conflict hypothesis, 121–122
Genetic–environmental interaction variance, 317, 317f
Genetics
    in agriculture, 3, 3f, 7–8, 8f
    applications of, 3–4, 10–11
    bacterial, 204–219
    basic concepts of, 2–7, 4, 11–12
    in biology, 4
    commercial applications of, 3–4
    divisions of, 5, 5f
    in evolution, 4, 9, 12
    future of, 10–11
    historical perspective on, 7–10
    importance of, 2–7
    in medicine, 3–4, 10–11
    model organisms in, 5–6, 6f, 7f
    in modern era, 10
    molecular, 4f, 5, 5f
    notation in. *See* Notation
    population, 5, 5f. *See also* Population genetics
    quantitative, 303–327. *See also* Quantitative genetics
    transmission, 5, 5f
    universality of, 4
    viral, 219–230
Genic balance system, 78
Genic interaction variance, 318
Genic sex determination, 77
Genome(s). *See also* Gene(s)
    of bacteria, 16f, 17, 206, 209f, 218
    chimpanzee vs. human, 153–154
    definition of, 4
    duplication of, 263, 386
    of eukaryotes, 16f, 17
    evolution of, 384–386, 386f
    human, unique genes in, 153–154
    sequencing of, 10, 10f, 11
    of viruses, 220f, 227
Genomewide association studies, 187, 187f, 323–324
Genomic imprinting, 120–122, 121f, 122t, 298
    epigenetics and, 122
Genotype
    definition of, 11–12, 46, 46t
    expression of. *See* Gene expression
    inheritance of, 46–47. *See also* Inheritance
Genotype–phenotype relationship, 11–12, 46–47. *See also* Gene expression
    continuous characteristics and, 124
    cytoplasmic inheritance and, 117–119, 119f
    environmental influences on, 123–125, 123f, 124f, 305–306
    expressivity and, 103, 103f
    gene interaction and, 106–115. *See also* Gene interaction
    genetic maternal effect and, 119–120, 120f
    genetic variation and, 338–339. *See also* Genetic variation
    heritability and, 316–324
    mutations and, 275–276
    penetrance and, 102f, 103
    polygenic inheritance and, 307–308, 308f, 309f
    quantitative traits and, 305–306, 305t, 306f
    sex influences on, 115–122, 122t
Genotypic frequency
    calculation of, 339
    Hardy–Weinberg law and, 341–342
    nonrandom mating and, 345–346
Genotypic ratios, 55–56, 56t
    observed vs. expected, 61–63, 63f
Georgiadis, Nicholas, 370
Germ-line mutations, 272f, 273
Germ-plasm theory, 8f, 9
Gierer, Alfred, 279
Gilbert, Walter, 10
Gilford, Hastings, 135
Globin genes, evolution of, 385, 386f
G_{M2} gangliosidase, in Tay-Sachs disease, 139
Goats, bearding of, 116, 116f
Goodness-of-fit chi-square test, 61–63, 62t, 63f, 172–173
Gordon, Leslie, 135
Green Revolution, 3, 3f
Grew, Nehemiah, 9
Grutzner, Frank, 73–74

Gynandromorphs, 257, 258f
Gyrase

Hair color, inheritance of, 43–44
    trinucleotide repeats and, 275, 275t
Haldane, John B. S., 10
Haploid cells, 19
Haploinsufficiency, 245–246, 246f
Haplotypes, 187
Hardy, Godfrey (G. H.), 341
Hardy–Weinberg equilibrium, 341
    definition of, 341
    genotypic frequencies at, 341–342
    testing for, 343–344
Hardy–Weinberg law, 341–345
    allelic frequencies and, 342–343
    extensions of, 343
    genotypic frequencies and, 341–342
    implications of, 342–343
Harris, Harry, 368
Heart disease
    congenital, 271
Helper T cells, in HIV infection, 229
Hemings, Sally, 88
Hemizygote
    definition of, 81
Hemophilia, 8, 140, 141f, 148t
Hen feathering, 116, 117f
Henking, Hermann, 75
Hereditary nonpolyposis colon cancer, 295
    faulty DNA repair in, 295t
Heredity. *See also* Inheritance
    chromosome theory of, 50, 75, 163, 170–171
    nondisjunction and, 82–83, 83f
    gene as fundamental unit of, 11
    principles of, 44–64
    sex influences on, 115–122, 122t
Heritability, 316–324
    broad-sense, 318
    calculation of, 318–320
        by degrees of relatedness, 320
        by elimination of variance components, 318–319
        by parent-offspring regression, 319–320, 319f
    definition of, 316
    environmental factors in, 321
    of intelligence, 322
    limitations of, 320–322
    narrow-sense, 318
    phenotypic variance and, 316–318
    population differences and, 321–322
    realized, 326
    selection response and, 324–327
    summary equation for, 318
Hermaphroditism, 75
    sequential, 77, 77f
Hershey, Alfred, 223
Heterogametic sex, 76
Heterokaryons, 189, 189f
Heterozygosity
    autosomal dominant traits and, 139, 139f, 143t
    autosomal recessive traits and, 138–139, 138f, 143t
    definition of, 46, 46t
    dominance and, 100–101, 101f

Heterozygosity (cont.)
  expected, 369
  inbreeding and, 345–347, 346f, 347f
  nonrandom mating and, 345–346, 346f, 347f
Heterozygote screening, 152
Hexosaminidase deficiency, in Tay-Sachs disease, 139
Hfr cells, 211–212, 212f, 213t
Hillmer, Alex, 161–162
Hiroshima, atomic bombing of, 289, 290, 290f
Histone(s), 17, 17f, 20
HIV (human immunodeficiency virus infection), 227–229, 228f, 371–372
Homogametic sex, 76
Homologous pairs, of chromosomes, 19, 20f, 32
  random separation of, 30
Homologous recombination
  in DNA repair, 294
Homologous traits, 380
Homozygosity
  autosomal dominant traits and, 139, 139f, 143t
  autosomal recessive traits and, 138–139, 138f, 143t
  definition of, 46, 46t
  dominance and, 100–103, 101f
  inbreeding and, 345–347, 346f, 347f
  nonrandom mating and, 345–347, 346f, 347f
  by state vs. descent, 345, 346f
Homunculus, 8, 9f
Honeybees
  epigenetics and, 122
Hooke, Robert, 8
Hopi Native Americans, albinism in, 1–2, 2f
Horizontal gene transfer, 218–219, 386
Hormones, in sex determination, 80–81
Hot spots, recombination, 191
Hrdlieka, Ales, 1
Hubby, John, 368
Human chorionic gonadotropin, in prenatal screening, 150
Human genetic studies, 135–154. *See also* Genetic studies
Human immunodeficiency virus, 227–229, 228f, 371–372
  evolutionary relationships of, 371–372, 372t
Huntington disease, 274t, 275
Hutchinson, Jonathan, 135
Hutchinson–Gilford progeria syndrome, 135–136, 135f, 136f
Huxley, Julian, 365
Hybrid breakdown, 374, 374t
Hybrid inviability, 374, 374t
Hybrid sterility, 374, 374t
Hybridization. *See also* Breeding
  allopolyploidy and, 260, 260f
  of plants, 9
  in situ, 190, 190f
  somatic-cell, 188–190, 189f
Hydroxylamine, as mutagen, 286, 286f
  familial, 139, 140f
Hypophosphatemia, 142
Hypostatic gene, 107

Immunodeficiency states, 227–229, 371–372
In situ hybridization
  in gene mapping, 190, 190f

In vitro fertilization, preimplantation genetic diagnosis and, 150–151
Inbreeding, 345–347, 346f, 347f
Inbreeding depression, 347
Incomplete dominance, 100–101, 101f, 101t, 103f
Incomplete linkage, 166–167, 169–170
Incomplete penetrance, 103, 103f
Incorporated errors, 282, 284, 284f
Independent assortment, 30, 56–57, 57f, 106, 162–163, 169
  chi-square test for, 172–174, 173f
  definition of, 56
  interchromosomal recombination and, 170. *See also* Recombination
  testcrosses for, 172–174, 173f
  vs. complete linkage, 165f, 166–167, 169–170
  vs. nonindependent assortment, 163, 164–165, 164f
  vs. segregation, 49t
Indian blue peafowl, feather color in, 85–86, 86f
Induced mutations, 281, 284–297. *See also* Mutations
Influenza virus, 229–230, 230f, 230t
In-frame deletions, 274, 279t
In-frame insertions, 274, 279t
Inheritance, 44–64. *See also* Heredity
  of acquired characteristics, 8, 9, 10t, 47
  of acquired traits, 47
  anticipation in, 122–123, 275
  blending, 9, 10t
  chromosome theory of, 50, 75, 163, 170–171
  nondisjunction and, 82–83, 83f
  codominance in, 101t, 102
  of continuous characters, 124
  cytoplasmic, 117–119, 118f, 119f, 122t
  of dominant traits, 48f, 49, 100–103, 101f, 101t, 137, 138f, 139, 139f, 141–142, 141f, 143t
  early concepts of, 7–10
  gene interactions and, 106–115
  of genotype vs. phenotype, 46–47
  incomplete dominance in, 100–101, 101f
  incomplete penetrance in, 103, 103f
  of linked genes, 161–191. *See also* Linkage; Recombination
  Mendelian, 10, 10t, 44–47
  in monohybrid crosses, 47–56. *See also* Monohybrid crosses
  polygenic, 307–308
  of quantitative characteristics, 53–54, 307–308, 308f
  of recessive traits, 49, 138–140, 138f, 141f, 143t
  segregation in, 49, 51f, 162–163
  sex-linked, 74, 81–90. *See also* Sex-linked traits
  studies of. *See* Genetic studies
  uniparental, 257
  of Y-linked traits, 142, 143t
Insertions, 273f, 274, 279t, 282–283, 283f
  in-frame, 274, 279t
Integrase, 227
Intelligence
  heritability of, 322
  inbreeding and, 347

Intercalating agents, as mutagens, 286–287, 287f
Interchromosomal recombination, 170
Interference, 182
Intergenic suppressor mutations, 277–288, 288f, 279t
Interkinesis, 27, 27t
Interphase
  in meiosis, 26–27, 27t, 28f
  in mitosis, 21–22, 21f, 23f, 24t. *See also* Cell cycle
Interrupted conjugation, in gene mapping, 213–215
Intrachromosomal recombination, 170
Intragenic mapping, 224–225, 225f, 226f
Intragenic suppressor mutations, 287, 289t
Inversions, 246–248, 246f–248f, 262t
  in evolution, 262–263
Ionizing radiation, 287–288. *See also* Radiation exposure
IQ
  heritability of, 322
  inbreeding and, 347
Isotopes, 276

Jacob, François, 214
Jacobsen syndrome, 274t
Japan, atomic bombing of, 289, 290, 290f
Jefferson, Thomas, 88
Jimson weed, aneuploidy in, 253, 254f
Johannsen, Wilhelm, 46, 307

Karpechenko, George, 260
Karyotypes, 240–241
  definition of, 240
  in genetic testing, 149
  human, 240–241, 240f
  preparation of, 240–241
Kinetochores, 20, 20f, 22, 24
Klinefelter syndrome, 79, 79f, 90, 254
Kucharski, Ryszard, 122

Language, genetic aspects of, 154
Leaf variegation, cytoplasmic inheritance and, 118–119, 119f
Leber hereditary optic neuropathy, 119
Lederberg, Joshua, 208–210, 220
Lederberg–Zinder experiment, 220–221, 221f
Leptotene, 26, 29f
Lesch–Nyhan syndrome, 148t
Lethal alleles, 100f, 103–104
Lethal mutations, 276, 279t
Levene, Phoebus Aaron, 273, 273f, 277
Lewontin, Richard, 368
Libraries. *See* DNA libraries
Li–Fraumeni syndrome, 296t
Limnaea peregra (snail), shell coiling in, 119–120, 120f
Linkage
  chi-square test for, 61–63, 62t, 63f, 172–174
  complete, 165f, 166–167, 169
  incomplete, 165f, 166–167, 169–170
  independent assortment and, 165f, 166–167, 169–170
  lod scores and, 186
  testcross for, 165–166, 165f
  three-gene, 176–183

Linkage analysis, 186–187. *See also* Gene mapping
   single-nucleotide polymorphisms in, 187
Linkage disequilibrium, 187
Linkage groups, 163
   in two-point crosses, 175–176
Linkage maps, 174–186. *See also* Gene mapping
Linked genes, 163–174
   crossing over with, 166–167, 167f, 169
   definition of, 163
   incomplete linkage of, 165f, 166–167, 169–170
      coupling in, 168, 169f
      notation for, 164
      predicting outcome of, 171–172
      recombination frequency for, 167
      in repulsion, 168, 169f
   recombination frequency for, 167
   testcrosses for, 165–167, 165f
Lipoproteins, in familial hypercholesterolemia, 139, 140f
Little, Clarence, 99–100
Loci, gene. *See* Gene loci
Lod scores, 186
Loss-of-function mutations, 276, 279t
Lymphocytes
   T, in HIV infection, 229
Lyon hypothesis, 89–90
Lyon, Mary, 89
Lysogenic life cycle, viral, 219, 220f
Lytic life cycle, viral, 219, 220f

M phase, in cell cycle, 21, 21f, 22–24, 23f, 24t. *See also* Mitosis
Maize
   recombination in, 170–171
Male-limited precocious puberty, 117, 117f
Male-pattern baldness, 161–162
Map unit (m.u.), 174
Mapping functions, 184f, 185
Maps. *See* Gene mapping
Markers, Y-linked, 87–88
Maternal age, aneuploidy and, 256, 256f
Maternal blood testing, 149–150
Mathematical models, in population genetics, 325
Mating
   assortative, 345–347
      in sympatric speciation, 378
   nonrandom, 345–347
Maximum likelihood approach, 382
Maximum parsimony approach, 382
McClintock, Barbara, 170–171, 170f
McClung, Clarence E., 75, 76
Mean, 310–311, 311f
Mechanical isolation, 374, 374t
Media, culture, 205f
Medicine, genetics in, 3–4, 10–11
Megaspores, 35, 36f
Megasporocytes, 35, 36f
Meiosis, 12, 25–37, 26f, 28f–33f
   in animals, 33–34, 34f
   cell division in, 26–27, 26f
   crossing over in, 27, 27t, 28f–32f, 29–30, 29f
   definition of, 26
   genetic consequences of, 31f

   genetic crosses and, 49–51
   genetic variation and, 29–30, 30f
   independent assortment in, 56–57, 57f
   overview of, 26–27, 28f–29f
   in plants, 35–37, 36f
   segregation in, 49, 51f, 162–163
   stages of, 26–27, 27t, 28f–29f
   vs. mitosis, 26, 31, 32f, 32t
Melanin, hair color and, 43–44
Melanocortin-1 receptor, 44
Mendel, Gregor, 9, 9f, 10, 44–45, 44f, 272
Mendelian inheritance, 10, 10t, 44–47. *See also* Inheritance
   first law of, 48f, 49, 49t. *See also* Segregation
   polygenic, 3307–308, 308f, 310f
   second law of, 48f, 49t, 56–57, 57f. *See also* Independent assortment
Mendelian population, 338
Meristic characteristics, 306
Merozygotes, 213
Messenger RNA. *See* mRNA (messenger RNA)
Metacentric chromosomes, 20, 20f, 240
Metaphase
   in meiosis, 26–27, 27t, 28f, 29f
   in mitosis, 22, 23f, 24t
Metaphase plate, 22
Methylation
   in genomic imprinting, 121, 122
Mice
   as model genetic organisms, 5, 7f, A10–A11
   yellow, 99–100, 99f, 100f, 103–104
Microcephalin, 154
Microsatellites
   variation in, 371
Microspores, 35, 36f
Microsporocytes, 35, 36f
Microtubules, spindle, 20, 20f, 22, 23f
Migration, 349–350
   allelic frequency and, 349–350, 350f, 358–359
Minimal media, 205
Mirabilis jalapa (four-o'-clocks), leaf variegation in, 118–119, 119f
miRNA. *See* MicroRNA (miRNA)
Mismatch repair, 290t, 291, 291f
Missense mutations, 275, 276, 279t
Mitochondrial diseases, 119
Mitochondrial DNA (mtDNA), 118
Mitosis, 12, 21–24, 21f, 24t
   as cell cycle phase, 21–24, 21f, 23f, 24t
   chromosome movement during, 22–24, 23f
   definition of, 26
   stages of, 21f, 22–24, 23f, 24t
   unequal crossing over in, mutations and, 243, 243f, 493, 493f
   vs. meiosis, 26, 31, 32f, 32t
Mitotic spindle, 20, 20f, 22, 23f. *See also under* Spindle
Mitton, Jeffrey, 344
MN blood group antigens, 102
Model genetic organisms, 5–6, 6f, 7f, A1–A13
   Arabidopsis thaliana, 5, 7f, A8–A9
   Caenorhabditis elegans, 5, 6f, A6–A7
   Drosophila melanogaster, 5, 6f, A2–A3
   Escherichia coli, 5, 6f, A4–A5
   Mus musculus, 5, 7f, A10–A11
   Saccharomyces cerevisiae, 5, 7f, A12–A13

Modified ratios, 111–112, 111t
Molecular clock, 384, 385f
Molecular evolution. *See also* Evolution
   DNA sequence variation and, 371–372
   molecular clock and, 385f
   protein variation and, 369
   rate of, 383–384
Molecular genetics, 5, 5f
Molecular markers, in gene mapping, 186
Molecular motors, 24
Molecular phylogenies, 380
Mollusks, sex determination in, 77, 77f
Monoecious organisms, 75
Monohybrid crosses, 47–56. *See also* Crosses; Inheritance
   F_1 generation in, 47f, 48
   F_2 generation in, 47f, 48–49
   F_3 generation in, 48f, 49
   P generation in, 47–48, 47f
   reciprocal, 48
Monopolins, 33f
Monosomy, 252, 262t
Monozygotic twins, 143, 144f
Morgan, 174
Morgan, Thomas Hunt, 10, 81–83, 81f, 83f, 160, 162, 174, 272
Mosaicism, 257, 258f
Mouse. *See* Mice
mtDNA. *See* Mitochondrial DNA (mtDNA)
Muller, Hermann, 287
Mullerian-inhibiting substance, 80
Mullis, Kary, 10
Multifactorial characteristics, 124
Multigene families, 385, 386f
Multiple alleles, 104, 104f, 105f
Multiple crossovers, 184–185, 184f
Multiple-loci crosses, 56–61
Multiplication rule, 52–53, 52f, 342
Mus musculus. *See* Mice
Mustard gas, as mutagen, 285
Mustard plant. *See Arabidopsis thaliana*
Mutagen(s), 284–285, 286f
   Ames test for, 288, 289f
Mutations, 12, 271–296. *See also specific genes*
   alkylating agents and, 286, 286f
   allelic frequencies and, 348–349, 348f, 358–359, 358t
   Ames test for, 288, 289f
   aneuploid, 252–257. *See also* Aneuploidy
   anticipation and, 122–123, 275
   base analogs and, 275–276, 275f
   base mispairing and, 281–282, 282f, 287–288
   base substitution, 273–276, 273f, 276f, 278f
   in cancer, 298–291, 289f
   causes of, 281–288
   chromosome, 239–263, 262t, 273. *See also* Chromosome mutations
   in cis configuration, 168, 169f, 226–227
   classification of, 272–275
   complementation tests for, 113, 226–227, 226f
   conditional, 276
   in coupling, 168, 169f
   deamination and, 284, 284f
   definition of, 272
   deletion, 273f, 274, 279t
   depurination and, 283–284, 284f

Mutations *(cont.)*
　DNA repair and, 290–295, 290t
　in evolution, 262–263
　expanding trinucleotide repeats and, 274–275, 274t, 275f, 279t
　experimental uses of, 272
　forward, 275, 277f, 279t
　frameshift, 274, 279t
　gain-of-function, 276, 279t
　gene, 273
　in gene mapping, 224–225, 226f
　in genetic analysis, 272
　germ-line, 272f, 273
　hydroxylating agents and, 276, 276f
　importance of, 272
　incorporated errors and, 282, 283f
　induced, 281, 284–287
　insertion, 283f, 284, 279t
　lethal, 276, 279t
　location of, 113
　loss-of-function, 276, 279t
　missense, 275, 276, 279t
　natural selection and, 358
　neutral, 276, 279t
　　genetic variation and, 369, 384
　nonsense, 275–276, 276f, 279t
　oxidative, 286, 287f
　phenotypic effects of, 275–276
　radiation-induced, 287–288, 287f, 289–290
　rates of, 280–281
　replication errors and, 281–283, 282f, 283f
　in repulsion, 168, 169f
　reverse, 275, 277f, 279t
　silent, 276, 276f, 279t
　somatic, 272–273, 272f
　SOS system and, 287–288
　spontaneous, 281–283
　　chemical changes and, 284–287, 285f, 286f
　　replication errors and, 282
　strand slippage and, 283, 283f
　study of, 288–290, 289f
　suppressor, 277–278, 277f, 278f, 279t
　tautomeric shifts and, 281, 282f
　tinman, 271
　in trans configuration, 168, 169f, 226–227
　transition, 273, 273f, 279t, 284
　translesion DNA polymerases and, 294–295
　transversion, 273, 273f, 279t
　unequal crossing over and, 243, 243f, 283, 283f
　up, 360
Myoclonic epilepsy of Unverricht–Lundborg type, 274t
Myotonic dystrophy, 274t

Nagasaki, atomic bombing of, 289, 290
Nail–patella syndrome, 185–186, 185f
Narrow-sense heritability, 318
National Bison Range, genetic drift in, 337–338
Native Americans, albinism in, 1–2, 2f
Natural selection, 324, 353–358
　allelic frequency and, 353–358, 354t, 355t, 357t, 358t, 359f, 369t
　directional, 356
　fitness and, 354
　general selection model and, 355

　mutation and, 358
　selection coefficient and, 354
Neel, James, 289, 347
Negative assortative mating, 345
Nematode. *See Caenorhabditis elegans* (nematode)
Neonatal screening, for genetic diseases, 151–152
Neural tube defects, 148t, 150
Neurofibromatosis, 249
Neurospora crassa (bread mold), 5
Neutral mutations, 276, 279t
　genetic variation and, 369, 384
Neutral-mutation hypothesis, 369, 384
Newborn screening, for genetic diseases, 151–152
Nilsson-Ehle, Herman, 307–308
Nitrogenous bases. *See* Base(s)
Nitrous acid, as mutagen, 286, 286f
Nodes, on phylogenetic tree, 380, 380f, 381f
Nondisjunction
　aneuploidy and, 252, 255f
　chromosome theory of inheritance and, 82–83, 83f
　definition of, 82–83
　Down syndrome and, 254
　maternal age and, 256
　mosaicism and, 257
　polyploidy and, 258, 258f
Nonhomologous end joining, 294
Nonindependent assortment, 163, 163f, 164f
Nonrandom mating, 345–347
Nonreciprocal translocations, 248–249, 262t
Nonrecombinant gametes, 165f, 166–167, 167f
Nonrecombinant progeny, 165f, 166–167, 168f
Nonsense mutations, 275–276, 276f, 279t
Nonsynonymous substitutions, 383, 383t, 384f
Normal distribution, 310, 310f
Notation
　for alleles, 48, 55
　for crosses, 49–50, 55
　for X-linked genes, 49, 55
Notch mutation, 245–246, 246f
Nuclear envelope, 16f, 17
Nuclear matrix, 18–19
Nucleotide(s). *See also* Base(s)
　deamination of, 284, 284f, 286f
　depurination of, mutations and, 283–284, 284f
Nucleotide substitutions, rates of, 382–383, 383t
Nucleotide-excision repair, 290t, 293–294, 293f
　in genetic disease, 295, 296t
Nucleus, 16f, 17
Nullisomy, 252, 262t

Obesity
　genetic factors in, adoption studies of, 146, 146f
Oncogenes, 227
Oocytes, 34, 34f
Oogenesis, 34, 34f
Oogonia, 34, 34f
Origin of replication, 21
Ornithorhynchus anatinus, sex determination in, 73–74
Osteogenesis imperfecta, 148t

Outcrossing, 345
Ovary, in plants, 35
Overdominance, 357
Ovum, 34, 34f
Oxidation
　mutations and, 286, 287f

P (parental) generation, in monohybrid crosses, 47–48, 47f
Pachytene, 27, 29f
Palindromic sequences, on Y chromosome, 87
Pangenesis, 8, 8f, 10t
Paracentric inversions, 246, 247f, 262t
Parental gametes, 165f, 166–167, 167f
Parental (P) generation, in monohybrid crosses, 47–48
Parental progeny, 165f, 166–167, 168f
Parsimony approach, for evolutionary relationships, 382
Parthenogenesis, 261
Patau syndrome, 256
Pauling, Linus, 384
Pavo cristatus (Indian blue peafowl), feather color in, 85–86, 86f
Peas, Mendel's experiments with, 45–49
Pedigree, definition of, 137
Pedigree analysis, 137–143
　autosomal dominant traits in, 139, 139f, 143t
　autosomal recessive traits in, 138–139, 138f, 143t
　proband in, 137, 138f
　symbols in, 137f
　X-linked traits in
　　dominant, 141–142, 141f, 143t
　　recessive, 139–140, 141f, 143t
　Y-linked traits in, 142, 143t
Penetrance
　definition of, 103
　incomplete, 103
Pentaploidy, 258. *See also* Polyploidy
Pentose sugars, 280
Pepper plant, fruit color in, 106–107, 107f
Pericentric inversions, 246, 248f, 262t
Petal color. *See* Flower color
Petri plates, 205, 205f
Phages. *See* Bacteriophage(s)
Pharmacology
　genetics and, 4
　recombinant DNA technology and, 547
Phenocopy, 124
Phenotype. *See also* Traits
　definition of, 12, 46, 46t
　expression of, 46–47, 106
　　gene interaction and, 106–115. *See also* Gene interaction
　factors affecting, 46–47
　genotype and, 11–12, 46–47. *See also* Genotype–phenotype relationship
　mutations and, 485–486
　novel, from gene interactions, 106–107, 107f
Phenotypic correlation, 327
Phenotypic ratios, 55–56, 56t
　from gene interaction, 111–112, 111t
　observed vs. expected, 61–63, 63f
Phenotypic variance, 316–318, 317f. *See also* Heritability

Phenotypic variation. See Genetic variation
Phenylketonuria (PKU), 123–124, 148t
Phenylthiocarbamide (PTC), taste sensitivity for, 365–366
Phylogenetic trees, 380–382, 380f–382f
Phylogeny, 380
Physa heterostroha, albinism in, 110, 111f
Physical maps, 174, 187–190. See also Gene mapping
    definition of, 174
Pigmentation. See Color/pigmentation
Pili, sex, 210, 211f
Pisum sativum (pea), Mendel's experiments with, 45–49
Plants
    alternation of generations in, 35, 35f
    aneuploidy in, 253, 254f
    breeding of, 3f, 7–8, 8f, 9, 260, 260f
        genetic correlations in, 327, 327t
        inbreeding and, 347
        quantitative genetics in, 303–304
    chloroplast DNA in, 118, 119. See also Chloroplast DNA (cpDNA)
    cytoplasmic inheritance in, 118–119, 119f
    flower color in
        inheritance of, 100–101, 101f, 102
        lethal alleles and, 101f, 102, 103–104
    flower length in, inheritance of, 315–316, 316f
    genetically engineered, 3, 3f, 7–8, 8f
    inbreeding in, 347
    life cycle of, 35, 35f
    Mendelian inheritance in, 45–49, 45f, 47f, 48f
    petal flower color, inheritance of, 100–101, 101f, 102
    polyploidy in, 258. See also Polyploidy
    sexual reproduction in, 35–37, 36f
Plaque, recombinant, 219, 220f, 224–225, 225f
Plasmid(s), 17
    bacterial, 206–207, 209f
    R, antibiotic resistance and, 216
Plating, 205
    replica, 206
Platypus, sex determination in, 73–74
Pleiotropy, 124
    genetic correlations and, 327
pol gene, 227, 229
Polar bodies, 34, 34f
Polydactyly, 103, 103f
Polygenic traits, 124, 307–309, 308f, 309f
    gene number determination for, 308–309
    inheritance of, 307–308
    statistical analysis of, 315–316
Polygeny, 124
Polymerase chain reaction (PCR)
    in preimplantation genetic diagnosis, 151
Polymorphic loci, proportion of, genetic variation and, 369, 369t
Polymorphisms, restriction fragment length. See Restriction fragment length polymorphisms (RFLPs)
Polynucleotide strands
    slippage of, 283, 283f
Polyploidy, 241, 241f, 252, 258–262, 262t
    allopolyploidy, 260, 260f
    in animals, 258, 261
    autopolyploidy, 258–259, 258f, 259f

    definition of, 252
    in evolution, 263
    in humans, 261
    significance of, 261–262
    in speciation, 378–379
Poly-X females, 79–80
Population(s), 310
    genetic structure of, 339–340
    Mendelian, 338
    migration of, 349–350
Population genetics, 5, 5f
    allelic frequency and, 339–340. See also Allelic frequency
    definition of, 338
    DNA sequence variation and, 371–372
    effective population size and, 352
    evolution and, 366–367
    founder effect and, 352
    genetic bottleneck and, 352–353
    genetic drift and, 350–353, 352f, 353f, 358–359, 358t
    genetic variation and, 5, 338–339. See also Genetic variation
    genome evolution and, 384–386
    genotypic frequency and, 339
    Hardy–Weinberg law and, 341–345
    mathematical models in, 339
    migration and, 349–350, 350f, 358–359, 358t
    mutations and, 348–349, 348f, 358–359, 358t
    natural selection and, 353–358, 358t
    nonrandom mating and, 345–347
    phylogenies and, 380
Population size
    effective, 352
    genetic drift and, 351–352
Position effect, 246
Positive assortative mating, 345
Postzygotic reproductive isolating mechanisms, 374, 374t
Prader–Willi syndrome, 121, 245t, 257
Precocious puberty, male-limited, 117, 117f
Preformationism, 8–9, 9f, 10t
Pregnancy-associated plasma protein, 150
Preimplantation genetic diagnosis, 150–151
Prenatal genetic testing, 148–151, 148t, 149f, 150f. See also Genetic testing
Prenatal sex selection, equality in, 75, 76f
Presymptomatic genetic testing, 152
Prezygotic reproductive isolating mechanisms, 374, 374t
Primary Down syndrome, 254. See also Down syndrome
Primary oocytes, 34, 34f
Primary spermatocyte, 33, 34f
Principle of independent assortment. See Independent assortment
Principle of segregation. See Segregation
Probability
    addition rule for, 52f, 53
    binomial expansion and, 53–54
    chi-square test and, 61–63, 62t, 63f
    definition of, 52
    multiplication rule for, 52–53, 52f
Probability method
    for dihybrid crosses, 57–58
    for monohybrid crosses, 52–55, 52f

Proband 137, 138f
Probes
    in in situ hybridization, 190, 190f
Proflavin, as mutagen, 286–287, 287f
Progeria, 135–136, 135f, 136f
Prokaryotes, 11, 16f, 17, 17f. See also Bacteria; Eubacteria
    cell reproduction in, 18, 19f
    cell structure in, 16f, 17, 17f
    chromosomes of, 18, 19f
    definition of, 17
    DNA in, 17, 17f
    genes of, 17
Prometaphase, 22, 23f, 24t
Proofreading
    in replication, 291
Prophages, 219
Prophase
    in meiosis, 26–27, 27t, 28f
    in mitosis, 22, 23f, 24t
Proportion of polymorphic loci, 369, 369t
Protein(s). See also Gene(s)
    evolution of, 384–386, 386f
    histone, 17, 17f, 20
    variation in. See Genetic variation
Proteomics, 11
Prototrophic bacteria, 205
Proviruses, 227, 228f
Pseudoautosomal regions, 76
Pseudodominance, 245
PTC, taste sensitivity for, 365–366
Puberty, male-limited precocious, 117, 117f
Punnett, Reginald C., 163
Punnett square, 51, 51f

Q banding, 241, 241f
Quad screen, 150
Qualitative genetics, 306
Qualitative traits, 304, 304f
Quantitative genetics, 303–327
    definition of, 303
Quantitative trait(s), 124, 304–316
    analytic methods for, 309–316. See also Statistical analysis
    genotype–phenotype relationship and, 305–306, 305t, 306f
    heritability of, 316–324. See also Heritability
    inheritance of, 53–54, 124, 307–308, 308f
    meristic, 306
    origin of, 303–304
    statistical analysis of, 309–316
    threshold, 306, 307f
    types of, 306
    vs. qualitative traits, 304, 304f
Quantitative trait loci (QTLs), 303–304, 322–324
    definition of, 303
    mapping of, 322–324, 323t

R banding, 241, 241f
R plasmids, antibiotic resistance and, 216
Rabbits, coat color in, 123, 124f
Radiation, ionizing, 287–288
Radiation exposure
    in Japan, 289, 290, 290f
    mutations and, 287–288, 287f, 289–290

Radiation exposure *(cont.)*
  in Russia, 289–290
Ratios, phenotypic/genotypic, 55–56, 55t, 56
  from gene interaction, 111–112, 111t
  observed vs. expected, 61–63, 63f
Reading frames, 274
Realized heritability, 326
Recessive epistasis, 107–108
  duplicate, 109–110
Recessive traits
  autosomal dominant, 139, 143t
  inheritance of, 49, 138–140, 138f, 141f, 143t
  X-linked, 139–140, 141f, 143t
Reciprocal crosses, 48
Reciprocal translocations, 249, 250f, 262t
Recombinant DNA technology, 3–4
Recombinant gametes, 165f, 166–167, 167f
Recombinant plaques, 219, 220f, 224–225, 224f
Recombinant progeny, 165f, 166–167, 167f
Recombination, 30, 161–187
  crossing over and, 27
  crossing-over and, 28f, 163f, 164f
  definition of, 162, 342
  hot spots for, 191
  independent assortment and, 30, 56–57, 57f, 106, 162–166, 163f
  interchromosomal, 170
  intrachromosomal, 170
  intragenic, 224–225, 226f
  inversions and, 246–248, 247f
  nonindependent assortment and, 163, 163f, 164f
  physical basis of, 170–171
  rate variation in, 191
  three-gene, 176–183
  two-gene, 163–176
Recombination frequencies, 167
  calculation of, 167, 180–181, 223–224
  gene mapping with, 174–175, 180–181
Red hair, inheritance of, 43–44
Regression, 313–314, 314f
Regression coefficient, 314, 314f
Reinforcement, 375
Replica plating, 206
Replication
  base pairing in. *See* Base(s)
  in cell cycle, 21, 22, 24, 24t, 25, 25f
  in eukaryotes, 19–24
  origin of, 21
  plasmid, 206–207, 208f
  proofreading in, 291
  spontaneous errors in, 281–283. *See also* Mutations
  viral, 227, 228f
Replication blocks, 287–288, 287f
Replication errors, 492
  mutations and, 281–283, 282f, 283f
Reproduction
  asexual, polyploidy and, 261
  cellular, 19–24. *See also* Cell cycle; Cell division
  sexual, 25–37. *See also* Meiosis; Sexual reproduction
Reproductive isolation
  mechanisms of, 373–374, 374t
    postzygotic, 374, 374t
    prezygotic, 373–374, 374t
  speciation and, 373–379
Repulsion (trans) configuration, 168, 169f
Response to selection, 324–327
Restriction fragment length polymorphisms (RFLPs), 186
  in evolutionary studies, 370
Retroviruses, 227, 228f, 229f
Reverse duplications, 242
Reverse mutations (reversions), 275, 277f, 279t
Reverse transcriptase, 227
Reverse transcription, 227, 284, 285f
Reversions, 275, 277f, 279t
Rhagoletis pomenella, evolution of, 378
Rickets, vitamin D-resistant, 142
rII mutations, in gene mapping, 224–226, 226f
RNA, 11, 12
RNA splicing. *See* Splicing
RNA viruses, 227–230, 228f–230f
Robertsonian translocations, 249, 249f, 251f
  aneuploidy and, 255
  in Down syndrome, 255
Robson, John, 285
Roca, Alfred, 371
Rocky Mountain bighorn sheep, 337–338, 371, 371f
Roesch, Luiz, 203–204
Rotman, Raquel, 223
Roundworms. *See Caenorhabditis elegans* (nematode)

S phase, of cell cycle, 21f, 22, 24t
Saccharomyces cerevisiae (yeast)
  duplication of, 263
  as model genetic organism, 5, 7f, A12–A13
Salmonella typhimurium
  in Ames test, 288
  transduction in, 220–222
Sample, 310
Sampling errors, 351
Sanger, Frederick, 10
Saunders, Edith Rebecca, 163
Schizosaccharomyces pombe. *See* Yeast
Schleiden, Matthias Jacob, 9
Schull, William, 347
Schwann, Theodor, 9
Screening
  for genetic diseases. *See also* Genetic testing
    postnatal, 151–152
    prenatal, 149–150
    maternal, 149–150
Second polar body, 34, 34f
Secondary oocyte, 34, 34f
Secondary spermatocyte, 33, 34f
Segmental duplications, 244
Segregation, 30, 48f, 49, 51f, 56–57, 57f, 162–163
  adjacent-1, 249, 250f
  adjacent-2, 249, 250f
  alternate, 249, 250f
  centromeric sequences in, 299
  chi-square test for, 61–63, 62t, 63f, 172–174
  independent assortment and, 30, 56–57, 57f, 162–163
  recombination and, 162–163, 164f. *See also* Recombination
  translocations and, 249, 250f

vs. independent assortment, 49t
Seidman, Christine, 271
Seidman, Jonathan, 271
Selection
  artificial, 325f. *See also* Breeding
  natural, 324, 353–358. *See also* Natural selection
Selection coefficient, 354
Selection differential, 325
Selection response, 324–327
  correlated responses and, 327
  limits to, 326–327, 326f
Separase, 32
Sequence alignments, in phylogenetic trees, 381–382
Sequential hermaphroditism, 77, 77f
Sex
  definition of, 74
  gamete size and, 74, 76f
  heredity and, 115–122, 122t
  heterogametic, 76
  homogametic, 76
Sex chromosomes
  aneuploidy of, 254
    in Klinefelter syndrome, 79, 79f, 90
    in Turner syndrome, 79, 79f, 90
  definition of, 75
  X, 81–90
    abnormal number of, 88–90
    discovery of, 75
    inactivation of, 88–90, 89f, 90f, 254
    in Klinefelter syndrome, 79, 79f, 90
    in sex determination, 75–77, 76f, 78–81. *See also* Sex determination
    structure of, 74f
    in triplo-X syndrome, 79–80
    in Turner syndrome, 79, 79f, 90
  Y, 86–88
    characteristics of, 87
    discovery of, 75
    evolution of, 86–87
    function of, 87
    genetic markers on, 80, 87–88
    in Klinefelter syndrome, 79, 79f, 90
    as male-determining gene, 80
    palindromic sequences on, 87
    in sex determination, 75–77, 76f, 78–81
    structure of, 74f
  Z, 76–77, 85–86, 86f
Sex determination, 73–81
  abnormalities in, 79–81, 79f
  chromosomal, 75–77, 78–81
    XX-XO, 76
    XX-XY, 76, 76f, 87–88
    ZZ-ZW, 76–77, 85–86
  in *Crepidula fornicata*, 77, 77f
  in *D. melanogaster*, 78, 79t
  definition of, 74
  environmental, 77–78, 77f
  genic, 77
  hormonal, 80–81
  in humans, 79–81
  sequential hermaphroditism in, 77, 77f
  Y gene in, 80, 80f
Sex pili, 210, 211f
Sex ratio, 75, 76f

genetic drift and, 351
Sex selection, equality in, 75, 76f
Sex-determining region Y gene, 80, 80f
Sexduction, 213
Sex-influenced traits, 115–117, 116f, 122t
Sex-limited traits, 115–117, 117f, 118f, 122t
Sex-linked traits, 74, 81–90, 122t
    chromosome theory of inheritance and, 82–83, 83f
    definition of, 81
    early studies of, 81–83
    inheritance of, 88
    nondisjunction and, 82–83, 83f
    recognition of, 88
    X-linked, 81–90
        color blindness as, 84–85, 84f
        in *D. melanogaster*, 81–83, 82f, 83f
        dosage compensation and, 88–90
        eye color as, 81–83, 82f, 83f
        inheritance of, 88
        X chromosome inactivation and, 88–90, 89f, 90f
    Y-linked, 81, 87–88
        inheritance of, 88
    Z-linked, 85–86, 86f
Sexual development
    precocious puberty in, 117, 117f
    sequential hermaphroditism in, 77, 77f
Sexual reproduction, 25–37
    in animals, 33–35, 34f
    in eukaryotes, 76f
    fertilization in, 26
    genetic variation and, 29–30
    meiosis and, 25–37. See also Meiosis
    in plants, 35–36, 35f, 36f
Sheep, bighorn, 337–338, 371, 371f
Shell coiling, genetic maternal effect and, 119–120, 120f
Short tandem repeats (microsatellites)
    variation in, 371
Shugosin, 32
Sickle-cell anemia, 148t
Silent mutations, 276, 276f, 279t
Single-nucleotide polymorphisms (SNPs), 187
Sister chromatids, 20f, 21
    counting of, 25
    separation of, 22–24, 23f, 24t, 25, 27, 30, 33
SIV_{cpz} virus, 228, 229f
Snails
    albinism in, 110, 111f
    shell coiling in, 119–120, 120f
Somatic mutations, 272–273, 272f
Somatic-cell hybridization, 188–190, 189f
SOS system, 287–288
Spartina anglica, evolution of, 378–379
Specialized transduction, 222
Speciation, 375–379
    allopatric, 375, 375f
    definition of, 375
    genetic differentiation and, 379
    polyploidy in, 378–379
    sympatric, 375, 377–378, 379f
Species
    biological species concept and, 373
    definitions of, 373
    reproductive isolation and, 373–379, 374t

Speech, genetic aspects of, 154
Sperm
    in animals, 33–34, 34f
    in plants, 35, 36
Spermatids, 34, 34f
Spermatocytes, 33, 34f
Spermatogenesis, 33–34, 34f
    vs. oogenesis, 34
Spermatogonia, 33, 34f
Spinal muscular atrophy, 274t
Spindle, mitotic, 20, 20f, 22, 23f
    centromeres and, 299
Spindle microtubules, 20, 20f, 22, 23f
Spindle-assembly checkpoint, 22
Spinocerebellar ataxia, 274t
Spontaneous mutations, 282–283
Sporophytes, 35, 35f
SRY gene, 80, 80f
Staining, chromosome, 241, 241f
Standard deviation, 311, 312f
Statistical analysis, 309–316
    correlation in, 312–313, 312f
    frequency distribution in, 309–310, 309f, 310f
    frequency in, 339–340
    mean in, 310–311, 311f
    normal distribution in, 310, 310f
    of polygenic traits, 315–316
    population in, 310
    of quantitative traits, 309–316
    regression, 313–315, 314f
    sample in, 310
    sampling error in, 351
    standard deviation in, 312, 312f
    variance in, 311f, 312
Stevens, Nettie, 75, 170
Strand slippage, 283, 283f
Submetacentric chromosomes, 20, 20f, 240
Suppressor mutations, 277–278, 279t
    intergenic, 277–278, 278f
    intragenic, 277, 277f
Sutton, Walter, 10, 50, 163, 170
Swine flu, 230, 230f
Symbols
    for alleles, 48, 55
    for crosses, 49–50, 55
    for X-linked genes, 49, 55
Sympatric speciation, 375, 377–378, 379f
Synapsis, 26
Synaptonemal complex, 27, 29f
Synonymous substitutions, 383, 383t, 384f

Tandem duplications, 242
Taster genes, 365–366
Tatum, Edward, 208–210
Tautomeric shifts, mutations and, 281, 282f
Tay–Sachs disease, 139, 148t, 152
Telocentric chromosomes, 20, 20f, 240
Telomere(s), 20–21, 20f
Telophase
    in meiosis, 27, 27t, 29f
    in mitosis, 23f, 24, 24t
Temperate phage, 219, 220f
Temperature-sensitive alleles, 123
Temporal isolation, 374, 374t
Testcrosses, 55, 165–166
    dihybrid, 59, 59f

    for independent assortment, 172–174, 173f
    with linked genes, 165–166, 165f
    monohybrid, 55. See also Crosses
    three-point, 176–183
    two-point, gene mapping with, 175–176
Testis
    in androgen insensitivity syndrome, 80
    spermatogenesis in, 33, 34f
Testosterone, in sex determination, 80
Tetrad, 27
Tetraploidy, 258. See also Polyploidy
Tetrasomy, 252, 262t
Three-point testcross, 176–183
    steps in, 182–183
Threshold characteristics, 306, 307f
tinman mutation, 271
Traits. See also Phenotype
    acquired, inheritance of, 8, 9, 10t, 47
    autosomal, dominant, 139, 143t
    continuous (quantitative), inheritance of, 124
    definition of, 46, 46t
    discontinuous, 124, 304, 304f
    dominant, 49
        autosomal, 139, 143t
        X-linked, 141–142, 141f, 143t
    heritability of, 316–324. See also Heritability
    homologous, 380–382
    meristic, 306
    multifactorial, 124
    pleiotropic, 124
    polygenic, 124, 307–309, 308f, 309f
        inheritance of, 307–308
        statistical analysis of, 315–316
    qualitative, 304
    quantitative, 304–309, 309–314. See also Quantitative trait(s)
    recessive, 49
        autosomal, 138–139, 138f, 143t
        X-linked, 139–140, 141f, 143t
    sex-influenced, 115–117, 116f, 122t
    sex-limited, 115–117, 117f, 118f, 122t
    sex-linked, 81–90. See also Sex-linked traits
    vs. characteristics, 46
    vs. genes, 11–12. See also Genotype-phenotype relationship
    X-linked, 81–90, 139–140, 141f, 143t. See also X-linked traits
    Y-linked, 81, 87–88, 142, 143t. See also Y-linked traits
    Z-linked, 85–86, 86f
Trans configuration, 168, 169f
Transcription, 12
Transducing phages, 220–221
Transductants, 221
Transduction, 208, 209f, 220–223, 222f
    cotransduction, 221
    in gene mapping, 221–223, 222f
    generalized, 220–221, 222f
    specialized, 222
Transformants, 217–218
Transformation
    in bacteria, 208, 209f, 217–218, 217f, 431–432
    in gene mapping, 217–218, 217f
Transforming principle, 274–275
Transfusions, ABO antigens and, 105, 105f
Transitions, 273, 273f, 274t, 284

Translation, 12
Translesion DNA polymerases, 294–295
Translocation(s), 248–251, 250f, 251f
    in Down syndrome, 254–256, 255f, 256f
    nonreciprocal, 248–249, 262t
    reciprocal, 249, 250f, 262t
    Robertsonian, 249, 249f, 251f
        aneuploidy and, 252, 255
        in Down syndrome, 255
Translocation carriers, 255, 256f
Transmission genetics, 5, 5f
Transversions, 273, 273f, 279t
Trichothiodystrophy, 295, 296t
Trinucleotide repeats, expanding, 274–275, 274t, 275f, 279t
Triploidy, 258. See also Polyploidy
Triplo-X syndrome, 79–80
Trisomy, 252
Trisomy 8, 256
Trisomy 13, 256
Trisomy 18, 256
Trisomy 21, 239–240, 254–256, 255f, 256f. See also Down syndrome
Triticum aestivum (wheat)
    kernel color in, inheritance of, 307–308, 308f
    polyploidy in, 261
Tubulin subunits, 22–24, 24f
Tumors. See Cancer
Turner syndrome, 79, 79f, 90, 254, 257
Twin(s)
    dizygotic, 143, 144f
    epigenetic effects in, 464
    monozygotic, 143, 144f
Twin studies, 144–145
    of asthma, 145, 145t
    concordance in, 144–145, 144t–146t
    of epilepsy, 144
Two-point testcrosses, gene mapping with, 175–176

Ultrasonography, fetal, 148, 149f
Ultraviolet light, as mutagen, 287–288, 287f
Unbalanced gametes, 259
Underdominance, 357
Unequal crossing over, mutations and, 243, 243f, 283, 283f
Uniparental disomy, 257
Uniparental inheritance, 257
U-tube experiment, 210, 210f

Variable expressivity, 103, 103f
Variance, 311, 311f
    definition of, 312
    environmental, 317
    genetic, 317
        additive, 317
        dominance, 317
    genetic-environmental interaction, 317, 317f
    genic interaction, 318
    phenotypic, 316–318, 317f
Variation. See Genetic variation
Vertebrates, genome of, duplication of, 263
Virulent phage, 219, 220f
Viruses, 18, 18f, 219–230. See also Bacteriophage(s)
    definition of, 219
    diversity of, 219, 219f
    evolution of, 228, 229f
    gene mapping in, 223–224, 224f
    genes of, 18, 18f
    genetic material in, 278–279
    in genetic studies, 204t
    genome of, 227
    proviruses and, 227, 228f
    retroviruses, 227, 228f, 229f
    RNA, 227–230, 228f–230f
    RNA in, 278–279
    in transduction, 208, 209f, 220–223, 222f
Vitamin D-resistant rickets, 142
Von Tschermak, Erich, 45
Vries, Hugo de, 45

Waardenburg syndrome, 137, 138f
Watson, James, 10
Weight, adoption studies of, 146, 146f
Weinberg, Wilhelm, 341
Weismann, August, 10
Wethington, Amy R., 110
Wheat
    kernel color in, inheritance of, 307–308, 308f
    polyploidy in, 261
Whole-genome duplication, 386
Wild-type alleles, 55
Wild-type bacteria, 205
Wilkins, Maurice, 10
Williams–Beuren syndrome, 245t, 246
Wilson, Edmund, 75, 170
Wobble, 281–282, 282f, 283f
Wolf–Hirschhorn syndrome, 245t, 246
Wollman, Elie, 214
Worm. See *Caenorhabditis elegans* (nematode)
Wright, Sewall, 10

X chromosome, 81–90
    abnormal number of, 88–90
    discovery of, 75
    inactivation of, 88–90, 89f, 90f, 254
    in Klinefelter syndrome, 79, 79f, 90
    in sex determination, 75–81, 76f
    structure of, 74f
    in triplo-X syndrome, 79
    in Turner syndrome, 79, 79f, 90
X^2 distribution, critical values for, 62t
X : A ratio, 78, 79t

Xenopus laevis (clawed frog), 5
Xeroderma pigmentosum, 295, 295f, 296t
xist gene, 90
X-linked genes
    dosage compensation for, 88–90
    notation for, 85
X-linked traits, 81–90
    color blindness as, 84–85, 84f
    in *D. melanogaster*, 81–83, 82f, 83f
    dominant, 141–142, 141f, 143t
    dosage compensation and, 88–90
    eye color as, 81–83, 82f, 83f
    inheritance of, 88, 139–140, 140f, 141f, 143t
    nondisjunction and, 82–83, 83f
    notation for, 85
    recessive, 139–140, 141f, 143t
    recognition of, 88
    X chromosome inactivation and, 88–90, 89f, 90f
X-ray(s). See also Radiation exposure
    mutations and, 287–288
XX-XO sex determination, 76, 76f
XX-XY sex determination, 76, 76f

Y chromosome, 80, 86–88
    characteristics of, 87–88
    discovery of, 75
    evolution of, 86–87
    genetic markers on, 87–88
    in Klinefelter syndrome, 79, 79f, 90
    as male-determining gene, 80
    palindromic sequences on, 87
    in sex determination, 75–81, 76f
    structure of, 74f
Yeast
    as model genetic organism, 5, 7f, A12–A13
Yellow mice, 99–100, 99f, 100f, 103
Y-linked markers, 87–88
Y-linked traits, 81, 87–88, 142, 143t
    inheritance of, 88
    notation for, 85
    recognition of, 88
Yule, George Udny, 307

Z chromosome, 76–77, 85–86, 86f
Zea mays (corn), 5
Zebrafish, *golden* mutation in, 5–6, 7f
Zinder, Norton, 220
Z-linked traits, 85–86, 86f
Zuckerandl, Emile, 384
Zygotene, 26, 29f
ZZ-ZW sex determination, 76–77